ANNUAL REVIEW OF ECOLOGY AND SYSTEMATICS

EDITORIAL COMMITTEE (1995)

Responsible for the organization of Volume 26
(Editorial Committee, 1993)

ANNUAL REVIEW OF ECOLOGY AND SYSTEMATICS

VOLUME 26, 1995

DAPHNE GAIL FAUTIN, *Editor*

University of Kansas

DOUGLAS J. FUTUYMA, *Associate Editor*

State University of New York at Stony Brook

FRANCES C. JAMES, *Associate Editor*

Florida State University

ANNUAL REVIEWS INC.   4139 EL CAMINO WAY   P.O. BOX 10139   PALO ALTO, CALIFORNIA 94303-0139

ANNUAL REVIEWS INC.
Palo Alto, California, USA

International Standard Serial Number: 0066–4162
International Standard Book Number: 0–8243–1426-3
Library of Congress Catalog Card Number: 71-135616

Annual Review and publication titles are registered trademarks of Annual Reviews Inc.

∞ The paper used in this publication meets the minimum requirements of
American National Standard for Information Sciences—Permanence of Paper
for Printed Library Materials, ANSI Z39.48-1984.

Annual Reviews Inc. and the Editors of its publications assume no responsibility for the
statements expressed by the contributors to this *Review*.

Typesetting by Kachina Typesetting Inc., Tempe, Arizona; John Olson, President;
Jeannie Kaarle, Typesetting Coordinator; and by the Annual Reviews Inc. Editorial Staff

PRINTED AND BOUND IN THE UNITED STATES OF AMERICA

PREFACE

SUSTAINABILITY — MORE THAN A BUZZ-WORD?

The eleven chapters in Volume 26 on the topic of environmental sustainability exemplify the broadened purview of the *Annual Review of Ecology and Systematics*, which now explicitly includes "applied" systematics and ecology, such as conservation biology and fisheries (see Preface to Volume 25 of *ARES*). What could be more applicable to us all than the sustainability of those resources on which human life on earth depends? The members of the Editorial Committee of *ARES* hope these chapters will promote linkages among disciplines that have contributions to make to the deliberations about our collective future.

Several chapters were written by authors invited to discuss sustainability from the perspectives of disciplines not typically included in *ARES*. In assembling this section, I have encountered one of the major problems faced by those attempting to integrate various sources of information bearing on this issue of ultimate gravity—communication. As becomes obvious from reading the chapters, the very term "sustainability" has a wide variety of applications and connotations. Topics and authors for the section were selected in part to demonstrate some of the ways in which this word is employed—a diversity that extends to the underlying assumptions and principles used in discussions of and proposals for sustainability. This section will have achieved one of its aims if readers come away recognizing some of these differences and the need to be explicit in their own usage and about the underpinnings of their positions.

This section illustrates disciplinary differences in style as well as in usage and opinion. The difficulty of recognizing, and then getting beyond, such differences may contribute to a lack of communication between disciplines (e.g. 8). A piece of writing in a style too different from that of a reader's discipline may be viewed as inaccessible or may not be taken seriously, because it does not conform to the customary formula for "real science." For example, natural scientists charge some social science with lacking rigor, but our emphasis on rigor has recently been questioned (5, 18). Authors who accepted the invitation (and challenge!) to write for this section are presumably among those most convinced of the necessity for cross-disciplinary communication to effect the changes needed to sustain human life on earth. Yet, editorial intervention was required to craft the contributions from fields outside the typical focus of *ARES* into a format at once representative of those disciplines—and so in their construction expressing the cross-disciplinary message of this section—and sufficiently familiar to be acceptable to *ARES* readers.

Biologists must understand the style as well as the information of the social sciences to participate fully in the debate about, and development of, a rational, scientifically based environmental policy. Extending even beyond the expanded horizons of *ARES* is essential for subjects that have policy implications because political, sociological, and economic concerns often override purely scientific considerations (e.g. 1, 21). One perspective on the recent collapse of the northwest Atlantic cod fishery is that economic interests and regulatory bureaucracy were unable or unwilling to respond to advice from scientists (10). As this introduction is being written in mid-1995, the US Congress is proposing to reduce the weight it gives in its deliberations to data from the natural sciences and to increase that given to social sciences (such as economics and politics). [How ironic, then, that this same Congress is proposing to eliminate support for research in the social sciences by the US National Science Foundation (11).] These chapters are intended to serve as bridges, in both style and substance, to a literature that deals with sustainability from perspectives that may be different from those to which we biological scientists are accustomed.

Among the principles addressed in this collection are some as seemingly incontrovertible as the value of biodiversity. Much of conservation is based on what most ecologists accept as a biological reality—the inherent value of diversity (see, for example, the chapter by Hartshorn). Nelson proposes that, regardless of scientific rationale, its prominence stems from underlying human religious or cultural values. Science is not unique in this regard; all human activities, he argues, including economics, are value-laden. From their perspective outside the discipline of economics, biologists may readily recognize certain premises of economics as more beliefs than objectively verifiable realities, but it may come as a surprise that outsiders to ecology can view biodiversity in similar light (for more on values, see, for example, 5).

Rather than purely an article of faith, biodiversity may be among those tenets classified by Hardin (8) as "default principles of science," which cannot be proven but are consistent with knowledge. The importance of biodiversity to ecology is related to the biological premises of evolutionary and functional uniqueness. The experience of environmental science is that the interaction of each piece of the biosphere with others contributes in some incompletely understood but unique fashion to produce the whole, so removal of any component of the system may have unforeseeable consequences for other components and the entire system. But as species become extinct at an ever-increasing rate (e.g. 13), it is unrealistic—and perhaps counter-productive—to insist that all be saved. The chapter by Humphries, Vane-Wright & Williams provides some guidance in the process of selecting taxa to conserve. It is likely that if biologists do not participate in this triage, it will be done by people with priorities other than the ecological, or there will be no selection at all.

The value that environmental scientists regard as inhering in habitats, organisms, and their variety is one of the more obvious contrasts with that of conventional economics, in which a resource acquires value only when put to direct human use. In this perspective trees not grown in a plantation or oceanic fish stocks acquire value only when harvested, their value as inventory, or "natural capital" having been ignored. Only recently have some people in economics, such as Goodland (7, p. 1), recognized that, from the perspective of the environment, "consumption of natural capital is liquidation." As discussed by Hartshorn, the participation of scientists in decision-making can provide the biological basis for rational linkage of conservation with development. Biologists can also address values of biological resources other than through direct consumption (e.g. 4, 15). Setting these values and having them recognized in the marketplace are problematic, according to Daly, but in late 1995, the World Bank announced it had made a first atttempt (16).

Hardin (8, p. 56) regards attitudes toward limits as one of the three "major ways in which ecology and economics differ." Although some economists no longer believe literally that natural resources are inexhaustible (17), operationally the conviction persists in the premise of "substitutability." As alluded to by Clark and explained more fully by Daly, this basic assumption of economics posits that one resource can substitute for another. For example, implements once made of metal (e.g. kitchen strainers) or wood (e.g. baskets for refuse) are now made of plastic because of the economic calculus that includes raw material availability, costs of fabrication and distribution, durability, and consumer preference. Substitutability is irreconcilable with biological uniqueness. Hoffman & Carroll point out that even options provided by "modern" technologies for crops are rather severely constrained because of limitations on flexibility imposed by the unique evolutionary history of each genome.

Substitutability is necessary to another core principle of conventional economics, as discussed by Nelson, Daly, and Clark—that of continual growth (e.g. 9). But it is antithetical to most biological systems, which exhibit dynamic equilibria. [It might be argued that the growth espoused by most economists is like organic diversity, which appears to have increased since the end of the Precambrian (2). However, enhanced taxonomic richness does not necessarily imply increased biomass, and so it might more properly be analogized to development, a concept that, according to Daly, should denote increased quality of human life, clearly distinct from quantitative growth.] Even the most benign agricultural practices, Buol emphasizes, remove nutrients until ultimately one is operationally exhausted, preventing further growth. The limiting nutrient can be replaced—at some financial cost and only by removal from another source—but there is no substitute for it.

Regulation, a major issue in sustainability, is especially problematic for common pool resources (e.g. 6). Regulation of oceanic fisheries, as discussed by Hilborn, Walters, & Ludwig, has had mixed success; not surprisingly to biologists, the success seems to correlate with the biology of the organisms. Nichols, Johnson & Williams trace the history of waterfowl management in the United States, which, through trial and error, appears to have resulted in a scientifically based, societally acceptable policy for management. Lessons for other common pool biological resources may be limited, for in the United States, waterfowl are exploited largely for sport, so harvest can be modified as necessary without threat to human livelihood. Some problems in exploitation of common pool resources and some case studies in distribution of water are examined by Becker & Ostrum, who conclude that various systems may be equally viable. However, the most successful models they discuss appear to be for subsistence-level or small-market agriculture and small populations.

Vandermeer assesses, in light of growing human populations, the ecological and economic potentials of various regimes proposed as alternatives to conventional agriculture. Such consideration of scale effects is rare in debates about sustainability, according to Hardin (8). Buol points out that distribution, and not merely size, of human populations affects the rate at which nutrients are depleted from the soil. As people are increasingly concentrated away from the source of their food, nutrients are exported in that food from rural soil to urban sewers and then to their ultimate repositories, distant from their sources. Similarly, commercial pelagic fisheries being established in the tropical Pacific remove the nutrients from a nutrient-limited ecosystem to distant areas. This differs from both traditional fisheries in the tropics where the nutrients are recycled within the local ecosystem, and the large-scale high latitude fisheries in which the resource is more widespread and nutrient limitation is less. Thus, ecologically, a fishery and a crop may be more similar than two fisheries, a distinction economists may not recognize. If such considerations are to be factored into policy, they must be put forth by ecologists, according to Clark and other authors in this volume as well as to those writing elsewhere (e.g. 3, 5, 19).

Another consequence of the increased urbanization that has accompanied human population growth is an obscuring of human reliance on natural resources (e.g. 8, 20). It is distressingly easy to find examples of ignorance of this linkage. *The Straits Times* newspaper of Singapore recently reprinted an article (14) on environmental pollution by an economist who asserted: "In the past, much of the economic growth and industrialisation in all Asean [Association of Southeast Asian Nations] states, except Singapore, depended upon the exploitation of natural resources, including forests, coastal zones, fertile agricultural lands and water resources." Few readers of *ARES* would contend that Singapore is independent of natural resources because its economy is based on industry and services rather than agriculture and extractive activities. Yet, apparently, some economists believe that to be true, and may even propose policy on that basis.

We hope this collection of chapters will reinforce the realization that not everyone shares information, definitions, and perspectives, even when ostensibly pursuing similar goals and using identical words. Beyond recognizing the diversity, we hope this special section will help readers of *ARES* to cross disciplinary boundaries by providing new perspectives, new information, and new literature. The disciplines represented by the readers of *ARES* are central to the concerns about how to achieve sustainability, for, without a firm grounding in ecological reality, "sustainability" is in danger of "becom[ing] a cant phrase, a 'greenwash' that will fade away like so many previous rhetorical colorings" (12, p. 1).

DAPHNE GAIL FAUTIN
EDITOR

Literature Cited

1. Arrow K, Bolin B, Costanza R, Dasgupta P, Folke C, Holling CS, et al. 1995. Economic growth, carrying capacity, and the environment. *Science* 268:520–21
2. Benton MJ. 1995. Diversification and extinction in the history of life. *Science* 268:52–58
3. Brady GL, Geets PCF. 1994. Sustainable development: the challenge of implementation. *Int. J. Sustain. Dev. World Ecol.* 1:189–97
4. Claridge G. 1994. Management of coastal ecosystems in eastern Sumatra: the case of Berbak Wildlife Reserve, Jambi Province. *Hydrobiologia* 285:287–302
5. Ehrlich PR. 1994. Ecological economics and the carrying capacity of earth. In *Investing in Natural Capital: The Ecological Economics Approach to Sustainability*, ed. A Jansson, M Hammer, C Folke, R Costanza, pp. 38–56. Washington, DC and Covelo, CA: Island
6. Farrow S. 1995. Extinction and market forces: two case studies. *Ecol. Econ.* 13:115–23
7. Goodland R. 1993. *The only true definition of environmental sustainability!!: an informal discussion.* Paper presented at the US Agency for Int. Dev. Environ. Forum, 15 September 1993
8. Hardin G. 1993. *Living Within Limits: Ecology, Economics, and Population Taboos.* New York and Oxford: Oxford Univ. Press
9. Hawken P. 1993. *The Ecology of Commerce: A Declaration of Sustainability.* New York: HarperCollins
10. Holmes B. 1994. Biologists sort the lessons of fisheries collapse. *Science* 264:1252–53
11. Lawler A. 1995. GOP plans would reshuffle science. *Science* 268:964, 65, 67
12. Lélé S. 1994. Sustainability, environmentalism, and science. *Pacific Inst. Rep.* Spring:1, 2, 5
13. McNeely JA, Miller KR, Reid WV, Mittermeier RA, Werner TB. 1990. *Conserving the World's Biological Diversity.* Gland, Switzerland, and Washington, DC: Int. Union Conserv. Nature & Nat. Resourc.

14. Manopimoke S. 1994. Asean beginning to address pollution problem. *Straits Times,* 31 October 1994, p. 26
15. Myers N. 1995. The world's forests: need for a policy appraisal. *Science* 268: 823–24
16. Passell P. 1995. The wealth of nations: a "greener" approach turns list upside down. *New York Times,* 20 Sept., B5, B12
17. Rosenberg AA, Fogarty MJ, Sissenwine MP, Beddington JR, Shepherd JG. 1993. Achieving sustainable uses of renewable resources. *Science* 262:828–29
18. Roush W. 1995. When rigor meets reality. *Science* 269:313–15
19. Schlesinger WH. 1989. The role of ecologists in the face of global change. *Ecology* 70:1
20. Tisdell C. 1994. Conservation, protected areas, and the global economic system: how debt, trade, exchange rates, inflation, and macroeconomic policy affect biological diversity. In *Protected Area Economics and Policy: Linking Conservation and Sustainable Development,* ed. M Munasinghe, J. McNeely, pp. 51–68. Washington DC: World Conserv. Union (IUCN)
21. Vincent JR. 1992. The tropical timber trade and sustainable development. *Science* 256:1651–55

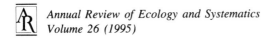
Annual Review of Ecology and Systematics
Volume 26 (1995)

CONTENTS

RELATED ARTICLES FROM OTHER *ANNUAL REVIEWS*

From the *Annual Review of Energy and the Environment,* Volume 20 (1995)

National Materials Flows and the Environment, IK Wenick, JH Ausubel
Atmospheric Emissions Inventories: Status and Prospects, JM Pacyna, TE Graedel
The Emerging Climate Change Regime, DM Bodansky

From the *Annual Review of Earth and Planetary Sciences,* Volume 23 (1995)

The Origin of Life in the Solar System: Current Issues, CF Chybam, GD McDonald

From the *Annual Review of Genetics,* Volume 29 (1995)

Conservation Genetics, R Frankham
Ecological Population Genetics, A Hoffman

From the *Annual Review of Phytopathology,* Volume 33 (1995)

The Impact of Molecular Characters on the Systematics of Filamentous Ascomycetes, GJ Samuels, KA Seifert

From the *Annual Review of Plant Physiology and Plant Molecular Biology,* Volume 46 (1995)

Plant Genomes: A Current Molecular Description, C Dean, R Schmidt
Cell Cycle Control, TW Jacobs
Physiological and Ecological Function within the Phytochrome Family, H Smith

For the convenience of readers, a detachable order form/envelope is bound into the back of this volume.

Annu. Rev. Ecol. Syst. 1995. 26:1–24

THE CONCEPT OF ENVIRONMENTAL SUSTAINABILITY[1]

Robert Goodland

S-5043, Environment Department, The World Bank, Washington, DC 20433

KEY WORDS: economic sustainability, development, economic development, natural
              capital, growth, limits of growth

ABSTRACT

This paper reviews the current status of the debate about the concept of environmental sustainability and discusses related aspects of growth, limits, scale, and substitutability. While the paths leading to environmental sustainability in each country or sector will differ, the goal remains constant. But this conceptualization is far from an academic exercise. Ensuring, within less than two human generations, that as many as 10 billions people are decently fed and housed without damaging the environment on which we all depend represents a monumental challenge.

INTRODUCTION

As soon as Prime Minister Gro Harlem Brundtland and her United Nations commission (105), in a brilliant feat, garnered almost worldwide political consensus on the urgent need for sustainability, many countries and institutions started to grapple with the same problem: Precisely what is sustainability, and, specifically, what does it mean for this particular sector, nation, or region? This paper outlines the concept of sustainability, then focuses on environmental sustainability (ES).

This paper seeks to define environmental sustainability partly by sharply distinguishing it from social sustainability and, to a lesser extent, from eco-

---

[1]NOTE: These personal opinions should in no way be construed as representing the official position of the World Bank Group.

1

nomic sustainability. These are contrasted in Figure 1. While overlap exists among the three, economic sustainability and ES have especially strong linkages. Defining each component of sustainability distinctly may help organize the action required to approach global sustainability in real life. Although this paper focuses more on the environmental aspects of sustainability, perhaps in the future, a general sustainability will come to be based on all three aspects—environmental, social, and economic.

Historically, economic theory has focused on efficiency of use of goods and, to a much lesser degree, on equity of distribution. Recent recognition of the pervasive economic significance of environmental conditions has forced two changes. First, the relatively new criterion of scale must now be added to the traditional criteria of allocation of resources and efficiency of use (30, 66). The "growth" debate emphasizes the scale of the growing human economic subsystem relative to the finite ecosystem. Ecologists and other biophysical scientists need to take more responsibility for leading the thinking on sustainable development and for seeing that efforts to achieve it are implemented promptly. Second, we must recognize that markets are almost invariably deficient as distributive mechanisms when natural resources are concerned.

Economic sustainability focuses on that portion of the natural resource base that provides physical inputs, both renewable (e.g. forests) and exhaustible (e.g. minerals), into the production process. ES adds consideration of the physical inputs into production, emphasizing environmental life-support systems without which neither production nor humanity could exist. These life-support systems include atmosphere, water, and soil—all of these need to be healthy, meaning that their environmental service capacity must be maintained. A healthy ozone shield, for example, prevents damage by ultraviolet b radiation to biota such as humans and crops. Continuous depletion or damage by human activities to irreplaceable and unsubstitutable environmental services would be incompatible with sustainability.

SOCIAL SUSTAINABILITY

The environment has now become a major constraint on human progress. Fundamentally important though social sustainability is, environmental sustainability or maintenance of life-support systems is a prerequisite for social sustainability. Redclift (74–76) claims that poverty reduction is the primary goal of sustainable development, even before environmental quality can be fully addressed. Poverty is increasing in the world in spite of global and national economic growth (see below). Poverty reduction has to come from qualitative development, from redistribution and sharing, from population stability, and from community sodality, rather than from throughput growth. Politicians will doubtless want the *impossible* goal of increasing throughput—

Social Sustainability	Economic Sustainability	Environmental Sustainability (ES)
Achieved only by systematic community participation and strong civil society. Cohesion of community, cultural identity, diversity, sodality, comity, tolerance, humility, compassion, patience, forbearance, fellowship, fraternity, institutions, love, pluralism, commonly accepted standards of honesty, laws, discipline, etc, constitute the part of social capital least subject to rigorous measurement, but for social sustainability. This "moral capital," as some call it, requires maintenance and replenishment by shared values and equal rights, and by community, religious and cultural interactions. Without this care it will depreciate as surely as will physical capital. Human capital—investments in education, health, and nutrition of individuals—is now accepted as part of economic development (102–104 108), but the creation of social capital as needed for social sustainability is not yet adequately recognized.	Economic capital should be stable. The widely accepted definition of economic sustainability "maintenance of capital", or keeping capital intact, has been used by accountants since the Middle Ages to enable merchant traders to know how much of their sales receipts they and their families could consume without reducing their ability to continue trading. Thus Hicks' (48) definition of income—"*the amount one can consume during a period and still be as well off at the end of the period*"—can define economic sustainability, as it devolves on consuming interest, rather than capital. We now need to extrapolate the definition of Hicksian income from sole focus on human-made capital and its surrogate (money) now to embrace the other three forms of capital (natural, social and human). Economics has rarely been concerned with natural capital (e.g., intact forests, healthy air) because until relatively recently it had not been scarce. This new scarcity, that of natural capital, arose because the scale of the human economic subsystem has now grown large relative to its supporting ecosystem (26–32). To the traditional economic criteria of allocation and efficiency must now be added a third, that of scale (28). The scale criterion would constrain throughput growth—the flow of material and energy (natural capital) from environmental sources to sinks, via the human economic subsystem. Economics values things in money terms, and is having major problems valuing natural capital, intangible, intergenerational, and especially common access resources, such as air. Because people and irreversibles are at stake, economics needs to use anticipation and the precautionary principle routinely, and should err on the side of caution in the face of uncertainty and risk.	Although ES is needed by humans and originated because of social concerns, ES itself seeks to improve human welfare by protecting the sources of raw materials used for human needs and ensuring that the sinks for human wastes are not exceeded, in order to prevent harm to humans. Humanity must learn to live within the limitations of the biophysical environment. ES means natural capital must be maintained, both as a provider of inputs ("sources"), and as a "sink" for wastes (23, 25, 72, 73, 74, 81, 83, 101). This means holding the scale of the human economic subsystem to within the biophysical limits of the overall ecosystem on which it depends. ES needs sustainable production and sustainable consumption. On the sink side, this translates into holding waste emissions within the assimilative capacity of the environment without impairing it. On the source side, harvest rates of renewables must be kept within regeneration rates. Non-renewables cannot be made fully sustainable, but quasi-ES can be approached for non-renewables by holding their depletion rates equal to the rate at which renewable substitutes can be created (38, 39). Ultimately, there can be no social sustainability without ES. ES supplies the conditions for social sustainability to be approached.

Figure 1. Comparison of social, economic and environmental sustainability

the flow of materials and energy from the sources of the environment, used by the human economy, and returned to environmental sinks as waste—by increasing consumption by all.

Countries truly sustaining themselves, rather than liquidating their resources, will be more peaceful than countries with unsustainable economies (41). Countries with unsustainable economies—those liquidating their own natural capital or those importing liquidated capital from other countries (e.g. Middle East oil or tropical timber 'mining') are more likely to wage war than are those with sustainable economies. When social sustainability has been clarified, possibly it will be relinked with ES, the whole contributing to sustainable development. Just as much of the world is not yet environmentally sustainable, neither is it socially sustainable. Disaggregation of social unsustainability will show what needs to be changed.

SUSTAINABILITY AND DEVELOPMENT

Sustainable development (SD) should integrate social, environmental, and economic sustainability and use these three to start to make development sustainable. The moment the term development is introduced, however, the discussion becomes quite different and more ambiguous. This paper is not focused on sustainable development, here assumed to be development that is socially, economically, and environmentally sustainable, or "development without throughput growth beyond environmental carrying capacity and which is socially sustainable" (27, 28, 30). World Wildlife Fund's (107) definition of sustainable development is similar: "Improvement in the quality of human life within the carrying capacity of supporting ecosystems." These definitions need to have the social aspects clarified, but they are less ambiguous than the Brundtland (105) definition: "development that meets the needs of the present without compromising the ability of future generations to meet their own needs."[2]

Part of the success of the Brundtland Commission's definition stems from its opacity (49), and the definition of sustainability in a growth context. But when the World Commission on Environment and Development (WCED) (106) reconvened five years later, calls for growth were striking for their absence. HRH Prince Charles commended WCED in their publication (106)

[2]This UN definition does not distinguish among the different concepts of growth and development. While development can and should go on indefinitely for all nations, throughput growth cannot. Sustainability will be achieved only when development supplants growth; when the scale of the human economy is kept within the capacity of the overall ecosystem on which it depends. If we acknowledge the finite nature of our planet, "sustainable growth" is an oxymoron (28, 29). Throughput growth has to be kept within carrying capacity or within the capacity of the environmental services of assimilation and regeneration.

for dropping their 1987 call for huge (5- to 10-fold) increases in economic growth.

This paper offers the case that ES does not allow economic growth, much less sustained economic growth. On the contrary, environmentally sustainable development implies sustainable levels of both production (sources), and consumption (sinks), rather than sustained economic growth. The priority for development should be improvement in human well-being—the reduction of poverty, illiteracy, hunger, disease, and inequity. While these development goals are fundamentally important, they are quite different from the goals of environmental sustainability, the unimpaired maintenance of human life-support systems—environmental sink and source capacities.

The need for sustainability arose from the recognition that the profligate, extravagant, and inequitable nature of current patterns of development, when projected into the not-too-distant future, leads to biophysical impossibilities. The transition to environmental sustainability is urgent because the deterioration of global life-support systems—which compose the environment—imposes a time limit. We do not have time to dream of creating more living space or more environment, such as colonizing the moon or building cities beneath the ocean. We must save the remnants of the only environment we have and allow time for and invest in the regeneration of what we have already damaged. We cannot "grow" into sustainability.

The tacit goal of economic development is to narrow the equity gap between the rich and the poor. Almost always this is taken to mean raising the bottom (i.e. enriching the poor), rather than lowering the top by redistribution (44). Only very recently has it been admitted that bringing the low-income countries up to the affluent levels of the countries of the Organization for Economic Cooperation and Development (OECD) in 40 or even 100 years is a totally unrealistic goal. Most politicians and most citizens have not yet realized that this goal is unrealistic. Most people would accept that it is desirable for southern low-income countries to be as rich as those in the northern hemisphere—and then leap to the false conclusion that it must therefore be possible. But if greater equality cannot be attained by growth alone, then sharing and population stability will be necessary.

Serageldin (78) makes the persuasive case that in low income countries achieving per capita income levels of $1,500 to $2,000 (rather than OECD's $21,000 average) is quite possible. Moreover, that level of income may provide 80% of the basic welfare provided by a $20,000 income—as measured by life expectancy, nutrition, education, and other aspects of social welfare. This tremendously encouraging case remains largely unknown, even in development circles. Its acceptance would greatly facilitate the transition to environmental sustainability. Colleagues working on the northern hemisphere's overconsumption should address the corollary not dealt with by Serageldin (78):

Can $21,000/capita countries cut their consumption by a factor of 10 and suffer "only" a 20% loss of basic welfare? If indeed both raising the bottom (low income rises to $2000) and lowering the top (OECD income declines to $16,000) prove feasible, that would be tremendously encouraging and would speed ES. But to accomplish the possible parts of the imperative of development, we must stop idolizing the impossible. The challenge to development specialists is to deepen this important argument.

Intergenerational and Intragenerational Sustainability

Most people in the world today are either impoverished or live barely above subsistence; the number of people living in poverty is increasing. Developing countries can never be as well off as today's OECD average. Future generations seem likely to be larger and poorer than today's generation. Sustainability includes an element of not harming the future (intergenerational equity), and some find the intergenerational equity component of sustainability to be its most important element (e.g. 105). If the world cannot move toward intragenerational sustainability during this generation, it will be that much more difficult to achieve intergenerational sustainability sometime in the future, for the capacity of environmental services will be lower in the future than it is today.

World population soars by 100 million people each year: Some of these people are OECD overconsumers, but most of them are poverty stricken. World population doubles in a single human generation—about 40 years. This makes achieving intergenerational equity difficult, although achieving intergenerational equity will probably reduce total population growth. Rather than focusing on the intergenerational equity concerns of ES, the stewardship approach of safeguarding life-support systems today seems preferable.

WHAT SHOULD BE SUSTAINED?

Environmental sustainability seeks to sustain global life-support systems indefinitely (this refers principally to those systems maintaining human life). Source capacities of the global ecosystem provide raw material inputs—food, water, air, energy; sink capacities assimilate outputs or wastes. These source and sink capacities are large but finite; sustainability requires that they be maintained rather than run down. Overuse of a capacity impairs its provision of life-support services. For example, accumulation of CFCs is damaging the capacity of the atmosphere to protect humans and other biota from harmful UVb radiation.

Protecting human life is the main reason anthropocentric humans seek environmental sustainability. Human life depends on other species for food, shelter, breathable air, plant pollination, waste assimilation, and other environmental life-support services. The huge instrumental value of nonhuman species to humans is grossly undervalued by economics. Nonhuman species of no present value to humans have intrinsic worth, but this consideration is almost entirely excluded in economics (exceptions are existence and option values). A question rarely posed by economists and not yet answered by any is: With how many other species is humanity willing to share the earth, or should all other species be sacrificed to make room for more and more of the single human species? Surely it is arrogant folly to extinguish a species just because we think it is useless today. The anthropocentric and ecocentric views are contrasted by Goodland & Daly (42).

Although biodiversity conservation is becoming a general ideal for nations and development agencies, there is no agreement on how much should be conserved, nor at what cost. Leaving aside the important fact that we have not yet learned to distinguish useful from nonuseful species, agreeing on how many other species to conserve is not central to the definition of environmental sustainability. Reserving habitat for other species to divide among themselves is important; let evolution select the mix of species, not us. But reserving a nonhuman habitat requires limiting the scale of the human habitat. "How much habitat should be conserved?", while an important question to ask, is moot; the answer is probably "no less than today's remnants." This brings us to the precautionary principle: In cases of uncertainty, sustainability mandates that we err on the side of prudence. Because survival of practically all the global life-support systems is uncertain, we should be very conservative in our estimate of various input and output capacities, and particularly of the role of unstudied, apparently "useless," species.

Many writers (15, 16, 19, 32, 40, 46, 52, 54, 60, 84) are convinced that the world is hurtling away from environmental sustainability, but economists have not reached consensus that the world is becoming less sustainable. What is not contestable is that the modes of production prevailing in most parts of the global economy are causing the exhaustion and dispersion of a one-time inheritance of natural capital—topsoil, groundwater, tropical forests, fisheries, and biodiversity. The rapid depletion of these essential resources, coupled with the degradation of land and atmospheric quality, shows that the human economy as currently configured is already inflicting serious damage on global supporting ecosystems, and future potential biophysical carrying capacities are probably being reduced (21, 22).

A HISTORY OF SUSTAINABILITY

A notion of economic sustainability was firmly embodied in the writings of JS Mill (61), and TR Malthus (57, 58). Mill (61) emphasized that environment ("Nature") needs to be protected from unfettered growth if we are to preserve human welfare before diminishing returns set in. Malthus emphasized the pressures of exponential population growth on the finite resource base. The modern neo-Malthusianism version is exemplified by Ehrlich & Ehrlich (33a–35) and by Hardin (45, 46). Daly's "Toward a Steady State Economy" (25, 26) and "Steady State Economics" (23) synthesized and extended these viewpoints on population and resources. Daly's "Steady State Economics" is a seminal work in which population and consumption pressures on environmental sources and sinks are clearly demonstrated—the flow of matter and energy from the environment, used by the human economy, and released back into the environment as wastes. Daly magisterially subsumes the issues of population and consumption factors into the single critical factor of *scale*.

Neither Mill nor Malthus is held in great esteem by most of today's economists, who are more likely to follow the technological optimism of David Ricardo (77). Ricardo believed that human ingenuity and scientific progress would postpone the time when population would overtake resources or "the niggardliness of nature." However, as poverty is increasing worldwide, that postponement seems to have ended.

The definition of environmental sustainability (Figure 2) hinges on distinguishing between throughput growth and development. The "Growth Debate" moved into the mainstream two decades after World War II. Boulding (12–14), Mishan (62, 63), and Daly (23–26), for example, seriously questioned the wisdom of infinite throughput growth on a finite earth. Throughput growth is defended by most economists, including Beckerman (7–9), who still rejects the concept of sustainability (10). "The Limits to Growth" (59) and "Beyond the Limits" (60) shook the convictions of the technological optimists. Meadows et al (59) concluded that "it is possible to alter these growth trends and establish a condition of ecological and economic stability that is sustainable into the future." Barney's (5) US Global 2000 Report (1980) amplified and clarified the limits argument. Large populations, their rapid growth and affluence, are unsustainable. The Ricardian tradition that still dominates conventional economics is exemplified by the Cornucopians Simon & Kahn in their 1984 response to the Global 2000 Report. Panayotou (70), Summers (89), and Fritsch et al (39) found growth compatible with sustainability and even necessary for it. The 1980 World Conservation Strategy (51) by the International Union to Conserve Nature and the World Wildlife Fund, and Clark & Munn's 1987 International Institute for Applied Systems Analysis report "Sustainable Development of the Biosphere" (17), reinforced these conclusions. Daly & Cobb's (32) prizewinning "For the Common Good" estimated that growth, at least in the United States, actually decreased people's well-being, and they outlined pragmatic operational methods to reverse environmental damage and reduce poverty. The growth debate and sustainability issues are usefully synthesized by Korten (54).

Few Nobel prizewinners in economics write on sustainability. Haavelmo &

Hansen (44) and Tinbergen & Hueting (90) repudiate throughput growth and urge the transition to sustainability. Solow's earlier writings (85) questioned the need for sustainability, but he is now modifying that position (87, 88). The World Bank adopted environmental sustainability in principle rather early on, in 1984, and now promotes it actively (2, 56, 65, 80–82). Major contributions to the sustainability debate were published as contributions to the 1992 UN Commission on Environment and Development conference in Rio de Janeiro, such as WCS's 1991 "Caring for the Earth: a strategy for sustainable living." An addendum to the Brundtland Commission (106) rectified and reversed the earlier (105) calls for "5- to 10-fold more growth," by placing the population issue higher on the agenda to achieve sustainability. Goodland, Daly, and El Serafy (43), supported by two Economics Nobelists (Tinbergen and Haavelmo), made the case that there are indeed limits, that the human economy has reached them in many places, that it is impossible to grow into sustainability, that source and sink capacities of the environment complement human-made capital (which cannot substitute for their environmental services), and that there is no way the southern hemisphere can ever catch up with the north's current consumerist life-style.

Since the late 1980s, a substantial corpus of literature on "Ecological Economics" [which now has a journal, society and textbook of the same name (19)], has espoused stronger types of sustainability (e.g. 3, 4, 18, 48, 50, 52, 64, 91, 94).

GROWTH COMPARED WITH DEVELOPMENT

The dictionary distinguishes between growth and development. "To grow" means "to increase in size by the assimilation or accretion of materials;" "to develop" means "to expand or realize the potentialities of; to bring to a fuller, greater or better state."

Growth implies quantitative physical or material increase; development implies qualitative improvement or at least change. Quantitative growth and qualitative improvement follow different laws. Our planet develops over time without growing. Our economy, a subsystem of the finite and nongrowing earth, must eventually adapt to a similar pattern of development without throughput growth. The time for such adaptation is now. Historically, an economy starts with quantitative throughput growth as infrastructure and industries are built, and eventually it matures into a pattern with less throughput growth but more qualitative development. While this pattern of evolution is encouraging, qualitative development needs to be distinguished from quantitative throughput growth if environmental sustainability is to be approached.

Development by the countries of the northern hemisphere must be used to free resources (the source and sink functions of the environment) for the growth and development so urgently needed by the poorer nations. Large-scale transfers to the poorer countries also will be required, especially as the impact of economic stability in northern countries may depress terms of trade and lower economic activity in developing countries. Higher prices for the exports of poorer countries, as well as debt relief, will be required. Most importantly, population stability is essential to reduce the need for growth everywhere. This includes both where population growth has the greatest impact (i.e. in the northern high-consuming nations) and where population growth is highest (i.e. in the southern, poor, low-consuming countries).

THE DEFINITION OF ENVIRONMENTAL SUSTAINABILITY

The definition of ES as the "maintenance of natural capital" constitutes the input/output rules in Figure 2.

The two fundamental environmental services—the source and sink functions—must be maintained unimpaired during the period over which sustainability is required (23, 27, 31). ES is a set of constraints on the four major activities regulating the scale of the human economic subsystem: the use of renewable and nonrenewable resources on the source side, and pollution and waste assimilation on the sink side. This short definition of ES is the most

---

1. *Output Rule:*
   Waste emissions from a project or action being considered should be kept within the assimilative capacity of the local environment without unacceptable degradation of its future waste absorptive capacity or other important services.
2. *Input Rule:*
   (a) *Renewables:* harvest rates of renewable resource inputs should be within regenerative capacities of the natural system that generates them.
   (b) *Nonrenewables:* depletion rates of nonrenewable resource inputs should be set below the rate at which renewable substitutes are developed by human invention and investment according to the Serafian quasi-sustainability rule (36–38). An easily calculable portion of the proceeds from liquidating nonrenewables should be allocated to research in pursuit of sustainable substitutes.
3. *Operational Principles:*
   (a) The scale (population × consumption per capita × technology) of the human economic subsystem should be limited to a level which, if not optimal, is at least within the carrying capacity and therefore sustainable.
   (b) Technological progress for sustainable development should be efficiency-increasing rather than throughput-increasing.
   (c) Renewable resources should be exploited on a profit-optimizing, sustained-yield, and fully sustainable basis.

Figure 2.   The definition of environmental sustainability

useful so far and is gaining adherents. The fundamental point to note about this definition is that ES is a natural science concept and obeys biophysical laws (Figure 2). This general definition seems to be robust irrespective of country, sector, or future epoch.

The paths needed by each nation to approach sustainability will not be the same. Although all countries need to follow the input/output rules, countries differ in the balance of attention between output and input that will be needed to achieve ES. For example, some countries or regions must concentrate more on controlling pollution (e.g. former centrally planned economies); some countries must pay more attention to bringing harvest rates of their renewable resources down to regeneration rates (e.g. tropical timber–exporting countries); some countries must bring their population to below carrying capacity; others must reduce their per capita consumption (e.g. all OECD countries).

There are compelling reasons why industrial countries should lead in devis-

Laws:

1. Neither growth in human population nor growth in the rates of resource consumption can be sustained.
2. The larger the population of a society and the larger its rates of consumption of resources, the more difficult it will be to transform the society to the condition of sustainability.
3. The response time of populations to changes in the total fertility rate is the length of time people live from their childbearing years to the end of life, or approximately 50 years.
4. The size of population that can be sustained (the carrying capacity) and the sustainable average standard of living are inversely related to one another.
5. Sustainability requires that the size of the population be less than or equal to the carrying capacity of the ecosystem for the desired standard of living.
6. The benefits of population growth and of growth in the rate of consumption of resources accrue to a few individuals; the costs are borne by all of society (the tragedy of the commons).
7. (Any) growth in the rate of consumption of a nonrenewable resource, such as a fossil fuel, causes a dramatic decrease in the life expectancy of the resource.
8. The time of expiration of nonrenewable resources, such as a fossil fuel, causes a dramatic decrease in the life expectancy of the resource.
9. When large efforts are made to improve the efficiency with which resources are used, the resulting savings are easily wiped out by the added resource needs that arise as a consequence of modest increases in population.
10. When rates of pollution exceed the natural cleansing capacity of the environment, it is easier to pollute that it is to clean up the environment.
11. Humans will always be dependent on agriculture so land and other renewable resources will always be essential.

Figure 3a.   Bartlett's Laws relating to sustainability and hypotheses about sustainability

Hypotheses:

1. For the 1994 average global standard of living, the 1994 population of the earth exceeds carrying capacity.
2. Increasing population size is the single greatest and most insidious threat to representative democracy.
3. The costs of programs to stop population growth are small compared to the costs of population growth.
4. The time required for a society to make a planned transition to sustainability increases with increases in the size of its population and the average per capita consumption of resources.
5. Social stability is a necessary, but not a sufficient, condition for sustainability. Social stability tends to be inversely related to population density.
6. The burden of the lowered standard of living that results from population growth and from the decline of resources falls most heavily upon the poor.
7. Environmental problems cannot be solved or ameliorated by increases in the rates of consumption of resources.
8. The environment cannot be enhanced or preserved through compromises.
9. By the time overpopulation and shortage of resources are obvious to most people, the carrying capacity has been exceeded. It is then too late to think about sustainability.

Figure 3b.   Bartlett's Sustainability Hypothesis (6)

ing paths toward sustainability. They have to adapt far more than do developing countries. If OECD countries cannot act first and lead the way, it is less likely that developing countries will choose to do so (95). Not only would it be enlightened self-interest for the north to act first, but it could also be viewed as a moral obligation. Second, developing countries are rightly pointing out (1, 11) that OECD countries have already consumed substantial amounts of environmental sink capacity (e.g. nearly all CFCs that are damaging the atmosphere were released by OECD countries) as well as source capacity (e.g. several species of great whales are extinct, and many stocks of fish and tropical timbers have been depleted below economically harvestable levels). Third, OECD countries can afford the transition to sustainability because they are richer. The rich would do themselves good by using the leeway they have for cutting overconsumption and waste (Figure 3a, 3b).

CAUSES OF UNSUSTAINABILITY

When the human economic subsystem was small, the regenerative and assimilative capacities of the environment appeared infinite. We are now painfully learning that environmental sources and sinks are finite. These capacities were very large, but the scale of the human economy has exceeded them. Source and sink capacities have now become limited. As economics deals only with scarcities, in the past source and sink capacities of the environment did not have to be taken into account. Conventional economists still hope or claim that economic growth is infinite or at least that we are not yet reaching limits to growth; hence the fierce recent repudiation of *Beyond the Limits of Growth* (60), and the welcome for Brundtland's call for "5- to 10-fold more growth" (105). The scale of the human economy is a function of throughput—the flow of materials and energy from the sources of the environment, used by the human economy, and then returned to environmental sinks as waste. Throughput growth is a function of population growth and consumption. Throughput growth translates into increased rates of resource extraction and pollution (use of sources and sinks). The scale of throughput has exceeded environmental capacities: That is the definition of unsustainability.

There is little admission yet that consumption above sufficiency is not an unmitigated good. The scale of the human economy has become unsustainable because it is living off inherited and finite capital (e.g. fossil fuels, fossil water); because we do not account for losses of natural capital (e.g. extinctions of species), nor do we admit the costs of environmental harm. The second reason for unsustainability is related to the first: government failure to admit that pollution and fast population growth are doing more harm than good.

THE TIME FOR ENVIRONMENTAL SUSTAINABILITY

Approaching sustainability is urgent. Consider that if release were halted today of all substances that damage the ozone shield, the ozone shield may need as much as one century to return to pre-CFC effectiveness. Every passing year means sustainability has to be achieved for an additional 100 million people. Though environmental sources and sinks have been providing humanity with their services for the last million years, and until recently have seemed vast and resilient, we have at last begun to exceed them and to damage them worldwide. Where environmental services are substitutable, the substitution achieved has been marginal. Most natural capital or environmental services cannot be substituted for, and their self-regenerating properties are slow and cannot be significantly hastened. That is why environmental sustainability has a time urgency.

Much of the resistance to accepting the necessity of a sustainability approach is that politicians have considered the consequences of doing so—controlling consumerism and waste, halting human population growth, and probably reducing population size, and relying on renewable energy—to be politically unacceptable. These are all felt to be politically damaging, so they are not put forward as much-needed societal goals. Instead, society calls for incremental progress in such disparate areas as enforcing the 'polluter pays' principle, support of women's reproductive health, educating girls, or clean technology. Important though these goals are, they are not enough, yet no one calls for the redistribution of resources from rich to poor. It is impossible for us to grow out of poverty and environmental degradation. It is precisely the nonsustainability of throughput growth beyond a certain scale that gives urgency to the concept of sustainability (27, 28). All forms of growth are unsustainable, whether in the number of trees, people, great whales, levels of atmospheric CO_2, or GNP.

The world will in the end become sustainable, one way or another. We can select the timing and nature of that transition and the levels of sustainability to be sought, or we can let depletion and pollution dictate the abruptness of the final inevitable transition. The former will be painful; the latter deadly. The longer we delay agreement on goals for levels of sustainability, the more the source and sink capacities will be damaged, the larger the number of people that will have to be accommodated on earth, and the more difficult the transition will be. For example, species extinctions are happening fast now, and they are accelerating. If that process continues for several decades, we will inevitably reach sustainability at a much poorer and less resilient level.

NATURAL CAPITAL AND SUSTAINABILITY

Of the four kinds of capital (natural, human, human-made, and social), environmental sustainability requires maintaining natural capital; understanding ES thus includes defining "natural capital" and "maintenance of resources" (or at least "non-declining levels of resources"). Natural capital—the natural environment—is defined as the stock of environmentally provided assets (such as soil, atmosphere, forests, water, wetlands), which provide a flow of useful goods or services; these can be renewable or nonrenewable, and marketed or nonmarketed. Sustainability means maintaining environmental assets, or at least not depleting them. "Income" is sustainable by the generally accepted Hicksian definition of income (47). Any consumption that is based on the depletion of natural capital is not income and should not be counted as such. Prevailing models of economic analysis tend to treat consumption of natural capital simply as income and therefore tend to promote patterns of economic activity that are unsustainable. Consumption of natural capital is liquidation, or disinvestment—the opposite of capital accumulation.

Now that the environment is so heavily used, the limiting factor for much economic development has become natural capital. For example, in marine fishing, fish have become limiting, rather than fishing boats. Timber is limited by remaining forests, not by sawmills; petroleum is limited by geological deposits and atmospheric capacity to absorb CO_2, not by refining capacity. As natural forests and fish populations become limiting, we begin to invest in plantation forests and fish ponds. This introduces an important hybrid category that combines natural and human-made capital—a category we may call "cultivated natural capital." This category is vital to human well-being, accounting for most of the food we eat, and a good deal of the wood and fibers we use. The fact that humanity has the capacity to "cultivate" natural capital dramatically expands the capacity of natural capital to deliver services. But cultivated natural capital (agriculture) is separable into human-made capital (e.g. tractors, diesel irrigation pumps, chemical fertilizers, biocides) and natural capital (e.g. topsoil, sunlight, rain). Eventually the natural capital proves limiting.

Natural Capital Is Now Scarce

In an era in which natural capital was considered infinite relative to the scale of human use, it may have been reasonable not to deduct natural capital consumption from gross receipts in calculating income. That era is now past. Environmental sustainability needs the conservative effort to maintain the traditional (Hicksian) meaning and measure of income now that natural capital is no longer a free good but is more and more the limiting factor in development. The difficulties in applying the concept arise mainly from operational problems of measurement and valuation of natural capital, as emphasized by Ahmad et al (2), Lutz (56), and El Serafy (37, 38).

Three Degrees of Environmental Sustainability

Sustainability can be divided into three degrees—weak, strong and absurdly strong—depending on how much substitution one thinks there is among the four types of capital (natural, human, human-made, and social) (32):

> *Weak environmental sustainability*: Weak ES is maintaining total capital intact without regard to the partitioning of that capital among the four kinds. This would imply that the various kinds of capital are more or less substitutes, at least within the boundaries of current levels of economic activity and resource endowment. Given current liquidation and gross inefficiencies in resource use, weak sustainability would be a vast improvement as a welcome first step but would by no means constitute ES. Weak sustainability is a necessary but not sufficient condition for ES. Weak sustainability is rejected by Beckerman (10), but the concept is finding some acceptance in economic circles. It means we could convert all or most of the world's natural capital into human-made capital or artifacts and still be as well off! For example, society would be better off, it is claimed by those espousing weak sustainability, by converting forests to houses, and oceanic fish stocks into nourished humans. Human capital—educated, skilled, experienced, and healthy people—is largely lost at the death of individuals, and so it must be renewed each generation, whereas social capital persists in the form of books, knowledge, art, family and community relations.
>
> *Strong environmental sustainability*: Strong ES requires maintaining separate kinds of capital. Thus, for natural capital, receipts from depleting oil should be invested in ensuring that energy will be available to future generations at least as plentifully as that enjoyed by the beneficiaries of today's oil consumption. This assumes that natural and human-made capital are not perfect substitutes. On the contrary, they are complements at least to some extent in most production func-

tions. A sawmill (human-made capital) is worthless without the complementary natural capital of a forest. The same logic would argue that if there are to be reductions in one kind of educational investments, they should be offset by increased investments in other kinds of education, not by investments in roads. Of the three degrees of sustainability, strong sustainability seems greatly preferable mainly because of the lack of substitutes for much natural capital, the fact that natural capital and not human-made capital is now limiting, and the need for prudence in the face of many irreversibilities and uncertainties. Pearce et al (71–73), Costanza (19), Costanza & Daly (20), Opschoor, Van der Straaten, van den Bergh, most ecologists, and most ecological economists prefer or are coming round to some version of strong sustainability.

Absurdly strong environmental sustainability: We would never deplete anything. Nonrenewable resources—absurdly—could not be used at all. All minerals would remain in the ground. For renewables, only net annual growth increments could be harvested in the form of the overmature portion of the stock. Some ecologists fear we may be reduced to this type of sustainability—harvesting only overmature growth increments of renewables, in which case this sustainability is better called "superstrong" sustainability (67).

There are tradeoffs between human-made capital and natural capital. Economic logic requires us to invest in the limiting factor, which now is often natural rather than human-made capital, which was previously limiting. Operationally, this translates into three concrete actions as noted in Figure 4.

SUSTAINABILITY AND SUBSTITUTABILITY

Conventional economics and technological optimists depend heavily on substitutability as the rule rather than the exception. Ecology has paid inadequate

---

1) FOSTER REGENERATION OF NATURAL CAPITAL:
   Encourage the growth of natural capital by reducing our current level of exploitation of it. For example, lengthen rotations (of forest cutting or arable crops) to permit full regeneration; limit catches (e.g. of fish) to prudently well within long-term sustained yield estimates.
2) RELIEVE PRESSURE ON NATURAL CAPITAL:
   Invest in projects to relieve pressure on natural capital stocks by expanding cultivated natural capital, such as tree plantations to relieve pressure on natural forests. Reducing pollution and waste provides more time for assimilative capacities to regenerate themselves.
3) IMPROVE EFFICIENCY IN USE OF NATURAL CAPITAL:
   Increase the end-use efficiency of products (such as improved cookstoves, solar heaters and cookers, wind pumps, solar pumps, manure rather than chemical fertilizer). Extend the life-cycle, durability, and recyclability of products to improve overall efficiency, as would taxing planned obsolescence and ephemerata.

Figure 4.   Rebuilding natural capital stocks

attention to the extent of substitutability between natural and human-made capital, yet it is central to the issue of sustainability. Substitutability is the ability to offset a diminished capacity of environmental source and sink services to provide healthy air, water, etc, and to absorb wastes. The importance of substitutability is that if it prevails, then there can be no limits, because if an environmental good is destroyed, it is argued, a substitute can replace it. When White Pine or sperm whales became scarce, there were acceptable substitutes. When easily gathered surficial oil flows were exhausted, drilling technology enabled very deep deposits to be tapped. In Europe, when the native forest was consumed, timber for houses was replaced with brick. If bricks did not substitute for timber, then timber was imported.

The realization that substitutability is the exception, rather than the rule, is not yet widespread, despite Ehrlich's warning (33, 34). However, once limits of imports cease to mask substitutability (e.g. US Pacific Northwest and British Columbia timber controversies show the limits of imports), then it becomes plain that most (but not all) forms of capital are more complementary or neutral and are less substitutable. Economists who hope that natural capital and human-made capital are substitutes claim that total capital (i. e. the sum of natural and human-made capital) can be maintained constant in some aggregate value sense. This reasoning, built on the questionable premise that human-made capital is substitutable for natural capital, means it is acceptable to divest natural capital (i.e. deplete environmental source or sink capacities) as long as an equivalent value has been invested in human-made capital. Even this weak sustainability is not required by national accounting rules. Indeed, our national accounts simply count natural capital liquidation as income (37, 38).

Unfortunately, that is also the way the world is being run at the micro or firm level; user costs are rarely calculated (53). We consume environmental source capacity by releasing many wastes (e.g. CO_2, CFCs, oxides of sulfur and nitrogen that make acid rain) into the air because we claim the investment in energy production and refrigeration (human-made capital) substitutes for healthy air or atmosphere. We extinguish species (depletion of biodiversity source capacity) by converting jungle to cattle ranches because the human-made capital (or strictly quasi-human made agriculture) is a substitute for the natural capital of biodiversity. Such "weak sustainability" has not yet been achieved, and it would be a great improvement were it attained. But, because human-made and natural capital are far from perfect substitutes, weak sustainability is a dangerous goal. It would be risky as an interim stage on the way to any reliable concept of sustainability.

Ecologists attach great importance to Baron Justus von Liebig's Law of the Minimum—the whole chain is only as strong as its weakest link. The factor in shortest supply is the limiting factor because factors are complements, not substitutes. If scarcity of phosphate is limiting the rate of photosynthesis, then photosynthesis would not be enhanced by increasing another factor such as nitrogen, light, water, or CO_2. If one wants faster photosynthesis, one must ascertain which factor is limiting and then invest in that one first, until it is no longer limiting. More nitrogen fertilizer cannot substitute for lack of phosphate, precisely because they are complements. Environmental sustainability is based on the conclusion that most natural capital is a complement for human-made capital, and not a substitute. Complementarity is profoundly unsettling for conventional economics because it means there are limits to growth, or limits to environmental source and sink capacities. Human-made capital is a very poor substitute for most environmental services. Substitution for some life-support systems is impossible.

A compelling argument that human-made capital is only a marginal substitute for natural capital is the reductio ad absurdum case in which all natural capital is liquidated into human-made capital. We might survive the loss of fossil fuels, but what would substitute for topsoil and breathable air? Only in science fiction could humanity survive by breathing bottled air from backpacks, and eating only hydroponic greenhouse food. If there is insufficient substitutability between natural capital and human-made capital, then throughput growth must be severely constrained and eventually cease. While new technology may postpone the transition from quantitative growth to qualitative development and environmental sustainability, current degradation shows that technology is inadequate. *"For natural life-support systems no practical substitutes are possible, and degradation may be irreversible. In such cases (and perhaps in others as well), compensation cannot be meaningfully specified."* (92, 93).

COMMON MISCONCEPTIONS ABOUT ENVIRONMENTAL SUSTAINABILITY

Is ES the Same as Sustained Yield?

There is a lively debate, especially in forestry and fishery circles, about whether environmental sustainability is "sustained yield" (S-Y). Clearly ES includes, but is far from limited to, sustained yield. ES is applied at the aggregate level to all the values of an ecosystem, not to just a few species of timber trees or fish. ES is akin to the simultaneous S-Y of many interrelated populations in an ecosystem. S-Y is ES restricted to a small fraction of the members of the ecosystem under consideration. S-Y is often used in forestry and fisheries to determine the optimal—most profitable—extraction rate of timber or fish. ES counts all the natural services of the sustained resource. S-Y counts only the service of the product extracted and ignores all other natural services. S-Y forestry counts only the timber value extracted; ES forestry counts all services, including protecting vulnerable ethnic-minority forest dwellers, biodiversity, genetic values, intrinsic as well as instrumental values, climate, wildlife, carbon balance, water source and water moderation values, ecosystem integrity in general (96), and of course, timber extracted. The relation between the two is that if S-Y is actually achieved, then the stock resource (e.g. the forest) will be nearer sustainability than if S-Y is not achieved. S-Y in tropical forestry is doubtful now (55) and will be more doubtful in the future as human population pressures intensify. But even were S-Y to be achieved, that resource is unlikely to have also attained ES. The optimal solution for a single variable, such as S-Y, usually (possibly inevitably) results in declining utility or declining natural capital sometime in the future, and therefore it is not sustainable.

Is ES Certain or Uncertain?

Environmental sustainability is a rather clear concept. However, there is much uncertainty about the details of its application. After scientists have spent centuries trying to estimate sustained yield for a few species of timber trees or fish, they are now questioning whether they can ever be successful (55), leaving aside whether humans would accept sustainable yield extraction rates once they were determined. Considering that ES is more complex than S-Y suggests a high degree of uncertainty. Today we are largely empirical in our assessment of assimilative capacity also. We allow a limit to be exceeded, often for years, before we muster the political will to start addressing the problem. Damage to the ozone shield was argued for years before CFC manufacturers agreed to phase CFCs out, and then they did so only when economic substitutes had been found. Even so, scientific understanding of biophysical linkages is weak, so there is much uncertainty, and hence a compelling need for the precautionary principle to prevail widely. Colleagues addressing ES should seek rough rather than precise indicators of sustainability so that we can move on. Better to be roughly right than precisely wrong.

Is ES More of a Concern for Developing Countries?

The countries of the northern hemisphere are responsible for the overwhelming share of global environmental damage today, and it is unlikely that poor countries will want to move toward sustainability if the north doesn't do so first. The northern hemisphere more than the southern can decrease global warming risks by reducing greenhouse gas emissions, for example. The north has to adapt to ES more than the south, and arguably before the south. The main exception is biodiversity, most of which is contained in tropical ecosystems. The north can afford to exert leadership on itself. But because developing economies depend to a much greater extent than do OECD economies on natural resources, especially renewables, the south has much to gain from reaching ES. In addition, because much tropical environmental damage is irreversible, it is either impossible or more expensive to rehabilitate tropical than temperate environments, so the south will gain more from a preventive approach than from emulating the short-sighted and expensive curative approach and similar mistakes of the north.

Does ES Imply Reversion to Autarky or the Stone Age?

As soon as society perceives that environmental sustainability means conservation of life-support systems, people will demand it on the grounds of welfare, equity, or economics. The poor suffer most from pollution and from the higher prices caused by depletion. The poor are least able to protect themselves against scarcities (e.g. of clean water, clean air) and pollution. Among the rich, only a few are acting on the message that affluence and overconsumption do not increase welfare. Much more education is needed for overconsumers to realize that rides in limousines are often slower as well as more polluting than those on the metro, and that eating three steaks a day reduces fitness. As the costs of overconsumption increase (sickness and decreased productivity, health costs, heart attack, stroke), this message will spread. Reducing waste means needing fewer land fills and trash incinerators, which would improve human welfare. The concept of sufficiency (doing more or enough with less) needs dissemination.

ES Involves Public Choice

ES as biophysical security is connected to welfare, and both are somewhat connected to economics, especially to efficiency of use. Public choice governs the rate at which society elects to approach ES voluntarily and purposefully, or, as at present, to recede from it. Society has the choice of an orderly transition to environmental sustainability on our terms, or of letting biophysical damage dictate the timing and speed of the transition. If society allows biophysical deterioration to make the transition to ES for us, the transition is likely to be unacceptably harsh for humans. That is why clarity and education are so important in the race to approach ES. Partly because recognition of the need for ES is so recent, political will and institutional capacity now have to catch up. There will be powerful losers when society decides to move toward ES and toward making polluters pay. Institutional strengthening therefore is a necessary condition for ES.

CONCLUSION

This paper reviews the current status of the debate about the concept of environmental sustainability and discusses related aspects of growth, limits, scale, and substitutability. While the paths leading to ES in each country or sector will differ, the goal remains constant. But this conceptualization is far from an academic exercise. The monumental challenge of ensuring, within less than two human generations, that as many as ten billion people are decently fed and housed without damaging the environment on which we all depend, means that the goal of environmental sustainability must be reached as soon as humanly possible.

ACKNOWLEDGMENTS

In addition to those from World Bank colleagues, comments on earlier drafts from Bill Adams, Jeroen van den Bergh, Don Ludwig, Raymond Mikesell, Johannes Opschoor, Michael Redclift, Jan van der Straaten, Salah El Serafy, Fulai Sheng, and Tom Tietenberg are warmly acknowledged. This paper honors Herman Daly, the leading sustainability theoretician, on whose work this paper is based. His enormous and generous contributions are gratefully acknowledged.

Literature Cited

1. Agarwal A, Narain S. 1991. Global warming in an unequal world. *Int. J. Sustainable Dev.* 1(1):98–104
2. Ahmad Y, El Serafy S, Lutz E, eds. 1989. Environmental accounting. Washington DC: World Bank. 100 pp.
3. Archibugi F, Nijkamp P, eds. 1989. *Economy and Ecology: Towards Sustainable Development.* Dordrecht: Kluwer. 348 pp.
4. Barbier E, ed. 1993. *Economics and Ecology: New Frontiers and Sustainable Development.* London: Chapman, Hall. 205 pp.
5. Barney GO, ed. 1980. *The Global 2000 Report to the President of the USA: Entering the 21st Century.* Harmondsworth: Penguin. 2 Vols.
6. Bartlett AA. 1994. *Reflections on Sustainability, Population Growth, and the Environment. Pop. Environ.* 16(1):5–35
7. Beckerman W. 1974. *In Defence of Economic Growth.* London: J. Cape. 287 pp.
8. Beckerman W. 1992. Economic growth: Whose growth? Whose environment? *World Dev.* 20:481–92
9. Beckerman W. 1992. *Economic Development: Conflict or Complementarity?* Washington DC: World Bank WPS 961. 42 pp.
10. Beckerman W. 1994. *"Sustainable development": Is it a useful concept? Environ. Values* 3:191–209
11. Beijing Declaration. 1991. Beijing ministerial declaration on environment and development. *Beijing, Peoples Republic of China, Ministerial Conference of (41 Ministers of) Developing Countries on Environment and Development,* 18–19 June 1991. 9 pp.
12. Boulding KE. 1966. The economics of the coming spaceship earth. In *Environmental Quality in a Growing Economy,* ed. H Jarret. Baltimore: Johns Hopkins Univ. Press
13. Boulding KE. 1968. *Beyond Economics.* Ann Arbor: Univ. Mich. 302 pp.
14. Boulding KE. 1992. *Towards a New Economics: Ecology and Distribution.* Aldershot, Hants.: Elgar. 344 pp.
15. Brown LB, et al. 1995. *State of the World: 1995.* Washington, DC: Worldwatch Inst. 255 pp.
16. Brown LB, Kane H. 1994. *Full House: Reassessing the Earth's Population Carrying Capacity.* New York: Norton. 261 pp.
17. Clark WC, Munn RE, eds. 1987. *Sustainable Development of the Biosphere.* Cambridge: Cambridge Univ. Press 491 pp.

18. Collard D, Pearce DW, Ulph D. eds. 1987. *Economics, Growth and Sustainable Environments.* New York: St. Martin's Press 205 pp.
19. Costanza R. ed. 1991. *Ecological Economics: The Science and Management of Sustainability.* New York: Columbia Univ. Press
20. Costanza R, Daly H. 1992. Natural capital and sustainable development. *Conserv. Biol.* 6:37–46 pp.
21. Daily G, Ehrlich P, Ehrlich A. 1994. Optimum population size. *Pop. Environ.* 15(6):469-475 pp.
22. Daily GC, Ehrlich PR. 1992. Population, sustainability and the earth's carrying capacity. *BioScience* 42(10):761–71
23. Daly H. 1977. *Steady State Economics.* Washington DC: Island Press. (1991 2nd ed.) 302 pp.
24. Daly HE. 1972. In defense of a steady-state economy. *Am. J. Agric. Econ.* 54(4):945–54 pp.
25. Daly HE, ed. 1973. *Toward a Steady State Economy.* San Francisco: Freeman
26. Daly HE. 1974. The economics of the steady state. *Am. Econ. Rev.* (May):15–21
27. Daly HE. 1988. On sustainable development and national accounts. In *Economics, Growth and Sustainable Environments,* ed. D Collard, DW Pearce, D Ulph. New York: St. Martin's Press 205 pp.
28. Daly HE. 1990. Toward some operational principles of sustainable development. *Ecol. Econ.* 2:1–6
29. Daly HE. 1990. Sustainable growth: an impossibility theorem. *Development* (SID) 3/4
30. Daly HE. 1992. Allocation, distribution, and scale: towards an economics that is efficient, just and sustainable. *Ecol. Econ.* 6(3):185–93
31. Daly HE. ed. 1980. *Economics, Ecology and Ethics: Essays Toward a Steady-State Economy.* San Francisco: Freeman
32. Daly HE, Cobb J. 1989. *For the Common Good.* Boston: Beacon. 492 pp.
33. Ehrlich P. 1989. The limits to substitutability: meta-resource depletion and a new economic-ecological paradigm. *Ecol. Econ.* 1:9–16
33a. Ehrlich P, Ehrlich A. 1989. Too many rich folks. *Populi* 16(3):3–29
34. Ehrlich P, Ehrlich A. 1989. How the rich can save the poor and themselves. *Pacific Asian J. Energy* 3:53–63
35. Ehrlich P, Ehrlich A. 1991. *Healing the Planet.* Boston: Addison-Wesley. 366 pp.
36. El Serafy, S. 1989. The proper calculation of income from depletable natural resources (25-39). In *Environmental Accounting for Sustainable Development,* YJ Ahmad et al. Washington, DC: World Bank/UNEP Symposium. 100 pp.
37. El Serafy S. 1991. The environment as capital (168-175) In *Ecological Economics,* ed. R. Costanza. New York: Columbia University Press 525 pp.
38. El Serafy S. 1993. Country macroeconomic work and natural resources. Washington, D.C.: The World Bank. Environment Working Paper No. 58:50 pp.
39. Fritsch B, Schmidheiny S, Seifritz W. 1994. Towards an ecologically sustainable growth society: physical foundations, economic transitions, and political constraints. Berlin, Springer 198 pp.
40. Goodland R. 1992. The case that the world has reached limits. *Population & Environment* 13(2):167-182 pp.
41. Goodland R. 1994. Environmental sustainability: imperative for peace. (19-46) In *Environment, Poverty, Conflict,* ed. N Graeger, D Smith, pp. 15–46. Oslo: Int. Peace Res. Inst. (PRIO): 125 pp.
42. Goodland R, Daly HE. 1993. *Poverty Alleviation Is Essential for Environmental Sustainability. Environ. Work. Pap. No. 42.* Washington, DC: World Bank. 34 pp.
43. Goodland, R, Daly HE, El Serafy S. 1992. *Population, Technology, Lifestyle: The Transition To Sustainability.* Washington DC: Islands. 154 pp.
44. Haavelmo T, Hansen S. 1992. On the strategy of trying to reduce economic inequality by expanding the scale of human activity. In *Population Technology Lifestyle: The Transition to Sustainability,* ed. R Goodland, R Daly, S El Serafy, pp. 38–51. Washington, DC: Island Press. 154 pp.
45. Hardin G. 1968. The tragedy of the commons. *Science* 162:1243–48
46. Hardin G. 1993. *Living Within Limits: Ecology, Economics and Population Taboos.* New York: Oxford Univ. Press 339 pp.
47. Hicks JR. 1946. *Value and Capital.* Oxford: Clarendon. 446 pp.
48. Hueting R. 1980. *New Scarcity and Economic Growth: More Welfare Through Less Production?* Amsterdam, North-Holland: 269 pp.
49. Hueting R. 1990. The Brundtland report: a matter of conflicting goals. *Ecol. Econ.* 2:109–17
50. International Institute of Applied Systems Analysis. 1992. *Science and Sustainability.* Laxenburg, Vienna: Int. Inst. Appl. Systems Anal. 317 pp.

51. International Union for the Conservation of Nature. 1980. *The World Conservation Strategy.* Gland, Switzerland: IUCN, WWF.
52. Jansson A, Hammer M, Folke C, Costanza R. 1994. *Investing in Natural Capital: The Ecological Economics Approach To Sustainability.* Washington DC: Island. 504 pp.
53. Kellenberg J, Daly H. 1994. *User Costs. Environ. Work. Pap.* Washington DC: World Bank
54. Korten D. C. 1991. Sustainable development. *World Policy J.* (Winter): 156
55. Ludwig D. 1993. Uncertainty, resource exploitation and conservation: Lessons from history. *Science* 260:17–53
56. Lutz E. ed. 1993. *Toward improved accounting for the environment. An UNSTAT-World Bank symposium.* Washington, DC: World Bank. 329 pp.
57. Malthus TR. 1964 [1836]. *Principles of Political Economy.* London: Kelley. 446 pp.
58. Malthus TR. 1970 [1878]. *An Essay on the Principle of Population.* Harmondsworth: Penguin. 291 pp.
59. Meadows D, Meadows D, Randers J, Behrens W. 1972. *The Limits to Growth.* New York: Universe. 205 pp.
60. Meadows D, Meadows D, Randers J. 1992. *Beyond the Limits: Global Collapse or a Sustainable Future.* Post Mills, VT: Chelsea Green. 300 pp.
61. Mill JS. 1900 [Rev. ed. 1848]. *Principles of Political Economy.* New York: Collier. 2 vols.
62. Mishan EJ. 1967. *The Costs of Economic Growth.* London: Staples. 190 pp.
63. Mishan EJ. 1977. *The Economic Growth Debate: An Assessment.* London: Allen & Unwin. 277 pp.
64. Netherlands. 1994. *The Environment: Towards a Sustainable Future.* Dordrecht: Kluwer Acad. 608 pp.
65. O'Connor J. 1995. *Monitoring environmental progress. World Bank Environment Dep. Tech. Paper.* Washington DC. v pp.
66. Opschoor JB, ed. 1992. *Environment, Economy, and Sustainable Development.* Groningen: Wolters-Noordhoff. 149 pp.
67. Opschoor JB. 1994. The environmental space and sustainable resource use. In *Sustainable Resource Management and Resource Use,* Netherlands Advisory Council for Res. on Nature and Environ., *Publ. RMNO 97.* 3:33–67
68. Opschoor JB, Reijnders L. 1991. Towards sustainable development indicators. In *In Search for Indicators of Sustainable Development,* ed. O. Kuik, M. Verbrugge, 2:1–27. Dordrecht: Kluwer
69. Opschoor JB, van der Straaten J. 1993. Sustainable development: an institutional approach. *Ecol. Econ.* 3:203–22
70. Panayotou T. 1993. *Green Markets: The Economics of Sustainable Development.* San Francisco: Int. Ctr. Econ. Growth. 169 pp.
71. Pearce DW, Barbier E, Markandya A. 1990. *Sustainable Development: Economics and Environment in the Third World.* Elgar: Aldershot. 217 pp.
72. Pearce DW, Markandya A, Barbier E. 1989. *Blueprint for a Green Economy.* London: Earthscan. 192 pp.
73. Pearce DW, Redclift M. eds. 1988. Sustainable development. *Futures* 20 (whole special issue)
74. Redclift MR. 1987. *Sustainable Development: Exploring the Contradictions.* London: Methuen. 221 pp.
75. Redclift MR. 1989. The meaning of sustainable development. *Geoforum* 23: 395–403
76. Redclift MR. 1994. Reflections on the 'sustainable development' debate. *Int. J. Sustainable Dev. World Ecol.* 1:3–21
77. Ricardo D. 1973 [1817]. *Principles of Political Economy and Taxation.* London: Dent. 300 pp.
78. Serageldin I. 1993. *Development Partners: Aid and Cooperation in the 1990s.* Stockholm: SIDA. 153 pp.
79. Serageldin I. 1993. Making development sustainable. *Finance Dev.* 30(4):6–10
80. Serageldin I, Daly H, Goodland R. 1995. The concept of sustainability. In *Sustainability and National Accounts,* ed. W van Dieren. Amsterdam: IMSA (for) The Club of Rome
81. Serageldin I, Steer A, eds. 1994. *Making Development Sustainable: From Concept to Action. ESD Occas. Paper 2.* Washington DC: World Bank. 40 pp.
82. Serageldin I, Steer A, eds. 1994. *Valuing the Earth. ESD Pap.* Washington DC: World Bank. 192 pp.
83. Simon JL, Kahn H. 1984. *The Resourceful Earth: A Response to Global 2000.* Oxford: Blackwell
84. Simonis U. E. 1990. Beyond growth: elements of sustainable development. Berlin, Edition Sigma 151 p.
85. Solow R. 1974. The economics of resources or the resources of economics. *Am. Econ. Rev.* 15:1–14
86. Solow R. 1991. See Toman
87. Solow R. 1993. An almost practical step toward sustainability. *Resources Policy* 19:162–72

88. Solow R. 1993. Sustainability: an economist's perspective. In *Selected Readings in Environmental Economics,* ed. R Dorfman, NS Dorfman, pp. 179–87. New York: Norton. 517 pp.

89. Summers L. 1992. Summers on sustainable growth. *Economist* 30 May. p. 91

90. Tinbergen J, Hueting R. 1992. GNP and market prices: wrong signals for sustainable economic success that mask environmental destruction. In *Population, Technology, Lifestyle: The Transition to Sustainability,* ed. R Goodland, H Daly, S El Serafy, pp. 52–62. Washington, DC: Island. 154 pp.

91. Tisdell C. 1992. *Environmental Economics: Policies for Environmental Management and Sustainable Development.* Aldershot, Hants: E. Elgar. 259 pp.

92. Toman MA. 1994. Economics and "sustainability": balancing trade-offs and imperatives. *Land Econ.* 70(4): 399–413

93. Toman M, Pezzey J, Krautkraemer J. 1993. Economic theory and "sustainability." *Univ. London, Economics Department, Discussion Paper 93-14.* 31 pp.

94. Turner R, ed. 1993. *Sustainable Environmental Economics and Management: Principles and Practice.* London: Belhaven. 389 pp.

95. von Weizsacker EU. 1992. *Sustainability: Why the North Must Act First.* Geneva, UN Acad. Environ. (June Workshop). 13 pp.

96. Westra L. 1994. *An Environmental Proposal for Ethics: The Principle of Integrity.* Lanham MD: Rowan & Littlefield. 237 pp.

97. World Bank. 1986. *Environmental Aspects of Bank Work.* Washington, DC: World Bank Operational Manual Statement OMS 2.36.

98. World Bank. 1990. *World Development Report.* Washington, DC: World Bank

99. World Bank. 1992. *World Development Report 1992: Development and the Environment.* New York: Oxford Univ. Press. 308 pp.

100. World Bank. 1991. *World Development Report.* Washington DC: World Bank

101. World Bank. 1993. *The World Bank and the Environment.* Washington, DC: World Bank. 193 pp.

102. World Bank. 1992. *Environmental Assessment Sourcebook.* Washington, DC: World Bank. 3 Vol.

103. World Bank. 1993. *World Development Report 1993: Investing in Health.* New York: Oxford Univ. Press 329 pp.

104. World Bank. 1995. *World Development Report: Infrastructure for Development.* Washington DC: World Bank. Oxford Univ. Press. 254 pp.

105. World Commission on Environment and Development. (The Brundtland Commission). 1987. *Our Common Future.* Oxford: Oxford Univ. Press

106. World Commission on Environment and Development. 1992. *Our Common Future Reconvened.* (22-24 April, Grosvenor Hotel, London). Geneva: Ctr. for Our Common Future. 32 pp.

107. World Wide Fund for Nature. 1993. *Sustainable Use of Natural Resources: Concepts, Issues and Criteria.* Gland, Switzerland. 32 pp.

Annu. Rev. Ecol. Syst. 1995. 26:25–44

SUSTAINABILITY OF SOIL USE

S. W. Buol

Soil Science Department, North Carolina State University, Box 7619, Raleigh, North Carolina 27695-7619

KEY WORDS: agroecosystems, erosion, fertilizer, food production, nutrient elements

ABSTRACT

The finite quantities of essential elements contained in soil determine its sustainable use for food production. Various soils differ substantially in their ability to provide essential elements. Methods of managing soil to facilitate production of food crops differ depending upon the type of crop grown and the characteristics of the soil. Historical evidence reveals that low rates of food crop production are possible for 100 or more years on some soils, while only one or two crops are obtainable on others. High rates of food production can be achieved only if the concentrations of essential elements in the soil are enhanced. When dedicated to food crop production, soil properties are altered by management practices to favor the requirements of the crop plant and discourage the growth of other vegetation. No soil can sustain the constant depletion of critical elements contained in the plant parts used as food products and transported to another location for consumption. This elemental depletion must be compensated for use to be sustainable.

INTRODUCTION

Humans have used soil longer than most other natural resources. Soil is such an integral component of natural and human-managed ecosystems that there exists a multitude of concepts relative to the nature and function of soil. The diversity of concepts of soil is compounded by the continuum of soil properties on the landscape, and geographical limitations of each person's scientific experience. The present identification of approximately 17,000 kinds of soil within the United States, and an undetermined number throughout the world, assures that diverse concepts develop among scientists who seldom compare

25

soils with entirely common properties (53). The social, economic, and political differences that often control the human use of soil further contribute to divergent concepts of sustainable soil use.

CHANGING CONCEPTS OF SOIL

Identifying and classifying soils is a continuing process (3). Early classification systems stressed agricultural uses and easily observed features of soil color and geologic association (12, 33, 48). With soil conceived as geologic material, such terms as residual soils, alluvial soils, laterite, moor soils, and marine soils were introduced.

About 100 years ago, a marked shift occurred in the concept that soils formed in response to soil-forming factors of climate and vegetation (10, 11). Glinka, a student of Dokuchaev, was instrumental in introducing this concept to the United States through CF Marbut's translation of his book (14). Such terms as "Podzol," "Chernozem," and "Solonetz" were introduced. Apparently independent of European scholars, the US Department of Agriculture began a federal soil survey program in 1899, and the practice of naming soils by place association in the United States was formalized (59).

Although several intermediate steps in soil classification were recorded, the most significant are summarized in the *Atlas of American Agriculture* (22) and the *USDA Yearbook of Agriculture* (1). Soil science was dominated by the concept of soil as an entity derived from soil-forming factors of parent material, climate, organisms, topography, and time (17). The limitations of classifying soils according to concepts of soil-forming processes controlled by factors external to the soil became apparent during attempts to revise the 1938 classification system in 1949 (37, 49).

Starting about 1950, achieving global recognition with the publication of *Soil Taxonomy* (41, 42), and continuing today (43), the development of the conceptual basis of recognizing and classifying soils has undergone significant changes. While concepts of soil genesis were employed to guide the structure of the taxonomic system, definitions of soils were made precise and quantitative. Only values that could be measured within the soil, or on samples taken from the soil, were used as classification criteria (39, 40). A completely new system of nomenclature formally eliminated such terms as lateritic, podzolic, etc that had proliferated without quantifiable identification. Perhaps the most significant aspect of Soil Taxonomy is its attempt to quantify and categorize dynamics of soil moisture and soil temperature regimes. This aspect of Soil Taxonomy clearly marks a conceptual understanding that each kind of soil is physically fixed to a location. For a soil to be sustainably used at a location, the management practices employed must address the soil properties both as they change each day of the year and as they are affected by extremes of

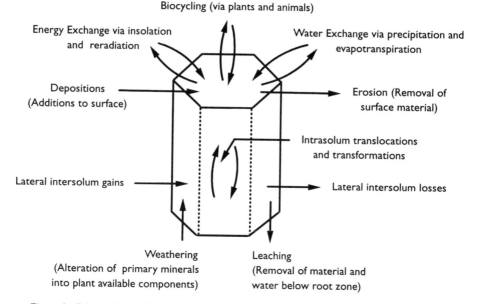

Biocycling (via plants and animals)

Energy Exchange via insolation and reradiation

Water Exchange via precipitation and evapotranspiration

Depositions (Additions to surface)

Erosion (Removal of surface material)

Intrasolum translocations and transformations

Lateral intersolum gains

Lateral intersolum losses

Weathering (Alteration of primary minerals into plant available components)

Leaching (Removal of material and water below root zone)

Figure 1 Schematic of soil as an open system (3).

weather occurring at frequencies of a decade or less. The paradigm shift from assuming soils to be static entities that came into being as the result of soil-forming factors to recognizing entities with real-time dynamics of temperature, moisture, and redox has profound significance to all ecological study, not just agroecology (15).

The use of dynamic soil properties to identify and classify soils should not be considered a completed task. Techniques for measuring and quantifying dynamic properties are time-consuming and expensive. Some individuals and organizations have not embraced the open-system, dynamic paradigm (Figure 1), citing limitations in methodology currently used to proxy for some unavailable long-term measurements. In many countries the required long-term data are being acquired in response to the need for a better understanding of whole-ecosystem dynamics. It is within the open-system concept of soil that sustainability of soil use is discussed in this chapter.

FUNCTIONS OF SOIL IN ECOSYSTEMS

Within any ecosystem, soil has four basic functions:

1. Soil must support the physical structure of plants so they can photosynthesize.

2. Soil must, in close synchrony with the air above it, have temperatures compatible with the physiology of the organisms within the ecosystem.
3. Soil must accept, retain, and emit water. Precipitation is sporadic in most parts of the earth while physiologic demands of plants are more nearly constant, subject, of course, to species-specific perturbations. Soil-stored water sustains plants during rainless periods.
4. Finally, soil is the exclusive source of several chemical elements required for both plant and animal (including human) physiology.

Physical Support

Physical support of plants is a fundamental requisite of soil. Relatively few soils used for food crop or forage fail this requisite. However, compaction of the soil surface invariably results if the surface is exposed to the direct impact of raindrops. Compacted surface conditions are frequently interpreted as responsible for poor vegetative growth without considering that the lack of vegetative cover to buffer the raindrop impact causes the compaction. Excessive animal or mechanical traffic, usually with vegetative removal, also compacts the soil surface. Some degree of surface compaction accompanies almost all seasonal crop production. Cultivators use a wide array of indigenous cultivation techniques that loosen the physical hardness of the surface soil to assure timely establishment of their planted seed. Left to natural processes, compacted areas are slow to revegetate and often undergo excessive erosion rates, until a fortuitous sequence of soil loosening, seed, moisture, and temperature reestablishes vegetation. Sustainable tree crop management on soil where rooting depth is restricted by physical or chemical soil properties is impaired by windfall losses (58). Grass or shrub vegetation, requiring less physical support, may occupy such sites.

Soil Temperature

Soil temperature is closely related to, but not the same as, the air temperature. The daily and annual dynamics of temperature are fixed properties, with associated probabilities, at any site. Natural plant growth and agronomic systems are confined by temperature dynamics. Daily soil temperature dynamics, especially in the surface soil, are modified by the vegetative cover, soil color, and soil texture (13). Plant breeding technology has adapted many food crops to allow their production in areas of temperate latitudes and with growing seasons colder and shorter than those to which the plants were initially adapted. Mulching to reduce high daily soil temperatures early in the growing season can reduce seedling damage in parts of the tropics with seasonal rainfall (19). Practices such as in greenhouse construction that alter soil temperature require substantial expense and have limited application except for high-value crops.

Soil Moisture

The ability of soils to supply soil moisture is closely related to weather patterns, i.e. climate. The probability of particular weather conditions at any site can be statistically quantified into a climatic representation and used as criteria to classify soil (26, 55–57).

Undesirable soil moisture conditions present some risk in every soil. Management or manipulation of the natural soil moisture regime to reduce the risks created by weather is of three forms—drainage, irrigation, or soil surface manipulation to affect infiltration.

Soils that have a high probability of being saturated for prolonged periods can be used for agriculture or forestry if drainage systems are installed (30). Drainage systems are installed to improve trafficability, to protect plants from excess water, and/or to control salinity. Most studies have found that changing land use from its natural vegetative cover to drained agricultural land use increases peak flow of runoff water, sediment loss, and nutrient loss (38). However, sediment and nutrient losses from drained cropland are usually small compared to those of cropland on well-drained soils because most poorly drained land is nearly level (38). Almost all agricultural drainage systems require transfer of excess water beyond ownership or jurisdictional boundaries; thus, legal and regulatory concerns are major sustainability issues.

Irrigation is one of the oldest soil-management technologies. When irrigation is not conducted correctly, salinization results. Irrigation technology is well understood (54), but its sustainability depends upon a reliable water source (usually off-site), and removal of accumulating salts often via a drainage system. Although irrigation is used on only about 10% of agricultural land (46) and has in places been sustainable since the beginning of recorded history, natural and legal disruptions of suitable water supply are continuing threats to sustainability.

Most soils are neither drained nor irrigated. Water infiltrates through the soil surface and redistributes within the soil in response to capillary and gravitational forces (Figure 2). Plant roots then extract the water in response to transpirational demand. Obviously, the frequency and amount of rainfall is the most critical variable for sustained soil use. Considerable differences exist among types of soil in their efficiency in using rainfall. Infiltration of rainfall into the soil is enhanced if soil is protected from direct raindrop impact. Undecomposed plant residues on the soil surface or complete live plant cover increases infiltration relative to that occurring in bare soil. Earthworm or insect burrows and root channels that open to the surface are effective conduits for infiltration. Mechanical disturbance to loosen compacted soil surfaces is also effective. However, if bare soil is not protected from the direct impact of

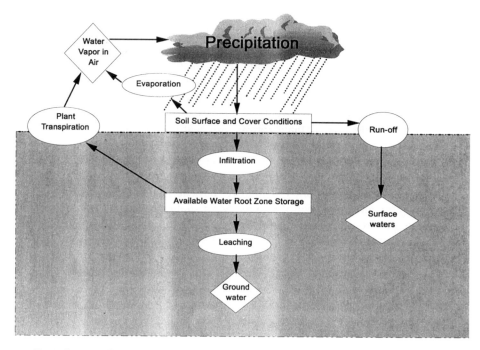

Figure 2 Flow of water as controlled by soil.

raindrops, infiltration rate will quickly decrease during the rain event, and runoff, with its erosion potential, will increase.

The amount of water retained from rapid leaching beyond the rooting depth of plants (available soil water) is dependent upon the volume of pores in the soil that are between about 0.01 mm and 0.0002 mm in diameter. Pores larger than about 0.01 mm in diameter empty of water under the force of gravity; pores smaller than about 0.0002 mm in diameter retain water at such a high capillary tension that it cannot be extracted by most plants. The amount of available water any soil can retain after wetting is controlled by the relative distribution of pore sizes. Sandy soils and clayey soils retain less available water per unit volume than do silty soils. In sand, a high proportion of the pores are too large, and in clayey soils a high proportion of the pores are too small.

Since little can be done to increase the amount of available waterholding capacity of a given soil, it is often advantageous to increase the rooting depth of crop plants, thereby increasing the amount of water available between rainfall events. If root penetration is restricted by hard, root-restricting layers, mechanical rupture by chisel or deep plowing is often used. In some soils root-restricting layers form in response to compaction immediately below the depth of cultivation, and some form of "chiseling" or "subsoiling" is required for continued use (9). In many soils, acid, infertile subsoil conditions limit downward root growth. This condition, known as aluminum toxicity, is not easily corrected unless calcium-containing materials such as ground limestone or gypsum are physically incorporated. Liming materials applied at the surface may take several years to neutralize subsoil acidity. Fortunately, many of the more acid soils also have low cation exchange capacity and the downward migration of calcium appears to occur more rapidly in such soils (4). Plant breeding to incorporate Al-tolerance in grain crops offers a partial alternative to lime application in some soils (16). Cultivars that can extend their roots into acid subsoils avail themselves of more plant-available water during drought than do acid-sensitive cultivars.

Soil Chemistry

Chemistry of the soil is seldom visible. Plants growing on soil deficient in certain elements display deficiency symptoms, but these are often overshadowed by insect and disease damage to the already weakened plants. All of the elements required for human and animal food enter the plant primarily through roots except carbon, which enters the plant through the stomata.

The presence of all the naturally occurring elements in physical mixture with organic compounds, and profuse inhabitation by microorganisms, assures that an almost infinite number of chemical reactions take place in the soil. It may never be possible to elucidate all of these reactions. Conceptually, it can be assumed that if a chemical reaction is possible at the temperatures and pressures ambient in soil, that reaction is taking place.

Three aspects of soil chemistry are germane to sustainable soil use. First, soil minerals are the only primary source of most elements required in food. Second, chemical reactions in the soil must decompose the organic compounds deposited in and on the soil by dead vegetation, microorganisms, and fauna. Third, the soil must provide the plant roots with essential elements as inorganic, ionic forms and do so rapidly enough to sustain desired growth rates (8).

Although it is impossible for plants to deplete a soil completely of a particular element, the total elemental content of soil provides an absolute maximum sustainability measurement. Total elemental analysis of soil requires that all the minerals be decomposed. However, the elemental amounts determined by total analysis are not representative of the amount of each element that will be available to plants. No absolute conversion of total elemental content to plant-available content can be made because the mineral forms differ among soils. However, available forms are seldom more than 10% of the total. Only

Table 1 Total and extractable elemental composition of some top-soil and subsoil layers in different soils.[a]

Soil	Layer (cm)	P (Kg ha^{-1})	K (kg ha^{-1})
Paleustalf[b]	0–18	700(16.8)[e]	1700(473)[e]
	42–60	320(3.6)	1800(237)
Ustipsamment[b]	0–18	200(27.8)	700(321)
	42–60	120(3.4)	500(78)
Haplustoll[b]	0–18	540(10.4)	8100(156)
	42–60	200(2.8)	2700(78)
Haplustox[c]	0–18	246(4)	NA[f] (78)
	42–60	212(2)	NA (23)
Hapludoll[d]	0–18	1,936(NA)	38,180(NA)
	42–60	1,672(NA)	37,018(NA)

[a] Assumed bulk density of 1 gm cm^{-3} (2M kg ha^{-1} 18cm^{-1}).
[b] Ibadon, Nigeria (24).
[c] Brazil (36).
[d] Wisconsin (23).
[e] Values in ( ) are extractable or exchangeable amounts.
[f] NA, not available

0.1 to 1% of the total elemental content is representative of the amount available to plants within a growing season. The unreliability of predicting plant growth from total elemental analysis caused soil scientists and agronomists to develop partial extractions, most often with dilute acids, and total elemental analyses of soil almost ceased in the 1930s. To obtain total elemental data representative of a wide range of soil types, publications of that era must be consulted (22, 23). However, total elemental data are sometimes reported for intensively studied areas such as the International Institute of Tropical Agriculture in Nigeria (24). Table 1 presents total and extractable elemental contents of P and K to illustrate the range in quantities that can be expected in various soils.

Partial extraction procedures, often called soil test values, do not directly measure the amount of a nutrient element available for plant uptake. Soil test procedures estimate the amount of available nutrient only through valid correlation with actual plant uptake. This has been effectively done with commonly grown grain crops. Almost all soil test procedures rely on short duration extractions in the laboratory, whereas actual nutrient element uptake takes place in the soil, with variable temperature and moisture conditions, physiologically contrasting plant species and cultivars, and contrasting growth rates and durations of maturation (31).

Air is the primary repository of nitrogen. However, nitrogen must be present in the soil as nitrate or ammonium for uptake by plant roots. Nitrogen availability in the soil often limits plant growth, but with 78% of the air composed of nitrogen, total depletion is not a concern for sustainability. The rate of nitrogen conversion from organic compounds into nitrate and/or ammonium forms available to plant roots is of most concern to high-yield crop production (44). The total soil content of P, K, Ca, Mg, and the other essential elements, the primary supply of which is in soil minerals, is finite in each soil. Sustained food production is ultimately limited by lack of one or more of these elements.

FUNCTIONS OF SOIL IN AGROECOSYSTEMS

Humans use only a small spectrum of the plant kingdom for food and fiber. During the course of human habitation, food and fiber plants have been strongly selected and more recently genetically altered or hybridized to better use space and production practices. Cultivars capable of high yields of edible grain or fruit per growing season have been developed. To accommodate mechanized harvest practices, cultivars with uniform ripening of grain or fruit have been selected. The crop plant is aggressively protected from weed competition by cultivation or selected use of herbicides. All agroecosystems contrast to natural ecosystems in two fundamental aspects: (*a*) products are removed from the site of growth, and (*b*) plant growth rates are expected to be much more rapid in the agroecosystem.

Elemental Export

Export of chemical elements from site of crop growth is a fundamental aspect of agroecosystems. Regardless of scale, human activity causes food to be transported before it is consumed. This may be on a local scale in the case of a shifting cultivator in the tropical rain forest, or on the large scale we see in the industralized world with its huge urban centers. Human food requires several chemical elements that originate only in soil minerals.

Representative amounts of some of the nutrient elements removed in crop and total tree harvest are presented in Table 2. At least three critical aspects must be considered when relating these data to sustainable soil use for food and fiber production. Food crops require the quantities of elements (listed in Table 2) in their short (approximately 100-day) growing season, while equivalent quantities are required by tree crops in 20 or more years. Thus, the concentration of soluble, ionic forms of the respective ions in contact with the plant root must be many times greater to produce the food crop successfully than to support tree or natural vegetation. The less extensive root system of food crops, as compared to trees, necessitates even higher concentrations within that smaller volume of soil.

When the values in Table 2 are compared with those in Table 1, it is readily apparent that, regardless of the native fertility of soil, continued export of

Table 2 Approximate amount of nutrient elements contained in various crops

Crop	Yield[a] kg ha^{-1}	N	P	K (kg ha^{-1})	Ca	Mg
Corn (grain)	9,416	151	26	37	18	22
(stover)	10,080	112	18	135	31	19
Rice (grain)	5,380	56	10	9	3	4
(straw)	5,610	34	6	65	10	6
Wheat (grain)	2,690	56	12	15	1	7
(straw)	3,360	22	3	33	7	3
Soybeans (grain)	2,690	168	18	52	8	8
Sugarcane (plant)	67,200	107	27	251	31	27
Cabbage	44,880	146	18	121	22	9
Onions	16,800	50	10	37	12	2
Peanuts (nuts)	2,800	101	6	15	1	3
Loblolly Pine (22 yrs)	84,000	135	11	64	85	23
Loblolly Pine (60 yrs)	234,000	344	31	231	513	80

[a] Total above ground values for loblolly (45, 47).

plants or plant parts will exhaust the elemental supply of any soil. Phosphorus is usually the most limiting of the soil-supplied elements, followed by potassium (18, 25). Nitrogen is often the most limiting element during any growing season, necessitating supplemental fertilization to attain high yield. However, soil nitrogen can be sustainably resupplied from the air at a slower rate by microbial fixation.

Although straw or stover may be cycled as plant residue, or even fed to cattle and recycled as manure, grain or seed contains the most significant amounts of the critical phosphorus. A high proportion of the grain becomes human food, the refuse of which is spatially concentrated near the house, the village, or the city, and not returned to the site from which it was extracted.

SOIL CHANGE UNDER HUMAN MANAGEMENT

There are so many systems of soil management that few generalities universally apply. Such diametrically opposite practices as irrigation and drainage are sometimes used in the same field the same year. Most crop management scenarios are employed to favor growth conditions for a desired food crop over the native vegetation. Management practices to achieve these conditions almost universally involve removing vegetation that competes for direct sunlight and water. Ridding a site of competing vegetation invariably involves cultivation to disrupt weed growth and to create physical conditions that assure good seed-to-soil contact for the crop plant.

In cultivation, the uppermost horizons of soil are physically mixed, and the resulting "plow layer" usually becomes more dense than the original horizons, but not so dense as to hinder seeding growth. Cultivated surfaces almost always decrease about 30% in carbon content compared to carbon content prior to cultivation (50). Initial reduction of soil carbon content is caused by loss of the unhumified plant remains that are instrumental in providing large pores in the soil (28). Soil carbon contents in agricultural use are controlled by crop-residue return rates (20, 32). Soil surfaces exposed to raindrop impact physically harden or "crust" and must be mechanically cultivated or mulched to improve infiltration. Microbial and faunal activity is changed by cultivation. Higher diurnal soil temperature fluctuations of unshaded soil and physical disturbance usually reduce earthworm and insect activity. Such soil alterations engendered by cultivation practices represent redirection of soil use and favor cultivated food crops over natural vegetation. Such changes are not inherently degenerate.

Sustainable soil use for crop production must be evaluated in terms of ability to sustain crop yield. The sequence of soil use and crop production upon initiation of cropping is very dependent upon the initial properties of the soil. A complete record of crop production at the Morrow plots, now a registered national landmark on the University of Illinois campus, offers one of the best records available for natively fertile, sustainable cropland use (29). In 1876 several experimental plots were established at the University of Illinois campus on a Flanagan silt loam (Aquic Argiudoll). Management details changed with time, reflecting changes in agricultural science, but the basic design remained unaltered. One set of plots has been planted to corn every year. One set of plots maintains a two-year rotation of corn alternating with oats, and more recently, corn alternating with soybeans. A third set of plots maintains a three-year rotation of corn, oats, and clover or alfalfa (legume). Manure, lime, phosphate fertilizer only, and complete fertilizer application, as recommended by soil test evaluation, have been introduced on subplots as these methods became standard practice among agronomists and farmers in the area. Table 3 averages corn yields for the four most recent years when all rotations were planted to corn, thereby assuring equal weather conditions in all plots. Clearly, both rotation schemes enhance yield of corn per hectare. However, in the three-year rotation, corn is harvested in only one third of the years. The legume

Table 3 Average subplot yields of corn in Morrow plots for 1961, 1967, 1973, and 1979 (29).

Treatment	Continuous corn	2-yr rotation corn-soybean	3-yr rotation corn-oats-legume[c]
		kg ha^{-1}	
No amendments	2671	3857	5299
MLP[a]	5293	7306	8209
MLNPK[b]	8165	8842	9450

[a] MLP—Manure, lime, P as rock phosphate and bonemeal added each year.
[b] MLNPK—Manure, lime and nitrogen, phosphate, potassium: fertilizer rate from soil test recommendation added each year.
[c] Legume—Clover, soybean or alfalfa plowed under for green manure every third year.

is used for green manure, so no income is realized during one third of the years. In the two-year rotation, corn is produced in only half of the years. Whether the yield gain per crop related to rotation is economically viable for the farmer depends upon the value obtained for the oats or soybeans. Total corn production per hectare over several years is thus reduced by rotation. Physically, all the systems have been sustainable for over 100 years, but economic realities favor the two-year rotation or continuous corn, depending upon the relative market value of corn and soybeans.

Perhaps the most interesting result of the Morrow plots is to compare the corn yield with no soil amendments for the first 10 years (1888–1897) of the experiment in the continuous corn treatment at 2,565 kg ha^{-1} with the last 10 years (1972–1981) at 2,991 kg ha^{-1}. Clearly, the naturally fertile Flanagan soil, quite representative of Central European and midwest United States loess-derived soil, can sustain a low level of continuous corn production for 100 years. The slight yield increase represented between these 10-year averages may be due to weather differences or perhaps modern hybrid seed, but is best viewed as insignificant. Substantial yield increases are achieved only when fertility is enhanced by additions of nitrogen, phosphorus, and potassium.

Quite a contrasting scenario of sustainable crop use of soils is obtained if long-term records are examined on naturally infertile soils like Yurimaguas (Typic Paleudult). In 1972 an experiment was started at Yurimaguas, Peru in the upper Amazon basin. Grain yields for a corn-soybean-rice rotation were compared with and without inputs of fertilizer and lime. The results are summarized in Figure 3. With 39 crops harvested between 1972 and 1991, it is clear that moderate yields are sustainable only with inputs of lime and fertilizer (51). Without fertilization, initial yields dropped to zero and remained there until an accidental fertilization took place in 1987. This additional fertility, also apparent after an interruption in planting during 1989, was quickly de-

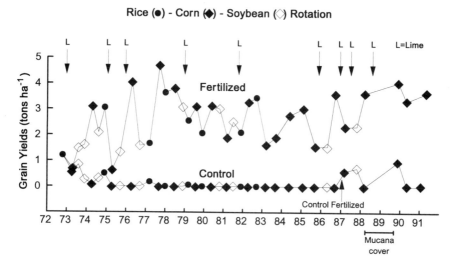

Rice (●) - Corn (◆) - Soybean (◇) Rotation

Figure 3   Grain yields in fertilized and unfertilized plots on a Paleudult soil in the Amazon basin (51).

pleted by one crop in 1990. Other experiments at the same Amazon jungle site demonstrated that intensive crop residue management extended non- fertilized production to only six crops, after which fertilizer was needed (34, 35).

A comparison of historical and present fertilizer use in natively fertile and infertile soils in the United States illustrates the soil management changes used to sustain agricultural production. Table 4 presents corn yields and fertilizer rates representative of farmer practices on a natively infertile Kandiudult in North Carolina and a natively fertile Argiudoll in Iowa in the early 1900s and approximately 60 years later. In both areas corn yields have more than tripled over that period of time. In the 1920s, small additions of fertilizer were routinely made on the Kandiudults in North Carolina, which are very similar to the Paleudults at Yurimaguas, while no fertilizer was used on the Argiudolls in Iowa, although yields were greater than those in North Carolina. By the 1980s yields remained higher in Iowa than in North Carolina. Also, more fertilizer was added annually to sustain Iowa's higher yield. By comparing the amounts of fertilizer used with the elemental requirements of corn (Table 2), it is apparent that fertilizer rates only slightly exceed annual crop extraction. Fertilizer rates in excess of grain extraction probably compensate for losses due to volatilization, leaching, mineral fixation, and erosion. Rates may also reflect farmer optimism that weather conditions may be good and higher than average yields may be obtained. Historical data such as these are poorly controlled but necessary to test such concepts as sustainability of soil use (27).

Table 4 Historical comparison of average farmer fertilization rates on an Ultisol and a Mollisol in the USA. (5) (Original data from soil survey reports and Agr. Extension Service data).

	Year	Kandiudult (N.C.)		Argiudoll (Iowa)	
		1925	1983	1919	1979
Corn yield	(kg ha^{-1})	2038	6899	2634	8153
Fertilizer rate	(kg ha^{-1})				
	N	36–53	134–177	0	168–202
	P	3–6	20	0	34–54
	K	6–11	75	0	75–111

MAINTAINING SUSTAINABLE USE OF SOILS

The levels of food production currently enjoyed with modern agriculture are possible only with considerable effort. A review of world soil resources and food production determined that labor-oriented agriculture without use of tractors, insecticides, herbicides, or chemical fertilizers could feed only 75% of the 1977 world population (6). Use of animal traction for cultivation, manure distribution, and maintenance of drainage and irrigation systems was included in Buringh & van Heemst's definition of labor-oriented agriculture. A companion study (7) estimates that with maximum use of available technology, food production could be increased 20 to 30 times the 1975 levels.

Figure 4 schematically illustrates the role of soil in the flow of essential food elements. The mineral pool, within the reach of crop plant roots, is the only source of most elements. Upon mineral weathering, the essential elements enter the available pool (see Figure 4) as inorganic ions in the soil water available for plant roots. Air provides carbon and nitrogen. While carbon, as CO_2, enters the plant from the air, nitrogen, as N_2, is usable only by certain microorganisms. These microorganisms, and their host plant, must be decomposed to release nitrate and ammonium into the plant available pool with the other essential elements. The energy of the sun drives the conversion of the inorganic forms of the essential elements taken from the soil by the plant roots into digestable organic forms within the crop plants. Once these substances are in the plant, the social and economic structure of society determines which plant parts are eaten as human or animal food and which are left as refuse at the site of plant growth. In all but the most primitive societies, the elements contained in the food are transported from the site of production and are a net loss to the productive soil. Unharvested parts of the plants remain in or on the soil. Their carbon is used as an energy source for microbes; approximately 75% is converted to CO_2, and the remainder is incorporated into microbial

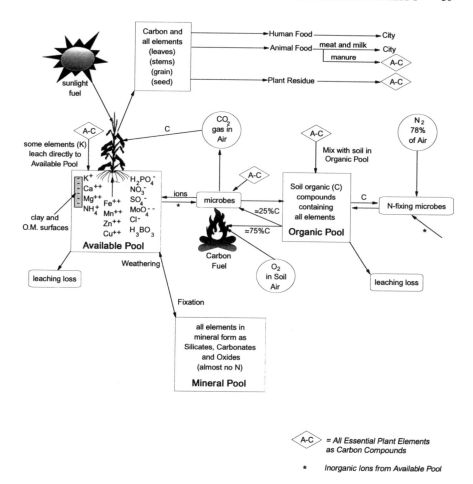

Figure 4   Flow of essential elements from the soil during food production.

tissue. Essential elements in the organic compounds are returned to the plant-available inorganic pool again to be used by plant roots and soil microbes. In some soils, leaching losses of soluble inorganic ions and some soluble organic compounds are possible as excess rainfall moves soil solution below the rooting depth.

When the plant-available inorganic pool of elements essential for plant growth is not maintained, the rate of plant growth declines, the soil surface is less well protected from the erosive energy of raindrops, and soil erosion rates increase. The process of erosion can be conceptualized as a natural process of exposing unweathered minerals (see Mineral Pool in Figure 4) to decomposition and exploitation of essential elements by plant roots. Concurrent with erosion on one part of the landscape is deposition on another part of the landscape, i.e. the floodplains. The many forms of erosion—landslides and gullies, in addition to surface sheet erosion—assure that most floodplain sediments are relatively fertile with minerals containing essential nutrient elements and plant residue from the eroded site. The location of earliest centers of civilization in floodplains must, in large part, be attributed to nutrient resupply to cultivated fields via deposition.

Soil erosion can be accelerated by human intervention, and sediment resulting from that erosion is not desirable from the present perspective of most societies. Sustainable soil use is threatened by erosion only when unsuitable soil material is exposed by that erosion (21). The open system concept of soil encompasses the reality that the surface of the soil is not a fixed point but moves up, by deposition, or down, by erosion, as soil constantly evolves with time (2).

Since the mid-1900s, growing populations in the United States have not required expanding the areas needed to grow food. This contrasts with the situation in the early part of the century (Figure 5). While many other changes have been made in farming practices—especially genetic improvement of crop

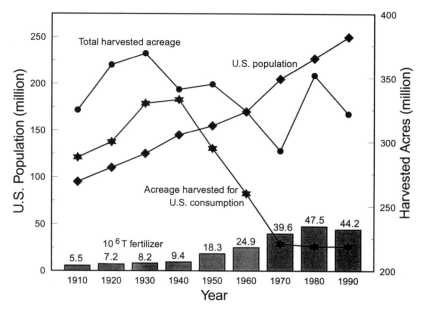

Figure 5 Harvested cropland, total and domestic use, fertilizer use, and population of United States each decade since 1910 (52).

species—the substantial increase in mineral fertilizer use has provided the plant-essential elements at a rate that soil minerals could not sustain. The high proportion of US cropland used to produce food for export reflects a political, social, and economic stability that fosters the infrastructure necessary to sustain this high rate of food production. Although the basic components illustrated in Figure 4 are maintained, with nutrient export to the cities and leaching losses compensated by fertilization, a multitude of apparently contrasting management practices are employed. The availability of alternative management options and the discretion of the farm owner-operator to select practices that best harmonize with soil, economic, and personal circumstances are together perhaps the most significant component of this demonstrated sustainability. Such conditions have not been present in many parts of the world.

SUMMARY

Soil is a dynamic volume of the earth's surface. There are numerous kinds of soil, each somewhat different from all others. Therefore, no one technology can assure sustainable soil use. Management options, harmonized with individual kinds of soils, are a fundamental requisite for sustainability. Although soils differ greatly in elemental composition, no soil can sustain nutrient removal without fertilization. The air contains an apparently unlimited supply of nitrogen that can be captured in the soil by microbial or industrial fixation, but the other essential elements—phosphorus, potassium, calcium, magnesium, etc—originate only in finite, albeit differing, quantities in each kind of soil. They must be replaced if food production and transport to urban centers are to be sustained.

To achieve sustainable and efficient food production, certain soil conditions must be altered to favor food crop production. Efficient use of the land area devoted to food production requires that the soil provide a high concentration of essential nutrients in an inorganic form usable by plants during a relatively short growing season. Modern technology has produced cultivars of food crops that are genetically capable of yields that far exceed those of only a few decades ago. This potential to obtain more human food per unit area devoted to food production cannot be achieved unless the soil contains an adequate quantity of the elements needed by that crop.

Without human intervention, soils support plants. Plants are transitory organisms that if left undisturbed upon completion of their life cycle return the elements they have extracted from the soil, to the soil, at the site of their growth. Humans are also transitory organisms that require elements extracted from the soil but unlike plants, they systematically transport the elements required in their metabolism and growth to centralized locations distant from the soil from which these elements were extracted. Sustainable soil use can be achieved only by replacing these elements in the soil at the site where the food crops are grown.

ACKNOWLEDGMENTS

Sincere appreciation is expressed to TL Grove, RJ McCracken, DD Richter, and TR Wentworth for their review suggestions.

Literature Cited

1. Baldwin M, Kellogg CE, Thorp J. 1938. Soil classification. In *Soils and Men, Yearbook of Agriculture,* pp 979-1001. Washington, DC: USDA. US Govt. Print. Off.

2. Buol SW. 1992. *Pedogenic-geomorphic concept for modeling.* In *Proc. First Soil Genesis Modeling Conference,* ed. WJ Waltman, ER Levine, JM Kimble. Lincoln, NE: USDA-SCS Nat. Soil Survey Center. 84 pp.

3. Buol SW, Hole FD, McCracken RJ. 1989. *Soil Genesis and Classification.* Ames, IA: Iowa State Univ. Press. 446 pp. 3rd ed.

4. Buol SW, Sanchez PA. 1986. Red soils in the Americas: morphology, classification and management. In *Proc. Int. Symp. on Red Soils,* ed. Inst. Soil Sci., Academia Sinica. pp 14-43. Beijing, China: Science Press; Elsevier, Amsterdam. 785 pp.

5. Buol SW, Sanchez PA, Weed SB, Kimble JM. 1990. Predicted impact of climatic warming on soil properties. In *Impact of Carbon Dioxide, Trace Gases, and Climate Change on Global Agriculture,* ed. BA Kimball, pp 71-82. *Am. Soc. Agron. Special Publ. No. 53.* Madison, WI. pp. 133.

6. Buringh P, van Heemst HDJ. 1977. *An estimation of world food production based on labour-oriented agriculture.* Wageningen, Netherlands: Ctr. World Food Market Res. Duivendaal. 46 pp.

7. Buringh P, van Heemst HDJ, Starling GJ. 1975. *Computation of absolute maximum food production of the world.* Dep. Soil Sci., Agric. Univ. Wageningen, The Netherlands: Duivendaal 10. 59 pp.

8. Corey RB. 1990. Physical-chemical aspects of nutrient availability. In *Soil Testing and Plant Analysis,* ed. DL Westerman, pp. 11-24. Madison, WI: Soil Sci. Soc. of Am. 784 pp.

9. Denton HP, Naderman GC, Buol SW, Nelson LA. 1986. Use of a technical soil classification system in evaluation of corn and soybean response to deep tillage. *Soil Sci. Soc. Am. J.* 50:1309–14

10. Dokuchaev VV. 1883. *Russian Chernozem in Collected Writings.* Vol. 3 (Transl. N. Kaner), Israel Prog. for Sci. Trans. Jerusalem, 1967. Available U.S. Dept. Commerce, Springfield, VA.

11. Dokuchaev VV. 1886. Report to the provincial zenstvo (local authority) of Nizhnii-Norgo-rod, No. 1. Main phases in the history of land assessment in European Russia, with classification of Russian soils. In *Collected Writings,* Vol. 4. Moscow: Acad. Sci., USSR. [1950]

12. Fallou FA. 1862. Pedologie oder allgemeine und besondere Bodenkunde. Dresden, Germany.

13. Geiger R. 1973. [1961] *The Climate Near the Ground.* Cambridge, MA: Harvard Univ. Press. 1961 (From German). 611 pp. 4th ed.

14. Glinka KD. 1927. *The Great Soil Groups of the World.* (Transl. CF Marbut) Ann Arbor, MI: Edwards Brothers. 235 pp.

15. Greenland DJ, Hayes MHB. 1981. Soil process. In *The Chemistry of Soil Process,* ed. DJ Greenland, MHB Hayes. New York: Wiley. 714 pp.

16. Gourley LM. 1993. Success in acid/low fertility soils in Columbia. In *Adaptation of Plants to Soil Stresses, Proc. Aug. 1-4, 1993 Worksh,* pp. 248-268. Lincoln, Nebraska: Univ. Neb., 347 pp.

17. Jenny H. 1941. *Factors of Soil Formation.* New York: McGraw-Hill. 281 pp.

18. Khasawneh FE, Sample EC, Kamprath EJ, eds. 1980. *The Role of Phosphorus*

in Agriculture. Madison, WI: Am. Soc. Agronomy. 910 pp.

19. Lal R, De Vleeschauwer O, Malafa Nganje R. 1980. Changes in properties of a newly cleared tropical Alfisol as affected by mulching. *Soil Sci. Soc. Am. J.* 44:829–33

20. Lal R, Mahboubi AA, Fausey NR. 1994. Long-term tillage and rotation effects on properties of a central Ohio soil. *Soil Sci. Soc. Am. J.* 58:517–22

21. Larson WE, Pierce FH, Dowdy RH. 1983. The threat of soil erosion to long-term crop production. *Science* 219:459–65

22. Marbut CF. 1935. Soils of the United States. In *USDA Atlas of American Agriculture,* Part 3, Advance Sheets No. 8. 98 pp.

23. Middleton HE, Slater CS, Byers HG. 1934. The physical and chemical characteristics of soils from the erosion experiment stations—second report. *Tech. Bull. No. 430, USDA.* Washington, DC: USGPO. 63 pp.

24. Moormann FR, Varley JA, Baker RM, Baker KF, Hughes JC, Brown G. 1981. Appendix 1: Profile locations, descriptions, and analyses. In *Characterization of Soils,* ed. DJ Greenland, pp. 284–421. Oxford: Oxford Univ. Press. pp. 260

25. Munson RD, ed. 1985. *Potassium in Agriculture.* Madison, WI: Am. Soc. Agron. 1223 pp.

26. Newhall F. 1972. Calculation of soil moisture regimes from the climatic record. *USDA-SCS. Rev. 4.* Washington, DC: USGPO

27. Norgaard RB. 1992. Environmental science as a social process. *Environ. Monitoring Assessment* 20:95–110

28. Nye PH, Greenland DJ. 1960. *The Soil Under Shifting Cultivation. Tech. Comm. No. 51.* Harpsenden: Commonwealth Bur. Soil. 156 pp.

29. Odell RT, Walter WM, Boone LV, Oldham MG. 1982. *The Morrow Plots: A Century of Learning. Agr. Expt. Bull. 775.* Univ. Ill. Urbana-Champaign, IL 22 pp.

30. Pavis GA, ed. 1987. *Farm drainage in the United States: History, Status and Prospects. USDA-ERS, Misc. 1455.* Washington, DC: USGPO

31. Peck TR, Soltanpour PN. 1990. The principles of soil testing. In *Soil Testing and Plant Analysis,* ed. RL Westerman, pp. 1-8. Madison, WI: Soil Sci. Soc. Am. 784 pp. 3rd ed.

32. Rassmussen PE, Parton WJ. 1994. Long-term effects of residue management in wheat-fallow: I. Inputs, yields, and soil organic matter. *Soil Sci. Soc. Am. J.* 58:523–30

33. Richthofen FF von. 1886. *Führer für Forschungsreisende.* Berlin

34. Sanchez PA. 1987. Management of acid soils in the humid tropics of Latin America. In *IBSRAM Management of acid tropical soils for sustainable agriculture: Proceedings of an IBSRAM [Int. Board for Soil Res. Manage.] Inaugural Worksh.,* pp. 63-107. Bangkok, Thailand

35. Sanchez PA, Benites JR. 1987. Low-input cropping for acid soils of the humid tropics. *Science* 238:1521-1527.

36. SCS-SMSS. 1986. *Tour Guide—Eighth International Soil Classification Workshop,* Brazil. Mimeo-Nat. Soil Survey Laboratory, USDA-Soil Conservation Service. Lincoln, Nebraska. pp. 285.

37. Simonson RW. 1959. Outline of a generalized theory of soil genesis. *Soil Sci. Soc. Am. Proc.* 23:152–56

38. Skaggs RW, Breve MA, Gilliam JW. 1994. Hydrologic and water quality impacts of agricultural drainage. *Crit. Rev. Environ. Sci. Technol.* 24:1–32

39. Smith GD. 1968. Soil classification in the United States. In *World Soil Resources Rep. 32,* pp. 6–24. Rome, Italy: FAO

40. Smith GD. 1986. *The Guy Smith Interviews: Rationale for Concepts in Soil Taxonomy. SMSS Tech. Monograph No. 11.* Ithaca, NY: Cornell Univ. 259 pp.

41. Soil Survey Staff. 1960. *Soil Classification, A Comprehensive System - 7th Approximation.* Washington, DC: USDA, USGPO. 265 pp.

42. Soil Survey Staff. 1975. *Soil Taxonomy. USDA Agr. Handbook No. 436.* Washington, DC; USGPO. 754 pp.

43. Soil Survey Staff. 1994. *Keys to Soil Taxonomy.* Washington, DC: USDA-SCS. 305 pp. 6th Ed.

44. Stevenson FJ. 1965. Origin and distribution of nitrogen in soil. In *Soil Nitrogen,* ed. WV Bartholomew, FE Clark, pp. 1–42. *Agron. Monograph No. 10.* Madison, WI: Am. Soc. Agron.

45. Switzer GL, Nelson LE, Hinesley LE. 1978. Effects of utilization on nutrient regimes and site productivity. In *Complete Tree Utilization of Southern Pine,* pp. 91–102. Madison, WI: For. Products Res. Soc. 361 pp.

46. Szabolcs I. 1990. Impact of climatic change on soil attributes: Influence on salinization and alkalization. In *Soils on a Warmer Earth,* ed. HW Scharpensel et al, pp. 61–69. *Developments in Soil Science 20.* Amsterdam: Elsevier

47. Tew DT, Morris LA, Allen HL, Wells CG. 1986. Estimates of nutrient removal, displacement and loss resulting

from harvest and site preparation of a *pinus taeda* plantation in the Piedmont of North Carolina. *For. Ecol. Manage.* 15:257–67

48. Thaer A. 1843. Grundsätze der rationellen Landwirtschaft. Cited in *Pedology*, JS Joffe, (1949, 2nd Ed.) New Brunswick, NJ: Pedology Publ. 5th Ed.

49. Thorp J, Smith GD. 1949. Higher categories of soil classification: order, suborder, and great soil groups. *Soil Sci.* 67:117–26

50. Troeh FR, Thompson LM. 1993. *Soil and Soil Fertility.* Oxford: Oxford Univ. Press. 462 pp. 5th Ed.

51. TROPSOILS. 1991. *Annual Report.* North Carolina State Univ. Soil Science Dept., Raleigh, NC.

52. US Bureau of the Census. 1910–1990. *Statistical Abstract of the United States.* Washington, DC: USGPO

53. USDA-SCS. 1990. *Soil Series of the United States, Including Puerto Rico and the U.S. Virgin Islands: Their Taxonomic Classification. Miscellaneous Publication Number 1483: 1990: 723-472/20319.* Washington, DC: USGPO. 459 pp.

54. US Salinity Laboratory Staff. 1954. *Diagnosis and Improvement of Saline and Alkali Soils. USDA, Agr. Handbook No. 60.* Riverside, CA. 160 pp.

55. Van Wambeke A. 1981. *Calculated Soil Moisture and Temperature Regimes of South America. Soil Cons. Serv. USDA-SMSS Tech. Mono. No. 2.* Ithaca, NY: Cornell Univ. Press

56. Van Wambeke A. 1982. *Calculated Soil Moisture and Temperature Regimes of Africa. Soil Cons. Serv. USDA-SMSS Tech. Mono. No. 3.* Ithaca, NY: Cornell Univ. Press

57. Van Wambeke A. 1985. *Calculated Soil Moisture and Temperature Regimes of Asia. Soil Cons. Serv. USDA-SMSS, Tech. Mono, No. 9.* Ithaca, NY: Cornell Univ. Press

58. Wilde SA. 1958. *Forest Soils.* New York: Ronald Press. 537 pp.

59. Whitney M. 1909. *Soils of the United States. U.S. Dept of Agric. Bur. Soils Bull. 55.* Washington, DC: USGPO

Annu. Rev. Ecol. Syst. 1995. 26:45–67

SUSTAINABLE EXPLOITATION OF RENEWABLE RESOURCES

R. Hilborn[1], *C.J. Walters*[2] *and D. Ludwig*[3]

[1]School of Fisheries WH-10, University of Washington, Seattle, Washington 98195,
[2]Fisheries Centre, University of British Columbia, Vancouver, British Columbia,
Canada, V6T 1Z4, [3]Departments of Mathematics and Zoology, University of British
Columbia, Vancouver, British Columbia, Canada V6T 1Z2

KEY WORDS: sustainable yield, harvesting, resource management, optimal exploitation,
             adaptive management

ABSTRACT

Sustainable exploitation of renewable resources depends on the existence of a
reproductive surplus, which is determined by the balance between births,
deaths, and somatic growth. The reproductive surplus differs spatially and
temporally as environmental conditions vary, and even in the absence of
exploitation, change is the rule and constancy is the exception. Sustainable
yields may be estimated by direct experimentation, observation of natural
systems, or deduction from biological understanding. Each of these approaches
has limitations, and for large-scale unique resources, the only way to determine
the response of the population to harvesting is by direct exploitation. To find
the maximum potential yield, the resource must be overexploited at some time,
or very similar resources must have been overexploited. Temporal changes in
environmental conditions mean that information on sustainability collected in
the past may have limited applicability in the future. The unregulated dynamic
of exploiters is to push the resources to overexploitation, and even when
regulated, exploiters have been very successful at modifying their behavior so
that regulations are less effective than anticipated. The most successful institutions at maintaining sustainability have been small-scale community or private ownership.

0066-4162/95/1120-0045$05.00

INTRODUCTION

In this chapter we begin by examining the biological basis for sustainable exploitation of renewable resources. We then consider problems in estimating the potential yield of a resource, the dynamics of the exploiters of the resource, and finally the institutional structures for managing the resources.

The report of the World Commission on Environment and Development (132) elevated concern about sustainable use of natural resources to high priority in the political and scientific agenda. The Ecological Society of America published an extensive research agenda to support the search for sustainable development (72). While the term "sustainable development" has been criticized as an oxymoron (25), the fields of fish, wildlife, and forest management have long histories of concern about sustainability, and much has been learned about the appropriate types of science and policy to support sustained exploitation. The purpose of this paper is to review the development and current state of knowledge in these fields and to suggest the relevance of this knowledge to the general problem of sustainable development.

Regulations to promote the conservation of salmon date from the fourteenth century (85), primarily because of the visibility of salmon in freshwater and the obvious direct impact of habitat destruction on such fish stocks. Problems of overexploitation were also recognized early in the development of fisheries; even in the Pacific northwest where the salmon runs seemed inexhaustible, the need for restrictions on catch was recognized soon after the fishery began to develop (4). The susceptibility of marine fishes to overexploitation was recognized much more slowly (87). Huxley (57, 58) was the champion of a biological school that believed that the fecundity of fishes was so large that fishing could not have an impact on the abundance of fish in the sea. Garstang (39), Hjort (52), and Russell (102) in Europe and Thompson (117, 118) in North America led the way in providing evidence that fishing not only could affect the abundance of marine fishes, but could reduce the abundance of certain fish stock so much that their potential yield was reduced.

By the 1950s the concept of maximum sustainable yield (MSY) was firmly entrenched in fisheries thinking as an objective of management and as an intrinsic property of most natural populations, and an elaborate set of methods had developed for calculating it (8, 66, 98). The simple view of MSY of the 1950s has gradually faded as we have come to recognize the complexity of society objectives, difficulty in estimating the productive potential of natural populations, and the problems in regulating the exploiters of the resources (66).

The histories of forestry and wildlife management show similar patterns. The concept of sustainable forestry dates from 1849 (32). The growth, survival, and economic calculations through which foresters arrive at "annual allowable cuts" and optimum "rotation ages" for harvesting forest tree stands are similar in principle to calculations fisheries biologists use for MSY assessments (18, 40). Wildlife ecologists have used concepts of annual reproductive surplus and logistic population growth to formulate policies for sustainable harvesting (14, 30, 31, 44, 70, 77, 78, 131).

THE BIOLOGICAL BASIS OF SUSTAINABILITY

The Principle of Reproductive Surplus

The biological basis for all sustainable harvesting is reproductive surplus. All natural populations are capable of net population growth under favorable conditions. Every single wheat seed can produce a plant that yields many seeds. When a few seeds per plant are retained for seeding next year's crop, the remainder is the reproductive surplus that can be harvested on a sustained basis. Similarly, a few seeds from mature trees can be grown in nurseries and planted to regenerate forests, or seeds from uncut trees can be a source of natural regeneration. Errington's (30) classic work on muskrats demonstrated one of the ecological mechanisms behind reproductive surplus in animals. Muskrat pairs that held good territories had high survival and reproductive success, and in general such pairs would produce surplus offspring that were forced out of their parents' home territories each year. Many of these dispersing muskrats could be harvested without affecting the parental breeding population, or some of the parents could be harvested and would be replaced by their own offspring. Fisheries scientists have long noted that net recruitment to many fish populations is independent of parental egg production over a wide range of parental abundance, implying very strong compensatory mortality (loss of "surplus" juveniles prior to recruitment) over this range (36, 83, 84).

The magnitude of the reproductive surplus varies greatly. Seeds planted in poor soils with poor moisture may not produce a reproductive surplus, while seeds planted in good soils with adequate amounts of moisture and nutrients may produce a very large reproductive surplus. A deer population in good habitat with few predators may be able to withstand substantial harvesting, while a population in poor habitat or with many predators may not be capable of any sustained yield. If one considers only population sizes, the reproductive surplus (and therefore the potential sustainable yield) depends on a simple balance of births and deaths. If the total biomass of the population is an important determinant of reproductive output, then the somatic growth of individuals in the population needs to be included as well.

The magnitude of the sustainable yield is determined by the reproductive surplus per reproductive unit, and the absolute size of the population. If one seed planted produces five seeds in the harvest, then the sustainable yield will be four times the population size. One must then determine how the rate of reproductive surplus depends on population size, that is, the extent of density dependence in the population. With most plant species, habitat is the dominant factor in limitation of population size, and as more and more seeds are deposited per unit area (be they wheat or trees), the reproductive surplus generated by each seed declines due to competition for space, moisture, and nutrients. There will exist an optimum number of seeds planted per unit of habitat to maximize the reproductive surplus.

Competition for resources must occur for most animal populations, but in the study of exploited fish populations, the relationship between population size and reproductive surplus has been the subject of much controversy. The debate began with Huxley (57, 58) arguing that since fish tend to have enormous numbers of eggs, the number of young fish surviving to reproductive age should be unaffected by the size of the adult population. He further argued that because of the potentially enormous reproductive surplus, fishing could not be expected to have any impact on fish populations. It is now widely accepted that fishing can have a major impact on fish populations, but the debate over the relationship between spawning stock abundance and reproductive surplus continues. In one major fisheries textbook of the 1980s, Gulland (46) summarized the data available: "more commonly the number of recruits is effectively independent of the adult stock size over most of the observed range of stock size." Myers et al (84) reviewed the data for 72 fish stocks and showed that for almost all stocks there was evidence that the reproductive surplus declined as adult population size was reduced.

The relationship between population size and reproductive surplus has been considerably less controversial in wildlife, where the much lower potential fecundity of birds and mammals has led theoreticians and empiricists to recognize the need to maintain an adequate number of breeding females in order to ensure a substantial reproductive surplus. But as noted by Gross (44), until recently there was wide confusion about whether the maximum surplus would occur at the largest breeding population size that the habitat could support (assuming perfect compensatory loss of all the annual surplus not harvested), or at some intermediate size, as is usually assumed in fisheries production theory. Forest ecologists have largely sidestepped the issue of reproductive surplus by assuming either that natural regeneration will always be adequate provided seed tree sources are not too distant from a harvested site (i.e. if a clear-cut site is not too large), or that recruitment will be assured through deliberate planting of seedlings. Forest managers concentrate instead on "silvicultural" practices to ensure adequate habitat for recruitment (physical conditions for seedling growth and survival, control of competition-succession interactions—37, 110).

The shape of the relationship between parental population size and reproductive surplus is one of the most critical features in sustainable harvesting.

If a population is not increasing, then in order to produce a reproductive surplus the population size must be reduced; the question then is how far to reduce it in order to maximize the total reproductive surplus (51). This question plagues everyone trying to determine the sustainable yield and optimal harvesting strategy for natural populations.

Sustainable harvesting requires more than a reproductive surplus. If the act of harvesting destroys the habitat or simplifies a population's genetic or spatial structure, the reproductive surplus may decline over time and could disappear. Some methods of agriculture and forestry deplete soil productivity or lead to erosion and habitat loss; some fishing methods such as dynamite fishing destroy habitat. The population structure of a species may depend on partially isolated populations that exchange individuals. The reproductive surplus in one portion of the population may be essential to recolonization of other habitats. Harvesting the reproductive surplus of the most productive population segments may contribute to the erosion of the spatial structure of the entire population (76, 99, 119).

One of the most pernicious problems in harvest management is that non-sustainable yields are available during the initial development of harvesting of a natural population that has a substantial accumulation of older animals and unproductive local subpopulations (99). This phenomenon leads to "fall down" in forestry, which refers to a drop in yield after an initial period of heavy exploitation. During the first rotation of cutting, each acre cut will be old-growth, but once the old-growth is exhausted, and the regenerated forest begins to be cut, the total harvest falls down from what was obtained from an acre of old-growth to that sustainable from a growing (but not maximum biomass) forest. Such a drop may occur even if a forest is regulated to achieve a balanced age structure (73).

The same problem occurs in exploiting long-lived fishes, where the yields obtainable from an unfished stock may exceed the sustainable yield by several orders of magnitude. For instance, a simple theory of fish exploitation based upon the logistic growth law implies that the optimum stock size for producing sustainable yield is at 50% of the unfished stock size. Thus the process of "fishing down" the stock will involve harvesting, on a one-time basis, 50% of the unfished stock. If the annual mortality rate of adult fish is 20% per year, this means that yield while fishing down will be 16 times the annual sustainable yield (51).

The large yields available during the early stages of forest and fishery development often lead to expectations of larger yields than are sustainable. The exploiting industries almost inevitably develop infrastructure to harvest and process the nonsustainable yield, and once that yield is gone they usually create great economic and political pressure to delay the "fall down." Any such delay is then likely to drive the stock to well below its most productive level, resulting in an even more severe fall down when the inevitable decline does come.

Historical Patterns in Biological Production

The simplest theories of sustainable harvesting treat population dynamics as unfolding in an unchanging environment, and these theories have traditionally tried to explain the changes in population abundance as a consequence of exploitation alone. For a long time, the logistic growth model was the starting point for analysis of harvesting of populations. This model assumes that surplus production depends on two parameters: an intrinsic rate of population growth when population density is low, and a "carrying capacity" population size at which the rate of growth becomes zero. If these two parameters were truly constant over time, the only explanation for population decline would be that the harvest was too high. Unfortunately, this view of the world is incompatible with what we know about natural variation: natural populations, even when unaffected by harvesting, often fluctuate dramatically. Such fluctuations imply that we cannot estimate sustainable production and harvest rates simply by examining the apparent impacts of historical harvesting.

Surveys of unexploited populations (51, 134) show that natural large-scale fluctuations are the rule, and that dramatic increases and declines in abundance have occurred in many populations as long as humans have observed them. Elton (29) documented the fluctuations in snowshoe hares, lynx, mice, voles, and lemmings. Waterfowl surveys show large fluctuations that cannot be attributed to harvesting alone (5). Some of the best documented examples are for fishes.

Caddy & Gulland (12) reviewed the time dynamics of exploited fish populations and found that most populations did not behave in the "steady state" as predicted by simple theory; they showed either cycles of high and low abundance or highly variable behavior the authors labeled "spasmodic." In investigations of the collapse of the California sardine population, Soutar & Isaacs (113, 114) used fish-scale depositions in anoxic sediments to track the abundance patterns of several species of fish off Santa Barbara for the last 2000 years; they found that sardines completely disappeared several times during the period examined. Similar patterns have been found in major pelagic fisheries in Peru (27) and South Africa (106). Indeed, while the collapse of the California sardine was long considered a classic example of fish stock collapse due to overfishing, the current interpretation of this fishery is much more ambiguous (67). Two conferences (65, 134) summarized what is known about long-term patterns in abundance of many fish stocks, and again variability and major change were the rule rather than the exception.

Temperature is perhaps the simplest variable to consider, for the reproductive surplus of almost any species is affected by it. The early life history stages of plants, birds, mammals, and fish are particularly sensitive to their thermal regimes, so we should expect that changes in temperature would affect changes in reproductive surplus. The impact of climate on fisheries has been shown to explain major changes in fish abundance (23, 41).

In addition to the changes in the climate discussed above, other aspects of environment also vary, and with them the potential reproductive surplus. In the late 1970s a number of major changes occurred in fish species of the northeastern Pacific (6, 35). The runs of sockeye and pink salmon to Alaska increased substantially, the abundance of Alaska pollock increased, the populations of king and tanner crab, fur seals, and many bird species declined. These changes appear to represent a shift in ecosystem structure, with a major readjustment in the reproductive surplus of many species. This particular ecosystem shift was noticed and documented because several species in the area were under long-term study (34, 48), and there have been major commercial fisheries for salmon, crab, and pollock. Such ecosystem shifts undoubtedly occur around the world.

Humans can affect the reproductive surplus of species in a variety of ways. Perhaps most obvious is the elimination of habitat. The rapid rate of extinction of species in the second half of the twentieth century is ascribed primarily to the loss of habitat (95). The clearing of temperate, and now tropical, forests (10, 97) for agriculture has been a major factor, as has land use for urbanization (120). The disappearance of Atlantic salmon over much of Europe and North America is largely due to habitat loss from stream blockage and pollution (85).

Introduction of new species, including pathogens, often has striking effects on populations and ecosystems and on the reproductive surplus. In the Great Lakes the inadvertent introduction of lamprey and zebra mussels, and the intentional introduction of Pacific salmon, have altered the reproductive potential of many (if not all) native species. The reproductive surplus of human populations has been repeatedly altered by the introduction of disease, whether bubonic plague into Europe or smallpox into the Americas.

Ecosystem Yield

Analyses of ecosystem energy flow and material cycling have long been used to make assessments and predictions of potential harvest; such ecosystem approaches have been articulated most precisely by Odum and colleagues (89), and they are a key part of recent thinking in ecosystem theory (64, 116) and ecological economics (21, 22, 38). Trophic assessments indicate that a high proportion of the world's primary production is already being appropriated in one way or another for human uses (121, 133), so we may be very near (or even beyond) global limits for total yields of natural products.

There has recently been a resurgence of interest in trophic analysis as a tool for applied ecology, stimulated by methodological developments such as the

ECOPATH trophic modeling system (15). The basic approach has been to use widely available, simple data on abundance, growth, and diet composition to make back-calculations of trophic flows (and likely natural limits to such flows). A key limitation of such calculations is that they are unlikely to provide credible predictions at the species level, even though such predictions are usually critical for economic evaluation of management alternatives (e.g. harvesting tons of low-valued species versus kilograms of high-valued ones that are often predators on the low-valued species).

ESTIMATING THE POTENTIAL YIELD

There are three basic methods to gain scientific understanding—experimentation, observation, and deduction. We are all familiar with the progress science has made with experimentation. For many scientists, the only real science is experimental science. Yet workers in fields such as astronomy must rely on observations of natural events. Charles Darwin is widely regarded as the greatest biological scientist who ever lived, yet he was almost exclusively an observer. Deduction, originally in the form of logic and mathematics and now increasingly as computation (numerical simulation), is also responsible for considerable scientific progress. All three methods have been employed to help understand the potential sustainable harvest from populations and ecosystems.

Experimental Methods

Agricultural experimental stations provide the model for experimental analysis of potential yield. Ideally we would like to compare alternative exploitation regimes on replicated sites under controlled circumstances. The basic requirements for an experimental approach are: 1) There must be several similar "systems"—the systems must be replicable, and 2) results must be available in a timely fashion so that they can be used—the systems must be evaluable. We must consider resources in the context of the spatial and temporal scales required for replication and evaluation.

Many of the major targets of modern sustainable harvesting such as whales and old-growth forest operate on very long time scales so that the potential for replication of experiments in time is limited. Other resources such as bluefin tuna and whales may exist as a few or single unique populations over global spatial scales, so there is no potential for replication in space. The agricultural experimental station model for understanding sustainability of these resources is clearly inappropriate. Other resources more amenable to learning by experimentation (71, 82) include forest management of key processes (7) and of rapidly growing trees, wildlife management of animals such as deer that exist in many discrete populations and have a short life span (43, 80), and many fisheries including freshwater lake management and management of salmon stocks where the life span is short and the potential for spatial replication exists (20, 122). Whole-lake experimentation is perhaps the most fully developed area for whole-ecosystem experimentation (108, 109). Many problems can be approached by a combination of methods. For instance, experimental methods can be used to evaluate certain components of hypotheses such as the reaction of plants to changes in CO_2 levels (115).

Understanding the potential for sustainable harvesting by experimentation is the natural approach for someone trained as a scientist: We would experimentally determine the shape of the reproductive surplus relationship to breeding population size, either by simultaneous experimentation with a number of "replicate" systems or temporally with the same system. The problem with this experiment is that it requires overexploitation of some of the systems at least at some times (driving it/them to low breeding population size) in order to determine the reproductive surplus at low population sizes. This creates two potential problems: If the resource is valuable, it may seem unwise to overexploit it deliberately. Secondly, if there are irreversible effects, or even long delays in recovery, the resource may be effectively destroyed. However, to understand the limits to exploitation, it is necessary to exceed the limits in some of the resources (51, 122, 127, 128).

Methods Based on Observation

The alternative to direct experimentation is learning from historical experience. If we have considerable experience with the same type of resource, we may know a great deal about the limits to sustainability. Because thousands of freshwater lakes have been managed in different ways, and hundred of stocks of Pacific salmon have been subjected to different harvests, we know a great deal about the response of these resources to exploitation. This type of analysis is analogous to the distinction between medical treatment trials and retrospective analysis. For a scientist, planned experiments with randomized treatments would be preferred, but we may still learn from repeated, unplanned experience.

The key conditions for good observational results are 1) a wide range of treatments have been applied, 2) the systems that have been treated were similar to begin with, and 3) the treatments produced differential responses. It is rare for any of these conditions to be met in observational data sets on natural resources. It is rare for management agencies to overexploit or underexploit resources deliberately. Thus, if systems are "well managed," the amount of contrast in treatments will be much less than would occur if the management regimes were experimental designs. Adequate replication (and measurement of replicate responses) of natural resource treatments are rare, and it is always possible that the lessons learned on one lake, river, etc will not apply to another system. Finally, the results of ecological experimentation or evaluation are frequently ambiguous and rarely cover all important response time scales.

In summary, learning by observation is very difficult with unique resources. If we want to learn how intensively a population can be exploited, we can find its limits only by exceeding them. WF Thompson (118), a pioneer of fisheries science, said "there is no way of knowing the strain a species will stand save by submitting it to one." This is perhaps the central problem with the concept of MSY: You cannot determine it without exceeding it. In experimental systems with replicates, you can exceed it in small experimental units where the cost of overexploitation is small.

Deductive Methods

Learning by observation is least helpful for "new" problems with which we have little experience. The issue is what to do when we are forced to extrapolate beyond the range of experience, and when experimentation is either physically impossible or economically impractical.

To make extrapolations, we are forced to rely at least to some degree on "general principles"—combining historical knowledge of the system with specific functional knowledge about key processes. Climate modeling is perhaps the largest-scale current example. Ecological examples include forest models based on individual tree simulations (110), population simulations by the International Whaling commission (60), and individual-based models (IBMs) for various fish populations (26).

There are fundamental logical and practical limits on our ability to make deductions about the behavior of macroscopic systems in nature by using a reductionist or functional approach (90, 116, 124, 129). These limits arise from factors ranging from inability to specify completely initial conditions, or to assess dependence of some predictions on initial conditions (chaotic dynamics), or to test model credibility based on historical data. These problems will necessarily force us to admit that several—and often many—alternative hypotheses are consistent with the data. Even in situations like the Laurentian Great Lakes (81) or the Columbia River Basin (69), where there have been profound human disturbances ranging from introduction of exotic species to overfishing to gross pollution, we are unable today to construct models that unequivocally apportion the blame for observed changes among those known disturbance agents.

Tracking Changes in Productivity

Experimentation, observation, and deduction can all be aimed at understanding the response of ecological systems to exploitation. But the primary presumption in adopting these approaches is that responses observed in the past will be repeated in the future if the same exploitation regime is applied. This assumption requires that the biological processes that give rise to the reproductive surplus relationship are stable in time. Earlier we discussed environmental and human factors that cause us to expect that reproductive surplus relationships will not be stable in time (123). This says, in effect, that our past experience about a system will become outdated, either suddenly with a regime shift, or perhaps slowly.

If changes are sudden, then after every shift we must discard our old understanding and start again with experimentation, observation, or deduction. If shifts occur often, we may not have learned about the new system before it changes. We have the same problem with smooth change; if the change happens faster than we can learn, then we can never quite achieve an optimal policy.

There are three possible solutions to the problem of changing production relationships. First, we can try to invest more resources in experimentation, observation, and deduction to learn faster, to try to beat the rates of natural change. Second, we can try to learn about and understand the dynamics of change. Finally, we can accept the limitations of our ability to learn about the systems and search for management policies that are robust to the changes.

We would argue that the first approach has very limited potential. Holling (54) argued that the apparent predominance of change in production relationships was a product of our mental time frame. Reductionist science has taught us to concentrate on detailed analysis at small space and time scales, but as Holling notes:

> They [fast changes] are increasingly caused by slow changes reflecting decadal accumulations of human influences on air and oceans and decadal to centuries transformations of landscapes. Those slow changes cause sudden changes in fast environmental variables that directly affect the health of people, productivity of renewable resources, and vitality of societies. Therefore analysis should focus on the interactions between slow phenomena and fast ones and monitoring should focus on long-term, slow changes in structural variables (Ref. 54, p. 553).

Events like El Niño were once complete surprises. As we have acquired more experience with them, we can incorporate the occasional El Niño event into our world view and anticipate how such events will affect the productivity of resources.

There is some prospect for designing highly robust policies for dealing with long-term variation in biological carrying capacities. For example, it is easily shown that the optimum annual harvest rate (proportion of population harvested each year) is independent of the carrying capacity parameter in most models for harvested populations; this rate depends, instead, only on the one parameter that defines the intrinsic rate of population increase. Thus if a fixed, time-invariant harvest rate is applied to a population with temporally varying carrying capacity, we expect the breeding population size to change "automatically" so as to track the carrying capacity change. The resulting harvest sequence will be near the theoretical optimum that could be achieved if the carrying capacity change were predictable, provided the change is not too rapid (91).

The Interaction Between Management Action and Knowledge

Some biological responses simply cannot be estimated or observed except through deliberate change in management policies. For instance, experience with Pacific salmon has shown that the relationship between number of fish allowed to spawn and the subsequent return of adult fish will show compensation, so that increasing number of spawners beyond some point will result in reduced rates of return, until the marginal benefit of additional spawners is zero. Biologists have conducted many studies to try to determine the factors that produce the compensatory mortality, and they have some idea of the "optimal" density for spawners per unit of habitat area. There has long been considerable discussion about the best number of fish to allow to spawn in the Fraser River sockeye salmon run, Canada's most important salmon run (100, 130). Yet biological knowledge of sockeye salmon and their habitat has not been sufficient to determine whether there would be benefits from increasing the number of fish allowed to spawn twofold or more. Walters & Hilborn (127) posed this as a problem in adaptive management, using management action to determine whether larger numbers of fish allowed to spawn would produce large adult returns. They argued that no small-scale data collection or experimentation could be performed to determine the impact of such a major change in number of fish allowed to spawn, since the response would likely involve large-scale colonization by fish of prehistoric spawning areas for which there are no productivity data (and little experience to predict how and when fish might disperse and colonize). The potential benefits were very large, while the costs in terms of foregone catch were also quite significant. In the 1980s and 1990s the escapements were roughly doubled on the Fraser River (by reducing harvest rates), and the number of adult fish produced also roughly doubled. A similar experiment was performed on the much smaller Rivers Inlet stock of sockeye salmon in British Columbia, where increased escapements did not result in larger adult returns (126).

The use of deliberate management changes as experimentation has come to be called "adaptive management." There has been much theoretical work and experience on when and how to do adaptive experiments (53, 59, 75, 122, 127, 128). This work involves explicit recognition that renewable resource decision-making is a problem in decision analysis, where the potential benefits and costs of alternative management actions are weighed, and the value of information learned in the process of management is incorporated in the long-term expected value of resource management. Three key issues in adaptive management have turned out to be: 1) the extent to which actions are reversible,

2) whether the system can be understood by small space and small time-scale experimentation, and 3) whether the rate of learning about the system is rapid enough to provide useful information for subsequent decisions.

Sainsbury (104, 105) describes the fisheries of NW Australia, where the abundance of high-value fishes had declined, and that of the low-value fishes had increased. He considered six hypotheses consistent with what was known about the biology of the ecosystem and the available data, but that resulted in dramatically different predictions about the consequences of different management actions. In such a circumstance there are two fundamental choices: either to take the action that appears to be best given your present knowledge, or to take an action that is expected to provide better ability to discriminate between alternative hypotheses at a later date. The second type of policy generally involves giving up short-term benefits (or taking higher short-term risks of overharvest) in favor of expected long-term benefits due to improved understanding.

The Limitations of Our Knowledge

Resource managers look to scientists to tell them the status of their resources and the potential yield. Ludwig et al (74) argued that such questions simply cannot be answered reliably in many, if not most, circumstances, because learning about natural resource systems is limited by 1) lack of replicates and controls, 2) lack of randomization in treatments in natural experiments, and 3) changes in underlying systems (49). They argued that managers pose the wrong questions—scientists cannot say very precisely how many fish are in the sea, or what the sustainable yield is, but they can design a monitoring and management system that will provide for long-term sustainable harvesting without severe risk to the stock.

The Scientific Committee of the International Whaling Commission (60) spent almost 10 years designing a harvest system that would meet the combined objectives of long-term sustainable exploitation with minimal risk to natural populations. The management system combined periodic surveys of the population abundance with very conservative quota-setting algorithms, demonstrated by simulation to be robust to a wide variety of alternative whale population structures and biological parameters. Despite the obvious difficulty in managing world-wide resources that are slow-growing, there is general consensus that the management procedures would work. However, the management procedures made no claim to be able to determine the abundance or long-term sustainable yield very reliably.

This highlights two perspectives on the appropriate direction for research in managing renewable resources. One school (107) suggests that intense detailed scientific research on the biological basis of the systems will provide improved understanding that in turn will lead to better management. A second school of thought (11, 125) argues that the space and time scales of many major systems are such that traditional scientific research will not provide additional useful improved understanding, and that improved design of the monitoring and management systems will provide greater benefits. Holling (54) & Lee (68) both argue that there is an important role for scientific research, but not if it is merely "disciplinary, reductionist and detached from people, policies and politics" (54). Holling argues that the needed research should be interdisciplinary, nonlinear, focused on the interaction between slow processes and fast ones, and should study cross-scale phenomena.

Scientific assessment blunders have played a major role in the collapse of some potentially sustainably harvested systems, particularly in fisheries. Early in the development of the Peru anchoveta fishery (once the world's largest fishery), leading fisheries scientists from around the world used a faulty equilibrium method for surplus production assessment to advise the Peruvian government that the fishery might be developed to a scale of 10–12 million tons annual catch (9, 45), a figure more than double what the fishery is likely actually to sustain (2). Similar problems have plagued assessment in other pelagic fisheries (92, 96). When Canada took over extended management jurisdiction (200-mile limit) of its east coast fish stocks in the late 1970s, after a period of intense fishing by foreign fleets, scientists overestimated the remaining abundance of cod off Newfoundland by over 200%, leading to a Canadian development policy that virtually destroyed the cod stock by 1991 (33, 56). Shrimp fisheries world-wide are now threatened by severe recruitment overfishing, justified in the past by arguments that there was no stock-recruit relationship (no limit to reproductive surplus) in shrimp (79).

Such blunders are not restricted to fisheries. There was much concern in the 1970s about overharvest versus habitat factors in causing declines in some North American waterfowl species, and there were calls for a large-scale experimental approach to the issue (86). But when this experiment came in 1979–1984, as the so-called stabilized duck hunting regulations evaluation (SDHRE), it involved the least informative possible experimental treatment—stabilized rather than variable harvest rates. Debate continues about whether waterfowl are being overharvested across North America (63).

British Columbia is one of North America's largest timber production areas, with an economy highly dependent on the forest industry. The provincial Ministry of Forests has recently announced a number of reductions in annual allowable cuts (and forestry employment) in order to correct for errors in sustainable harvest estimation made during the 1970s. These errors mainly involved overestimates in the "operable" (productive and economically accessible) forest land area (rather than production rates per unit area), due partly to underestimates of land withdrawals from the production base for purposes of environmental protection (parks, reserves, etc.)

THE DYNAMICS OF EXPLOITERS

Perhaps the biggest failure in natural resource management has been the widespread neglect of the dynamics of the exploiters. The tradition in fisheries, wildlife, and forestry has been to train managers in the biological understanding of fish, wildlife, and trees, yet once involved as managers, they soon discover that they are managing people much more than fish, wildlife, or trees.

Gordon (42) was the first to examine the economic behavior of fishermen and explain why most fisheries were unprofitable and overexploited. This work formed the quantitative basis for understanding the tragedy of the commons (47), and it has led to an extensive literature in both economics and fisheries (19). The economic theory, which assumes that individuals are unregulated and that they attempt to maximize individual profitability, suggests that the abundance of a resource will be depleted to a point where it is not profitable for anyone to expand the fishing effort. In simple theory this is known as the "bionomic equilibrium." Whether the stock is biologically overexploited at the bionomic equilibrium depends upon the relationship among catch rate and abundance, price of the product, and costs of fishing. If price is high or costs low, or if catch rate does not decline as rapidly as abundance, the normal outcome is biological overexploitation. In the extreme, when the animals are easy to capture and their reproductive rate is low, the bionomic equilibrium may result in extinction of the resource (17). Gordon's theory of economic behavior in open access resources has been validated by numerous empirical studies and is now widely accepted not only in fisheries, but in all open-access resource use including wildlife, forestry, and water rights.

Long before the theory of open access fisheries was developed, societies recognized the need to regulate resources to prevent overexploitation. Pacific Islanders adopted many social mechanisms to reduce exploitation rates, including property rights and closed areas and times invoked by tribal customs (61, 62). Maine lobstermen have long engaged in self-regulation by a combination of social agreement and territoriality (1). Indeed forms of self-regulation are common in many traditional fisheries. However, regulation of commercial and recreational fisheries has generally required direct intervention. Regulations to prevent overexploitation of salmon in England date back to at least 1285 (85), and closed seasons and gear restriction were common by the fourteenth century. In the twentieth century, various forms of regulation are accepted as common practice including seasons, gear restrictions, daily catch limits on individuals or vessels, and more commonly total catch limits.

Exploiters rarely are passive in the face of regulation. When fisheries managers attempt to limit overcapitalization by limiting the number of vessels in a fishery, the vessels are built longer, wider, and with bigger engines. When the length of vessels is limited, the new boats are built wider. When total size or tonnage is limited, additional electronics are added. It has proven almost impossible to prevent fishermen from increasing their harvesting power if it is in their individual interests to do so. The only fishery management systems that have proven effective in preventing increased capitalization in catching power are those with some form of property rights that changed the economic incentives for individual fishermen or firms (93, 94).

INSTITUTIONAL APPROACHES TO ACHIEVING SUSTAINABILITY

The most successful institution for promoting sustainable exploitation of fish, wildlife, and forests has been private ownership (16). The use of private property rights in traditional tropical fisheries (61), private forests around the world, private game parks in Africa and Europe (55), and private freshwater fishing preserves illustrates that long-term sustainable exploitation of renewable resources is possible. Private ownership does not guarantee sustainable exploitation; the owners of private resources will use the resource in such a way as to maximize their own return from the properties. Under private ownership, an old-growth forest of native trees is likely to be harvested quickly and replaced by regenerating forests, often with different species mixes. Private fishing reserves almost always replace native fishes with more "desirable" species.

It may be in the interest of private property owners to overexploit or drive to extinction species that have low reproductive rates (17, 19), and in times of financial crisis private property owners may be led into depletion of their own resources. On a longer time scale, private property owners have little economic incentive to maintain genetic diversity, soil structure or depth, or other attributes with a rate of change that is small compared with the discount rate of money.

Private ownership of fish resources in industrial countries is unusual, and the majority of institutions currently in place involve public ownership or administration. The dominant management institution has been the state or national fisheries agency (50, 88). For fish stocks that cross international boundaries or are found on the high seas, international agencies such as the International Pacific Halibut Commission (IPHC), International Pacific Salmon Fisheries Commission (IPSFC), International Commission for North Atlantic Fisheries (ICNAF), and the International Commission for South East Atlantic Fisheries (ICSEAF) have been formed. IPHC, IPSFC, and ICSEAF can all be considered successful in that the biological status of the major fish stock under their jurisdiction is healthy. ICNAF has been considerably less successful. Generally, the commissions with a smaller number of member states have been better able to implement restrictions, whereas agencies with many members that rely on consensus are obviously less able to take tough measures. There is a strong negative correlation between institutional complexity and the health of the stocks. Small countries or organizations that have few players (New Zealand, Iceland, IPHC) have much better track records than do more complex institutions such as the US and European fisheries agencies.

Regardless of institutional structure, as the need to preserve or rebuild fish stocks has become more apparent, two divergent approaches have been tried (112). The first approach is to reduce catches through gear restrictions, closed times, closed areas, and total catch quotas. While the responses of fishermen often thwart these measures due to increased efficiency or political pressure to prevent reductions in catch (103), the biological health of fisheries such as the Pacific halibut and salmon in British Columbia and Alaska shows that these methods can work. Despite the natural fluctuations in these stocks, the management agencies have used the traditional tools of fisheries management to assure that the reproductive potential of the stocks has not been threatened.

The second approach recognizes the difficulty in restricting fishermen, and instead attempts to maintain fish stocks by artificial propagation of juveniles in hatcheries and releasing these fish into the wild. Large-scale releases of salmon in the Pacific northwest and cod in the Atlantic began in the nineteenth century and have been almost totally ineffective at maintaining natural populations (112).

Fisheries institutions are attempting to correct the imperfections we have listed in two ways. The first is using different forms of property rights to change incentives for fishermen. Limited entry and area licensing are now common in industrialized countries, and these remedy some of the problems of economic inefficiency. The second attempts to account for intrinsic scientific uncertainty by using cautious management, often called the "precautionary principle" (13). Many management agencies are now reluctant to allow high fishing mortalities or to reduce spawning stocks to low abundance.

SUMMARY

The key lessons learned from the study of sustainable exploitation of fish, wildlife, and forests are: 1) The historical record shows that biological overexploitation is almost universal at some point in the development of a resource, and even when biological overexploitation is avoided, economic overexploitation is the norm; 2) to avoid overexploitation there must be deliberate willingness to forego attempts at maximizing yield; 3) we have the knowledge (from plenty of historical experience with overexploitation) to design management systems that will provide long-term sustainable harvest even when tracking unpredictable environmental changes, but 4) institutionally, we are generally unable to control exploiters well enough to make the changes necessary to track changing biological productivity and biological understanding. Successful management in the future will rest not so much on better science as on the implementation of better institutional arrangements for controlling exploiters and creating incentives for them to behave more wisely.

While some (3, 101) have challenged our assertion (74) that overexploitation is nearly inevitable, few fisheries that have been exploited for long periods have not gone through periods of both biological and economic overexploitation. The two North American fisheries commonly cited as well managed—Pacific halibut and Pacific salmon (3)—both went through periods of biological overexploitation in their development and are now economically overexploited.

The experience in wildlife management shows many similarities in understanding dynamics, particularly for large-scale systems such as caribou, moose, North American waterfowl, and East African wildlife. The difficulties of determining the potential harvest with large-scale, long-lived resources that occur in changing ecosystems are not affected by the taxonomic group. However, wildlife management has several advantages, including the considerable potential for replication and transfer of historical experience, and the economic pressure on wildlife resources is generally less than on commercial fisheries.

In forestry, the time scales are very large, but replication is possible. It is reasonably easy to establish what appears to be sustainability on a short time horizon, but it is hard to determine long-term sustainability with respect to nutrient cycling, soil loss, and climate change. However, historical experience shows that the economic pressure for forest overexploitation is severe.

Science has an important role to play in advancing the understanding of sustainability, but we agree with Butterworth (11) who argues that most science has not been directed toward this objective. The existing models of funding for science have led to the greatest portion of the funds being spent on basic research on spatial and temporal scales that are very unlikely to produce results useful for sustainable management—more research should concentrate at the appropriate scales where the important processes act (54, 68, 74).

Ehrlich & Daily (28) have suggested that fisheries are not typical of the problems facing sustainable development, and that the lessons we listed above have limited application to issues such as sustaining soil, freshwater, forests, atmospheric composition, and biodiversity. We believe that the lessons learned from fish, wildlife, and forests are quite applicable to these problems, and that much of the disagreement between authors on this subject is less than may appear. Ehrlich & Daily discuss three components in scientific understanding (24): 1) problem perception, 2) mechanistic understanding, and 3) strategic assessment. Using this framework, fisheries management went through problem perception in the late nineteenth and early twentieth centuries, the mechanistic understanding evolved from about 1900 to the 1950s when the modern theory of fishing was largely fully developed, and the problem of strategic assessment has been the main focus from the 1950s to the present. Considerable uncertainty remains about some of the mechanisms, especially the relative importance of environment and harvesting. The strategic assessments are frequently full of uncertainty, and as our scientific understanding has developed, we have begun to recognize that the uncertainty is considerably larger than previously admitted. The more we learn, the more we must admit we don't know. As society moves toward trying to find optimum levels of release for substances like atmospheric pollutants, we will undoubtedly discover that tracking atmospheric sustainability is as difficult as tracking sustainability in fish, forests, and wildlife.

ACKNOWLEDGMENTS

This research was supported by grants from Washington Sea Grant Program and the National Sciences and Engineering Research Council of Canada under grants A9239 and A5869.

Literature Cited

1. Acheson JA. 1972. Territories of the lobsterman. *Nat. Hist.* 81:60–69
2. Aguero M. 1987. A bioeconomic model of the Peruvian pelagic fishery. In *The Peruvian Anchoveta and Its Upwelling Ecosystem: Three Decades of Change,* ed. D Pauly, I Tsukayama, pp 307–24. ICLARM Stud. Rev. 15
3. Aron W, Fluharty D, McCaughran D, Roos JF. 1993. Fisheries management. *Science* 261:813
4. Babcock JP. 1902. Report of the Commissioner of Fisheries of British Columbia. *Can. Govt. Rep.* Ottawa
5. Barker RJ, Sauer JR. 1991. Modelling population change from time series data. See Ref. 78, pp. 182–94
6. Beamish RJ, McFarlane GA (Eds). 1989. Effects of ocean variability on recruitment and an evaluation of parameters used in stock assessment models. *Can. Spec. Publ. Fish. Aquat. Sci. No. 108.*
7. Beier C, Rasmussen L. 1994. Effects of whole-ecosystem manipulations on ecosystem internal processes. *TREE* 9:218–23
8. Beverton RH, Holt SJ. 1957. On the dynamics of exploited fish populations. *UK Min. Agric. Fish., Fish. Invest. (Ser. 2) No. 19*
9. Boerma, LK, Gulland J. 1973. Stock assessment of the Peru anchoveta (*Engraulis ringens*) and management of the fishery. *J. Fish. Res. Board Can.* 30: 2226–35
10. Botkin DB, Talbot LM. 1992. Biological diversity and forests. In *Managing the Worlds's Forests, Looking for Balance Between Conservation and Development,* ed. NP Sharma, pp. 47–74. Dubuque, IA: Kendall/Hunt
11. Butterworth DS. 1989. The Benguela ecology programme: successful and appropriate? *S. Afr. J. Sci.* 85:633–43
12. Caddy JF, Gulland JA. 1983. Historical patterns of fish stocks. *Mar. Policy 7:* 267-78
13. Cameron J, Abouchar J. 1991. The precautionary principle: a fundamental principle of law and policy for the protection of the global environment. *B.C. Int. Comp. Law Rev.* 14:1–27
14. Caughley G, Sinclair ARE. 1994. *Wildlife Ecology and Management.* Oxford: Blackwell Sci.

15. Christensen V, Pauly D. 1993. *Trophic Models of Aquatic Ecosystems.* Manila, Philippines: ICLARM

16. Christy FT JR. 1969. Fisheries goals and the rights of property. *Trans. Am. Fish. Soc.* 98:369–78

17. Clark CW. 1973. The economics of overexploitation. *Science* 181:630–34

18. Clark CW. 1976. *Mathematical Bioeconomics: The Optimal Management of Renewable Resources.* New York: Wiley-Intersci.

19. Clark CW. 1985. *Bioeconomic Modelling and Fisheries Management.* New York: Wiley

20. Collie JS, Walters CJ. 1993. Models that 'learn' to distinguish among alternative hypotheses. *Fish. Res.* 18: 259–75

21. Costanza R, ed. 1991. *Ecological Economics: The Science and Management of Sustainability.* New York: Columbia Univ. Press

22. Costanza R, Cornwell L. 1992. The 4P approach to dealing with scientific uncertainty. *Environment* 34:12–20,42

23. Cushing DH. 1982. *Climate and Fisheries.* London: Academic

24. Daily GC, Ehrlich PR. 1992. Population, sustainability, and earth's carrying capacity. *BioScience* 42:761–71

25. Daly HE. 1990. Towards some operational principles of sustainable development. *Ecol. Econ.* 2:1,5

26. DeAngelis DL, Gross LJ, eds. 1992. *Individual-Based Models and Approaches in Ecology.* New York: Chapman & Hall

27. De Vries TJ, Pearcy WG. 1982. Fish debris in sediments of the upwelling zone off central Peru: a late Quaternary record. *Deep-Sea Res.* 28:87–109

28. Ehrlich PR, Daily GC. 1993. Science and the management of natural resources. *Ecol. Applic.* 3:558–60

29. Elton CS. 1942. *Voles, Mice, and Lemmings; Problems in Population Dynamics.* Oxford: Clarendon

30. Errington PL. 1946. Predation and vertebrate populations. *Q. Rev. Biol.* 21: 221–45

31. Errington PL. 1963. *Muskrat Populations.* Ames, IA: Iowa State Univ. Press

32. Faustman M. 1968. *On the determination of the value which forest land and immature stands poses for forestry.* [Orig. publ. 1849.] Transl. in M. Gane, ed. *Inst. Pap. 42,* Commonwealth For. Inst., Oxford Univ.

33. Finlayson AC. 1994. *Fishing for truth: a sociological analysis of northern cod stock assessments from 1977 to 1990. Inst. Soc. Econ. Res., Soc. & Econ. Stud.* No. 52, Memorial Univ. Newfoundland, St. John's, Canada

34. Francis RC, Hare SR. 1994. Decadal-scale regime shifts in the large marine ecosystems of the North-East Pacific: a case for historical science. *Fish. Oceanogr.* 3:279–91

35. Francis RC, Sibley TH. 1991. Climate change and fisheries: what are the real issues? *Northwest Environ. J.* 7:295–307

36. Frank KT, Leggett WC. 1994. Fisheries ecology in the context of ecological and evolutionary theory. *Annu. Rev. Ecol. Syst.* 25:401–22

37. Franklin JF, Perry DA, Schowaiter TD, Harmon ME, McKee A, Spies TA. 1989. Importance of ecological diversity in maintaining long-term site productivity. In *Maintaining the Long-Term Productivity of Pacific Northwest Forest Ecosystems,* ed. DA Perry, pp. 82–97. Portland, OR: Timber

38. Folke C, Kaberger T., eds. 1991. *Linking the Natural Environment and the Economy: Essays from the Eco-eco Group.* Dordrecht, Netherlands: Kluwer Acad.

39. Garstang W. 1900. The impoverishment of the sea—a critical summary of the experimental and statistical evidence bearing upon the alleged depletion of the trawling grounds. *J. Mar. Biol. Assoc. UK* 6:1–69

40. Getz W, Haight RG. 1989. *Population harvesting: Demographic Models of Fish, Forest, and Animal Resources.* Princeton, NJ: Princeton Univ. Press

41. Glantz MH. 1992. *Climate Variability, Climate Change, and Fisheries.* Cambridge: Cambridge Univ. Press

42. Gordon HS. 1954. The economic theory of a common property resource: the fishery. *J. Polit. Econ.* 62:124–42

43. Gratson MW, Unsworth JW, Zager P, Kruck L. 1993. Initial experiences with adaptive resource management for determining appropriate antlerless elk harvest rates in Idaho. *Trans. N. Am. Wildl. Nat. Resour. Conf.* 58:610–19

44. Gross JE. 1969. Optimum yield in deer and elk populations. *Trans. N. Am. Wildl. Nat. Resour. Conf.* 35:372–86

45. Gulland JA. 1968. Population dynamics of the Peruvian anchoveta. *FAO Fish. Tech. Paper 72*

46. Gulland JA. 1983. *Fish Stock Assessment: A Manual of Basic Methods.* New York: Wiley

47. Hardin. G. 1968. The tragedy of the commons. *Science* 162:1243–48

48. Hare SR, Francis RC. 1995. Climatic change and salmon production in the northeast Pacific Ocean. In *Ocean Cli-*

mate and Northern Fish Populations, ed. RJ Beamish. Can. Spec. Publ. Fish. Aquat. Sci. 121: In press

49. Hilborn R, Ludwig D. 1993. The limits of applied ecological research. Ecol. Applic. 3:550–52

50. Hilborn R, Pikitch EK, Francis RC. 1993. Current trends in including risk and uncertainty in stock assessment and harvest decisions. Can. J. Fish. Aquat. Sci. 50:874–80

51. Hilborn R, Walters CJ. 1992. Quantitative Fisheries Stock Assessment: Choice, Dynamics and Uncertainty. New York: Chapman & Hall

52. Hjort J. 1914. Fluctuations in the great fisheries of northern Europe. Rapports, Cons. Perm. Int. Explor. Mer, 20

53. Holling CS ed. 1978. Adaptive Environmental Assessment and Management. Chichester England: Wiley

54. Holling CS. 1993. Investing in research for sustainability. Ecol. Applic. 3:552–55

55. Hudson, RJ, Drew KR, Baskin LM. eds. 1989. Wildlife Production Systems. Cambridge: Cambridge Univ. Press

56. Hutchings JA, Myers RA. 1994. What can be learned from the collapse of a renewable resource? Atlantic cod, Gadus morhua, of Newfoundland and Labrador. Can. J. Fish. Aquat. Sci. 51: 2126–46

57. Huxley TH. 1881. The herring. Nature 23:607–13

58. Huxley TH. 1884. Inaugural address. Fisheries Exhibition Lit. 4:1–22

59. Imperial MT, Hennessee T, Robadue Jr. D. 1993. The evolution of adaptive management for estuarine ecosystems: the national estuary program and its precursors. Ocean Coast. Manage. 20:147–80

60. International Whaling Commission. 1988. Comprehensive assessment workshop on management. Rep. Int. Whal. Commn. 38

61. Johannes RE. 1978. Traditional marine conservation methods in Oceana and their demise. Annu. Rev. Ecol. Syst. 9:349–64

62. Johannes RE. 1981. Words of the Lagoon. Berkeley: Univ. Calif. Press

63. Johnson DH, Owen M. 1991. World waterfowl populations: status and dynamics. See Ref. 78, pp. 635–52

64. Jorgensen SE, Patten BC, Straskraba M. 1992. Ecosystems emerging: toward an ecology of complex systems in a complex future. Ecol. Modelling 62:1–27

65. Kawasaki TS, Tanaka S, Toba Y, Taniguchi A. eds. 1991. Long-Term Variability of Pelagic Fish Populations and Their Environment. Oxford: Pergamon. 402 pp.

66. Larkin PA. 1977. An epitaph for the concept of maximum sustained yield. Trans. Am. Fish. Soc. 106:1–11

67. Lasker R, MacCall A. 1983. New ideas on the fluctuations of the clupeoid stock off California. Proc. Joint. Oceanogr. Assembly, 1982. General Symposia, pp. 110–20. Ottawa: Can. Natl. Comm./SCOT

68. Lee KN. 1993. Greed, scale mismatch, and learning. Ecol. Applic. 3:560–64

69. Lee KN. 1993. Compass and Gyroscope: Integrating Science and Politics for the Environment. Washington, DC: Island

70. Leopold AS. 1955. Too many deer. Sci. Am. 139:101–8

71. Likens GE. 1985. An experimental approach for the study of ecosystems. J. Ecol. 73:381–96

72. Lubchenco J, Olson AM, Brubaker LB, Carpenter SR, Holland MM, et al. 1991. The Sustainable Biosphere Initiative: an ecological research agenda. Ecology 72: 371–412

73. Ludwig D. 1993. Forest management strategies that account for short-term and long-term consequences. Can. J. For. Res. 23:563–72

74. Ludwig D, Hilborn R, Walters C. 1993. Uncertainty, resource exploitation, and conservation: lessons from history. Science 260:17,36

75. McAllister MK, Peterman RM. 1992. Experimental design in the management of fisheries: a review. N. Am. J. Fish. Mgmt. 12:1–18

76. McCall AD. 1990. Dynamic Geography of Marine Fish Populations. Seattle: Univ. Wash. Press

77. McCullough DR. 1990. Detecting density dependence: filtering the baby from the bath water. Trans. N. Am. Wildl. Nat. Resour. Conf. 55:534–43

78. McCullough DR, Barrett RH, eds. 1991. Wildlife 2001: Populations. London: Elsevier Appl. Sci.

79. McGuire TR. 1991. Science and the destruction of a shrimp fleet. Maritime Anth. Stud. 4:32–55

80. McNab J. 1983. Wildlife management as scientific experimentation. Wildl. Soc. Bull. 11:397–401

81. Milliman SR, Grima AP, Walters CJ. 1987. Policy making within an adaptive management framework, with an application to lake trout (Salvelinus namaycush) management. Can. J. Fish. Aquat. Sci. 44 (Suppl. 2):425–30

82. Mooney HA, Medina E, Shindler DW, Schulze EW and Walker BH. Editors.

1990. *Ecosystem Experiments.* Sci. Comm. on Prob. of Environ. ed. Vol. 45. New York: Wiley

83. Myers RA, Barrowman NJ. 1994. Is fish recruitment related to spawner abundance? Int. Council Explor. Sea G:37

84. Myers RA, Rosenberg AA, Mace PM, Barrowman N, Restrepo VR. 1994. In search of thresholds for recruitment overfishing. *ICES J. Mar. Sci.* 51:191

85. Netboy A. 1968. *The Atlantic Salmon: A Vanishing Species?* Boston: Houghton Mifflin

86. Nichols AD, Johnson FA. 1989. Evaluation and experimentation with duck management strategies. *Trans. N. Am. Wildl. Nat. Resour. Conf.* 54:566–93

87. Nielsen LA. 1976. The evolution of fisheries management philosophy. *Mar. Fish. Rev.* 38:15–22

88. O'Boyle R. 1993. Fisheries management organizations: a study of uncertainty. *Can. Spec. Publ. Fish. Aquat. Sci.* 120:423–36

89. Odum HT. 1982. *Systems Ecology.* New York: Wiley

90. Oreskes N, Shrader-Frechette K, Belitz K. 1994. Verification, validation, and confirmation of numerical models in the earth sciences. *Science* 263:641–46

91. Parma A, Deriso R. 1990. Experimental harvesting of cyclic stocks in the face of alternative recruitment hypotheses. *Can. J. Fish. Aquat. Sci.* 47:595–610

92. Pauly D. 1994. On the sex of fish and the gender of scientists. *Fish & Fisheries Series No. 14,* London: Chapman & Hall

93. Pearse PH. 1980. Property rights and the regulation of commercial fisheries. *J. Bus. Admin.* 11:185–209

94. Pearse PH. 1993. Fishing rights and fishing policy: the development of property rights as instruments of fisheries management. In *The State of the World's Fisheries Resources, Proc. World Fisheries Conf.,* Athens, pp. 76–90. New Delhi: Oxford & IBH Publ.

95. Peters RL, Lovejoy TE. 1990. Terrestrial fauna. See Ref. 120, pp. 353–70

96. Pitcher TJ, Hart PJB. 1982. *Fisheries Ecology.* London: Chapman & Hall

97. Repetto R, Gillis M. Eds. 1988. *Public Policies and the Misuse of Forest Resources.* Cambridge: Cambridge Univ. Press

98. Ricker WE. 1958. *Handbook of Computations for Biological Statistics of Fish Populations. Bull. Fish. Res. Board Can. No. 119*

99. Ricker WE. 1973. Two mechanisms that make it impossible to maintain peak period yields from Pacific salmon and other fishes. *J. Fish. Res. Board. Can.* 30:1275–86

100. Ricker WE. 1987. Effects of the fishery and of obstacles to migration on the abundance of Fraser River sockeye salmon (*Oncorhynchus nerka*). *Can. Tech. Rep. Fish. Aquat. Sci. No 1522*

101. Rosenberg AA, Fogarty MJ, Sissenwine MP, Beddington JR, Shepherd JG. 1993. Achieving sustainable use of renewable resources. *Science* 262:828–29

102. Russell ES. 1931. Some theoretical considerations on the "Overfishing" problem. *J. Cons. Cons. Int. Explor. Mer.* 6:3–27

103. Saetersdahl G. 1980. A review of past management of some pelagic stocks and its effectiveness. In *The Assessment and Management of Pelagic Fish Stocks,* ed. A Saville, pp. 505–515. *Cons. Int. Expl. Mer 177*

104. Sainsbury KJ. 1988. The ecological basis of multispecies fisheries, and management of a demersal fishery in tropical Australia. In *Fish Population Dynamics,* ed. J. Gulland, pps 349–82. Chichester: Wiley

105. Sainsbury KJ. 1991. Application of an experimental approach to management of a tropical multispecies fishery with highly uncertain dynamics. *ICES Mar. Sci. Symp.* 193:301–20

106. Shackleton LY. 1987. A comparative study of fossil fish scales from three upwelling regions. *S. Afr. J. Mar. Sci.* 5:79–84

107. Shannon LV, Shackleton LY, Siegfried WR. 1988. The Benguela Ecology Programme: the first five years. *S. Afr. J. Sci.* 84:472–75

108. Schindler DW. 1987. Detecting ecosystem responses to anthropogenic stress. *Can. J. Fish. Aquat. Sci.* 44(Suppl. 1):6–25

109. Schindler DW, Mills KH, Malley DF, Findlay DL, Shearer JA, Davies IL, et al. 1985. Long-term ecosystem stress: the effects of years of experimental acidification on a small lake. *Science* 228:1395–1401

110. Shugart HH. 1984. *A Theory of Forest Dynamics.* New York: Springer-Verlag

111. Sissenwine, M.P. 1984. The uncertain environment of fish harvesters, fishery scientists, and fishery managers. *Mar. Res. Econ.* 1:1–30

112. Smith TD. 1994. *Scaling Fisheries: The Science of Measuring the Effects of Fishing, 1855–1955.* Cambridge: Cambridge Univ. Press

113. Soutar A, Isaacs JD. 1969. History of fish populations inferred from fish

scales in anaerobic sediments off California. *Calif. Coop. Oceanic Fish. Invest. Rep.* 13:63–70

114. Soutar A, Isaacs JD. 1974. Abundance of pelagic fish during the 19th and 20th centuries as recorded in anaerobic sediments of the Californias. *Fish. Bull.* 72:257–73

115. Strain BR. 1991. Available technologies for field experimentation with elevated CO_2 in global change research. See Ref. 82, pp. 245–61

116. Straskraba M. 1993. Ecotechnology as a new means for environmental management. *Ecol. Eng.* 2:311–31

117. Thompson WF. 1919. The scientific investigation of marine fisheries, as related to the work of the Fish and Game Commission in southern California. *Fish. Bull. (California)* 2:3–27

118. Thompson WF. 1922. The marine fisheries, the state and the biologist. *Sci. Mon.* 15:542–50

119. Tuck GN, Possingham HP. 1994. Optimal harvesting strategies for a metapopulation. *Bull. Math. Biol.* 56:107–27

120. Turner BL III, Clark WC, Kates RW, Richards JF, Mathews JT, Meyer WB, eds. 1990. *The Earth As Transformed by Human Action: Global and Regional Changes in the Biosphere Over the Past 300 Years.* Cambridge: Cambridge Univ. Press

121. Vitousek PM, Ehrlich PR, Matson PA. 1986. Human appropriation of the products of photosynthesis. *BioScience* 36:368–73

122. Walters CJ. 1986. *Adaptive Management of Renewable Resources.* New York: MacMillan

123. Walters CJ. 1987. Nonstationarity of production relationships in exploited populations. *Can. J. Fish. Aquat. Sci.* 44 (Suppl. 2):156–65

124. Walters CJ. 1993. Dynamic models and large scale field experiments in environmental impact assessment and management. *Aust. J. Ecol.* 18:53–61

125. Walters CJ, Collie JS. 1988. Is research on environmental factors useful to fisheries management? *Can. J. Fish. Aquat. Sci.* 45:1848–54

126. Walters CJ, Goruk RD, Radford D. 1993. Rivers Inlet sockeye salmon: an experiment in adaptive management. *North Am. J. Fish. Mgmt.* 13:253–62

127. Walters CJ, Hilborn R. 1976. Adaptive control of fishing systems. *J. Fish. Res. Board Can.* 33:145–59

128. Walters, CJ, Hilborn. R. 1978. Ecological optimization and adaptive management. *Annu. Rev. Ecol. Syst.* 8:157–88

129. Walters CJ, Holling CS. 1990. Large-scale management experiments and learning by doing. *Ecology* 71:2060–68

130. Ward FJ, Larkin PA. 1964. Cyclic dominance in Adams River sockeye salmon. *Prog. Rept. 11, Int. Pac. Salmon Fish. Commn.* New Westminster, Canada

131. Watt KEF. 1968. *Ecology and Resource Management: A Quantitative Approach.* New York: McGraw-Hill

132. World Commission on Environment and Development. 1987. *Our Common Future.* Oxford: Oxford Univ. Press

133. Wright DH. 1990. Human impacts on energy flow through natural ecosystems, and implications for species endangerment. *Ambio* 19:189–94

134. Wyatt T, Larrañeta MG, eds. 1988. Long-term changes in marine fish populations. Vigo:[sk.n.], 1988 (Bayona: imprinta REAL). 45

Annu. Rev. Ecol. Syst. 1995. 26:69–92

CAN WE SUSTAIN THE BIOLOGICAL BASIS OF AGRICULTURE?

Carol A. Hoffman and C. Ronald Carroll

Institute of Ecology, University of Georgia, Athens, Georgia 30602-2202

KEY WORDS: rhizosphere, microorganism, transgenic plants, mycorrhizae, sustainability

ABSTRACT

Large areas of the tropics are inherently marginal for general agriculture, and inappropriate management is decreasing productivity on even high-quality soils. For improving the biological basis of long-term agriculture sustainability, especially on marginal lands with low fertility and depleted soil organic matter, we must improve management practices. These practices include increasing soil organic matter and water-stable aggregates by using cover crops during fallow, finding better matches between crops and local environment, enhancing microbial activities in the rhizosphere, and selecting for more beneficial VA mycorrhizal species and better crop-VAM matches. New crop varieties, including transgenic crops, with improved pest and disease resistances and improved root characteristics to increase the beneficial interactions in the rhizosphere could increase productivity with low environmental cost. Analysis of environmental risks associated with release of transgenic plants, including possible risks from hybridization with wild relatives, must be considered as part of any implementation plan.

INTRODUCTION AND RATIONALE

In this review we argue that the biological basis of agriculture can be sustained, but that this will require both managing the agroecosystem and changing the biological players. Specifically, we focus on three key areas: 1) managing plant rhizosphere processes, 2) the potential and limitations for modifying these

69

0066-4162/95/1120-0069$05.00

processes, and 3) the contributions and limits of biotechnology in improving productivity from transgenic plants. While we recognize that the ecological basis for sustainable crop production has many dimensions, we have restricted this review to crop plants and crop-microorganism interactions. Furthermore, the processes that lead to sustainable crop production in the biophysical sense are only part of a much larger equation for sustainable agriculture that must also include social and economic processes (7, 99).

The need for sustainable crop production is most acute, and the biophysical barriers to be surmounted most formidable, where lands are only marginally suitable for agriculture. In this review, we emphasize those lands that are marginal primarily due to low inherent fertility and low soil organic matter.

In the tropics, marginal lands include highly weathered, low fertility Oxisols and Ultisols that constitute about 43% of land area (106) and low base saturated soils, mostly in arid regions, another 17% (105). The remaining 40% of variably fertile tropical lands constitutes the primary agricultural base in tropical regions. Within this base of variably fertile lands, locally high levels of salt and aluminum, micronutrient deficiencies, low pH, poor physical structure, and low levels of available phosphorus may further limit agricultural potential. In temperate North America, marginal lands include large areas of shallow and/or eroded soils, such as the eroded piedmont and sandy coastal plain of the southeast, the hilly Palouse of Washington state, parts of California's Central Valley, and regions along the Mississippi and Missouri rivers.

Not only are many lands inherently low quality for general agricultural use, but vast areas have been degraded by human activities. The World Resources Institute estimates that "An area approximately the size of China and India combined has suffered moderate to extreme soil degradation caused mainly by agricultural activities, deforestation, and overgrazing in the past 45 years..." (127). Over the past decade, approximately one half of the area of tropical forest lost each year expands the base of productive agriculture; the other half simply replaces agricultural land that is worn out and abandoned. Houghton observes (55, p. 311), "If agriculture could be made sustainable throughout the tropics, total agricultural area could continue to grow at current rates while, at the same time, rates of deforestation could be reduced by approximately 50%."

In these marginal lands, sustainable crop production will require build-up and maintenance of soil organic matter, formation of water–stable soil aggregates, increased microbial transformation of nutrients in the rhizosphere, selection of locally adapted crop and microbial ecotypes, and improved resistance of crops to pests and disease. On marginal lands, agrochemicals and tillage should be reduced for economic and ecological reasons and fertility should be improved and maintained by greater emphasis on ecological processes. Thus, two key areas of research include the ecological management of rhizosphere processes, especially the interactions between microorganisms and plant roots, and enhanced genotype-environment matching with crops through the use of native genetic resources and cautious application of biotechnology. The general goal in this research agenda should be to find the appropriate trade-off between increased yield and long-term yield stability through the substitution of management for external inputs.

THE DETERMINANTS OF SOIL FERTILITY

Soil fertility is best defined with respect to particular crops. Soils that are generally of low quality for agriculture may be quite good for particular crops. For example, the high sulfur-containing clays (cat clays) in many tropical coastal wet areas have extremely low pH when oxidized upon exposure to air. However, when cat clays remain submerged and the sulfur is in a reduced state, as in paddy rice production, they are less acidic and can be very productive. The physical cracking and swelling characteristics of heavy textured vertisols limit their use for many crops, yet they represent some of the most productive cottonlands in central Texas and parts of the tropics.

SOIL ORGANIC MATTER AND SOIL BIOLOGY

Soil organic matter (SOM), especially the more labile components, plays the key role in maintaining soil fertility and structure in many soils. SOM is chemically heterogeneous, variously consisting of dead roots, sloughed off root cells, exudates from living roots, microbial and invertebrate biomass, fungal hyphae, mucilages and polysaccharides, among other components. The more labile components, especially within the rhizoplane (on or close to the surface of roots), serve as a ready source of energy-rich carbon compounds necessary to support high levels of microbial activity. This microbial pool activity contributes to plant nutrition through decomposition, ammonification, nitrogen fixation, and solubilizing phosphorus (11). Together with mucilages and polysaccharides from roots and microbes and the network of fungal hyphae, soil organic matter is essential to the formation of water-stable aggregates, the fundamental building blocks of soil structure.

In regions with long, warm, wet growing seasons, maintaining high levels of soil organic matter is difficult. In these regions, bare fallow periods and clean cultivation practices, while perhaps useful for pest and disease control, can result in the rapid depletion of SOM, greatly reduced microbial activity, and, where clay content is high, dense soil structure. Management of SOM is clearly a key factor for the maintenance and improvement of agricultural productivity as well as for the restoration of degraded and abandoned agricultural lands. While SOM in the rhizosphere can be maintained by bringing in plant residues from off-site, in most cases, SOM will need to be maintained by managing crop and fallow vegetation and soil biology on-site. In the short term, opportunities exist for improving the management of SOM by using naturally occurring microorganisms and crop mixtures. In the longer term, we believe that plant breeding and biotechnology could improve the way we manage plant-microbe interactions to improve soil organic matter.

MANAGING THE RHIZOSPHERE

Most attempts to modify the rhizosphere have been limited to the legume-rhizobia association, either in a legume crop or in legume cover/cereal rotations. When legume/cereal rotations have been in place for several years, total nitrogen in the production field is often similar to conventional cereal production systems using fertilizer. However, the partitioning and chemical form of the nitrogen differ. In rotation systems, the largest fractions of nitrogen are generally in microbial and weed biomass. In conventional systems, proportionally more nitrogen ends up in the cereal crop. Long-term average yields usually slightly favor the conventional system, but when fertilizer costs are included, the net economic yield often favors the cover crop rotation system (84). Because many recent reviews discuss rhizobia and rotation systems with respect to nitrogen management in temperate agriculture, we instead focus on free-living microorganisms in the rhizosphere and mutualist mycorrhizas in marginal agricultural lands.

We believe that three broad areas of investigation will be particularly important contributors to the biological basis for the sustainability of agroecosystems. These are the role of root exudates in microorganism-mediated plant nutrient uptake, the role of the microorganism community in suppression of crop root disease, and the role of microorganisms, especially mycorrhizae, in the restoration of degraded agricultural lands. Significant research advances in these areas will depend on the collaboration of plant breeders, molecular geneticists, and ecologists.

THE ROLE OF ROOT EXUDATES IN MICROORGANISM-MEDIATED PLANT NUTRIENT UPTAKE

Predator-prey interactions among rhizosphere microorganisms play a key role in plant nutrient dynamics. The by-products of bacterial grazing by protozoans and nematodes are important sources of ammonium nitrogen, phosphorus, and other plant nutrients (10, 11, 23, 71). Bacterial grazing also turns over soil organic carbon (12, 13, 71). Thus, the rapid turn-over of microbial biomass is an important source of plant nutrients. Furthermore, the high metabolic activity of the microbial community is important to the release of biological compounds that help form the binding material for water–stable soil aggregates.

Most of the energy to the rhizosphere microbial community is supplied by plants roots in the form of dead roots, sloughed-off cell materials, and exudates from living roots. A surprising amount of photosynthate is exuded from roots, actively and passively, as energy-rich sugars. We measured in situ partitioning of root and rhizosphere respiration in wheat seedlings (8). Of the total rhizosphere respiration, 40.6% was from root respiration and 59.4% was from rhizo-microbial respiration. Thus, over half of the total rhizosphere respiration was based on root exudate as the substrate.

Although root exudates are important contributors to microbial metabolism, several important questions remain to be answered before the relevance of root exudation to crop productivity can be asserted. First, exudation has largely been studied in modern lines of agricultural crops, mostly cereals and pasture grasses. Because traditional lines and wild ancestors have not been studied, we don't know how agricultural selection may have modified carbon flow from roots into the rhizosphere. We do know that some crops and their wild relatives have different patterns of resource allocation. For example, wild relatives and tetraploid cultivars of wheat transport proportionately more assimilates to roots after anthesis than do hexaploid cultivars (25).

Second, we do not know the complete genetic controls for root exudation; thus, the basis for modification of root exudation through genetic engineering is unknown.

Third, roots may contain allelochemicals used in weed, disease, or herbivore suppression. The extent to which these allelochemicals are present in the rhizosphere from root exudate or from root material decomposition is not well known. But the consequences of these allelochemicals for plant nutrient uptake could be significant if they affect microbial interactions. For example, release of an allelochemical could shock rhizoplane bacteria into temporary stasis. Grazing by bacterial predators would continue to release ammonium nitrogen but, because the bacterial community was in stasis, less nitrogen would be taken up by bacteria and a larger fraction would be available to the plant.

Fourth, some chemicals in the exudate may act as signals that increase bacterial activity without significantly contributing to the nutritional base. We have been investigating the distribution of cucurbitacins in the roots and rhizosphere of wild and cultivated squash (*Cucurbita* spp.). Concentrations equivalent to that found in *Cucurbita andreana* roots increase the growth of the beneficial bacterium *Pseudomonas fluorescens*.

A general goal for plant breeders and molecular geneticists would be to achieve improved uptake from the pool of microbial nitrogen through modification of the root exudation process so that the exudation process could subsidize a plant's nitrogen budget through the use of surplus photosynthate.

But many plant breeders have a different opinion. Evans (26) has suggested that because the application of nitrogen fertilizer and irrigation can decrease the need for large root systems (104), breeders could increase yield by minimizing investment in roots when ample nutrients are supplied.

THE ROLE OF THE MICROBIAL COMMUNITY IN SUPPRESSION OF CROP ROOT DISEASE

Some rhizoplane bacteria, especially pseudomonads, confer protection against root pathogens. As populations of beneficial bacteria increase and the disease incidence declines, the soil is said to be suppressive to those diseases in decline. Pseudomonads suppress bacterial and fungal pathogens directly by releasing antibiotics and indirectly by chelating iron, an essential mineral with limited mobility. Chelation is accomplished by siderophores, fluorescent pigments with high Fe^{+++} affinity.

Suppressive soils and pseudomonads have been best studied with respect to wheat "take all" (*Gaeumannomyces graminis* var. *tritici*). For example, strains of *Pseudomonas fluorescens* and *P. aureofaciens* produce phenazine antibiotics that protect wheat roots against take-all disease (94, 113, 114). Mutant strains lacking phenazine production were ineffective against take-all. Similarly, pseudomonads are less effective against take-all in iron-rich culture media.

There is considerable scientific and commercial interest in genetically engineering microorganisms to produce desirable agricultural traits. Pseudomonads with high affinity for roots may be particularly useful carriers of introduced traits, such as delta-endotoxin genes from *Bacillus thuringiensis* against insects, or as phenotypes with negative traits deleted, as with the "ice-minus" strain of *Pseudomonas syringae* that confers protection against frost damage. The genetic bases for siderophore pigments, the membrane receptor, and the genes for antibiotic synthesis have been identified and in some cases cloned (17).

Molecular geneticists have used various techniques to ensure that the introduced gene is stable and remains within the GEM strain. For example, the delta-endotoxin Bt gene was introduced into a *Pseudomonas fluorescens* strain through first cloning the gene into a non-self transposable vector (i.e. a defective transposon) (88). There are several significant concerns over the release of genetically engineered microorganisms (GEMs), mostly related to unknown environmental consequences, containment problems, and possible horizontal transmission into other microorganisms (116). Once released, GEMs may generally be much more difficult to contain than genetically modified macroorganisms, such as higher plants, because of their long-distance dispersal potential and rapid population growth. Containment of the introduced gene in the target genome or even within the target species may also be problematic. Techniques for producing GEMs are becoming widely known, and the careless use of plasmid vectors in countries with poor environmental regulation could result in horizontal transmission of undesirable traits into nontarget microorganisms.

More emphasis should be placed on methods, including genetic modification, to modify the environment of particular rhizospheres to favor naturally occurring beneficial microorganisms. We maintain that modifying crop plants to produce more root exudate or better signal compounds in order to increase the population of antibiotic-producing wild type pseudomonads by 25% is ecologically safer than releasing a GEM that produces 25% more antibiotic.

Beneficial strains of microorganisms are effective against particular pathogens; usually the level of protection is strongly dependent on the type of crop and the soil in which it is grown. Planting material can be inoculated with the appropriate beneficial microorganism to provide a temporal advantage over the pathogen, but the augmentation may not be effective and usually must be repeated with every planting. A more general solution to stable pathogen control might be found if the ecological basis for the long-term stability of suppressive soils were better understood.

MANAGING MICROORGANISMS THAT HAVE MULTIPLE BENEFICIAL ROLES: THE SPECIAL CASE OF MYCORRHIZAS

Mutualistic associations between roots and fungi, known as mycorrhizas, are extremely widespread. In agriculture, the most important mycorrhizas are the zygomycetous fungi (Order Glomales), generally referred to as vesicular-arbuscular mycorrhizae (VAM) or arbuscular mycorrhizae (AM) (83). Of the common crop plant families, only Brassicaceae, Chenopodiaceae, and Polygonaceae typically lack mycorrhizas (31, 117).

VAM play important multiple roles in plant nutrition, disease resistance, rhizosphere microbial processes, competition with weeds, and soil structure (100). Table 1 is a synoptic list of some of the more important ways that VAM may contribute to the biotic base of sustainable agriculture.

Crops may lose their VAM association if they are intercropped with nonhost crops, if long bare fallow periods are part of the cropping cycle, or if the soil is seriously disrupted through intensive tillage or misuse of agrochemicals, particularly heavy applications of phosphorus-rich fertilizer (59). The species of VAM may shift during continuous cropping monoculture from those that are good mutualists to VAM species that provide less benefit to the plant but are good competitors. In continuous cropped maize, the shift is often from effective *Gigaspora/Acaulospora/Glomus* species to less beneficial species of

Table 1   A synoptic list of traits sometimes found in vesicular-arbuscular mycorrhizal associations. The inclusion of a trait does not imply general occurrence, only that it has been established in at least one study.

Increased uptake of soil nutrients (74)

Secondary improvement of *Rhizobium*-based nitrogen fixation through improved plant nutrition (3)

Increased uptake of ammonium nitrogen (4)

Hyphal proliferation in organic-rich microsites (112)

Shift to less effective VAM mutualist species over time in continuous monoculture cropping (60)

Shift to less effective VAM mutualist species in high fertilizer regimes (59)

Enhanced contribution in tropical crop mixtures (109)

More effective mutualist species (e.g. *Gigaspora* spp.) more common in low input than in conventional maize or soybean (20)

Possible competitive advantage to crops against nonmycorrhizal weeds through direct antagonism by the fungi (30)

Improve fecundity of plants with asymptomatic root pathogens (86)

Provide resistance against root pathogens and tolerance to root disease damage (18)

Resistance and/or tolerance to parasitic nematodes (28)

Binding of stable macro-aggregates in soil structure (81)

Glomus, e.g. *G. intraradix* (60, 61). Also, *Gigaspora* abundance is associated with increases in water–stable soil macroaggregates (81). Some of the conflicting information about the benefits provided by VAM to crops could be due to the preponderance of VAM studies using species of *Glomus* extracted from continuous or highly fertilized crops (e.g. 41). In at least one case, allelochemicals from the VAM inhibit the nonhost in plant competition experiments (30); it would be useful to test a range of VAM species to see which produced the most effective allelochemicals.

Polycultures of two or more crop species planted in close proximity provide a mixture of VAM hosts with differing VAM species relations. Polyculture cropping may maintain a more effective community of VAM mutualists and thereby may contribute to the phenomenon of transitive overyielding observed in many intercropped systems (120). Within-row polycultures of maize and beans have higher rates of VAM infection and higher yields of both species than when maize and beans are intercropped by rows (109). The trade-off between potential competition resulting from the overlap of roots between crop species and possible benefits accruing from VAM needs investigation (see 61 for discussion of VAM in yield declines in continuous monocultures).

Thompson (115) suggests important ways to enhance VAM-crop associations. The use of VAM-building hosts in the cropping and fallow cover periods to select for more effective mutualist species seems especially worthy of further investigation.

There is potential for improving VAM-crop mutualisms through breeding. Genetic variation in the VAM association has been demonstrated for wheat, pearl millet, sorghum, maize, cowpea, chickpea, and peanut (115). Wheat illustrates the complex interactions of soil-crop genotype-VAM genotype that are likely to exist. Hetrick et al (48) used mixed cultures of *Glomus* spp. to test the response of ancestral, old, and new wheat varieties. They found a strong genetic basis for differences in VAM dependence among cultivars and a growth depression associated with certain cultivar-VAM combinations. Hetrick at al (49) demonstrated a consistent dependency on VAM for wheat cultivars released before 1950, and they suggest that modern breeding has reduced the dependency of wheat on VAM. Koide et al (69) compared wild and domestic oat and reported little response by the wild oat to VAM inoculation. *G. intraradix* was used as the VAM species, an unfortunate choice as we noted above, and their results might have been different if a more effective mutualist species had been used.

The considerable genetic variation that occurs among VAM and crop host plants can affect the effectiveness of the mutualism. Colonization rates and some benefits from VAM appear to behave as heritable traits among plant cultivars of peanut (66), and therefore the basis exists for improving VAM associations through breeding. Breeding for improved VAM association should select plants for appropriate root architecture (60, 108). Variation in root architecture occurs within species, and VAM colonization levels are strongly correlated with root "coarseness." The role of root exudate quantity and composition on the VAM association is also a potentially important research direction.

Could plant growth–promoting bacteria also enhance colonization by VAM (28)? Many pseudomonads that enhance mycotrophy occur in ectomycorrhizal associations and appear to encourage some species of ectomycorrhizal fungi over others (21, 22). Although *Pseudomonas fluorescens* greatly increases in the presence of VAM association (68), the potential to enhance pseudomonad density, or the consequences of high pseudomonad density on VAM associations, are both unknown.

RHIZOSPHERE PROCESSES: A NEW AVENUE FOR PLANT BREEDING AND BIOTECHNOLOGY

Traditionally, plant breeders have concentrated on selecting for yield characteristics and disease resistance. Transgenic crops have incorporated many traits including disease resistance, longer shelf life for the harvest, and tolerance to herbicides (Table 2). Plant breeders and the biotechnology industry have not made root properties or the rhizosphere community a focus for crop improvement (90). We believe breeding and genetic engineering research could contribute significantly to sustainable agriculture in three important new areas: root architecture, root exudate production, and mutualistic association between roots and microorganisms. Within crop species considerable variation in root architecture and exudate production and composition exists. There is also considerable variation in the plant-microbial mutualism due to differences among lines within one microbial species. The relative contributions of heritable genetic differences and phenotypic plasticity are unknown.

THE BIOLOGICAL RESOURCE BASE OF AGRICULTURE

In order for plant breeding to assume a major role in the development of environmentally sound agricultural systems, the biological characteristics of crops must be known. Yet cutting-edge training in breeding emphasizes biotechnology, and the classically trained germplasm expert is increasingly rare. Although crop genetic diversity is preserved in germplasm repositories (85), without the in-depth knowledge provided by germplasm experts about the biological variability of crop organisms and the genetic materials available for crop improvement, agricultural sustainability as measured by varietal improvement may lag.

Germplasm stocks can have a significant positive impact on productivity, but obstacles to finding the desired characteristic or to understanding its expression in a matrix of correlated characters may make incorporation through breeding difficult. For example, incorporation of germplasm from wild species has greatly improved tomato production. Genetic resistance against nematodes (*Meloidogyne incognita*) was bred into *Lycopersicon esculentum* from an accession of the wild species *Lycopersicon peruvianum* (102). To date, this gene has never been found again in any accession of this species, not even in the original seed stock cited as the source of the gene (103). In this case, a rare gene in a single accession from a wild species provided a unique contribution to crop productivity.

Similarly, *Lycopersicon cheesmanii* is salt tolerant (101). Barriers to hybridization between this species and cultivated tomato have prevented its use. However, if the genes controlling salt tolerance are identified, it may be possible to develop a salt tolerant tomato (63), which would have broad application to areas such as California, coastal Peru, and Ecuador where irrigation has led to salt accumulation in soils.

The causes, mechanisms, or pathways for some crop characteristics may be unknown. For example, wild relatives of wheat have higher net CO_2 exchange rates per unit leaf area (CER) than do modern hexaploid wheats (26), and yield improvement by increasing CER is theoretically possible. However, the high yield of modern cultivars is positively correlated with size of the flag leaf (leaf subtending the infructescence), and the flag leaf in wild relatives is small (small leaves are better adapted to the dry environments in which the wild species grow). Breeding for increased CER could actually decrease yield of modern wheat if leaf size decreased concurrently. Correlated characters can interfere with breeding objectives, when the reason for the correlation is not clear. Breeding for a single culm in grain crops could increase yield if resources were reallocated to seeds instead of multiple stems. However, the plant could become more vulnerable to lodging or any factor that damaged the meristem (25). Evans stated, "crop yield is the integrated end product of a great variety of processes, and focusing on any one of these, however important, is likely to have counter-intuitive effects, even when supported by quite comprehensive simulation models." Plant breeders who rely upon genetic transformation should be mindful of this; manipulation of crop plants on a gene-by-gene basis is likely to present surprises.

BIOTECHNOLOGY/GENETIC ENGINEERING AS A TOOL FOR CROP IMPROVEMENT

Although theoretically any gene can be transferred between species using genetic engineering techniques, several limitations currently exist: 1) Only single gene traits have been successfully manipulated; 2) some major crops cannot be succesfully transformed or regenerated from calli; 3) gene insertion cannot be directed to particular chromosomes (62); 4) inserted genes may not all be equally expressed or may have limited expression (27, 53); and 5) transgenes may become inactivated in certain environments (27).

TRAITS CURRENTLY UNDER DEVELOPMENT AND THEIR POTENTIAL CONTRIBUTION TO SUSTAINABILITY

Biotechnology offers the potential for the development of crop varieties, crop symbionts, or biocontrol agents that increase food security and environmental quality. To this we offer one caveat. If the seduction of biotechnology is the ability to modify crops by the integration of a single characteristic from any

Table 2 Trait and crop for field testing permits issued by USDA from 1992–1994. In several cases it was not possible to determine the specific product that was produced by the transgene.

Trait	Number of permits	Specific product	Crop
Herbicide tolerance	107	Bromoxynil tolerant	cotton
		Sulfonylurea tolerant	soybean, tobacco, cotton
		Glyphosate tolerant	soybean, cotton, tomato, tobacco, corn, rapeseed, wheat, beet, lettuce
		Phosphinothricin	corn, soybean, alfalfa, rice, beet, peanut, wheat, rapeseed
Virus resistance	99	Cucumber mosaic virus (CMV)	squash, melon, corn, cucumber, tomato
		Papaya ringspot virus (PRV)	papaya, melon, squash, plum
		Tobacco etch potyvirus (TEV)	tobacco
		Tobacco mosaic virus (TMV)	tomato, corn
		Tomato yellow leaf curl virus (TYLCV)	tomato
		Potato leaf roll virus (PLRV)	potato
		Potato virus X (PVX)	potato
		Potato virus Y (PVY)	potato, tobacco
		Alfalfa mosaic virus (AMV)	alfalfa
		Maize chlorotic dwarf virus (MCDV)	corn
		Maize chlorotic mottle virus (MCMV)	
		Maize dwarf mosaic virus (MDMV)	
		Watermelon mosaic virus 2 (WMV2)	melon, squash
		Zucchini yellow mosaic virus (ZYMC)	watermelon, cucumber
			melon, squash
		Galleria mellonella	watermelon, cucumber
		Cecropin production	potato
		Beet curly top virus (BCTV)	potato, apple
		Vibrio cholera	tobacco
		Soybean mosaic virus (SBMV)	alfalfa
			soybean, melon

Trait	No.	Target/agent	Crops
Fungal resistance	11	Tomato spotted wilt virus (TSWV)	tomato, lettuce, peanut
		Barley yellow dwarf virus (BYDV)	potato, barley
		Barley yellow mosaic virus (BYMV)	potato
		Beet necrotic yellow vein (BNYVV)	beet
		Fire blight	apple
		Rhizoctonia	potato, tobacco
		Phytophera	tobacco, cucumber
			melon
		Erwinia carotovora	potato
		Botrytis cinerea	tobacco, tomato
		Alternaria	tomato
		Mildew	squash
		Downy mildew	lettuce
		Verticillium dahliae	potato
Insect resistance	49	Bt t (coleopteran)	potato, cotton, alfalfa, eggplant
		Bt k (lepidopteran)	cotton, amalanchier, tobacco, tomato, corn, walnut, rapeseed, cranberry
		wheat germ agglutinin	corn
		insecticidal protein	cotton, corn
Plant quality traits			
Delayed ripening	16	low polygalacturonase	tomato
Oil profile alteration	18	High laurate	rapeseed
Increased solids	8		potato
Cold tolerance	3		tomato
Seed storage protein	8		potato
Miscellaneous			
Heavy metals sequestered			soybean, corn, sunflower
Pigment			potato
Industrial enzymes			petunia
Nutritional quality		methionine protein	rapeseed
Fatty acid alterations			corn
			rapeseed

species on earth, then the fulfillment of this goal relies upon the knowledge about and availability of these traits in global biodiversity. Yet even the crop germplasm in the US National Germplasm System is largely undocumented (98), and biodiversity in native habitats is rapidly being destroyed (125). We discuss the positive developments in biotechnology and the problems that could diminish these contributions.

Traits for transgenic crops in advanced stages of development fall into four primary categories (Table 2). To date, most of the field testing permits issued by the USDA are for herbicide-tolerant crops.

HERBICIDE TOLERANCE Developers of herbicide-tolerant (HT) crops (52) claim that incorporation of these varieties into agroecosystems will increase sustainability by, 1) requiring fewer applications of chemicals, 2) substituting more environmentally benign herbicides (e.g glyphosate) for more persistent and toxic herbicides (e.g. 2,4-D), and 3) allowing better timing of herbicide application to deliver chemicals (more along the lines of IPM). For example, in Iowa in 1985, approximately 48% of acres planted to corn were treated with a pre-plant herbicide, 36% received a pre-emergence herbicide, and 35% received a post-emergence herbicide (126). Currently, Pioneer is field-testing a glyphosate-tolerant corn, and DeKalb has produced HT soybeans that could be used to decrease reliance on 2,4-D and atrazine. Cotton, which receives heavy chemical applications, has been engineered for tolerance to both glyphosate and bromoxynil. Given that competition from weeds reduces yields approximately 9.5% (14) to 11% (92), minimal usage of herbicides on HT crops could increase sustainability not only by decreasing chemical usage but also by increasing harvestable yields per acre. More production on less land could mean that marginal lands could be taken out of production and converted to more sustainable types of land use. Alternatively, if herbicides were substituted for mechanical weed control methods, then decreased soil erosion could increase topsoil retention and decrease sediment input to aquatic systems.

INSECT RESISTANCE (IR) Insects reduce crop productivity by 12–14% (95, 14). Total agricultural output could increase significantly without additional land clearing if insect damage decreased. Incorporation of *Bacillus thuringiensis* (Bt) endotoxins (47), cowpea trypsin inhibitors (50), or protease inhibitors (51) could decrease losses to insects.

Improved control of trait expression could mitigate some problems (loss of resistance or impacts on non-target organisms) associated with transgenic IR crops (36, 78). For example, corn plants transformed for Bt expression with a special promoter (maize phosphoenolpyruvate carboxylase or PEPC) produce Bt toxin only in pollen and photosynthetic parts of the plant, whereas other promoters (such as CaMV 35S) cause Bt toxin production in all plant tissues

(70). Similarly in rice, a viral coat protein against rice tungro disease is targeted for expression in phloem because the viral vector, the green leafhopper, is primarily a phloem feeder (76).

FUNGAL AND VIRAL RESISTANCE   Given that fungicides comprise 21% of the world market for chemical pesticides (67), incorporation of fungal resistance should benefit food production and environmental quality. Similarly, because virus control often relies upon pesticides to control the insect vector, viral resistant plants should diminish environmental costs. Traditional plant breeding has successfully improved crop performance (77), but genetic transformation might increase the number of crops protected.

RISKS ASSOCIATED WITH GENETICALLY ENGINEERED CROP PLANTS

Genetically engineered crops are not a panacea for sustainability. For example, availability of HT varieties may actually increase herbicide use (34). In addition, weeds may evolve herbicide tolerance due to increased use of "relatively safe" herbicides. This might foster the development of crop varieties tolerant of more toxic herbicides. Also, overuse of herbicides may lead to cross-tolerance (39). At least two species of weeds in Australian wheat fields have evolved cross-resistance to a wide variety of herbicides.

Insects may also evolve resistance to genetically engineered IR varieties (1, 37). Widespread incorporation of Bt toxins into many crops could cause loss of effectiveness. Several strategies might increase resistance durability such as incorporation of multiple Bt genes or multi-lining (37).

Genetically engineered crops might generate additional risks (54). Transgenic crops might become pests by invading natural communities (124). In field tests, oilseed rape with transgenes for antibiotic resistance or herbicide tolerance was no more invasive than were unmodified plants in the absence of the selective agent (15). However, invasiveness must be assessed for a particular genotype under the appropriate selective regime; some traits pose greater risk than others (65, 116).

Depending upon the location in which a crop grows, cross-compatible wild relatives may grow nearby. If crop pollen fertilizes wild relatives, the hybrids may express the engineered trait. This could lead to the development of more serious weeds or might change the dynamics of plant and animal communities adjacent to fields (2, 16, 54). Several crops in the United States have cross-compatible wild relatives (Table 3), and the inclusion of selectively advantageous traits into their genome could make them more abundant, perhaps enhancing their weediness. In theory, agricultural sustainability is protected from this event by enforcement of the Plant Pest Act and Plant Protection Act, implemented by the USDA. However, the current case involving transformed squash shows that legal protection is weak.

A variety of yellow crookneck squash (*Cucurbita pepo* var. ZW-20) has been engineered for resistance to watermelon mosaic virus 2 and zucchini yellow mosaic virus. The Upjohn/Asgrow Seed Company petitioned USDA-APHIS to give the variety nonregulated status, a requirement for commercialization. This crop has compatible wild relatives in the United States, including *C. pepo* var. *ovifera*, which is a weed of soybean fields in the midwest (6, 44, 75, 89). Hybridization between ZW-20 and wild *C. pepo* could produce a plant with greater fitness if viruses exert selection on wild populations. In the USDA finding, the only data concerning viral risk assessment were compiled by the Asgrow Company. In addition to the conflict of interest issue, the data (119) were inadequate to make a judgment; it appears that 14 plants (no sample size shown) were surveyed. Because viral proteins were not found in these plants, it was inferred that transfer of viral resistance characteristics would not present additional environmental problems. Despite public comments from ecologists that enhanced weediness may occur and that the data are insufficient to make a final determination (Rissler, Ellstrand in 29, H. Wilson, personal communication), USDA recommended deregulation for ZW-20 squash (80). The APHIS assessment states that the squash "is unlikely to increase the weediness potential for any other cultivated plant or native wild species with which it can interbreed." Instead of careful experiments accurately assessing risk, armchair interpretation of limited data has directed the finding.

Studies of pollen flow, pollen competition, and hybrid fitness are beginning with sunflower (LH Rieseberg, A Snow, unpublished) and could be useful if transgenic hybrids are developed for commercial release. The particular transgenic trait could substantially affect any estimates of hybrid persistence and environmental risk. Concerns about transgenic hybrids with wild species may be even more important in other parts of the world where numerous wild progenitors are present (24). Wild *Cucurbita* species (probably *C. ecuadorensis*) are a major problem along the northwest coast of Ecuador. During wet years (e.g. as influenced by El Niño events), wild *Cucurbita* vines flourish and frequently smother areas of young regenerating forest.

The critical issue in determining risk from transgenic plants is the selective value of the individual. Clearly, possession of a trait that frees an organism from the constraints of a previously limiting factor can lead to ecological release (107). However, selection does not operate on a single factor, but on the individual. If a transgenic crop or a crop-wild hybrid possesses characteristics that confer disadvantages in the particular environment, then the transgene may still be lost from the population. For example, imagine a hybrid between cultivated tomato with a Bt transgene and its wild relative. Even if the Bt gene confers an advantage to the hybrid by protection against herbivores,

Table 3   Wild relatives of US crops located in the US. Weedy status is indicated by *.

Crop	Wild relatives
Wheat	None (73)
Corn	*Tripsacum floridanum* (123)
	Tripsacum dactyloides
Soybeans	None
Cotton	*Gossypium tomentosum* (32, 56)
	Gossypium thurberi (43)
Sorghum	*Sorghum halepense** (40)
	*Sorghum bicolor**
Rapeseed	*Brassica campestris* (96)
	*Brassica nigra**
	*Brassica rapa**
Barley	*Hordeum arizonicum* (42)
	Hordeum brachyantherum
	Hordeum bulbosum
	Hordeum depressum
	Hordeum jubatum
	Hordeum nodosum
	Hordeum leporinum
	*Secale cereale**
Oats	*Avena fatua**
	Avena sterilis
Rice	*Oryza sativa** (9)
Sunflower	*Helianthus deserticola* (46)
	H. anomalus
	H. niveus
	*H. debilis**
	H. praecox
	H. petiolaris
	H. neglectus
	*H. bolanderi**
	H. paradoxus
	H. argophyllus
	H. annuus ssp. *jaegeri**
	H. annuus ssp. *lenticularis**
	H. annuus ssp. *texanus**
	H. annuus spp. *annuus**
Peanuts	None (111)
Sugarbeets	*Beta vulgaris* (19)
Beans, dry	*Phaseolus coccineus* (110)
	Phaseolus lunatus
Potatoes	None (45)
Grapes	*Vitis rotundifolia* (93)
Sugarcane	*Miscanthus japonicus* (38)
	*Sorghum halepense** (97)
	*Sorghum bicolor**
Tobacco	*Nicotiana glauca** (35)

Table 3   (*continued*)

Crop	Wild relatives
Rye	*Secale cereale*
Tomato	*Lycopersicon esculentum* var. *cerasiforme** (102)
Green peas	*Pisum elatius* (5)
	Pisum humile
Watermelon	None
Peas (dry)	*Pisum elatius* (5)
	Pisum humile
Cabbage	*Brassica campestris* (Nishi et al 1961 in 96)
	*Brassica nigra** (Nishi et al 1970 in 96)
	*Brassica juncea** (Kakizaki 1925 in 96)
	*Raphanus sativus** (79)
Cucumber	None (87)
Sweet Potatoes	*Ipomoea batatas* (126)
Cauliflower	*Brassica campestris* (Nishi et al 1961 in 96)
	*Brassica nigra** (Nishi et al 1970 in 96)
	*Brassica juncea** (Kakizaki 1925 in 96)
	*Raphanus sativus** (79)
Green peppers	*Capsicum annuum* var. *minus* (57)
Green beans	*Phaseolus coccineus* (110)
	Phaseolus lunatus
Pumpkin	*Cucurbita texana* (123)
	Cucurbita okeechobeensis
	Cucurbita foetidissima
	Cucurbita palmata
	Cucurbita digitata
Carrots	*Daucus carota**
Alfalfa	*Medicago sativa* (17a)

the hybrid could still be disadvantaged in its natural environment if the hybrid genome also makes the hybrid less drought tolerant, interferes with dispersal, or prevents seed dormancy (33). Even if possession of the transgene reduces fitness of the hybrid, the transgene still may become established in wild populations if backcrossing to wild-type individuals produces an Fn that possesses all characteristics of the wild type plus the transgene. In addition, selective advantage of the hybrid is not necessary for establishment of the transgene in natural populations because of scale-dependent evolutionary forces, such as recurrent migration or the spatial and temporal structure of the wild populations (72). Risk management and mitigation of gene impact on the environment may be the means of dealing with inevitable movement of the transgene beyond its intended limits (64).

Transgenic plants may also have an impact on non-target soil organisms. Incorporation of biomass from plants expressing one or more Bt endotoxins into soil could influence soil biota (58) or biogeochemical cycles (118). Although impact studies on soil organisms are incomplete, Bt toxin residues remained in the soil for at least 30 days after leaf material incorporation into soil (91). However, current methodologies accurately measure only extractable Bt (91); larger amounts of Bt endotoxin from plant matter may bind to clay (121) or soil organic components (82) and may exert effects over a longer time period. Methodologies for assessing the impact of synthetic organic chemicals, allelochemicals, or extracellular enzymes on soil organisms may be applicable to transgenic plant products (82).

CONCLUSIONS

Large areas of the tropics are inherently marginal for general agriculture. Badly degraded tropical lands that no longer can support productive agriculture represent an area similar in size to the tropical land area currently supporting agriculture. Shallow and eroded soils in parts of the temperate zone are also marginal for general agriculture, and some lands, once highly fertile, are losing their productivity through inappropriate management. The need to find more sustainable agricultural practices is most acute on these marginal lands to stop the process of degradation.

For improving the biological basis for long-term agricultural sustainability, especially on marginal lands with low fertility and depleted soil organic matter, we need to place greater emphasis on management of renewable resources within the crop field. These management practices should aim to increase soil organic matter and water-stable aggregates by using cover crops during fallow, finding better matches between crops and the local environment, enhancing microbial activities in the rhizosphere, and selecting for more beneficial VA mycorrhizal species and better crop-VAM matches. We believe that transgenic crops have a place in this comprehensive management strategy. New crop varieties with improved pest and disease resistances (obtained through biotechnology) and improved root characteristics to increase the beneficial interactions in the rhizosphere (currently derived from traditional plant breeding) would help realize the goal of high productivity with low environmental cost. Genetic engineering to enhance herbicide resistance might be useful in an ideal world, but given current political realities, the potential for ecological risk, and the economic pressures faced by farmers on marginal lands, there is too much risk to endorse the use of these transgenic plants.

Literature Cited

1. Altman DW, Wilson FD, Benedict JH, Gould F. 1992. Biopesticides and resistance. *Science* 255:903–4
2. Andow D. 1994. Community response to transgenic plant release: using mathematical theory to predict effects of transgenic plants. *Mol. Ecol.* 3:65–70
3. Bagyaraj DJ. 1984. Biological interactions with VA mycorrhizal fungi. In *VA Mycorrhiza*, ed. C Powell, DJ Bagyaraj, pp. 131–53. Boca Raton, FL: CRC. 234 pp.
4. Barea JM, Azcon-Aguilar C, Azcon R. 1987. Vesicular-arbuscular mycorrhiza improve both symbiotic N_2 fixation and N uptake from soil as assessed with a ^{15}N technique under field conditions. *New Phytol.* 106:717–25
5. Ben-Ze'ev N, Zohary D. 1973. Species relationships in the genus *Pisum* L. *Isr. J. Bot.* 22:73–91
6. Bridges DC, Baumann PA. 1992. *Weeds Causing Losses in the United States.* Champaign, IL: Weed Soc. Am. 404 pp.
7. Carroll CR, Vandermeer JH, Rossett PM, ed. 1990. *Agroecology.* New York: McGraw-Hill. 641 pp.
8. Cheng W, Coleman DC, Carroll CR, Hoffman CA. 1993. In situ measurement of root respiration and soluble carbon concentrations in the rhizosphere. *Soil Biol. Biochem.* 25:1189–96
9. Chu Y-E, Oka HI. 1970. Introgression across isolating barriers in the wild and cultivated *Oryza* species. *Evolution* 24: 344–55
10. Clarholm M. 1985. Interaction of bacteria, protozoa and plants, leading to mineralization of soil nitrogen. *Soil Biol. Biochem.* 17:181–87
11. Clarholm M. 1994. The microbial loop in soil. In *Beyond the Biomass,* ed. K Ritz, J Dighton, KE Giller, pp. 221–38. New York: Wiley
12. Coleman DC. 1994. Compositional analysis of microbial communities: Is there room in the middle? In *Beyond the Biomass,* ed. K Ritz, J Dighton, KE Giller, pp. 201–20. New York: Wiley
13. Coleman DC, Anderson RV, Cole CV, Elliott ET, Woods L, et al. 1978. Trophic interactions in soils as they affect energy and nutrient dynamics: IV. Flows of metabolic and biomass carbon. *Microb. Ecol.* 4:373–80
14. Cramer HH. 1967. *Plant Protection and World Crop Production.* Berlin: Bayer/Leverkusen. 542 pp.
15. Crawley MJ, Hails RS, Rees M, Kohn D, Buxton J. 1993. Ecology of transgenic oilseed rape in natural habitats. *Nature* 363:620–23
16. Dale PJ. 1994. The impact of hybrids between genetically modified crop plants and their related species: general considerations. *Mol. Ecol.* 3:31–36
17. Davison J. 1988. Plant beneficial bacteria. *Bio/Technology* 6:282–86
17a. Deam CC. 1940. *Flora of Indiana.* Dep. Conserv., Div. For. Indianapolis, IN. 1236 pp.
18. Dehne HW. 1982. Interactions between vesicular arbuscular mycorrhiza fungi and plant pathogens. *Phytopathology* 72:1115–18
19. Doney DL, Whitney ED. 1990. Genetic enhancement in *Beta* for disease resistance using wild relatives: a strong case for the value of genetic conservation. *Econ. Bot.* 44:445–51
20. Douds DD Jr, Janke RR, Peters SE. 1993. VAM fungus spore populations and colonization of roots of maize and soybean under conventional and low-input sustainable agriculture. *Agric. Ecosys. Environ.* 43:325–35
21. Dupponis R. 1991. Some mechanisms involved in growth stimulation of ectomycorrhizal fungi by bacteria. *Can. J. Bot.* 68:2148–52
22. Dupponis R, Garbaye J. 1991. Mycorrhization helper bacteria associated with the Douglas fir *Laccaria laccata* symbiosis: effects in vitro and in glasshouse conditions. *Ann. Soc. For.* 48: 239–51
23. Elliot ET, Coleman DC, Cole CV. 1979. The influence of amoebae on the uptake of nitrogen by plants in gnotobiotic soil. In *The Soil-Root Interface,* ed. JL Harley, S Russell, pp. 221–29. London/New York: Academic. 448 pp.
24. Ellstrand NC, Hoffman CA. 1990. Hybridization as an avenue of escape for engineered genes. *BioScience* 40:438–42
25. Evans LT. 1993. *Crop Evolution, Adaptation, and Yield.* Cambridge/New York: Cambridge Univ. Press. 500 pp.
26. Evans LT, Dunstone RL. 1970. Some physiological aspects of evolution in wheat. *Aust. J. Biol. Sci.* 23:725–41
27. Finnegan J, McElroy D. 1994. Transgene inactivation: Plants fight back! *Bio/Technology* 12:883–88
28. Fitter AH, Garbaye J. 1994. Interactions between mycorrhizal fungi and other soil organisms. In *Management of Mycorrhizas in Agriculture, Horticulture*

and Forestry, ed. AD Robson, LK Abbott, N Malajcuk, pp. 123–32. Netherlands: Kluwer Acad.

29. Fox JL. 1994. USDA likely to okay Asgrow's engineered squash. Bio/Technology 12:761–62

30. Francis R, Read DJ. 1994. The contribution of mycorrhizal fungi to the determination of plant community structure. In Management of Mycorrhizas in Agriculture, Horticulture and Forestry, ed. AD Robson, LK Abbott, N Malajcuk, pp. 11–25. Netherlands: Kluwer Acad.

31. Gerdemann JW. 1975. Vesicular-arbuscular mycorrhizae. In The Development and Function of Roots, ed. JG Torrey, DT Clarkson, pp. 575–91. New York: Academic

32. Gerstel GU. 1956. Segregation in the new allopolyploids of Gossypium. I. The R_1 locus in certain New World–wild American hexaploids. Genetics 41:31–44

33. Gliddon C. 1994. The impact of hybrids between genetically modified crop plants and their related species: biological models and theoretical perspectives. Mol. Ecol. 3:41–44

34. Goldberg R, Rissler J, Shand H, Hassebrook C. 1990. Biotechnology's Bitter Harvest. Rep. Biotechnol. Work. Group. 73 pp.

35. Goodspeed TW. 1954. The genus Nicotiana. Chron. Bot. 16:1–536

36. Gould F. 1988. Evolutionary biology and genetically engineered crops. BioScience 38:26–33

37. Gould F. 1991. The evolutionary potential of crop pests. Am. Sci. 79:496–507

38. Grassl CO. 1980. Breeding Andropogoneae at the generic level for biomass. Sugarcane Breed. Newsl. 43:41–57

39. Gressel J. 1988. Multiple resistances to wheat selective herbicides. New challenges to molecular biology. Oxford Surv. Plant Mol. Cell. Biol. 5:195–203

40. Hadley HH. 1958. Chromosome numbers, fertility and rhizome expression of hybrids between grain sorghum and johnsongrass. Agron. J. 50:278–82

41. Hamel C, Furlan V, Smith DL. 1992. Mycorrhizal effects on interspecific plant competition and nitrogen transfer in legume-grass mixtures. Crop Sci. 32: 991–96

42. Harlan JR. 1968. On the origin of barley. USDA Handbk. 338:9–31

43. Harland SC, Atteck OM. 1941. The genetics of cotton. XVIII. Transference of genes from diploid North American wild cottons (Gossypium thurberi Tod., G. armourianum Kearney, and G. aridum comb.nov. Skovsted) to tetraploid New World cottons (Gossypium barbadense L. and G. hirsutum L.). J. Genet. 42:1–47

44. Harrison S, Oliver LR, Bell D. 1977. Control of Texas gourd in soybeans. Proc. South. Weed Sci. Soc. 30:46

45. Hawkes JG. 1958. Potatoes. I. Taxonomy, cytology and crossability. In Manual of Plant Breeding, Vol. III. Breeding Tubers and Root Crops, ed. H Kappert, W Rudolf, pp. 1–43. Berlin: FRG

46. Heiser CB. 1978. Taxonomy of Helianthus and the origin of domesticated sunflower. In Sunflower Science and Technology, ed. JF Carter, 19:31–53. Madison, WI: Am. Soc. Agron. Crop Sci. Soc. Am.

47. Hernstadt C, Soares GG, Wilcox ER, Edwards DL. 1986. A new strain of Bacillus thuringiensis with activity against coleopteran insects. Bio/Technology 4:305–8

48. Hetrick BAD, Wilson GWT, Cox TS. 1992. Mycorrhizal dependence of modern wheat varieties, landraces, and ancestors. Can. J. Bot. 70:2032–40

49. Hetrick BAD, Wilson GWT, Cox TS. 1993. Mycorrhizal dependence of modern wheat cultivars and ancestors: a synthesis. Can. J. Bot. 71:512–18

50. Hilder VA, Gatehouse AMR, Boulter D. 1990. Genetic engineering of crops for insect resistance using genes of plant origin. In Genetic Engineering of Crop Plants, ed. GW Lycett, D Grierson. pp. 51–66. London: Butterworths

51. Hilder VA, Gatehouse AMR, Boulter D. 1993. Transgenic plants conferring insect tolerance: protease inhibitor approach. In Transgenic Plants, Vol. 1, Engineering and Utilization, ed. S Kung, R Wu, pp. 317–38. San Diego: Academic. 383 pp.

52. Hinchee MAW, Padgette SR, Kishore GM, Delannay X, Fraley RT. 1993. Herbicide-tolerant crops. In Transgenic Plants, Vol. 1, Engineering and Utilization, ed. S. Kung, R Wu, pp. 243–63. San Diego: Academic. 383 pp.

53. Hobbs SA, Kpodar P, DeLong CM. 1990. The effect of T-DNA copy number, position, and methylation on reporter gene expression in tobacco transformants. Plant Mol. Biol. 15: 851–64

54. Hoffman CA. 1990. Ecological risks of genetic engineering of crop plants. BioScience 40:434–37

55. Houghton RA. 1994. The worldwide extent of land-use change. BioScience 44:305–13

56. Hutchinson JB, Silow RA, Stephens SG.

1947. *The Evolution of* Gossypium. London: Oxford Univ. Press. 160 pp.

57. IBPGR. 1983. *Genetic Resources of* Capsicum. *AGPG/IBPGR/82/12.* Rome: IBPGR Secretariat. 49 pp.

58. Jepson PC, Croft BA, Pratt GE. 1994. Test systems to determine the ecological risks posed by toxin release from *Bacillus thuringiensis* genes in crop plants. *Mol. Ecol.* 3:81–89

59. Johnson NC. 1993. Can fertilization of soil select less mutualistic mycorrhizae? *Ecol. Appl.* 3:749–57

60. Johnson NC, Pfleger FL. 1992. Vesicular-arbuscular mycorrhizae and cultural stress. In *Mycorrhizae in Sustainable Agriculture, Am. Soc. Agron. Special Publ. No. 54.* pp. 71–98

61. Johnson NC, Copeland PJ, Crookston RK, Pfleger FL. 1992. Mycorrhizae: possible explanation for yield decline with continuous corn and soybean. *Agron. J.* 84:387–90

62. Jones JD, Dunsmuir P, Benbrook J. 1985. High level expression of introduced chaemeric genes in regenerated tranformed plants. *EMBO J.* 4:2411–18

63. Jones RA. 1987. Genetic advances in salt tolerance. In *Tomato Biotechnology,* ed. DJ Nevins, RA Jones, pp. 125–37. New York: Liss

64. Kareiva P, Morris W, Jacobi CM. 1994. Studying and managing the risk of cross-fertilization between transgenic crops and wild relatives. *Mol. Ecol.* 3:15–21

65. Keeler KH, Turner CE. 1991. Management of transgenic plants in the environment. In *Risk Assessment in Genetic Engineering,* ed. M Levin, H Strauss, pp. 189–218. New York: McGraw Hill

66. Kesava Rao PS, Tilak KVBR, Arunachalam V. 1990. Genetic variation for VA mycorrhizal-dependent phosphate mobilization in groundnut (*Arachis hypogaea* L.). *Plant Soil* 122:137–42

67. Klassen W. 1993. Pest management and biologically based technologies: a look to the future. In *Pest Management: Biologically Based Technologies,* ed. RD Lumsden, JL Vaughn, pp. 410–22. Beltville, MD: Am Chem Soc. 435 pp.

68. Klyuchnikov AA, Kozhevin PA. 1990. Dynamics of *Pseudomonas fluorescens* and *Azospirillum brasiliense* populations during the formation of the vesicular-arbuscular mycorrhiza. *Microbiology* 59:449–52

69. Koide R, Mingguang L, Lewis J, Irby C. 1988. Role of mycorrhizal infection in the growth and reproduction of wild vs. cultivated plants. *Oecologia* 77:537–43

70. Koziel MG, Beland GL, Bowman C, Carozzi NB, Crenshaw R, Crossland L, et al. 1993. Field performance of elite transgenic maize plants expressing an insecticidal protein derived from *Bacillus thuringiensis. Bio/Technology* 11:194–200

71. Kuikman PJ, Jansen AG, VanVeen JA, Zehnder AJB. 1990. Protozoan predation and the turnover of soil organic carbon and nitrogen in the presence of plants. *Biol. Fertil. Soils* 10:22–28

72. Linder CR, Schmitt J. 1994. Assessing the risks of transgene escape through time and crop-wild hybrid persistence. *Mol. Ecol.* 3:23–30

73. Maan SS. 1987. Interspecific and intergeneric hybridization in wheat. In *Wheat and Wheat Improvement,* ed. EG Heyne, 13:453–71. Madison, WI: Am. Assoc. Agron. Crop Sci. Soc. Am.

74. Marschner H, Dell B. 1994. Nutrient uptake in mycorrhizal symbiosis. In *Management of Mycorrhizas in Agriculture, Horticulture and Forestry,* ed. AD Robson, LK Abbott, N Malajcuk, pp. 89–102. Netherlands: Kluwer Acad. Publ.

75. McCormick LL. 1987. Weed survey-southern states. *South. Weed Sci. Soc. Res. Rep.* 30:184–215

76. McCouch SR, Ronald P, Kyle MM. 1993. Contributions of biotechnology to crop improvement. In *Crop Improvement for Sustainable Agriculture,* ed. MB Callaway, CA Francis, pp. 157–91. Lincoln: Univ. Neb. 261 pp.

77. McGarvey PB, Kaper JM. 1993. Transgenic plants for conferring virus tolerance: satellite approach. In *Transgenic Plants,* Vol. 1, *Engineering and Utilization,* ed. S Kung, R Wu, pp. 277–96. San Diego: Academic. 383 pp.

78. McGaughy WH, Whalon ME. 1992. Managing insect resistance to *Bacillus thuringiensis* toxins. *Science* 258:1451–55

79. McNaughton IH. 1973. Synthesis and sterility of *Raphanobrassica. Euphytica* 22:70–88

80. Medley TL. 1994. Availability of determination of nonregulated status for virus resistant squash. *Fed. Reg.* 59 (238):64189

81. Miller RM, Jastrow JD. 1992. The role of mycorrhizal fungi in soil conservation. In *Mycorrhizae in Sustainable Agriculture. Am. Soc. Agron. Special Publ. No. 54.* pp. 29–43

82. Morra MJ. 1994. Assessing the impact of transgenic plant products on soil organisms. *Mol. Ecol.* 3:53–55

83. Morton JB, Benny GL. 1990. Revised

classification of arbuscular mycorrhial fungi (Zygomycetes): a new order, Glomales, two suborders, Glomineae and Gigasporineae and two new families Acaulosporaceae and Gigasporaceae with an emendation of Glomaceae. *Mycotaxon* 37:471–91

84. National Research Council. 1989. *Alternative Agriculture.* Washington: Natl. Acad. 448 pp.
85. National Research Council. 1991. *Managing Global Genetic Resources: The U.S. National Plant Germplasm System.* Washington: Natl. Acad. 171 pp.
86. Newsham KK, Fitter AH, Watkinson AR. 1994. Root pathogenic and arbuscular mycorrhizal fungi determine fecundity of asymptomatic plants in the field. *J. Ecol.* 82:805–14
87. Nijs APM den, Custers JBM. 1990. Introducing resistances into cucumber by interspecific hybridization. In *Biology and Utilization of the Cucurbitaceae,* ed. DM Bates, RW Robinson, C Jeffrey, pp. 382–96. Ithaca, NY: Comstock, Cornell Univ. Press. 1st ed. (Facsimile).
88. Obukowicz MG, Perlak FJ, Kusano-Kretzmer K, Mayer EJ, Bolten SL, Watrud LS. 1987. IS50L as a non-self transposable vector used to integrate the *Bacillus thuringiensis* delta-endotoxin gene into the chromosome of root-colonizing Pseudomonads. *Gene* 51:991–96
89. Oliver L, Harrison S, McClelland M. 1983. Germination of Texas gourd (*Cucurbita texana*) and its control in soybeans (*Glycine max*). *Weed Sci.* 31:700–6
90. O'Toole JC, Bland WL. 1987. Genotypic variation in crop plant root systems. *Adv. Agron.* 41:91–145
91. Palm CJ, Donegan K, Harris D, Seidler RJ. 1994. Quantification in soil of *Bacillus thuringiensis* var. *kurstaki* endotoxin from transgenic plants. *Mol. Ecol.* 3:145 -151
92. Parker ML, Fryer JD. 1975. Weed control problems causing major reductions in world food supplies. *FAO Plant Protection Bull.* 23:83–94
93. Patel GI, Olmo HP. 1955. Cytogenetics of *Vitis.* I. The hybrid *V. vinifera* X *V. rotundifolia. Am. J. Bot.* 42:141–59
94. Pierson LS III, Thomashow LS. 1992. Cloning and heterologous expression of the phenazine biosynthetic locus from *Pseudomonas aureofaciens* 30-84. *Mol. Plant-Microbe Interactions* 5:330–39
95. Pimentel D. 1976. World food crisis: energy and pests. *Bull. Entomol. Soc. Am.* 22:20–26

96. Prakash S, Hinata K. 1980. Taxonomy, cytogenetics and origin of crop Brassicas, a review. *Opera Bot.* 55:1–57
97. Price S. 1957. Cytological studies in *Saccharum* and allied genera. III. Chromosome numbers in interspecific hybrids. *Bot. Gaz.* 118:146–59
98. Qualset CO. 1991. Plant biotechnology, plant breeding, population biology and genetic resources. In *Agricultural Biotechnology at the Crossroads, Biological, Social and Institutional Concerns,* ed. JF MacDonald, pp. 81–90. Ithaca: Natl. Agric. Biotech. Council. 307 pp.
99. Ragland J, Lal R, eds. 1993. *Technologies for Sustainable Agriculture in the Tropics. Am. Soc. Agron. Special Publ. No. 56.* 313 pp.
100. Read DJ, Lewis DH, Fitter AH, Alexander IJ. eds. 1992. *Mycorrhizas in Ecosystems.* Wallingford, UK: CAB Int.
101. Rick CM . 1973. Potential genetic resources in tomato species: clues from observations in native habitats. In *Genes, Enzymes, and Populations,* ed. A. Srb, pp. 255–69. New York: Plenum. 359 pp.
102. Rick CM. 1979. Biosystematic studies in *Lycopersicon* and closely related species of *Solanum.* In *Biology and Taxonomy of the Solanaceae,* ed. JG Hawkes, RN Lester, AD Skelding, pp. 667–77. New York: Academic
103. Rick CM. 1992. Source of Mi gene. *Rep. Tomato Genet. Coop.* 42:33
104. Rufty TW, Raper CD, Huber SC. 1984. Alterations in internal partitioning of carbon in soybean plants in response to nitrogen stress. *Can. J. Bot.* 62:501–8
105. Sanchez PA. 1976. *Properties and Management of Soils in the Tropics.* New York: Wiley. 618 pp.
106. Sanchez PA, Salinas JG. 1981. Low-input technology for managing oxisols and ultisols in tropical America. *Adv. Agron.* 34:279–406
107. Schmitt J, Linder CR. 1994. Will transgenes lead to ecological release? *Mol. Ecol.* 3:71–74
108. Schwab SM. 1987. Considerations of vesicular-arbuscular mycorrhiza physiology in breeding for enhanced mineral uptake by plants. In *Genetic Aspects of Plant Mineral Nutrition by Plants,* ed. EW Gabelman, BC Loughman, pp. 603–15. Dordrecht: Martinus Nijhoff
109. Sieverding E. 1991. *Vesicular-Arbuscular Mycorrhiza Management in Tropical Agroecosystems.* Berlin: Germany: Hartmit Bremer Verlag. 371 pp.
110. Smartt J. 1976. Comparative evolution of pulse crops. *Euphytica* 25:139–43

111. Smartt J. 1978. The evolution of pulse crops. *Econ. Bot.* 32:185–98
112. St. John TV, Coleman DC, Read CPP. 1983. Growth and spatial distribution of nutrient-absorbing organs: selective exploitation of soil heterogeneity. *Plant Soil* 71:474–93
113. Thomashow LS, Weller DM, Bonsall RF, Pierson LS III. 1990. Production of the antibiotic phenazine-1-carboxylic acid by fluorescent *Pseudomonas* species in the rhizosphere of wheat. *Appl. Environ. Microbiol.* 56:908–12
114. Thomashow LS, Weller DM. 1988. Role of a phenazine antibiotic from *Pseudomonas fluorescens* in biological control of *Gaeumannomyces graminis* var. *tritici. J. Bact.* 170:3499–508
115. Thompson JP. 1994. What is the potential for management of mycorrhizas in agriculture? In *Management of Mycorrhizas in Agriculture, Horticulture and Forestry,* ed. AD Robson, LK Abbott, N Malajcuk, pp. 191–200. Netherlands: Kluwer Acad. Publ.
116. Tiedje JM, Colwell RK, Grossman YL, Hodson RE, Lenski RE, et al. 1989. The planned introduction of genetically engineered organisms: ecological considerations and recommendations. *Ecology* 70:298–315
117. Trappe JM. 1987. Phylogenetic and ecological aspects of mycotrophy in the angiosperms from an evolutionary standpoint. In *Ecophysiology of VA Mycorrhizal Plants,* ed. GR Safir, pp. 5–25. Boca Raton, FL: CRC Press
118. Trevors JT, Kuikman P, Watson B. 1994. Transgenic plants and biogeochemical cycles. *Mol. Ecol.* 3:57–64
119. USDA. 1994. *APHIS-USDA Petition 92–204–01 for Determination of Nonregulated Status for ZW-20 Squash: Environmental Assessment and Finding of No Significant Impact*
120. Vandermeer JH. 1990. Intercropping. In *Agroecology,* ed. CR Carroll, JH Vandermeer, PM Rossett, pp. 481–516. New York: McGraw Hill. pp.
121. Venkateswerlu G, Stotsky G. 1992. Binding of the protoxin and toxin proteins of *Bacillus thuringiensis* subsp. *kurstaki* on clay minerals. *Curr. Microbiol.* 25:225–33
122. Whitaker TW, Bemis WP. 1964. Evolution in the genus *Cucurbita. Evolution* 18:553–59
123. Wilkes HG. 1990. Teosinte and other wild relatives of maize. In *Recent Advances in the Conservation and Utilization of Genetic Resources: Proceedings of the Global Maize Germplasm Workshop.* pp. 70–80. Mexico, DF: CIMMYT
124. Williamson M. 1994. Community response to transgenic plant release: predictions from British experience of invasive plants and feral crop plants. *Mol. Ecol.* 3:75–79
125. Wilson EO. 1992. *The Diversity of Life.* New York: Norton. 424 pp.
126. Wintersteen W, Hartzler R. 1987. Pesticides used in Iowa crop production in 1985. *Coop. Extension Serv. Rep. Pm-1288,* Iowa State Univ., Ames, IA
127. World Resources. 1992. *World Resources 1992–93. A Report by the World Resources Inst.,* p. 111. Oxford: Oxford Univ. Press.

Annu. Rev. Ecol. Syst. 1995. 26:93–111

MEASURING BIODIVERSITY VALUE FOR CONSERVATION

Christopher J. Humphries, Paul H. Williams, and Richard I. Vane-Wright

Biogeography and Conservation Laboratory, The Natural History Museum, Cromwell Road, London SW7 5BD, United Kingdom

KEY WORDS: biodiversity, conservation, biogeography, character richness, character combinations, natural hierarchy, option value, species richness, surrogacy, taxic diversity.

ABSTRACT

Practical approaches to measuring biodiversity are reviewed in relation to the present debate on systematic approaches to conservation, to fulfil the goal of representativeness: to identify and include the broadest possible sample of components that make up the biota of a given region. Rather than adapting earlier measures that had been developed for other purposes, the most recent measures result from a fresh look at what exactly is of value to conservationists. Although debate will continue as to where precisely these values lie, more of the discussion has been devoted to ways of estimating values in the absence of ideal information. We discuss the current principles by assuming that the currency of biodiversity is characters, that models of character distribution among organisms are required for comparisons of character diversity, and that character diversity measures can be calculated using taxonomic and environmental surrogates.

INTRODUCTION

Over the last two decades almost all arguments about nature conservation have involved the issue of biological diversity and ways to preserve it (23, 31, 48, 64, 107). These discussions culminated in the 1992 Convention on Biological Diversity and its implementation (37). The conservation of biodiversity is a vast undertaking, requiring the mobilization of existing data, huge amounts of new information, and the monitoring and management of wildlife on an un-

93

precedented scale. From this newly articulated desire to preserve and protect all of nature's variety has arisen a battery of scientific methods to devise a calculus of diversity (57) that can be put to practical use. This account is an attempt to review the key principles that have emerged to date. We describe the various efforts that have been made to provide effective measures of biodiversity within the context of a general strategy for conservation, and how these have been combined with analytical methods of reserve selection for the design of networks of conservation areas capable of satisfying representation goals.

Goals and Representativeness

Extinction of organisms has been increasing at rates that now far exceed the geological background rate (89, 90). Analyses of geological extinction rates have indicated that the present diversity crisis has moved from the abstract to the concrete (71) and that many parts of the globe are losing organisms through pollution, conversion of natural habitats, and environmental degradation due to human overpopulation (87). That conservation intervention is necessary is hardly a new idea; it has been part of environmental policy for more than a century (4). The issues of measuring and conserving wholesale diversity for assessing "ecosystem health" have emerged only relatively recently (70). The future of restoration ecology will depend in large measure on the extent to which the variety of extant organisms is represented in well-managed reserve systems (5).

McNeely et al (61) considered biodiversity to be "an umbrella term for the degree of nature's variety, including both the number and frequency of ecosystems, species or genes in a given assemblage." This can be interpreted as a hierarchy of genes within populations, populations within species and higher taxa, and taxa within functioning ecosystems. Genes and taxa represent the material products of evolution, a process that has been taking place over the last 3.5 billion years, involving millions of diversification events (71). Consequently, organisms represent different combinations of shared and novel genetic systems: While some parts of the genome, such as ribosomal RNA, are as old as life itself and shared by all organisms, the formation of new genetic material has occurred throughout evolution. This historical hierarchy of life is reflected in the taxonomic hierarchy, from kingdoms down to species and infraspecific units (30, 99). The significance of taxonomic relationships for biodiversity measurement has been interpreted in a number of different ways which are discussed in more detail below.

In addition to the wealth of diversity, amounting to estimates of 5–15 million extant species (32, 58, 93, 94, 121) which can be represented in the taxonomic hierarchy, historical relationships are also clearly detectable in biogeography, with distinctive geographical and ecological distributions of genes, species, and higher taxa evident at global, regional, and local levels (10, 11, 38, 52, 65, 66). As with genealogies, which can be difficult to detect because of homoplasy, geographical patterns are often complicated by stochastic processes of chance dispersals and introductions into non-native habitats (for review see 63). However, we concur with Croizat (11) that life and Earth evolved together to form distinctive patterns that can be uncovered through biogeographic analysis. Biological diversity is thus characterized by taxonomic turnover at all geographical scales, with the result that location of the components of biodiversity is absolutely critical to effective network design for in situ conservation.

In a practical strategy for conserving biodiversity, the minimum survey requirements include analysis of geographical distributions (preferably estimating taxonomic relationships of taxa) and analyzing these relational patterns (42). The aim is to identify particular areas that collectively include all taxa under consideration and then manage these areas as networks of biodiversity reserves, of both natural and semi-natural systems (108), to sustain viable populations of all component taxa. In this way it is hoped to preserve as wide a range of ecosystems, taxa, genes, and other biological features as possible. This strategy recognizes the basic currency of biodiversity as characters, models the distribution of this currency among species, and then compares the spatial distribution of species on various geographical scales. Margules & Austin (55) have noted that new data bases, even for well-known organisms, need to be constructed if accurate ecological and geographical surveys are going to improve on the ad hoc information generally available from existing biological collections. In view of the fact that inventories of only about 1.4 - to 1.8 million of the estimated 5–15 million extant species have been produced (94), it is most likely for many areas of the world that surrogates for character diversity at levels higher than species will have to be used (28, 33). In order of increasing remoteness, these will move to higher taxa, species assemblages, land classes, landscapes, and even whole ecosystems (114). To relate this scale of surrogacy to characters, differential weighting of taxa, based on phylogenetic models for predicting character diversity, provides a quantitative scheme for estimating biodiversity value at its most fundamental level.

Difference in Diversity

A simple count of species is still perhaps the most common approach used by both ecologists and systematists for measuring diversity to yield species-richness values for assemblages in areas (56, 77, 88). Species-richness measures depend on the size of the area being sampled (9). For comparisons between areas, accurate identifications of the taxa are considered fundamental because the species of each area are treated as having equal weight, whatever the range occupied. For decades this approach has been used to compare patterns of diversity at global, regional, and local scales (39). However crude such a measure might be, species richness is the one most widely used for a variety of purposes. For example, extinction rates are calculated either in absolute terms as a rate loss or as an estimate of a ratio of species lost in relation to amounts of transformed habitat (39). Each species constitutes a unit (89, 90) such that loss of one species means that the world is poorer by one unit species or a defined area is poorer by one locally extinct taxon.

Regarding species as equally represented in different areas under comparison is considered insufficient by ecologists, who often want to know about species, beyond their presence or absence, including such information as commonness or rarity (33). Besides the number of species in a particular area, it is also necessary to know how individuals are apportioned within it (54, 77), for example the Shannon index (48, 54, 76, 96). In a more elaborate system (see 7, 59), Whittaker (109–111) suggested that at least three measures are required to determine diversity: alpha diversity (species richness of standard site samples), beta diversity (differentiation between samples along habitat gradients), and gamma diversity for a geographic area (differentiation between areas at larger scales). Various modifications to these concepts have resulted in considerable variation in precise definition of alpha and beta diversity in the modern literature (52).

Over the last five years biologists and economists have begun to question seriously whether we can afford to treat all species equally. The panda logo of the World Wide Fund (WWF) is a reminder that human beings have always intuitively considered that some organisms are more important than others and should receive higher priorities in assessments and choices among conservation areas. One could ask whether one species of panda is equivalent to one species of rat. We suppose that the WWF chose the panda as a symbol because it is a rare, threatened, and large charismatic animal guaranteed to evoke interest and draw large quantities of charitable funds. Rats are more important in the economy of nature, but they lack appeal for more than a few people.

The 1994 Chambers Dictionary defines diversity as "the state of being diverse; difference; dissimilarity; variety." Thus, when choosing to prioritize one area for conservation against another, the one containing the most unlike taxa would be more diverse than the one containing similar taxa. For example, if comparing one area containing two sibling species of daisy with another area containing a columbine and a daisy, the latter area would be more diverse and could warrant a higher priority than the former area (see 41). Such a decision is justified in terms of the number of character differences or higher taxa being represented by the component taxa of each area (100, 120). Thus, as a basic rule of thumb, the taxonomic hierarchy can be used for an inclusive measure of diversity, including genetic diversity (44). However, there is a problem now that the taxonomic hierarchy has come to mean a variety of different things, ranging from traditional classifications simply aggregating species into genera and families, to highly corroborated phylogenetic hypotheses with detailed data on tree topology and branch lengths (25).

CLADOGRAM MEASURES OF TAXONOMIC DIFFERENCE

The use of phylogenetic history in assessing diversity has been interpreted in different and sometimes conflicting ways (47, 48). For example, Erwin (19, and 8) valued areas for conservation that could cater for future evolution in so-called "species-dynamo" areas. Erwin suggested that in areas with many closely related species, members of rapidly evolving clades ("evolutionary fronts") make greater contributions to future diversity than do species-poor lineages. There are a number of problems with this approach if we accept that areas with a greater number of different taxa are more diverse than those that contain actively evolving clades. In terms of characters, or degrees of difference between taxa, choosing "dynamo areas" could result in selection for low character diversity, and such a choice further makes the unwarranted assumption that we know something about future patterns of evolution (112, 116).

There are a number of ways to justify the choice of one area over another. In contrast to Erwin's approach, for example, one could assume that evolution of characters was clock-like, either implicitly or explicitly. In this model, character richness could be maximized for a restricted set of organisms by choosing representatives of earliest diverging taxa. Vane-Wright et al (in 57, 100) were the first to provide an explicit quantitative method, *root weight*, which provided an absolute measure that scored basal clades with higher weights as compared to the trivial clades higher in rooted trees. A similar approach, the *phylogenetic diversity* method, using weighted and unweighted binary ranking procedures to give diversity scores to terminal taxa in cladograms (43, 67, 117), appeared shortly thereafter. By assigning weights in terms of the number of taxa from a particular node and the position of the node in a tree, Nixon & Wheeler's (67) two procedures performed in a manner similar to root weight by giving highest weights to early diverging taxa. Stiassny & de Pinna (92) noted that basal taxa within lineages of freshwater fishes (i.e. groups that are species poor when compared to their sister groups), are particularly vulnerable to environmental pressures (notably due to range-size rarity). Because known freshwater groups and 25% of all vertebrates fall within this category, the authors considered loss of such taxa would cause disproportionate character loss. Contrary to Erwin, they suggest that all threatened basal lineages should receive higher weights in biodiversity assessments than their species-rich sister clades.

The root weight and phylogenetic diversity methods (99, 154) are both capable of selecting early-diverging and basal taxa by using information about the position of all taxa within a clade to assign absolute scores to each terminal taxon relative to its position in the branching topology. Williams et al (119) showed that such approaches had consequences that might seem undesirable. For example, considering all vertebrates using root weight or higher taxon richness (119, 120) would preferentially favour selection of basal groups of fishes as first choice subsets. The implications of considering the whole tree of life would give a perverse result of selecting several lineages of bacteria, while eukaryotes could be represented by one protist (116). Williams et al (119) proposed new ways for assessing taxonomic diversity of faunas and floras based on pairwise differences between taxa in cladograms. They noted that diversity considered as relative genealogical divergence, even in a restricted sense, was open to different interpretations.

For the first concept, their higher taxon richness measure was implemented to search for a flora or fauna in which species diverged closest to the root of the cladogram, to capture areas richest in highest taxa. However, seen from another perspective, different floras and faunas might contain samples of species that could be highly clumped or regularly dispersed across a cladogram. In the second concept, regular dispersion of species (terminal taxa) across a fully resolved pectinate cladogram would show more of a different kind of diversity than a clumped selection of an equal number of terminal taxa within the cladogram. Williams et al (119) were thus seeking to define diversity in terms of a pattern of relationship. Subsequently, the higher taxon measure selecting for early diverging taxa has been shown to increase the likelihood of capturing character-rich biotas, while the dispersion measure capturing the most regularly dispersed faunas and floras samples the richest combination of characters (25, 116). As diversity was assessed by the numbers of nodes between each pairwise comparison, the advantage of these methods was that any available branching diagram could be used for large numbers of species in the absence of branch length data.

SAMPLES OF CHARACTERS

Branch Lengths

The earliest reaction to Vane-Wright et al (in 57, 100), by Altschul & Lipman (4), suggested that in addition to nodes on cladograms, lengths of branches should also be taken into consideration in biodiversity measures. Vane-Wright et al (100) and Williams et al (119) emphasized that the best estimates of genealogy should be used, and they concentrated on node-counting measures because of the lack of reliable branch-length information for most taxa. Whereas cladograms are estimates of the most robustly supported relationships between taxa, branching diagrams interpreted as phylogenetic trees that incorporate branch length information, require more exacting interpretative models of character distributions (112) to calculate and predict accurately the pattern of diversity value.

Characters cannot all be counted directly, but their distribution among organisms can be predicted from their genealogy (22) by using a model of how character distributions may be related to the genealogical pattern. Different biotas can then be compared by using a taxonomic measure to predict or model differences of representation in these underlying characters. It is important to note that different classes of characters show very different patterns of distribution among organisms. If one class of characters is considered to be of real value, then it becomes important to identify that class of characters appropriate for diversity comparisons. Special cases have been made for interspecific genetic diversity (39), ecological and functional characters (86, 102), morphological diversity, and phenotypic characters (116). Attention has tended to converge onto valuing interspecific genetic diversity (23–25, 78, 91, 103–105), albeit in some instances to suit particular kinds of data such as molecular sequences and genetic distance data (12, 13).

"Phylogenetic diversity" has been used as a term most frequently by Faith (20, 25) to measure feature richness by using a spanning-subtree length technique. Unfortunately, the term, "phylogenetic diversity" had been used by Nixon & Wheeler (67) for their node-counting measure, and the term "feature richness" was already in use by conservationists to include cultural sites, geological structures etc. Nonetheless, Faith correctly recognized the most appropriate general measure of what we shall call "character richness." Phylogenetic diversity is intended to identify floras and faunas with the greatest representation of individual genetic or morphological characters (20). It is based on genealogical classifications of organisms that use measures of character differences as estimates of character distance along all branches (see also 20, 21, 26, 91, 103–105). These approaches show that phylogenetic pattern has the potential to quantify and estimate biodiversity at a range of resolutions, from branching patterns in tree topologies to variation among characters, to cater for almost all available information within any kind of classification or topology. Faith (25) characterized this range at three levels; existing hierarchies that may or may not reflect phylogenies, classifications concordant with phylogenies, and well-corroborated phylogenetic estimates with both topological and branch-length information.

Models of Character Change

Predicting character distributions requires a well-corroborated estimate of the phylogenetic tree and selection of an explicit process model as to how character changes are distributed on phylogenetic trees. At its simplest there are three cladistic models (116), although several variations have been described (25 for example). The most popular of these is the empirical model, which assumes any sample of characters to be representative of the entire universe of valued characters. More contentious are anagenetic or cladogenetic models. The clock-like anagenetic model is most appropriate for selectively neutral character changes, so that actual branch lengths on an ultrametric tree are assumed to be representative of character diversity. The cladogenetic model on the other hand suggests that character change is under stabilising selection and is primarily associated with the nodes of trees (43, 117).

If we had complete knowledge of the character universe, this could be used directly in diversity measures (78, 91, 103, 119). However, complete knowledge is never available. The empirical model uses counts of characters from a sample of character differences among organisms, and this is used for the branch lengths (23). Even though most characters are unsampled (74, 75), the empirical model assumes the sample to be unbiased (116, 117).

The clock model is popular among molecular biologists as the number of character changes along branches is related to duration (anagenesis). Clock models have been challenged (36, 98) because base substitutions can occur at different rates in highly constrained sequences (1, 2). In efforts to provide generalized models for all hierarchies, Faith (25) has also inferred equal branch lengths from taxonomic name hierarchies, which must be considered dubious given the arbitrariness of conventional taxonomic naming and ranking (117), but in the absence of other information, this is arguably better than nothing.

The cladogenetic model of character evolution is that favored by punctuated equilibrium theory (18) for character changes primarily associated with clado-genesis. In this model, characters important in conservation are those associated with nodes on trees. The debate about the relevance of cladogenetic models versus clock models has been equated to the debate about morphological versus molecular characters in systematics. Williams et al (116) suggest that different character suites are likely to follow different processes and models of change. This is what challenges the "universal" empirical model favored (at least implicitly) by Faith. Because different models will give different results from diversity measures, it focuses attention on choice of those characters and models most appropriate to perceived biodiversity value.

Although Faith (25) implies that only his methods (20–25) consider character diversity, we believe that all phylogenetic weighting methods devised to date make one standard assumption: that the underlying topology reflects character diversity. Williams et al (119, see also 112, 116) had already made the key point that choice of the underlying model is critical in terms of being able to justify the assumptions. The stated goal of protecting diversity has been equated by Weitzman (103–105) and Faith (20) with protecting as much character diversity (i.e. richness) as possible, and both maximum likelihood approaches (103, 105) and parsimony methods (20) have been employed in efforts to maximize feature diversity.

In summary, of the methods described to date, those appropriate for assessing character diversity fulfil May's (57) request for a calculus of biodiversity. In discussing a range of measures, Williams et al (119) identified three criteria for measuring diversity of subsets of terminals on cladograms: taxon richness, higher taxon diversity, and dispersion across the classification (regularity).

However, higher taxon diversity and dispersion are incompatible. Clarification of diversity value as characters (20) and of how this can be related to taxonomy by evolutionary models (116) indicates that the distinction between higher taxon diversity and dispersion corresponds to measures of character richness and character combinations, respectively (see below). Thus, after selecting for an appropriate currency (characters) and a particular evolutionary model that will predict the distribution of characters over cladograms and trees, the remaining consideration is to decide whether greatest value resides within individual characters or in combinations of characters. However, before this issue is considered some discussion about characters as the currency of biodiversity is necessary.

VALUES, CURRENCIES, AND MEASURES OF BIODIVERSITY

Norton (69) suggests that because there can never be one truly objective measure of biodiversity, measures appropriate for restricted purposes are the only ones available. Measures of biological diversity, like all measures, imply a set of values (39, 61, 84, 114, 121), and the idea that genetic diversity is perceived by many biologists to be the basic currency of biological diversity seems to satisfy the widest range of interests (17). Although it is a reductionist argument (48), such a view implies that expressible genetic resources, however measured (116), are the important building blocks of the biosphere, and that all other levels within the hierarchy, even species (but also land classes and ecosystems), are the vessels containing, or surrogates representing, the basic currency (114).

Genetic resources can be assessed in myriad ways, raising the question of what constitutes an operational definition of diversity. It is clear that in the context of conservation biology, diversity is generally viewed as an anthropocentric value system. For example, IUCN/UNEP/WWF (44, 45, 61, 84) stresses that the preservation of genetic diversity in agriculture, forests, and fisheries is both a matter for future insurance and investment against environmental change. Consequently, option value can be an important justification for conservation because we are always working in ignorance or uncertainty about which characters will be valuable in the future. In the practical occupation of choosing in situ conservation areas there may be a general-purpose (if not universal) currency of diversity that can satisfy many sectoral interests, can be measured in particular groups of organisms or geographical areas, and can be integrated with the larger-scale subdivisions of the biosphere, including habitats, vegetation zones, and ecosystems.

Option Value

Although we consider that areas can be assessed by degrees of character difference and that this justifies the need for differential weighting of one area versus another, it is important to recognize that there are other opinions. While there remain disagreements on what warrants adequate representation, we will also disagree on which aspects of the taxonomic system should be used in measures of diversity. While arguments have been put forward for weighting, others have been proposed against it. To suggest that, because we lack knowledge of the future values of species, all species should be considered equally important (121) depends on where the fundamental currency of biodiversity is seen to lie (25).

The issue of whether we should weight or not has been clarified by using the taxonomic hierarchy to evaluate species through "option value" (20, 25, 103, 104). Option value (39, 106) reflects the willingness of a risk averse society to pay a premium, on top of the value itself, to guarantee access to resources of uncertain future supply. Haneman (40) suggests that option value has two interpretations: It may be akin to a risk premium arising from uncertainty as to future potential value should a protected area be preserved. Or, it may refer to the irreversibility of destruction and to the need for information. According to McNeely (60), in conservation: "option value is a means of assigning a value to risk aversion in the face of uncertainty." Norton (68) rejects the use of option value and argues that it means identifying species, guessing what uses they might have, and estimating the likelihood that discoveries about them will be made at a future date. In a simple sense, anticipated future-use value of species is considered impossible to apply (16). Those making other non-weighting arguments have been similarly sceptical about attempting to assign current value to species (6), arguing that taxonomic weighting refers to maintenance of future options (97). Wilson (121) favors equal weighting because all species are indeed "irreplaceable resources to humanity," in terms of commodity, amenity, and also morality (25, 68, 121). Reid (83), however, identifies option value of biodiversity with maximizing human capacity to adapt to changing ecological conditions, which in turn requires maximizing the rest of life's capacity to adapt to change. The key to this suggestion is that if we accept that option value reflects capacity to respond to environmental change, we cannot operate at the level of species but should instead focus on biodiversity value at the fundamental level at which the objects for future use reside: expressible and heritable characters (114).

Equating option value with characters satisfies the requirement of IUCN (44), insofar as protecting subsets of taxa with as many features as possible would be the best safety net for keeping options open for future generations. As Faith (20, 21, 25) points out, characters can be seen as the fundamental "currency" with option value to the future. As Williams (114) puts it, biodiversity conservation must then focus on maximizing the amount of currency (counted as the number of different characters) to be held within the conservation bank (the set of secure species, ecosystems, or areas). If characters are seen as the basic unit of currency with equal weights applied to them (because of uncertainty), the vessels in which they occur (individuals, species, ecosystems, or areas) naturally may have different values because they contribute different numbers or combinations of complementary or novel characters.

Character Richness

If diversity value is interpreted to mean that it is the separate characters that are useful, as is the case for making pharmaceuticals or imparting disease-resistance, then the biodiversity measure that can identify the subset of organisms with the greatest number of character states would be chosen. Character richness for subsets of organisms can be maximized on metric trees with branch lengths scaled to character changes, by choosing a set of organisms within the floras or faunas of the areas under consideration linked by the maximum spanning subtree length (20, 21, 27, 112, 113). Ultrametric trees with clock-like character changes show maximum diversity in those areas that are richest in earliest diverging taxa. Williams et al (119) produced a measure for this when branch length data were lacking. Spanning-subtree length is also a consistent approach (20, 21, 27), but the branch lengths implied by ultrametric trees must be added (114).

Character Combinations

Within ecosystems, organisms are integrated by functional relationships. These may depend on combinations of characters. Thus an alternative interpretation of option value would be to assess the richness of different combinations of characters. To do this it is necessary to search for subsets of organisms that are most evenly spaced or distributed over the topology of the phylogenetic tree. This regular pattern will represent taxa at all levels within the hierarchy and therefore the largest number of diagnostic character combinations for subsets of a given size (112, 118).

Williams et al (112, 118, 119) devised the cladistic dispersion measure, which favors sets of organisms with the largest and most even distribution of nodes between them. The logical basis of this criterion is that the more evenly spread choices maximize the degree to which subtrees represent the overall shape of the classification, and thereby the most dispersed subtrees are more representative of the variety of subgroups with their different character combinations (112). However, this measure has no clearly justified balance between regularity, numbers of organisms, and subtree length (21, 119).

Taxonomic dispersion has subsequently been developed by Faith & Walker (25, 27, 29), who have drawn from a family of p-median procedures used in operations research (53, 95) to locate objects in regular patterns on networks (such as a metric tree). The general model represents different combinations of characters as "discs" of differing diameter along the branches of metric trees (21, 27). If particular unique combinations of characters come and go over time, any point on the tree may be the center of distribution for a variety of character state combinations. The distance that can be travelled along the tree while still finding that combination defines the combination radius. All points are equally likely to be centers for a combination, and combinations can have any radius with equal probability. When searching for hyperdiverse sets of terminals, their continuous p-median measure maximizes representation of all historical character combinations by maximizing intersection of all discs anywhere on the tree with the selected terminals, and by minimizing the sum or average distances from all points on the tree to their nearest selected terminal among the p-set of terminals. The discrete p-median measure maximizes extant character combinations by maximizing intersection of discs centered on unselected terminals with the selected terminals, by minimizing the sum or average distances from all unselected terminals on a tree to the nearest selected terminal. Williams (114) suggests that the discrete version theoretically may be more appropriate in the context of conservation, because this seeks to maximize the representation of only those character combinations that actually exist as modern species (terminals on a tree).

Surrogacy

Representing character richness and character combinations has all the signs of providing a neat theoretical solution, but problems of application remain. Practical use of taxonomic measures recognizes that all characters of organisms can never be counted directly, which means that surrogate approaches are inevitable (28, 114). It is possible to recognize a scale of surrogacy within which suitable indirect measures reflecting character richness and character combinations can be chosen. The use of surrogates can exploit a predictive relationship between the surrogate variable and the target variable to reduce costs, but in all cases the surrogate must be demonstrated to predict diversity (51, 70).

In the absence of ideal knowledge of all characters, taxonomic diversity measures of the kind described above are capable of estimating option value for all groups of organisms and sectoral interests, such as crop cultivars, from phylogenetic information. This has been done for Old World fruit bats by Mickleburgh et al (62), and for more than 30 other groups of plants, mammals, and invertebrates of global, regional, and local interest (118).

In some ways, the original weighting debate is turning full circle as, for large numbers of taxa at least, species richness is now seen to be a good surrogate for character richness. As Williams (114) points out, for this to fail would require that more species-rich faunas and floras become progressively taxonomically clumped as the number of taxa sampled increases. This result provides a stronger theoretical basis for using species-richness in diversity value measurement.

To ensure adequate representation from as wide a sample of organisms as possible, data sets for different species groups have been combined, and this has become an issue because incongruent patterns of species distribution are known at global (35, 115), national (72, 79, 80, 101), and local scales (85). Currently, various ways of combining data sets are being explored, for example, by summation of data sets (101) in a manner akin to the "total evidence" approach in systematics (15, 46). The predictive relationship depends on the notion that included diversity is indicative of total diversity, although it is unclear under what conditions predictor relationships for pattern by summation are better than when using one particular component group. There have been a number of claims from sectoral studies that well-chosen indicator groups are predictors of overall diversity [e.g. mycalesine satyrid butterflies as predictors of other groups in Madagascar (49, 50)]. These are only expected to work if there is a high level of nestedness (14, 73) or orderedness (85) in the distributions of the different species groups within each analysis.

Williams (114) has mentioned that, in principle, indicator groups could be selected and used in combination with a taxonomic diversity measure, but it remains to be seen whether adding taxonomic weight to species richness of one indicator group is a better predictor of total character richness than is species richness of the indicator group on its own. Kremen & Lees (personal communication) are trying to test this directly in Madagascar, using mycalesines as the target indicator taxon, through comparison with species richness for a number of other groups. Williams (114) predicts that such patterns might be predicted when there are strong repeating patterns in genealogy and space that could be explained by a vicariant model of biogeography (66), although it is quite possible that dispersals could obscure such patterns.

Any higher taxon, in theory, can be a surrogate for species. On one hand, this greatly reduces dependence on extrapolation between groups of the indicator approach, while on the other, it avoids some pitfalls of environmental surrogates. Williams (114) indicates that the idea behind such an approach is that mapping 1000 genera or families represents more of diversity than mapping 1000 species and may incur little or no extra data-gathering cost. Gaston & Williams (34, see also 115, 118) have used higher taxon richness as a surrogate for species richness, but when taxa are monophyletic, it may be possible to use higher taxa as surrogates for character richness using the tree-based measures (112).

A further surrogate approach, environmental diversity, has been described by Faith & Walker (27, 28) as an expression of environmental pattern so that environmental variation can be seen as a continuum rather than being partitioned into clusters. Faith & Walker relate environmental variables to species distributions using a standard ecological continuum model. Thus, because they can estimate environmental space from environmental data directly or indirectly by using some indicator group, the corresponding surrogate-measure, "environmental diversity," makes best-possible use of either kind of data. They conclude that the "arbitrariness" of the attribute method can be replaced with a robust surrogate "pattern" approach that is flexible and avoids unwarranted assumptions (28). While this approach is very promising, and maybe the best available for many poorly sampled areas of the world, it too usually has to depend on very simple and unrealistic assumptions about even and regular species distributions in niche space, and assumptions that the distribution of species among areas is at or near equilibrium with governing environmental factors (114).

We consider that, compared with most other approaches, taxon-based surrogates retain identities of attributes (taxa) for each area unit under consideration. By retaining the identity of taxa across areas, and hence some knowledge of spatial turnover (118), faunas and floras within networks can be assessed efficiently using complementarity (100). Critically, this conclusion does not suggest that option value in terms of characters has to be scored using mathematically complex taxonomic measures, but option value supports the use of less direct approaches that have been identified within the usual three tiers of genes, species, and ecosystems. These range from direct counts of characters, to progressively more remote surrogates in species, higher taxa, ordinations of species assemblages, land forms, landscapes and ecosystems (114).

SUMMARY AND CONCLUSIONS

Biodiversity can be assessed from a wide range of viewpoints for conservation, but some basic principles are emerging that are common to many of the evaluation methods that have been proposed. Biodiversity value can be identified with option value and character richness (or combinations) and can be estimated using phylogenetic models or by using surrogates such as species-richness, assemblage-richness, or land-class richness. Combined with the ability to assess areas in terms of their complementarity (turnover) and degrees of rarity (endemism), this approach provides the minimum framework for an appropriate system of relative measurement (99). To date, there are very few studies that accommodate all of these principles into one system, although the picture is changing.

A more general framework for the assessment of biodiversity must be able to use available information from a range of levels, and it is clear that some basic issues have yet to be resolved. However, for all of these methods it has been shown that there is an initial requirement to recognize attributes, whether genes, characters, species, higher taxa, or land classes. At the most fundamental level, characters can be considered as attributes with equal option value.

Secondly, these attributes have to be scored for geographical analyses at the appropriate spatial scales, from small pastoral units, to large standardised regional grid cells (81).

The third requirement is an estimate of the pattern as a summary of relationships among the attributes or parcels of attributes that need to be conserved. The thrust of this paper has been to review how phylogenetic pattern can be used to predict this underlying distribution of characters for biodiversity (20, 25, 100, 117, 119). In the case of grid cells as areas with species as parcels of character attributes, the phylogenetic relationships of the species provide the underlying pattern that can be used, in turn, to determine levels of biodiversity representation (e.g. character richness) and relative option value.

The fourth requirement is to provide a reliable model of how the pattern of conserved areas should provide information about diversity at the level of the attributes. Still being explored is the question of which model provides a robust explanation of how variation in the attributes is predicted by the pattern, and there remain some unresolved issues. For example, if unwarranted assumptions are made by one model for linking characters and option value to the phylogenetic pattern, then the resulting diversity scores may be even less defensible than when simply viewing all species as equal. Faith (25), for example, criticized Weitzman's (103) approach as being hard to justify because it requires assumptions of clock-like rates of evolution both for derivation of the pattern and for the model linking pattern to characters. We suspect that when empirical branch length data are not considered reliable, the cladogenetic model may prove to be more predictive than simple species richness or the clock model.

ACKNOWLEDGMENTS

We should like to thank Rüdiger Wehner, permanent Fellow, and Wolf Lepenies, Director, of the Wissenschaftskolleg zu Berlin for inviting us to participate in the class of 1993–1994, where many of the ideas presented here were discussed throughout the year with colleagues, other fellows, and guests: Katrina Brown, Erhard Denninger, Ashok Desai, Dan Faith, Kevin Gaston,

Werner Greuter, Gudrun Henne, Ian Kitching, Kathleen MacKinnon, Chris Margules, Jeff McNeely, Sandy Mitchell, Beatrice Murray, Norman Myers, Nick (A.O.) Nicholls, Bob Pressey, Eduardo Rabossi, Gustav Ranis, Tony Rebelo, Wolfgang Streeck, Josef Settele, and Campbell Smith.

Literature Cited

1. Albert VA, Backlund A, Bremer K. 1994. DNA characters and cladistics: the optimisation of functional history. In *Models in Phylogeny Reconstruction*, ed. RW Scotland, DJ Siebert, DM Williams, pp. 249–72. Oxford: Clarendon Press

2. Albert VA, Mishler BD, Chase MW. 1992. Character-state weighting for restriction site data in phylogenetic reconstruction, with an example from chloroplast DNA. In *Molecular Systematics of Plants,* ed. PS Soltis, DE Soltis, JJ Doyle, pp. 369–403. New York: Chapman & Hall

3. Altschul SF, Lipman DJ. 1990. Equal animals. *Nature* 348:493–94

4. Anderson D, Grove R. 1987. *Conservation in Africa*. Cambridge, UK: Cambridge Univ. Press

5. Austin MP, Margules CR. 1986. Assessing representativeness. In *Wildlife Conservation Evaluation*. ed. MB Usher, pp. 45–67. London: Chapman & Hall

6. Aylward B. 1992. Valuing the environment. In *Global Biodiversity*, ed. B Groombridge, pp. 407–25. London: Chapman & Hall

7. Bond WR. 1989. Describing and conserving biotic diversity. In *Biotic Diversity in Southern Africa: Concepts and Conservation,* ed. BJ Huntley, pp. 2–18. Cape Town: Oxford Univ. Press

8. Brooks DR, Mayden RL, Mclennan DA. 1992. Phylogeny and biodiversity: conserving our evolutionary legacy. *TREE* 7:55–9

9. Brown JH. 1988. Species diversity. In *Analytical Biogeography,* ed. AA Myers, PS Giller, pp. 57–89. London: Chapman & Hall

10. Croizat L. 1958. *Panbiogeography*. Caracas: Publ. by the author

11. Croizat L. 1964. *Space, Time, Form: The Biological Synthesis*. Caracas: Publ. by the author

12. Crozier RH. 1992. Genetic diversity and the agony of choice. *Biol. Conserv.* 61: 11–15

13. Crozier RH, Kusmierski RM. 1995. Genetic distances and the setting of conservation priorities. *Conserv. Biol.* In press

14. Cutler A. 1991. Nested faunas and extinction in fragmented habitats. *Conserv. Biol.* 5:496–505

15. Donoghue M. 1995. Approaches to parsimony: "Consensus" or "total evidence." *Annu. Rev. Ecol. Syst.* 26: In press

16. Ehrenfeld D. 1988. Why put value on biodiversity? In *Biodiversity,* ed. EO Wilson, pp. 212–16. Washington, DC: Natl. Acad. Press

17. Ehrlich PR, Wilson EO. 1991. Biodiversity studies: science and policy. *Science* 253:758–61

18. Eldredge N, Gould SJ. 1972. Punctuated equilibria: an alternative to phyletic gradualism. In *Models in Paleobiology,* ed. TJM Schopf, pp. 82–115. San Francisco: Freeman Cooper

19. Erwin TL. 1991. An evolutionary basis for conservation strategies. *Science* 253: 750–52

20. Faith DP. 1992. Conservation evaluation and phylogenetic diversity. *Biol. Conserv.* 61:1–10

21. Faith DP. 1992. Systematics and conservation: on predicting the feature diversity of subsets of taxa. *Cladistics* 8:361–73

22. Faith DP. 1993. Biodiversity and systematics: the use and misuse of divergence information in assessing taxonomic diversity. *Pacific Conserv. Biol.* 1:53–57

23. Faith DP. 1994. Genetic diversity and taxonomic priorities for conservation. *Biol. Conserv.* 68:69–74

24. Faith DP. 1994. Phylogenetic diversity: a general framework for the prediction of feature diversity. See Ref. 31, pp. 251–68

25. Faith DP. 1994. Phylogenetic pattern and the quantification of organismal biodiversity. *Phil. Trans. R. Soc. Lond. B* 345:45–58

26. Faith DP, Cawsey EM. *Phylorep, software for phylogenetic representativeness.* Publ. by authors

27. Faith DP, Walker PA. 1993. *Diversity Software Package and Reference and User's Guide.* Canberra: CSIRO, Div. Wildlife and Ecol.

28. Faith DP, Walker PA. 1995. Environmental diversity: how to make best possible use of surrogate data for assessing the relative biodiversity of sets of areas. *Biodiv. Conserv.* In press

29. Faith DP, Walker PA. 1995. Hotspots and fire-stations: on the use of biotic and evironmental data to estimate the relative biodiversity of different sets of areas. *Biodiv. Lett.* In press

30. Fernholm B, Bremer K, Jornvall H. 1989. *The Hierarchy of Life; Molecules and Morphology in Phylogenetic Analysis.* Amsterdam, New York, Oxford: Excerpta Medica

31. Forey PL, Humphries CJ, Vane-Wright RI, eds. 1994. *Systematics and Conservation Evaluation.* Oxford: Clarendon

32. Gaston KJ. 1991. The magnitude of global insect species richness. *Conserv. Biol.* 5:283–96

33. Gaston KJ. 1994. *Rarity.* London: Chapman & Hall

34. Gaston KJ, Williams PH. 1993. Mapping the world's species—the higher taxon approach. *Biodiv. Lett.* 1:2–8

35. Gaston, KJ, Williams PH, Humphries CJ. Large scale patterns of diversity: spatial variation in family richness. *Proc. Roy. Soc. Series B* Submitted

36. Gillespie JH. 1991. *The Causes of Molecular Evolution.* Oxford: Oxford Univ. Press

37. Glowka L, Burhenne-Guilmin F, Synge H, McNeely JA, Gündling L. 1994. *A Guide to the Convention on Biological Diversity.* Gland: IUCN

38. Grehan JR. 1993. Conservation biogeography and the biodiversity crisis: a global problem in space/time. *Biodiv. Lett.* 1:134–40

39. Groombridge B. 1992. *Global biodiversity: Status of the Earth's Living Resources.* London: Chapman & Hall

40. Haneman, WM. 1989. Information and the concept of option value. *J. Environ. Econ.* 16:23–37.

41. Harper JL, Hawksworth DL. 1994. Biodiversity: measurement and estimation. *Phil. Trans. Roy. Soc. B* 345:5–12

42. Humphries CJ, Vane-Wright RI, Williams PH. 1991. Biodiversity reserves: setting new priorities for the conservation of wildlife. *Parks* 2:34–38

43. Humphries CJ, Williams PH. 1994. Cladograms and trees in biodiversity. In *Models in Phylogenetic Construction,* ed. RW Scotland, DM Siebert, D Williams, pp. 335–52. Oxford: Clarendon

44. IUCN, UNEP, WWF. 1980. *World Conservation Strategy, Living Resource Conservation for Sustainable Development.* Gland, Switzerland: IUCN, UNEP, WWF

45. IUCN, UNEP, WWF. 1991. *Caring for the Earth, a Strategy for Sustainable Living.* Gland: IUCN

46. Kluge AG. 1989. A concern for evidence and a phylogenetic hypothesis of relationships among Epicrates (Boidae, Serpentes). *Syst. Zool.* 38:7–25

47. Krajewski C. 1991. Phylogeny and diversity. *Science* 254:918–19

48. Krajewski C. 1994. Phylogenetic measures of biodiversity: a comparison and critique. *Biol. Conserv.* 69:33–39

49. Kremen C. 1992. Assessing the indicator properties of species assemblages for natural areas monitoring. *Ecol. Applicat.* 2:203–17

50. Kremen C. 1994. Biological inventory using target taxa: a case study of the butterflies of Madagascar. *Ecol. Applicat.* 4:407–22

51. Landres PB, Verner J, Thomas JW. 1988. Ecological uses of vertebrate indicator species: a critique. *Conserv. Biol.* 2:316–28

52. Latham RE, Ricklefs RE. 1993. Continental comparisons of temperate-zone tree species diversity. In *Species Diversity: Historical and Geographical Perspectives,* ed. RE Ricklefs, D Schluter, pp. 294-314. Chicago: Univ. Chicago Press

53. Love RF, Morris JG, Wesolowsky GO. 1988. *Facilities Location. Models and Methods.* London: North-Holland

54. Magurran AE. 1988. *Ecological Diversity and Its Measurement.* London: Croom Helm

55. Margules CR, Austin MP. 1994. Biological models for monitoring species decline: the construction and use of data bases. *Phil. Trans. R. Soc. Lond. Ser. B* 344:69–75 CK

56. May RM. 1981. Patterns in multispecies communities. In *Theoretical Ecology,* ed. RM May, pp. 197–227. Sunderland, MA: Sinauer Assoc.

57. May RM. 1990. Taxonomy as destiny. *Nature* 347:129–30
58. May RM. 1992. How many species inhabit the earth? *Sci. Am.* October:18–24
59. McNaughton SJ. 1994. Conservation goals and the configuration of biodiversity. See Ref. 31, pp. 41–62
60. McNeely JA. 1988. *Economics and Biological Diversity.* Gland: IUCN
61. McNeely JA, Miller KR, Reid WV, Mittermeier RA, Werner TB. 1990. *Conserving the World's Biodiversity.* Washington, DC: IUCN, WRI, CI, WWF and WB
62. Mickleburgh SP, Hutson AM, Racey PA. 1992. *Old World Fruit Bats, an Action Plan for Their Conservation.* Gland: IUCN
63. Myers AA, Giller PS. 1988. *Analytical biogeography; An Integrated Approach to the Study of Animal and Plant Distributions.* London & New York: Chapman & Hall
64. Myers N. 1979. *The Sinking Ark: A New Look at the Problem of Disappearing Species.* Oxford: Pergammon
65. Nelson G, Ladiges PY. 1990. Biodiversity and biogeography. *J. Biogeogr.* 17:559–60
66. Nelson G, Platnick N. 1981. *Systematics and Biogeography: Cladistics and Vicariance.* New York: Columbia Univ. Press
67. Nixon KC, Wheeler QD. 1992. Measures of phylogenetic diversity. In *Extinction and Phylogeny,* ed. MJ Novacek, QD Wheeler, pp. 216–34. New York: Columbia Univ. Press
68. Norton BG. 1988. The constancy of Leopold's land ethic. *Conserv. Biol.* 2:93–102
69. Norton BG. 1994. On what we should save: the role of culture in determining conservation targets. See Ref. 31 pp. 23–39
70. Noss RF. 1990. Indicators for monitoring biodiversity. *Conserv. Biol.* 4:355–64
71. Novacek MJ, Wheeler QD. 1992. Introduction: Extinct taxa: Accounting for 99.9...% of the Earth's biota. In *Extinction and Phylogeny,* ed. MJ Novacek, QD Wheeler, pp. 1–16. New York: Columbia Univ. Press
72. Oosterbroek P. 1994. Biodiversity of the Mediterranean region. See Ref. 31, pp 289-307
73. Patterson BD, Atmar W. 1986. Nested subsets and the structure of insular mammalian faunas and archipelagoes. *Biol. J. Linn. Soc.* 28:65–82
74. Patterson C. 1994. Null or minimal models. In *Models in Phylogeny Reconstruction,* ed. RW Scotland, DJ Siebert, DM Williams, pp. 173-92. Oxford: Clarendon
75. Patterson C, Williams DM, Humphries CJ. 1993. Congruence between molecular and morphological phylogenies. *Annu. Rev. Ecol. Syst.* 24:153–88
76. Peet RK. 1974. The measurement of species diversity. *Annu. Rev. Ecol. Syst.* 5:285–307
77. Pielou EC. 1967. The use of information theory in the study of the diversity of biological populations. *Proc. 5th Berkeley Symposium on Mathematics and Statistical Probability* 4:163–77
78. Polasky S, Solow A, Broadus J. 1993. Searching for uncertain benefits and the conservation of biological diversity. *Environ. Res. Econ.* 3:171–81
79. Prendergast JR, Quinn RM, Lawton JH, Eversham BC, Gibbons DW. 1993. Rare species, the coincidence of diversity hotspots and conservation strategies. *Nature* 365:335–37
80. Prendergast JR, Wood SN, Lawton JH, Eversham BC. 1993. Correcting for variation in recording effort in analyses of diversity hotspots. *Biodiv. Lett.* 1:39–53
81. Pressey RL, Possingham HP, Margules CR. 1995. Optimality in reserve selection algorithms: When does it matter and how much? *Biol. Conserv.* In press
82. Raup DM. 1988. Diversity crises in the geological past. In *Biodiversity,* ed. EO Wilson, pp. 51–57. Washington, DC: Natl. Acad. Sci. Press
83. Reid WV. 1994. Setting objectives for conservation planning. See Ref. 31, pp. 1–13
84. Reid W, Miller KR. 1989. *Keeping Options Alive: The Scientific Basis for Preserving Biodiversity.* Washington, DC: WRI
85. Ryti RT. 1992. Effect of focal taxon on the selection of nature reserves. *Ecol. Applic.* 2:404–10
86. Schulze E-D, Mooney HA. 1993. *Biodiversity and Ecosystem Function.* Berlin: Springer-Verlag
87. Sisk TD, Launer AE, Switky KR, Ehrlich PR. 1994. Identifying extinction threats: global analyses of the distribution of biodiversity and the expansion of the human enterprise. *BioScience* 44:592–604
88. Smith EP, van Belle G. 1984. Nonparametric estimation of species richness. *Biometrics* 40:119–29
89. Smith FDM, May RM, Pellew R, Johnson TH, Walter KR. 1993. Estimating extinction rates. *Nature* 364:494–96
90. Smith FDM, May RM, Pellew R, Johnson YH, Walter KR. 1993. How

much do we know about the current extinction rate? *TREE* 8:375–78

91. Solow AR, Broadus JM, Tonring N. 1993. On the measurement of biological diversity. *J. Environ. Econ. Manage.* 24:60–68

92. Stiassny MLJ, de Pinna MCC. 1994. Basal taxa and the role of cladistic patterns in the evaluation of conservation priorities: a view from freshwater. See Ref. 31, pp. 235–49

93. Stork NE. 1995. How many species are there? *Biodiv. Conserv.* In press

94. Stork NE. 1994. Inventories of biodiversity: more than a question of numbers. See Ref. 31, pp. 81–100

95. Tansel BC, Francis RL, Lowe TJ. 1983. Location on networks: a survey. Part 1: the p-center and p-median problems. *Manage. Sci.* 29:482–97

96. Taylor LR. 1978. Bates, Williams, Hutchinson—a variety of diversities. *Symp. R. Ent. Soc. Lond.* No 9:1–18

97. Tisdell, 1990. Economics and the debate about the preservation of species, crop varieties, and genetic diversity. *Ecol. Econ.* 2:77–90.

98. Usher MB. 1986. *Wildlife Conservation Evaluation.* London: Chapman & Hall

99. Vane-Wright RI. 1994. Systematics and the conservation of biodiversity: global, national and local perspectives. In *Perspectives on Insect Conservation,* ed. KJ Gaston, TR New, MJ Samways, pp. 197–211. Andover, UK: Intercept

100. Vane-Wright RI, Humphries CJ, Williams PH. 1991. What to protect? Systematics and the agony of choice. *Biol. Conserv.* 55:235–54

101. Vane-Wright RI, Smith CR, Kitching IJ. 1994. Systematic assessment of taxic diversity by summation. See Ref. 31, pp. 309–26

102. Walker BH. 1992. Biodiversity and ecological redundancy. *Conserv. Biol.* 6:18–23

103. Weitzman ML. 1992. On diversity. *Q. J. Econ.* 107:363–406

104. Weitzman ML. 1992. Diversity functions. *Discussion paper number 1610.* Harvard University, MA: Harvard Inst. of Econ. Res.

105. Weitzman ML. 1993. What to preserve? An application of diversity theory to crane preservation. *Q. J. Econ.* 108:157–83

106. Weisbrod, BA. 1964. Collective-consumption services of individual-consumption goods. *Q. J. Econ.* 78: 471–77

107. Western D. 1992. The biodiversity crisis: a challenge for biology. *Oikos* 63: 29–38

108. Westman WE. 1990. Managing for biodiversity: unresolved science and policy questions. *BioScience* 40:26–33

109. Whittaker RH. 1960. Vegetation of the Siskiyou mountains, Oregon and California. *Ecol. Monogr.* 30:279–338

110. Whittaker RH. 1972. Evolution and the measurement of species diversity. *Taxon* 21:213–51

111. Whittaker RH. 1975. *Communities and Ecosystems.* New York: Macmillan

112. Williams PH. 1993. Choosing conservation areas: using taxonomy to measure more of biodiversity. In *Int. Symp. on Biodiversity and Conservation,* ed. T-Y Moon, pp. 194–227. Seoul: Korean Entomol. Inst.

113. Williams PH. 1994. *Using Worldmap. Priority Areas for Biodiversity.* Version 3.0. London: The author

114. Williams PH. 1995. Comparing character diversity using biological surrogates. In *Priority Areas Analysis: Systematic Methods for Conserving Biodiversity,* ed. CJ Humphries, CR Margules, RL Pressey, RI Vane-Wright. Oxford: Oxford Univ. Press In preparation

115. Williams PH, Gaston KJ. 1994. Measuring more of biodiversity: Can higher-taxon richness predict wholesale species richness? *Biol. Conserv.* 67:211–17

116. Williams PH, Gaston KJ, Humphries CJ. 1995. Do conservationists and molecular biologists value differences between organisms in the same way? *Biodiv. Lett.* 2:67–78

117. Williams PH, Humphries CJ. 1994. Biodiversity, taxonomic relatedness, and endemism in conservation. See Ref. 31, pp. 269–87

118. Williams PH, Humphries CJ, Gaston KJ. 1994. Centres of seed plant diversity: the family way. *Proc. R. Soc. Lond. Ser. B.* 256:67–70

119. Williams PH, Humphries CJ, Vane-Wright RI. 1991. Measuring biodiversity: taxonomic relatedness for conservation priorities. *Aust. Syst. Bot.* 4: 665–9.

120. Williams PH, Vane-Wright RI, Humphries CJ. 1993. Measuring biodiversity for choosing conservation areas. In *Hymenoptera and Biodiversity,* ed. J LaSalle, ID Gauld, pp. 309–8. Wallingford: CAB Int.

121. Wilson EO. 1992. *The Diversity of Life.* London: Penguin

122. WRI, IUCN, UNEP. 1992. *Global Diversity Strategy: Guidelines for Action to Save, Study and Use the Earth's Biotic Wealth Sustainably and Equitably.* New York: WRI

Annu. Rev. Ecol. Syst. 1995. 26:113–33

HUMAN ECOLOGY AND RESOURCE SUSTAINABILITY: The Importance of Institutional Diversity

C. Dustin Becker and Elinor Ostrom

Workshop in Political Theory and Policy Analysis, Indiana University, 513 North Park, Bloomington, Indiana 47408-3895

KEY WORDS: common-pool resources, institutions, complexity, spatial and temporal scales, design principles

---

ABSTRACT

We define the concept of a common-pool resource based on two attributes: the difficulty of excluding beneficiaries and the subtractability of use. We present similarities and differences among common-pool resources in regard to their ecological and institutional significance. The design principles that characterize long-surviving, delicately balanced resource systems governed by local rules systems are presented, as is a synthesis of the research on factors affecting institutional change. More complex biological resources are a greater challenge to the design of sustainable institutions, but the same general principles appear to carry over to more complex systems. We present initial findings from pilot studies in Uganda related to the effects of institutions on forest conditions.

INTRODUCTION

Aristotle asked how we as political animals could design and create institutions that would assure our survival with "some measure of good life within it." Today, numerous writings on sustainability reflect a shared awareness that natural resources are becoming increasingly scarce. Abnormally high extinction rates, deforestation rates, a thinning ozone layer, and increasing carbon dioxide and water pollution levels are but a few of the vital signs that humanity,

113

0066-4162/95/1120-0113$05.00

in its aggregate, has gone beyond a sustainable relationship with Earth's natural resources.

The original scope—that of a city—of Aristotle's question is still relevant. Some of the worst environmental problems on the globe are found in urban slums where local residents have little authority to design and create their own institutions. In addition, the realm of some environmental problems has greatly expanded from the size of a Greek city to the extent of the Earth. Aristotle's question, though, at its heart, remains the same. What multitiered social arrangements would best allow the future billions of us to have the comforts of basic human needs: clean water and air, nutritious foods, shelter, and human dignity? While there is no single answer to this question, general principles can be derived from both ecology and social science that are usable by diverse individuals in multiple settings to create and recreate institutions that will help humans assign more realistic valuations to ecological goods and services, and thereby be more likely to manage these resources in a sustainable manner.

A key challenge is establishing a common ground in the fractured academic world of the natural and social sciences. The discipline of human ecology tries to do this. The concept of common-pool resources and the question of how to govern and manage them, for example, have recently been addressed by many human ecologists (7, 10, 14, 47). In this paper, we define the concept of a common-pool resource based on two attributes: the difficulty of excluding beneficiaries and the subtractability of use. We present similarities and differences among common-pool resources in regard to their ecological and institutional significance. The design principles that characterize long-surviving delicately balanced resource systems governed by local rules systems are presented, as is a synthesis of the research on factors affecting institutional change. More complex biological resources are a greater challenge to the design of sustainable institutions, but the same general principles appear to carry over to more complex systems. We present initial findings from pilot studies in Uganda related to the effects of institutions on forest conditions.

RETHINKING THE TRAGEDY OF THE COMMONS

A decade ago, when the National Academy of Sciences first established a panel to study common-property institutions,[1] many scientists interested in

[1]This was the Panel on Common Property Resources. For publications, see National Research Council (52) and Bromley et al (14). A flurry of books and dissertations have been generated as the result of the Panel's initial activities and findings (7, 10, 12, 13, 23, 34, 38, 39, 45, 47, 54, 56, 57, 60, 63, 64, 69, 73, 74, 85, 86, 89). The International Association for the Study of Common Property, which was formed after the NAS panel finished its activities, will have had its fifth meeting in Bodø, Norway, May 24–28, 1995.

natural resource policy problems presumed that the users of common-pool resources were helplessly caught in a tragedy of the commons. Scientists assumed that users were destined to continue overharvesting unless external solutions were imposed on them. The solutions to be imposed were frequently presented as "the only way" to reduce externalities and increase efficiency. One proposed solution was control of natural resources by a central government agency. The second favored solution was the imposition of private property. Something had to be wrong with the theories, the interpretation of the theories, or the policy prescriptions if solutions as different as state control and market control were both proposed as the only way to manage natural resources efficiently.

By clarifying terms and conducting careful empirical research, researchers have identified a wide diversity of institutional arrangements that individuals have used to overcome tragedy of the commons scenarios. The dominant theories of a decade ago have not been proved wrong; rather, their claim to universal applicability has been successfully challenged. Both experimental and field research readily establishes that when those using resources whose legal status is open access are constrained by diverse factors to act independently, the predictions derived from the "tragedy of the commons" (31), the Prisoners' Dilemma game (32), and the logic of collective action (55; see also 68) are empirically supported (41, 60). Where valued resources are left to be open access, one can expect conflict, overuse, and the potential of destruction.

Common-Pool Resources

Scientific progress is made across the biological and social sciences when crucial attributes are identified that generate predictions about the behavior of the objects under study. Entities that share these crucial attributes differ in other respects that may also be important in providing a full explanation of system behavior. For social scientists interested in questions of resource governance and management, two attributes of resources are the first to be identified in efforts to understand how institutions interact with resources to produce incentives leading toward destruction or sustainability of a resource system. These attributes are 1) the *difficulty of exclusion*, and 2) the *subtractability of benefits* consumed by one person from those available to others (60).

EXCLUSION   The goods and events that individuals value differ in terms of how easy or costly it is to exclude or limit potential beneficiaries (users) from consuming them once they are provided by nature or through the activities of other individuals. Fencing and packaging are physical means of excluding potential beneficiaries from goods. To be effective, however, fencing and packaging must be backed by property rights that are feasible to defend (in an economic and legal sense). It follows that the feasibility of excluding or limiting use by potential beneficiaries is derived both from the physical attributes of the goods and from the institutions used in a particular jurisdiction (60).

Excluding or limiting potential beneficiaries from using a common-pool resource is a nontrivial problem for many reasons. In some cases, the problem is the sheer size of a resource. For example, the total cost of "fencing" an inshore fishery, let alone an entire ocean, is prohibitive. In other cases, the additional benefits from exclusion, or placing restrictions on use, are calculated to be less than the additional costs from instituting a mechanism to control use. In still other cases, basic constitutional or legal considerations prevent exclusion or limiting use (60). Knowing that a resource is one that is difficult to exclude, one can predict that "free riding" behavior will occur. *Free riding* is a term used to describe situations where some individuals "free ride" on the efforts of other individuals to provide either the good itself or the set of rules (and their monitoring and enforcement) that would enable individuals to achieve a sustainable, long-term utilization pattern in relationship to a resource (55, 68).

SUBTRACTABILITY   The goods and events that individuals value also differ in terms of the degree of subtractability of one person's use from that available to be used by others. If one fisherman lands a ton of fish, that ton is not available for others. On the other hand, one person's enjoyment of a sunset does not subtract from others' enjoyment of a sunset. Information is the extreme case of a good that is not subtractable. Most natural resources, on the other hand, are characterized by subtractable uses.

Goods characterized by problems of exclusion without any subtractability are considered to be public goods (60, 68). Institutions well-adapted for providing public goods are unlikely to solve the overharvesting and potential destruction problems faced in coping with common-pool resources characterized by problems of exclusion and subtractability. Common-pool resources include both natural resources and artifactual facilities designed by humans. Besides sharing two attributes in common, this broad set of resource systems differs on many other attributes.

A common-pool resource creates the conditions for the existence of a stock that may be quantified in terms of resource units. This stock may be the source of one or more flows of resource units over time. Examples of common-pool resources and their resource units include: 1) a groundwater basin and acre-feet of water, 2) a fishing ground and tons of fish, 3) an oil field and barrels of oil, 4) computer facilities and processing time, and 5) parking garages and parking spaces. It is the resource units that are subtractable from a resource. The fish (oil, water) being harvested are a flow, appropriated from a stock of fish (oil, water) (60).

The distinction between the resource stock and the flow of resource units is especially useful in connection with renewable resources that are predictable enough so that one can define a regeneration rate. As long as the rate of appropriation of resource units from a common-pool resource does not exceed the regeneration rate, the resource stock will not be exhausted. When a resource has no natural regeneration (an exhaustible resource), then any appropriation rate will eventually lead to exhaustion.

Subtractability and difficulties of excluding beneficiaries help identify two of the core problems anyone trying to develop institutions leading to sustainability must solve. First, the boundaries that define or include individuals who are authorized to access, harvest, manage, exclude, or sell rights to the use of a resource need to be well-defined (see 71). Second, rules allocating harvesting rights and duties must be devised to keep total use within the bounds of sustainable use. Subtractability and exclusion are not, however, all of the important attributes that affect problems of designing institutions to sustain a resource.

OTHER ATTRIBUTES OF COMMON-POOL RESOURCES   The challenge of devising workable rules, however, is also affected by other attributes that differentiate among types of common-pool resources. Schlager et al (70), for example, discuss the importance of the degree of mobility of resource units and the presence or absence of storage. Mobility refers to the spatial movement of resource units, such as flowing water or migratory fish. Water in a lake, or lobsters, are relatively stationary when contrasted to water in a river, or salmon. Storage refers to the capacity to store and retain harvested units. The amount of storage in an irrigation system is close to zero on most run-of-the-river irrigation systems, but it can be extremely high in a conjunctive use system, depending on groundwater basins for storage or a large surface irrigation system with many dams along the course. Schlager et al (70) illustrate how these attributes affect the severity of the problems resource users face, the ease with which they can solve problems, and the type of institutional arrangements most frequently used to solve these problems. The effects of mobility and storage are due to the impact of these physical attributes on:

1. the information users have about their common-pool resources and the problems they are experiencing;
2. the likelihood that users will be able to capture the benefits that issue from their efforts to solve problems; and
3. their assurance about the behavior of other users (70, p. 297).

Whether the resource units must be used within a resource system or are exportable also affects the incentives of users and the problems of regulation. Fish and water can be exported. Once resource units are removed from a common-pool resource, they are similar to other private goods, but their marketability may be one of the incentives that leads to overharvesting. Parking space and computer processing time must be used by multiple individuals within the same entity. Resource units that must be used in a shared facility can be affected by the time and type of use of the other units being used simultaneously and sequentially. Other attributes, such as the asymmetry of interests among participants, are also important (81).

Whether the common-pool resource is physical or biological also makes a considerable difference in the ability of institutions to use and care for a resource sustainably. Institutions adapted for single product physical resources have an easier allocation problem to solve than the set of allocation rules required for the sustained use of biological resources, especially multispecies ones like fisheries or diverse tropical forest ecosystems. Compared to the sharing of a physical resource, the appropriation of complex biological resources has many more ecological variables influencing what resources can be used, when they can be used, and to what extent they may be extracted, while still maintaining the integrity of the resource base. Complex biological resources typically have more uncertainty associated with them than do single physical resources. According to Mayr (46), emergence, the origination of unsuspected qualities or properties at higher levels of integration in complex hierarchical systems, is vastly more important in living than in inanimate systems. We discuss this in more detail later.

DESIGN PRINCIPLES AND ROBUST INSTITUTIONS

As part of an extended effort to study common-pool resources and the institutional arrangements that enhance the capacity of individuals to use these resources in a sustainable way over long periods of time, we have identified a diversity of common-pool resources and related institutions that have been used intensively by humans (56, 59, 60, 90, 91). These institutions are considered to be robust in the sense that both the resource systems and the institutions have survived for long periods of time. The day-to-day operational rules of these systems have undergone change, but these changes have occurred within a set of collective-choice and constitutional-choice rules (75). Some of the robust systems that have been studied in some detail include those of the Swiss Alpine meadows (26, 27, 53); Japanese mountain areas (48); 1000-year-old Spanish irrigation systems (28, 42); California groundwater basins (12); and indigenous irrigation systems in the Philippines (78).

Robust institutions tend to be characterized by most of the design principles listed in Table 1. Clearly defining the boundaries (Principle 1) helps to identify who should receive benefits and pay costs. Equating benefits and costs (Principle 2) is considered a fair procedure in most social systems. Decisions by local users to establish harvesting and protection rules (Principle 3) enable those with the most information and stake in a system to have a major voice in regulating use. The first three principles together help solve core problems associated with free riding and subtractability of use. Rules made to solve these problems are not, however, self-enforcing. Thus, monitoring (Principle 4), graduated sanctioning (Principle 5), and conflict-resolution mechanisms (Principle 6) provide ongoing mechanisms for invoking and interpreting rules and finding ways of assigning sanctions that increase common knowledge and agreement. Recognizing the formal rights of users to do the above (Principle 7) prevents those who want to evade local systems from claiming a lack of

Table 1 Design principles derived from studies of long-enduring institutions for governing sustainable resources

1. *Clearly Defined Boundaries*
   The boundaries of the resource system (e.g. groundwater basin or forest) and the individuals or households with rights to harvest resource products are clearly defined.
2. *Proportional Equivalence Between Benefits and Costs*
   Rules specifying the amount of resource products that a user is allocated are related to local conditions and to rules requiring labor, materials, and/or money inputs.
3. *Collective-Choice Arrangements*
   Most individuals affected by harvesting and protection rules are included in the group who can modify these rules.
4. *Monitoring*
   Monitors, who actively audit physical conditions and user behavior, are at least partially accountable to the users and/or are the users themselves.
5. *Graduated Sanctions*
   Users who violate rules are likely to receive graduated sanctions (depending on the seriousness and context of the offense) from other users, from officials accountable to these users, or from both.
6. *Conflict-Resolution Mechanisms*
   Users and their officials have rapid access to low-cost, local arenas to resolve conflict among users or between users and officials.
7. *Minimal Recognition of Rights to Organize*
   The rights of users to devise their own institutions are not challenged by external governmental authorities, and users have long-term tenure rights to the resource.
For resources that are parts of larger systems:
8. *Nested Enterprises*
   Appropriation, provision, monitoring, enforcement, conflict resolution, and governance activities are organized in multiple layers of nested enterprises.

Source: Ostrom (56, p. 90).

legitimacy. In addition, nesting a set of local institutions into a broader network of medium- to larger-scale institutions helps to ensure that larger-scale problems are addressed as well as those that are smaller. Institutions that have failed to sustain resources tend to be characterized by very few of these design principles, and those that are characterized by some, but not most, of the principles are fragile.

The design principles are articulated in Table 1 in a general language. The specific ways that individual users have crafted rules to meet these principles vary. Successful, long-enduring irrigation institutions, for example, have developed different ways of meeting the second design principle of achieving congruence or proportionality between the costs of building and maintaining irrigation systems and the distribution of benefits. Three examples below illustrate the diversity of specific rules that meet the second design principle.

The *zanjeras* of Northern Philippines are self-organized systems in which farmers obtain use-rights to previously unirrigated land from a large landowner by building a canal that irrigates the landowner's land and that of a *zanjera*. At the time of land allocation, each farmer who agrees to abide by the rules of the system receives a bundle of rights and duties in the form of *atars*. Each *atar* defines three parcels of land located in the head, middle, and tail sections of the service area where the holder grows his or her crops. Responsibilities for construction and maintenance are allocated by *atars*, as are voting rights. In the rainy seasons, water is allocated freely. In a dry year, water may be allocated only to the parcels located in the head and middle portions. Thus, everyone receives water in plentiful and scarce times in rough proportion to the amount of *atars* they possess. *Atars* may be sold to others with the permission of the irrigation association, and they are inheritable (see 18, 78, 84).

When the *Thulo Kulo* irrigation system was first constructed in 1928, 27 households contributed to a fund to construct the canal and received shares proportionate to the amount they invested. Since then, the system has been expanded by selling additional shares. Measurement and diversion weirs or gates are installed at key locations so that water is automatically allocated to each farmer according to the proportion of shares owned. Routine monitoring and maintenance is allocated to work teams so that everyone participates in proportion to their share ownership, but emergency repairs require labor input from all shareholders regardless of the size of their share (see 43, 44). Similar self-organized systems exist throughout Nepal (37, 61, 67, 76, 93, 94).

In 1435, 84 irrigators served by two interrelated canals in Valençia gathered at the monastery of St. Francis to draw up and approve formal regulations to specify who had rights to water from these canals, how the water would be shared in good and bad years, and how responsibilities for maintenance would be shared. The modern *Huerta* of Valençia, composed of these plus six additional canals, now serves about 16,000 ha and 15,000 farmers. The right to water inheres in the land itself and cannot be bought and sold independently of the land. Rights to water are approximately proportionate to the amount of land, as are obligations to contribute to the cost of monitoring and maintenance activities (see 42).

Five hundred years after the irrigators of Valençia devised a system to allocate water rights, modern water users dependent upon the groundwater basins underlying Los Angeles County also started meeting to discuss how to allocate water rights among municipal, industrial, and agricultural users (12, 56). They all faced the potential destruction of their valuable groundwater basins if they continued the pumping race that unsettled property rights had encouraged during the prior decades. Over a 40-year period, groundwater pumpers in four adjacent basins established private water associations that invested heavily in obtaining accurate information about the geologic and hydrologic characteristics of the basins. Further, water producers bargained in the shadow of equity courts to define water shares based on historical use patterns and to develop proportionate reductions in authorized, annual water withdrawals. The costs of monitoring these systems over time and enforcing the agreements are paid for by water users in proportion to the amount of water they produce. Participants also established limited, special-purpose water districts to manage other aspects of the groundwater basins not covered by limiting the quantity of withdrawals. The institutional arrangements developed by water users meet all eight of the design principles listed in Table 1 (56, p. 180). Several of these agreements have now been in practice for half a decade and show no signs of losing their overall effectiveness in protecting the sustainability of the groundwater basins over time.

These four systems differ substantially. The *zanjeras* are ways landless laborers can acquire use-rights to land and water. These could be called communal systems. The *Thulo Kulo* system comes as close to allocating private and separable property rights to water as is feasible in an irrigation system. This might be called a private or market solution. The *Huerta* of Valençia has maintained centuries-old land and water rights that forbid the separation of water rights from the land being served. The Valençian system differs from both "communal" and "private property" systems because water rights are firmly attached to ownership of land. The California groundwater systems privatized water rights (the flow), and a vigorous market for these water rights ensued. At the same time, limited-purpose, public, local jurisdictions assumed responsibility for managing the basins (the resource system or stock) by assessing pump taxes and undertaking replenishment programs. Underlying these strong differences, however, is the basic design principle that the costs of constructing, operating, and maintaining these systems are roughly proportional to the benefits that participants obtain. Since those who are users of these systems have devised their rules over time using trial and error methods, one should not presume that there was a conscious overall plan to develop institutions that met the design principles. Rather, the design principles are an effort of careful observers to identify commonalities that help to account for sustainability of fragile resources over very long periods of time.

Differences like the ones illustrated here lead us to stress the importance of design principles rather than specific institutional solutions to common-pool resource problems. The contribution of social and biological scientists to the study of sustainable resource systems can be substantial if general theoretical principles are identified rather than searching for particular institutional solutions that are prescribed as universal solutions. Universal solutions can tend to become slogans, such as "nature reserves," "privatization," "individual transferable quotas," "integrated rural development projects," and "joint management schemes." There are successful and unsuccessful examples of all of these types of programs that differ from one another in important institutional and ecological variables.

Slogans may mask important underlying principles involved in many of the successful efforts to utilize individual transferable quotas or joint management schemes rather than providing useful guides for reform. Strict privatization of water rights is not a feasible option within the broad institutional framework of many countries. Nor is it easy to see how one can accomplish this in a run-of-the-river system. The only institutional arrangements that we know where water users own strict quantitative shares to water are those involving storage—either groundwater or constructed surface storage. Even in groundwater basins, one can privatize the flow of the resource (once good scientific information is available about long-term climate, geologic, and hydrological characteristics). But the basin itself cannot be parcelled out and must be used jointly. Nor can a dam be parcelled out—especially one that generates hydroelectric power, recreation, and flood control benefits in addition to water supply. Thus, flow characteristics may be allocated in a different way than stock and facility characteristics. On the other hand, authorizing the beneficiaries of a common-pool resource to participate in the design of their own systems—design principles 3 and 7 combined—is a feasible reform within the broad institutional framework of most countries and most ecological settings.

FACTORS AFFECTING INSTITUTIONAL VARIETY

Not only is a substantial variety of rules used to reduce the cost of externalities from unregulated use of natural resources, but neighboring systems that appear to face similar situations frequently adopt different solutions. Within a few miles of Valençia is Alicante, where irrigators long ago built a surface dam and adopted rules separating water from the land. The weekly water market in Alicante has operated for centuries. Adjacent to *Thulo Kulo* is *Raj Kulo*, where the allocation of water (and labor responsibilities) is according to the amount of land owned. Near the *zanjeras* are many irrigation systems with quite different rules for distributing water and input responsibilities. Near the water basins in Los Angeles County are basins in Orange, San Bernardino, and Riverside Counties where pumpers have refused to undertake litigation to clarify their water rights (12).

The variety of rules selected by local users who appear to face similar circumstances raises the question of whether the choice of institutional arrangements is an evolutionary process involving selection for more efficient institutions over time. In an important article, Alchian (2) demonstrated how the pressure of competitive markets would select surviving firms that used profit-maximizing strategies whether they had chosen these strategies self-consciously or not. Some advocates of market orders (72, 83, 88) have argued that individuals will slowly establish new and more efficient institutions through a series of spontaneous individual decisions. The improved group outcome is conceptualized as an unintended result of individual learning and adjusting behavior over time. It is not quite clear, however, what selection principle is at work outside of competitive markets.

Others, including Knight (35) and Ostrom (56, 58), point out that changes in rules usually occur within a meta set of rules at a collective choice or constitutional level and within settings that vary in terms of pressure for survival or excellence. The meta set of rules may assign different advantages to various participants in the rule-changing process. Those with the most voice in collective-choice processes may refuse to support a change if they do not benefit themselves from the change in rules. This can occur even when the aggregate benefit of a rule change is large. Thus, to explain a change in rules, one needs to analyze not only the status quo distribution of costs and benefits but also the distributional effects of proposed rules (40) and how these relate to the meta rules used for making and changing rules.

To explain institutional change, one needs to analyze the relationships between variables characterizing the resource, the community of individuals involved, and the meta rules for making and changing rules. Sufficient theoretical and empirical research has been conducted on this and the closely related theory of collective action to enable one to specify important variables and the direction of their impact. The following variables appear to be conducive to the selection of norms, rules, and property rights that reduce externalities (6, 7, 19, 38–40, 48, 49, 56):

1. Accurate information about the condition of the resource and expected flow of benefits and costs are available at low cost.

2. Participants are relatively homogeneous in regard to asset structure, information, and preferences.

3. Participants share a common understanding about the potential benefits and risks associated with the continuance of the status quo as contrasted with changes in norms and rules that they could feasibly adopt.

4. Participants share generalized norms of reciprocity and trust that can be used as initial social capital.

5. The group using the resource is relatively small and stable.

6. Participants do not discount the future at a high rate.

7. Participants have the autonomy to make many of their own operational rules, which if made legitimately, will be supported and potentially enforced by external authorities.

8. Participants use collective-choice rules that fall between the extremes of unanimity or control by a few (or even bare majority), and thus they avoid high transaction or high deprivation costs.

9. Participants can develop relatively accurate and low-cost monitoring and sanctioning arrangements.

Many of these variables are, in turn, affected by the type of larger regime in which users are embedded. If the larger regime facilitates local self-organization by providing accurate information about natural resource systems, providing arenas in which participants can engage in discovery and conflict-resolution processes, and providing mechanisms to back up local monitoring and sanctioning efforts, the probability of participants adopting more effective rules over time is higher than in regimes that ignore resource problems or presume that all decisions about governance and management need to be made by central authorities.

THE CHALLENGES OF COMMON-POOL BIOLOGICAL RESOURCES

Biological resources, both single species and multispecies systems, have a set of uncertainties that stem from being part of an interacting community of organisms. A physical resource like water may have many competing uses, but whether a replenishable amount is used in farming or municipal water supply does not affect the long-term sustainability of the resource. In contrast, the harvest of individuals from living populations in terrestrial or aquatic ecosystems can affect the future availability of both the particular species and others with which it interacts. For example, a community timber-harvesting program that is modeled to be sustainable could eliminate the sustainable harvest of medicinal herbs in a forest by altering the sunlight and moisture in the understory. The ecological and economic trade-offs associated with multispecies systems are extremely complex and thus require long-term trial-and-error social experiments to arrive at optimal, or even feasible, use patterns

(20). Imposing rigid private property or central authority on a multispecies resource use system that has evolved over centuries may not only adversely affect the very human groups who have been responsible for successfully husbanding resources but may also adversely affect the ecological function of the resource systems (11, 22, 24, 25).

One principle that today is grossly inappropriate for complex systems, but still guides allocation of biological resources, is the concept of maximum sustained yield (MSY). This is not to say that we should completely abandon attempts to manage some single-species resources at MSY, but the concept fails in most dynamic systems. Plant and animal populations that have very predictable regenerative properties within a complex community can be managed in a sustained-yield fashion, but they are more an exception than the rule. A greater number of species have chaotic population patterns (92). When recruitment and yield are predictable, allocation rules and institutions may be fairly simple, understandable, and trusted by users. For example, the plaice fishery in the North Sea has a long history of MSY, and it has a predictable stock-recruitment pattern, showing only a six-fold variation during three decades (9). In contrast, the North Sea haddock varies by 500-fold in its abundance at recruitment and has recently been overfished to the point of requiring closure of the fishery (36).

Setting the MSY of a living resource is challenging because environmental conditions fluctuate, causing values of MSY to fluctuate. Climatic, demographic, and environmental stochasticity can greatly alter the sustainability of a set of species being managed at MSY. On a closed system, mixed-species game ranch in Kenya, when predators were reduced, the number of wildebeests, Thomson's gazelles, Grant's gazelles, and hartebeests increased. Sex ratios were purposely skewed toward females to increase productivity, and the populations were culled in a sustainable proportion to their reproductive potential. Conservative but complex biological models were employed to select the animals for culling. For a few years, the system produced an economically profitable and biologically sustainable supply of meat to specialty markets in Nairobi. However, the ecological sustainability of the closed system collapsed during a drought in 1984, when ungulate numbers and diversity plummeted, requiring managers to open the system to acquire more stock to keep the business going (82). Getting the biology right is not easy!

To complicate matters further, misinterpretation of economic models and self-interest have favored overexploitation of many common-pool resources, especially fisheries. For decades, and probably still for some today, there has been a misconceived notion that maximum economic yield (total revenue minus total cost) usually falls below MSY (30). Clark (17) showed that discounting the future (the preference to make one dollar today rather than two dollars in the future) encourages fishing far beyond MSY under many reasonable conditions. Many current-day institutions encourage users to discount the future, strongly leading to strategies that are neither ecologically nor economically sustainable.

Societies driven to the efficiency level that favors overcapitalization and large corporate exploitation have failed to use diverse biological resources sustainably, largely because resource developers have banked on substitutability for sustainability. So long as any species of sea organism will do for a commodity such as "fish meal," or so long as natural capital can be reinvested in any form of capital, biological resources can be, by definition, economically sustaining (80; but see 33). Trees can be cut and sold, made into pulp and paper, and earnings may be reinvested in other sectors. From this perspective, economic sustainability is operationally defined as exploiting nature's capital whenever it is efficient to do so, and switching to other forms of capital when nature's value drops below some threshold. Substitution as currently practiced has serious ecological limits, because eventually all the components get used up. If the switches among resources were made within the realm of biological capital only, and with the goal of sustaining each resource in the pool of potential commodities, we would stand a better chance of sustaining all of the resources. Each form of natural capital (species) is sustainable only when its use is equal to its replenishment as dictated by interactions with other species and disturbance regimes that affect the system.

To achieve ecologically sustainable use of diverse biological resources, either we need reliable models that adequately predict complex demographics, effects of species interactions, and disturbance stochasticity, or we need to be very conservative in our exploitation of any one resource in the array. In practice, the latter approach is the more attainable, but it is certainly not the current paradigm in commercial resource management.

Human societies with a long history of interdependence with multispecies resources should evolve institutions that optimize the turnover and use of extractables without compromising the functional aspects of the ecosystem or the future availability of important products (8). In other words, their rule systems should maintain the diversity that sustains the multitude of useful species. What sort of social arrangements would do this? Opportunistic substitution would be sustainable at low human densities, so until the resources were scarce, one would predict little in the way of restraints on use. Excavations of ancient food middens of native people living on the Aleutian Islands indicate that exploitation of certain species in the Pacific kelp bed ecosystem caused the biological community to flip back and forth between two major equilibria. There were periods when sea otter bones and fish dominated the middens, and periods when mainly sea urchin skeletons dominated the refuse piles (77).

Under conditions where diverse and scarce resources are in demand (consumptively or nonconsumptively), one would predict institutions with some form of strict protection of the habitat that supports the scarce biota. Sacred forests in Africa (29) and Shaman's gardens in Amazonia (66) have been interpreted as institutionalized insurance against overexploitation of rare biological resources, especially medicinal or ceremonial species. Natural area protection at local, national, and international scales indicates widespread attempts to buffer biological resources from overutilization (15).

Rules that favor the use of a little bit of everything, but overutilization of nothing, would be expected along with complex social systems for resource allocation and monitoring. In the modern-day Mayan community of Chunhuhub in the forests of Quintana Roo, Mexico, people still speak of "Yuntzilob," the forest deities who bring disaster on those who overuse biological resources. According to Anderson (3), many of the well-educated youth of this community today see the ecological importance of these beliefs.

Subsistence societies around the world have evolved institutions that sustain themselves and the complex biological natural resources upon which they depend for survival (66). In the Amazon, Tukano Indians living along the blackwater tributaries have strict rules against deforestation along the river margins. These riparian areas are reserved for the feeding grounds of fish. The ecological basis of this rule becomes apparent in the relative lack of fish in flooded areas that have been deforested (16), and in the poor agricultural yields on the low-nutrient blackwater flood plains. The relatively high dietary protein levels and population densities of the Tukano also support the idea that the rule against deforestation has a direct advantage to the individual rule makers in terms of Darwinian fitness.

Throughout Amazonia, Amerindians practice a form of agriculture that is considered to be benign in terms of ecosystem functions such as nutrient cycling, depletion of species, and gas exchange (87). Many Amerindians maintain small, species-diverse garden plots of less than an acre, similar in size to a natural gap produced by treefall. They combine perennial tree crops with natural forest regeneration in their method of swidden cultivation (21). Their fallow swiddens are often more species rich than the mature forest. Semi-nomadic Kayapo Indians in Brazil take edible tubers, fruit trees, and medicinal plants from the forest interior and plant them along their trails and adjacent to their campsites, thus changing the distributions of useful resources, without depleting original stocks (65).

Currently, the ecologically sustainable systems of traditional societies are in direct conflict with the dominant product orientation of global-market enterprises such as livestock ranching, timber harvesting, and monoculture farming (50, 51, 79). The sustainability of complex multispecies resource systems is also endangered by the decay of the indigenous taboos and rules that were so adaptive in the first place (4). Population growth has put a strain on traditional institutions. With more people demanding land, some modern Mayan communities have modified traditional rules for fallow periods, shortening them to allow more families the opportunity to farm (3). People can self-organize on a local level to maintain complex biological systems, but they must have a long-term, adaptive relationship with the resources to do so.

CURRENT RESEARCH ON HUMAN INSTITUTIONS AND FOREST ECOLOGY

Future research will need to include theoretical and empirical studies that specifically address how heterogeneity of participants, multispecies or multiproduct resource systems, and long time horizons affect the selection and performance of institutions. We are currently developing a theoretical and empirical program of research—the International Forestry Resources and Institutions (IFRI) research program (see 62). The IFRI research program combines an effort to examine how diverse institutional arrangements perform within similar and across different ecological zones with an effort to monitor and understand human-ecological systems interactions over long periods of time. The theoretical questions relate in part to whether the design principles derived from studying robust institutions of smaller and somewhat simpler common-pool resources (see Table 1) are applicable in designing institutions to sustain complex forest ecologies. Further, an important set of questions relate to the conditions under which local communities will overcome severe collective-action problems to design, monitor, and enforce their own local institutions and how well diverse types of local, regional, and national institutions will perform in different types of forest ecologies (see 49).

Our current research program is now beginning a long-term operational phase after a two-year design, pretest, and pilot phase. Three collaborating research centers in Bolivia, Nepal, and Uganda have been established, and initial research has also been undertaken in India. In an IFRI study of two adjacent and similar-sized parcels of forested land in Uganda, institutions explained major differences in the physical and biological condition of the forest. Namungo forest was family-owned, was well-monitored, and had clear rule structures. Collective and individual action by local forest consumers could be used to negotiate and modify rules for exploitation of many subsistence products in Namungo's forest. Commercial use of Namungo forest was forbidden, and sanctions were clear. In contrast, Lwamunda forest, as public property, was poorly monitored, lacked any rules negotiated by local users, had unclear boundaries, and had poorly specified sanctions. Based on the design principles discussed above (Table 1), Lwamunda was expected to show more "open-access" utilization and more degradation than Namungo forest. In 30 random plots of 300 m^2 made in each forest, plots degraded by timber milling, charcoal-making, and commercial exploitation of fuelwood were significantly more prevalent in Lwamunda than in Namungo forest. A different set of tree species dominated each forest area, the depleted species clearly reflecting recent utilization preferences (5). Rules and their effectiveness at the local level are critical to the sustainability of complex biological resources.

Initial findings from an in-depth study of five forests in the Almora district of the Kumaon region of Uttar Pradesh in India indicate that the expected strong relationship between population density (number of people per hectare of forest) and lower forest density (cubic meters of tree biomass per hectare) is not present in the first five self-organized communities sampled in this region (1). Further, Agrawal finds that somewhat larger communities are better able to raise the needed resources to hire forest guards to monitor the use of communal forests. Thus, the presumption in collective-action theory that very small communities are better able to overcome collective-action problems may not hold when the type of collective action required involves the mobilization of substantial resources.

CONCLUSION

Now that nature's capital including water, old-growth forests, and fishery stocks is becoming scarcer relative to growing stocks of human-made capital, investments in protecting nature's capital and the efficiency of its use are becoming more central to long-term economic sustainability. To achieve long-term economic sustainability, we need more than ever before a combination of institutions that restrain shortsighted and selfish behavior and that make rules based on flexible and cautious models of the ecology of complex biological systems. We are entering an age when we need to reduce our losses of natural capital, and one substitution after another is no longer a risk we can afford to take. Flipping from complex equilibria to simpler ones to eke out sufficient food in the short run will produce long-run scarcities of essential biological resources. Coevolution via cautious trial-and-error exploitation is a better strategy than use of rigid MSY models for setting harvest levels of multispecies resources. Extinction roulette with poor management of multispecies resources is not a prudent strategy for the twenty-first century.

ACKNOWLEDGMENTS

This article results from work on Social and Ecological System Linkages of the Property Rights Program of the Beijer International Institute of Ecological Economics, The Royal Swedish Academy of Science, and the International Forestry Resources and Institutions (IFRI) research program at the Workshop in Political Theory and Policy Analysis, Indiana University, supported by the Forests, Trees and People Program of the Food and Agriculture Organization of the United Nations. Becker is an Assistant Professor at the School of Public and Environmental Affairs and a Research Associate at the Workshop in

Political Theory and Policy Analysis. Ostrom is the Arthur F. Bentley Professor of Political Science and the Co-Director of the Workshop in Political Theory and Policy Analysis.

Literature Cited

1. Agrawal A. 1994. *Small is beautiful, but could larger be better? A comparative analysis of five village forest institutions in the Indian Middle Himalayas.* Presented at 1st Annu. Meet. Int. For. Resourc. Inst. Network, Oxford For. Inst., Oxford, India, Dec. 15–18
2. Alchian A. 1950. Uncertainty, evolution, and economic theory. *J. Polit. Econ.* 58:211–21
3. Anderson EN. 1992. *Can Ejidos work? Forest management in a Maya community.* Presented at Int. Assoc. for the Study of Common Property Meet., Washington, DC, Sept.
4. Atran S. 1993. Itza Maya tropical agroforestry. *Curr. Anthropol.* 34(5):633–700
5. Becker CD, Banana A, Gombya-Ssembajjwe W. Early detection of tropical forest degradation: an IFRI pilot study in Uganda. *Environ. Conserv.* Forthcoming
6. Benjamin P, Lam WF, Ostrom E, Shivakoti G. 1994. *Institutions, Incentives, and Irrigation in Nepal. Decentralization: Finance & Management Project Report.* Burlington, VT: Assoc. in Rural Dev.
7. Berkes F, ed. 1989. *Common Property Resources: Ecology and Community-based Sustainable Development.* London: Belhaven
8. Berkes F, Folke C. 1994. *Linking social and ecological systems for resilience and sustainability.* Presented at Worksh. on Property Rights and the Performance of Natural Resource Systems, Swedish Acad. Sci., Beijer Intl. Inst. of Ecological Econ., Stockholm
9. Beverton RJH. 1962. Long-term dynamics of certain North Sea fish populations. In *The Exploration of Natural Animal Populations*, ed. ED LeCren, MW Holdgate, pp. 242–59. Oxford: Blackwell
10. Blaikie PM, Brookfield H. 1987. *Land Degradation and Society.* London: Methuen
11. Blewett RA. 1994. *Property rights as a cause of the tragedy of the commons: institutional change and the pastoral Maasai in Kenya.* Presented at Public Choice Soc. Meet., Austin, Texas, April 8–10
12. Blomquist W. 1992. *Dividing the Waters: Governing Groundwater in Southern California.* San Francisco: ICS
13. Bromley DW. 1991. *Environment and Economy: Property Rights and Public Policy.* Oxford: Blackwell
14. Bromley DW, Feeny D, McKean M, Peters P, Gilles J, et al., eds. 1992. *Making the Commons Work: Theory, Practice, and Policy.* San Francisco: ICS
15. Castañeda J. 1992. Union of forest Ejidos and communities of Oaxaca. In *Development or Destruction: The Conversion of Tropical Forest to Pasture in Latin America*, ed. TE Downing, SB Hecht, HA Pearson, and C Garcia-Downing, pp. 333–34. Boulder, CO: Westview
16. Chernela JM. 1989. Managing rivers of hunger: the Tukano of Brazil. *Adv. Econ. Bot.* 7:238–48
17. Clark CW. 1990. *Mathematical Bioeconomics: The Optimal Management of Renewable Resources.* New York: Wiley
18. Coward EW, Jr. 1979. Principles of social organization in an indigenous irrigation system. *Hum. Organ.* 38(1):28–36
19. Curtis D. 1991. *Beyond Government: Organizations for Common Benefit.* London: Macmillan
20. Dove MR. 1993. A revisionist view of tropical deforestation and development. *Environ. Conserv.* 20:17–24
21. Dufour DL. 1990. Use of tropical rainforests by native Amazonians. *BioScience* 40(9):652–59
22. Folke C, Holling CS, Perrings C. 1994.

Biodiversity, ecosystems and human welfare. Work. Pap. No. 49. Stockholm: Swedish Acad. Sci., Beijer Intl. Inst. Ecological Econ.

23. Fortmann L, Bruce JW, eds. 1988. *Whose Trees? Proprietary Dimensions of Forestry.* Boulder, CO: Westview

24. Gadgil M, Berkes F, Folke C. 1993. Indigenous knowledge for biodiversity conservation. *Ambio* 22(2-3):151–56

25. Gadgil M, Rao PRS. 1994. *On designing a system of positive incentives to conserve biodiversity for the ecosystem people of India.* Presented at Worksh. on Design Principles, Swedish Acad. Sci., Beijer Int. Inst. Ecol. Econ., Stockholm

26. Glaser [Picht] C. 1987. *Common property regimes in Swiss alpine meadows.* Presented at Conf. on Advances in Compar. Institutional Anal., Inter–University Ctr. of Postgrad. Stud., Dubrovnik, Yugoslavia

27. Glaser [Picht] C. 1987. *Field notes to CPR colleagues.* Bloomington: Indiana Univ., Worksh. in Polit. Theory & Policy Anal.

28. Glick TF. 1970. *Irrigation and Society in Medieval Valencia.* Cambridge, MA: Harvard Univ. Press, Belknap Press

29. Gombya-Ssembajjwe W. 1994. *Sacred forests in modern Ganda society.* Presented at FAO For. Work. Group Meet. on Common Property, Oxford For. Inst., Oxford, UK, Dec. 15–18

30. Gordon HS. 1954. The economic theory of a common property resource: the fishery. *J. Polit. Econ.* 62:124–42

31. Hardin G. 1968. The tragedy of the commons. *Science* 162:1243–48

32. Hardin R. 1982. *Collective Action.* Baltimore, MD: Johns Hopkins Univ. Press

33. Jansson A, Hammer M, Folke C, Costanza R. 1994. *Investing in Natural Capital.* Washington, DC: Island

34. Keohane RO, Ostrom E, eds. 1995. *Local Commons and Global Interdependence: Heterogeneity and Cooperation in Two Domains.* London: Sage

35. Knight J. 1992. *Institutions and Social Conflict.* Cambridge, MA: Cambridge Univ. Press

36. Krebs CJ. 1994. *Ecology.* New York: Harper Collins. 4th ed.

37. Laitos R, et al. 1986. *Rapid Appraisal of Nepal Irrigation Systems. Water Management Synthesis Rep. No. 43.* Fort Collins: Colorado State Univ.

38. Lam WF. 1994. *Institutions, engineering infrastructure, and performance in the governance and management of irrigation systems: the case of Nepal.* PhD thesis. Indiana Univ., Bloomington

39. Lee M. 1994. *Institutional analysis, public policy, and the possibility of collective action in common pool resources: a dynamic game theoretic approach.* PhD thesis. Indiana Univ., Bloomington

40. Libecap GD. 1989. *Contracting for Property Rights.* Cambridge, MA: Cambridge Univ. Press

41. Libecap GD. 1995. The conditions for successful collective action. In *Local Commons and Global Interdependence: Heterogeneity and Cooperation in Two Domains,* ed. R Keohane, E Ostrom, pp. 161–90. London: Sage

42. Maass A, Anderson RL. 1986. *...and the Desert Shall Rejoice: Conflict, Growth and Justice in Arid Environments.* Malabar, FL: RE Krieger

43. Martin EG. 1986. *Resource mobilization, water allocation, and farmer organization in hill irrigation systems in Nepal.* PhD thesis. Cornell Univ.

44. Martin EG, Yoder R. 1983. The Chherlung Thulo Kulo: a case study of a farmer-managed irrigation system. In *Water Management in Nepal: Proceedings of the Seminar on Water Management Issues,* July 31–Aug. 2, Appendix I, 203–17. Kathmandu, Nepal: Ministry of Agric., Agric. Projects Serv. Ctr., Agric. Dev. Council

45. Martin F. 1989/1992. *Common-Pool Resources and Collective Action: A Bibliography.* Vols. 1 and 2. Bloomington: Indiana Univ., Worksh. in Polit. Theory & Policy Anal.

46. Mayr E. 1982. *The Growth of Biological Thought.* Cambridge, MA: Belknap

47. McCay BJ, Acheson JM. 1987. *The Question of the Commons: The Culture and Ecology of Communal Resources.* Tucson: Univ. Ariz. Press

48. McKean MA. 1992. Management of traditional common lands (Iriaichi) in Japan. In *Making the Commons Work: Theory, Practice, and Policy,* ed. DW Bromley, D Feeny, MA McKean, P Peters, JL Gilles, et al, pp. 63–98. San Francisco: ICS

49. McKean MA, Ostrom E. 1995. Common property regimes in the forest: just a relic from the past? *Unasylva* 46:3–15

50. Moran E. 1983. *The Dilemma of Amazonian Development.* Boulder, CO: Westview

51. Moran E. 1993. *Through Amazonia Eyes: The Human Ecology of Amazonian Populations.* Iowa City: Univ. Iowa Press

52. National Research Council. 1986. *Proceedings of the Conference on Common Property Resource Management.* Washington, DC: Natl. Acad.

53. Netting R McC. 1981. *Balancing on an Alp.* New York: Cambridge Univ. Press
54. Netting R McC. 1993. *Smallholders, Householders: Farm Families and the Ecology of Intensive, Sustainable Agriculture.* Stanford, CA: Stanford Univ. Press
55. Olson M. 1965. *The Logic of Collective Action: Public Goods and the Theory of Groups.* Cambridge, MA: Harvard Univ. Press
56. Ostrom E. 1990. *Governing the Commons: The Evolution of Institutions for Collective Action.* New York: Cambridge Univ. Press
57. Ostrom E. 1992. *Crafting Institutions for Self-Governing Irrigation Systems.* San Francisco: ICS
58. Ostrom E. 1994. Constituting social capital and collective action. *J. Theoret. Polit.* 6(4):527–62
59. Ostrom E, Gardner R, 1993. Coping with asymmetries in the commons: self-governing irrigation systems can work. *J. Econ. Perspect.* 7(4):93–112
60. Ostrom E, Gardner R, Walker J. 1994. *Rules, Games, and Common-Pool Resources.* Ann Arbor: Univ. Mich. Press
61. Ostrom E, Lam WF, Lee M. 1994. The performance of self-governing irrigation systems in Nepal. *Hum. Syst. Manage.* 13(3):197–207
62. Ostrom E, Wertime MB. 1994. *International forestry resources and institutions (IFRI) research strategy.* Work. pap. Bloomington: Indiana Univ., Worksh. in Polit. Theory & Policy Anal.
63. Ostrom V, Feeny D, Picht H, eds. 1993. *Rethinking Institutional Analysis and Development: Issues, Alternatives, and Choices.* San Francisco: ICS. 2nd ed.
64. Pinkerton E. 1989. *Co-operative Management of Local Fisheries: New Directions for Improved Management and Community Development.* Vancouver: Univ. British Columbia Press
65. Posey DA. 1985. Indigenous management of tropical forest ecosystems: the case of the Kayapo Indians of the Brazilian Amazon. *Agrofor. Syst.* 3:139–58
66. Posey DA. 1993. Indigenous knowledge in the conservation of world forests. In *World Forests for the Future,* ed. K Ramakrishna, GM Woodwell, pp. 59–77. New Haven, CT: Yale Univ. Press
67. Pradhan P. 1989. *Increasing Agricultural Production in Nepal: Role of Low-Cost Irrigation Development through Farmer Participation.* Kathmandu, Nepal: Int. Irrigation Manage. Inst.
68. Sandler T. 1992. *Collective Action.* Ann Arbor: Univ. Mich. Press
69. Schlager E. 1994. Fishers' institutional responses to common-pool resource dilemmas. In *Rules, Games, and Common-Pool Resources,* ed. E Ostrom, R Gardner, J Walker, pp. 247–65. Ann Arbor: Univ. Mich. Press
70. Schlager E, Blomquist W, Tang SY. 1994. Mobile flows, storage, and self-organized institutions for governing common-pool resources. *Land Econ.* 70(3):294–317
71. Schlager E, Ostrom E. 1992. Property-rights regimes and natural resources: a conceptual analysis. *Land Econ.* 68(3):249–62
72. Schotter A. 1981. *The Economic Theory of Social Institutions.* Cambridge, MA: Cambridge Univ. Press
73. Sengupta N. 1991. *Managing Common Property: Irrigation in India and the Philippines.* London: Sage
74. Sengupta N. 1993. *User-Friendly Irrigation Designs.* New Delhi: Sage
75. Shepsle KA. 1989. Studying institutions: some lessons from the rational choice approach. *J. Theoret. Polit.* 1:131–49
76. Shivakoti GP. 1991. *Organizational effectiveness of user and non-user controlled irrigation systems in Nepal.* PhD thesis. Mich. State Univ., Dep. Resource Dev.
77. Simenstad CA, Estes JA, Kenyon KW. 1978. Aleuts, sea otters, and alternate stable-state communities. *Science* 200:403–11
78. Siy RY Jr. 1982. *Community Resource Management: Lessons from the Zanjera.* Quezon City: Univ. Philippines Press
79. Skole DL, Chomentowski WH, Salas WA, Nobre AB. 1994. Physical and human dimensions of deforestation in Amazonia. *BioScience* 44(5):314–21
80. Solow R. 1992. *An almost practical step toward sustainability.* Transcript of an invited lecture. Washington, DC: Resources for the Future
81. Sparling EW. 1990. Asymmetry of incentives and information: the problem of watercourse maintenance. In *Social, Economic, and Institutional Issues in Third World Irrigation,* ed. R Sampath, RA Young, pp. 195–213. Boulder, CO: Westview
82. Stelfox JB. 1985. *Mixed species game ranching.* PhD thesis. Univ. Alberta, Edmonton, Alberta, Canada
83. Sugden R. 1986. *The Economics of Rights, Cooperation and Welfare.* London: Basil Blackwell
84. Svendsen M, Small L. 1990. Farmers' perspective on irrigation performance. *Irrigation and Drainage Systems* 4:385–402

85. Tang SY. 1992. *Institutions and Collective Action: Self-Governance in Irrigation.* San Francisco: ICS

86. Thomson JT. 1992. *A Framework for Analyzing Institutional Incentives in Community Forestry.* Rome: Food and Agric. Organ. of the United Nations

87. Uhl C. 1987. Factors controlling succession following slash and burn agriculture in Amazonia. *Ecology* 75: 377–407

88. von Hayek FA. 1967. Notes on the evolution of rules of conduct. In *Studies in Philosophy, Politics, and Economics,* ed. FA von Hayek, pp. 66–81. Chicago: Univ. Chicago Press

89. Wade R. 1988. *Village Republics: Economic Conditions for Collective Action in South India.* New York: Cambridge Univ. Press

90. Weissing F, Ostrom E. 1991. Irrigation institutions and the games irrigators play: rule enforcement without guards. In *Game Equilibrium Models II: Methods, Morals, and Markets,* ed. R Selten, pp. 188–262. Berlin: Springer-Verlag

91. Weissing F, Ostrom E. 1993. Irrigation institutions and the games irrigators play: rule enforcement on government- and farmer-managed systems. In *Games in Hierarchies and Networks: Analytical and Empirical Approaches to the Study of Governance Institutions,* ed. FW Scharpf, pp. 387–428. Frankfurt: Campus Verlag/Boulder, CO: Westview

92. Wilson JA, French J, Kleban P, McKay SR, Townsend R. 1991. Chaotic dynamics in a multiple species fishery: a model of community predation. *Ecol. Modelling* 58:303–22

93. Yoder RD. 1986. *The performance of farmer-managed irrigation systems in the hills of Nepal.* PhD thesis. Cornell Univ.

94. Yoder RD. 1994. *Locally Managed Irrigation Systems.* Columbo: Int. Irrigation Manage. Inst.

Annu. Rev. Ecol. Syst. 1995. 26:135–54

SUSTAINABILITY, EFFICIENCY, AND GOD: Economic Values and the Sustainability Debate

Robert H. Nelson

School of Public Affairs, University of Maryland, College Park, Maryland 20742-1821

KEY WORDS: growth, ecology, idea of progress, economics profession, religion

---

ABSTRACT

Economics is not only a technical subject; it also reflects a strong set of values. The values embedded in the economic way of thinking are often at odds with the way of thinking of biologists, ecologists, and other physical scientists. Economists value nature in terms of its benefits for human consumption and its usefulness in promoting economic growth. Growth is so critical because it can alleviate material scarcity in the world, and poverty is the true source of evil behavior. A recent dissident group of "ecological economists" argues that, rather than growth, a more appropriate goal is sustainability. The conflicting values implicit in mainstream economics and in ecological economics partly reflect deep underlying theological differences. These differences can be traced back to old messages of the Judeo-Christian tradition, now being manifested in secular form.

INTRODUCTION

Economic language is often the currency of contemporary policy debate (71). To say that a policy is economically "efficient" is to make a strong claim for the social legitimacy of that policy, to administer in a secular age what in an earlier time would have been said to be the blessing of God (28). Economists have occupied many important positions in government (67, 84). Key domestic

135

0066-4162/95/1120-0135$05.00

agencies such as the Congressional Budget Office and the Office of Management and Budget, and international organizations such as the World Bank, traditionally have employed many economists on their policy staffs. In such positions, economists can exert a significant influence on government decisions concerning a wide range of issues (31), including environmental and natural resource policies that have a major bearing on current sustainability debates.

This chapter examines the value perspective that economists bring to the sustainability issue. Members of the economics profession are by no means monolithic in their views on sustainability, but the value system embedded in the economic approach to public policy tends to lead to a certain way of thinking about environmental and natural resource problems (15, 22, 87). This economic outlook differs substantially from the perspectives of many biologists and other physical scientists.

The subject of sustainability, to be sure, has not been a central concern in the economic literature. Until recently, only a few economists such as Daly (16) had addressed the issue in the terms of the current debate. More recently, in response to wide public interest, some leading figures in mainstream economics have sought to show how sustainability can be understood in economic terms (82). However, it is fair to say that most economists today are still skeptical about whether the concept of sustainability is a useful guide for social action or whether it merits substantial professional attention for policy-making purposes.

Economists as Advocates for a Value System

Economists today increasingly recognize that in social policy matters it is almost impossible to be "value-neutral" (78). In giving policy advice, economists—or any social scientists—will inevitably reflect a set of intellectual constructs and presuppositions that at some level become a matter of faith (88). From the progressive era early in this century, however, economists argued that their policy efforts were those of the value-neutral expert (59). The governing process included two separate realms: objective professional expertise and subjective value systems. The social value judgments should be made by politicians in the democratic process, setting the broad course for society. Economists, along with other expert professionals, should frame the options for consideration and, once politicians had chosen among these options, should assist in the technical tasks of implementation (65). But economists should not seek to press their own values (so the reigning economic orthodoxy went for many years), because that would be an improper role for a professional.

Economists began entering government in large numbers in the New Deal years (1930s) and during World War II (early 1940s). They found, in many cases, that they could not reconcile their actual experiences in the policy world with the norms of proper professional conduct that they had been taught.

Indeed, well before the current discussions of sustainability, many leading policy economists were led to abandon the previous claims to value neutrality. In 1968, Kaysen (37) observed that "the role of the economist in policy formation in these areas is almost diametrically opposite to that envisaged in the formal theory of policy making. . . . He functions primarily as a propagandist of values, not as a technician supplying data for the pre-existing preferences of the policy makers."

Schultze (75) said that "political values permeate every aspect of the decision-making process in the majority of federal domestic programs. There is no simple division of labor in which the 'politicians' achieve consensus on an agreed-on set of objectives while the 'analysts' design and evaluate—from efficiency and effectiveness criteria—alternative means of achieving those objectives." Within this framework, Schultze contends, an economist should act in the policy role as "a partisan advocate for efficiency" (76), which in practice would mean efficiency as determined through the specific value-lens of economic analysis (9).

The experience of economists in government also showed that most of the details of higher economic theory were too far removed from the real world to have practical consequences for policy decisions (80). The statistical results of econometrics were typically too fragile to be relied upon in setting policies affecting millions of people. Instead, a few key ideas—"just common-sense economics ... the kind of basic analytical framework that we all sort of got in Econ. 101," as one former economist with the President's Council of Economic Advisers put it (1)—were the greatest source of economic influence.

Cairncross (12) wrote of his long experience as a policy adviser to the British government that it was the economic "way of thinking" that had the greatest impact in the policy-making process. In general, as Weiss (90) wrote, the key policy role of the social sciences was that they provided "the intellectual background of concepts, orientations and intellectual generalizations that inform policy." This economic way of thinking and the value system it represents now put those in the mainstream of the economics profession at odds with many of those most concerned with the sustainability of the world future (64).

In more recent years, some economists have found that even the higher reaches of economic theory are not truly separated from implicit value elements. Extending the thinking of philosopher Richard Rorty (72) into the economic realm, McCloskey (53) has argued that economic theorizing with its scientific claims "promises knowledge free from doubt, free from metaphysics, morals and personal conviction. What it is able to deliver renames as scientific methodology ... the economic scientist's metaphysics, morals and personal convictions"—and, it might be added, in some cases religion as well.

The economist's claims to scientific objectivity are, so McCloskey and others (41) have said, a rhetorical device, not to be taken literally. If not by

explicit calculation, economists are effectively seeking to exclude non-econo-
mists from the debate and to stake a claim for political power based on
scientific authority. Yet, in regarding the views of economists on sustainability
and other issues, it is the value system as much as the technical analysis that
drives the conclusions (21, 74).

The Economic Way of Thinking

The value system of economics begins with the fact that economics is a social
science, and thus it is about the interactions of people and their welfare.
Animals, plants, the physical state of the world, and other material conditions
do not enter into consideration, except in so far as they provide a backdrop to
human well-being. It is in this sense similar to the biblical view that human
beings alone are made in the image of God, and that God created the world
out of nothing for his enjoyment and for human use (91).

That is not to say that sustaining nature necessarily commands a low priority
in the economic value system. If people derive much pleasure ("utility," to use
the economist term) from nature, then preservation of natural conditions may
be a high social priority. But it is the fact that people benefit from or choose
to protect nature for reasons of their own doing, not the intrinsic necessity of
sustaining elements of nature per se, that counts in economic thinking (24).

Another key feature of the economic way of thinking is that the factors
entering into human welfare are regarded as substitutable. Thus, no one good
or service—no one biological or natural system—has any automatic claims.
If any one item is not available for human use in the future, economists expect
that people will be able to obtain a suitable substitute. By the essence of its
method, economics is concerned with tradeoffs. Given that item A costs so
much, economists ask whether more of it will add or detract from social well
being, recognizing that producing or maintaining A requires giving up the
"opportunity cost" of other items that could be obtained by society for the
same expenditure of resources (96).

Another way of saying this is that the economic way of thinking rejects the
idea that some things are literally "priceless" (38). Many people will say that
preserving a species, saving human lives, or some other goal is beyond any
consideration of costs, but economists regard such assertions as a rhetorical
and political device. These assertions are claims on resources by partisans of
particular causes, rather than a reasonable basis for priority-setting by society.
Adding up all the claims made for "priceless" objects, or even for perfectly
realizing one goal, economists suggest, could well exceed the total social
resources available (69). It is thus not only objectionable in principle, but it
may also be physically impossible to realize such demands.

Another important element of the economic way of thinking is that well-
being is derived from consumption. The economic world is divided into acts

of production and acts of consumption, and it is only the latter that enter into the "utility functions" by which economists rank one consumption set—and, in the aggregate over all people, one social outcome—relative to another. In policy-making, this outlook translates into opposition by economists to preserving for their own sake particular industries, jobs, communities, and other portions of the physical infrastructure of production. Economists have similarly opposed the many proposals that society should choose a social infrastructure of production—a set of laws, regulations, and other institutional mechanisms—on the basis of the morality of each mechanism.

Economists thus oppose policies to curb speculation, arguing that speculative practices provide socially useful incentives in the market to conserve resources today in order to provide for greater total production in the long run. The wage level is part of the arrangements for production and thus is not itself an item of consumption. Hence, most economists oppose the government interference with market wages that many people have sought on social equity grounds such as the "just wage" arguments of the past or the more recent "comparable worth" claims. Similarly, economists argue that the best way to control pollution is to allow the market system to operate, requiring the creation of formal rights to pollute the environment that could either be sold by the government (i.e. polluters would pay a tax) or transferred into private ownership for market trading (42, 73, 83). Economists have rejected the value objections made by some environmentalists that such a policy would be unethical, amounting to the official sanctioning by government of immoral behavior—analogous to issuing permits for a form of antisocial, if not criminal, activity against nature.

In general, the economic way of thinking argues that public policy should be determined by the end of achieving efficient use of resources to maximize production and consumption, not by the moral desirability of the physical methods and social institutions used to achieve this end.

In considering whether current social consumption should be reduced, as some have said, for reasons of sustainability, the economic way of thinking thus finds that there is only one sound reason for doing so. It might be desirable to reduce current consumption, if this reduction will allow for increases in future consumption—either through greater present investment or reduced depletion of existing natural resources. The possibility that consumption should be reduced because the act of consumption is not good for the soul, or is not what actually makes people happy, has no place within the economic value system (64).

To be sure, some economists have found tensions within their own thinking with regard to the presence of these strong value elements. Thus, if society wants to declare that preserving a wilderness should be accomplished, regardless of the costs, or that consumption should be reduced as an ascetic act of self-denial, then economists should as value-free scientists defer to this choice. However, as noted above, economists in practice act as strong advocates for the values embedded in the economic way of thinking.

In recent years, some economists have sought to reconcile this tension by introducing a new concept of "existence value" (44). The existence value is the amount that a person would be willing to pay simply for the knowledge that a wilderness, an endangered species, a distinctive forest ecology, or some other object exists in the world. In this way, the traditional necessary link in economics between individual benefit and actual consumption would be broken. A person could be said to derive benefit from many things that he or she did not expect ever to experience or consume directly. Economists then propose to calculate and aggregate these individual benefits of existence values in order to derive estimates of total social benefits and costs of particular policy proposals (57).

An active debate has broken out within the economics literature over the desirability and validity of introducing existence values into the repertoire of economic analysis (33). The objections are many, but two practical concerns are particularly telling. First, existence value is sufficiently nebulous that its calculation is subject to very wide ranges of estimates (79), and the credibility of the calculations is often doubtful. Second, existence value is not, in principle, limited to wilderness, endangered species, and other objects in nature, but could be attributed to jobs, dams (a powerful symbol of progress for earlier generations), highways, and indeed virtually any object that is invested with symbolic significance by some member of society. To try to measure in dollar terms the economic magnitude of individual and aggregate social benefits derived from all these symbolic associations would greatly complicate the practice of policy economics. The introduction of existence value, so the critics argue, seeks to expand the scope of the values reflected in economic analysis but in the end threatens to undermine the clarity of the economic way of thinking.

Economic Values and Sustainability

The subject of sustainability is not altogether a new one for economists. Mill, one of the great economists of the nineteenth century, addressed it famously in the following terms (56):

> The preceding chapters comprise the general theory of the economical progress of society, in the sense in which those terms are commonly understood; the progress of capital, of population, and of the productive arts. But in contemplating any progressive movement, not in its nature unlimited, the mind is not satisfied with merely tracing the laws of the movement; it cannot but ask the further question, to what goal? Towards what ultimate point is society tending by its industrial progress? When the progress ceases, in what condition are we to expect that it will leave mankind?
>
> I cannot therefore regard the stationary state of capital and wealth with the

unaffected aversion so generally manifested towards it by political economists of the old school. I am inclined to believe that it would be, on the whole, a very considerable improvement on our present condition. I confess I am not charmed with the ideal of life held out by those who think that the normal state of human beings is that of struggling to get on; that the trampling, crashing, elbowing, and treading on each other's heels, which form the existing type of social life, are the most desirable lot of human kind, or anything but the disagreeable symptoms of one of the phases of industrial progress.... The best state for human nature is that in which, while no one is poor, no one desires to be richer, nor has any reason to fear being thrust back by the efforts of others to push themselves forward.

Until recently, few other economists sought to consider what a sustainable world might be and what it might mean for social arrangements (63). However, in a 1991 lecture to the Woods Hole Oceanographic Institution, Solow (82) applied the economist's value lens on the world to the subject of sustainability. His conclusions are of particular interest as an illustration of the application of traditional economic values to this subject by a leading contemporary economist (a Nobel prize winner in 1987).

First, Solow acknowledged that, like most economists, he started off skeptically. Even where an effort had been made to develop "carefully thought out definitions and discussions" of sustainability, the fact was that "they all turn out to be vague." Indeed, Solow was of the opinion that "sustainability is an essentially vague concept, and it would be wrong to think of it as being precise." If there were a meaning, it belonged to the realm of ethics rather than science: "It says something about a moral obligation that we are supposed to have for future generations." As long as it is understood as a declaration of a broad social value, sustainability "is not at all useless."

Solow then observed that sustainability cannot literally mean "to leave the world as we found it in detail"—something not only physically "unfeasible" but also "when you think about it not even desirable." He stated that, instead, sustainability must be understood in the terms of "an obligation to conduct ourselves so that we leave to the future the option or the capacity to be as well off as we are." Thus, society is morally obligated to act to ensure that the social welfare of future generations will be at least at the level of the present generation.

As noted above, economic thinking considers that there is no reason in principle why any one form of consumption should automatically trump other possible ways of attaining well-being. As Solow said at Woods Hole, "What about nature? ... I think that we ought, in our policy choices, to embody our desire for unspoiled nature as a component of well-being. But we have to recognize that different amenities really are, to some extent, substitutable for one another." If people will feel happier going to baseball games (now and in the future) than visiting wilderness areas, then building baseball stadiums should command a higher government priority. As Solow elaborated, "sustainability doesn't require that any *particular* species of owl or any *particular* species of fish or any *particular* tract of forest be preserved." Unless a species contributes instrumentally to future production and consumption (perhaps by the use of its genetic code), sustainability offers no grounds for the Endangered Species Act, as economists think about the matter.

Solow does not believe that the welfare of future generations can be entrusted simply to the workings of the market. Active government policy intervention might be necessary. Indeed, he urged his audience to think about what government policy measures might be needed to ensure "sustainability as a matter of distributional equity between the present and the future." In this framework, sustainability becomes "a problem about savings and investment. It becomes a problem about the choice between current consumption and providing for the future." And government, as economists have believed at least since Keynes, can play a major role in determining levels of total social consumption and investment. Thus, the issue of sustainability becomes a part of macroeconomics.

To be sure, by this definition, sustainability had never been much of a problem in the modern era and would not be unless future developments cause a drastic reversal of the economic trends of the nineteenth and twentieth centuries. As Solow observed, "you could make a good case that our ancestors, who were considerably poorer than we are, ... were probably excessively generous in providing for us." In other words, past generations saved and invested so much, sacrificing their own consumption for our benefit, that they ironically ended up accomplishing a massive redistribution of income from a relatively poorer group of people in those days to a relatively richer group today. Based on this past precedent, and given the general value presumption in favor of a more equal distribution of income, the current generation perhaps should be looking to increase its consumption, to redistribute income from the richer people expected in the future to those of us who are less well off today.

This economic way of thinking about sustainability obviously is not what many people who now express concern about the issue have in mind. Solow's concerns reflected in part the long-standing interest of economists in the determinants of economic growth. Indeed, early in his career, Solow had been a leading developer of several aspects of "growth theory," which addressed much the same policy questions of determining appropriate levels of investment and consumption over the long run (81).

In policy-making circles, economists have been particularly prominent as advocates for sustained economic growth. The Council of Economic Advisers, the leading vehicle for transmitting the views of professional economists to the US government, was created expressly for the purpose of maintaining full employment and setting the economy on a path of long-run sustained growth (23). The pursuit of growth is one of the principal elements in the economic value system, one often in conflict with the views of those most concerned today about sustainability.

The Value of Growth

Economic values, some people are sure to think, leave nothing sacred, reduce everything to crass material terms. It is said that "an economist is someone who knows the price of everything and the value of nothing." The economic value system is at odds with important religious traditions, which consider parts of life as transcendent, above the daily routine of production and consumption. When economists come to apply their economic way of thinking to marriage, for example, it is treated as a contractual relationship to be negotiated to serve the individual advantage and convenience of each party (7).

Indeed, thinking economically does not come naturally to most people. The economics profession expects that instilling the way of thinking of economics will require long and intensive training, typically requiring many years of graduate school. This is not a matter of the analytical complexity of the subject. As noted above, the practice of policy economics depends mostly on a firm grasp of a small number of fairly elementary principles. Rather, absent regular reinforcement, the policy analyst is likely to slip into modes of thought grounded in value traditions that inject a greater element of the sacred—the priceless—into the affairs of mankind.

Indeed, a lifelong professional commitment to the practice of economic values depends on a strict discipline that might be described as requiring a certain religious zeal of its own (25). Not all economists are comfortable describing the matter in such terms, but many have observed that the value system of economics, like most value systems, shares important qualities with religion (28). In the case of economics, the theological elements remain implicit, as are almost all the important values that underpin the economic way of thinking. At the heart of the religious side of economics is a conviction of the powerful value gains of economic growth. Economists might be said to be the "priesthood" for a secular religion of growth (85).

In a survey of leading American economists, Baumol (6) was asked not long ago to explain why he had decided to enter the profession. He replied "I believe deeply, with Shaw, that there are few crimes more heinous than poverty. Shaw, as usual, exaggerated when he told us that money is the root of all evil. But he did not exaggerate by much." The source of evil, as Baumol sees it, is poverty, and poverty can be solved by growth. In finding the solution for evil, economists are addressing a subject that has also been central to the history of religion. Economists are, in effect, expressing a secular faith. This "economic theology" might be regarded as one belief system within the larger "religion of progress," as it has been described, that has characterized much of the thinking of the modern age (11, 45).

Marxism, socialism, capitalism, virtually all the major systems of economic thought of the past 200 years, are particular branches of this modern religion of progress (60). Schumpeter (77) once wrote of "the gospel of Marx." These secular religions differed on the specific details of how economic growth would be realized—they sometimes even fought wars with one another over the details of economic interpretation, much as Christians warred over the details of interpreting the Bible—but they found no disagreement that satisfying all real material needs would greatly transform the world for the better. For them, the explanation for why people cheat, lie, steal, and otherwise behave badly is the pressure of material deprivation. In other words, poverty is the original sin, and the road to secular salvation is economic growth that eventually ends scarcity and banishes evil.

As the most influential economist of the twentieth century, Keynes (40) in 1930 predicted that, with existing rates of economic growth, the world would have all the material goods it needed within 100 years. Like many ordinary people, Keynes regarded the economic value system as a crass and lower species of morality that should be abandoned as soon as sufficient material advance made this possible. A sustained commitment to economic progress would mean that "we shall be able to rid ourselves of many of the pseudo-moral principles which have hag-ridden us for two hundred years, by which we have exalted some of the most distasteful of human qualities into the position of the highest virtues. We shall be able to afford to dare to assess the money-motive at its true [base] value." If people kept the faith and bore with the current situation, the maintenance of rapid economic growth would fairly soon lead mankind "out of the tunnel of economic necessity into daylight"—to a new heaven on earth. It is probably fair to say that the sustainability concerns of today were never taken as a serious problem, if considered at all, by Keynes.

In America the economics profession emerged during the progressive era, as part of the broader progressive aim to create the necessary social instruments for the scientific management of society. Progressivism has been described by historians as "a secular Great Awakening" that sparked "a moral fervor that had all the earmarks of a religious revival" (13, 29). The message of this secular religion was yet another economic theology, described by historians as the "gospel of efficiency" (34). Waldo (89) would observe that, in the progressive era, "it is yet amazing what a position of dominance 'efficiency' assumed, how it waxed until it had assimilated or overshadowed other values, how men and events came to be degraded or exalted according to what was assumed to be its dictate."

In progressive religion, efficient and inefficient become in essence moral categories, the test of whether an action serves an ultimate purpose. It is similar to the distinction between good and evil in Judeo-Christian religion. Efficiency could be so exalted because, if economic growth is the road to secular salvation, the valid test of whether an action contributes to the salvation of humanity is whether it is efficient.

It is no coincidence that the emergence of sustainability as an issue comes at a time when faith in economic progress is waning. Indeed, to declare that an action is sustainable seems today to be serving a function similar to declaring it efficient in the past. In neither case is it meant to be a precise statement about the consequences of the action (70). Pezzey (68) surveyed the definitions given in his review of writings on the subject of sustainable development, and found more than 50 concepts. In practice, virtually every group in society today seems to think that there are valid grounds for regarding its activities as a genuine key to realizing a sustainable future. For example, reflecting the all-purpose term of approval that sustainability has become, the President of the National Coal Association (46) recently declared that "in reality, our 250-year supply of coal is the only domestic source of energy that meets the definition of 'sustainability.' ... Without a doubt, the catalyst for a stable U.S. economy under sustainable development is an abundant and secure energy supply," led by coal.

In current discussions, to say that an action is sustainable is, in essence, to declare that it is socially legitimate. Terms such as "providence" in the medieval era, "natural law" in the Enlightenment, "efficiency" in the progressive era, and now "sustainability" at the end of the century, tell us more about our basic value systems—the gods we worship and who must bless our actions—than they do about the character of any specific action so described.

The shift from efficiency to sustainability no doubt reflects in part the moral disappointments of the twentieth century, relative to the hopes for economic progress that were widely shared at the beginning. The economic advances promised were, on the whole, realized in the western world. Yet, contrary to the basic assumption of all the various branches of the religion of progress, a whole new order of material abundance did not seem to change the basic moral and spiritual condition of the world. Indeed, the record of the twentieth century would be filled with world wars, genocides, prison camps, nuclear bombs, and many other dismal objects and events. Instead of leading to a secular salvation, science and technology seem to have magnified the powers of destruction, causing some economists to fear that economic efficiency was making the arrival of a terrible new hell on earth as likely as heaven on earth (8).

Predictions of Environmental Collapse

On the failure of the twentieth century to live up to the early hopes, there are few dissenters. Yet most economists continue to regard the continued pursuit of economic growth as an appropriate policy goal, even while they have lowered their expectations for the consequences. Perhaps heaven on earth will not be reached by this route, but one can still reasonably hope, most economists believe, for a better future (4, 86).

Some economists outside the mainstream, however, have recently banded together in a new subfield of "ecological economics" that is prepared to reach more radical conclusions (14, 43). In favoring the application of a new criterion of sustainability to public policies of all kinds, ecological economists in some cases argue that the whole world faces serious economic and social disorders of various kinds, if there are not drastic changes in basic social and economic arrangements to curb or even reverse growth (10, 27).

This concern, to be sure, is at least as old as the beginnings of the industrial revolution. In perhaps the most famous answer of all, Malthus (50), a leading economist of his time, contended that population growth was sure to outrun available food supplies. Later in the nineteenth century, Jevons (36), another prominent figure in the history of economic thought, argued that coal was certain to run out in England, and no other energy substitute would be available to sustain the existing standard of living. When such concerns emerged again in the United States in the late 1960s and early 1970s, it was at first physical scientists who typically pressed the case (20).

More recently, the United Nations Educational, Scientific and Cultural Organization (UNESCO) published a set of explorations from the field of ecological economics on the subject of sustainable development. As Mayor (52) summarized the overall conclusion: "unless development is distinguished from economic growth, the turn-off towards sustainable development will be missed." Time is running short to avert grave environmental and social damages to the very fabric of the earth because "too many warning signs have already been ignored suggesting that, in North and South alike, we are moving in the wrong direction and that there may be few, if any, short-cuts back."

The Report of the World Commission on Environment and Development (93) in 1987 indicated that the total world economy might have to grow five to ten times its current size. Without such growth, the poor of the world would be unable to come up to the living standards of existing developed nations, leaving the world with unacceptable long-run inequalities. The editors of the UNESCO report found (27), however, that an attempt to achieve "anything remotely resembling" this magnitude of economic increase would "simply speed us from today's long-run unsustainability to imminent collapse" of the world environment. It would be essential to achieve a future "pattern of development without throughput growth." To avoid the permanent maintenance of large disparities in income between rich and poor countries, the scale of economic activity of rich countries would have to decline (27): "[E]cological constraints are real and more growth for the poor must be balanced by negative throughput growth for the rich." It would also be necessary to accomplish major transfers of income from rich to poor countries as well as to shift the relative magnitudes of production. Other major institutional changes, including sharp curtailments in world trade, would also be necessary conditions to achieving a sustainable future (18).

The Mainstream Economic Critique

In applying the lens of economic analysis to such forecasts of looming environmental destruction, mainstream economists have responded with considerable skepticism (5). They point out that the prices of most minerals and other natural resources have shown a fairly consistent trend of decline for about a hundred years (3, 94). In the short run, at least, rather than shortages, an excess of food and minerals and associated employment losses and other disruptive economic transitional effects have been the greatest policy concerns of many world governments (92).

The existing trends in environmental degradation are less favorable, but awareness of the environmental problem is more recent, and institutional adaptation can be expected to occur slowly. As the economic way of thinking sets the framework of analysis, the "sink capacity" of the environment has, in effect, been treated as a free good in a large commons (32). The development of regulatory and pricing mechanisms to bring access to the commons under control is only about 25 years old in economically advanced nations and has just begun in most less developed countries.

Reflecting this economic perspective, the World Bank (92) states that "the environmental debate has rightly shifted away from concern about *physical limits* to growth toward concern about incentives for *human behavior* and policies that can overcome *market and policy failures*." As economists explain, "the reason some resources—water, forests, and clean air—are under siege while others—metals, minerals, and energy—are not is that the scarcity of the latter is reflected in market prices and so the forces of substitution, technical progress, and structural change are strong."

In the 1980s, deregulation of oil and gas ended the "energy crisis" of the 1970s, after some unsatisfactory earlier attempts by governments to apply price controls and other regulatory mechanisms. Current anxieties about an environmental crisis and the policy responses of many world governments to environmental problems, many economists think, are in a category similar to that of the 1970s mishandling of energy problems. If the political hurdles can be overcome, environmental problems should be amenable to the same types of pricing and market solutions as the energy crisis (2, 30, 66).

In the economic way of thinking, more economic growth rather than less will be the answer to a large class of environmental problems. Higher incomes both create stronger public demands for environmental amenities and help to bring about a more sophisticated political process that will respond effectively and rapidly to growing public demands for environmental amenities. On the whole, experience has shown that the higher the income of a country around the world, the lower the level of air, water, and other pollution and in general the higher the quality of the environment for human use (92). The quality of the environment is yet another example of the general economist view that growth will be the answer to the problems of the world.

Underlying Moral Elements

There is no way in principle to resolve the question of whether existing rates of economic growth can continue for the foreseeable future without creating unacceptable environmental stresses. Economists can point to the 200-year history of mistaken predictions of food, energy, timber, and other dire crises sure to occur in the near future. Yet, the fact that these predictions essentially all proved wrong does not guarantee that they must always be erroneous. In the end, it comes down to a matter of judgment, based on the weight of the evidence available, and which risk seems greater—the risk of making major sacrifices today that prove to be needless, or the risk of not taking precautions, and then later generations possibly suffering the consequences.

Moreover, complicating the matter, as in most controversies involving economists, further powerful value elements underlie the discussion, revolving around the merits or lack of such in the value system implicit in the mainstream economic view (17, 19, 26, 95). The value system of most economists regards the environment as a factor of production. Labor and capital have long received attention from economists as key factors of production. Land and natural resources were also recognized many years ago as significant factors, although regarded as playing a declining role in a modern economy. It is only recently, however, that economists have come to regard the sink capacity of the environment as yet another input that must be allocated among industries through the same supply and demand mechanisms and incentives that control the use of any factor of production.

In thus treating the environment as a "commodity," the economic way of thinking is offensive to many religious traditions. Many could be given, but consider one example, the tradition of Protestantism. In the sixteenth century, one of the founders of the Protestant faith, Calvin (39), stated that God "brought forth living beings and inanimate things of every kind, that in a wonderful series he distinguished an innumerable variety of things, that he endowed each kind with its own nature, assigned functions, appointed places and stations." As a result, it is possible for us to enjoy "a slight taste of the divine from contemplation of the universe."

Calvin then, and other Protestants up to the present day, regarded nature as a manifestation of the presence of God in the universe, a source of religious enlightenment and spiritual inspiration (61). A wilderness area is a cathedral of sorts, because it allows a person to derive spiritual inspiration by coming into the close presence of the divine in the world (58). To put a wilderness area to use—to regard it as a factor of production—is to deface a church. The same ethic may extend to many other aspects of nature (54). Nature must be preserved because it is a part of "the Creation." To put a price on nature is to put a price on something that belongs to God.

To say that human beings can remake the world in the place of God, that they can redo the creation, would have been declared a heresy 500 years ago. Today, a journal of environmental opinion merely observes that with their values "economics, and economists, are traditional enemies of the environment" (48).

The Biblical Treatment of Sustainability

The very subject of the sustainability of current society also is hardly a "value-neutral" question. Implicit in the mainstream economic devotion to eliminating poverty through growth and progress is an essentially Christian way of regarding the world: The record of all human history is a gradual advance from a degraded condition to a future in which happiness and spiritual contentment will reign. The question of the sustainability of society also has powerful religious overtones and comes up many times in the Bible. Genesis, Chapter 6 in the King James version, notes that "men began to multiply on the face of the earth, and daughters were born unto them." We learn shortly thereafter that God, looking down on the spread of mankind over the earth, was mightily displeased with this and other elements of his Creation—that "the wickedness of man was great in the earth." Indeed, God's displeasure was so great that he resolved to "destroy man whom I have created from the face of the earth; both man, and beast, and the creeping thing, and the fowls of the air." It was, to use a more contemporary language, a negative verdict on sustainability.

In a recent translation of the Bible, the same verse is given in present-day English. God is said to be displeased with the fact that "now a population explosion took place upon the earth." As a result of this and other signs that human beings are failing to fulfill his intentions, he resolves to "cover the earth with a flood and destroy every living being." He recants later only to the extent of allowing Noah to save two of every species, as the Endangered Species Act today seeks to sustain the animal heritage of the earth in the face of the unsustainable spread of population and economic development.

After the Creation, the question of sustainability appears for the first time, in the Garden of Eden, where Adam and Eve lived in harmony and bliss but could not sustain this condition; instead they were cast into a world of pain and suffering when they succumbed to the temptations of the devil. Later books of the Old Testament are filled with other places and societies that, owing to their wickedness, suffer the wrath of God. This divine retribution usually takes the form of an environmental disaster—if not a great flood, then famine, drought, pestilence, or other natural calamity. The greatest environmental threats with respect to the spread of greenhouse gases and resulting global warming are today seen—perhaps coincidentally, perhaps not—in terms of many of the same consequences: the onset of flooding, famine, drought, pestilence, and other natural catastrophes.

In our secular age, people are not likely to speak in mainstream policy circles of the "wickedness" of mankind. Yet, among radical members of the environmental movement, who express the strongest doubts about the sustainability of our current civilization, there is a strong sense of current human depravity. Brower, perhaps the most prominent environmentalist of the past 50 years, argued in his standard "sermon" that "We're hooked. We're addicted. We're committing grand larceny against our children. Ours is a chain letter economy …. When [such] rampant growth happens in an individual, we call it cancer" (55). Foreman, the founder of the radical environmental organization Earth First, views human beings as the "cancer of the earth" (49).

These, to be sure, are extreme views. Yet, large numbers of people today do believe that the moral condition of the world is bad and getting worse. In secular circles, while people no longer typically believe in divine retribution, they often do have a sense that some form of disaster might be a consequence of the many transgressions of human beings against one another, against other species, and against the earth. These expectations of punishment seem, in many cases, to take the same forms as the expectations of God's imposition of a severe justice in the Bible—the arrival of environmental calamities.

Perhaps what we are seeing in current discussions of "sustainability" is the reappearance in secular form of an old biblical message of great power in the history of western civilization. The biblical messages were, of course, delivered by priests, ministers, and other clergy. Today, the discussions of sustainability are carried on mostly by biological, physical, and social scientists. But, as many commentators have noted, scientists in the modern age have, in many respects, taken the places of the priesthoods of old (51).

The Bible, of course, is filled with messages of hope and redemption as well as of the wrath of God being visited upon the earth. And today as well, both messages can be found within the broad field of economics. The mainstream economic view of the world holds out a path to heaven on earth through further economic growth; a minority economic prophesy warns that current human failings will bring on great natural calamities and perhaps hell on earth.

This is not, of course, to say that the conflicting visions of economists—which themselves mirror divisions of opinion covering much wider groups in society—are matters of values and morality alone. Far from it. But it would also be most naive to think that economics is a matter of objective facts and scientific laws alone, unaffected by moral judgments.

CONCLUSION

The positivist philosophy that dominated so many fields in the twentieth century was responsible for the idea that value questions and scientific questions are essentially separate domains. Thus, government was portrayed by progressive political theorists as consisting of two strictly separate domains of "politics" and "administration." Economists portrayed their efforts as taking exogenously given social values and then determining the policy actions that would serve these values in the most efficient way possible. Philosophers for many years neglected ethical subjects and confined their studies to linguistics and other narrowly drawn topics that were particularly amenable to specific analytical methods. Only theology sought to maintain a grand world view, but theological studies ranked low in the academy, mirrored in governing circles by the virtual exclusion of religious considerations as legitimate elements of the public policy debate.

All this has been breaking down in the final quarter of the twentieth century. The old formal dichotomies of fact and value, politics and administration, science and religion are collapsing (35). In matters of governance, perhaps the one greatest contribution to this breakdown was Lindblom's (47) classic article in the field of public administration on "The Science of 'Muddling Through.'" As Lindblom pointed out, the political process could seldom supply the values in advance to guide professional administrators. Instead, politics and administration, objective and subjective elements, were thoroughly interwoven in the making of government policy. The values of society were not set in advance; instead, the values could only be realized after the fact. It was through the very process of making administrative decisions that society often discovered what it believed about its own values.

Today, basic questions of growth, of efficiency, and of sustainability and others that society confronts are also inseparable mixtures of value and scientific elements (62). As Lindblom said, we mostly muddle through in trying to deal with this blend, to uncover which is which, and to decide how the two should be combined, as be combined they must. Indeed, few philosophers now would say that science is "scientific" in the sense used in the first half of the twentieth century. Whenever science is applied to real world questions, it supplies its own lens on the world, which in the end might aptly be described as its own theology.

What is most remarkable, and is still not adequately appreciated, is the degree to which our current economic policy debates owe their assumptions and moral perspectives to the Judeo-Christian heritage (60). If they are outwardly secular, it takes only a slight probing below the surface to find major biblical elements. Sustainability is perhaps a new word, but there is no more value-charged question in the writings of the Bible. To ask whether a society is sustainable is to ask whether its people are living according to God's commands.

Literature Cited

1. Allen WR. 1977. Economics, economists, and economic policy: modern American experiences. *Hist. Polit. Econ.* 9:48–88
2. Anderson TL, Leal DR. 1991. *Free Market Environmentalism.* Boulder, CO: Westview
3. Barnett HJ, Morse C. 1963. *Scarcity and Growth: The Economics of Natural Resource Availability.* Baltimore: Johns Hopkins Univ. Press
4. Bast JL, Hill PJ, Rue RC. 1994. *EcoSanity: A Common-Sense Guide to Environmentalism.* Lanham, MD: Madison
5. Baumol WJ. 1986. On the possibility of continuing expansion of finite resources. *Kyklos* 39:167–79
6. Baumol WJ. 1992. On my attitudes: sociopolitical and methodological. In *Eminent Economists: Their Life Philosophies,* ed. Michael Szenberg, pp. 51–59. New York: Cambridge Univ. Press
7. Becker G. 1991. *A Treatise on the Family.* Cambridge, MA: Harvard Univ. Press
8. Boulding KE. 1970. *Beyond Economics: Essays on Society, Religion, and Ethics.* Ann Arbor: Univ. Mich. Press
9. Bromley DW. 1990. The ideology of efficiency: searching for a theory of policy analysis. *J. Environ. Econ. Manage.* 19:86–107
10. Brown LR, et al. 1991. *State of the World: A Worldwatch Institute Report on Progress Toward a Sustainable Society.* New York: WW Norton
11. Bury JB. 1932. *The Idea of Progress: An Inquiry into its Origin and Growth.* New York: Macmillan
12. Cairncross A. 1985. Economics in theory and practice. *Am. Econ. Rev.* 75:1–14
13. Callahan RE. 1962. *Education and the Cult of Efficiency: A Study of the Social Forces that Have Shaped the Administration of the Public Schools.* Chicago: Univ. Chicago Press
14. Costanza R. ed. 1991. *Ecological Economics: The Science and Management of Sustainability.* New York: Columbia Univ. Press
15. Cropper ML, Oates WE. 1992. Environmental economics: a survey. *J. of Econ. Lit.* 30:675–740
16. Daly HE. 1977. *Steady-State Economics: The Economics of Biophysical Equilibrium and Moral Growth.* San Francisco: WH Freeman
17. Daly HE, Cobb JB. 1989. *For the Common Good: Redirecting the Economy toward Community, the Environment and a Sustainable Future.* Boston: Beacon
18. Daly HE, Townsend KN, eds. 1993. *Valuing the Earth: Economics, Ecology, Ethics.* Cambridge, MA: MIT Press
19. Devall B, Sessions S. 1985. *Deep Ecology.* Salt Lake City: Peregrine Smith
20. Ehrlich PR. 1968. *The Population Bomb.* New York: Ballantine
21. Etzioni A. 1988. *The Moral Dimension: Toward a New Economics.* New York: Free Press
22. Fisher AC. 1981. *Resource and Environmental Economics.* New York: Cambridge Univ. Press
23. Flash ES. 1965. *Economic Advice and Presidential Leadership: The Council of Economic Advisors.* New York: Columbia Univ. Press
24. Freeman AM. 1979. *The Benefits of Environmental Improvement.* Baltimore: Johns Hopkins Univ. Press
25. Galbraith JK. 1958. *The Affluent Society.* Boston: Houghton Mifflin
26. Gillroy JM, Wade M, eds. 1992. *The Moral Dimensions of Public Policy: Beyond the Market Mechanism.* Pittsburgh: Univ. Pittsburgh Press
27. Goodland R, Daly H, El Serafy S, von Droste B. 1991. Introduction. In *Environmentally Sustainable Economic Development: Building on Brundtland,* eds. R Goodland, et al. New York: UN Educ., Sci. and Cultural Org.
28. Goodwin CD. 1989. Doing good and spreading the gospel (economic). In *The*

Spread of Economic Ideas, ed. D. Colander, AW Coats, pp. 157–73. New York: Cambridge Univ. Press
29. Haber S. 1964. *Efficiency and Uplift: Scientific Management in the Progressive Era, 1890–1920*. Chicago: Univ. Chicago Press
30. Hahn RW, Stavins RN. 1991. Incentive-based environmental regulation: a new era from an old idea? *Ecology Law Q.* 18:1–42.
31. Harberger AC. 1993. The search for relevance in economics. *Am. Econ. Rev.* 83:1–16
32. Hardin G. 1968. The tragedy of the commons. *Science* 162:1243–48
33. Hausman JA, ed. 1993. *Contingent Valuation: A Critical Assessment*. New York: North Holland
34. Hays SP. 1959. *Conservation and the Gospel of Efficiency: The Progressive Conservation Movement, 1890–1920*. Cambridge, MA: Harvard Univ. Press
35. Heisenberg W. 1958. *Physics and Philosophy: The Revolution in Modern Science*. New York: Harper
36. Jevons WS. 1965 [1906]. *The Coal Question: An Inquiry Concerning the Progress of the Nation and the Probable Exhaustion of our Mines*. New York: Augustus Kelley
37. Kaysen C. 1968. Model-makers and decision makers: economists in the public policy process. *The Public Interest* Summer, 1968, pp. 80–95
38. Kelman S. 1981. Cost-benefit analysis: an ethical critique. *Regulation* 5(1):33–40
39. Kerr HT, ed. 1989. *Calvin's Institutes: A New Compendium*. Louisville, KY: Westminster/John Knox
40. Keynes JM. 1930. Economic possibilities for our grandchildren. In *Essays in Persuasion*, ed. JM Keynes, pp. 358–73. New York: Norton
41. Klamer A, McCloskey D, Solow R, eds. 1988. *The Consequences of Economic Rhetoric*. New York: Cambridge Univ. Press
42. Kneese AV, Schultze CL. 1975. *Pollution, Prices and Public Policy*. Washington, DC: Brookings Inst.
43. Korten DC. 1991–1992. Sustainable development. *World Policy J.* Winter, 1991–1992. 9:157–90
44. Krutilla JV. 1967. Conservation reconsidered. *Am. Econ. Rev.* 57:777–86
45. Lasch C. 1991. *The True and Only Heaven: Progress and its Critics*. New York: Norton
46. Lawson R. 1994. Coal, electricity and a sustainable future. *Coal Voice* Fall, 1994, p. 38

47. Lindblom CE. 1959. The science of "muddling through." *Public Admin. Rev.* 19:79–88
48. Lindler B. 1988. Making economics less dismal. *High Country News* 20(19):10
49. Looney DS. 1991. Protection or provocateur? (interview with Foreman). *Sports Illustrated* May 27 issue, pp. 54–58
50. Malthus T. 1798. *An Essay on the Principle of Population*. Ann Arbor: Univ. Mich. Press
51. Manuel FE, Manuel FP. 1979. *Utopian Thought in the Western World*. Cambridge, MA: Harvard Univ. Press
52. Mayor F. 1991. Foreword. In *Environmentally Sustainable Economic Development: Building on Brundtland*, eds. R. Goodland, H Daly, S El Serafy, and B von Droste, p. 3. New York: UN Educ., Sci., Cultural Org.
53. McCloskey DN. 1985. *The Rhetoric of Economics*. Madison: Univ. Wisc. Press
54. McKibben B. 1989. *The End of Nature*. New York: Random House
55. McPhee JA. 1971. *Encounters with the Archdruid*. New York: Farrar, Straus & Giroux
56. Mill JS. 1987. [1848] *Principles of Political Economy*. New York: Augustus M Kelley
57. Mitchell RT, Carson RT. 1989. *Using Surveys to Value Public Goods: The Contingent Valuation Method*. Washington, DC: Resources for the Future
58. Nash R. 1967. *Wilderness and the American Mind*. New Haven: Yale Univ. Press
59. Nelson RH. 1987. The economics profession and the making of public policy. *J. Econ. Lit.* 25:49–91
60. Nelson RH. 1993. *Reaching for Heaven on Earth: The Theological Meaning of Economics*. Lanham, MD: Rowman & Littlefield
61. Nelson RH. 1993. Environmental Calvinism: The Judeo-Christian roots of eco-theology. In *Taking the Environment Seriously*, ed. RE Meiners, B Yandle, pp. 233–55. Lanham, MD: Rowman & Littlefield
62. Nelson RH. 1995. *Public Lands and Private Rights: The Failure of Scientific Management*. Lanham, MD: Rowman & Littlefield
63. Norgaard RB, Howarth RB. 1991. Sustainability and discounting the future. In *Ecological Economics: The Science and Management of Sustainability*, ed. R. Costanza, pp. 88–101. New York: Columbia Univ. Press
64. Norton BG. 1991. Thoreau's insect analogies: or why environmentalists

hate mainstream economists. *Environ. Ethics* 13:235–51

65. Nourse EG. 1953. *Economics in the Public Service: Administrative Aspects of the Employment Act.* New York: Harcourt Brace

66. Oates W, ed. 1992. *The Economics of the Environment.* Brookfield, VT: Edward Elgar

67. Pechman JA. 1989. The United States. In *The Role of the Economist in Government: An International Perspective,* ed. JA Pechman, pp. 111–124. New York: Harvester Wheatsheaf

68. Pezzey J. 1989. *Economic Analysis of Sustainable Growth and Sustainable Development. Work. Pap. No. 15.* Washington, DC: World Bank Environ. Dep.

69. Portney PR. 1990. EPA, the evolution of federal regulation. In *Public Policies for the Environment,* ed. P. Portney, pp. 7–25. Washington, DC: Resources for the Future

70. Redclift M. 1987. *Sustainable Development: Exploring the Contradictions.* London: Methuen

71. Rhoads SE. 1985. *The Economist's View of the World: Government, Markets and Public Policy.* New York: Cambridge Univ. Press

72. Rorty R. 1979. *Philosophy and the Mirror of Nature.* Princeton, NJ: Princeton Univ. Press

73. Ruff LE. 1970. The economic common sense of pollution. *The Public Interest* 19:69–85

74. Sagoff M. 1988. *The Economy of the Earth: Philosophy, Law and the Environment.* New York: Cambridge Univ. Press

75. Schultze CL. 1968. *The Politics and Economics of Public Spending.* Washington, DC: Brookings Inst.

76. Schultze CL. 1982. The role and responsibilities of the economist in government. *Am. Econ. Rev.* 72:62–66

77. Schumpeter JA. 1942. *Capitalism, Socialism and Democracy.* New York: Harper

78. Sen A. 1987. *On Ethics and Economics.* New York: Basil Blackwell

79. Shavell S. 1993. Contingent valuation of the nonuse value of natural resources: implications for public policy and the liability system. In *Contingent Valuation: A Critical Assessment,* ed. J. Hausman. New York: North Holland

80. Simon HA. 1986. The failure of armchair economics. *Challenge* November/December:18–23

81. Solow RM. 1970. *Growth Theory: An Exposition.* New York: Oxford Univ. Press

82. Solow RM. 1991. *Sustainability: An Economist's Perspective.* Woods Hole, MA: Woods Hole Oceanographic Inst.

83. Stavins R, ed. 1988. *Project '88, Harnessing Market Forces to Protect our Environment: Incentives for the New President.* Washington, DC: US Govt. Print. Off.

84. Stein H. 1985. The Washington economics industry. *Am. Econ. Rev.* 76:1–9

85. Stigler GL. 1982. *The Economist as Preacher and Other Essays.* Chicago: Univ. Chicago Press

86. Stroup RL. 1991. *Progressive Environmentalism: A Pro-Human, Pro-Science, Pro-Free Enterprise Agenda for Change.* Dallas: Natl. Ctr. for Policy Anal.

87. Teitenberg T. 1992. *Environmental and Natural Resource Economics.* New York: Harper Collins

88. Tribe LH. 1972. Policy science: analysis or ideology? *Philos. Public Affairs* 2: 66–110

89. Waldo D. 1948. *The Administrative State: A Study of the Political Theory of American Public Administration.* New York: Ronald

90. Weiss CH. 1977. Research for policy's sake: the enlightenment function of social research. *Policy Anal.* 3:531–45

91. White L. 1967. The historical roots of our ecological crisis. *Science* 155:1203–07

92. World Bank. 1992. *World Development Report 1992: Development and the Environment.* New York: Oxford Univ. Press

93. World Commission on Environment and Development. 1987. *Our Common Future.* New York: Oxford Univ. Press

94. World Resources Institute. 1994. *World Resources 1994–95: People and the Environment.* New York: Oxford Univ. Press

95. Worster D. 1993. *The Wealth of Nature.* New York: Oxford Univ. Press

96. Zerbe RO, Dively DD. 1994. *Benefit-Cost Analysis: In Theory and Practice.* New York: Harper Collins

Annu. Rev. Ecol. Syst. 1995. 26:155–75

ECOLOGICAL BASIS FOR SUSTAINABLE DEVELOPMENT IN TROPICAL FORESTS

Gary S. Hartshorn
World Wildlife Fund, 1250 24th Street NW, Washington, DC 20037

KEY WORDS: tropical forest management, community-based forestry, sustainable tropical
              forestry, tropical forest ecology, tropical forests

---

ABSTRACT

Unless sustainable development becomes much more prevalent in tropical forests, appreciable areas of unprotected tropical forests will not survive far into the twenty-first century. Sustainable tropical forestry must integrate forest conservation and economic development. Key ecological factors discussed here include: reproduction, natural regeneration, growth, ecosystem functions, and biodiversity conservation. Four models of sustainable tropical forestry are described: 1) Industrial timber production based on the PORTICO company in Costa Rica that owns and manages its production forests and makes a substantive investment in research. 2) Community-based timber production using the Yánesha Forestry Cooperative in the Peruvian Amazon as an example of local empowerment over the protection and use of forests. This Coop has a local processing facility that enables most of the timber to be marketed, and it uses an innovative strip-cut management system that promotes excellent natural regeneration of native tree species. 3) Community-based production of nontimber forest products depends on local rights of access or tenure to tropical forests. However, more information is needed on harvestable levels and management techniques as local preferences move from subsistence uses to commercial production. 4) Locally controlled nature tourism is touted as the most benign use of tropical forests, but local communities receive minimal economic returns and have little say in prioritizing development objectives.

155

INTRODUCTION

Sustainable development in tropical forests is a high profile, often contentious global issue. National and local owners of tropical forests believe it is their prerogative—indeed right—to use their forest resources to meet development objectives. Environmentalists argue that tropical forests are unique global resources that should be protected from any or all development activities. The development–protection dichotomy reaches extremes with regard to tropical forests, involving issues such as species protection, concessionary policy, multinational involvement, indigenous peoples, local access and/or tenure, timber vs. nontimber forest products, and sustainable development, inter alia. Sustainable tropical forestry based on natural forests has the potential to conserve far more biodiversity than does plantation forestry.

In the context of tropical forests, the phrases "sustainable development" and "sustainability" have generated much discussion (18, 28, 41, 59, 88, 96). Books are devoted to sustainable forestry (5, 108). Critics point out that, in the case of forests, sustainability of development cannot be proven until the second or third rotation. This may be a valid criticism, but it ignores the importance of process, i.e. working toward more sustainable forestry.

Unprotected tropical forests are seriously threatened unless in the near future they are brought under sustainable management regimes. Despite national systems of protected areas, boycotts and threatened bans on tropical timber, and public concern about tropical deforestation, nevertheless, loss of tropical forests averaged 154,000 km^2 per year during the 1980s (125). Unless there is much greater effort and commitment to finding sustainable ways to use tropical forests, few unprotected tropical forests will survive far into the twenty-first century.

This chapter focuses on the status of ecological efforts to promote and test sustainable development in tropical forests. To put it simply: How can tropical forests be used without destroying them? Forest policy and socioeconomic aspects of forest use are treated only peripherally when they are regarded as relevant to this discussion. The classic literature on tropical silviculture is not reviewed here; readers interested in tropical forest management should consult the traditional literature (10, 34–36) as well as more recent holistic perspectives (20, 97, 126).

ECOLOGICAL BASIS AND CRITERIA FOR SUSTAINABLE FORESTRY

Several key ecological factors are critical to sustainable tropical forestry: reproduction and genetics; natural regeneration; growth; seed dispersal; seedling establishment; light regimes; ecological processes; and ecosystem functions. Biodiversity issues to be addressed include the effects of harvesting not only on the focal species, but on other dependent or independent species in the same ecosystem.

Reproduction

Focal species must be able to reproduce successfully. In the context of sustainable tropical forestry, however, there is an appalling dearth of scientific information on the reproductive biology of commercial timber species in natural forests (68). Key reproductive features such as phenology, breeding system, pollinator, seed disperser, and seed viability are known for few tropical forest species that are the source of timber or of nontimber forest products (11). The exceptions to this generalization are tropical forest species (e.g. *Bactris gasipaes, Tectona grandis, Theobroma cacao*) domesticated for plantation culture.

Understanding the reproductive biology of a focal species in natural forest may permit sustainable harvest, maintain dependent species, and minimize risk of extinction. But the harvesting of prereproductive individuals or all adults may drastically lower or abrogate seed production. Repetitive harvesting of all near-adult individuals dooms the population to extinction, analogous to harvesting a plantation. Ignorance of seasonality of reproduction can also have serious consequences. In the Amazon Basin, for example, mahogany (*Swietenia macrophylla*) is traditionally harvested from natural forests in the dry season, just before fruits mature and disperse the winged seeds at the end of the dry season (80). If a few seed trees were left or if mahogany harvesting could be delayed to later in the dry season or the beginning of the rains, far more viable seed would be naturally available for maintaining the population.

The concept of keystone species (111) in tropical forests has not been adequately explored, nor has its relevance to sustainable tropical forestry been evaluated. A keystone species provides critical food resources (usually fruit) that support populations of users (e.g. frugivore guild) during seasonal scarcity. In Manu, Peru, palms are the principal food source for monkeys during the annual period of food scarcity (112). Figs, with some individuals always in fruit, also often play a keystone role for the frugivore guild (122).

One should not assume that keystone species fulfill their eponymous function every year. Supra-annual fluctuations in seasonality, such as those associated with the El Niño-Southern Oscillation (ENSO) phenomenon, may have community-wide effects on phenology and food availability. The best documentation comes from the Barro Colorado Island research facility in Panama's Lake Gatún, where an ENSO-related weak dry season (e.g. 1958, 1970) precluded flowering and fruiting of several tree and liana species. The scarcity of fleshy fruits the following rainy season triggered famine among the frugivore guild (38, 66). Fluctuations in the seasonality of rainfall may be exacerbated by global climate change (53).

Reproductive capacity, whether sexual or vegetative, needs to be considered in assessing the threat of extinction as well as its relevance to any harvesting regime. The extreme example is monocarpic species that grow for many years before synchronously flowering and fruiting and then die (e.g. bamboos, talipot palm *Corypha umbraculifera, Tachigali* spp.). Overharvesting prereproductive individuals of monocarpic species can have devastating effects on the population. Massive, natural die-offs of postreproductive individuals can also have important conservation consequences, such as the bamboo die-off in central China that forced pandas to forage at lower, more accessible elevations, increasing their susceptibility to poaching and capture (106).

As suggested above, vegetative reproduction (67, 109) may be an important factor in sustainability. The Brazilian sassafras tree (*Ocotea odorifera*) is a primary source of safrol oil used in the manufacture of two chemicals, heliotropin (used as a fixative in the fragrance industry) and piperonyl butoxide (an enhancer of some insecticides like pyrethrins). Because *O. odorifera* occurs in the highly endangered Atlantic forests of Brazil, IBAMA (National Environmental Agency) lists it as an endangered species. Like many members of the Lauraceae, *O. odorifera* coppices vigorously from cut stumps, whether in primary forest, secondary forest, or even open pasture. Not only does stump-sprouting capacity offer an excellent opportunity for sustainable management of sassafras trees, it also suggests that in the dwindling Atlantic forests of Brazil, this species may well have the rare capacity to survive deforestation.

Natural Regeneration

Natural regeneration of commercial species has long been a principal focus and limitation of tropical forest management systems (19, 39, 43, 69, 105, 120). Though experts agree that adequate regeneration is a fundamental criterion of sustainability, the question is, how much is adequate? Sampling the number of seedlings or saplings of the focal species may be a poor predictor of future stocking of harvestable individuals. In natural habitats like tropical forests, most plant populations go through a fairly restrictive bottleneck in their life cycle, usually in the seed or seedling stage (45, 58). A reproductive plant may produce copious seeds of low viability, resulting in very few seedlings. More commonly, viable seeds require specific microsite conditions such as moisture or heat (e.g. *Ochroma lagopus* seed germination is favored by high temperatures) or chemical or physical scarification through a disperser's gut (119). An extreme example of the latter is *Calvaria*, a Sapotaceae tree endemic to Mauritius. Only large trees now exist and they fruit abundantly, but the seeds do not germinate. Seeds that pass through domestic turkeys do germinate, suggesting that this tree species may have depended on the dodo (*Raphus cucullatus*), extinct since 1681 (76).

In neotropical forests, high densities (> $10/m^2$) of seedlings are sufficiently uncommon to be noticeable. Tropical forest species with abundant seedlings seem to have very tiny seeds of short viability (e.g. *Vochysia* spp.) or large, chemically protected seeds (e.g. *Pentaclethra macroloba*). Unusually dense carpets (> $100/m^2$) of seedlings under the parental crown (e.g. *Ampelocera macrocarpa,* some Sapotaceae) suggest that important seed dispersers/predators are missing from the forest habitat (64).

A contrasting pattern of supra-annual, synchronous mast fruiting is well known for the dipterocarp forests of Southeast Asia (124). After skipping several years, numerous species among several genera of the dominant tree family, Dipterocarpaceae, synchronize fruit maturation, which has the effect of satiating seed predators. Many seed eaters such as hornbills (Bucerotidae) and wild boar (*Sus scrofa*) migrate as they track mature fruit. Due to the relatively long interval between mast crops by canopy dipterocarps, adequate regeneration of the commercial species is a key prelogging requirement of the Malayan Uniform System (7, 10). Unfortunately, logging companies do not always respect the regeneration requirement.

Once forest seedlings have exhausted their seed reserves, they come under fierce competition for light, nutrients, and moisture and are subject as well to attack by herbivores and parasites. The net result is usually such a drastic reduction in numbers that, by the sapling stage of a plant's life cycle, numbers of juveniles in the population may be inadequate to replace harvested adults. The forest managers' preoccupation with inadequate regeneration of focal species has prompted considerable investment in enrichment planting.

Because of low density of adults and poor natural regeneration, selectively logged forests have been enriched by line-planting of high value species, such as mahogany (*Swietenia macrophylla*). To provide the more favorable light regime that mahogany requires, broad lanes (ca. 2 m wide) are cut through the forest understory for planting nursery-grown seedlings. Provided there is an acceptable match of species to site, enrichment planting is usually successful in establishing vigorous saplings. However, enrichment planting is seldom viable economically, because of the cost of frequent weeding to favor the planted trees.

In contrast to the long tradition of species-based tropical silviculture, research on tropical forest dynamics prompted a more holistic, community-level perspective on the role of tree falls (gaps) in natural regeneration of native tree species (113). Several studies of neotropical forest dynamics [for trees > 10 cm diameter at breast height (dbh)] indicate high turnover rates of 75–150 years (51), tree mortality rates of 1–2% per year (51), stand half-life of 38–50 years (73), and high species-dependence on gaps for successful establishment

(17, 30, 48). Researchers over the past three decades have illuminated the central role of disturbance in the dynamics and regeneration not only of tropical forests, but also of many other biomes as well (92). The focus on natural regeneration of trees in tropical forests contributed substantially to the development of the intermediate disturbance hypothesis for maintaining high species diversity (27, 63). Frequent, stochastic disturbance is theorized to preempt competitive exclusion of gap opportunists from complex communities (25). The role of disturbance in community dynamics, natural regeneration, and maintenance of diversity continues to attract much research (31).

As concern mounted in the late 1970s over tropical deforestation, I realized that the revisionary thinking about gap dynamics might have potential to improve tropical forest management. I proposed simulating gap dynamics by clear-cutting long, narrow strips in tropical forests suitable for production forestry (46, 49). Because approximately half of the native tree species in the La Selva (Costa Rica) forest depend on gaps for successful establishment, I reasoned that simulating gaps could stimulate natural regeneration of many tree species. The La Selva research shows that a higher proportion of canopy tree species require gaps than do subcanopy or understory tree species (50). Also, many valuable tropical timber species are dependent on gaps (e.g. *Cedrela odorata, Cedrelinga catenaeformis, Swietenia macrophylla*). A commercial application of the gap-based model of tropical forest management is described below in the Sustainable Forestry Models section.

Growth Rates

Extremely slow growth rates of many tropical trees in primary forests are perceived as a major impediment to sustainable tropical forestry. Growth rates ranging from 0 to 10 mm of annual diameter increment mean a tree must be many decades old before it attains harvestable size. The combination of slow growth rates of valuable timber species in natural forests (73) and inadequate natural regeneration of valuable timber species discouraged many tropical foresters from seriously considering production forestry as a viable use of complex tropical forests (cf 72). Yet these growth rates seemed incongruous with the dynamic filling of gaps in the forest canopy. Gap-dependent, light-demanding tree species can attain the canopy (30–40 m) in 10–20 years, demonstrating not only rapid height growth, but appreciable diameter increment as well (74). Gap species are prime candidates for plantation forestry and enrichment planting, too.

Though favorable light regimes in gaps, strips, or second growth areas often promote rapid growth of trees, climbing vines and scandent shrubs grow even faster. These dependent plants can quickly overtop gaps as well as trees. Even canopy emergents can be festooned with lianas that often drastically reduce the seed production of large trees. Robust lianas interlinking canopy tree crowns not only may reduce growth, they may also complicate the selective felling of trees. Thus, "climber control" is an integral part of preharvesting and silvicultural treatments of natural forests (6). Cutting of large lianas 6–18 months before tree felling is a common prescription to facilitate the felling and to minimize damage to forest structure. Similarly, the periodic control of woody climbers in simulated gaps and young secondary forest is essential to promoting good growth of young trees.

Lack of ecological information about liana species is a major constraint to sustainable development of nontimber forest products. With the notable exception of the more important rattan species in southeast Asia, even less is known about the ecology of lianas than of canopy trees in most tropical forests (101). The fact that some lianas (e.g. *Chondrodendron* spp., *Uncaria* spp.) have extraordinary pharmaceutical properties has resulted in the uncontrolled harvest of these species (J. Duke, personal communication); such harvesting, combined with lack of cultivation, may threaten their survival.

Biodiversity Conservation

In most tropical countries, production forestry and biodiversity conservation are considered inimical (or incompatible), and the responsible national agencies are usually in different ministries (13). However, if sustainable tropical forestry could maintain natural forest habitats, harvesting timber or nontimber forest products could be compatible with biodiversity conservation objectives (86, 87, 98). Most developing countries in the natural range of tropical forests have 2–10 times as much area of tropical forests remaining as has been designated in national parks and equivalent reserves (125). [However, most countries with tropical forests also have less than the recommended 10% of the national territory in protected areas (60).] Thus, the development of sustainable production systems for timber and/or nontimber forest products that maintain natural forests could complement the national system of protected areas for the conservation of biodiversity.

Selective logging of high-value timber usually has serious repercussions on forest biodiversity (32). Logging roads provide access for hunters, poachers, miners, farmers, and colonists. Even subsistence hunting can decimate populations of larger vertebrates (65). Enforcement of regulations is often so weak or nonexistent that even though logging per se does not destroy the forest, the accompanying defaunation and deforestation tend to have disastrous effects on biodiversity (102, 103).

Ethnobiologists are finding that indigenous tribes manage forest succession to enhance abundance of useful plants (9). They may plant or favor medicinal plants and fruit trees in their garden plots so that preferred species are available or attract game animals. Indeed, there is growing evidence that indigenous people have had a more profound effect on primary forest composition and structure than was previously thought (e.g. 44).

Efforts to design and test sustainable development models for tropical forest products have been almost entirely entrepreneurial. While recent research has tended to focus on timber vs. nontimber uses (91), one of the great challenges for sustainable tropical forestry is to integrate timber and nontimber forest production. The community-based management of forest products such as bushmeat, medicinal plants, fibers for basket weaving, and edible fruits could diversify the resource base and add value to forest products. Most community-based initiatives currently focusing on nontimber forest products tend to target limited markets, with modest economic benefits. Generating significant economic benefits for local communities is a major problem with most ecotourism initiatives as well (see below).

Ecosystem Functions

In addition to providing renewable commodities and supporting biodiversity, tropical forests have several important ecosystem functions. Tropical forests protect watersheds, reduce soil erosion, and stabilize local and global climate, inter alia. Though these key ecosystem functions can be achieved to a substantial degree by tree plantations, they are most efficiently met by intact natural forests with their multiple layers of vegetation and full complement of wildlife—both vertebrate and invertebrate (99). Yet these critical ecosystem functions are ignored in national development planning and by local landowners.

Catchment forests are critically important to many hydrologic functions. On front ranges—where they are often called cloud forests—they intercept moisture-laden clouds, adding appreciable condensation drip to normal precipitation. The characteristic abundance of epiphytes, particularly mosses, in cloud forests greatly increases the surface area for moisture condensation. Because many important watershed catchments are at cooler, mid-elevations (ca. 1000–3000 m), organic matter accumulates on the soil surface and in the upper soil layer (79). Not only do epiphyte-laden catchment forests intercept much moisture, but also the abundant organic matter on and in the forest soil functions like a gigantic sponge in absorbing the heavy rainfall. Catchment forests are particularly effective in modulating stream flow, reducing peak flows during high rainfall periods, and releasing groundwater during seasonal drought. The maintenance and protection of forests in watersheds feeding reservoirs, not surprisingly, should be a top priority; however, many developing countries continue to ignore the critical hydrologic functions of forest catchments until drought or flooding brings the lesson home.

Forest cover protects soil from the erosive forces of rainfall and surface runoff. Deforested slopes may suffer soil erosion two orders of magnitude greater than that in comparable sites with forest protection (115). Even on steep slopes, forests are amazingly tenacious, not only minimizing soil erosion, but also reducing the incidence of landslips and slides. Mechanized logging causes considerable soil disturbance (78). Though reforestation can lessen soil degradation as well as rehabilitate soil productivity, much degradation occurs during the landscape conversion process and in the early establishment phase of plantations. Landscape and national level classification of land-use potential is fundamentally important to identifying and zoning landscapes unsuitable for agriculture (sensu lato) that should remain in protective forests. Human population growth, inequitable land tenure and access, failed agrarian reform, and national colonization and development programs force rural farmers onto steeper slopes and farther upstream in important watersheds.

The cutting and burning of tropical forests contribute 20–25% of the greenhouse gases accumulating in our planet's atmosphere; hence, these are significant factors in global climate change. The substantial areas of unprotected forests in forest-rich countries, and the national efforts to convert these forests to pasture or agriculture, suggest that tropical deforestation will continue to be an important factor in global climate change. Sustainable development justifications for protecting tropical forests are a high priority for development assistance agencies and international conservation organizations.

An ecosystem function often overlooked in sustainable tropical forestry initiatives is the role of mycorrhizae (61). There is some evidence that logging may depress mycorrhizal inoculation potential for seedlings (2). Obligate mycorrhizal associates may be essential to sustainable forest management systems (62).

SUSTAINABLE FORESTRY MODELS

The very few commercial forestry operations that have been certified as sustainable sources of tropical timber indicate the paucity of successful models of sustainable forest management. In fact, much has been written about sustainable forestry issues, but there are few documented sustainable models. Almost by definition, industrial harvesting of tropical timber is not sustainable, because it is essentially a mining operation to extract only the most valuable timber (83, 85, 116). Even with national laws and regulations requiring an investment in postharvest silviculture to ensure future timber harvests, such investment seldom happens. That is why the most interesting models of sustainable tropical forestry are small-scale initiatives at the community or small-landowner level (89, 100). Scaling up these small, local pilot projects to commercially significant levels is a great challenge (40). Because alternatives to timber [or nontimber forest products (NTFPs), as they are popularly termed] are generally not of interest to timber companies, a great array of NTFPs are attracting considerable interest as a source for local community development. Finally, nature-based tourism (or ecotourism) attempts to capitalize on the burgeoning interest in tropical forests and wildlife.

Industrial Timber Production

Concessionary arrangements, which are the dominant mode for producing industrial timber from tropical forests, have been severely criticized for policy and practices antithetical to sustainable tropical forestry (71). Generally, a government grants a timber-harvesting concession to a commercial company for a moderate period (one to a few decades). There are many things wrong with the concessionary approach to tropical forestry, such as little or no incentive for the concessionaire to invest in or practice sustainable forestry, a mining approach for valuable timbers, token royalty charges for stumpage, and ignorance or abuse of local rights and tenure (75).

The Costa Rican company, PORTICO S.A., has made an unusual commitment to sustainable forestry by managing its own forests on a sustainable basis. Through enlightened leadership in the mid-1980s, PORTICO used a debt-for-equity swap with a US bank to purchase several thousand hectares of production forests in northeast Costa Rica, to buy and upgrade two sawmills, and to expand and modernize their hardwood door manufacturing plant. By 1988 PORTICO was exporting 65,000 hardwood doors, primarily to the United States.

Because most of Costa Rica is privately owned in relatively small holdings, PORTICO bought several properties, mostly in the 100–500 ha range. A novel approach was to purchase only the forested portions of these properties, leaving the seller's working farm not only intact, but with an infusion of capital. Thus, the purchase of production forests by PORTICO did not expel local people. In fact, many local residents were able to find off-farm employment with PORTICO forestry operations.

Carapa nicaraguensis (Meliaceae) is the principal timber tree used by PORTICO, marketed under the trade name "royal mahogany." Trees of this species dominate the swamp forests of northeast Costa Rica, where they can attain 200 cm dbh and a height of 45 m, with commercial volumes averaging 40 m^3/ha. Furthermore, *C. nicaraguensis* (formerly called *C. guianensis* in Central America) has prolific natural regeneration of seedlings (81). Thus, it is an ideal candidate for sustainable forest management.

PORTICO developed a comprehensive forest management plan for each property, which is managed as a production unit. These management plans are based on detailed stand inventories of timber volume by species, diameter-class distribution, and regeneration status. Early harvesting was done by contract loggers, who caused considerable damage to the residual stand by indiscriminate felling and skidding. A switch to employee logging coupled with field training in directional felling and minimal skidding of logs greatly reduced logging damage. This is vitally important to the long-term sustainable production of timber from such forests.

PORTICO has made a major commitment to research and to monitoring in its production forests (12). It has long-term control plots (each 1–4 ha) to monitor natural growth and mortality in unlogged or managed forests. These complement many other plots in which the effects of felling and skidding damage are assessed and experimental silvicultural treatments such as thinning of competing individuals are monitored. Research results support the management plan (R. Peralta, PORTICO Director of Research, personal communication). An interesting internal analysis indicates that PORTICO's investment in research approximately doubles the cost of its timber; however, the vertical integration of production enables the higher costs of raw material to be absorbed and covered by the export sale of the final product (L. Torres, PORTICO President, personal communication).

The conservation of biodiversity is an integral component of PORTICO's forest management (23). The company actively patrols its properties and effectively prohibits hunting in the forests. As part of company-sponsored biodiversity inventories, we found more wildlife in recently logged, managed forests than in the adjoining Barra del Colorado Wildlife Sanctuary, presumably due to better protection from hunting in the former (G. Hartshorn, unpublished). The company has a formal collaborative agreement with the National Biodiversity Institute (INBio) for biodiversity inventory and monitoring research. PORTICO is also responsive to recommendations for better integration of production forestry and biodiversity conservation. For example, when we recommended that the company exclude all *Dipteryx panamensis* trees from harvesting because this tree's fruits are a prime food for endangered green macaws (*Ara ambigua*), PORTICO immediately banned logging or damage to all mature *Dipteryx* trees in its production forests.

Community-Based Timber Production

The involvement of local communities in sustainable forestry traditionally focused on nontimber forest products (see below). Only in the past decade have some communities embarked upon direct involvement in the sustainable production of tropical timber (8, 89). One innovative approach is the Yánesha Forestry Cooperative (COFYAL) in the Peruvian Amazon (52). COFYAL was created through the joint efforts of USAID and the Peruvian government to bring sustainable development to the Palcazú Valley in the Central Selva region. Originally conceived as a traditional rural development project to assist colonists with agriculture and livestock, the USAID-funded environmental assessment of the proposed project concluded that conventional approaches would fail. The recommended alternative of sustainable forestry was accepted.

A contemporaneous social soundness analysis of the proposed rural development project noted that the Palcazú Valley is home to the last native communities of Amuesha Indians, an Arawakan indigenous group compressed into
the central and southern part of the lower Palcazú Valley (110). In the loan
agreement with the Peruvian government, USAID required that, prior to dis-
bursing loan funds, all native community land claims be officially recognized
and legally titled. The Peruvian government complied, recognizing 11 native
communities of Amuesha Indians in the Palcazú Valley. When the Palcazú
project started in 1984, the Amuesha comprised about 60% of the Valley's
5000 inhabitants. Initial surveys of land-use capability and location of production forests in the lower valley indicated that the extensive native communities
held most of the remaining forests suitable for production forestry.

In part because Peruvian law recognized only agricultural cooperatives, it
took two years to obtain official status for the Yánesha Forestry Cooperative
(COFYAL), founded by 70 individual Amueshas and five native communities.
(Yánesha is what the Amuesha call themselves, whereas the latter name is how
they are known in Peru and in the anthropologic literature.) The objectives of
COFYAL, the first forestry cooperative in South America, are to: 1) provide
a source of employment for members of the native communities; 2) manage
the communities' natural forests for sustained yield of forest products; and 3)
protect the cultural integrity of the Yánesha people.

All forest management and timber production is under the control of
COFYAL. COFYAL adopted the strip-cut technique for managing complex
tropical forests (49, 55). To improve use of timber from the strip cuts, technical
advisors to the project designed complementary local processing technologies
to be managed by the local landowners (114). On land ceded by the Shiringa-
mazú native community, the cooperative has two sawmills, a bank of 44
pressure caps for preservation of poles and posts, a portable charcoal kiln, and
the appropriate supporting equipment. Local processing, in combination with
access to national and regional markets, enables COFYAL to process and sell
most of the wood harvested from the production strips.

The strip-cut technique is the ecological cornerstone of COFYAL's sustainable forestry initiative. Strip cuts promote outstanding natural regeneration of
hundreds of native tree species (54). Even with occasional thinning, only a
handful of tree species have been lost from the regenerating strips. The strips
may enhance biodiversity conservation by providing more appropriate regeneration sites for rare tree species; for example, I found four seedlings of
Minquartia guianensis on two demonstration strips, yet I could not locate a
parent tree within 500 m. Early growth rates in combination with silvicultural
release suggest that a projected 30–40 year rotation may be attainable.

Though COFYAL's progress and success have been severely compromised
by the 1988 pullout of USAID, occasionally lengthy shutdowns due to national

conflicts and local outbreaks of terrorism, and lack of financial and technical support, still the cooperative survives. The resiliency of COFYAL epitomizes the importance of a transdisciplinary approach to sustainable tropical forestry, where the activities must be ecologically sound, economically viable, socially responsible, and politically acceptable.

Community-Based Production of Nontimber Forest Products

The diversity of nontimber forest products (NTFPs) spans the spectrum of tropical biodiversity: bamboo, bark, bird nests, bushmeat, dyes, eggs, feathers, fiber, fish, fodder, fruits, fuelwood, gums, heart of palm, hides, honey, lac, latex, leaves, live animals, medicines, mushrooms, nuts, oils, ornamentals, rattans, resins, spices, thatch, and vines. A remarkable array of NTFPs have been harvested for centuries or millenia by traditional hunter-gatherers (e.g. 3, 16, 57, 82, 90, 117, 118). This last half-century has seen NTFPs become important commercial commodities as well (42, 84). With few exceptions (e.g. rattan), NTFPs are secondary in economic importance to timber. Though commercial logging activities may destroy some NTFPs, the purported lower economic value of the NTFPs plus their more intensive labor requirements and unsuitability for mechanized extraction of NTFPs often relegate or assign them to local communities.

Over the past decade there has been an increase of research and development interest in NTFPs, with particular emphasis on bringing greater economic benefits to local people. Initially, research efforts focused on valuation of NTFPs (29, 91). The rapid and substantive involvement of cultural anthropologists has broadened the NTFP issue to address indigenous rights, tenure, and local empowerment (26, 75, 93). Thus, the control, use, and marketing of NTFPs have become critical factors in the local empowerment of forest dwellers, ethnic groups, and local communities (4, 21).

Government designation of vast extractive reserves in Brazil and Colombia has received much attention (37, 104, 107). Brazil's extractive reserves total over 20,000,000 ha in the Amazon Basin, primarily benefiting local rubber tappers who maintain an extensive network of forest trails to tap wild rubber trees (*Hevea brasiliensis*) during the rainy season. Rubber tappers collect Brazil nuts during the dry season. The development of local processing capacity and more direct access to markets are key factors in improving the economic returns to local communities (26).

Before World War II, forest dwellers in Southeast Asia had vast areas of primary forest from which to collect NTFPs, but those who mechanized logging of the valuable dipterocarp forests ignored or abused local people while they opened up logged-over areas for shifting cultivation and promoted human migration. Depletion of and reduced access to NTFPs fueled greater community involvement in forest protection. Joint community management has become national policy in India (94, 95). Local communities have successfully rehabilitated degraded forests and improved watershed protection in northern Thailand (24). Even in forest-rich countries like Indonesia, local communities are protecting their traditional forests and NTFPs (1). Scarcity of forest land and severe watershed degradation in the Philippines is opening the door for community coordination and initiatives to maintain the few remaining forests and to rehabilitate the extensive degraded landscapes with NTFP species (121).

Despite the surge of interest in NTFPs and many new projects to promote NTFPs as the economic basis for local community development, little ecological information is available about the focal species (cf 47, 70). What are sustainable harvest levels? Where are the population bottlenecks? How can populations be enhanced? How can NTFP species be managed in rebuilding forests? Many of these fundamental ecological questions need to be answered if sustainable commercial production of NTFPs is to benefit rural communities.

Locally Controlled Nature Tourism

Nature-based tourism (or ecotourism, as it is popularly known) is usually portrayed as the most benign economic use of natural habitats, simply because no products are removed (123). Ecotourism is defined as travel to relatively undisturbed or uncontaminated natural areas with the specific objective of studying, admiring, and enjoying the scenery and wild plants and animals, as well as any cultural manifestations (both past and present) found in those areas (22). Nature-based tourism has boomed in countries such as Thailand, Kenya, and Costa Rica, while many developing countries are promoting tourism as an attractive development option (15). In the rush to cash in on the global tourism boom, however, the involvement and interests of local communities tend to be overlooked or overridden by outside economic powers. Though proponents of ecotourism have focused on the potential benefits for protected areas, ecotourism does not have to be restricted to formally designated national parks and equivalent reserves. Hence, tropical forests with intact wildlife or scenic features are potentially attractive to tourists and could more easily generate economic returns to local communities than do traditional rural development projects.

Ecotourism brings economic benefits in four ways: 1) It is a growth industry; 2) the market comes to the resource; 3) tourism helps diversify the economy; and 4) it stimulates economic growth in rural areas (14). However, the environmental impacts of ecotourism, or more importantly, significant growth in ecotourism, are poorly known and largely unstudied. For example, what is the area's carrying capacity for tourists? How can infrastructure mitigate the environmental impacts of increased tourism? What are the environmental tradeoffs of growth in ecotourism? For popular sites, how can increasing visits and higher expectations be managed?

Local entrepreneurs can play a leadership role in meeting the increasing needs and expectations of ecotourists, as well as diversifying the attractions and opportunities for tourists (33). The importance of local initiatives is demonstrated by world-class nature attractions like Costa Rica's Monteverde Cloud Forest Reserve, Nepal's Royal Chitwan National Park, and Thailand's Khao Yai National Park. Active community involvement in diversifying lodging choices, promoting local crafts and artisans, providing music festivals adjacent to scenic areas, offering local nature guides, etc, not only strengthens the local economy, but often has a synergistic effect on local support for the focal natural resources (56).

Another area in which nature-based tourism offers potential synergies is with research. The recent surge in safer, less rigorous access to forest canopies is helping to protect several tropical forests while providing researchers with better and easier access to the last terrestrial frontier of tropical biology. In northeast Peru, the Amazon Center for Environmental Education and Research (ACEER) has constructed a 450 m walkway through the tropical forest canopy. Though nature tourists and educational groups are the primary users of the walkway, researchers are finding it attractive, too. For example, the beaver-tailed lizard, *Uracentron flaviceps*, was presumed to be extremely rare because only a few specimens exist in the world's herpetological collections. It turns out to be quite common in the ACEER forest along the canopy walkway.

CONCLUSIONS

Sustainable development is critical to the survival of most unprotected tropical forests. The physical, political, and/or legal owners of tropical forests insist that these sometimes vast natural resources be used for development objectives. However, traditional development for pastures or agriculture is synonymous with deforestation. Sustainable development in tropical forests is a new paradigm that attempts to integrate forest conservation and economic development.

Biological and ecological factors critical to sustainable tropical forestry include: reproduction, natural regeneration, growth, and ecosystem functions. Though a few plant species successfully maintain populations through vegetative reproduction, the production of seeds and seedlings is fundamental to most sustainable harvesting systems. Inadequate natural regeneration long constrained attempts to manage complex tropical forests; however, recent advances in our understanding of gap dependency offer new hope for successful establishment and growth of preferred species. The selective harvesting of timber causes considerable damage to tropical forests, as well as opening the forests for uncontrolled hunting and land clearing. Tropical forests have important but little appreciated ecological functions in watershed and soil protection, local rainfall regimes, and global climate change.

Four models of sustainable tropical forestry are discussed: industrial timber production, community-based timber production, community-based production of nontimber forest products, and locally controlled nature tourism. More has been written about the merits, definition, and criteria for sustainability than has been devoted to documenting sustainable models of tropical forestry. Many of the more interesting models involve local communities, whether they focus on timber, nontimber forest products, or ecotourism. Industrial timber production from tropical forests is largely through government concessions and almost by definition nonsustainable, because of the short time period and minimal investment in future harvests.

The Costa Rican door manufacturing company PORTICO is unusual in many respects: It owns its own production forests; it manages its forests on a sustainable basis; it invests in directed research; it effectively protects biodiversity; and its wood utilization is vertically integrated. PORTICO's primary timber species, *Carapa nicaraguensis,* dominates its forests, and produces abundant seedlings and adequate growth.

The Yánesha Forestry Cooperative in the Peruvian Amazon is an innovative approach to community-based timber production. The Coop adopted the strip-cut forest management technique that promotes excellent natural regeneration of hundreds of native tree species. Local processing of timber for sawnwood, preserved poles, and charcoal enables the Coop to use nearly all of the native timbers and to generate attractive economic returns.

Nontimber forest products traditionally have been key to the subsistence of local communities. Much effort is now going into determining commercial potential, sustainable harvesting levels, and market opportunities for several nontimber forest products. These forest commodities are increasingly important in strengthening local control of and access to tropical forests. Much more research is needed on nontimber forest products for them to play a more important role in sustaining tropical forests.

Nature-based tourism (or ecotourism) is considered the most benign use of tropical forests because no products are removed. National and local booms in ecotourism tend to ignore or marginalize local communities from decision-making, as well as from receiving appropriate economic returns. The most attractive sites face problems in determining how many visitors can be handled and how to mitigate the environmental impact of increased tourism. Synergies between ecotourism and field research are beginning to occur, particularly with the rapid development of safer techniques to access the forest canopy.

Literature Cited

1. Abdoellah O, Lahjie AB, Wangsadidjaja SS, Hadikusumah H, Iskandar J, Sukmananto B. 1993. Communities and forest management in East Kalimantan: pathway to environmental stability. *Res. Network. Rep. No. 3*, Univ. Calif., Berkeley

2. Alexander I, Ahmad N, See LS. 1992. The role of mycorrhizas in the regeneration of some Malaysian trees. *Phil. Trans. R. Soc. Lond. B* 335:379–88

3. Anderson AB. 1988. Use and management of native forest dominated by acai palm (*Euterpe oleracea* Mart.) in the Amazon estuary. *Adv. Econ. Bot.* 6:144–54

4. Anderson AB, May PH, Balick MJ. 1991. *The Subsidy From Nature: Palm Forests, Peasantry, and Development on an Amazon Frontier.* New York: Columbia Univ. Press. 233 pp.

5. Aplet GH, Johnson N, Olson JT, Sample VA, eds. 1993. *Defining Sustainable Forestry.* Washington, DC: Island. 328 pp.

6. Appanah S, Appanah F, Putz FE. 1984. Climber abundance in virgin dipterocarp forest and the effect of pre-felling climber cutting on logging damage. *Malay. For.* 47:335–42

7. Appanah S, Weinland G, Bossel H, Kreiger H. 1989. Are tropical rain forests non-renewable? An enquiry through modelling. *J. Trop. For. Sci.* 2:331–48

8. Ascher W. 1994. *Communities and Sustainable Forestry in Developing Countries.* Ctr. Trop. Conserv., Duke Univ. Durham, NC. 33 pp.

9. Balee W, Gely A. 1989. Managed forest succession in Amazonia: the Ka'apor case. In *Resource Management in Amazonia: Indigenous and Folk Strategies,* ed. DA Posey, W Balee, pp. 129–58. New York: NY Bot. Gard.

10. Baur GN. 1968. *The Ecological Basis of Rainforest Management.* Sydney, Australia: New South Wales Govt. Printer. 499 pp.

11. Bawa KS, Hadley M. eds. 1990. *Reproductive Ecology of Tropical Forest Plants.* Paris: UNESCO/Man and the Biosphere. 421 pp.

12. Bianchi-Sweron H, Valerio-Garita J, Simula M. 1993. *Industria Forestal Sostenible: Estudio de Caso Sobre Portico S.A., Costa Rica.* INDUFOR Proyecto INEFAN-Int. Trop. Timber Organ. PD 155/91, Helsinki, Finland

13. Blockhus JM, Dillenbeck MR, Sayer JA, Wegge P, eds. 1992. *Conserving Biological Diversity in Managed Tropical Forests: Proceedings of a Workshop held at the IUCN General Assembly, Perth, Australia, 30 Nov.–1 Dec. 1990.* Gland, Switzerland: Int. Union Conserv. Nature & Nat. Resourc. (Int. Union Conserv. Nature). 244 pp.

14. Boo E. 1990. *Ecotourism: The Potentials and Pitfalls.* Vols. 1, 2. Washington, DC: World Wildlife Fund. 71 pp. 172 pp.

15. Boo E. 1992. *The Ecotourism Boom: Planning for Development and Management.* Washington, DC: World Wildlife Fund. 14 pp.

16. Boom BM. 1989. Use of plant resources by the Chacobo. In *Resource Management in Amazonia: Indigenous and Folk Strategies,* ed. DA Posey, W Balee, pp. 78–96. New York: NY Bot. Gard.

17. Brandani A, Hartshorn GS, Orians GH. 1988. Internal heterogeneity of gaps and species richness in Costa Rican tropical wet forest. *J. Trop. Ecol.* 4–99–119

18. Budowski G. 1988. Is sustainable harvest possible in the tropics? *Am. For.* 94:34–37, 80–81

19. Burgess PF. 1972. Studies on the regeneration of the hill forests of the Malay peninsula. *Malay. For.* 35:103–21

20. Buschbacher RJ. 1990. Natural forest management in the humid tropics: ecological, social, and economic considerations. *Ambio* 19:253–58

21. Cabarle BJ. 1991. Community forestry and the social ecology of development. *Grassroots Dev.* 15:3–9

22. Ceballos-Lascurain H. 1987. *Estudio de Prefactibilidad Socioeconomica del Turismo Ecologico y Anteprovecto Arquitectonico y Urbanistico del Centro de Turismo Ecologico de Sian Ka'an, Quintana Roo.* Mexico City: SEDUE

23. Cheney K. 1991. Costa Rica: good logging. *Third World Week January* 4:37–38

24. Chuntanaparb L, Chamchong C, Hoamuangkaew W, Ongprasert P, Limchoowong S, et al. 1993. Community allies: forest co-management in Thailand. *Res. Netw. Rep. No. 2,* Univ. Calif., Berkeley

25. Clark DA. 1986. Regeneration of canopy trees in tropical wet forests. *Trends Ecol. Evol.* 1:150–54

26. Clay J. 1992. Some general principles and strategies for developing markets in North America and Europe for non-timber forest products: lessons from Cul-

tural Survival Enterprises, 1989–1990. *Adv. Econ. Bot.* 9:101–6

27. Connell JH. 1978. Diversity in tropical rain forests and coral reefs. *Science* 199: 1302–10

28. D'Silva E, Appanah S. 1993. *Forestry Management for Sustainable Development.* World Bank, Econ. Dev. Inst., Washington, DC. 46 pp.

29. de Beer JH, McDermott MJ. 1989. *The Economic Value of Non-Timber Forest Products in Southeast Asia.* Amsterdam: Netherlands Comm. for IUCN. 175 pp.

30. Denslow JS. 1980. Gap partitioning among tropical rainforest trees. *Biotropica* 12(suppl.):47–55

31. Denslow JS. 1987. Tropical rain forest gaps and tree species diversity. *Annu. Rev. Ecol. Syst.* 18:431–51

32. Dykstra DP, Heinrich R. 1992. Sustaining tropical forests through environmentally sound harvesting practices. *Unasylva* 43:9–15

33. Eber S. ed. 1992. *Beyond the Green Horizon: Principles for Sustainable Tourism.* Surrey, England: World Wildlife Fund. 54 pp.

34. Food and Agriculture Organization. 1989. *Management of Tropical Moist Forests in Africa. FAO Forestry Paper 88.* Rome: FAO. 165 pp.

35. Food and Agriculture Organization. 1989. *Review of Forest Management Systems of Tropical Asia: Case-Studies of Natural Forest Management for Timber Production in India, Malaysia and the Philippines. FAO Forestry Paper 89.* Rome: FAO. 228 pp.

36. Food and Agriculture Organization. 1993. *Management and Conservation of Closed Forests in Tropical America. FAO Forestry Paper 101.* Rome: FAO. 141 pp.

37. Fearnside PM. 1989. Extractive reserves in Brazilian Amazonia. *Bioscience* 39: 387–93

38. Foster RB. 1982. Famine on Barro Colorado Island. In *The Ecology of a Tropical Forest: Seasonal Rhythms and Long-Term Changes,* ed. EG Leigh Jr, AS Rand, DM Windsor, pp. 201–12. Washington, DC: Smithsonian Inst. Press

39. Fox JED. 1976. Constraints on the natural regeneration of tropical moist forest. *For. Ecol. Manage.* 1:37–65

40. Fox J. 1992. The problems of scale in community resource management. *Enviro. Manage.* 16:289–97

41. Gane M. 1992. Sustainable forestry. *Commonwealth For. Rev.* 71:83–90

42. Godoy RA, Bawa KS. 1993. The economic value and sustainable harvest of plants and animals from the tropical forest: assumptions, hypotheses, and methods. *Econ. Bot.* 47:215–19

43. Gomez-Pompa A, Vasquez-Yanes C, Guevara S. 1972. The tropical rain forest: a nonrenewable resource. *Science* 117:762–65

44. Gomez-Pompa A, Flores JS, Sosa V. 1987. The 'Pet-Kot': a man-made tropical forest of the Maya. *Interciencia* 15: 10–15

45. Gomez-Pompa A, Whitmore TC, Hadley M, eds. 1991. *Rain Forest Regeneration and Management.* Paris: UNESCO. 457 pp.

46. Gorchov DL, Cornejo F, Ascorra C, Jaramillo M. 1993. The role of seed dispersal in the natural regeneration of rain forest after strip-cutting in the Peruvian Amazon. In *Frugivory and Seed Dispersal: Ecological and Evolutionary Aspects,* ed. TH Fleming, AJ Estrada, 107/108:339–49. Oxford, OH: Miami Univ.

47. Hall P, Bawa K. 1993. Methods to assess the impact of extraction of non-timber tropical forest products on plant populations. *Econ. Bot.* 47:234–47

48. Hartshorn GS. 1980. Neotropical forest dynamics. *Biotropica* 12(suppl.):23–30

49. Hartshorn GS. 1989. Application of gap theory to tropical forest management: natural regeneration on strip clear cuts in the Peruvian Amazon. *Ecology* 70: 567–69

50. Hartshorn GS. 1989. Gap-phase dynamics and tropical tree species richness. In *Tropical Forests: Botanical Dynamics, Speciation, and Diversity,* ed. LB Holm-Nielsen, IC Nielsen, H Balsev. pp. 65–73. London: Academic

51. Hartshorn GS. 1990. An overview of neotropical forest dynamics. In *Four Neotropical Forests,* ed. AH Gentry. pp. 585–99. New Haven, CT: Yale Univ. Press

52. Hartshorn GS. 1990. Natural forest management by the Yanesha Forestry Cooperative in Peruvian Amazonia. In *Alternatives to Deforestation: Steps Toward Sustainable Use of the Amazonian Rain Forest,* ed. AB Anderson, pp. 128–38. New York: Columbia Univ. Press

53. Hartshorn GS. 1992. Possible effects of global warming on the biological diversity in tropical forests. In *Global Warming and Biological Diversity,* ed. RL Peters, TE Lovejoy, pp. 137–46. New Haven: Yale Univ. Press. 386 pp.

54. Hartshorn GS, Pariona-Arias W. 1993. Ecologically sustainable forest management in the Peruvian Amazon. In *Perspectives on Biodiversity: Case Studies of Genetic Resource Conservation and*

Development, ed. C Potter, J Cohen, pp. 151–66. Washington, DC: AAAS

55. Hartshorn GS, Simeone R, Tosi JA Jr. 1987. Sustained yield management of tropical forests: a synopsis of the Palcazu development project in the Central Selva of the Peruvian Amazon. In *Management of the Forests of Tropical America: Prospects and Technologies,* ed. JC Figueroa-Colon, FH Wadsworth, S Branham. pp. 235–43. Rio Piedras, Puerto Rico: Inst. Trop. For.

56. Healy RG. 1994. *Ecotourism, Handicrafts and the Management of Protected Areas in Developing Countries.* Ctr. Trop. Conserv., Duke Univ., Durham, NC. 28 pp.

57. Hladik CM, Bahuchet S, de Garine I, eds. 1990. *Food and Nutrition in the African Rain Forest.* Paris: UNESCO/MAB. 96 pp.

58. Hubbell SP, Foster RB. 1990. The fate of juvenile trees in a neotropical forest: implications for the natural maintenance of tropical tree diversity. In *Reproductive Ecology of Tropical Forest Plants,* ed. KS Bawa, M Hadley, pp. 317–41. Paris: UNESCO. 421 pp.

59. ITTO. 1990. *ITTO Guidelines for the Sustainable Management of Natural Tropical Forests. ITTO Technical Series No. 5.* Yokohama, Japan: ITTO. 18 pp.

60. IUCN. 1994. *1993 United Nations List of National Parks and Protected Areas.* Gland, Switzerland: IUCN. 313 pp.

61. Janos DP. 1988. Mycorrhiza applications in tropical forestry: Are temperate-zone approaches appropriate? In *Trees and Mycorrhiza,* ed. FSP Ng, pp. 133–88. Kuala Lumpur, Malaysia: For. Res. Inst. Malaysia. 305 pp.

62. Janos DP. 1995. Mycorrhizas, succession, and the rehabilitation of deforested lands in the humid tropics. In *Fungi and Environmental Change,* ed. JC Frankland, N Magan, GM Gadd. Cambridge, UK: Cambridge Univ. Press. In press

63. Janzen DH. 1970. Herbivores and the number of tree species in tropical forests. *Am. Nat.* 104:501–28

64. Janzen DH, Martin PS. 1982. Neotropical anachronisms: the fruits the Gomphotheres ate. *Science* 215:19–27

65. Johns AD. 1992. Species conservation in managed tropical forests. In *Tropical Deforestation and Species Extinctions,* ed. TC Whitmore, JA Sayer. pp. 15–53. London: Chapman & Hall

66. Kauffmann JH. 1962. *Ecology and Social Behavior of the Coati, Nasua narica,* on Barro Colorado Island, Panama. Univ. Calif. Publ. Zool. 60

67. Khan ML, Tripathi RS. 1986. Tree regeneration in a disturbed sub-tropical wet hill forest of northeast India: effect of stump diameter and height on sprouting of four tree species. *For. Ecol. Manage.* 17:199–209

68. Koshoo TN. 1990. Reproductive ecology of tropical forest plants: concluding remarks. In *Reproductive Ecology of Tropical Forest Plants,* ed. KS Bawa, M Hadley, pp. 403–411. Paris: UNESCO. 421 pp.

69. Kio PRO. 1976. What future for natural regeneration of tropical high forest? An appraisal with examples from Nigeria and Uganda. *Commonwealth For. Rev.* 55:309–18

70. LaFrankie JV. 1994. Population dynamics of some tropical trees that yield non-timber forest products. *Econ. Bot.* 48:301–9

71. Lamprecht H. 1988. Constraints and opportunities of technical assistance in forestry in the humid tropics. *Allgemeine Forst- und Jagdzeitung* 159:103–7

72. Leslie AJ. 1977. When theory and practice contradict. *Unasylva* 29:2–17

73. Lieberman D, Lieberman M. 1987. Forest tree growth and dynamics at La Selva, Costa Rica (1960–1982). *J. Trop. Ecol.* 3:347–58

74. Lieberman D, Lieberman M, Hartshorn G, Peralta R. 1985. Growth rates and age-size relationships of tropical wet forest trees in Costa Rica. *J. Trop. Ecol.* 1:97–109

75. Lynch OJ, Alcorn JB. 1994. Tenurial rights and community-based conservation. In *Natural Connections: Perspectives in Community-Based Conservation,* ed. D Western, RM Wright, pp. 373–92. Washington, DC: Island. 581 pp.

76. Mabberley DJ. 1983. *Tropical Rain Forest Ecology.* New York: Chapman & Hall. 156 pp.

77. Maini JS. 1992. Sustainable development of forests. *Unasylva* 43:3–8

78. Malmer A, Grip H. 1990. Soil disturbance and loss of infiltrability caused by mechanized and manual extraction of tropical rainforest in Sabah, Malaysia. *For. Ecol. Manage.* 38:1–12

79. Marrs RH, Proctor J, Heaney A, Mountford MD. 1988. Changes in soil nitrogen-mineralization and nitrification along an altitudinal transect in tropical rain forest in Costa Rica. *J. Ecol.* 76:466–82

80. Martini AMZ, Rosa NdeA, Uhl C. 1994. An attempt to predict which Amazonian tree species may be threatened by logging activities. *Environ. Conserv.* 21:152–62

81. McHargue LA, Hartshorn GS. 1983. Seed and seedling ecology of carapa guianensis. *Turrialba* 33:399–404
82. Myers N. 1988. Tropical forests: much more than stocks of wood. *J. Trop. Ecol.* 4:209–21
83. Nadarajah T. 1994. *The Sustainability of Papua New Guinea's Forest Resource. NRI Discussion Paper No. 76.* Soc. Stud. Div., Natl. Res. Inst., Papua New Guinea. 70 pp.
84. Nepstad DC, Schwartzman S, eds. 1992. *Non-Timber Products, from Tropical Forests: Evaluation of a Conservation and Development Strategy.* Vol. 9. New York: NY Bot. Gard. 164 pp.
85. OTA. 1984. *Technologies to Sustain Tropical Forest Resources.* Washington, DC: US Congress Off. Technol. Assessment. 344 pp.
86. OTA. 1992. *Combined Summaries: Technologies to Sustain Tropical Forest Resources and Biological Diversity.* Washington, DC: US Congress Off. Technol. Assessment. 88 pp.
87. Panayotou T, Ashton PS. 1992. *Not by Timber Alone: Economics and Ecology for Sustaining Tropical Forests.* Washington, DC: Island. 282 pp.
88. Pearce F. 1994. Are Sarawak's forests sustainable? *New Scientist* 26 Nov:28–32
89. Perl MA, Kiernan MJ, McCaffrey D, Buschbacher RJ, Batmanian GJ. 1991. *Views from the Forest: Natural Forest Management Initiatives in Latin America.* Washington, DC: World Wildlife Fund. 30 pp.
90. Peters CM. 1990. Population ecology and management of forest fruit trees in Peruvian Amazonia. In *Alternatives to Deforestation: Steps Toward Sustainable Use of the Amazon Rain Forest.* ed. AB Anderson, pp. 86–98. New York: Columbia Univ. Press. 281 pp.
91. Peters CM, Gentry AH, Mendelsohn RO. 1989. Valuation of an Amazonian rain forest. *Nature* 339:655–56
92. Pickett STA, White PS, eds. 1985. *The Ecology of Natural Disturbance and Patch Dynamics.* Orlando, FL: Academic
93. Poffenberger M. ed. 1990. *Keepers of the Forest: Land Management Alternatives in Southeast Asia.* West Hartford, CT: Kumarian. 289 pp.
94. Poffenberger M, ed. 1992. *Sustaining Southeast Asia's Forests. Research Network Report No. 1.* Univ. Calif., Berkeley
95. Poffenberger M. 1994. The resurgence of community forest management in eastern India. In *Natural Connections:*
    Perspectives in Community-Based Conservation, ed. D Western, RM Wright, pp. 53–79. Washington, DC: Island. 581 pp.
96. Poore D, Burgess P, Palmer J, Rietbergen S, Synnott T. 1989. *No Timber Without Trees: Sustainability in the Tropical Forest.* London: Earthscan. 252 pp.
97. Poore D, Sayer J. 1991. *The Management of Tropical Moist Forest Lands: Ecological Guidelines.* Gland, Switzerland: IUCN-World Conservation Union. 69 pp. 2nd ed.
98. Primack RB, Hall P. 1992. Biodiversity and forest change in Malaysian Borneo. *BioScience* 42:829–37
99. Primack RB, Lovejoy TE. eds. 1995. *Ecology, Conservation and Management of Southeast Asian Forests.* New Haven, CT: Yale Univ. Press. 292 pp.
100. Primack RB, Tieh F. 1994. Long-term timber harvesting in Bornean forests: the Yong Khow case. *J. Trop. For. Sci.* 7:262–79
101. Putz FE, Mooney HA. eds. 1991. *The Biology of Vines.* Cambridge: Cambridge Univ. Press
102. Redford KH. 1992. The empty forest. *BioScience* 42:412–22
103. Robinson JG, Redford KH. 1994. Community-based approaches to wildlife conservation in neotropical forests. In *Natural Connections: Perspectives in Community-Based Conservation,* ed. D Western, RM Wright, pp. 300–19
104. Salafsky N, Dugelby BL, Terborgh JW. 1992. *Can Extractive Reserves Save the Rain Forest?* Ctr. Trop. Conserv., Duke Univ. Durham NC. 24 pp.
105. Saulei SM. 1984. Natural regeneration following clear-fell logging operations in the Gogol Valley, Papua New Guinea. *Ambio* 13:351–54
106. Schaller G. 1993. *The Last Panda.* Chicago: Univ. Chicago Press. 291 pp.
107. Schwartzman S. 1989. Extractive reserves: the rubber tappers' strategy for sustainable development. In *Fragile Lands of Latin America: Strategies for Sustainable Development,* ed. JO Browder, pp. 150–65. Boulder, CO: Westview
108. Sharma NP. ed. 1992. *Managing the World's Forests: Look for Balance Between Conservation and Development.* Dubuque, IA: Kendall/Hunt. 605 pp.
109. Stocker GC. 1981. Regeneration of a north Queensland rainforest following felling and burning. *Biotropica* 13:86–92
110. Stocks A, Hartshorn G. 1993. The Palcazu project: forest management and

native Yanesha communities. *J. Sustain. For.* 1:111–35

111. Terborgh J. 1986. Keystone plant resources in the tropical forest. In *Conservation Biology,* ed. ME Souls, pp. 330–44. Sunderland, MA: Sinauer

112. Terborgh J. 1986. Community aspects of frugivory in tropical forests. In *Frugivores and Seed Dispersal,* ed. A Estrada, TH Fleming, pp. 371–84. Boston: Dr. W Junk

113. Tomlinson PB, Zimmermann MH, eds. 1978. *Tropical Trees as Living Systems.* New York: Cambridge Univ. Press. 675 pp.

114. Tosi JA. 1982. *Sustained Yield Management of Natural Forests: Forestry Sub-Project, Central Selva Resources Management Project, Palcazu Valley, Peru.* Lima, Peru: USAID. 68 pp.

115. Tosi-Olin J. 1980. *Estudio Ecologico Integral de las Zonas de Afectacion del Proyecto Arenal.* San Jose, Costa Rica: Centro Cientifico Tropical. 617 pp.

116. Uhl C, Guimaraes-Vieira IC. 1989. Ecological impacts of selective logging in the Brazilian Amazon: a case study from the Paragominas region of the state of Para. *Biotropica* 21:98–106

117. van Beusekom CF, van Goor CP, Schmidt P, eds. 1987. *Wise Utilization of Tropical Rain Forest Lands. Tropenbos Scientific Series 1.* Ede: the Netherlands. 154 pp.

118. van Schaik C, Kramer R, Shyamsundar P, Salafsky N. 1992. *Biodiversity of Tropical Rain Forests: Ecology and Economics of an Elusive Resource.* Ctr. Trop. Conserv. Duke Univ., Durham, NC. 36 pp.

119. Vasquez-Yanes C, Orozco-Segovia A. 1984. Ecophysiology of seed germination in the tropical humid forest of the world: a review. In *Physiological Ecology of Plants of the Wet Tropics,* ed. E Medina, HA Mooney, C Vasquez-Yanes, pp. 38–49. The Hague: Dr W Junk

120. Viana VM. 1990. Seed and seedling availability as a basis for management of natural forest regeneration. In *Alternatives to Deforestation in Amazonia: Steps Toward Sustainable Use of the Amazon Rain Forest,* ed. A Anderson, pp. 99–115. New York: Columbia Univ. Press. 281 pp.

121. Walpole P, Braganza G, Burtkenley-Ong J, Tengco GJ, Wijanco E. 1993. *Upland Philippine Communities: Guardians of the Final Forest Frontier. Res. Netw. Rep. No. 4.* Univ. Calif., Berkeley

122. Wheelwright NT, Haber WA, Murray KG, Guindon C. 1984. Tropical fruit-eating birds and their food plants: a survey of a Costa Rican lower montane forest. *Biotropica* 16:173–92

123. Whelan T. 1991. Ecotourism and its role in sustainable development. In *Nature Tourism: Managing for the Environment,* ed. T Whelan, pp. 3–22. Washington, DC: Island

124. Whitmore TC. 1984. *Tropical Rain Forests of the Far East.* Oxford: Clarendon. 352 pp. 2nd ed.

125. WRI. 1994. *World Resources 1994–95.* New York: Oxford Univ. Press. 400 pp.

126. Wyatt-Smith J. 1987. *The Management of Tropical Moist Forest for the Sustained Production of Timber: Some Issues. IUCN/Int. Inst. Environ. Dev. Tropical Forestry Policy Paper 4.* Gland, Switzerland

Annu. Rev. Ecol. Syst. 1995. 26:177–99

MANAGING NORTH AMERICAN WATERFOWL IN THE FACE OF UNCERTAINTY[1]

James D. Nichols

National Biological Service, Patuxent Environmental Science Center, Laurel, Maryland 20708

Fred A. Johnson

Office of Migratory Bird Management, Patuxent Environmental Science Center, Laurel, Maryland 20708

Byron K. Williams

Vermont Cooperative Fish and Wildlife Research Unit, School of Natural Resources, University of Vermont, Burlington, Vermont 05405

KEY WORDS: waterfowl (Anatidae), hunting, adaptive management, population dynamics, North America

---

ABSTRACT

Informed management of waterfowl (or any animal population) requires management goals and objectives, the ability to implement management actions, periodic information about population and goal-related variables, and knowledge of effects of management actions on population and goal-related variables. In North America, international treaties mandate a primary objective of protecting migratory bird populations, with a secondary objective of providing hunting opportunity in a manner compatible with such protection. Through the years, annual establishment of hunting regulations and acquisition and management of habitat have been the primary management actions taken by federal agencies. Various information-gathering programs were established and, by

---

[1]The US government has the right to retain a nonexclusive, royalty-free license in and to any copyright covering this paper.

the 1960s, had developed into arguably the best monitoring system in the world for continentally distributed animal populations. Retrospective analyses using estimates from this monitoring system have been used to investigate effects of management actions on waterfowl population and harvest dynamics, but key relationships are still characterized by uncertainty. We recommend actively adaptive management as an approach that can meet short-term harvest objectives, while reducing uncertainty and ensuring sustainable populations over the long-term.

INTRODUCTION

The term "waterfowl" refers collectively to members of the family Anatidae—ducks, geese, and swans. Waterfowl have been referred to as "the most prominent and economically important group of migratory birds in North America" (24). The breeding distributions of the 45 species of waterfowl that are native breeders in North America range from the southern United States to Alaska and the Canadian arctic (7). As suggested by the term "waterfowl," wetlands are an essential habitat component for these species throughout their ranges. Most North American species are migratory, breeding in the northern United States and Canada during the spring and summer and migrating along traditional pathways to wintering grounds in the United States and, for some species, Mexico and even Central and South America. The prairies of north-central United States and south-central Canada are an extremely important breeding area for many duck species, whereas many goose and swan species breed farther north in Alaska and the Canadian arctic.

Waterfowl hunting and associated management efforts have a long history in North America. This history has been closely linked with scientific investigations of waterfowl ecology that have guided waterfowl management over the years. Monitoring programs were established for the purpose of estimating key demographic parameters for waterfowl populations (16, 59, 82), and biological understanding has been a cornerstone of many programs for the management of waterfowl habitats and the annual setting of waterfowl hunting regulations in North America. Few examples exist of a more successful long-term collaboration between wildlife research and management.

Despite this success, there remains substantial uncertainty about the effects of management on waterfowl populations. For example, waterfowl harvest management in North America continues to be limited by a less than complete understanding about the ecological relationships linking biological processes to harvest mortality (3, 64, 67). Key relationships in the process of reproduction are yet to be fully understood, knowledge about the ecology of migration and patterns of movement is incomplete, and the role and importance of randomness in the environment awaits more comprehensive assessment.

This uncertainty extends to North American waterfowl populations and their sustainability (their ability to persist indefinitely, or at least into the forseeable future). The current health of North American waterfowl populations and their habitats varies from one species and location to another. Some arctic-nesting goose populations are increasing rapidly in abundance to the point where they are damaging habitat on the breeding grounds. Some duck populations have exhibited substantial declines, and breeding and wintering habitat for many duck species continues to be destroyed and degraded by agriculture and other human activities. Demands for high levels of sport harvest continue, and subsistence harvest (by indigenous peoples) is not well regulated or well monitored.

We believe that adaptive management (39, 41, 88, 89) offers the best approach to dealing with the various sources of uncertainty in future management efforts for North American waterfowl (50). In general terms, adaptive management involves (i) the choice of actions, taking into account uncertainty as to their consequences; (ii) monitoring and assessment of population dynamics; and (iii) use of the monitoring and assessment information in future decision-making. By this accounting, any management scenario that monitors the status of resources and tailors decisions accordingly can be described as adaptive. However, we use adaptive management more formally to represent a systematic process of using information generated by management actions to improve biological understanding and inform future decision-making. Passive and active forms of adaptive management are distinguished by the use of management actions to acquire useful information. In particular, we use the phrase adaptive waterfowl management in what follows to mean the active pursuit of information as an objective of the decision-making process. Thus delimited, adaptive waterfowl management can be described as an approach to dealing with the "dual control problem" of simultaneously pursuing harvest and conservation objectives on the one hand and the objective of improved understanding about population dynamics on the other (89).

Our purpose in this paper is first to review the history and evolution of waterfowl management in North America and then to describe the adaptive process that we propose for future management. We begin by presenting a conceptual basis for animal population management. We proceed to a brief historical review of the evolution of waterfowl management in North America, focusing on the close relationship between management approaches and ecological research on waterfowl populations. Following this review, we discuss some of the specific lessons learned from the North American experience and then conclude with a description of the proposed adaptive approach to waterfowl management.

CONCEPTUAL BASIS FOR ANIMAL POPULATION MANAGEMENT

We believe that there are four fundamental requirements for the informed management of any animal population. First, the manager must develop explicit goals or objectives (e.g. these might involve harvest and population size). Second, the manager must have the ability to implement management actions that are relevant to the attainment of goals (e.g. actions might involve hunting regulations and protection/management of wetlands). Third, the manager must develop a program to gather information on important state variables (e.g. population size) and goal-related variables (e.g. harvest) for the managed population. Fourth, the manager must have a hypothesis or model about the effects of management actions on state and goal-related variables.

Given these four components, informed management can be implemented as an iterative process. Periodically, the information-gathering program provides an estimate of system state and goal-related variable(s). This information is used in conjunction with the model(s) of the harvested population to decide what management action is best with respect to the specified management objectives. Conceptually, the management of an animal population is a simple and straightforward process, but numerous problems typically arise as it is implemented.

A SHORT HISTORY OF NORTH AMERICAN WATERFOWL MANAGEMENT

The following review is abbreviated and selective. It reflects the bias of our own experiences, which have been in the United States and have emphasized ducks.

Before 1930

Prior to the mid-1800s waterfowl were extremely abundant in North America (19, 71, 75), and hunting occurred throughout the year by both market and recreational hunters (19, 60, 75). Waterfowl appear to have been perceived as an infinite natural resource, meriting no management intervention.

The late 1800s and early 1900s were characterized by declining waterfowl numbers and the growing recognition that protection from excessive hunting and habitat loss was needed (5, 19, 28, 70). In the United States, the federal government was granted authority to implement management actions in the form of hunting regulations (5, 19, 86) and land acquisition and protection (31, 77). The objective of such management, as stated in the Migratory Bird Treaty Act of 1918, was to protect migratory birds, and it was further specified that other objectives such as hunting would be permitted only to the extent that they were compatible with protection (19, 86).

Research on waterfowl distribution and migratory habits was begun (5, 12, 19, 31), but little information accrued on waterfowl abundance (annual estimates of system state) or about effects of management actions on waterfowl populations. The existing (albeit limited) knowledge of waterfowl abundance and population ecology led to the establishment of a closed season for all species during the primary breeding season as well as total protection from hunting for some species existing in very low numbers.

1930–1950

Waterfowl numbers were low during much of this period, and the US government carried out responsibilities under the Migratory Bird Treaty Act, primarily through acquisition and management of waterfowl habitat and annual setting of hunting regulations (5, 17, 19, 28, 31). Decisions about the timing of the hunting season and where to purchase wetlands were aided by information on bird migration and distribution provided by waterfowl banding and winter survey data (5, 54, 59, 82). Winter survey data provided annual assessments of population status that were used in the development of hunting regulations (16), and band recovery data were used to provide indications of hunting intensity (53).

Regulations were restricted when populations were in decline (63, 86), indicating a clear effort to change management actions in response to changes in system state. There were no explicit models relating management actions to subsequent changes in population size. Instead, management actions were guided by the common-sense ideas that increased hunting mortality could lead to reductions in waterfowl abundance and that habitat acquisition and improvement could lead to increases in abundance.

1951–1975

From 1950–1975, the various data collection programs required to provide information for waterfowl management matured into probably the best such system for any continentally distributed animal population(s) in the world (59, 82). These programs reflect a productive collaboration between research and management and include: aerial breeding ground surveys (Figure 1) providing estimates of pond numbers, and estimates of adult population size and indices of brood numbers for prairie-nesting waterfowl species (5, 33, 59); a harvest survey consisting of a mail questionnaire survey and a parts (duck wings and goose tails) collection survey, providing estimates of the waterfowl harvest by species, sex, and age (5, 10, 58, 59); an operational banding program, data from which are used to estimate harvest rates and annual survival rates (9, 36);

Figure 1 Strata and transects for the North American Waterfowl Breeding Population and Habitat Survey (59).

and a winter survey providing indices of waterfowl numbers and distribution on the wintering grounds (59, 82).

Research using data from these surveys found evidence of a strong positive relationship between May and July pond numbers on prairie breeding areas and fall age ratio of mallards (*Anas platyrhynchos*), permitting prediction of numbers of young mallards in the fall using aerial survey estimates of numbers of ponds in the spring and summer (1, 18). Field research provided evidence of the ability of specific land management practices to increase waterfowl reproductive rates (6, 20). Regression-based estimates of the positive, linear relationship between hunting and overall mortality of mallards (36) were used to predict the total mortality, and then the total population size, expected to result from imposition of hunting regulations leading to specific band recovery rates (29).

Prior to this period, band recovery and survey data had led to the conclusion (54) that North American waterfowl followed four major flyways (migration paths and their associated wintering grounds; Figure 2). A flyway council system based on these geographic units was developed whereby state and federal (US, Canadian, Mexican) representatives were given a major role in

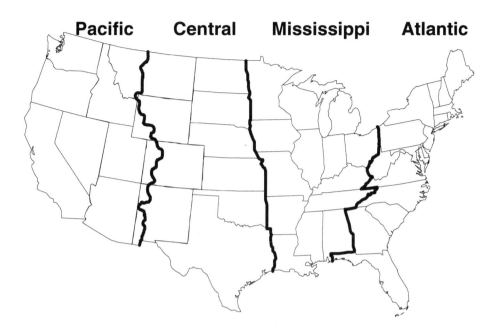

Pacific   Central   Mississippi   Atlantic

Figure 2   Administrative waterfowl flyways in the United States (5).

the coordination of flyway-specific management activities, including development of annual hunting regulations (43). The establishment of the flyway councils was accompanied by an increase in the complexity and geographic variation of hunting regulations (63, 86).

Some important waterfowl populations exhibited substantial fluctuations during this period. Each summer, waterfowl managers and researchers considered current population sizes and habitat conditions, together with the dual goals of hunting opportunity and healthy waterfowl populations, and developed a desired population size for the following spring. Population models were used to derive specific numerical harvest objectives termed "harvestable surplus" (16) or "allowable harvest" (29). Band recovery data were used to estimate the direct effect of hunting on waterfowl mortality and thus on waterfowl population dynamics (29, 36). Armed with objectives, timely information about system state, and estimated effects of management on system state, North American waterfowl managers varied regulations in direct response to population fluctuations. Indeed, except for the disagreements among different interest groups about population goals, waterfowl management in North America during the 1960s and early 1970s was viewed as an ideal example of scientific management of animal populations.

1976–1993

During the early 1970s, researchers recognized that previous methods for estimating annual survival rates from waterfowl band recovery data required very unrealistic assumptions, so they developed more reasonable estimation methods (9, 79, 80). More importantly, previous inferences about the positive, linear relationship between hunting and overall mortality rate (the "additive" mortality hypothesis) were shown by Anderson & Burnham (3) to be an inevitable consequence of sampling covariation between estimators; they thus destroyed all evidence of this relationship. New tests of the relationship between hunting and overall mortality for mallards, using reasonable statistical approaches, supported the "compensatory" mortality hypothesis, that for a certain range of hunting mortality rates, changes in hunting mortality were compensated by changes in nonhunting mortality such that overall mortality remained unchanged (3). Subsequent retrospective analyses of data from North American ducks have provided a mixture of inferences about the effects of hunting on overall mortality and population dynamics (64, 67, 86), whereas the few studies of geese have supported the additive mortality hypothesis (26, 34, 76).

The relationship between hunting and overall mortality rates is central to reasonable management, and the two extreme hypotheses describing this relationship (completely compensatory, completely additive) lead to very different management strategies (2, 92). The results of initial analyses supporting the compensatory mortality hypothesis (3) led to relatively liberal hunting regulations during the late 1970s. Regulations were experimentally stabilized in both the United States and Canada during 1979–1984 (83) at relatively liberal levels in an effort to investigate effects of environmental variation on population parameters. Duck populations were generally low during the 1980s (Figure 3), and regulations following the stabilized period were restrictive, reflecting a "risk-aversive conservatism" in the face of uncertainty about effects of hunting on waterfowl populations (86).

Research on the relationship between hunting mortality rate and the various components of hunting regulations led to the conclusion that major changes in regulations produced the intended changes in waterfowl harvest rates, but that there was little evidence of the effectiveness of so-called special regulations designed to fine-tune harvest management (64, 68). Continued research on the relationship between habitat management and waterfowl population dynamics (21, 22) led to the development of explicit models relating different types of habitat management to mallard reproductive rates (and hence population status) in prairie breeding areas of the United States (14, 15, 49).

In response to concern over low populations of several important duck species and continuing high rates of wetland habitat loss and degradation,

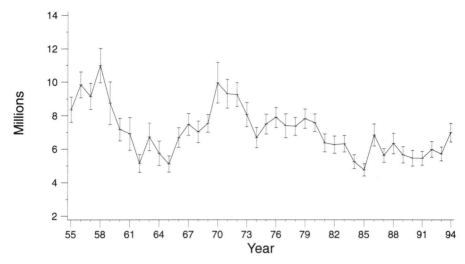

Figure 3   Breeding population estimates and 95% confidence limits for mallards in North America.

representatives of Environment Canada and the US Department of the Interior signed the North American Waterfowl Management Plan in 1986 (24). This document reflects the recognition that recovery and perpetuation of North American waterfowl populations depend on restoring wetlands and their associated ecosystems throughout the continent. The Plan lists explicit habitat objectives and numerical population goals for many waterfowl species (24, 87).

The last two decades have been characterized by low populations of several duck species, by large and increasing populations of several goose species, and by uncertainty about effects of management actions. Research results continued to influence management decisions, and the inconclusive nature of many such results was a major source of management uncertainty. During this period, management goals were periodically presented, and the data collection programs continued to provide useful information about waterfowl population status. The importance of quality habitat to waterfowl populations was generally accepted, as reflected in the North American Waterfowl Management Plan.

LESSONS LEARNED

We believe that waterfowl management in North America has been successful in many respects. Most waterfowl populations remain healthy, millions of hectares of habitat have been purchased or are under conservation easement, and sport harvest is carefully monitored and regulated. A number of important reasons explain this success: Various pieces of legislation vested management jurisdiction with federal governments and curtailed commercial hunting; treaties provided for international cooperation in migratory bird management; funding mechanisms, including the sale of hunting permits, facilitated the purchase and management of important habitats; large-scale monitoring and research programs were implemented, and for the most part, harvest levels were conservative.

Our overall positive assessment notwithstanding, we believe that substantial room for improvement exists in North American waterfowl management. In the following discussion we point to some of the lessons that emerge from our examination of the history of waterfowl management in North America in hopes that they may be useful in our future management efforts to ensure sustainable waterfowl populations. We have organized this discussion into two general categories: (i) management objectives and the process of making decisions, and (ii) the relationship between management actions and population status.

Management Objectives and Decision-Making

The treaties between the United States and Canada and the United States and Mexico provide only a broad mandate for migratory bird conservation. The establishment of specific management objectives is the responsibility of those federal, state, and provincial agencies vested with management authority. The mandated objectives of federal waterfowl management are clear in specifying protection and conservation of migratory bird populations first and sport hunting second. However, the existence of dual, potentially antagonistic, objectives leaves much room for discussion and argument.

For example, in years of favorable habitat conditions, there will typically be many different sets of hunting regulations (and resultant harvest rates) that permit sport harvest and that should also result in population growth. Complete specification of this management problem requires the assignment of "weights" reflecting the relative importance of the population and harvest objectives. The flyway council system permits input to decisions about hunting regulations from federal and state government agencies and from private organizations ranging from private hunting groups to strict protectionist organizations. We believe that much of the controversy that has arisen in the development of duck hunting regulations over the years has resulted from the different relative weights or values placed on these two objectives by various agencies and interest groups.

We believe further progress in waterfowl management will require clearly stated objectives that identify measurable responses. We fully appreciate that development of explicit objectives will be extremely difficult. However, in the absence of explicit objectives, it is not possible objectively to compare alternative management choices or to gauge management performance. Objectives also help define and bound the extent of the ecological, social, and economic models that are necessary for evaluating alternative management strategies. Ultimately, consensus on specific objectives may not be possible, but managers will nonetheless benefit from a better understanding of the nature and breadth of the desires of the various resource-user groups.

The operational aspects of decision-making in waterfowl management, particularly those involving harvest regulation, have been well documented (8, 17, 28, 29, 85). Currently, regulations governing sport harvest are promulgated annually in Canada, the United States, and Mexico in an elaborate process that is designed to elicit input from state, provincial, and federal conservation agencies. No such formal process exists for decisions regarding habitat management, although joint ventures under the North American Waterfowl Management Plan provide some opportunity for review and coordination of activities. Rather than discuss the details of these processes, we believe it is more instructive to focus on the conceptual aspects of making waterfowl management decisions.

Both science and management make use of statistical inference, but the different objectives of these two endeavors may lead to different statistical perspectives and approaches. One such difference in perspective is related to the treatment of Type I and II error rates. In science there is a strong bias against Type I errors, in which a null hypothesis (the hypothesis of no difference) is mistakenly rejected. Thus, the investigator typically assigns a low probability (e.g. 0.05) for Type I errors, despite the fact that lower probabilities of Type I errors produce higher probabilities for Type II errors (failures to reject false null hypotheses and, hence, to detect real differences). This tendency, when applied in waterfowl harvest management, has sometimes placed the burden of proof on those charged with resource maintenance, rather than on those seeking higher levels of exploitation (67, 86). In many instances, we probably should be more concerned with Type II errors, where a real response to management or the environment goes undetected. We should assess risks associated with the two types of errors and establish error rates accordingly, rather than relying on the traditional error rates used in the scientific literature (37).

Decision-making in the face of uncertainty implies risk, and we should make efforts to evaluate the risks associated with alternative decisions. Although management at the federal level has focused on the risk of declines in waterfowl abundance, such management has not dealt with risk in a consistent manner. Perhaps the greatest need for the sustainable management of waterfowl, as with other renewable resources, is a strategic plan for coping with the inherent uncertainty and risk in the decision-making process (38, 88). Methods of Bayesian inference and decision may be helpful in addressing this and other problems faced by waterfowl managers (42, 96).

Investigating and Modeling the Relationship Between Management Actions and Population Status

In our earlier outline of the conceptual basis for management, we specified the need for two kinds of information: (i) periodic information about variables related to both the state of the managed system (e.g. population size) and management goals (e.g. number of birds harvested), and (ii) information about the relationship between management actions and population status. The data-collection programs implemented for North American waterfowl populations (5, 17, 59, 82) meet the first need.

In response to the second information need, we have often estimated key relationships using retrospective studies that compute measures of association between historical changes in relevant demographic variables and management actions (66). These studies lack replication and random assignment of treatments to experimental units, two of the key features of manipulative experimentation (25). Inferences from these retrospective studies are thus weak, admitting alternative explanations for observed changes in population response variables (64–66, 68).

Some characteristics of waterfowl harvest management in North America have been especially detrimental to efforts to understand effects of management actions through retrospective analyses. For example, our ability to draw inferences from retrospective studies has been limited by the historical tendency to manage for population stability by liberalizing hunting regulations when waterfowl were abundant and restricting regulations when waterfowl populations were low (50, 67). This harvest strategy has produced a large-scale confounding of environmental, density-dependent, and harvest effects (Figure 4). We agree with the many investigators who have recommended either experimental hunting seasons or an adaptive management approach as means of drawing stronger inferences (3, 4, 50, 64–68, 95).

In many cases, the sheer number and complexity of management actions has overwhelmed managers' abilities to evaluate their effects (68). Complexity of hunting regulations (86) has contributed greatly to uncertainty about management effects, while it is not clear that it has increased hunter satisfaction. We recommend restricting the number of management options to a relatively small number of very different alternatives, because the costs of learning about the effects of many small, "fine-tuning" changes in regulations likely would be prohibitive, especially when viewed in the context of the relatively small gains that might arise from their use (68).

Both the implementation of management decisions regarding migratory birds and the evaluation of effects of these actions are made more difficult by

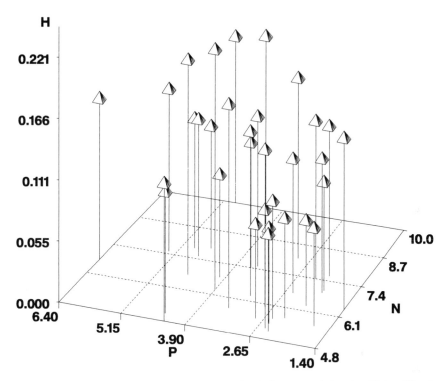

Figure 4   Harvest rates (*H*) of adult male mallards in relation to population size (*N*, in millions) and pond numbers (*P*, in millions) in Prairie Canada.

problems of geographic scale. Waterfowl populations rarely form discrete units, and heterogeneity among populations sharing study area can make inference about management effects difficult (44, 46). We recommend that experimentation or adaptive management consider scale problems in the implementation of different management treatments, trying to direct studied management actions at specific groups of birds, to the extent possible.

In addition to key relationships that have received (usually retrospective) study, some important relationships have received little study under any approach, and these merit additional attention. Possible density-dependence of reproductive (1, 23, 47, 51) and survival (1, 3, 47, 64, 67) rates in duck populations has received insufficient attention, despite its pivotal importance in defining population responses to management.

Decisions about habitat acquisition and management require information about the relationship between habitat characteristics and waterfowl survival and reproduction. Some investigations of the relationship between specific habitat characteristics and components of reproductive rate have been completed (20, 21, 47), and results have been incorporated in population modeling efforts (13, 14, 15, 49). The relationship between habitat and survival probability, however, has not been well-studied.

We have devoted inadequate attention to functional relationships involving humans (57), such as the relationship between hunting regulations and hunter participation and the positive influence of hunter numbers on waterfowl population status through provision of funds for habitat acquisition and management. Miller & Hay (61) and Hochbaum & Walters (40) provide good examples of considering waterfowl hunters in modeling efforts.

Our "knowledge" of effects of management actions on waterfowl populations can be encoded in models, and these models can then be used to consider consequences of alternative actions. Uncertainty associated with these models translates directly into uncertainty about the appropriateness of management decisions. We believe that the degree of uncertainty about management effects has strongly influenced the degree to which explicit population models have been used to guide waterfowl management during the last three decades. During the 1960s, when the relationship between hunting regulations and population status was believed to be known with high certainty, explicit population models played an important role in management decisions (29). Results of Anderson & Burnham (3) did not support these models, and explicit population models have seen only limited use in guiding harvest management actions during the last two decades (95).

Although uncertainty of key functional relationships is a legitimate and important problem, we do not believe that it should be used as a rationale for not engaging in model development and use. We believe that explicit models are important in providing a clear basis for management decisions and making predictions that can be used as the basis for future learning.

We believe that mechanistic models may make better choices for management applications than are phenomenological models (empirical models that simply describe an observed statistical relationship, without reference to the mechanism responsible for the relationship). Mechanistic models have a greater probability of providing accurate predictions outside the range of conditions experienced during model development. This conclusion emerges from a consideration of model forms that characterize the relationship between harvest rate and annual survival rate (50).

We also believe that models incorporating spatial dynamics of waterfowl populations will become increasingly important as issues of population distribution and harvest allocation come to the forefront in discussions of waterfowl management objectives. Some work has been completed on estimating area-specific rates of survival and movement for waterfowl (35, 78) and on incorporating such estimates into population models describing spatial dynamics (52, 90).

Finally, we repeat the recommendation of Conroy (11) that models subjected to few or no validation efforts should not be used unquestioned as the basis for management decisions. Uncertainty about key relationships leads us to advocate the approach of considering multiple models (rather than a single most-probable model) in the development of management strategies and then assessing relative credibility by comparing competing predictions with subsequent observations.

MANAGING IN THE FACE OF UNCERTAINTY

Components of Uncertainty in Waterfowl Management

There are at least four identifiable attributes of waterfowl biology that generate uncertainty and motivate the need to account for uncertainty in waterfowl management. We identify these sources of uncertainty using the terminology of operations research and decision theory in order to emphasize that the specific problems of waterfowl management fall within a broad class of problems associated with management of stochastic systems.

The first source of uncertainty is uncontrollable (and possibly unrecognized) *environmental variation,* which influences biological processes and induces stochasticity in population dynamics (1, 2, 16, 18, 19, 29, 47, 66, 71, 72). For example, weather variables and habitat conditions on breeding and wintering grounds can influence reproductive rates (1, 2, 18, 30, 32, 47, 51, 72, 74), survival rates (47), and migration and distribution patterns (45, 69).

The second source of uncertainty is limited knowledge about underlying biological mechanisms and about relationships between management actions and population status, identified in what follows as *structural uncertainty.* An example is the management of waterfowl harvests, for which there is a substantial lack of agreement as to which hypothesis (i.e. "additive" or "compensatory") best describes the relationship between harvest rate and annual survival rate (3, 64, 67). Because of the cost associated with such uncertainty (50), it is important to seek its elimination as a management goal, along with other traditional harvest goals.

Uncertainty about population status, referred to as *partial observability,* reflects imprecision in the monitoring of a biological system. Such uncertainty imposes limits on harvest management, even if one understands with certainty the underlying biological mechanisms and has total control over harvest rates. Partial observability limits the ability to recognize the need for protection, or to respond to utilization opportunities when they occur.

Partial controllability expresses the fact that management decisions only partially control the actual magnitude of the corresponding action [e.g. harvest regulations control actual harvest rates (and harvest effects) only within certain limits of precision]. The inability to specify harvest rates accurately can limit both short-term management performance and the reduction of structural uncertainty, irrespective of monitoring precision.

The uncertainty factors listed above are operative on any biotic resource subject to management. Indeed, in recent years, management of renewable resources has increasingly recognized the need to account for uncertainty (39, 62, 88, 93). It is most fortunate that by now a powerful statistical and mathematical theory is available for the treatment of uncertainty in dynamic systems (84), and computer software is being developed for assessment of uncertain systems (55, 56, 94).

A Systematic Approach to Adaptive Waterfowl Management

Adaptive management was defined in the Introduction. A technical specification of adaptive waterfowl management involves the following components:

MANAGEMENT OPTIONS   An array of potential management actions must be available for decision-making at each decision point in some relevant time frame. For example, adaptive harvest management might include a range of potential regulations from "restrictive" to "liberal", with the proviso that (i) the regulations represent realistic alternatives, and (ii) they include enough variation to elicit differential population responses. We use a_t to represent the management action in year t, and $\underline{A}$ to represent a sequence $\{a_1, \ldots, a_T\}$ of actions over a time frame T. The sequence $\underline{A}$, sometimes called a management strategy or policy (91), might consist of a series of decisions about land management on wildlife refuges, along with the annual setting of harvest regulations. The management of waterfowl ultimately consists of policy choices in accordance with management objectives, recognizing that the action specified at a particular time should be tailored to population and habitat conditions.

MODEL SET   An adaptive approach recognizes a collection of alternative biological mechanisms for population dynamics, with uncertainty as to which is most appropriate for the population under consideration. These are represented by dynamic population models, each model predicting population responses to management as functions of initial population status, environmental conditions, and management actions. Population dynamics are expressed by

$$N_{t+1} = N_t + G_i\,(N_t, a_t, \underline{e}_t, z_t) \qquad\qquad 1.$$

for model m_i, where $\underline{e}_t$ are time-varying environmental or habitat conditions, the random variable z_t represents a white noise process, and $G_i\,(N_t, a_t, \underline{e}_t, z_t)$ is the net population growth from t to $t+1$. It sometimes is useful to express environmental or habitat conditions similarly,

$$\underline{e}_{t+1} = \underline{e}_t + \underline{E}_i(\underline{e}_t, a_t, \underline{z}_t), \qquad\qquad 2.$$

and a generic representation includes both population and environmental state variables in a single-state transition equation:

$$\underline{x}_{t+1} = \underline{x}_t + \underline{F}_i(\underline{x}_t, a_t, \underline{z}_t), \qquad\qquad 3.$$

recognizing that a_t can influence either population status or habitat conditions or both.

MONITORING PROGRAM   To assess the state of the system and to gauge model performance in tracking population dynamics, some level of population monitoring is required. Let y_t represent data that are recorded about the population at time t, with the value y_t stochastically dependent on the system state $\underline{x}_t$:

$$y_t = \underline{g}(\underline{x}_t) + \varepsilon_t, \qquad\qquad 4.$$

and with the random variable ε_t independent of $\underline{z}_t$. Monitoring data accumulate through time, and each year additional data are added to an extant database Y_t:

$$Y_{t+1} = \{Y_t, \underline{y}_{t+1}\}. \qquad\qquad 5.$$

In general, the more sophisticated and precise the program for monitoring population status (i.e. the smaller the variances in ε_t), the easier it is to resolve uncertainties about biological mechanisms and thus improve the management of waterfowl.

MEASURES OF UNCERTAINTY   Key to an adaptive approach is the tracking of the confidence (or equivalently, the uncertainty) associated with each population model under consideration. Here we use $p_i(t)$ to represent the likelihood at time t that model m_i is the most appropriate for describing population dynamics. This notation indicates that the likelihoods vary among models, and the likelihoods change through time as the population responds to management actions. Variation in the likelihood values through time is based on the comparison of monitoring data and model predictions and therefore is informed by monitoring data: $p_i(t) = p(m_i \mid Y_t)$.

OBJECTIVE FUNCTION   An objective function is a formal expression of management objectives and is needed to compare and evaluate different management policies. The function provides a measure of the effect of different management policies and thereby permits identification of optimal policies. For example, a useful objective function for harvest regulation might include the total predicted harvest over the timeframe, as influenced by regulatory strategies and model likelihood values. An objective function for habitat management might include both resource benefits and possible management costs. We use $V(\underline{A} \mid Y_0, \underline{p}_0)$ to denote the value of the objective function, conditional on accumulated monitoring information and current likelihood values. For example, an objective based on total accumulated harvest might be:

$$\underline{V}(\underline{A} \mid Y_0, \underline{p}_0) = \sum_i p_0(i) E\left(\sum_{t=0}^{T} [H_i (N_t, a_t) \mid Y_0]\right), \qquad 6.$$

with $E[\Sigma_t H_i(N_t, a_t) \mid Y_0]$ the total expected harvest for model m_i, given the current data Y_0 and action a_t at time t.

With these components, the adaptive management of waterfowl can be expressed in terms of dynamic optimization. Thus, waterfowl managers seek a policy $\underline{A}$ over the timeframe $\underline{T}$ that maximizes $V(\underline{A} \mid Y_0, \underline{p}_0)$ subject to:

$$\underline{x}_{t+1} = \underline{x}_t + \underline{F}_i(\underline{x}_t, a_t, \underline{z}_t) \qquad 7.$$

$$\underline{y}_{t+1} = \underline{g}(\underline{x}_t) + \underline{\varepsilon}_t, \qquad 8.$$

recognizing that particular actions a_t in $\underline{A}$ at each point in time are dependent on accumulated monitoring information and the model likelihoods at that time.

Expressing the adaptive management problem in this way allows us to use the theory and methods of dynamic estimation and optimization (84), particularly the procedures for analysis of Markov decision processes (73, 94).

Advantages and Limitations in Managing Waterfowl Adaptively

An important advantage to using an adaptive approach to dealing with uncertainty in waterfowl management is that it requires making explicit the factors entering into the decision-making process, thus reducing ambiguities. Other advantages accrue because of the dynamic nature of an adaptive approach, with an accounting for population changes through time and for future consequences of present actions. A dynamic framework involving an extended management timeframe requires of management that it be future-oriented, balancing the current benefits of resource use against future benefits accruing to resource conservation and sustainability. Another benefit of an adaptive approach is that it establishes a framework to include nonsportsmen and others with strong conservation interests, without excluding those who engage in the sport hunting of warterfowl. Finally, adaptive management provides a framework in which managers and researchers can work cooperatively on issues that are important to each group. Value is ascribed to information and understanding to the extent that they contribute to the goals of resource management, so that biological monitoring, assessment, and research are recognized as contributing to improved management.

Recognizing the many advantages of an adaptive approach to waterfowl management, it also is useful to recognize that there are some potentially important limitations as well. One such concern involves identification of the models to be used (50). It is not likely that management can be informative of population dynamics, if the models under consideration are inadequate. But the recognition of, and agreement on, reasonable candidate models can be problematic and likely require considerable effort, creativity, and goodwill among stakeholders. Biological relationships controlling population dynamics also can change through time. If the rate of change in key relationships is similar to the rate of learning through adaptive management, then learning essentially becomes impossible. The potential for this problem is real. For example, evidence from banding assessments suggests compensatory patterns of hunting mortality for mallards in the 1970s (3), but additive effects in the 1980s (81). Despite this potential limitation, adaptive management should still be preferable to static approaches to management when key biological relationships change over time. Beyond these technical problems, full implementation of adaptive waterfowl management requires agreement among stakeholders about objectives, constraints, model sets, and management options, as well as an institutional environment conducive to objective management. The necessary cooperation among groups can be developed only through participation and interaction. We are hopeful that such cooperation will be achieved and believe that adaptive management provides a framework for rational waterfowl management that meets the needs for change and offers an excellent opportunity to achieve sustainability.

Literature Cited

1. Anderson DR. 1975. Population ecology of the mallard. V. Temporal and geographic estimates of survival, recovery, and harvest rates. *US Fish Wildl. Serv. Res. Publ. 125.* 110 pp.
2. Anderson DR. 1975. Optimal exploitation strategies for an animal population in a Markovian environment: a theory and an example. *Ecology* 56: 1281–97
3. Anderson DR, Burnham KP. 1976. Population ecology of the mallard. VI. The effect of exploitation on survival. *US Fish Wildl. Serv. Res. Publ. 128.* 66 pp.
4. Anderson DR, Burnham KP, Nichols JD, Conroy MJ. 1987. The need for experiments to understand population dynamics of American black ducks. *Wildl. Soc. Bull.* 15:282–84
5. Anderson DR, Henny CJ. 1972. Population ecology of the mallard. I. A review of previous studies and the distribution and migration from breeding areas. *US Fish Wildl. Serv. Res. Publ. 105.* 166 pp.
6. Balser DS, Dill HH, Nelson HK. 1968. Effect of predator reduction on waterfowl nesting success. *J. Wildl. Manage.* 32:669–82
7. Bellrose FC. 1976. *Ducks, Geese & Swans of North America.* Harrisburg, PA: Stackpole. 543 pp.
8. Blohm RJ. 1989. Introduction to har-

vest—understanding surveys and season setting. *Proc. Int. Waterfowl Symp.* 6: 118–33

9. Brownie C, Anderson DR, Burnham KP, Robson DS. 1978. Statistical inference from band recovery data: a handbook. *US Fish Wildl. Serv. Res. Publ. 130.* 212 pp.

10. Carney SM. 1992. *Species, Age and Sex Identification of Ducks Using Wing Plumage.* Washington DC: US Fish Wildl. Serv. 144 pp.

11. Conroy MJ. 1993. The use of models in natural resource management: prediction, not prescription. *Trans. N. Am. Wildl. Nat. Res. Conf.* 58:509–19

12. Cooke WW. 1906. Distribution and migration of North American ducks, geese, and swans. *USDA Biol. Surv. Bull. 26.* 90 pp.

13. Cowardin LM, Johnson DH. 1979. Mathematics and mallard management. *J. Wildl. Manage.* 43:18–35

14. Cowardin LM, Johnson DH, Frank AM, Klett AT. 1983. Simulating results of management actions on mallard production. *Trans. N. Am. Wildl. Nat. Res. Conf.* 48:257–72

15. Cowardin LM, Johnson DH, Shaffer TL, Sparling DW. 1988. Applications of a simulation model to decisions in mallard management. *US Fish Wildl. Serv. Tech. Rep. No. 17.* 28 pp.

16. Crissey WF. 1957. Forecasting waterfowl harvest by flyways. *Trans. N. Am. Wildl. Conf.* 22:256–68

17. Crissey WF. 1963. Exploitation of migratory waterfowl populations in North America. In *Proc. European Meeting on Wildfowl Conservation, 1st,* St. Andrews, Scotland, pp. 105–22. London: Nature Conservancy

18. Crissey WF. 1969. Prairie potholes from a continental viewpoint. Saskatoon Wetlands Seminar. *Can. Wildl. Serv. Rep. Ser.* 6:161–71

19. Day AM. 1949. *North American Waterfowl.* Harrisburg, PA: Stackpole. 329 pp.

20. Duebbert HF, Kantrud HA. 1974. Upland duck nesting related to land use and predator reduction. *J. Wildl. Manage.* 38:257–65

21. Duebbert HF, Lokemoen JT. 1976. Duck nesting in fields of undisturbed grass-legume cover. *J. Wildl. Manage.* 40:39–49

22. Duebbert HF, Lokemoen JT. 1980. High duck nesting success in a predator-reduced environment. *J. Wildl. Manage.* 44:428–37

23. Dzubin A. 1969. Comments on carrying capacity of small ponds for ducks and

possible effects of density on mallard production. Saskatoon Wetlands Seminar. *Can. Wildl. Serv. Rep. Ser.* 6:138–60

24. Environment Canada and US Department of the Interior. 1986. *North American Waterfowl Management Plan.* Washington, DC: US Fish Wildl. Serv. 19 pp.

25. Fisher RA. 1971. *The Design of Experiments.* New York: Hafner. 248 pp.

26. Francis CM, Richards MH, Cooke F, Rockwell RF. 1992. Long-term changes in survival rates of lesser snow geese. *Ecology* 73:1346–62

27. Deleted in proof.

28. Gabrielson IN. 1941. What is behind the waterfowl regulations. *Report of the Special Senate Committee on the Conservation of Wildlife Resources,* 77th Congress. Washington, DC. 34 pp.

29. Geis AD, Martinson RK, Anderson DR. 1969. Establishing hunting regulations and allowable harvest of mallards in the United States. *J. Wildl. Manage.* 33:848–59

30. Hammond MC, Johnson DH. 1984. Effects of weather on breeding ducks in North Dakota. *US Fish Wildl. Serv. Tech. Rep. No. 1.* 17 pp.

31. Hawkins AS. 1984. The U.S. response. In *Flyways,* ed. AS Hawkins, RC Hanson, HK Nelson, HM Reeves, pp. 2–9. Washington, DC: US Dep. Interior.

32. Heitmeyer ME, Fredrickson LH. 1981. Do wetland conditions in the Mississippi Delta hardwoods influence mallard recruitment. *Trans. N. Am. Wildl. Nat. Res. Conf.* 46:44–57

33. Henny CJ, Anderson DR, Pospahala RS. 1972. Aerial surveys of waterfowl production in North America. 1955–71. *US Fish Wildl. Serv. Spec. Sci. Rep., Wildl. 160.* 48 pp.

34. Hestbeck JB. 1994. Survival of Canada geese banded in winter in the Atlantic Flyway. *J. Wildl. Manage.* 58:748–56

35. Hestbeck JB, Nichols JD, Malecki RA. 1991. Estimates of movement and site fidelity using mark-resight data of wintering Canada Geese. *Ecology* 72:523–33

36. Hickey JJ. 1952. Survival studies of banded birds. *US Fish Wildl. Serv. Spec. Sci. Rep., Wildl. 15.* 177 pp.

37. Hilborn R. 1992. Can fisheries agencies learn from experience? *Fisheries* 17:6–14

38. Hilborn R, Pikitch EK, Francis RC. 1993. Current trends in including risk and uncertainty in stock assessment and

harvest decisions. *Can. J. Fish. Aquat. Sci.* 50:874–80

39. Hilborn R, Walters CJ. 1992. *Quantitative Fisheries Stock Assessment.* New York: Chapman & Hall. 570 pp.

40. Hochbaum GS, Walters CJ. 1984. Components of hunting mortality in ducks: a management analysis. *Can. Wildl. Serv. Occas. Pap. 52.* 27 pp.

41. Holling CS, ed. 1978. *Adaptive Environmental Assessment and Management.* New York: Wiley. 363 pp.

42. Iversen GR. 1984. Bayesian statistical inference. *Quant. Soc. Sci. Appl. Ser. 07–043.* Newbury Park, CA: Sage. 79 pp.

43. Jahn LR, Kabat C. 1984. Origin and role. In *Flyways,* ed. AS Hawkins, RC Hanson, HK Nelson, HM Reeves, pp. 374–83. Washington, DC: US Dep. Interior

44. Johnson DH, Burnham KP, Nichols JD. 1986. The role of heterogeneity in animal population dynamics. *Proc. Int. Biometrics Conf., 13th.* 15 pp.

45. Johnson DH, Grier JW. 1988. Determinants of breeding distributions of ducks. *Wildl. Monogr. 100.* 37 pp.

46. Johnson DH, Nichols JD, Conroy MJ, Cowardin LM. 1988. Some considerations in modeling the mallard life cycle. In *Waterfowl in Winter,* ed. MW Weller, pp. 9–20. Minneapolis, MN: Univ. Minn. Press

47. Johnson DH, Nichols JD, Schwartz MD. 1992. Population dynamics of breeding waterfowl. In *Ecology and Management of Breeding Waterfowl,* ed. BDJ Batt, AD Afton, M G Anderson, CD Ankney, DH Johnson, et al, pp. 446–85. Minneapolis, MN: Univ. Minn. Press

48. Deleted in proof.

49. Johnson DH, Sparling DW, Cowardin LM. 1987. A model of the productivity of the mallard duck. *Ecol. Model.* 38: 257–75

50. Johnson FA, Williams BK, Nichols JD, Hines JD, Kendall WL, et al. 1993. Developing an adaptive management strategy for harvesting waterfowl in North America. *Trans. N. Am. Wildl. Nat. Res. Conf.* 58:565–83

51. Kaminski RM, Gluesing EA. 1987. Density- and habitat-related recruitment in mallards. *J. Wildl. Manage.* 51:141–48.

52. Koford RR, Sauer JR, Johnson DH, Nichols JD, Samuel MD. 1992. A stochastic population model of mid-continent mallards. In *Wildlife 2001: Populations,* ed. DR McCullough, RH Barrett, pp. 170–81. New York: Elsevier

53. Lincoln FC. 1930. Calculating waterfowl abundance on the basis of banding returns. *USDA Circ. 118.* 4 pp.

54. Lincoln FC. 1935. The waterfowl flyways of North America. *USDA Circular 342.* 12 pp.

55. Lubow B. 1993. *Stochastic Dynamic Programming (SDP): User's Guide.* Fort Collins, CO: Colo. Coop. Fish & Wildl. Res. Unit. 119 pp.

56. Lubow B. 1995. SDP: Generalized software for solving stochastic dynamic optimization problems. *J. Wildl. Manage.* In press

57. Ludwig D, Hilborn R, Walters C. 1993. Uncertainty, resource exploitation, and conservation: lessons from history. *Science* 260:17,36

58. Martin EM, Carney SM. 1977. Population ecology of the mallard. IV. A review of duck hunting regulations, activity, and success, with special reference to the mallard. *US Fish Wildl. Serv. Res. Publ. 130.* 137 pp.

59. Martin FW, Pospahala RS, Nichols JD. 1979. Assessment and population management of North American migratory birds. In *Environmental Biomonitoring, Assessment, Prediction, and Management—Certain Case Studies and Related Quantitative Issues,* ed. J Cairns Jr, GP Patil, WE Walters, pp. 187–239. Fairland, MD: Int. Coop. Publ. House

60. Meanley B. 1982. *Waterfowl of the Chesapeake Bay Country.* Centreville, MD: Tidewater. 210 pp.

61. Miller JR, Hay MJ. 1981. Determinants of hunter participation: duck hunting in the Mississippi Flyway. *Am. J. Agric. Econ.* 63:677–84

62. Murphy DD, Noon BD. 1991. Coping with uncertainty in wildlife biology. *J. Wildl. Manage.* 55:773–82

63. Nelson HK, Bartonek JC. 1990. History of goose management in North America. *Trans. N. Am. Wildl. Nat. Res. Conf.* 55:286–92

64. Nichols JD. 1991. Responses of North American duck populations to exploitation. In *Bird Population Studies: Their Relevance to Conservation and Management,* ed. CM Perrins, J-D Lebreton, G Hirons, pp. 498–525. Oxford, UK: Oxford Univ. Press

65. Nichols JD. 1991. Science, population ecology, and the management of the American black duck. *J. Wildl. Manage.* 55:790–99

66. Nichols JD. 1991. Extensive monitoring programmes viewed as long-term population studies: the case of North American waterfowl. *Ibis* 133: Suppl. 1:88–98

67. Nichols JD, Conroy MJ, Anderson DR,

Burnham KP. 1984. Compensatory mortality in waterfowl populations: a review of the evidence and implications for research and management. *Trans. N. Am. Wildl. Nat. Res. Conf.* 49:535–54

68. Nichols JD, Johnson FA. 1989. Evaluation and experimentation with duck management strategies. *Trans. N. Am. Wildl. Nat. Res. Conf.* 54: 566–93

69. Nichols JD, Reinecke KJ, Hines JE. 1983. Factors affecting the distribution of mallards wintering in the Mississippi Alluvial Valley. *Auk* 100:932–46

70. Oberholser HC. 1917. The great plains waterfowl breeding grounds and their protection. In *USDA Yearbook*, pp. 197–204. Washington, DC: US Govt. Print. Off.

71. Phillips JC, Lincoln FC. 1930. *American Waterfowl*. Boston, MA: Houghton Mifflin. 312 pp.

72. Pospahala RS, Anderson DR, Henny CJ. 1974. Population ecology of the mallard. II. Breeding habitat conditions, size of the breeding population, and production indices. *US Fish Wildl. Serv. Res. Publ. 115.* 73 pp.

73. Puterman ML. 1994. *Markov Decision Processes: Discrete Stochastic Dynamic Programming*. New York: Wiley. 649 pp.

74. Raveling DG, Heitmeyer ME. 1989. Relationships of population size and recruitment of pintails to habitat conditions and harvest. *J. Wildl. Manage.* 53:1088–1103

75. Reeves HM. 1966. Influence of hunting regulations on wood duck population levels. In *Wood Duck Management and Research: A Symposium*, ed. JB Trefethen, pp. 163–81. Washington, DC: Wildl. Manage. Inst.

76. Rexstad EA. 1992. Effect of hunting on annual survival of Canada geese in Utah. *J. Wildl. Manage.* 56:297–305

77. Salyer JC II, Gillett FG. 1964. Federal refuges. In *Waterfowl Tomorrow*, ed. JP Linduska, pp. 497–508. Washington, DC: US Dept. Interior

78. Schwarz CJ. 1993. Estimating migration rates using ring recoveries. In *The Use of Marked Individuals in the Study of Bird Population Dynamics: Models, Methods, and Software*, ed. J-D Lebreton, PM North, pp. 255–64. Berlin: Birkhauser Verlag

79. Seber GAF. 1970. Estimating time-specific survival and reporting rates for adult birds from band returns. *Biometrika* 57:313–18

80. Seber GAF. 1972. Estimating survival rates from bird-band returns. *J. Wildl. Manage.* 36:405–13

81. Smith GW, Reynolds RE. 1992. Hunting and mallard survival, 1979–88. *J. Wildl. Manage.* 56:306–16

82. Smith RI, Blohm RJ, Kelly ST, Reynolds RE. 1989. Review of data bases for managing duck harvests. *Trans. N. Am. Wildl. Nat. Res. Conf.* 54:537–44

83. Sparrowe RD, Patterson JH. 1987. Conclusions and recommendations from studies under stabilized duck hunting regulations: management implications and future directions. *Trans. N. Am. Wildl. Nat. Res. Conf.* 52:320–26

84. Stengel RF. 1994. *Optimal Control and Estimation*. New York: Dover. 639 pp.

85. Trost R, Dickson K, Zavaleta D. 1993. Harvesting waterfowl on a sustained yield basis: the North American perspective. In *Waterfowl and Wetland Conservation in the 1990s—A Global Perspective. Int. Waterfowl and Wetlands Research. Bur. Spec. Pub. 26,* ed. M Moser, RC Prentice, J Van Vessem, pp. 106–12. Slimbridge, UK

86. US Department of the Interior. 1988. *Supplemental Environmental Impact Statement: Issuance of Annual Regulations Permitting the Sport Hunting of Migratory Birds*. Washington, DC: US Fish Wildl. Serv. 340 pp.

87. US Department of the Interior, Environment Canada, and Secretaria de Desarrollo Social Mexico. 1994. *1994 Update to the North American Waterfowl Management Plan. Expanding the Commitment*. Washington, DC: US Fish Wildl. Serv. 30 pp.

88. Walters CJ. 1986. *Adaptive Management of Renewable Resources*. New York: MacMillan. 374 pp.

89. Walters CJ, Hilborn R. 1978. Ecological optimization and adaptive management. *Annu. Rev. Ecol. Syst.* 9:157–88

90. Walters CJ, Hilborn R, Oguss E, Peterman RM, Stander JM. 1974. Development of a simulation model of mallard duck populations. *Can. Wildl. Serv. Occas. Pap. 20.* 35 pp.

91. Williams BK. 1982. Optimal stochastic control in natural resource management: framework and examples. *Ecol. Model.* 16:275–97

92. Williams BK. 1988. MARKOV: a methodology for the solution of infinite time horizon Markov decision processes. *Appl. Stochastic Models Data Anal.* 4: 253–71

93. Williams BK. 1989. Review of dynamic optimization methods in renewable

natural resource management. *Nat. Res. Model.* 3:137–216

94. Williams BK. 1995. Adaptive optimization and the harvest of biological populations. *Biometrics.* In review

95. Williams BK, Nichols JD. 1990. Modeling and the management of migratory birds. *Nat. Res. Model.* 4:273–311

96. Winkler RL. 1972. *Introduction to Bayesian Inference and Decision.* New York: Holt, Rinehart, & Winston. 563 pp.

Annu. Rev. Ecol. Syst. 1995. 26:201–24

THE ECOLOGICAL BASIS OF ALTERNATIVE AGRICULTURE

John Vandermeer

Department of Biology, University of Michigan, Ann Arbor, Michigan 48109

KEY WORDS: agroecosystems, alternative agriculture, sustainability, soils, pest control,
            multiple cropping

ABSTRACT

The critique of modern agriculture has spawned a host of alternatives, collectively known as the alternative agriculture movement. Its critics have been fierce, its proponents zealous. Making sense of the movement is similar to making sense of the original critique—always eclectic, sometimes contradictory, too often romantic, now and then nonsensical, and occasionally brilliant. This review discusses definitions of the alternative agriculture movement, substitutes for pest control, soil management, integration of all aspects of the farming operation, and the problem of conversion of one form to another.

INTRODUCTION

The critique of modern agriculture (33, 162) has most naturally spawned an alternative, or rather a host of alternatives, collectively known as the alternative agriculture movement (5, 19, 20, 103, 126). Its critics have been fierce, its proponents zealous. Making sense of the movement is similar to making sense of the original critique—always eclectic, sometimes contradictory, too often romantic, now and then nonsensical, and occasionally brilliant.

Ultimately the nature of "alternative agriculture" is not all that clear. Even the name used is as diverse as the proponents—"alternative," "sustainable," "holistic," "ecological," and "organic" agriculture, plus such all-encompassing titles as "permaculture" and "low input sustainable agriculture" (LISA). Concern has been voiced over the proper articulation of the concept (27, 31, 34, 51, 56, 68, 91, 103, 107, 110, 112, 164), and while all participants in this

201

0066-4162/95/1120-0201$05.00

debate incorporate a critique of the modern system, each has his or her own emphasis.

For example, Poincelot (138, p. 2) emphasizes resource conservation—"the main future threat to agriculture is a diminishing resource base." The goal of sustainable agriculture is "the elimination of agriculture's consumption and pollution of limited resources." At an even more general level, Widdowson (198, p. 1) notes that "Holistic agriculture is concerned in obtaining a correct grouping in farming systems which are in themselves sustainable.... The organization should be such as to give the fullest use of land resources and the best utilization of available labour and capital...."

The political development of the alternative program, in its broadest context, has been analyzed by Beus & Dunlap (17–19), who refer to a loose coalition of interest groups that share a fundamental criticism of modern industrial agriculture as the "externalities/alternatives" or "ex/al" coalition. They also note that this coalition "is viewed with considerable contempt by many in the traditional agricultural establishment" (citing 28, Vol. 19:432). Even more broadly, MacRae et al (111), in an ambitious attempt not only to define sustainable agriculture but to set out a program for its massive implementation in Canada, note that sustainability includes "nutrient and water cycles, energy flows, beneficial soil organisms, natural pest controls, and the humane treatment of animals, ... [to] ensure the well-being of rural communities, and to produce food that is nutritious and uncontaminated with products that might harm human and livestock health."

An alternative definitional scheme separates "sustainable" from "alternative," allowing that all agriculture is probably less sustainable than might be desired, and that alternative forms are always welcome when they will increase sustainability (132, 133). Such a scheme may help to defuse the tension that sometimes arises when modern agriculture is forced to face the problems it has created.

The debate about the definition of alternative agriculture remains vigorous. I suspect it is too early to be worried about formal definitions that might guide scientific work. It is better simply to leave the definitional debate to active discourse until the problem and its solutions are better understood (27, 32, 51), as suggested by the National Research Council, which characterized alternative farming practices as "... not a well-defined set of practices or management techniques. Rather, they are a range of technological and management options used on farms striving to reduce costs, protect health and environmental quality, and enhance beneficial biological interactions and natural processes" (126).

Much of what is relevant for the ecologist can be conveniently codified in a simple three-part classification—pest management, soil management (including water management), and integration (or what might be more marketably codified as integrated pest management, integrated soil management, and integrated farm management—the I3 program). Each of these includes a variety of techniques, all of which have some ecological basis. Many are based on well-known and popular ecological assumptions, whereas others are based on popular but discredited ideas. Some have been clearly demonstrated to be efficacious; others are assumed to be so without evidence. Many have been thoroughly studied from an ecological point of view; others await such serious study. My purpose here is to review the ecological bases of the various proposals and to assess the state of verified ecological knowledge about them.

SUBSTITUTING FOR PEST CONTROL

The key problem on which the post–World War II agricultural revolution focused was that of pests: The methodology for controlling pests, much like the methodology in warfare (149), was total annihilation. This was achievable through the use of new poisons that sprang ultimately from war research. That this technique could threaten human health and cause environmental damage was not generally recognized until Rachel Carson's *Silent Spring* was published in 1962; only in 1978 were the general ecological and evolutionary mechanisms behind those problems formulated for the lay public. The most important of these mechanisms was the now well-known pesticide treadmill (179)—pest resistance, pest resurgence, and the formation of secondary pest outbreaks—which led farmers into spraying ever greater quantities of pesticides (179).

In recognition of the pesticide treadmill, the problem of pest control became formulated as a program—integrated pest management (IPM). This formulation derived from a systematization of what had been a variety of issues. The key ideas, originally articulated by vandenBosch (179), were extremely simple: 1) Don't spray poisons unless it is necessary, and 2) manage the ecosystem in such a way that it doesn't become necessary. These principles remain at the foundation of almost all functioning IPM programs (97, 184).

Pests have always been problems, and their control a goal of production. Control through biological and cultural practices was diverse and universal (78). The ecological study of agricultural pest situations has become common for insect pests (88, 140, 144, 148, 184), for crop diseases (59, 121, 185, 189, 201), and for weeds (54, 84, 101). Application of ecological principles to solve the problems created by these pests has met with mixed success.

Application of ecological theory to biological control practice for insect populations has been an extremely active field since the explosion of theoretical ecology in the 1970s, with most theory based either on the Lotka Volterra predator-prey equations or the Nicholson-Baily model (15, 123). Both formal and informal theory have focused on the relatively obvious idea that a successful biological control agent must not itself go extinct if it is to perform its function for more than a short period of time. Thus, while practitioners seek

natural enemies that are efficient at killing the pest, the emphasis from ecological theory has been on agents that form stable equilibria with their prey,
but at prey densities that are below the economic threshold (15, 69). This view
has its detractors (123, 124) who note that evidence from actual case studies
of biological control do not suggest that concern over the population stability
of the introduced enemy and its pest has ever really been a component of any
real biological control, and that subtler modes of population regulation (such
as through a habitat mosaic) may provide the same practical effect as that of
formal population equilibrium (122). Debate continues over the interpretation
of data (71, 122).

In addition to the focus on stability of the predator-prey interaction in a
biological control system, a general consensus seems to have evolved regarding
two assumptions: 1) The control agent must be a specialist, and 2) the probability of success is far greater in a perennial crop system than an annual crop
system (159). Several authors have challenged this point of view (123, 135).
While it seems to be the case that most successful programs have been in
perennial systems (94), and most successful introductions of biological control
agents have been specialists, nothing in ecological theory convincingly suggests that such should always be the case. Recently it has been suggested that
the combination of a specialist to deal with pest outbreaks, and a generalist to
deal with pests during normal times, would frequently be the optimum strategy
(65).

The biological control literature also seems to have settled on a series of
enemy "qualities" to be sought in the search for a biological control agent.
These include synchrony of activity between control agent and pest, effectiveness at low prey density, reproductive capacity greater than that of the pest,
dispersal ability greater than that of the pest, ease of management, high searching ability, requirement of only a few pest individuals to complete the life
cycle, and climatic similarity in needs of control agent and pest (75, 123, 194).
While most of these qualities are sensible either from the point of view of
qualitative ecological insight or formal ecological theory, none has been subjected to the same detailed empirical examination that Murdoch and colleagues
have applied to the question of equilibrium and monophagy (122), and thus
these must be judged as tentative hypotheses still. Furthermore, other forms
of biological control, such as mass liberation and cultural control for conservation of natural enemies (11), are well known only from specific examples.
All of this suggests that the underlying ecological principles involved in
biological control are either so obvious they hardly need ecological study, or
they remain at the level of interesting speculations that may or may not turn
out to be useful or true upon detailed study.

Application of ecological principles to questions of cultural control of insect
pests has been equally vigorous as, if somewhat more eclectic than, biological

control. These techniques have included the management of vegetation texture (9, 11, 144), various forms of tillage (76), sanitation, careful planning of planting and harvesting dates, resistant varieties, fertilizer manipulations, and others (11). The ecological principles involved in these cultural control methods are many and varied, but many of them fall within the general category of plant apparency theory (52). The general idea is to reduce the plant's effective apparency with regard to the pest in question.

The ecology of diseases in agroecosystems has followed a distinct trajectory, despite the obvious theoretical parallel to predator prey theory (8). Most theory has treated the disease system, at least superficially, much like a classical epidemiological problem (96), after a simple and approximate classification of diseases into the mono- versus poly-cyclic categories (single versus multiple infections during a single cropping season) (188, 202). Subsequent to initial infection the progress of the disease is usually thought of as exponential-like or logistic-like, and ecological questions revolve around the nature of the initial infection and its epidemiological buildup (58, 121).

As in the case of most of the biological control literature, plant disease has been largely thought of as a two-dimensional problem, the crop and the disease. Recently it has been suggested that disease dynamics are fundamentally three-dimensional, with the crop, the disease organism, and the so-called aegricorpus (the diseased tissue of the crop) (105, 121). Clearly it is the aegricorpus that increases according to the logistic or exponential pattern, while the dynamics of the other two dimensions are controlled elsewhere—the crop by socioeconomic and ecological forces, and the disease by landscape aspects of the crop itself. This three-dimensional approach has not yet been completely explored but would seem to hold great potential. It is, after all, well known that the formulation of multidimensional systems in one or two dimensions frequently results in an overly simplified, and often quite mistaken, statement of the problem (98). With three or more dimensions, questions of indirect effects and higher-order interactions become relevant (1, 180, 181, 197), and complicated, sometimes with enigmatic, dynamic behavior emerging (70, 151, 183).

Weeds have plagued agriculture for centuries. The homology between the farmer's perception of weed problems and the classic ecological notions of plant competition is obvious (182, 196). The formulation of plant competition as response and effect (61) has been suggested as a focus on weed problems and their management (84). Such a focus emphasizes the dual role of the crop as recipient and promoter of competition, and leads quite naturally to a management philosophy that seeks to control the effect of weeds by manipulating the ecosystem (54, 84, 101). Precedent for such management strategies can be found in many traditional agricultural systems. For example, the *mal monte* (bad weeds) and *buen monte* (good weeds) of Mexican farmers recognizes that the vegetation community as a whole must be managed to promote those aspects that are beneficial (35). A great deal of attention has been paid to the use of cover crops (a good weed) to control weeds (47, 79).

SOIL MANAGEMENT

The components of soil management can be divided into three general categories: 1) physical structure, 2) nutrient dynamics, and 3) biota. The three are inextricably interrelated—physical structure influences the nature of the microbiota, which determines the nutrient dynamics, which feed back on the microbiota, which partially determines the macrobiota, which in turn has an important effect on soil physical structure. Modern agriculture is thought to have provoked significant changes in all three components, generally sacrificing long-term utility for short-term improvement in nutrient status. The effects range from obvious and spectacular to subtle and not well documented. At the obvious end of the spectrum is the massive soil erosion occurring throughout the contemporary world (26, 41, 162) and codified in popular mythology (64). At the subtle end of the spectrum is a great deal of circumstantial evidence suggesting that modern management techniques have negatively affected soil structure, long-term nutrient availability, and the integrity of soil biota.

A more functional approach to defining categories of soil science is the model of van Noordwijk & De Willigen (177) in which the salient components of the relation between applied material (chemical fertilizer, organic matter, or some sort of contaminant) and dry matter production can be conveniently illustrated (Figure 1). In quadrat I the material applied is transferred to the soil, a process in which some amount of the material is inevitably lost (e.g. the shaded area in quadrat I would likely be very large for a tropical ultisol but low for a temperate mollisol with high humic content). This quadrat is the usual focal point of soil chemistry, physics, or biology. Quadrat II illustrates the process of nutrient uptake by plants and involves all the factors usually thought of as root ecology. The shape of the uptake curve depends on root biomass density, microbial actions (e.g. mycorrhizal infections or pathogenic infections), and physical and chemical factors associated with rhizosphere dynamics. Quadrat III is largely concerned with plant physiology. Quadrat IV represents, of course, the consequence of composing the functions of the first three quadrats, and it maps the data normally obtained in simple fertilizer trials. This model enables one to tease apart the various ecological factors that might be affecting the ultimate curve of dry matter production as a function of nutrient input.

While a soil's "texture" is a property of the physical nature of the soil, its "structure" is a consequence of the physical nature interacting with the biota. The physical nature combines with various gums and resins of biological origin

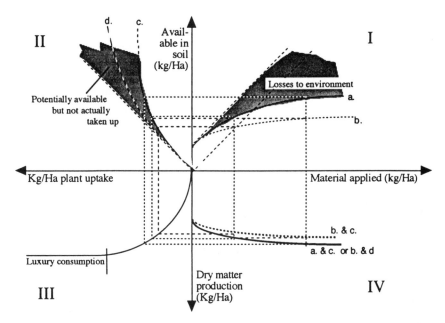

Figure 1   The theoretical formulation of van Noordwijk & Willigen (177). In quadrat I the applied material (e.g. nitrate) becomes available in the soil solution, with certain losses to the environment (shaded area). In quadrat II the material available in the soil is taken up by the plant at slower (*c*) or faster (*d*) rates, depending on ecological conditions in the rhizosphere and the physiology of the plant itself. In quadrat III the absorbed material is converted to plant tissue. Thus the process of translating material applied to dry matter production is separated into three component processes.

to produce an aggregate structure (a natural aggregate is referred to as a *ped*), which dictates a certain natural pore space distribution, which, in turn, is modified by living macrobiota such as arthropods, earthworms, and growing roots (25). Altering the biota can have dramatic effects on this structure in easily imagined ways, and it is generally acknowledged that, for a variety of reasons, one might speak of good versus bad soil structure with respect to a specific crop (e.g. enough pore space must be available for adequate infiltration of water into the soil, to say nothing of retaining proper hydrostatic pressure to allow the crops to take up what water is available, although too much pore space may result in high levels of evaporation and general loss of moisture). Furthermore, pore space is dialectically related to root growth in the sense that growing roots not only create new pore spaces, but many roots grow through existing voids, some of which were created by formerly growing roots (178, 191).

That conventional farming has had an important effect on physical structure of soils seems well documented (106, 143). In particular, a variety of studies have found greater soil aggregate stability when alternative methods are used (83, 152), possibly because of greater amounts of soil organic matter (127, 139) and certainly due, in part, to changes in plowing technology (29).

Soil macroporosity is a major determinant of the structural quality of the soil, contributing not only to crop growth but also to general environmental quality (23). In an experimental "integrated farm," soil macroporosity and the proportion of pores influenced by soil biota were significantly higher than in a conventional farm (30), but both the details of the physical structure itself and the role of soil biota in forming that structure depended strongly on the details of the cropping system (21). In another experimental system, macroporosity was higher under the integrated system, but total root density was higher in the conventional system (154). Referring back to the van Noordwijk-de Willigen theory (Figure 1) we note the potential compensatory nature of these two results. Curve b refers to availability (i.e. the ordinate in quadrat I), reduced because of the lower microporosity of the conventional system (presumably because of higher hydrostatic pressure within the soil), and curve b combined with curve c results in a dry matter production curve lower than that with the higher macroporosity, curve a. But if root biomass increases at the same time the macroporosity decreases, that could compensate for the expectation in decreased biomass production, as suggested by the curve $a\&c$ or $b\&d$ in Figure 1.

Elevating the nutrient status of soils has been a centerpiece of the modern model ever since Leibig invented the law of the minimum and Lawes & Gilbert initiated the long-term experiments at Rothamstead. The customary treatment of nutrients in terrestrial ecosystems is dual: first an analysis of how the nutrients get into and move through the soil solution to arrive at the rhizosphere, and second, the uptake within the rhizosphere, quadrats I and II in the Van Noordwijk and de Willingen model (Figure 1). Unfortunately, the commonly used term "availability" in the context of nutrients is not as clear as it might initially seem. The term might be sensibly applied either to that which makes its way to the soil solution to be a potential target of utilization by the crops (quadrat I of Figure 1) or to that which makes its way from the soil solution pool to be actually incorporated into the crop (quadrat II of Figure 1). Furthermore, the notion that there might be a meaningful definition of availability of a single nutrient is questionable in the face of modern knowledge of nutrient interactions (48). The alternative program tends to look at nutrients as a "system" that can be "in balance" or not, something metaphorically similar to ecosystem "health." Many traditional farmers the world over have held this point of view (170), and many alternative agricultural research workers have taken inspiration from them (6). It is perhaps ironic that modern reductionist science is now arriving at the same point of view—that an examination of nutrients one by one, codified in the sacrosanct philosophy of Justus von Leibig, is misleading and that nutrient absorption is an affair of interaction.

The major source of nitrogen in most soils is nitrate, available largely due to the activity of microorganisms (81, 128). Ammonium cannot be stored in great quantities in cells because of its toxicity. Nitrate is stored in cell vacuoles but must be reduced to ammonium before being processed into amino acids (i.e. the constructive phase of the nitrogen cycle is the reverse of the decomposition phase—nitrate to ammonium to amino acid to protein). The basic processing equipment for nitrogen is distributed differently in legumes than in other plants (128). The major source of nitrogen in more general terms is the atmosphere, and biological nitrogen fixation is a dominant feature of terrestrial ecosystems (134). Most of this process occurs through the symbiosis of legumes and *Rhizobium*, although a significant potential exists through the action of free-living nitrogen-fixing bacteria, such as *Azotobacter* and *Azospirillum* (128). While it is generally thought that having a large concentration of ammonium ions in the soil solution is desirable, too little is understood about free amino acids and larger organic molecules to reach a final judgment about what may ultimately be the best source of nitrogen (119).

Another recent topic of seemingly immense ecological importance is the so-called "added nitrogen interaction" effect (13, 153). When inorganic nitrogen is added to an agroecosystem, it seems sometimes to stimulate the use of native nitrogen. This is thought to be due to one or more of the following: (i) increased net mineralization of soil N with subsequent consumption of the mineralized N by the plant, (ii) increased root exploration, and/or (iii) pool-substitution in which added nitrogen stands proxy for native inorganic soil nitrogen that would otherwise have been removed from that pool (153). This sort of indirect interaction is exemplary of the complexities that exist in the agroecosystem.

Soil biota is involved in many other aspects of soil biology besides biological nitrogen fixation (166). Of particular interest to alternative agriculture is the role of earthworms in promoting the physical and chemical structure of soil (73, 74, 86, 114) as well as other indirect effects such as promoting nodulation (46, 163), in dispersal of mycorrhizal fungi (66), and even in suppression of diseases (167), although worms occasionally act as disease dispersers also (171). Other components of soil fauna are also implicated in various structural, chemical, or biological aspects of soil function (24, 62, 108, 117, 145, 156). Recent research has convincingly shown that low-input systems generally harbor a richer soil biota (42, 155, 193, 204), especially earthworms (16).

The soil factor that transcends all three of the general categories and is intimately involved in the general character of quadrats I and II of Figure 1 is organic matter (169). A major determinant of soil structure through its action of creating soil ped structure, soil organic matter is sometimes the principal of creating soil ped structure, soil organic matter is sometimes the principal force determining cation exchange capacity. Soil organic matter obviously derives from the soil biota as well as partially determining the nature of the latter, and its breakdown supplies the soil with mineral nutrients. The formation of soil organic matter at the most general level is nothing more than the consequences of decomposition, with a succession of heterotrophic organisms from primary microflora, which attack the fundamental components, "succeeded by secondary, tertiary, and so forth, microflora that thrive on the cells and by-products [of the previous successional stages]" (39). This process is immensely complicated, as evidenced by the many published flow charts illustrating just the "simplest" processes involved in both managed and unmanaged systems (e.g. 39). From the point of view of alternative agriculture, two issues seem of primary importance: the basic partitioning of organic matter during the decomposition process, and the nitrogen-to-carbon ratio.

Regarding the basic partitioning of organic matter, it is now well known, if perhaps oversimplified, that organic matter decomposes into two distinct fractions, the "fast" fraction, or the active organic matter, and the "slow" fraction. For example, after a single year, in a woodland ecosystem in England, 99% of the sugars and 75% of the cellulose in the litter had decomposed, but only 25% of the waxes and 10% of the phenols (116). The slow fraction ultimately contributes to the physical structure of the soil while the fast fraction is the main supplier of mineral nutrients. Different types of organic matter may very well contain different fractions of fast and slow raw materials and thus ultimately have different effects on the soil, although the details as to why this happens remain elusive (85).

The raw material that goes into the decomposition system has a highly variable chemistry. This chemistry is conveniently characterized by the carbon:nutrient ratio (166), the most important one of which is the carbon:nitrogen ratio. If organic matter contains abundant carbon relative to nitrogen, soil microorganisms are not able to grow fast enough to decompose the material efficiently, and they eventually must use nitrogen already in the soil. The application of organic matter may thus result in an initial decrease in nitrogen availability for the crops due to the needs of the microorganisms, the bane of the beginning organic gardener. C:N ratios vary enormously, from 250 for sawdust, to 90 for cornstalks, to 20 for legumes, to 11 for humus, to 4 for bacteria (44). As a rule of thumb, a C:N ratio less than 20 generates a net release of nitrogen for plant growth, while a C:N ratio greater than 20 involves immobilization of nitrogen, therefore making less N available for plant growth. The C:N ratio changes over time as the carbon is utilized in respiration and the nitrogen is recycled. For plant residues, it has been estimated that about 20–40% of the carbon is actually incorporated into biological material. The rest is released as CO_2 (4), raising some question about the role of soils and organic matter therein to the global CO_2 problem (67, 82). The rate of decomposition of various organic materials, and thus the C:N ratio at a given time, depends on O_2, moisture content, temperature, pH, substrate specificity, and available minerals (38, 93). It is also strongly influenced by management factors such as tillage and inorganic fertilizer applications (141, 158). The nature of the material itself also affects the decomposition rate (80). In a recent study, it was estimated that agriculture without supplementary fertilization was economical for 65 years on temperate prairie and for six years in a tropical semiarid thorn forest, while an extremely nutrient-poor Amazonian soil showed no potential for agriculture beyond the three-year lifespan of the forest litter mat (169). On the other hand, long-term experiments in England (139 years; 2) and Denmark (90 years; 152) clearly show the benefits of organic matter for long-term soil structure and crop yields.

The provisioning of organic matter in the form of animal manure is an old procedure and is thought to have considerable potential for adding alternative organic matter to cropping systems (22, 36). It is commonly employed in traditional agriculture in underdeveloped countries (5, 14), although it is sometimes viewed as only a supplement to inorganic fertilizers (120). In at least one study, farmyard manure was shown to provide a greater increase in soil nitrogen than did green manure (137).

There has been much interest in the role of green manure in agroecosystems. The effects of living mulches are difficult to summarize since their purpose is multiple—they are frequently employed in an attempt to control weeds (53) or insects (160), or simply to add organic matter to the system (45). While most research concentrates on the role of living mulch as having one or the other of these effects, researchers in alternative agriculture routinely recognize the multiple potential of living mulch (40, 79, 182). Unfortunately there is little in the way of comprehensive ecological theory to enable us to predict when and where a cover crop will function well.

Mulching has long been a mainstay of organic agriculture. To decrease bulk for ease in transport and to reduce the C:N ratio of applied material, it is customary to compost the mulch before application. Most alternative agriculturalists regard compost as a central component of the ultimate sustainable agricultural system. Recently some well-deserved attention has been focused on the biological processes involved in the composting process (92, 175). The practical problem is that compost should be "biomature" when it is applied to the field (203), but the more theoretical issue is that the decomposition processes involved in composting are just as complicated as the decomposition processes in soils (115, 166), and our ecological knowledge of them remains empirical, eclectic, and superficial.

That soil physical structure is dependent on humus is clear. Recently it has come to light that humic substances also affect plant metabolism directly (190).

Work at the physiological level has shown that low molecular weight humic substances stimulated the production of ATPase on plasma membranes and tonoplast (113). Sodium humate also apparently interferes with the protein ion carriers on the cell membrane (63), and humic acid is capable of altering the lipid matrix of the cell membrane (150). Thus, the well-established role of humus in the chemical and physical structure of soil is not the only concern. It may very well have an important direct physiological role also. It should be noted that these studies refer specifically to low molecular weight humic acids, that is, molecules that are capable of penetrating the cell wall. They are thus not likely the same fraction of humus that contributes to cation exchange capacity and soil structure, but rather may be a component of later stages of decomposition.

The IPM program of alternative pest control strategists has been a key organizing principle for proponents of alternative agriculture. Curiously, a parallel pattern did not evolve with regard to the other major issue in alternative agriculture, soil. It is not difficult to imagine a program of ISM (Integrated Soil Management), and a general outline of such a program is easily visualizable (apply fertilizer to the soil only when necessary, manage the cropping system to minimize the need to apply nutrients from the outside), perhaps based on the generalized notion of "soil quality," although that is still a somewhat slippery concept (133, 165).

INTEGRATION

Probably the idea most promoted by advocates of an alternative agriculture is an integration of all aspects of the farming operation. This may take various forms, depending on the advocate or practitioner. One persistent idea is that an agroecosystem should mimic the functioning of nonmanaged ecosystems (49, 129), with tight nutrient cycling, vertical structure, and the preservation of biodiversity (130). One manifestation of this idea is the incorporation of trees into the agroecosystem, something that has been part of traditional agroecosystems in tropical regions for some time (60, 87, 109). Another manifestation has been the attempt to use ecological principles as part of the design criterion, thus replacing what had become a strictly economic decision-making process with one that includes ecological ideas also. For example, the natural biodiversity of most unmanaged ecosystems has been used as a rationale to suggest that multiple cropping is generally a laudable goal (this is discussed below), or that the famous diversity-stability hypothesis of theoretical ecology might be related to promoting stability through diversity in agroecosystems (186). Many of these ideas suffer from a lack of rigorous definition in the field of ecology itself, and thus their application to the design of agroecosystems is perhaps premature.

Multiple cropping (55, 57, 90, 199, 200) or one of its many variants has been often cited as an important integrative component of the alternative agriculture agenda. It is often observed that more traditional forms of agriculture, especially in the tropics, include some form of multiple cropping at their cores, and it is thought that the abandonment of this procedure was specifically to accommodate the methods of modern conventional agriculture (i.e. application of chemicals and mechanization). If this were true, the ecological benefits of multiple cropping would be lost in the modern system, and it would make sense to argue for their reintegration in the alternative program.

Exactly what are the ecological benefits thought to accrue from the practice of multiple cropping? The hypothesized ecological benefits have been divided into two categories (182), reduced competition (or the competitive production principle) and facilitation. In the case of the competitive production principle, it is thought that two different species occupying the same space will use all necessary resources more efficiently than a single species occupying that same space, much as is sometimes believed to happen in natural ecosystems. In the case of facilitation, it is thought that one crop species has some sort of positive effect on the other species. Theoretical details of each of these two categories can be found in Vandermeer (182).

The evidence for the competitive production principle is scant. For the most part the enormous literature on multiple cropping (see bibliographies in 55, 90, 131, 182) is not of use, since experiments do not generally accommodate the data required to answer the key question, "Does the intercrop yield better than a combination of the two optimal monocultures?" At one extreme a review of experiments with dialell mixtures of grasses (172) found only very weak evidence that interspecific competition, on average, was smaller than intraspecific competition. On the contrary, the overall pattern of the data suggested something more akin to the neutral competition hypothesis (77, 157), in which inter and intraspecific competition are not significantly different.

At the other extreme, mixtures of legumes and nonlegumes frequently seem to provide evidence of intercrop advantage (72, 125, 173). Yet even here it is not really clear what mechanism is involved. On the one hand, it could be a case of the competitive production principle, that if the legume and nonlegume are tapping different pools of nitrogen, interspecific competition for that nutrient is thereby reduced (161). On the other hand, the legume could be facilitating the growth of the nonlegume by supplying it with extra nitrogen. Very few experiments have addressed this problem (142), and the general literature is not amenable to differentiating between these two alternatives (182). That the legume may simply sequester nitrogen that would otherwise leach out of the soil and later release it to the nonlegume has been suggested in some cases (3).

The general conclusion seems to be that, although theoretically two ecologically different crops could fit into an environment more efficiently than an equal biomass of the same crop, accumulated evidence offers little support for that idea. It would seem to be equally logical, especially in light of recent theory in plant ecology (37, 77, 157), to suggest that different species of annual crops are more or less interchangeable ecologically, and the only advantage of intercropping is either from a socioeconomic point of view, perhaps from the long-term point of view of soil conservation, or from some sort of special facilitative effect characteristic of the particular combination involved. Perhaps the competitive production principle will turn out to be the competitive neutrality principle.

Nevertheless, the alternative mechanism for intercrop advantage, the facilitative effect (182), while certainly not ubiquitous, seems well-established in particular cases. For example, a vast literature exists, both strictly empirical (118, 144) and mechanistic (56, 89, 146, 173), on how intercropping creates the sort of vegetative texture that controls specialist pests. In all of these cases, the second crop acts as a facilitator of the first crop by somehow controlling potential pests, whether by resource concentration or by means of enemies (7, 146, 148).

Ultimately, ecological principles speak to many other integrative aspects of a potential alternative agriculture program. Topics such as multiple interacting components and complex dynamics (the problem of viewing as simple that which is complex) or the various integrative problems associated with conversion need to be explored from an ecological point of view: Much recent ecological theory may very well be relevant. Additionally, the interface of a socioeconomic framework with ecology, while always acknowledged, needs to be more explicit. A host of integrative topics more traditionally associated with alternative agriculture, such as agroforestry, conservation tillage, the incorporation of animals, risk reduction, biodiversity, and system design, have been excluded from this review because of space limitations.

A recent formulation seeks to combine many of the ecological and social forces under a single umbrella, the "syndromes of production" (10). Elaborating on the loosely formulated notion that the agroecosystem represents alternative sets of management techniques and ecological relations (174), Andow & Hidaka (10) compare the traditional *shizen* system of rice production to the now contemporary Japanese system and suggest that it represents a distinct group of management techniques and, by implication, ecological relations. The syndromes of production focus may provide a useful way of directing future research.

THE PROBLEM OF CONVERSION

Several studies have compared actually existing alternative systems with their modern counterparts. This comparison is always suspect because the matching of farms is, in principle, very difficult. The alternative farms are alternative for a large number of possible reasons, and the fact of their alternativeness may not be the determining feature of their success or failure. For example, if alternative farms just happen to develop on the best soils and those that were on the bad soils just happen to develop the use of modern inputs, the compensating factor of the good soils is what makes the alternativeness work. Or if alternative farmers are mainly marginalized farmers of a Third World country, and their alternativeness is a consequence of a depressed socioeconomic condition and marginal lands, their having yields lower than yields of a modern highly capitalized farm would hardly be of interest.

Acknowledging the potential importance of these factors in possibly creating spurious results, it is nevertheless of interest to examine those comparative studies. In India, comparing seven farm pairs, each one alternative and one conventional farm, no difference was encountered in the critical factors of yields or economic performance (176), although differences in other factors were striking such as less dependence on external inputs for the alternative farms. This finding reflects a general conclusion that functioning low-input systems do not perform more poorly than conventional high-input ones (12, 50, 104, 126), and so it has provided political fodder for those who seek to promote alternative agriculture.

On the other hand, attempts at setting up experimental low-input systems side-by-side with conventional ones have generally failed. Even in cases where multiple factors of low input are engaged, the conversion from conventional to low-input sustainable agriculture has been something of a failure thus far. Performance (yield or profitability) is almost always reduced in the alternative system. Sometimes, lowered yields are compensated by reduced input costs (192, 195), but frequently they are not (95, 102, 168). We are thus left with an uncomfortable paradox—that real functioning alternative systems work well, but attempts to convert conventional ones have failed. This suggests something about the inadequacy of our systematic ecological knowledge of agroecosystems.

Cuba represents a fascinating experiment currently in progress. Faced with a virtual elimination of all of its high technology inputs from the former Soviet Union and the Eastern Block since 1989, Cuba has been forced to transform its agriculture wholesale. All pest control is with biological control or IPM, soil amendments are almost entirely organic, weed management is based on managing plant competition, mycorrhizae substitute for imported phosphorous, and vermiculture makes mounds of worm compost from farm and city waste (43, 136, 147, 187). The transformation is still in its infancy, but Cuba represents a microcosm (if 11 million people can be called a microcosm) worthy of considerable study for those interested in alternative agriculture. Results thus far suggest: 1) Some components are dramatic failures (e.g. the Voisson cattle management system), 2) some components have seen great success (e.g. IPM), and 3) some problems have arisen that were never before acknowledged as problems of an alternative model (e.g. new forms of labor management).

CONCLUSIONS

In the end what can be said about the ecological basis of alternative agriculture? Given our current state of understanding with respect to agroecosystems, general principles may be more pragmatic than ecological. First, the curative/prophylactic, or responsive/preventative paradigm (184) seems like a generally good rule of thumb as we slowly work toward a better agriculture. The programmatic conclusion is to seek solutions that can deal with extant problems as they arise (responsive technologies) but also to develop solutions that ultimately prevent those problems from arising in the first place (preventative technologies). I rather expect ecological principles will have more to do with the latter than the former.

Second, a first approximation to understanding a system as complicated as an agroecological one may be embodied in the functional composition model developed by van Noordwijk & de Willigen (177) (Figure 1). Frequently the ultimate agroecosystem is far too multidimensional to be dealt with in this fashion, but as an initial epistemological tool, the functional composition approach allows us to explore a level of complexity that is somewhere between the one-dimensional approach of conventional agriculture and the hopelessly grand approach of systems ecology. We can acknowledge complexity but try to represent it as a lower-dimensional system. Sometimes this will be possible, as it seems to be in the case of Van Noordwijk & de Willigen's model applied to soil fertility. However, at other times we will perhaps have to be more creative. It is always a danger that one might fall into the reductionist's trap, which is exactly the opposite of what most alternative agriculturalists intend. But it is important to distinguish between reductionism as a research tactic and reductionism as a philosophy—the latter is the folly of modernism, the former is a necessity of science (99).

Third, in pursuit of the elusive sustainable system, the syndrome of a production model (10) is likely to be a useful metaphor in the future. Indeed, the failure of single-shot attempts at conversion coupled with the proverbial high performance of actually functioning low-input systems suggests the utility of the syndromes of production model—i.e. conventional agriculture and sustainable agriculture represent alternative "adaptive peaks" in the landscape, and one may not be able to climb the alternative peak without temporary loss of yield and/or profitability. On the other hand, one can always strive to jump from one peak to the other.

Fourth, we must remember that the problems faced by real farmers in today's world are mainly economic and political (100)—falling prices for their products in the face of rising prices for inputs in the developed world, lack of land tenure security and disappearing markets in the underdeveloped world. Sustainable ecological techniques will remain irrelevant until these problems are solved.

Literature Cited

1. Abrams PA. 1991. Strengths of indirect effects generated by optimal foraging. *Oikos* 62:1676–176

2. AFRC. 1991. *Rothamsted Experimental Station: Guide to the Classical Field Experiments.* Harpenden: AFRC Inst. Arable Crops Res.

3. Agamuthu P, Broughton WJ. 1985. Nutrient cycling within the developing oil-palm-legume ecosystem. *Agric. Ecosys. Environ.* 13:111–23

4. Alexander M. 1977. *Introduction to Soil Microbiology.* New York: John Wiley. 2nd ed.

5. Altieri MA. 1987. *Agroecology: The Scientific Basis of Alternative Agriculture.* Boulder, CO: Westview

6. Altieri MA. 1990. Why study traditional agriculture? In *Agroecology*, ed. CR Carroll, JH Vandermeer, P Rosset, p. 551–64. New York: McGraw-Hill

7. Altieri MA, Letourneau DK. 1984. Vegetation diversity and insect pest outbreaks. *CRC Crit. Rev. Plant Sci.* 2:131–69

8. Anderson RM, May RM. 1991. *Infectious Diseases of Humans: Dynamics and Control.* Oxford: Oxford Univ. Press

9. Andow DA. 1991. Vegetational diversity and arthropod population response. *Annu. Rev. Entomol.* 36:561–86

10. Andow DA, Hidaka K. 1989. Experimental natural history of sustainable agriculture: syndromes of production. *Agric. Ecosyst. Environ.* 27: 447–62

11. Andow DA, Rosset PM. 1990. Integrated pest management. In *Agroecology*, ed. CR Carroll, JH Vandermeer, PM Rosset. New York: McGraw-Hill

12. Andrews RW, Peters SE, Janke RR, Sahs WW. 1990. Converting to sustainable farming systems. In *Sustainable Agriculture in Temperate Zones*, ed. CA Francis, CB Flora, LD King, pp. 281–314. New York: Wiley

13. Azam F, Simmons FW, Mulvaney RL. 1994. The effect of inorganic nitrogen on the added nitrogen interaction of soils in incubation experiments. *Biol. Fertil. Soils* 18:103–8

14. Babu SC, Subramanian SR, Rajasekeran B. 1991. Fertilizer and organic manure use under uncertainty: policy comparisons for irrigated and dryland farming systems in South India. *Agric. Syst.* 35: 89–102

15. Beddington JR, Free CA, Lawton JH. 1978. Characteristics of successful enemies in models of biological control of insect pests. *Nature* 273:513–19

16. Berry EC, Karlen DL. 1993. Comparison of alternative farming systems. II. Earthworm population density and species diversity. *Am. J. Altern. Agric.* 8: 21–26

17. Beus CE, Dunlap RE. 1993. Agricultural policy debates: examining the alternative and conventional perspectives. *Am. J. Alt. Agric.* 8:98–106

18. Beus CE, Dunlap RE. 1990. Conventional vs. alternative agriculture: the paradigmatic roots of the debate. *Rural Sociol.* 55:590–616

19. Beus CE, Dunlap RE. 1991. Assessing alternative vs. conventional agricultural paradigms: a proposed scale. *Rural Sociol.* 56:432–60

20. Boeringa R, ed. 1980. Alternative methods of agriculture. *Agric. Environ.* 5:1–199

21. Boersma OH, Kooistra MJ. 1994. Differences in soil structure of silt loam Typic Fluvaquents under various agricultural management practices. *Agric. Ecosyst. Environ.* 51:21–42

22. Boschi S, Tano F. 1994. Effects of cattle manure and components of pig slurry

on maize growth and production. *Eur. J. Agron.* 3:235–41

23. Bouma J. 1991. Influence of soil macroporosity on environmental quality. *Adv. Agron.* 46:1–37

24. Bouwman LA, Zwart KB. 1994. The ecology of bacterivorous protozoans and nematodes in arable soil. *Agric. Ecosyst. Environ.* 51:145–60

25. Brewer R. 1964. *Fabric and Mineral Analysis of Soils.* New York: Wiley

26. Brosten D. 1988. How much can we lose? *Agrichem. Age* 32:7–8

27. Brown BJ, Hanson ME, Liverman DM, Merideth RW Jr. 1987. Global sustainability: toward definition. *Environ. Manag.* 11:713–19

28. Browne WP. 1988. *Private Interests, Public Policy, American Agriculture.* Lawrence: Univ. Press Kans.

29. Bruce RR, Langdale GW, Dillard AL. 1990. Tillage and crop rotation effects on characteristics of a sandy surface soil. *Soil Sci. Soc. Am. J.* 54:1744–47

30. Brussard L. 1994. An appraisal of the Dutch programme on soil ecology of arable farming systems (1985–1992). *Agric. Ecosyst. Environ.* 51:1–6

31. Buttel FH. 1990. Social relations and the growth of modern agriculture. In *Agroecology*, ed. CR Carroll, JH Vandermeer, P Rosset, pp. 113–145. New York: McGraw-Hill

32. Carroll CR, Vandermeer JH, Rosset P, eds. 1990. *Agroecology.* New York: Mc-Graw-Hill

33. Carson R. 1962. *Silent Spring.* New York: Houghton-Mifflen

34. Carter HW. 1989. Agricultural sustainability: an overview and research assessment. *Calif. Agric.* 43:16–18, 37

35. Chacon JC, Gliessman SR. 1982. Use of the "non-weed" concept in traditional tropical agroecosystems of southeastern Mexico. *Agroecosystems* 8:1–11

36. Chambers BJ, Smith KA. 1992. Soil mineral nitrogen arising from organic manure applications. *Aspects Appl. Biol.* 30:135–43

37. Chesson PL, Warner RR. 1981. Environmental variability promotes coexistence in lottery competitive systems. *Am. Nat.* 117:923–43

38. Clark MD, Gilmore JT. 1983. The effect of temperature on decomposition at optimum and saturated soil water contents. *Soil Sci. Soc. Am. J.* 47:927–29

39. Coleman DC, Cole CV, Elliott ET. 1984. Decomposition, organic matter turnover, nutrient dynamics in agroecosystems. In *Agricultural Ecosystems: Unifying Concepts*, ed. R Lowrance, BR Stinner, G House, pp. 83–104. New York: Wiley

40. Costello MJ. 1994. Broccoli growth, yield and level of aphid infestation in leguminous living mulches. *Biol. Agric. Hortic.* 10:207–22

41. Crosson PR, Stout A. 1983. *Productivty Effects of Cropland Erosion in the United States.* Baltimore, MD: Johns Hopkins Univ. Press

42. Didden WAM, Marinissen JCY, Vreeken-Buijs MJ, Burgers SLGE, de Fluiter R, et al. 1994. Soil meso- and macrofauna in two agricultural systems: factors affecting population dynamics and evaluation of their role in carbon and nitrogen dynamics. *Agric. Ecosyst. Environ.* 51:171–86

43. Dlott J, Perfecto I, Rosset P, Burkham J, Monterrey J, Vandermeer JH. 1993. Management of insect pests and weed. *Agric. Hum. Values* X:9–15

44. Donahue RL, Miller RW, Shickluna UC. 1977. *Soils: An Introduction to Soils and Plant Growth.* Englewood Cliffs, NJ: Prentice-Hall

45. Dou Z, Fox RH. 1994. The contribution of nitrogen from legume cover crops double-cropped with winter whet to tilled and non-tilled maize. *Eur. J. Agron.* 3:93–100

46. Doube BM, Ryder MH, Davoren CW, Stephens PM. 1994. Enhanced root nodulation of subterranean clover (*Trifolium subterraneum*) by *Rhizobium leguminosarium* biovar *trifolii* in the presence of the earthworm *Aporrectodea trapezoides* (Lumbricidae). *Biol. Fertil. Soils* 18:169–74

47. Dyck E, Liebman M, Erich MS. 1995. Crop-weed interference as influenced by a leguminous or synthetic fertilizer nitrogen source: I. Doublecropping experiments with crimson clover, sweet corn, lambsquarters. *Agric. Ecosys. Environ.* In press

48. Evangelou VP, Wang J, Phillips RE. 1994. New developments and perspectives on soil potassium quantity/intensity relationships. *Adv. Agron.* 52:173–227

49. Ewel JJ. 1986. Designing agricultural ecosystems for the humid tropics. *Annu. Rev. Ecol. Syst.* 17:245–71

50. Faeth P, Repetto R, Kroll K, Dai Q, Helmers G. 1991. *Paying the Farm Bill: US Agricultural Policy and the Transition to Sustainable Agriculture.* Washington DC: World Resourc. Inst.

51. Farshad A, Zinck JA. 1993. Seeking agricultural sustainability. *Agric. Ecosys. Environ.* 47:1–12

52. Feeny PP. 1976. Plant apparency and

chemical defense. *Rec. Adv. Phytochem.* 10:1–40

53. Fischer A, Burrill L. 1993. Managing interference in a sweet corn-white clover living mulch system. *Am. J. Alt. Agric.* 8:51–56

54. Forcella F, Eradat-Osskoui K, Wagner SW. 1993. Application of weed seedbank ecology to low-input crop management. *Ecol. Applic.* 3:74–83

55. Francis CA. 1986. *Multiple Cropping: Practices and Potentials.* New York: Macmillan

56. Francis CA, Youngberg G. 1990. Sustainable agriculture—an overview. In *Sustainable Agriculture in Temperate Zones,* ed. CA Francis, CB Flora, LD King, pp. 1–23. New York: Wiley

57. Francis CA, Prager M, Laing DR, Flor CA. 1978. Genotype X environment interactions in bush bean cultivars in monoculture and associated with maize. *Crop Sci.* 18:237–46

58. Fry WE. 1982. *Principles of Plant Disease Management,* New York: Academic

59. Fry WE, Leonard KJ, eds. 1989 *Plant Disease Epidemiology.* New York: McGraw-Hill

60. Getahun A, Reshid K, Munyua H. 1991. *Agroforestry for Development in Kenya: An Annotated Bibliography.* Nairobi, Kenya: Int. Council Res. Agrofor. (ICRAF)

61. Goldberg DE. 1990. Components of resource competition in plant communities. In *Perspectives on Plant Competition,* ed. JB Grace, D Tilman, pp. 27–49. New York: Academic

62. Griffiths BS, Young IM. 1994. The effects of soil structure on protozoa in a clay-loam soil. *Eur. J. Soil Sci.* 45:285–92

63. Guminski S, Sulej J, Glabiszewski J. 1983. Influence of sodium humate on the uptake of some ions by tomato seedlings. *Acta Soc. Bot. Poloniae.* 52:149–64

64. Guthrie W. 1940. *Dust Bowl Ballads.* Boston: Rounder Records

65. Haila Y, Levins R. 1992. *Humanity and Nature: Ecology, Science and Society.* London: Pluto

66. Harinikumar KM, Bagyaraj DJ. 1994. Potential of earthworms, ants, millipedes, termites for dissemination of vesicular-arbuscular mycorrhizal fungi in soil. *Biol. Fertil. Soils* 18:115–18

67. Harrison KG, Broecker WS, Bonani G. 1993. The effect of changing land use on soil radiocarbon. *Science* 262:725–26

68. Hassebrook C. 1990. Developing a so-

cially sustainable agriculture. *Am. J. Alt. Agric.* 5:50, 96

69. Hassell MP. 1978. *The Dynamics of Arthropod Predator-Prey Systems.* Princeton, NJ: Princeton Univ. Press

70. Hastings A, Hom CL, Ellner S, Turchin P, Godfray HCJ. 1993. Chaos in ecology: Is mother nature a strange attractor? *Annu. Rev. Ecol. Syst.* 24:1–33

71. Hawkins BA, Thomas MB, Hochberg ME. 1993. Refuge theory and biological control. *Science* 262:1429–32

72. Haynes RJ. 1980. Competitive aspects of the grass-legume association. *Adv. Agron.* 33:227–61

73. Hindell RP, McKenzie BM, Tisdall JM, Silvapulle MJ. 1994. Relationships between casts of geophagous earthworms (Lumbricidae, Oligochaeta) and matric potential. I. Cast production, water content, bulk density. *Biol. Fertil. Soils* 18:119–26

74. Hindell RP, McKenzie BM, Tisdall JM, Silvapulle MJ. 1994. Relationships between casts of geophagous earthworms (Lumbricidae, Oligochaeta) and matric potential. II. Clay dispersion from casts. *Biol. Fertil. Soils* 18:127–31

75. Horn DJ. 1988. *Ecological Approach to Pest Management.* New York: Guilford

76. House GJ, Crossley DA Jr. 1987. Legume cover cropping, no-tillage practices, soil arthropods: ecological interactions and agronomic significance. In *The Role of Legumes in Conservation Tillage,* ed. JF Power. Soil Cons. Soc. Am.

77. Hubbell SP, Foster RB. 1986. Biology, chance, history and the structure of tropical rain forest tree communities. In *Community Ecology,* ed. J Diamond, TJ Case, pp. 314–29. New York: Harper & Row

78. Huffaker CB, Messenger PS, eds. 1976. *Theory and Practice of Biological Control.* New York: Academic

79. Ingels C, Van Horn M, Bugg RL, Miller PR. 1994. Selecting the right cover crop gives multiple benefits. *Calif. Agric.* 48:43–48

80. Janzen HH, Kucey RMN. 1988. C N and S mineralization of crop residues as influenced by crop species and nutrient regime. *Plant Soil* 106:35–41

81. Jarrell WM. 1990. Nitrogen in agroecosystems. In *Agroecology,* ed. CR Carroll, JH Vandermeer, P Rosset, pp. 385–412. New York: McGraw-Hill

82. Jenkinson DS, Adams DE, Wild A. 1991. Model estimates of CO_2 emissions from soil in response to global warming. *Nature* 351:304–6

83. Jordahl JL, Karlen DL. 1993. Compari-

son of alternative farming systems. III. Soil aggregate stability. *Am. J. Alt. Agric.* 8:27–33

84. Jordan N. 1993. Prospects for weed control through crop interference. *Ecol. Applic.* 3:84–91

85. Juma NG. 1994. A conceptual framework to link carbon and nitrogen cycling to soil structure formation. *Agric. Ecosyst. Environ.* 51:257–67

86. Kang BT, Akinnifesi FK, Pleysier JL. 1994. Effect of agroforestry woody species on earthworm activity and physicochemical properties of worm casts. *Biol. Fertil. Soils* 18:193–99

87. Kang BT, Reynolds L, Atta-Krah AN. 1990. Alley farming. *Adv. Agron.* 43:315–59

88. Kareiva P. 1983. Influence of vegetation texture on herbivore populations: resource concentration and herbivore movement. In *Variable Plants and Herbivores in Natural and Managed Systems,* ed. RF Denno, MS McClure, pp. 259–89. New York: Academic

89. Kareiva P. 1986. Trivial movement and foraging by crop colonizers. In *Ecological Theory and IPM Practice,* ed. M Kogan. New York: Wiley

90. Kass DC. 1978. Polyculture cropping systems: review and analysis. *Cornell Int. Agric. Bull.* 32:1–69

91. Keeney DR. 1989. Toward a sustainable agriculture: need for clarification of concepts and terminology. *Am. J. Alt. Agric.* 4:101–5

92. Kirchman H, Witter E. 1992. Composition of fresh, aerobic and anaerobic farm animal dungs. *Bioresource Technol.* 40:137–42

93. Kowalenko CG, Ivarson KC, Cameron DR. 1978. Effect of moisture content, temperature, nitrogen fertilzation on carbon dioxide evolution from field soils. *Soil Biol. Biochem.* 10:417–23

94. Laing JE, Hamai J. 1976. Biological control of insect pests and weeds by imported parasites, predators, pathogens. In *Theory and Practice of Biological Control,* ed. CB Huffaker, PS Messenger, pp. 686–744. New York: Academic

95. Lanini WT, Zalom F, Marois J, Ferris H. 1994. Researchers find short-term insect problems, long-term weed problems. *Calif. Agric.* 48:27–33

96. Leonard KJ, Mundt CC. 1984. Methods for estimating epidemiological effects of quantitative resistance to plant diseases. *Theor. Appl. Genet.* 67:219–30

97. Levins R. 1986. Perspectives on IPM: from an industrial to an ecological model. In *Ecological Theory and Integrated Pest Management,* ed. M Kogan. New York: Wiley

98. Levins R, Vandermeer JH. 1989. The agroecosystem embedded in a complex ecological community. In *Agroecology,* ed. CR Carroll, JH Vandermeer, P Rosset, pp. 341–62. New York: McGraw-Hill

99. Levins R, Lewontin R. 1985. *The Dialectical Biologist.* Cambridge: Harvard Univ. Press

100. Lewontin RC. 1982. Agricultural research and the penetration of capital. *Sci. People* Jan-Feb:12–17

101. Liebman M, Dyck E. 1993. Crop rotation and intercropping strategies for weed management. *Ecol. Applic.* 3:92–122

102. Liebman M, Rowe RJ, Corson S, Marra MC, Honeycutt CW, Murphy BA. 1993. Agronomic and economic performance of conventional vs. reduced input bean cropping systems. *J. Prod. Agric.* 6:369–77

103. Lockeretz W. 1988. Open questions in sustainable agriculture. *Am. J. Alt. Agric.* 3:174–81

104. Lockeretz W, Shearer G, Kohl DH. 1981. Organic farming in the corn belt. *Science* 211:540–47

105. Loegering WQ. 1984. Genetics of the pathogen-host association. In *The Cereal Rusts,* ed. WR Bushnell, AP Roelfs, 1:165–92. Orlando, FL: Academic

106. Logsdon SD, Radke JK, Karlen DL. 1993. Comparison of alternative farming systems. I. Infiltration techniques. *Am. J. Alt. Agric.* 8:15–20

107. Lowrance R, Hendrix PF, Odum EP. 1985. A hierarchical approach to sustainable agriculture. *Am. J. Alt. Agric.* 1:169–73

108. Lussenhop J. 1992. Mechanisms of microarthoropod-microbial interactions in soil. *Adv. Ecol. Res.* 23:1–33

109. MacDicken KG, Vergara NT. 1990. *Agroforestry: Classification and Management.* New York: Wiley

110. MacRae RJ, Hill SB, Mehuys GR, Henning J. 1990. Farm-scale agronomic and economic conversion to sustainable agriculture. *Adv. Agron.* 43:155–98

111. MacRae S, Hill B, Henning J, Bentley AJ. 1990. Policies, programs, regulations to support the transition to sustainable agriculture in Canada. *Am. J. Alt. Agric.* 5:76–92

112. Madden JP. 1990. The economics of sustainable low-input farming systems. In *Sustainable Agriculture in Temperate Zones,* ed. CA Francis, CB Flora, LD King, pp. 315–41. New York: Wiley

113. Maggioni A, Varanini Z, Pinton R, De

Biase MG. 1992. Humic substances affect transport properties of root membranes. In *Humus, Its Structure and Role in Agriculture and Environment Kubat Proc. 10th Symp. Humus et Planta,* Prague, August 19–23, 1991. Amsterdam: Elsevier

114. Marinissen JCY. 1994. Earthworm populations and stability of soil structure in a silt loam soil of a recently reclaimed polder in the Netherlands. *Agric. Ecosyst. Environ.* 51:75–87

115. Mathur SP, Owen G, Dinel H, Schnitzer M. 1993. Determination of compost biomaturity. I. Literature review. *Biol. Agric. Hortic.* 10:65–85

116. Minderman G. 1968. Addition, decomposition and accumulation of organic matter in forests. *J. Ecol.* 56: 355–62

117. Moore JC. 1994. Impact of agricultural practices on soil food web structure: theory and application. *Agric. Ecosyst. Environ.* 51:239–47

118. Moreno RA, Mora LE. 1984. Cropping pattern and soil management influence on plant diseases: II. Bean rust epidemiology. *Turrialba* 34:41–45

119. Mori S, Mishionwa Y, Uchino H. 1979. Nitrogen absorption by plant roots from culture medium where organic and inorganic nitrogen coexist. I. Effect of pretreatment nitrogens on the absorption of treatment nitrogen. *Soil Sci. Plant Nutr.* 25:39–50

120. Motavalli PP, Singh RP, Anders MM. 1994. Perception and management of farmyard manure in the semi-arid tropics of India. *Agric. Syst.* 46:189–204

121. Mundt CC. 1989. Disease dynamics in agroecosystems. In *Agroecology.*, ed. CR Carroll, JH Vandermeer, P Rosset, pp. 263–99. New York: McGraw-Hill

122. Murdoch WW. 1994. Population regulation in theory and practice. *Ecology* 75:271–87

123. Murdoch WW, Chesson J, Chesson PL. 1985. Biological control in theory and practice. *Am. Nat.* 125:L344–66

124. Murdoch WW, Reeve JD, Huffaker CB, Kennett CE. 1984. Biological control of scale insects and ecological theory. *Am. Nat.* 123:371–92

125. Nair PKR, Patel UK, Singh RP, Kaushik MK. 1979. Evaluation of legume intercropping in conservation of fertilizer nitrogen in maize culture. *J. Agric. Sci.* 93:189–94

126. National Research Council. 1989. *Alternative Agriculture.* Washington, DC: Natl. Acad.

127. Oades JM. 1984. Soil organic matter and structural stability: mechanisms and implications for management. *Plant Soil* 76:319–37

128. Oaks A. 1992. A re-evaluation of nitrogen assimilation in roots. *Bioscience* 42: 103–11

129. Oldeman RDA. 1983. The design of ecologically sound agroforests. In *Plant Research and Agroforestry*, ed. PA Huxley, pp. 173–207. Nairobi, Kenya: Int. Ctr. Res. Agrofor. (ICRAF)

130. Paoletti MG, Pimentel D. 1992. Biotic diversity in agroecosystems. *Agric. Ecosys. Environ.* 40:1–355

131. Papendick RI, Sanchez A, Triplett GB, eds. 1976. *Multiple Cropping. Am. Soc. Agron. Spec. Pub. 27*

132. Par JF, Hornick SB. 1992. Agricultural use of organic amendments: a historical perspective. *Am. J. Alt. Agric.* 7:181–89

133. Parr JF, Papendick RI, Hornick SB, Meyer RE. 1992. Soil quality: attributes and relationship to alternative and sustainable agriculture. *Am. J. Alt. Agric.* 7:5–11

134. Peoples MB, Herridge DF. 1990. Nitrogen fixation by legumes in tropical and subtropical agriculture. *Adv. Agron.* 44: 155–223

135. Perfecto I. 1991. Ants (Hymenoptera: Formicidae) as natural control agents of pests in irrigated maize in Nicaragua. *J. Econ. Entomol.* 84:65–70

136. Perfecto I. 1994. The transformation of Cuban agriculture after the Cold War. *Am. J. Alt. Agric.* 9:98–108

137. Persson J, Kirchmann H. 1994. Carbon and nitrogen in arable soils as affected by supply of N fertilizers and organic manures. *Agric. Ecosys. Environ.* 51: 249–55

138. Poincelot RP. 1986. *Toward a More Sustainable Agriculture.* Westport, CT: AVI

139. Pojasok T, Kay BD. 1990. Assessment of a combination of wet sieving and turbidimetry to characterize the structural stability of moist aggregates. *Can. J. Soil Sci.* 70:33–42

140. Power AG, Kareiva P. 1990. Herbivorous insects in agroecosystems. In *Agroecology*, ed. CR Carroll, JH Vandermeer, P Rosset, pp. 301–27. New York: McGraw-Hill

141. Rasmussen PE, Collins HP. 1991. Long term impacts of tillage, fertilizer and crop residue on soil organic matter in temperate semiarid regions. *Adv. Agron.* 45:93–134

142. Reeves M. 1992. *Nitrogen dynamics in a maize bean intercrop in Costa Rica.* PhD thesis. Univ. Mich., Ann Arbor

143. Reganold JP, Elliot LF, Unger YL. 1987. Long-term effects of organic and

conventional farming on soil erosion. *Nature* 330:370–72

144. Risch SJ, Andow D, Altieri MA. 1983. Agroecosystem diversity and pest control: data, tentative conclusions, new research directions. *Environ. Entomol.* 12:625–29

145. Rogers SL, Burns RG. 1994. Changes in aggregate stability, nutrient status, indigenous microbial populations, seedling emergence, following inoculation of soil with *Nostoc muscorum. Biol. Fertil. Soils* 18:209–15

146. Root RB. 1973. Organization of a plant-arthropod association in simple and diverse habitats: the fauna of collards (*Brassica oleracea*). *Ecol. Monogr.* 43:94–125

147. Rosset P, Benjamin M. 1994. *The Greening of the Revolution: Cuba's Experiment with Organic Agriculture.* Melborne: Ocean

148. Russell EP. 1989. Enemies hypothesis: a review of the effect of vegetational diversity on predatory insects and parasitoids. *Environ. Entomol.* 18:590–99

149. Russell EP. 1993. *War on insects: Warfare, insecticides and environmental change in the US, 1870–1945.* PhD thesis. Univ. Mich., Ann Arbor

150. Samson GS, Visser A. 1989. Surface-active effects of humic acids on potato cell membrane properties. *Soil Biol. Biochem.*, 21:343–47

151. Schaffer WM, Kot M. 1986. Chaos in ecological systems: the coals that Newscastle forgot. *Trends Ecol. Evol.* 1:58–63

152. Schjønning P, Christensen BT, Carstensen B. 1994. Physical and chemical properties of a sandy loam receiving animal manure, mineral fertilizer or no fertilizer for 90 years. *Eur. J. Soil Sci.* 45:257–68

153. Schnier HF. 1994. Nitrogen-15 recovery fraction in flooded tropical rice as affected by added nitrogen interaction. *Eur. J. Agron.* 3:161–67

154. Schoonderbeek D, Schoute JFT. 1994. Root and root-soil contact of winter wheat in relation to soil macroporosity. *Agric. Ecosyst. Environ.* 51:89–98

155. Scow KM, Somasco O, Gunapala N, Lau S, Venette R, et al. 1994. Transition from conventional to low-input agriculture changes soil fertility and biology. *Calif. Agric.* 48:20–26

156. Siepel H. 1994. Life-history tactics of soil microarthropods. *Biol. Fertil. Soils* 18:263–78

157. Silvertown J, Law R. 1987. Do plants need niches? Some recent developments in plant community ecology. *Trends Ecol. Evol.* 2:24–26

158. Simard RR, Angers DA, Lapierre C. 1994. Soil organic matter quality as influenced by tillage, lime, phosphorus. *Biol. Fertil. Soils* 18:13–18

159. Simmonds FJ, Bennett FD. 1977. *Biological Control of Agricultural Pests. Proc. XV Int. Congr. Entomol.* 1976

160. Smith JG. 1976. Influence of crop background on aphids and other phytophagous insects on brussels sprouts. *Ann. Appl. Biol.* 83:1–13

161. Snaydon RW, Harris PM. 1979. Interactions belowground—the use of nutrients and water. In *Proc. Int. Worksh. Intercropping,* ed. RW Willey, pp. 188–201. Hyderabad, India: ICRISAT

162. Soule J, Carré D, Jackson W. 1990. Ecological impact of modern agriculture. In *Agroecology,* CR Carroll, JH Vandermeer, P Rosset, pp. 165–188. New York: McGraw-Hill

163. Stephens PM, Davoren CW, Ryder MH, Doube BM. 1994. Influence of the earthworm *Aporrectodea trapezoides* (Lumbricidae) on the colonization of alfalfa (*Medicago sativa* L.) roots by *Rhizobium meliloti* L5–30R and the survival of *R. meliloti* L5–30R in soil. *Biol Fertil. Soils* 18:63–70

164. Stinner BR, House GJ. 1989. The search for sustainable agroecosystems. *J. Soil Water Conserv.* 44:111–16

165. Stork NE, Eggleton P. 1992. Invertebrates as determinants and indicators of soil quality. *Am. J. Alt. Agric.* 7:38–47

166. Swift MJ, Heal OW, Anderson JM. 1979. Decomposition in terrestrial ecosystems. In *Studies in Ecology.* Vol. 5. Oxford: Blackwell Sci.

167. Szczech M, Rondomanski W, Brzeski MW, Smolinska U, Kotowski JF. 1993. Suppressive effect of a commercial earthworm compost on some root infecting pathogens of cabbage and tomato. *Biol. Agric. Hortic.* 10:47–52

168. Temple SR, Somasco OA, Kirk M, Friedman D. 1994. Conventional, low-input and organic farming systems compared. *Calif. Agric.* 48:14–19

169. Tiessen H, Cuievas E, Chacon P. 1994. The role of soil organic matter in sustaining soil fertility. *Nature* 371:783–85

170. Toledo VM. 1990. The ecological rationality of peasant production. In *Agroecology and Small Farm Development.*, ed. M Altieri, SB Hecht, pp. 53–60. Boca Raton, FL: CRC

171. Toyota K, Kimura M. 1994. Earthworms disseminate a soil-borne plant pathogen, *Fusarium oxysporum* f. sp. *raphani. Biol. Fertil. Soils* 18:32–36

172. Trenbath BR. 1974. Biomass productivity of mixtures. *Adv. Agron.* 26:177–210
173. Trenbath BR. 1976. Plant interactions in mixed crop communities. In *Multiple Cropping.* ed. RI Papendick, A Sanchez, GB Triplett, pp. 129–170. *Am. Soc. Agron. Spec. Publ. 27*
174. Trenbath BR, Conway GR, Craig IA. 1990. Threats to sustainability in intensified agricultural systems: analysis and implications for management. In *Agroecology: Researching the Ecological Basis for Sustainable Agriculture,* ed. SR Gliessman, pp. 337–65. New York: Springer-Verlag
175. Ulén B. 1993. Losses of nutrients through leaching and surface runoff from manure-containing composts. *Biol. Agric. Hortic.* 10:29–37
176. van der Werf E. 1993. Agronomic and economic potential of sustainable agriculture in South India. *Am. J. Altern. Agric.* 4:185–91
177. van Noordwijk M, de Willigen P. 1986. Qualitative root ecology as element of soil fertility theory. *Neth. J. Agric. Sci.* 34:273–82
178. van Noordwijk M, Kooistra MJ, Boone FR, Veen BW, Schoonderbeek D. 1993. Root-soil contact of maize, as measured by thin-section technique; I. Validity of the method. *Plant Soil* 139:109–18
179. vandenBosch R. 1978. *The Pesticide Conspiracy.* New York: Doubleday
180. Vandermeer JH. 1969. The competitive structure of communities: an experimental approach using protozoa. *Ecology* 50:362–71
181. Vandermeer JH. 1980. Indirect mutualism: variations on a theme by Stephen Levine. *Am. Nat.* 116:441–48
182. Vandermeer JH. 1989. *The Ecology of Intercropping.* Cambridge UK: Cambridge Univ. Press
183. Vandermeer JH. 1993. Loose coupling of predator prey cycles: entrainment, chaos, intermittency in the classic MacArthur consumer-resource equations. *Am. Nat.* 141:687–716
184. Vandermeer JH, Andow DA. 1986. Prophylactic and responsive components of an integrated pest management program. *J. Econ. Entomol.* 79:299–302
185. Vandermeer JH, Power A. 1990. Epidemiology of the corn stunt system in Central America: I. The basic model and some prospects for cultural control. *Ecol. Model.* 52:235–48
186. Vandermeer JH, Schultz B. 1990. Variability, stability, risk in intercropping. In *Agroecology: Researching the Ecological Basis for Sustainable Agriculture,* ed. SR Gliessman, pp. 205–29. New York: Springer-Verlag
187. Vandermeer JH, Carney J, Gesper P, Perfecto I, Rosset P. 1993. Cuba and the dilemma of modern agriculture. *Agric. Hum. Values* X:3–8
188. Vanderplank JE. 1963. *Plant Diseases: Epidemics and Control.* New York: Academic
189. Vanderplank JE. 1982. *Host-Pathogen Interactions in Plant Disease.* New York: Academic
190. Vaughan D, Malcom RE. 1985. *Soil Organic Matter and Biological Activity.* Dordrecht: Martinus Nijhoff/Junk
191. Veen BW, van Noordwijk M, de Willigen P, Boone FR, Kooistra MJ. 1992. Root-soil contact of maize, as measured by a thin section technique; III. Effects on shoot growth, nitrate and water uptake efficiency. *Plant Soil* 139:131–38
192. Vereijken P. 1989. Research on integrated arable farming and organic mixed farming in The Netherlands. In *Current Status of Integrated Arable Farming Systems Research in Western Europe,* ed. P Vereijken, DJ Roye, pp. 41–50. *IOBC/WPRS Bull. 1989/XII/5*
193. Vreeken-Buijs MJ, Geurs M, de Ruiter PC, Brussaard L. 1994. Microarthropod biomass-C dynamics in the belowground food webs of two arable farming systems. *Agric. Ecosys. Environ.* 51: 161–70
194. Waage JK, Hassell MP. 1982. Parasitoids as biological control agents: a fundamental approach. *Parasitology* 84: 241–68
195. Weijnands FC, Vereijken P. 1992. Region-wise development of prototypes of integrated arable farming and outdoor horticulture. *Neth. J. Agric. Sci.* 40:225–38
196. Weiner J. 1990. Plant population ecology in agriculture. In *Agroecology.*, ed. CR Carroll, JH Vandermeer, P Rosset, pp. 235–62. New York McGraw-Hill
197. Werner EE. 1992. Individual behavior and higher-order species interactions. *Am. Nat.* 140:S5–S32
198. Widdowson RW. 1987. *Towards Holistic Agriculture: A Scientific Approach.* Oxford: Pergamon
199. Willey RW. 1979. Intercropping—its importance and its research need. Part I. Competition and yield advantages. *Field Crop Abstr.* 32:1–10
200. Willey RW. 1979a. Intercropping—its importance and its research need. Part II. Agronomic relationships. *Field Crop Abstr.* 32:73–85
201. Wolfe MS. 1985. The current status and

prospects of multiline cultivars and variety mixtures for disease resistance. *Annu. Rev. Phytopathol.* 23:251–73

202. Zadoks JC, Schein RD. 1979. *Epidemiology and Plant Disease Management.* New York: Oxford Univ. Press

203. Zucconi F, Forte M, Monaco A, De Bertoldi M. 1981. Biological evaluation of compost maturity. *Biocycle*, July/Aug. 27–29

204. Zwart KB, Burgers SLGE, Bloem J, Bouwman LA, Brussaard L, et al. 1994. Population dynamics in the belowground food webs in two different agricultural systems. *Agric. Ecosyst. Environ.* 512:187–98

Annu. Rev. Ecol. Syst. 1995. 26:225–48

ECONOMIC DEVELOPMENT VS. SUSTAINABLE SOCIETIES:
Reflections on the Players in a Crucial Contest

John G. Clark

Department of History and Environmental Studies Program, University of Kansas, Lawrence, Kansas 66045

KEY WORDS: sustainable development, environmental degradation, natural capital, ecological sustainability, commons systems

---

ABSTRACT

The World Commission on Environment and Development adopted and legitimated the idea of sustainable development in its report *Our Common Future*. Without substantiation, WCED claimed that economic growth and environmental protection were compatible. The 1992 UN Conference on Environment and Development at Rio de Janeiro adopted the idea, without further testing, as its intellectual core. Since 1992, the United Nations, the United States, and many other nations have created agencies to track progress toward sustainable development. In favor of the idea are individuals, including, it appears, most economists, who advocate centralization, internationalization, and rapid economic development. The opposition consists principally of people from academic disciplines, especially ecologists and humanists. This group does not communicate effectively, but if it did, it might agree that: economic development and environmental protection are not compatible; insistence by economists that all natural resources be given a dollar value is useless, if not harmful; biodiversity has intrinsic value; sustainable development weakens local autonomy; and social welfare is a key component of environmental health.

To strengthen the defense of ecosystems, ecologists, humanists, and others should apply their knowledge to practical environmental problems. By making their knowledge accessible in local political arenas, they will concurrently

225

shore up the ability of local units to protect their environments and speak with force in larger political arenas. All proponents of environmental health must become advocates of environmental justice.

INTRODUCTION

The idea of sustainable development has been defined by scores of individuals. Organization men—a recognizable elite who may work for General Motors, the United Nations, the World Bank, or a university—dominate the debate over its meaning. Scientists and economists are the leading participants, although historians, philosophers, and assorted others also contribute. In this debate, scientists, especially ecologists, are best informed and, of all the academics, save the economists, are most committed to applying their knowledge to policy. Economists come across as the least flexible; those who do open their minds to new ways of understanding the globe are generally ignored by their colleagues. Individuals in the so-called social sciences pursue their revisionist and counter-revisionist squabbles with no apparent impact on politics or policy. Most historians are content to write history that prepares no one for action, while only a few philosophers care whether their societies adhere to an environmental ethic.

The meanings of sustainable development weigh less heavily on my mind than do the operational contexts within which economic growth occurs. People with power, regardless of their understanding of nature, make decisions every day that collectively affect the health of the planet. Politicians, government officials, bankers, developers, and facilitators rule in each nation. Few of these notables know or care to know the meanings of sustainable development, a term that lacks clarity (88). I begin by discussing the origins of the idea of sustainable development, then identify those interests most enamored of it, arguing that the idea's strongest advocates ignore ecology and equity. I search for ideas that aim at ecological health and social welfare and that embrace applied values. Among the most attractive to me are those ascribing high value to local democratic control over community decisions. In 1987, the World Commission on Environment and Development (WCED) adopted the term "sustainable development" as the informing concept of its report *Our Common Future* (154). WCED fostered the unsubstantiated idea that economic development and environmental protection were compatible—indeed, that economic development nurtured environmental sustainability. In chapters on food security, species and ecosystems, energy, and other topics, WCED employed sustainable development as a mystical goal rather than as a concrete objective that might be achieved by particular actions. Economic growth at 3–6% would, WCED promised, safeguard ecosystems and further human welfare (154, pp. 50–52).

Prior to the UN Conference on Environment and Development (UNCED) at Rio de Janeiro in June 1992, the concept of sustainable development had attracted only modest criticism from a few individuals (11, 17, 101, 106, 135). But UNCED's preparatory committees ignored those commentaries, adopting the idea without qualification. This untested notion, infiltrating virtually every discussion, elicited widespread acceptance among the government representatives who crafted the documents discussed at Rio. This reductionist concept appeared in the title of the key Rio document *Agenda 21. A Blueprint for Action for Global Sustainable Development into the 21st Century* (138) and countless times in the body of the report.

One *Agenda 21* recommendation (p. 275) called for the creation of a UN Commission on Sustainable Development, which was approved by the General Assembly in 1992. Each nation adopting *Agenda 21* promised an equivalent organization. President Clinton formed the President's Commission on Sustainable Development (PCSD) in 1993; its members tour the United States promoting the idea of sustainable development. PCSD promises harmony between environmental protection and economic development but is unable to present data supporting the value of the idea (39, 92).

Evidence surrounding us clearly indicates that people engaged in economic development at local, national, and global levels are indifferent to or ignorant of the concept of sustainable development. Look outside! Everywhere you see fields asphalted, top soil flowing into storm sewers or contaminated by chemicals, wetlands drained, seashores developed, air and water polluted, grasslands and forests ravaged .... These are common sights in the First World (developed countries). Restraint and careful planning are difficult to discern; uninhibited growth predominates.

Lesser developed countries (LDCs, South, Third World) suffer great degradation of their environment because of low standards of welfare. Infrastructural inadequacies force billions to wash with, cook with, and drink water contaminated by human, animal, and industrial wastes. In Brazil, millions of citizens are condemned to life in *favelas* (slums). One, Rocinha in Rio, contains as many as 300,000 residents. Built on a steep and eroding mountain slope, Rocinha's people rely on gravity to move their wastes down the mountain side, and they function without access to health or welfare services (JG Clark, personal observation). Yet, the Brazilian government and corporate planners scheme to crisscross the magnificent Pantanal wetlands and plains with a canal and to cut down a vast stretch of the diminishing Atlantic rain forest north of Salvador for a luxury resort complex.

Daily, millions of individual decisions, based on conventional economic understanding of returns on investment, unleash activity that further degrades the environment, including the land of tribal and rural folk in Cambodia, Nigeria, Kenya, Brazil, and other nations (3, 33, 88). Sustainable development, say WCED (154) and other growth promoters, promises the resolution of environmental crises and poverty without any sacrifice in standards of living among affluent people. No wonder sustainable development captured the allegiance of prosperous decision-makers in every country.

The idea of sustainable development offered by WCED and UNCED owes more to economists than to scientists or humanists. With modest differences in emphasis, such early advocates of self-sustained growth and the modernization of the Third World as Hirschman (68), Rostow (124), and Black (24) identified several prerequisites of sustained development: availability of long-term development capital, stable industrial labor force, urbanization, technological capabilities, risk-taking entrepreneurs and innovators, large-scale production units, and commercial agriculture. As nations achieved these, economic growth and modernization followed. Emulating the experience of the First World spelled success for the Third World.

In collateral studies, social scientists (6, 20, 76) advanced as further prerequisites the democratization, bureaucratization, and rationalization of political structures. To achieve this, modernizing elites in the LDCs would reshape national institutions, many of colonial origin, into the image of American or European institutions. Having democratized themselves, the modernizing elite would, with massive aid from the First World, reshape their economies so as to fulfill the requirements for self-sustained economic growth. All segments of the population would partake of economic betterment. In none of this did the environment appear as more than the housing for valuable resources.

WCED (154) and *Agenda 21* (138) failed to raise these unfulfilled ideas to a new level of meaning. So deeply attuned were First and Third World national elites to the ritualistic formulas of economic development that they could do no more than intone past prayers for augmented growth in GNP per capita. Concomitantly, national elites campaigned successfully for expanded international trade. True, WCED wrote with feeling about meeting the essential needs of the poor, conserving and enhancing the resource base, and avoiding further environmental degradation. Both WCED and UNCED underlined the need to merge environmental considerations with economic goals in decision-making (why not economic considerations with environmental goals?). Strengthening democracy received attention (154, p. 41, p. 57, pp. 60–61, p. 63).

"At a minimum," according to WCED, "sustainable development must not endanger the natural systems that support life on Earth: the atmosphere, the waters, the soils, the living beings" (154, p. 45), necessarily implying "harmony among human beings and between humanity and nature" over many generations (154, p. 65). Some argue that these words are no more than descriptive of the state of the world. In this best of all possible worlds, things are even getting better, or so Simon (127–129) assures us.

Simon's cornucopianism, in its modern guise, can be traced to resource economists like Hotelling (74), who developed theories of optimal rates of resource depletion. This happy concept, as I understand it, assured humanity that at the very moment a resource was fully depleted, it would no longer be desired. The utility of the depleted resource, or its social purpose, would be fulfilled by a new resource created by human technological ingenuity. Optimality theory adheres strictly to supply and demand forces and ignores ecological consequences.

One cornucopian, Barnett (18), blasted Meadows et al (98) for asserting the finite capacity of the Earth to provide resources and to absorb waste. The environment, promised Barnett, had nothing to fear from economic growth (18, p. 146–47). The cornucopians, politely described as superoptimists by Tisdell (136), depicted a world in wonderful condition. They ignored pollution and ran roughshod over reasonable estimates of the globe's carrying capacity. However, they made eminently good sense compared to a bizarre prognostication by Ausubel (15) that dismissed weather as increasingly irrelevant to a global population soon to live under domes.

The WCED interpretation of sustainable development required, however, a profound shift in philosophical and scientific content from the meaning assigned those words by a prior user. At the 1981 meeting of the General Assembly of the International Union for the Conservation of Nature and Natural Resources (IUCN), delegates approved a "World Conservation Strategy" (WCS) that employed the term "sustainable development" at least twice (81). By sustainable development, IUCN clearly meant the "ecologically sound use of natural resources" or the "sustainable utilization of natural resources" (81). Believing that environmental protection and economic development could be harmonized, the IUCN called on nations and development agencies to apply WCS. Unfortunately, no takers appeared. Nonetheless, IUCN employed the concept of "sustainable development" as if it read "ecologically sustainable development."

Champions of sustainable development such as WCED subordinated sustainability to development. They were unaware of, or unsympathetic to, individuals who viewed the globe as subject to natural laws that limit growth and who advocated social ideas that question the desirability of short-term, self-sustained growth (see 26, 40–42, 46, 62, 98, 108, 126).

Boulding (26) first, and then Schumacher (126) and Daly (40–42), indicted conventional market-driven economics for ignoring humanity's dependence upon the natural world. In a striking metaphor, Boulding likened the people of Earth to travelers on a spaceship who must forever recycle their wastes to survive (26). Daly condemned growthmania and its supporting cornucopian economics, advancing a case for a steady-state economy in which capital stock (natural capital, in Boulding's terms) would be maintained, undepleted, for the use of successive generations (40, 108). Cautioning economists that failure to acknowledge the first and second laws of thermodynamics imperiled all life, these scholars emphasized that we live in a condition of high entropy compared with preindustrial peoples of the past and present.

In a book sponsored by IUCN, Dasman (46, pp. 27–35) argued the relevance of ecological concepts to all development. Perhaps reflecting the conservative mind-set of IUCN, Dasman refrained from insisting that economic development defer to biological life systems. Implicitly, this work suggested the possibility of melding conventional economic goals and ecological principles, but the matter of priority was not pursued. Meadows et al (98) argued for the primacy of ecological values, presenting two arguments: 1) Sometime soon we will run out of many natural resources, and 2) an expanding population and rising per capita consumption will soon produce too much waste for the world to absorb. In 1974, Barnett (18) attacked Meadows et al on the first count but ignored the second. Meadows et al's error regarding the first point— by some decades or centuries—delighted the cornucopians but should not obscure the accuracy of the second point. In a finite system, "constraints ... can act to stop exponential growth" (98, p. 156). These might be sociopolitical restraints on birth rates and family size. Experts in environmental health suggest the readiness of new viruses and mutated pests to pounce on humanity and our sources of sustenance (99, p. 516).

The ultimate restraint in a world of oxygen breathers and water drinkers must be biological. Does not the word "sustainable" imply limits to growth? Organisms almost invariably expand their populations beyond the limits of their resources, at which point the autophage effect is triggered and communities begin to devour life-sustaining resources at a suicidal rate (62). Dramatic population decline ensues. Until humanity developed industrial technologies that raised food production per capita and muted the impact of epidemic diseases, the autophage effect, or its political economy equivalent, the Malthusian principle, acted to slow human population growth. Hanson (62, p. 10) predicted the resurrection of the autophage effect as consumption and waste exploded. The system would either crash, Hanson reasoned, or a new economic order based on ecological principles would emerge. Hanson may have been too optimistic.

WCED's version of sustainable development owes more to the Hotelling (74) belief in the technological quick-fix and Rostow's (124) prescription for self-sustained growth than to the reasoning of Boulding (26) or Hanson (62). Rostow's (124) book, after all, was subtitled *A Non-Communist Manifesto*. To Rostow, Marxist economic determinism and Communist tyranny threatened free people. Pursuing Rostow's stages of growth started nations along the road to a democratic-controlled capitalism (124, pp. 151, pp. 164–65). As Vernadsky chanted during the heady days of Allied victory in 1945, "[O]ur democratic ideals are in tune with the elemental geological processes, with the laws of nature, with the noosphere.[1] Therefore we may face the future with confidence. It is in our hands. We will not let it go" (142, p. 12).

The premonitory essays of Boulding (26), Daly (40–42), Ehrlich (53, 54), and Commoner (34) bore little immediate fruit. Between 1974 and 1978, Environmental Conservation,[2] for years alone among scholarly biological journals in its energetic advocacy of ecological sustainability and biodiversity, printed eight articles suggesting the unsustainability and environmentally degrading consequences of human activity. In 1974, Ehrlich (54) attributed worrisome ecosystem collapse to human population pressures. The impact of fossil-fuel burning on the environment, depleting the ozone layer, causing acid deposition, and threatening global warming formed the substance of several articles (25, 84, 121, 155), while others focused on the vulnerability of biophysical life support systems (89, 134, 137).

Ignoring both increasing pollution and exacerbated maldistribution of wealth within and between nations, national leadership elites relied on growth to cure recession and inflation during the 1970s and 1980s. They believed that stimulating growth required only the unleashing of market-driven economic forces. Structural adjustments such as privatization, deregulation of business, abandonment of health and welfare subsidies, and internationalization of economies —strategies advocated by the World Bank (153)—would protect accumulated wealth and open new opportunities for investors. Their theory that something might trickle down to the poor formed an essential part of the political agenda, attesting in part to the failure of WCED and UNCED to fashion an idea of sustainability that encompassed equity, justice, and ecological health.

WCED (154) and UN rhetoric (138) included all the elements that seemed to herald a paradigm shift in attitudes regarding the environment. Efficiency and technology, they said, would produce more output with less energy input. Cultural and ecological awareness would preserve biodiversity. Population and ecosystem potentials would be harmonized. Agricultural production would benefit from a redefinition of land use and rational national agricultural policies. Environmental factors would be fully integrated with economic development decisions at all political levels. Equity and democracy would prevail. Nevertheless, hundreds of pages in *Our Common Future* (154) and *Agenda 21* (138) consistently relied on economic growth and the transfer of new and additional funds and technologies from the North to the South to create the false promise of a viable future. Rates of economic growth, insisted WCED, must be accelerated in order to achieve sustainable development.

Sustainable development of the WCED variety means business-as-usual.

---

[1]Noosphere is defined as a new state of the biosphere, the realm of "freely thinking humanity."
[2]Formerly, this was the *Biological Conservation,* 1968–1974, which was published in collaboration by IUCN, UNEP, WWF, and the International Association for Ecology.

This satisfies business interests, government officialdom around the world, powerful international institutions such as the World Bank, the International Monetary Fund, UN Food and Agricultural Organization, and national elites whose internationalized assets are relatively safe from the erratic fluctuations of their own economies. "Promoting Sustainable Development" is an avowed goal of General Motors Corporation (57, p. 21). GM consistently opposed environmental regulation. It now touts its fuel economy improvements since 1974, though it has always opposed federal standards and any rise in them (57, p. 14). Meanwhile, in Sao Paulo, Brazil, GM's huge plant refuses to comply with Sao Paulo State water quality standards. Employing thousands of people, GM threatens to move if coerced.[3]

Like GM, the World Bank brandishes the term "sustainable development" whenever possible. The World Bank (153) contends that growth is the key to eradicating poverty and protecting the environment. The growth myth, the core of WCED's sustainable development, relies on the application of structural adjustments in the LDCs that increase user fees on water, sanitation, electricity, and other utilities, build airports, canals, hydroelectric facilities, and highways, privatize public services, and internationalize economies.

For 50 years, the World Bank has pursued the same policies, yet as many as 1.7 billion people lack access to safe drinking water and sanitation services. According to Korten (85), a Manilan nongovernmental organization (NGO) leader, and Rich (122), an Environmental Defense Fund lawyer, the World Bank perpetuates the myth of trickle-down welfare. Confounding the Bank's claims, a UN Development Programme report in 1992 (139, p. 3, pp. 34–35) showed that the richest 20% already receive 83% of world income while the poorest 20% receive less than 2%. The rich allow very little wealth to escape their hands.

Pressures from NGOs may have somewhat diminished the Bank's readiness to fund environmentally damaging projects (14), but endemic distrust of the World Bank persists among environmental NGOs. To participate in the fiftieth anniversary of the World Bank in 1994, US and LDC NGOs sponsored a "50 years is enough" campaign, dedicated to the dismantling of the Bank. However, the Bank appears impervious to change, perhaps now more entrenched than ever as a consequence of UNCEDs acceptance of the Global Environmental Facility, managed by World Bank and UNDP, as the primary funding agency for Rio's sustainable development initiatives.

General Motors, the World Bank, and the International Chamber of Commerce (78, 79) regard sustainable development favorably because it poses no threat to conventional economic development. Other interpreters of sustainable

---

[3]*Author interview with the environmental compliance manager of a British thread firm in Sao Paulo, Brazil, July 1994.

development, primarily economists, government officials, and employees of international organizations, defended the workability of the idea within the parameters established by WCED and UNCED. Internationalizing predilections and the presumed congruence between environmental protection and economic development provide the intellectual cement for this group.

The internationalizers (29–31, 56, 75, 91, 94, 95, 119) advocate some variety of "one world," a supranational system that integrates national and international systems, perhaps via a global council or world forum. Typically, the internationalists promote global free trade and such UN-sponsored conventions as the treaties on ozone, climate change, and biodiversity. Claims are made that governments will then "secure their citizens future" (56, p. 49) by reducing poverty and protecting the environment. Others claim that international planning has proven the mutual dependence of development and environmental management (30, p. 300; 91, p. 198).

Not everyone applauded the internationalization of economies and polities. If global international relations had entered a new stage in its evolution, as Caldwell believed (31), who would benefit and who would lose? Three recent volumes (58, 82, 125) offer criticism of globalization, each drawing heavily from the experience of communities, indigenous or not, managing ecosystems in common. Chapters in Sachs (125) focus on the erosion of local control and the degradation of local ecologies when environmental management shifts to national and global arenas of power. Johnston & Burton (83, p. 213) assert that "Development processes ... and governments, all deny human rights," an indictment pervasive in chapters of *Who Pays the Price?* (3, 4, 32, 132) that investigate the impacts on indigenous people of economic development and the absence of a protective net of civil and environmental rights. Sustainable development, lacking a land ethic, indifferent to the carrying capacity of local ecosystems, and rooted in a materialist world view, as Worster (157, p. 153–55) itemized the idea's flaws, cannot protect ecologies from exogenous forces.

Grumbine (58) advanced the idea that local ecological integrity depends upon a defensive bulwark consisting of science, values, and common sense. Biology can identify ecosystems and the communities within them, but culture determines how ecologies are treated. Contributors to Grumbine (58) offered the principles of conservation biology as the most solid defense against ecosystem destruction. But those principles—coevolutionary processes that produce diversity in natural communities and that depend upon size and other factors to preserve genetic pools—imply parallel value judgments. Soulé (131, p. 42–44) defined those values as: "Diversity of organisms is good .... Ecological complexity is good .... Evolution is good."

Hardin (64) no longer assumes, as he did in "Tragedy of the Commons" (63), that commons were unmanaged. Nor is it assumed any longer that native peoples acted as ecological saints. As early as 5000 BC, people in the basin of Mexico overexploited their resources (73). Living unsustainably is the sin not only of modern societies. Hoopes (73) and Ludwig et al (90) remind us that sustainability has rarely occurred in human history.

At a generalized level, people involved in common property systems participate closely in its operation, cooperate to ward off external threats to their resources, and normally recognize the desirability of a steady-state society. If success attends these efforts, the system at some moment may be described as sustainable. To biologists, such an achievement suggests the protection of biodiversity. But are/were these people conservationists?

Traditional peoples and conservation biologists may, Redford & Stearman (118) alert us, define conservation differently. The former aim at preservation of general habitat characteristics, not the conservation of particular species. Indeed, traditional peoples may wish to improve their positions by increasing the flow of natural products from their domains (1, 5, 21). In short, they may desire to develop their common property. Raising production can place culture, and perhaps the environment, at risk. Is this their prerogative?

Among a large number of analyses of common property systems, several impressed me. The editors of *The Ecologist* (50) charged nations, WCED, and the World Bank, among others, with subverting local control over environments. *The Ecologist* employed the term "enclosure" to describe the historical process of transforming all resources into commodities that destroyed common property systems and marginalized their people. Bahuguna (16) and Guha (60) document this by recounting the story of the Chipko (tree-hugger) movement in India, a local defense against deforestation with Buddhist and Ghandian roots. From India as well came the story of the invasion of traditional fisheries by modern trawlers of capitalist venturers. Kurien (86) documented the cooperative and individualist responses of the fisherfolk to high technology operations driven by the demand for protein of an exploding population. After the fishing stock suffered severe decline, political agitation produced controls that prompted the trawlers to move elsewhere. However, local use of the outboard motor also caused a reduction of the catch. Moreover, the local population and its demand for fish were growing. Vulnerability characterized this traditional fishing society.

Chipko's activists and Kurien's fisherfolk satisfied Vivian's (143) definition of people engaged in common resource systems who managed for sustainable development. Vivian's understanding of that term, however, incorporating economic equity, justice, and resource rights into a relatively stable situation, diverged radically from the intention of WCED. By introducing civil and resource rights into the equation, Vivian (143), Bahuguna (16), Guha (60), Ostrom & Becker in this volume, and Kurien (86) move the discourse into a realm that directly challenges the power of governing elites—a realm insufficiently explored by biologists and others dedicated to ecosystem conservation.

Biologists like Alcorn (5) and Bennett (21) would safeguard the indigenes [the means] to preserve the forests [the ends]. Vivian (143) and Kurien (86) would defend the basic rights of traditional peoples [the ends] by preserving their habitats [the means]. Encroachment upon such rights destroys livelihoods and cultures. *The Ecologist* (50) sides with Vivian and Kurien as does, I believe, Wilson. Writing of the stewardship of nature, Wilson (151, p. 7) expressed the belief that "people and nature must be mutually self-reinforcing." As these views form part of the perimeter defense against the invasion of common property systems by outside economic interests, an integrative effort is required. Biologists should be more explicit about rights when they attempt to design ecosystem management plans. The intrusion of nonlocal capital, like tourism, under the guise of sustainable development, displaces and disempowers people.

The relevance of common property systems to today's world is vigorously proclaimed by Berkes & Farvar (22, p. 13–14) and by Berkes (23) as providing democratic and reasonable ways to manage property and sustain ecologies. However, not a few traditional resource systems are run by authoritarian elites who enrich themselves by cooperating with outside economic and political interests. The presumed efficiency of common property systems is a weak argument because of their small output. Grounding the defense in biodiversity has protected very little. Claims on behalf of the excellent stewardship exercised over habitats by traditional peoples afford no protection to Brazilian Indians (3, 132). To preserve these traditional systems requires acknowledgment in the highest law of the land of the prior rights of traditional peoples to their resources. Achieving this, however, guarantees neither enforcement nor sustainable use.

The disciplinary preoccupations burdening economists and ecologists are as likely to damage as to strengthen the case for common property systems or even the case for strong, self-determining local communities. Economists present a clearer danger both to local power and ecological sustainability. The narrow economic definitions of product and property as carrying a monetary value cannot encompass or be relevant to many cultures. The economist Pearce (109, 110) would impose a universal monetary value on the various uses traditional peoples make of the flora, fauna, and minerals, although these have no cash value to the users. Undeterred by this, economists, whether ecologically sensitive or not, enter numbers and values in their ledgers.

For their part, ecologists (see 149) may insist that biodiversity, as the single most important prerequisite for global health and welfare, should be defined as the common heritage of mankind. Raven (117), a politically astute biologist, advocated an international treaty to protect biodiversity, asserting the equivalence of biodiversity and global sustainability and the necessity for population stability. While Raven and Pearce (109, 110) recognized that local management of these resources infrequently benefits local people, they opted to shift the role of protection to national and global authorities. In contrast, Vivian (143) would protect the locals in their rights to these resources. In the Raven and Pearce scenario, a pharmaceutical firm might negotiate with a government for the privilege of seeking medicinally valuable species in a tropical rain forest. Vivian would insist upon direct negotiations between local people and the firm with the government protecting its citizens against unfair tactics. Danke (43) described and analyzed a process in the United States whereby civil units negotiate directly with firms to safeguard local ecosystems and might even offer economic rewards to firms that pledge to adopt sustainable practices.

Debate waxes and wanes in local, national, and international forums targeting environmental protection and economic development. A constant, however, is the unwillingness of governments to empower their citizens or to institute systems of social justice. In the First World, recent analyses (54a, 86a) suggest that democracy functionally operates only at a certain and rising income level and even then benefits some races or ethnic groups more than others.

The crisis shaking the world is not environmental but political. Without functional democratic societies in which social justice is embedded, business-as-usual will prevail and environmental destruction will persist. Sustainable development á la WCED promotes normal economic growth as the remedy for environmental degradation. Thus, this idea can contain no ecological or moral content. To be effective in dispelling the beliefs and moderating the behavior of those espousing sustainable development necessitates thorough democratization, accurate scientific and technical information, and the effective commingling of the two. An idea of sustainability in which ecological health takes precedence over economic development must be anchored in ecological principles and social justice. Notwithstanding the routineness of most of their work, ecologists have advanced our perceptions of sustainability further than most scholars; historian Worster's (157) work relies heavily on both broad and focused ecological studies. Ecologists seem less susceptible to abstractness and romanticism than do many social scientists and humanists. The most exciting students of Earth's biophysical status learn from one another.

Biologists (particularly ecologists), atmospheric specialists, and experts in public health should make their knowledge accessible and useful in local and state forums. Scientists must cease hiding behind walls of objectivity and become advocates of policies that remain consistent with their scientific knowledge. Some (e.g. Ehrlich and Wilson) have been involved for decades, but mostly in national and international arenas. Communities, counties, and states—perhaps less alluring stages—urgently need good applied science to defend their environments. The science departments of major state universities could form ad hoc oversight teams to scrutinize the actions of state environmental agencies and to offer constructive criticism.

For decades, ecologists such as Holdgate, Kassas & White (69) have presented scholarly warnings of intensifying global environmental deterioration. Malone (97), in 1976, described scientists as linking humanity to the natural biosphere. Scientists, well aware of the steady contamination of air, water, and soils, must become sophisticated about such critical social issues as population, food security, and access to natural resources. Malone urged scientists to illuminate issues relevant to environmental policy. He praised scientists who preferred "the advocacy role" (97, p. 87). Slobodkin (130) warned that ecology will lose its relevance unless its research is driven by practical questions. Frustrated by the failure of politicians to act with common sense on the basis of the best available data, Myers (102) asserted that biologists were as competent as economists to engage in policymaking.

But normal science, conducted by thoroughly institutionalized practitioners in the laboratories of government, corporations, and universities, emphasizes uncertainty and, therefore, the need for more research money. Two decades ago, Dasman (46, p. vii) suggested that need for effective application of ecological principles exceeded the need for more research. The values and strategies enunciated by two recent and influential statements on the role of science in protecting the planet diverge dramatically from the arguments of scientists (12, 13, 46, 100–102) who assert that current knowledge is capable of retarding, if not reversing, the further deterioration of soils, water, and forests.

The Ecological Society of America (ESA) (49) and the US National Research Council (140) both fashioned powerful cases for massive infusions of research money. ESA's Sustainable Biosphere initiative, launched in 1988, accepted as articles of faith reliance on investigator-initiated research and the retention of disciplinary integrity, each of which conforms to the research structure of universities. Complexities and uncertainties, rather than the availability of data useful now in defending the biosphere, were stressed to justify funding aimed at global change, biological diversity, and sustainable ecological systems (49).

Congruent with the ESA message were urgent calls by Wilson (150, 151), Holloway (71, p. 12), Resser, Lubchenko & Levin (120), and the US National Research Council (140, pp. 37–40) for the emergency recruitment of systematists to conduct national inventories of species, especially in remaining tropical rain forests. Ludwig and coauthors (90, pp. 17–18) objected to the simplistic assumptions that ignorance is the problem (19, pp. 640–42) and that more of such research would yield remedies to resource exploitation. The wild species of Brazil's Pantanal and Atlantic rain forests are plundered and destroyed by squatters and poachers who seek survival and by ranchers, resort developers, and industrialists who seek profits. The Mississippi River between Baton Rouge and New Orleans may be the most toxic stretch of water in the United States. In none of these cases is ignorance the explanation.

Scientists lack the fortitude to engage head-on with life-threatening biological and physical trends. As currently organized, ecology is unlikely to break through conventions. True, ESA admonishes its members to consider means to influence environmental policymaking, but this is a remark hidden at the tail end of a report on funding priorities (49, pp. 401–405). Nonetheless, ESA desires a sustainable biosphere (49, pp. 373–77). Ecologists possess dynamic potential to nudge polities in the proper directions if they would loosen the self-imposed bonds of scientific objectivity. Two decades ago, the botanist Poore (114, pp. 244–45), along with Malone (97), admitted the need to couple values with political skill in convincing society of the vulnerability of its natural capital.

Boulding's (26) recognition of the subsidies humanity draws from nature (10) inspired him to characterize natural capital as intrinsically valuable, of incomparably higher order than conventional capital stock. Economists Costanza (36–38) and Daly (40–42), collaboratively and individually, adopted and refined the idea of natural capital (38). A minimum necessary condition for sustainability, they wrote (36, pp. 16–17), "is the maintenance of the total natural capital stock at or above the current level." As the qualitative enhancement of life depended upon natural capital, these inventive individuals sought like Pearce (109, 110) to create an accounting system that captured the dollar value of natural capital. Costanza and Daly, while employing the term "sustainable development," were not wedded to it. Daly recognized natural capital as an inevitable limitation on development. But unable to carry the economic argument further, he had recourse to moral argument. Daly, then, defined as moral imperatives population control by the Third World and reduction of consumption by the First World (41). Why remain bound by an accounting mentality that blocks consideration of alternative standards of valuation? The idea of sustainability discussed by Goodland (this volume) bears small philosophical resemblance to that of WCED. Goodland considers environmental sustainability a rigorous concept. Is it not more than that ... is it not a good, a right?

Economists are able to explain as rational the actions of people who elect to deplete natural resources, including natural capital, and destroy ecosystems. They apply a theory of discounting in which a good is worth more now than later—a bird in the hand is worth two in the bush. Conservation is anathema. By employing the Hotelling (74) idea that supply and demand for a resource will reach zero simultaneously, economists offer technology as the substitute for the depleted resource. To an extent, Costanza (38), Daly (40), and Pearce (110) diverge from such economic wisdom. But they insist that every resource carries a dollar value. Dasgupta (45), an economist, approached a concept of natural capital, but by insisting that natural capital be priced in a market, he was unable to pursue the idea to its logical and ecological roots. Warford (145) also pinned his hopes for sustainable development on market forces, albeit constrained by some regulation.

The monetization of natural capital often assumes the form of cost-benefit analysis (17) and "willingness-to-pay" (WTP) (110). In analyzing the tradeoffs between resource use and environmental protection, Barbier advanced a theory of natural resource scarcity as a costing methodology (17, pp. 92–115). But this theory has practical application to a very limited number of substitutable resources. How does one valuate air scarcity or the death of a lake or the deforestation of vast areas? Pearce's (110, pp. ix–x; 3, 12, 21–23) WTP assigns value on the basis of the preference of individuals for environmental enhancement. Each consumer calculates the cost or benefit of preserving, say, the Grand Canyon. If the costs exceed the benefits, the resource is expendable. Happily, the Grand Canyon's preservation value for the United States, in Pearce's valuation, came to $7.4 billion while the costs totaled but $3 billion. With a net preservation advantage of $4 billion, or less than the cost of decommissioning a nuclear plant, there is every reason to protect the Canyon.

Most scientists and humanists find such accounting useless, if not pernicious. Although McNeeley (93) doubted the assertion that biological resources were beyond value, other scientists rejected economic rationality as of dubious benefit to the environment. Robinson (123) dismissed as impossible the valuation of the atmosphere but accepted levying penalties on polluters. Ehrenfield (51) and *The Ecologist* (50, pp. 111–28), seconded by several others (2, 61, 107, 116), viewed biodiversity as beyond economic valuation. Commodification and pricing deny intrinsic value and permit extinction or contamination without regard to future consequences. Economic utilitarianism precludes the simple recognition that nature's subsidies are unquantifiable. The methods of economists are, wrote Wilson (152, p. 305), "elastic and poorly calibrated. They have no sure way of valuating the ecosystem services that species provide singly and in combination...." The indifference of economists to biological and cultural values needlessly marginalizes their environmental role. Daly & Cobb's efforts (41) to meld economics and values, consonant with ideas found in Worster (157) and Wilson (152, pp. 319–22), carry inherently greater potential to do environmental good than do accounting and indexing schemes. Few ecologists or humanists consider integration with economics useful in the struggle to protect ecological and human health.

Wilson (152, pp. 322–29) used the term "sustainable development." Worster concluded that the "smooth words of sustainable development may lead us into quicksand" (157, p. 144). Yet, they aimed at similar goals, each upholding the pre-eminent importance of biological over material wealth, of natural capital over human capital. Admitting to the uncertainties of measuring the carrying capacities of ecosystems, they embraced as a moral imperative the rights of future generations to an undiminished natural capital. WCED's (154) undeviating materialism contrasts sharply with this practical eco-centrism.

Aware of the degree to which science is hostage to politics and appropriations, a few, but too few, ecologists (see 28) advocate more science directly applicable to ecological sustainability. But ecology, potentially the most important of the life sciences, languishes insufficiently connected to the real world, or so Peters believes (112, pp. xi–xii, pp. 1–3). My reading of Pimm (113) and Pomeroy & Alberts (115) lends credibility to Peters' assessment. While writing (113, 115) of ecosystem perspectives, population resilience, community structure, extinction, habitat stability—a conventional menu of issues—the sole purpose seemed to be the validation of ecological biology. The *Annual Review of Ecology and Systematics* 1992 special section on global environmental change (10) offered little of obvious use to policymakers. In 1990, the special issue of *Ambio* (8) on sustainable development failed to discuss the concept but did promote research dedicated to raising agricultural production. The environmental benefits of Ecological Society of America's appearances before Congress, as applauded by Holland (70), remain unclear to me.

Voices are raised championing an environmental ethic that marshals both ecological and welfare values in defense of the natural world. Disagreement about how much nature should be saved is often linked to the meaning of agricultural sustainability. Deep ecologists such as A Naess (103, p. 30) proclaim the equality of all life, while agroecologists such as Altieri (7, pp. 166–167, pp. 196–198) advocate an agriculture without chemicals or machines. To some, such views may seem unconnected to agricultural systems forced to feed a steadily rising population. Altieri's visions fail to elicit support from agriculturalists, say from India, who seek production with the least unsustainability (66, pp. 274–76)

In this crucial debate over agricultural sustainability, science and technology are identified as both culprit and savior. Perlas (111) and others (65, 71, 72, 100) view specialized agriculture as grossly inefficient, toxic, and genetically suicidal, but centralist and technocratic resolutions are approved by DuPuis & Gesler (48) and others (27, 35), who maintain that sustainable farming in the Third World necessitates an increase in crop yields through new technologies.

The debate over sustainable agriculture and resources runs afoul of population growth. In Latin America 62% of the people live in poverty, providing an ever growing petri dish for infectious disease organisms. At the International Conference on Population and Development in Cairo, September, 1994, heads of state spoke eloquently about raising standards of living. But two poor Indian women delegates, acutely aware of their lack of power, bemoaned their inability to return home with a hopeful message about aid for family planning (133a, p. 16). Population presses sharply against sustainable uses of air, water, and land, not only in the Third World, but most critically there.

In a world in which not all ecosystems can be saved, the question is who will decide (144). Conservation biologists assume they should make the decisions, but a more just claim can be made by those who live with or near the biodiversity to be saved. The agendas of conservation biologists and inhabitants may not be congruent. Nor can agreement among the inhabitants be assumed (above 86). Western (146, 147) lamented the peripheral role of biology in conservation decision-making. One reason may be that biologists such as Westman (148) too often ignore the needs of residents. In an article "What is ecosystem management?" Grumbine (59) allowed for accommodation to human use if it did not weaken the integrity of the area. That may be unacceptable to inhabitants who can demonstrate prior rights.

A thoughtful essay by Ehrenfeld (52, pp. 247–50) resonated with concern that internationalization threatens local distinctiveness and autonomy. As resources—air, water, soils—become scarce because of pollution and overuse, powerful economic and political organizations will, as Thapa & Weber (133) observed, intensify their efforts to monopolize them. In the Philippines, inaccessibility to good land, runaway population, and rampant government corruption drove one third of the population into the uplands, where they were condemned to permanent poverty due to steep slopes, thin soils, erosion, and low crop-yields (47). Poor states, rarely democratic, offer to foreign capital valuable concessions, many of which come at the expense of workers and poor people.

The struggle for water intensifies throughout the world. Scientists insist that governments adopt river basins as the nucleus of management systems. Newson (105) argues that the water sustainability thus gained must be consonant with social justice. The imperfect allocation of water and resulting stresses must be linked to increasing demand (100, pp. 18–29), but inequities in control and distribution are equally significant. Tens of thousands of communities lack safe water and suffer attendant illnesses. Leadership elites channel development funds into projects that yield their clients more immediate returns than would water or sanitation systems.

We have known for over 30 years that excessive burning of fossil fuels and forests produces serious air pollution and may alter the climate. National governments negotiated international treaties to reduce acid deposition, ozone depletion, and global warming. These new political arrangements protect the economic interests of the Western industrial systems that produced the pollution. In 1987, Hildyard (67) condemned the World Resources Institute (WRI) for assigning to the poor and to the rich equal blame for deforestation. Undaunted, WRI's report of 1990–1991 (156, pp. 3–17) shifted the blame for global warming to the poor nations.

Centralized international regimes such as the UN Framework Convention for Climate Change establish program priorities on the basis of economic interests that are national and international, not local. During the 1980s, the negligence of Union Carbide in India and Exxon in Alaska caused life-destroying disasters. Today both companies operate as if they had done no harm. Meanwhile, as several studies convincingly demonstrate, people in Nepal and sub-Saharan Africa desperately search for firewood because they are too poor to consider more efficient woodstoves or kerosene (44, 55, 77, 133, 141). These people are not ignorant of the damage they do. They would accept sound ecological management if accompanied by equity and justice. Rarely do discussions of climate change descend to such mundane levels.

During the next 20 years, the poor will overwhelm the world's cities, particularly those in the Third World. During the formative years of the industrial world, its cities outdid all human habitats in ecological destruction. Today, such Third World cities as Cairo, Sao Paulo, and Mexico City have joined the ranks of heavily polluted and polluting mega-cities. Plans offered in the First World for sustainable cities enjoin urban leaders to encroach less on nature, increase the density of buildings, restructure transportation, reduce energy use, improve urban ventilation, and on and on in a vein that seems thoroughly disconnected with the reality of a Delhi as described by Varshney & Aggarwal (141, p. 39). Although P Naess (104) pessimistically assessed the willingness of urban residents to accept environmental regulations, the ambient air quality of most American cities has improved since 1970.

Sustainability requires the simultaneous application of ecological principles and social justice. An argument that rejects equity must be considered as morally reprehensible as an argument that rejects ecological sustainability. Holdgate (69), Malone (97), and Wilson (152) condemned anthropocentrism while applying ethical values to conservation. These scientists, along with nonscientists such as Sachs (125) and Worster (157), fault contemporary economic theory for its unalloyed application of anthropocentric and instrumentalist values. They agree that ecological sustainability and the economic development promoted by WCED are incompatible. Modern economic theory nurtures greed.

Economic elites in each nation seek common ground and profit. UNCED's preparatory process afforded a comfortable habitat in which national governments drafted documents that committed their governments to nothing. Powerful interests, banks and corporations, the providers of jobs and wages in Sao Paulo and Bhopal sensed the harmlessness of the whole exercise. Thus the ready acceptance of WCED's concept of sustainable development at and after Rio.

What have those opposing an unsustainable future to offer? Science and morality! Neither have fared well when pitted against the appetites of economic development. Identifying modernity, materialism, and/or capitalism as the hydraheaded panzer leading the "fanatical drive against the earth" (157, p. 219) reflects at best the beginning of an initial step. The political arena will determine the outcome, and politically such words carry no meaning and call forth no volunteers. Life scientists and their peers—historians, philosophers, and other creative individuals—must humanize their knowledge. Environmental scientists have much to offer in seeking justice, equity, and democracy, in striving for the common good. We can be taught by examples from the natural world that sustainability is not a totally unrealistic goal.

Ecologists, historians, and others can function symbiotically in local and democratic communities. This alliance can moderate development and nurture sustainability if sound and realistic data are presented. Only if growth compulsions and legitimating ideas are held in abeyance at critical junctures can a semblance of sustainability be achieved. Democratic communities must make this fight, ecosystem by ecosystem, issue by issue. This may be what Presidents Jefferson, Lincoln, and Wilson had in mind when each, in his own words, exhorted humanity to sustain democracy as its last best hope.

\9Any *Annual Review* chapter, as well as any article cited in an *Annual Review* chapter, may be purchased from the Annual Reviews Preprints and Reprints service. 1–800–347–8007; 415–259–5017; email: arpr@class.org

---

Literature Cited

1. Abate T. 1992. Into the northern Philippines rainforest. *BioScience* 42:246–51
2. Adams JGU. 1991. On being economical with the environment. *Global Ecol. Biogeogr. Lett.* I:161–63
3. Albert B. 1992. Indian lands, environmental policy and military geopolitics in the development of the Brazilian Amazon: the case of the Yanomami. *Dev. Change* 23:71–99
4. Albert B. 1994. Bold miners and Yanomami Indians in the Brazilian Amazon: the Hashimu Massacre. In *Who Pays the Price? The Sociocultural Context of Environmental Crises,* ed. BR Johnston, pp. 47–55. Washington, DC: Island. 249 pp.
5. Alcorn JB. 1993. Indigenous people and conservation. *Conserv. Biol.* 7:424–25
6. Almond GA, Coleman JS. 1960. *The Politics of the Developing Area.* Princeton, NJ: Princeton Univ. Press. 591 pp.
7. Altieri MA. 1991. Increasing biodiversity to improve insect pest management in agro-ecosystems. In *The Biodiversity of Microorganisms and Invertebrates: Its Role in Sustainable Agriculture,* ed. DL Hawksworth, pp. 165–92
8. *Ambio.* 1990. Science of Sustainable Development. XIX (Special Issue)
9. Anderson AB, May PH, Balick MJ. 1991. *The Subsidy from Nature: Palm Forests, Peasantry, and Development of an Amazon Frontier.* New York: Columbia Univ. Press. 233 pp.
10. *Annual Review of Ecology and Systematics.* 1992. Special section on global environmental change. *Annu. Rev. Ecol. Syst.* 23:1–235
11. Archibugi F, Nijkamp P, eds. 1989. *Economy and Ecology: Towards Sustainable Development.* Dordrecht; Kluwer. 343 pp.
12. Asibey EOA. 1975. Blackfly and the environment. *Environ. Conserv.* 2:25–28
13. Asibey EOA. 1977. The blackfly dilemma. *Environ. Conserv.* 4:291–95
14. Aufderheide P, Rich B. 1988. Environmental reform and the multilateral banks: the greening of development lending. *World Policy J.* V:301–21

15. Ausubel J. 1991. Does climate still matter? *Nature* 350:649-52
16. Bahuguna S. 1988. Chipko: The people's movement with a hope for the survival of humankind. *ifda dossier* 63: 3-14
17. Barbier E. 1989. *Economics, Natural Resource Scarcity and Development: Conventional and Alternative Views.* London: Earthscan. 256 pp.
18. Barnett HJ. 1974. Economic growth and environmental quality are compatible. *Policy Sci.* 5:137-47
19. Baron J, Galvin KA. 1990. Future directions of ecosystem science. *BioScience* 40:640-42
20. Bendix R. 1967. Tradition and modernity reconsidered. *Compar. Stud. Soc. Hist.* IX:292-346
21. Bennett BC. 1992. Plants and people of the amazonian rainforest. *BioScience* 42:597-607
22. Berkes F, ed. 1989. *Common Property Resources: Ecology and Community-based Sustainable Development.* London: Belhaven. 302 pp.
23. Berkes F, Farvar MT. 1989. Introduction and overview. In *Common Property Resources,* ed. F Berkes, pp. 1-17. London: Belhaven
24. Black C. 1966. *The Dynamics of Modernization: A Study in Comparative History.* New York: Harper & Row. 206 pp.
25. Bormann FH. 1974. Acid rain and the environmental future. *Environ. Conserv.* 1:270
26. Boulding K. 1966. The economics of the coming spaceship Earth. In *Environmental Quality in a Growing Economy,* ed. H Jarett, pp. 1-14
27. Caesar K. 1990. Developments in crop research for the Third World. *Ambio* XIX:353-57
28. Cairns J. 1993. Environmental science and resource management in the 21st century: scientific perspective. *Environ. Toxicol. Chem.* 12:1321-29
29. Caldwell L. 1984. Political aspects of ecologically sustainable development. *Environ. Conserv.* 11:299-308
30. Caldwell L. 1990. *International Environmental Policy: Emergence and Dimensions.* Durham, NC: Duke Univ. Press. 460 pp. 2nd ed.
31. Caldwell L. 1991. Globalizing environmentalism: threshold of a new phase in international relations. *Soc. Nat. Resourc.* 4:259-72
32. Chance N. 1994. Contested terrain: a social history of human environmental relations in arctic Alaska. In *Who Pays the Price?* ed. BR Johnston, pp. 170-86
33. Chokor BA. 1991. Government policy and environmental protection in the developing world: the example of Nigeria. *Environ. Manage.* 17:15-30
34. Commoner B. 1971. *The Closing Circle: Man, Nature, and Technology.* New York: Knopf. 326 pp.
35. Conway G, Barbier EB. 1990. *After the Green Revolution; Sustainable Agriculture for Development.* London: Earthscan. 205 pp.
36. Costanza R, ed. 1991. *Ecological Economics: The Science and Management of Sustainability.* New York: Columbia Univ. Press
37. Costanza R, Daly HE, Bartholomew JA. 1991. Goals, agenda, and policy recommendations for ecological economics. In *Ecological Economics,* ed. R. Costanza, pp. 1-20. New York: Columbia Univ. Press
38. Costanza R, Daly HE. 1992. Natural capital and sustainable development. *Conserv. Biol.* 6:37-46
39. D'Addio T. 1994. U.S. response to the UN Commission on Sustainable Development. *Renewable Resources J.* 12:16-18
40. Daly HE. 1974. Steady-state economics versus growthmania: a critique of the orthodox conceptions of growth, wants, scarcity, and efficiency. *Policy Sci.* 5: 149-67
41. Daly HE. 1990. Toward some operational principles of sustainable development. *Ecol. Econ.* 2:1-6
42. Daly HE, Cobb JB. 1989. *For the Common Good: Redirecting the Economy Toward Community, the Environment, and a Sustainable Future.* Boston: Beacon. 482 pp.
43. Daneke GA. 1993. Integrating social and environmental costs into high-tech industrial development planning: experiences and expectations in the United States. In *Technology Policy. Towards an Integration of Social and Ecological Concerns,* ed. G Aichholzar, G Schienstack. pp. 205-21. Vienna: Inst. Adv. Stud.
44. Dang H. 1993. Fuel substitution in sub-Saharan Africa. *Environ. Manage.* 17: 285-83
45. Dasgupta P. 1990. The environment as a commodity. *Oxford Rev. Econ. Policy* 6:51-67
46. Dasman RF, Milton, JP, Freeman PH, eds. 1973. *Ecological Principles for Economic Development.* London: Wiley. 252 pp.
47. Dixon JA, Fallon LA. 1989. *The Concept of Sustainability: Origins, Extensions, and Usefulness for Policy. En-*

vironmental *Department. Division Work. Pap. No. 1989–1.* World Bank: Policy & Res. Div.

48. DuPuis EM, Geisler C. 1988. Biotechnology and the small farm. *BioScience* 38:406–11

49. Ecological Society of America's Committee for a Research Agenda for the 1990s. 1991. The sustainable biosphere initiative. *Ecology* 72:371–412

50. *The Ecologist.* 1993. *Whose Common Future? Reclaiming the Commons.* Philadelphia: New Soc. 216 pp.

51. Ehrenfeld D. 1988. Why put a value on biodiversity? In *Biodiversity,* ed. ED Wilson, FM Peter, pp. 212–16. Washington, DC: Natl. Acad. Press

52. Ehrenfeld D. 1989. Hard times for diversity. In *Conservation for the Twenty-first Century,* ed. D Western, MC Pearl, pp. 242–50. New York: Oxford Univ. Press

53. Ehrlich PR. 1968. *The Population Bomb: Population Control or Race to Oblivion.* New York: Ballantine. 201 pp.

54. Ehrlich PR. 1974. Human population and environmental problems. *Environ. Conserv.* 1:15–19

54a. Elshtain, JB. 1995. *Democracy on Trial.* New York: Basic. 153 pp.

55. Fox J. 1984. Firewood consumption in Nepal. *Environ. Manage.* 3:243–50

56. French HF. 1992. *After the Summit: The Future of Environmental Governance. Worldwatch Paper 107.* Washington, DC: Worldwatch Inst. 62 pp.

57. General Motors Public Interest Report. 1994.

57a. Goodland R. 1995. The concept of environmental sustainability. *Annu. Rev. Ecol. Syst.* 26:1–24

58. Grumbine RE, ed. 1994. *Environmental Policy and Biodiversity.* Washington, DC: Island. 415 pp.

59. Grumbine RE. 1994. What is ecosystem management? *Conserv. Biol.* 8:27–38

60. Guha R. 1995. The malign encounter: the chipko movement and competing visions of Nature. In *Who Will Save the Forest?,* ed. T Banuri, FA Marglin, pp. 80–113. London: Zed. 195 pp.

61. Hanemann WM. 1988. Economics and the preservation of biodiversity. In *Biodiversity,* ed. ED Wilson, FM Peters, pp. 193–99. Washington, DC: Natl. Acad. Press

62. Hanson JA. 1977. Towards an ecologically-based economic philosophy. *Environ. Conserv.* 4:3–10

63. Hardin G. 1968. The tragedy of the commons. *Science* 162:1243–48

64. Hardin G. 1991. The tragedy of the unmanaged commons: population and the disguises of providence. In *Commons Without Tragedy: Protecting the Environment from Overpopulation—A New Approach,.* ed. RV Andelson, pp. 162–85. London

65. Hargrove TR, Cabanella VL, Coffman WR. 1988. Twenty years of rice breeding: the role of semidwarf varieties in rice breeding for Asian farmers and the effects on cytoplasmic diversity. *BioScience* 38:675–81

66. Hawksworth DL, ed. 1991. *The Biodiversity of Microorganisms and Invertebrates: Its Role in Sustainable Agriculture.* Wallingford, UK: CAB Int. 302 pp.

67. Hildyard N. 1987. Tropical forests: a plan for action. *The Ecologist* 17:129–33

68. Hirschman AO. 1958. *The Strategy of Economic Development.* New Haven: Yale Univ. Press. 217 pp.

69. Holdgate MW, Kassas M, White GF. 1982. World environmental trends between 1972 and 1982. *Environ. Conserv.* 9:11–29

70. Holland MM. 1994. Contributions to science policy from the Ecological Society of America: a 10-Year Retrospective. *Bull. Ecol. Soc. Am.* 75:103–12

71. Holloway J. 1991. Biodiversity and tropical agriculture: a biogeographic view. *Outlook on Agric.* 20:9–13

72. Holloway JD, Stork NE. 1991. The dimensions of biodiversity: the use of invertebrates as indicators of human impact. In *The Biodiversity of Microorganisms and Invertebrates,* ed. DL Hawksworth, pp. 37–61. Paris

73. Hoopes J. 1994. *Human ecological impacts in the Basin of Mexico.* Pap. presented Rockefeller Program Semin. on Environ. Univ. Kans., Lawrence

74. Hotelling H. 1931. The economics of exhaustible resources. *J. Polit. Econ.* 39:137–75

75. Housman R. 1994. Reconciling trade and the environment. Lessons from the North American Free Trade Agreement. *Environ. Trade Ser.* 3. Geneva: UNEP. 65 pp.

76. Huntington SP. 1966, Political modernization: America vs Europe. *World Polit.* 18:378–414

77. Hyman EL. 1994. Fuel substitution and efficient woodstoves: Are they the answers to the fuelwood supply problem in Northern Nigeria? *Environ. Manage.* 18:23–32

78. International Chamber of Commerce. 1991. The Business Charter for Sustainable Development. Paris

79. International Chamber of Commerce

Report on the UNCED Conference and Associated Events. 1992. *Document 210/410.* Paris

80. Deleted in proof

81. International Union for the Conservation of Nature. 1981. World conservation strategy. *Bulletin* 12

82. Johnston BR. ed. 1994. *Who Pays the Price? The Sociocultural Context of Environmental Crises.* Washington, DC: Island. 249 pp.

83. Johnston BR, Burton Gregory. 1994. Human environmental rights issues and the multinational corporation: industrial development in the free trade zones. In *Who Pays the Price?*, ed. ER Johnston, pp. 206–15. Washington, DC: Island

84. Johnston HS. 1974. Pollution of the atmosphere. *Environ. Conserv.* 1:163–76

85. Korten DC. 1991–1992. Sustainable development. A review essay. *World Policy J.* IX:157–90

86. Kurien J. 1991. *Ruining the commons and responses of the commoners: coastal overfishing and fishermen's actions in Kerala State, India.* Discuss. Pap. 23. Geneva: UN Res. Inst. for Soc. Dev.

86a. Lasch C. 1995. *The Revolt of the Elites and the Betrayal of Democracy.* New York: WW Norton. 276 pp.

87. Lé-Lé SM. 1991. Sustainable development: a critical review. *World Dev.* 19: 607–21

88. Loita Maimina Enkuyio Conservation Trust Co. 1994. Forest of the lost child: indigenous cultures biodiversity and the CBD. *Ecoforum* 19:7–6

89. Lovins A. 1976. Long-term constraints on human activity. *Environ. Conserv.* 3:3–14

90. Ludwig D, Hilborn R. Walters C. 1993. Uncertainty, resource exploitation, and conservation: lessons from history. *Science* 260:17–36

91. McCormick J. 1989. *Reclaiming Paradise: The Global Environmental Movement.* Bloomington, IN: Ind. Univ. Press. 259 pp.

92. Macilwain C. 1994. Clinton's green think tank under a cloud. *Nature* 370: 239

93. McNeely JA. 1998. *Economics and Biodiversity: Developing and Using Economic Incentives to Conserve Biological Diversity.* Gland, Switzerland: IUCN. 236 pp.

94. MacNeil J. 1939. Our 'common future', sustaining the momentum. In *Economy and Ecology*, ed. F Archibugi, P Nijkamp, pp. 15–25. Dordrecht: Kluwer

95. MacNeil J, Winsemius P, Yakushiji T.

1991. *Beyond Interdependence. The Making of the World's Economy and the Earth's Ecology.* New York: Oxford Univ. Press. 159 pp.

96. Majumdar SK, ed. *Conservation and Resource Management.* Easton, PA: Penn. Acad. Sci. 444 pp.

97. Malone TF. 1976. The role of scientists in achieving a better environment. *Environ. Conserv.* 3:81–89

98. Meadows D, Meadows D, Randers J, Behrens WW III. 1972. *The Limits to Growth. A Report for the Club of Rome's Project on the Predicament of Mankind.* New York: Universe. 205 pp.

99. Miller JA. 1989. Diseases for our future: global ecology and emerging viruses. *BioScience* 39:509–17

100. Myers N. 1988. *Natural resource systems and human exploitation systems: physiobiotic and ecological linkages.* Environmental Dep. Work. Pap. No. 12. World Bank: Policy Plan. Res. Staff

101. Myers N. 1989. The environmental basis of sustainable development. In *Environmental Management and Economic Development* ed. G Schramm, J Warford, pp. 57–68. Baltimore: Johns Hopkins Univ. Press for the World Bank

102. Myers N. 1991. Biologists as policymakers. *Environ. Conserv.* 18:6

103. Naess A. 1988. Deep ecology and ultimate premises. *The Ecologist* 13:128–31

104. Naess P. 1993. Can urban development be made environmentally sound? *J. Environ. Manage. Planning* 36:309–33

105. Newson M. 1992. Water and sustainable development. *J. Environ. Plan. Manage.* 35:175–83

106. Norgaard R. 1988. Sustainable development: a co-evolutionary view. *Futures* 20:606–20

107. Norton B. 1988. Commodity, amenity, and morality: the limits of quantification in valuing biodiversity. In *Biodiversity*, ed. ED Wilson, FM Peter. pp. 200–5. Washington, DC: Natl. Acad. Press

108. Olson M, Landsberg H, eds. 1973. *The No Growth Society.* New York: Norton. 259 pp.

109. Pearce D. 1988. Economics, equity and sustainable development. *Futures* 20: 598–605

110. Pearce D. 1993. *Economic Values and the Natural World.* Cambridge, MA: MIT Press. 129 pp.

111. Perlas N. 1988. The sustainable agriculture movement. *Orion* 35–47

112. Peters RH. 1991. *Critique of Ecology.* New York: Cambridge Univ. Press. 166 pp.

113. Pimm SL. 1991. *The Balance of Nature? Ecological Issues in the Conservation*

of Species and Communities. Chicago: Univ. Chicago Press. 434 pp.

114. Poore D. 1975. Conservation and development. *Environ. Conserv.* 2:243–46

115. Pomeroy LR, Alberts JJ, eds. 1988. *Concepts of Ecosystem Ecology: A Comparative View.* New York: Springer-Verlag. 384 pp.

116. Randall A. 1988. What mainstream economics has to say about the value of biodiversity. In *Biodiversity,* ed. ED Wilson, FM Peter, pp. 217–23. Washington, DC: Natl. Acad. Press

117. Raven PA. 1990. The politics of preserving biodiversity. *BioScience* 40: 769–74

118. Redford KH, Stearman AM. 1993. Forest dwelling native Amazonians and the conservation of biodiversity. *Conserv. Biol.* 7:248–55

119. Repetto R. 1994. *Trade and Sustainable Development.* Geneva: UNEP. 45 pp.

120. Resser PG, Lubchenco J, Levin S. 1991. Biological research priorities—a sustainable biosphere. *BioScience* 41:625–27

121. Revelle RR, Shapero DC. 1979. Energy and climate. *Environ. Conserv.* 5:81–91

122. Rich B. 1994. *Mortgaging the Earth: The World Bank, Environmental Impoverishment, and the Crisis of Development.* Boston: Beacon. 376 pp.

123. Robinson JM. 1997. The atmosphere as a resource. In *Conservation and Resource Management,* ed. SK Majumdar, pp. 68–82. Easton, PA: Penn. Acad. Sci.

124. Rostow WW. 1971. *The Stages of Economic Growth.* London: Cambridge Univ. Press. 178 pp. 2nd ed.

125. Sachs W, ed. 1993. *Global Ecology: A New Arena of Political Conflict.* London: Zed Books. 262 pp.

126. Schumacher EF. 1973. *Small is Beautiful.* New York: Harper & Row. 290 pp.

127. Simon JL. 1983. Life on earth is getting better, not worse. *The Futurist* XVII:7–15

128. Simon JL. 1990. *Population Matters: People, Resources, Environment, and Immigration.* New Brunswick, NJ: Transaction. 577 pp.

129. Simon JL, Kahn H, eds. 1984. *The Resourceful Earth: A Response to Global 2000.* New York: Blackwell. 595 pp.

130. Slobodkin LB. 1988. Intellectual problems of applied ecology. *BioScience* 38: 337–42

131. Soulé ME. 1994. What is Conservation Biology? In *Environmental Policy and Biodiversity,* ed. RE Grumbine, pp. 35–53. Washington, DC: Island

132. Sponsel L. 1994. The Yanomami holocaust continues. In *Who Pays the Price?,* ed. BR Johnston, pp. 37–46. Washington, DC: Island

133. Thapa GB, Weber KE. 1990. Actors and factors in deforestation in "Tropical Asia." *Environ. Conserv.* 17:19–27

133a. *The Earth Times.* September 24, 1994

134. Thompson PA. 1975. The collection, maintenance, and environmental importance of the genetic resources of wild plants. *Environ. Conserv.* 2:223–28

135. Tisdell C. 1988. Sustainable development: differing perspectives of ecologists and economists, and relevance to LDCs. *World Dev.* 16:373–84

136. Tisdell C. 1990. *Natural Resources, Growth, and Development.* New York: Praeger. 136 pp.

137. Trout SA, Delistraty DA. 1977. Food resources of the oceans: an outline of status and potentials. *Environ. Conserv.* 4:243–52

138. UN Convention on Environment and Development. 1992. *Agenda 21: The United Nations Programme of Action from Rio.* New York: UN. 294 pp.

139. UN Development Programme. 1992. *Human Development Report 1992.* New York: Oxford Univ. Press. 216 pp.

140. US National Research Council. 1992. *Conserving Biodiversity. A Research Agenda for Development Agencies.* Washington, DC: Natl. Acad. Press. 127 pp.

141. Varshney CK, Aggarwal M. 1992. Ozone pollution in the urban atmosphere of Delhi. Part B: Urban environment. *Atmos. Environ.* 26B-3:291–300

142. Vernadsky WI. 1945. The biosphere and the noosphere. *Am. Sci.* 33:1–12

143. Vivian J. 1991. *Greening at the Grassroots: People's Participation in Sustainable Development. Discussion Paper 22.* UN Res. Inst. Soc. Dev., Geneva

144. Walker BH. 1992. Biodiversity and ecological redundancy. *Conserv. Biol.* 6: 18–23

145. Warford J. 1989. Economic development and environmental protection. *Nat. Resourc. For.* 13:233–41

146. Western D. 1991. Biology and conservation: making the relevant connection. *Conserv. Biol.* 5:431–33

147. Western D, Pearl MC, eds. 1989. *Conservation for the Twenty-first Century.* New York: Oxford Univ. Press 365 pp.

148. Westman WE. 1990. Managing for biodiversity. *BioScience* 40:26–33

149. Wilson ED, Peter FM, eds. 1988. *Biodiversity.* Washington, DC: Natl. Acad. Press. 521 pp.

150. Wilson ED. 1988. The current state of biological diversity. In *Biodiversity,* ed. ED Wilson, FM Peter, pp. 3–18. Washington, DC: Natl. Acad. Press

151. Wilson ED. 1989. Conservation: the next hundred years. In *Conservation,* ed. D Western, M Pearl, pp. 7:6–7. New York: Oxford Univ. Press

152. Wilson ED. 1992. *The Diversity of Life.* Cambridge: Belknap Press. 424 pp.

153. World Bank. 1994. *World Bank Development Report.* New York: Oxford Univ. Press.

154. World Commission on Environment and Development. 1987. *Our Common Future.* New York: Oxford Univ. Press. 406 pp.

155. World Meteorological Organization 1976. Statement on modification of the ozone layer due to human activities. *Environ. Conserv.* 3:63–70

156. World Resources Institute. 1990. *World Resources 1990–91: A Guide to the Global Environment.* New York: Oxford Univ. Press. 783 pp.

157. Worster D. 1993. *The Wealth of Nature. Environmental History and the Ecological Imagination.* New York: Oxford Univ. Press. 255 pp.

158. Yudelman SW. 1987. *Hopeful Openings: A Study of Five Women's Development Organizations in Latin America and the Caribbean.* West Hartford, CT: Kumarian. 127 pp.

Annu. Rev. Ecol. Syst. 1995. 26:249–68

VESTIGIALIZATION AND LOSS OF NONFUNCTIONAL CHARACTERS

Daniel W. Fong

Department of Biology, The American University, Washington, DC 20016-8007

Thomas C. Kane

Department of Biological Sciences, University of Cincinnati, Cincinnati, Ohio 45221

David C. Culver

Department of Biology, The American University, Washington, DC 20016-8007

KEY WORDS: reduction, flightless, eyeless, mutation, pleiotropy

ABSTRACT

Reduction and total loss of characters are common evolutionary phenomena. Vestigialization of any morphological, physiological, or behavioral feature can be expected upon relaxation of selection on the trait. Direct selection of vestigialization is rarely documented. Most explanations of evolutionary reductions invoke indirect selection through energy economy or antagonistic pleiotropy arguments, while some invoke the effects of accumulation of neutral mutations. A few documented cases of trade-offs between fitness and wing reduction or pesticide resistance in some insects, and between fitness and resistance to phages or antibiotics in bacteria suggest that indirect selection is a plausible mechanism for evolutionary reductions. Expression of presumably useless genes suggests that neutral mutation arguments require a longer time than is available for the observed reductions. Rapid decay of useless behaviors may require explanations in terms of trade-offs among neural pathways for information processing.

249

0066-4162/95/1120-0249$05.00

INTRODUCTION

> Rudimentary, atrophied, or aborted organs. Organs or parts in this strange condition, bearing the stamp of inutility, are extremely common throughout nature.
>
> Charles Darwin 1859, p. 418

Darwin drew heavily on examples of new features, such as the vertebrate eye, to support his theory of natural selection (21); yet, as the quotation indicates, he was also impressed with the pervasive nature of rudiments or vestiges. Vestigialization begins when a trait is rendered nonfunctional, or becomes a selective liability outright, due to shifts in the environment. The feature then atrophies over time. Eventually it will persist at a simplified stage in which further reduction is maladaptive, or it will disappear entirely. The intermediate, reduced structure is the vestige. JBS Haldane (cited in 43, p. 61) allowed that probably for every case of progressive evolution in the sense of descendants being more complex in structure and behavior than their ancestors, there have been ten cases of regressive evolution. A cursory examination of introductory biology texts yields such examples as simplified morphology of parasitic animals, reduced number of digits in horses, loss of limbs in snakes, the vermiform appendix and coccyx of humans, vestigial wings of flightless birds, eye and pigment loss of cave-dwelling organisms, etc.

The vestigialization and loss of a structure can be of evolutionary significance. Rudiments of portions of the primitive reptilian jaw have evolved into critical components of the modern mammalian inner ear (79). In the salamander family Plethodontidae, the loss of lungs apparently freed the hyobranchial apparatus from serving as a buccal respiratory pump, thus allowing for the evolution of a projectile tongue, a specialized feeding mechanism that contributed to the extensive radiation of the family (54, 101).

Causes of Vestigialization

Modern evolutionary biology has emphasized the appearance of new structural or functional complexes (15). Theories of vestigialization and regressive evolution have assumed minor importance (see 6, 26, 69, 73, 109), except for evolutionary biologists studying cave-dwelling organisms (4, 17, 20, 107).

Vestigial or missing structures may be signs of evolutionary trade-offs. The evolution of elaborated extra-optic sensory structures of cave animals may require eye and pigment reduction due to energy economy in resource-poor environments such as caves (67, 68, 88). Eye and pigment loss frees up energy required for elaboration of extra-optic sensors. Similarly, compensatory sensory system trade-offs could be the most plausible explanation for eye reduction in cave animals (40). In the broader context, metabolic efficiency has received indirect support as the cause of flightlessness in some island-dwelling birds (59). More generally, character loss may typically be a consequence of indirect selection (38, p. 41). At the molecular level no feature may be selectively neutral, because nonessential messages will produce "noise" in the biochemical pathways of an organism, and natural selection will eliminate unused features and the biochemical messages they produce (73).

Vestigialization and loss of structures may have nothing to do with adaptation, either directly or indirectly. Eye and pigment reduction in cave-dwelling animals may be the result of the relaxation of selection, particularly stabilizing selection (107). In essence, structures lose complexity due to the accumulation through genetic drift of selectively neutral mutations. Whether this process can result in the observed rates of loss has been questioned (4, 55), but plausible genetic models of eye and pigment loss by neutral mutation and drift can produce the observed rates (11, 17). Rates of evolution of even presumably adaptive characters often could be accounted for, in principle, by drift (56). Such nonadaptive explanations are of particular interest because they echo the question of the relative commonness of adaptive and nonadaptive evolution (36, 61, 99). Convergence is often cited as one of the strongest pieces of evidence for natural selection in evolution. If reductions occurring independently among diverse taxa, such as eye loss in cave animals, are simple consequences of accumulating neutral mutations, convergence may also reflect nonadaptive evolution (see also 80).

The interesting aspect, and the rub, of evolutionary reductions is not that they are too difficult but rather that they are too easy to explain in theory. Distinguishing among various theories of regressive evolution is hampered by lack of empirical information and by experimental limitations posed by many of the organisms in question. Little is known of the genetic basis for many of the characters in question or of the energetic cost of their development and maintenance. Estimates of rates of evolutionary change are rarely attempted. Further, little is known about the relationship between character reduction and its effects on fitness. We review here some empirical case studies and theoretical perspectives on evolutionary reductions, evaluating recent progress and suggesting potential avenues for future work.

REDUCTION AND LOSS OF DIGITS AND LIMBS IN TETRAPODS

Reduction and loss of digits and limbs is a repeated theme in many lines of tetrapods, including amphibians, reptiles, birds, and mammals (45). Reduction of digits and limbs generally follows an orderly sequence, proceeding proximad from distal elements in amphibians (1) and reptiles (33). Analyses of patterns of limb formation in many tetrapods, especially amphibians (see 63, 87), have greatly clarified the developmental processes underlying digit and

limb reduction. Such structural changes can proceed rapidly under weak selection (45); otherwise, our understanding of the ultimate evolutionary causes
of digit and limb reduction is far from complete. Simplification of bony
elements in the limbs and skull of some salamanders may be a consequence
of responses to selection for miniaturization (37). The reduction of limbs in
squamate reptiles was likely preceded by elongation of the body (33). Whereas
limb reduction may have occurred multiple times in lizards such as skinks and
legless lizards, leg loss may have occurred only once in snakes (33, 74). Instead
of providing a catalog of limb reduction and loss in tetrapods, we concentrate
on one case for which sufficient data allow a critical review: the evolution of
flightlessness in birds.

Flightless Birds

Loss of sustained flight has occurred repeatedly in birds and is associated with
reductions in the size of skeletal elements in the wings and of the flight muscles
(28). Flightlessness is found among 26 families in 17 orders of birds (53). In
reviewing the distribution of extant and recently extinct (within 15,000 years)
flightless species in ten orders, Roff (78) estimated that flightlessness has
evolved independently at least 30 times.

Most considerations of flightlessness in birds have emphasized flightless
birds on islands. The argument is that, in the absence of predators, there is
relaxed selection to maintain flight for predator avoidance, and that reduction
of wings along with pectoral muscle mass results in energy conservation (see
59). The association between flightlessness and insularity is statistically significant for rails, in which all 17 flightless species, as compared to only 18 of
105 volant species, are found on islands. Each of these 17 species probably
represents an independent transition to flightlessness (78). The basal metabolic
rates of several flightless rail species are significantly lower than their volant
congeners, and basal metabolic rate is inversely correlated with pectoral muscle
mass relative to total mass (59). This seems to support the idea that flightlessness evolved as a response, manifested as reductions in wing size and pectoral
muscle mass, to selection for energy conservation through lowered metabolic
rate. The critical and unproven assumption is that lowered metabolic rate
results in increased fitness in rails.

The association of ratites with islands is suggested by the fact that 25 of the
41 species are/were from New Zealand and Madagascar. If the ratites had
evolved from a single flightless ancestor, their higher frequency of occurrence
on islands than on continents might not indicate that insularity is necessary
for the evolution of flightlessness (78), but rather that flightless species adapted
well to the islands, or that flightlessness led to a high speciation rate with a
higher proportion persisting on predator-free islands than on continents.

Excluding rails and ratites, Roff (78) estimated 15 transitions to flightlessness for the other birds: 8 transitions involve insular species, 4 involve coastal species, and 3 involve species of inland waters. The coastal transitions include penguins, the great auk, steamer ducks, and a few extinct California coast ducks, while 4 of the insular transitions include species, such as the Galapagos cormorant, that forage off the coast of the islands. For at least the extant species in both groups, the shortened stout wings play an important role in propulsion during dives; in addition, steamer ducks use their wings as oars for propulsion on the water (see 51–53). Not surprisingly, steamer ducks have high relative pectoral muscle mass (59). Thus wing reductions in these birds may be the direct result of selection for enhanced locomotory performance in water at the expense of aerial flight capability. The three inland water transitions occur in three species of grebes living in isolated low-latitude mountain lakes. Flightlessness in grebes, which are foot-propelled divers, is associated with structural reduction and loss of locomotory capacity of their wings; energy conservation is again the mechanism invoked for their wing reduction (50). Constancy and high productivity of the habitat are proposed as reasons for relaxation of selection for maintenance of aerial flight on steamer ducks (53) and flightless grebes (50).

Two generalizations emerge from the bird data. First, there was relaxed selection for maintenance of flight due to release from predation on islands, release from migration due to constancy and high productivity of the habitats, or both. Second, there are at least two patterns of reduction. In the penguins and steamer ducks, for example, the smaller wings probably resulted from direct selection for improved performance in an aquatic medium. In other words, flightlessness is misleading in these birds, they just fly under water. In the rails, grebes, and ratites, the causes of flightlessness and wing reduction are not as clear. The only evidence for the energy economy argument is the observed lower metabolic rate of a few flightless rails compared to volant congeners with larger flight muscles, yet the lowered metabolic rate could be a consequence of, rather than the reason for, flightlessness. Data on intraspecific variation in wing size and flight muscle mass and in fitness components, such as egg or clutch size, are needed.

LUNG LOSS IN PLETHODONTID SALAMANDERS

It is generally accepted that lunglessness in plethodontid salamanders is a derived condition evolved from lunged ambystomatid-like ancestors (24). Assuming cool, swift, oxygen-rich, upland Appalachian streams in the late Mesozoic as ancestral habitats, lung loss in plethodontids is hypothesized as an adaptation for increased ballast, and thus decreased risk of downstream drift, in concert with increased reliance on cutaneous instead of pulmonary respiration (105, also see 5, 7). According to this hypothesis, selection for maintenance of lungs in ancestral plethodontids was weak because cutaneous respiration was sufficient, and the presence of lungs was maladaptive due to buoyancy. The first point is supported by the fact that both lunged and lungless salamanders show low energy demands as reflected by metabolic rates (27), and that even in lunged salamanders, cutaneous respiration accounts for a large percentage of total oxygen consumption at low temperatures (102, 103). The second point is problematical. Although larvae of *Ambystoma maculatum* with reduced lungs are less buoyant and thus drift for significantly shorter distances compared to larvae with larger lungs (7), the selective advantage is unclear. Reduced downstream drift may increase larval survival and population stability for species with lengthy larval stages (5), but whether the larval duration of ancestral plethodontids was short or lengthy is debatable (83).

Alternatively, plethodontids could have a terrestrial origin because cool, oxygen-rich, upland Appalachian streams may not have existed in the late Mesozoic—much of the area was purported to be a peneplane, and the climate was subtropical (82). In such a scenario, lunglessness is hypothesized to be a consequence of reduced buccal volume due to selection of narrower heads (82), or of the evolution of terrestrial courtship and mating (72). The first hypothesis is based on the observation that in lunged ambystomatids narrower-headed species show a greater reliance on cutaneous respiration than do wider-headed species (103, 104) because they have lower capacity for pulmonary respiration due to reduced tidal volume of the buccal pump. Selection of narrower heads in ambystomatid-like ancestral plethodontids resulted in less reliance on pulmonary respiration, although the selective advantage of narrower heads is unclear. This hypothesis assumes weak or no selection for maintenance of lungs and assumes that the presence of lungs was not maladaptive per se. The second hypothesis assumes strong selection for shifting courtship and mating activities of ancestral plethodontids from water to land, in order to escape the costs of migration from land to water and competition for mates at breeding ponds. The energetic demand of courtship and mating activities in slow-flowing to stagnant, relatively hypoxic breeding ponds required pulmonary respiration to supplement cutaneous respiration. Shifting such activities to land resulted in energetically less expensive breeding activity in an environment with greater oxygen supply, with a consequent relaxation of selection for pulmonary supplement to cutaneous respiration. This hypothesis also assumes weak or no selection for maintenance of lungs and that lungs were not maladaptive per se in ancestral plethodontids. Thus, if plethodontids did originate on land, their lunglessness is essentially a case of the evolutionary loss of a nonfunctional character. The already low energy demand of salamanders as a group also suggests that energy economy arguments may have little significance as the ultimate cause of lung loss.

SUBTERRANEAN ANIMALS

Cave-Dwelling Organisms

Among the most distinctive features of animals from caves and other subsurface habitats are the reduction and loss of eyes and pigment. These have occurred independently across many taxa, such as arachnids, insects, crustaceans, fish, and salamanders, with literally hundreds of documented cases (see reviews in 17, 34, 98). The ubiquity of these reductions is underscored by the estimated 50,000 to 100,000 obligate cave-dwelling species in the world (18), most of which are eyeless.

Wilkens (107), working with the characin fish *Astyanax fasciatus,* argued that neutral mutation is responsible for eye and pigment loss in cave animals. The argument is that without light the presence or absence of eyes and pigment has no effect on fitness. Most mutations affecting a complex system such as an eye are likely to be degenerative (17), and ultimately some of these mutations will be fixed in the population by genetic drift. Others (e.g. 4, 55) have objected that the process of neutral mutation and genetic drift is too slow to account for rates of eye and pigment loss; *A. fasciatus* populations probably have been isolated in caves for fewer than 10,000 generations. Nonetheless, there are plausible scenarios for eye loss resulting from mutations at a subset of loci of a polygenic system (17), or from high rates of mutation from functional to nonfunctional alleles (11). F_1 hybrids between *A. fasciatus* populations have larger eyes than either parental population (106), indicating that eye reduction involves different genes in different populations. Data on protein electrophoretic variation also support the hypothesis of independent isolation in different caves (3). Wilkens (108) pointed out that there should be a relaxation of stabilizing selection accompanying isolation in caves. He argued that the increased variability of eyes in recently isolated populations supports this view, but he did not attempt to measure stabilizing selection directly. Culver et al (19) measured stabilizing selection in a cave-dwelling population of the amphipod *Gammarus minus* and failed to find any reduction in stabilizing selection relative to spring-dwelling populations. However, stabilizing selection was rare in both habitats.

Although neutral mutation is probably an important force in eye and pigment loss, it is not the only factor. In particular, energy economy and pleiotropy can be invoked to couple the reduction in eyes and pigment with the increase in extra-optic sensory structures (20, 29, 67, 88). There is evidence of direct selection for small eyes in different cave-dwelling populations of *G. minus,* even when selection for extra-optic sensory structures is taken into account (41). In these cases, instead of energy economy, there may be an evolutionary trade-off at the level of neurological connections in the brain. Compensatory innervation of parts of the optic processing areas of the brain by projections of neurons from olfactory or tactile sensory organs are observed in mutant anophthalmic mice (42) and in cave fish (100). Finally, the standardized rate (see 56) of increase of extra-optic sensors (antennae in particular) is nearly an order of magnitude less than the standardized rate of loss of eye characters in *G. minus* (20). The estimated rates of reduction for two eye size characters are 2.4×10^{-5} and 1.5×10^{-5}, whereas rates of increase for six antenna size-characters ranged from 8.9×10^{-7} to 1.1×10^{-6} (20, p. 179). There are of course other factors that may be involved, such as differences in the number of loci involved, but the consistent differences in rates of change of optic and extra-optic sensory structures suggest that both selection and neutral mutation are important in the evolution of reduced eyes of *G. minus*. Extra-optic sensory structures increase in size as a result of directional selection. Optic structures decrease in size as a result of directional selection and neutral mutation.

Mole Rats

The eyes of the mole rat *Spalax ehrenbergi* are minute and covered by skin, and they are operationally blind, as light flashes evoke no action potential along the optic pathway (see review in 62). The atrophied eyes of the mole rat, however, do seem to function in photoperiodic entrainment of circadian activity and thermoregulatory rhythms. The mole rat also shows severe reduction in the thalamic and tectal areas of the brain, which discern form and motion, but this is accompanied by the hypertrophy of structures that serve photoperiodic functions (14). It is suggested that such a highly specific set of reductions and hypertrophies in the brain indicate that eye reduction in mole rats, which, in effect, eased a costly metabolic burden, cannot be the simple consequence of relaxed selection for vision, but rather, it may be part of a highly complex set of adaptations to the underground environment (14). Nevo (62) pointed out that because of the tight coupling of photoperiodism and reproductive cycles in mammals, even atrophied eyes are important in photoperiod detection, so it is not surprising that there are no completely eyeless subterranean mammals.

INSECTS

Not surprisingly, a taxonomically diverse group such as insects exhibits many cases of character loss and vestigialization. For example, adult mayflies (Ephemeroptera) spend their brief existence mating and do not feed, and they have nonfunctional vestigial mouthparts (25). Nasute soldiers have evolved independently in two phyletic lines of termites (26). Instead of using their mandibles for defense, as is common in primitive taxa, these soldiers are convergent for use of a "squirt gun," a prolonged portion of the head that ejects irritating fluids; and they are also convergent for nonfunctional vestigial mandibles (26). While these examples further illustrate the pervasive nature of vestigialization, we concentrate on two cases that can shed light on the causes of character reduction. These are the loss of flight and the loss of hearing.

Flightlessness and Wing Reduction

The loss of flight and of wings from winged flight–capable ancestors has evolved numerous times in the insects (see review in 77). The incidence of flightlessness is correlated with the mode of metamorphosis. Flightlessness is almost nonexistent in the hemimetabolous orders with naiads as immatures, is uncommon among holometabolous orders with larvae as immatures, and is common in the paurometabolous orders with nymphs as immatures. Roff (77) suggested that, relative to orders with the other two types of metamorphosis, flightlessness is more likely to evolve in paurometabolous insects in which the immatures are more mobile and have feeding habits similar to those of adults. In other words, there is relaxed selection for maintenance of flight for dispersal. The incidence of flightlessness is positively correlated with latitude and altitude, and is high in woodland, desert, ocean surface, and aquatic habitats, while it is low in stream, river, and pond margins. The data suggest that habitat persistence is conducive to the evolution of flightlessness (22, 77, 78), which also implies relaxed selection for the maintenance of flight.

The incidence of flightlessness is higher among females than males, which Roff (77) suggested allows more resources for egg production. The hypothesis is that fecundity selection is constrained by selection to maintain flight but can proceed when selection for flight is relaxed, and thus flightlessness is indirectly adaptive. This trade-off hypothesis is supported by data from wing dimorphic insects. Among 22 wing-dimorphic species, the short-winged morph is more fecund and can reproduce sooner than the long-winged morph (75). The flightless short-winged morph of the cricket *Gryllus rubens* is also more fecund than the long-winged morph and is more efficient at converting assimilated nutrients into biomass, a potential result of lower respiration costs by vestigial flight muscles (60). Finally, in the crickets *Teleogryllus oceanicus* and *Gryllus firmus,* experimental removal of wings upon metamorphosis increases egg production (76). Wing dimorphism is also found in water striders (93). In some species flightless females have shorter preoviposition periods and produce more eggs over the initial reproductive stages (92). The case with water striders, however, is complicated by recent evidence indicating that wing dimorphism and short-wings may be ancestral conditions in some genera, and that there exist two types of short-wing morphs that develop through different morph determination mechanisms (2).

These insect data are illuminating in two respects. First, trade-offs between flight maintenance and fitness components are observed. Second, there is evidence for a physiological mechanism underlying the observed trade-offs. Thus, when selection for flight is relaxed, flightlessness and associated wing and especially muscle reduction in insects probably evolved as correlated responses, through energy economy, to selection for enhanced fitness components.

Loss of Hearing

Tympanal ears have evolved repeatedly in insects (32). In some groups, especially noctuid moths, tympanal ears are tuned to between 20 kHz to 50 kHz and primarily function as detectors of foraging bats. The echolocation calls of bats trigger evasive aerial maneuvers by the insect (89). That insect hearing is an effective strategy against bat predation is underscored by the evidence that some bats shift echolocation frequencies above or below the most sensitive frequencies of tympanal ears (84).

Female gypsy moths (*Lymantria dispar*) are winged but are not exposed to predation by bats because they do not fly and are less sensitive to high frequency sound compared to males that fly (8). Noctuid moths endemic to bat-free islands of French Polynesia (a region that bats apparently never colonized) show reduced sensitivity to frequencies above 35 kHz compared to moths in bat-inhabited areas, and deafness in these moths is probably derived from ancestors with full auditory capacity (31).

Praying mantises show widespread sexual dimorphism in ultrasound sensitivity and in ear morphology (110). It is always the females that show reduced hearing, ranging from mild loss to complete deafness. Except in cases of mild loss, hearing reduction is associated with ears that are structurally different from those of the male. These mantises are also dimorphic in wing length, which is highly correlated with dimorphism in auditory sensitivity. Short-winged flightless individuals have reduced hearing compared to functionally winged mantises of the same species and mantises usually fly at night. There are also strong indications that wing reduction and hearing loss in female mantises are derived conditions (110).

In all these cases predation by echolocating bats is probably the selective agent behind the maintenance of functional tympanal ears. Full to partial deafness is always associated with release from bat predation. Although the commonness of this phenomenon among noctuid moths is unknown, it is common in the Mantodea. Structural dimorphism, which reflects differences in hearing sensitivity, occurs among 63 of 183 genera of mantises suitable for anatomical comparison between sexes, and hearing sensitivity dimorphism should be more widespread because it is not always associated with structural dimorphism. These 63 dimorphic genera span 3 families and 11 subfamilies within the large Mantidae (110). Thus, hearing loss associated with release from bat predation has likely evolved multiple times within the Mandotea and certainly within the insects. Assuming that full auditory acuity or fully developed tympanal ears in the absence of bat predation is not maladaptive per se, reduction of these features is another indication of the prevalence of vestigialization of nonfunctional characters.

The dimorphism of wing length in mantises is identical to the patterns of wing reduction discussed in the previous section. It is probable that flightlessness and wing reduction in the female mantises also can lead to an increase in fitness components. This result, however, confounds the interpretation of the evolution of hearing loss; i.e. do hearing loss and associated simplified tympanal structures also represent energy savings that contribute to fitness gains? We suggest that a phylogenetic analysis may permit the teasing apart of the two phenomena.

REVERSAL OF RESISTANCE

Much applied research has concentrated on the evolution of resistance to antibiotics in bacteria and to pesticides in agricultural pests and arthropod disease vectors, because of severe public health and economic consequences. Studies of the evolution of resistance are germane to our focus because reversal of resistance in the absence of the antibiotic or pesticide is analogous to the evolution of vestigialization and loss of nonfunctional characters.

Important pest control agents include insecticidal protein toxins derived from *Bacillus thuringiensis,* a common soil bacterium (see review in 96). These proteins act by binding to the brush border membrane of the insect midgut epithelium causing swelling and lysis of the cells. Resistance has recently evolved in the field in the diamondback moth *Plutella xylostella* and in nine insect species selected for resistance in the laboratory. The primary mechanism of resistance is reduced binding of the toxin to the midgut membrane. Strains of *P. xylostella* selected for resistance at levels 25 to 2800 times that of unselected strains show reversal of resistance to levels of unselected strains were not exposed to *B. thuringiensis* for many generations. Restoration of binding of the toxin to the midgut is the mechanism of reversal of resistance (97). Rapid reversal to sensitivity at pre-selection levels is also observed in the cockroach *Blattella germanica* resistant to pyrethroids (12) and in other insects (see review in 58).

Adverse pleiotropic effects on fitness-correlated life-history characters resulting from evolution of resistance has been hypothesized (64, 71, 86) and has been observed in a few cases (9, 35, 57). For example, strains of *P. xylostella* resistant to *B. thuringiensis* show reduction in survival, egg hatching, and fecundity compared to sensitive strains (35). That such trade-offs between fitness and resistance are observed infrequently may result, with continued exposure to the insecticides, from selection to ameliorate the adverse effects on fitness (see 9, 81).

Amelioration of fitness costs associated with the evolution of resistance is observed in strains of *Escherichia coli* resistant to the virus T4 (47, 48) and in strains of *Bacillus subtilis* resistant to the antibiotic rifampicin (13). *Escherichia coli* resistance to T4 is accompanied by reduction of competitive fitness compared to sensitive strains. The competitive fitnesses of resistant strains grown in the absence of T4 can be restored to the level of sensitive strains, but they do not revert to sensitivity. Restoration of fitness resulted from selection directly for genetic changes that counter the adverse pleiotropic effects of resistance. In the case of *B. subtilis,* adverse pleiotropic effects of resistance were ameliorated through two mechanisms. The first is through replacement of mutations conferring resistance at high fitness cost with ones at lower cost. The second is through selection for modifiers at other loci that compensate for the fitness cost of the original mutations. These results can shed light on the energy economy and pleiotropy hypotheses of regressive evolution. Lenski (48) notes that trade-offs due to competition for limited energy or material should intensify with further adaptation, while trade-offs due to disruption of genetic integration should diminish. Even genetic trade-offs that persist can be difficult to detect, depending on the relative frequencies of loci that acquire resources and those that allocate resources to the traits involved in the trade-off (39).

VESTIGIAL BEHAVIOR

Decay of behavioral traits presumed to be selectively neutral, though rare, has been documented. Vestigial behavior can be especially significant in the study of mechanisms of vestigialization because a "useless" behavior can be elicited only under appropriate experimental conditions. Such a trait is not expressed in natural settings and thus is not subject to selection directly, nor can any pleiotropic effects or energetic costs of its expression be subject to selection. Two plausible explanations of the decay of such behavior involve the accumulation of mutations underlying the genetic basis for the behavior or trade-offs at the level of neural circuitry.

Decay of Sexual Behavior

All-female strains of *Drosophila mercatorum,* selected for parthenogenesis from bisexual wild ancestors, were maintained in the laboratory for over 20 years without males, thus rendering female mating behavior a nonfunctional trait (10). The parthenogenetic strains, upon exposure to males, showed greatly reduced mating propensity, measured as mating speed, compared to females of sexual strains. Because female mating behavior is elicited only in the presence of males, this trait was not expressed while the parthenogenetic strains were maintained without males and thus was not subject to selection. Carson et al (10) suggested accumulation of neutral mutations as the genetic mechanism for these reductions.

Behavior of the Cave Fish Astyanax

The cave and epigean forms of the characin fish *Astyanax fasciatus* have been subjects of intensive studies (see 107). Cave forms differ from the epigean form in several behavioral features that have been demonstrated, through crosses among populations, to be under polygenic control. Specifically, cave forms show reduced intensity or complexity of aggressive behavior (66), schooling behavior (65, 66), fright reaction to alarm substances (30), and feeding behavior (85). Wilkens (107) ascribed accumulation of neutral mutations as the mechanism for these reductions.

Decay of Egg-Rejection Behavior

Egg-rejection behavior in the Village Weaver, *Ploceus cucullatus,* a colonial nesting bird from Africa, had probably evolved to counter brood parasitism by the Didric Cuckoo, *Chrysococcyx caprius* (see 16). Village Weavers were introduced to Hispaniola in the eighteenth century, where they were not exposed to brood parasites until the arrival of the Shiny Cowbird, *Molothrus bonariensis,* in the early 1970s. Hispaniola Village Weavers are less discriminatory and reject a lower percentage of foreign eggs than do their African counterparts (16). The low level of rejection behavior probably is not an adaptation to parasitism by the Cowbird because of the recent contact between the species, and intraspecific brood parasitism is unknown in this population. Cruz & Wiley (16) concluded that egg-rejection behavior is selectively neutral in the absence of brood parasitism, and that the low level of rejection observed in Hispaniola Village Weavers represents the vestige of a nonfunctional trait.

VESTIGIAL GENES

Regal (73) noted that a biological molecule may be used in a variety of physiological systems and thus the same symbol (molecule) may encode different messages, providing, in the cybernetic sense, the potential for noise in the system. He argued that natural selection will favor the reduction or elimination of unused features because this will reduce or eliminate erroneous messages in physiological pathways, a process he called "streamlining evolution." He noted that this could be accomplished by either eliminating the genetic material from the genome or by "turning off" sections of the DNA. Li (49), in a similar context, suggested that pseudogenes and other forms of "nonfunctional" DNA may remain in the genome as long as they do not impede the cellular physiology.

How much streamlining of the genome actually occurs is an open question. The discovery in the late 1970s of noncoding intervening sequences (introns) within the protein coding regions of the genome of many eukaryotes has led to a debate over their origin and significance. Doolittle (23), among others, argued that introns arose early in evolution, perhaps as parts of inefficiently organized pre-cellular genomes, and he suggested that their absence in modern prokaryotes might be a consequence of selection for more efficient organization and expression in rapidly growing cells. This hypothesis is in the spirit of streamlining at the molecular level. More recently, however, Doolittle and colleagues (94) have concluded that introns arose late in evolution as insertions in the structural genes of eukaryotes. Because most introns have no known function, yet their messages are transcribed and must then be spliced out prior to translation, a late origin for introns is contrary to the hypothesis of streamlining evolution.

The occasional appearance of atavistic features indicates that the genes for reduced or lost morphological characters are often retained in the genome. Chimeras of chick epithelium and mouse molar mesenchyme can result in tooth development in which enamel matrix proteins are secreted by the chick epithelium (44). The opsin gene, which encodes a visual pigment protein, is actively transcribed in the early development of the blind cave-dwelling characin *Astyanax fasciatus* (46), and sequence data suggest that this gene is capable of producing a functional protein (111). Thus, toothlessness in birds and eye regression in *A. fasciatus* appear to be consequences of disruptions of developmental or regulatory pathways rather than disruptions of specific protein-coding regions of the genome.

Maintenance of expression of a "useless" gene, such as the opsin gene in *A. fasciatus,* may reflect the recency of cave isolation and eye reduction in these populations (111). Moles and mole rats, however, have a long evolutionary history in subterranean habitats and, presumably, a long evolutionary history of eye reduction, yet they may still retain genes for three and one lens-specific crystallin proteins, respectively (70). With evolutionary times for eye degeneration to occur of 25 million years for the mole rat and 45 million years for the mole, the crystallin genes should, in the absence of selective constraints, be expected to have accumulated enough mutations to have been rendered silent (70). Their continued expression is an enigma.

The consequences of vestigialization at the molecular level are unclear. Some structural genes involved in the development of vestigial structures have been turned off, such as the gene for enamel protein in birds. Others, such as the crystallin protein genes of blind burrowing mammals, continue to be expressed, opposing the noise suppression hypothesis. Further, the hypothesis of a more recent origin for introns is counter to the view of streamlining at the molecular level. The difficulty is that the lack of a known function for introns is not conclusive proof that they are useless. Likewise, the mere fact that crystallin genes have been conserved in moles and mole rats is suggestive of some unknown selective advantage (70). Pseudogenes that occur in many multi-gene families, produce no protein, and appear to have no other phenotypic consequences would seem to be examples of the outcome of streamlining evolution. Even here the issue is not clearcut. Many pseudogenes exhibit rates and patterns of mutation that suggest that these nonfunctional sequences are selectively neutral (see 49), supporting the idea of streamlining evolution. Recent studies on some *Drosophila* pseudogenes (95), however, have revealed slower evolutionary rates, and patterns of codon and substitution bias that indicate as yet unknown evolutionary constraints on these sequences. How vestigialization is manifested at the molecular level requires a more thorough understanding of the functioning of genetic and developmental systems.

CONCLUSIONS

Although we have concentrated only on selected case studies of vestigial characters, there should be no question that evolutionary reduction and loss of characters is common. In addition, we have not even discussed any case of character reduction in plants, such as the loss of photosynthetic ability in parasitic plants (e.g. Indian Pipe), and there should be many examples. Generally if there is relaxation of stabilizing or directional selection on a character, whether morphological, behavioral, or physiological, vestigialization can be expected. A focus on vestigialization can provide a different perspective on evolutionary phenomena. Recent debates on the flight capabilities of *Archaeopteryx* have centered on whether its small flight muscles could have generated sufficient power for sustained flight (see 90). An indicator of aerodynamic function in birds is vane asymmetry of the flight feathers, with the central shaft being closer to the leading edge, which is necessary to generate lift (see 91). Vane asymmetry of *Archaeopteryx lithographica* is not different from that of modern flightless birds but is significantly lower (more symmetrical) than that of flying birds, suggesting that *A. lithographica* was incapable of flight (91). The interesting point, from our perspective, is the regression of vane asymmetry to the ancestral state in the flightless birds. Although wings and feathers are smaller in flightless birds and may represent energy savings, variation in the position of the central shaft in the feather probably does not reflect differences in energy expenditure. The selective advantage of a more symmetrical feather in flightless birds is also unclear. The answers may come from examinations of pathways of feather development or from studies of vane symmetry and thermal properties of feathers. Our point is that many more interesting examples of vestigialization in nature may be revealed from such a perspective.

Understanding of the causes of vestigialization and loss of characters is incomplete. If the reduction of a character is obviously adaptive, then vestigialization is of interest only as another example of evolution through natural selection. Adaptive reduction in such a direct sense is exemplified by the short and stout wings used for swimming by penguins and steamer ducks. Even when selection does operate on a vestigial character, as in the case of direct selection against eyes in the cave-dwelling amphipod *Gammarus minus,* precisely how selection works can still be unknown. However, most examples of evolutionary reduction are of interest because they resist explanations as adaptations per se. Most explanations of character reduction invoke indirect selection in terms of energy economy or antagonistic pleiotropy arguments, although what is meant by energy in such a context is usually unstated, and few, if any, such arguments are framed as testable hypotheses. Furthermore, such arguments generally rest on the assumption that trade-offs among populations also operate within populations. In this light, the documented trade-off between fitness components and wing size and wing muscle mass in the cricket *Gryllus rubens* is significant, especially because a physiological mechanism underlying the trade-off can be inferred. The many cases of documented trade-offs between fitness components and flightlessness in other insects, and especially between fitness and resistance to pesticides, indicate that this is a plausible mechanism for character reduction; the more general implication is that pleiotropy is widespread. More analyses of trade-offs between fitness components and vestigial structures within populations are needed to ascertain whether it is a general explanation for character reduction.

It is more difficult to explain vestigial behavior in terms of energy economy and pleiotropy, because the normal behavior is not expressed and the vestigial behavior is elicited only under experimental conditions. Indeed, in all cases of vestigial behavior the investigators invoked the accumulation of neutral mutations to explain the behavioral decay. The neutral mutation argument is attractive in its simplicity, yet it is difficult to test. The expression of presumably useless genes, either naturally as in the case of opsins in the cave fish and in mole rats, or experimentally induced as in the case of enamel proteins in birds, suggests that very long evolutionary times are needed for mutation to destroy the genetic basis of a character. Yet the decay of the egg-rejection behavior of the Village Weaver, and especially the sexual behavior of *Drosophila mercatorum,* have proceeded rapidly. Trade-offs involving the neural circuitry responsible for the processing of information connected to the vestigialized behavior and other neural functions may explain this phenomenon.

Finally, we note that studies of vestigial characters generally make three assumptions. The first is that the reduced state is derived, and the second is that the character is nonfunctional. Although these are usually valid assumptions, the cases of the ancestral short-winged condition in water striders and of the photoperiodic function of the vestigial eyes in mole rats indicate caution must be exercised. The third assumption generally equates uselessness with selective neutrality per se. That this assumption is not always valid is reflected by the debate over whether the rates of evolution of pseudogenes are constrained by selection.

Literature Cited

1. Alberch P, Gale EA. 1985. A developmental analysis of an evolutionary trend: digital reduction in amphibians. *Evolution* 39:8–23
2. Andersen NM. 1993. The evolution of wing polymorphism in water striders (Gerridae): a phylogenetic approach. *Oikos* 67:433–43
3. Avise JC, Selander RK. 1972. Evolutionary genetics of cave-dwelling fishes of the genus *Astyanax. Evolution* 26:1–19
4. Barr TC. 1968. Cave ecology and the evolution of troglobites. *Evol. Biol.* 2:35–102
5. Beachy CK, Bruce RC. 1992. Lunglessness in plethodontid salamanders is consistent with the hypothesis of a mountain stream origin: a response to Ruben and Boucot. *Am. Nat.* 139:839–47
6. Brace CL. 1963. Structural reduction in evolution. *Am. Nat.* 97:39–49
7. Bruce RC, Beachy CK, Lenzo PG, Pronych SP, Wassersug RJ. 1994. Effects of lung reduction on rheotactic performance in amphibian larvae. *J. Exp. Zool.* 268:377–80
8. Cardone B, Fullard JH. 1988. Auditory characteristics and sexual dimorphism in the gypsy moth. *Physiol. Entomol.* 13:9–14
9. Carrière Y, Deland JP, Roff DA, Vincent C. 1994. Life-history costs associated with the evolution of insecticide resistance. *Proc. R. Soc. Lond. B* 258:35–40
10. Carson HL, Chang LS, Lyttle TW. 1982. Decay of female sexual behavior under parthenogenesis. *Science* 218:68–70
11. Chakraborty R, Nei M. 1974. Dynamics of gene differentiation between incompletely isolated populations of unequal sizes. *Theor. Pop. Biol.* 5:460–69
12. Cochran DG. 1993. Decline of pyrethroid resistance in the absence of selection pressure in a population of German cockroaches (Dictyoptera: Blattellidae). *J. Econ. Entomol.* 86:1639–44
13. Cohan FM, King EC, Zawadzki P. 1994. Amelioration of the deleterious pleiotropic effects of an adaptive mutation in *Bacillus subtilis. Evolution* 48:81–95
14. Cooper HM, Herbin M, Nevo E. 1993. Ocular regression conceals adaptive progression of the visual system in a blind subterranean mammal. *Nature* 361:156–59
15. Cracraft J. 1990. The origin of evolutionary novelties: pattern and process at different hierarchical levels. In *Evolutionary Innovations*, ed. MH Nitecki, pp. 21–44. Chicago: Univ. Chicago Press. 304 pp.
16. Cruz A, Wiley JW. 1989. The decline of an adaptation in the absence of a presumed selection pressure. *Evolution* 43:55–62
17. Culver DC. 1982. *Cave Life: Evolution and Ecology*. Cambridge: Harvard Univ. Press. 189 pp.
18. Culver DC, Holsinger JR. 1992. How many species of troglobites are there? *Bull.* 54:79–80
19. Culver DC, Jernigan RW, O'Connell J, Kane TC. 1994. The geometry of natural selection in cave and spring populations of the amphipod *Gammarus minus* Say (Crustacea: Amphipoda). *Biol. J. Linn. Soc.* 52:49–67
20. Culver DC, Kane TC, Fong DW. 1995. *Adaptation and Natural Selection in Caves: The Evolution of* Gammarus mi-

nus. Cambridge: Harvard Univ. Press. 240 pp.

21. Darwin C. 1859. *On the Origin of Species.* London: Murray. 502 pp. 1st ed. (Facsimile).

22. Denno RF, Roderick GK, Olmstead KL, Dobel HG. 1991. Density-related migration in planthoppers (Homoptera: Delphacidae): the role of habitat persistence. *Am. Nat.* 138:1513–41

23. Doolittle WF. 1987. The origin and function of intervening sequences in DNA: a review. *Am. Nat.* 130:915–28

24. Duellman WE, Trueb L. 1986. *Biology of Amphibians.* New York: McGraw-Hill. 670 pp.

25. Edmunds GF. 1972. Biogeography and evolution of Ephemeroptera. *Annu. Rev. Entomol.* 17:21–43

26. Emerson AE. 1961. Vestigial characters of termites and processes of regressive evolution. *Evolution* 15:115–31

27. Feder ME. 1976. Lunglessness, body size, and metabolic rate in salamanders. *Physiol. Zool.* 49:398–406

28. Feduccia A. 1980. *The Age of Birds.* Cambridge: Harvard Univ. Press. 196 pp.

29. Fong DW. 1989. Morphological evolution of the amphipod *Gammarus minus* in caves: quantitative genetic analysis. *Am. Midl. Nat.* 121:361–78

30. Fricke D. 1988. Reaction to alarm substance in cave populations of *Astyanax mexicanus* (Characidae, Pisces). *Ethology* 76:305–8

31. Fullard JH. 1994. Auditory changes in noctuid moths endemic to a bat-free habitat. *J. Evol. Biol.* 7:435–45

32. Fullard JH, Yack JE. 1993. The evolutionary biology of insect hearing. *Trends Ecol. Evol.* 8:248–52

33. Gans C. 1975. Tetrapod limblessness: evolution and functional corollaries. *Am. Zool.* 15:455–67

34. Ginet R, Decou V. 1977. *Initiation à la Biologie et à l'Ecologie Souterraines.* Paris: J-P Delarge. 345 pp.

35. Groeters FR, Tabashnik BE, Finson N, Johnson MW. 1994. Fitness costs of resistance to *Bacillus thuringiensis* in the diamondback moth (*Plutella xylostella*). *Evolution* 48:197–201

36. Haldane JBS. 1933. The part played by recurrent mutation in evolution. *Am. Nat.* 67:5–19

37. Hanken J, Wake DB. 1993. Miniaturization of body size: organismal consequences and evolutionary significance. *Annu. Rev. Ecol. Syst.* 24:501–19

38. Harvey PH, Pagel MD. 1991. *The Comparative Method in Evolutionary Biology.* New York: Oxford Univ. Press. 239 pp.

39. Houle D. 1991. Genetic covariance of fitness correlates: What genetic correlations are made of and why it matters. *Evolution* 45:630–45

40. Jones R, Culver DC. 1989. Evidence for selection on sensory structures in a cave population of *Gammarus minus* Say (Amphipoda). *Evolution* 43:688–93

41. Jones R, Culver DC, Kane TC. 1992. Are parallel morphologies of cave organisms the result of similar selection pressures? *Evolution* 46:353–65

42. Katz MJ, Lasek RJ, Kaiserman-Abramof IR. 1981. Ontophyletics of the nervous system: eyeless mutants illustrate how ontogenetic buffer mechanisms channel evolution. *Proc. Natl. Acad. Sci.* 78:397–401

43. Kimura M. 1983. *The Neutral Theory of Molecular Evolution.* Cambridge: Cambridge Univ. Press. 367 pp.

44. Kollar EJ, Fisher C. 1980. Tooth induction in chick epithelium: expression of quiescent genes for enamel synthesis. *Science* 207:993–95

45. Lande R. 1978. Evolutionary mechanisms of limb loss in tetrapods. *Evolution* 32:73–92

46. Langecker TG, Schmale H, Wilkens H. 1993. Transcription of the opsin gene in degenerate eyes of cave-dwelling *Astyanax fasciatus* (Teleostei, Characidae) and of its conspecific epigean ancestor during early ontogeny. *Cell Tissue Res.* 273:183–92

47. Lenski RE. 1988. Experimental studies of pleiotropy and epistasis in *Escherichia coli.* I. Variation in competitive fitness among mutants resistant to virus T4. *Evolution* 42:425–32

48. Lenski RE. 1988. Experimental studies of pleiotropy and epistasis in *Escherichia coli.* II. Compensation for maladaptive effects associated with resistance to virus T4. *Evolution* 42:433–40

49. Li W-H. 1983. Evolution of duplicate genes and pseudogenes. In *Evolution of Genes and Proteins,* ed. M Nei, RK Koehn, pp. 14–37. Sunderland: Sinauer Assoc. 331 pp.

50. Livezey BC. 1989. Flightlessness in grebes (Aves, Podicipedidae): its independent evolution in three genera. *Evolution* 43:29–54

51. Livezey BC, Humphrey PS. 1983. Mechanics of steaming in steamer-ducks. *Auk* 100:485–88

52. Livezey BC, Humphrey PS. 1984. Diving behaviour of steamer ducks *Tachyeres* spp. *Ibis* 126:257–60

53. Livezey BC, Humphrey PS. 1986. Flightlessness in steamer-ducks (Anatidae: Tachyeres): its morphological bases and probable evolution. *Evolution* 40:540–58

54. Lombard RE, Wake DB. 1986. Tongue evolution in the lungless salamanders, family Plethodontidae. IV. Phylogeny of plethodontid salamanders and the evolution of feeding dynamics. *Syst. Zool.* 35:532–51

55. Ludwig W. 1942. Zur evolutorischen Erklärung der Höhlentiermerkmale durch Allelelimination. *Biol. Zentral.* 62:447–82

56. Lynch M. 1990. The rate of morphological evolution in mammals from the standpoint of neutral expectation. *Am. Nat.* 136:727–41

57. McKenzie JA. 1993. Measuring fitness and intergenic interactions: the evolution of resistance to diazinon in *Lucilia cuprina. Genetica* 90:227–37

58. McKenzie JA, Batterham P. 1994. The genetic, molecular and phenotypic consequences of selection for insecticide resistance. *Trends Ecol. Evol.* 9:166–69

59. McNab BK. 1994. Energy conservation and the evolution of flightlessness in birds. *Am. Nat.* 144:628–42

60. Mole S, Zera AJ. 1993. Differential allocation of resources underlies the dispersal-reproduction trade-off in the wing-dimorphic cricket, *Gryllus rubens. Oecologia* 93:121–27

61. Muller HJ. 1949. The Darwinian and modern conceptions of natural selection. *Proc. Am. Philos. Soc.* 93:459–70

62. Nevo E. 1991. Evolutionary theory and processes of active speciation and adaptive radiation in subterranean mole rats, *Spalax ehrenbergi* superspecies, in Israel. *Evol. Biol.* 25:1–125

63. Oster GF, Shubin N, Murray JD, Alberch P. 1988. Evolution and morphogenetic rules and the shape of the vertebrate limb in ontogeny and phylogeny. *Evolution* 42:862–84

64. Parker MA. 1990. The pleiotropy theory for polymorphisms of disease resistance genes in plants. *Evolution* 44:1872–75

65. Parzefall J. 1983. Field observations in epigean and cave populations of the Mexican characid *Astyanax mexicanus* (Pisces, Characidae). *Mem. Biospeol.* 10:171–76

66. Parzefall J. 1985. On the heredity of behavior patterns in cave animals and their epigean relatives. *NSS Bull.* 47:128–35

67. Poulson TL. 1963. Cave adaptation in amblyopsid fishes. *Am. Midl. Nat.* 70:257–90

68. Poulson TL, White WB. 1969. The cave environment. *Science* 165:971–81

69. Prout T. 1964. Observations on structural reduction in evolution. *Am. Nat.* 98:239–49

70. Quax-Jeuken Y, Bruisten S, Bloemendal H, de Jong WW. 1985. Evolution of crystallins: expression of lens-specific proteins in the blind mammals mole (*Talpa europaea*) and mole rat (*Spalax ehrenbergi*). *Mol. Biol. Evol.* 2:279–88

71. Rausher MD. 1992. Natural selection and the evolution of plant-insect interactions. In *Insect Chemical Ecology: An Evolutionary Approach*, ed. BD Roitberg, MB Isman, pp. 22–88. New York: Chapman & Hall. 359 pp.

72. Reagan NL, Verrell PA. 1991. The evolution of plethodontid salamanders: Did terrestrial mating facilitate lunglessness? *Am. Nat.* 138:1307–13

73. Regal PJ. 1977. Evolutionary loss of useless features: Is it molecular noise suppression? *Am. Nat.* 111:123–33

74. Rieppel O. 1988. A review of the origin of snakes. *Evol. Biol.* 22:37–130

75. Roff DA. 1986. The evolution of wing dimorphism in insects. *Evolution* 40:1009–20

76. Roff DA. 1989. Exaptation and the evolution of dealation in insects. *J. Evol. Biol.* 2:109–23

77. Roff DA. 1990. The evolution of flightlessness in insects. *Ecol. Monogr.* 60:389–421

78. Roff DA. 1994. The evolution of flightlessness: Is history important? *Evol. Ecol.* 8:639–57

79. Romer AS. 1966. *Vertebrate Paleontology*. Chicago: Univ. Chicago Press. 468 pp.

80. Romero A. 1985. Can evolution regress? *NSS Bull.* 47:86–88

81. Roush RT, McKenzie JA. 1987. Ecological genetics of insecticide and acaricide resistance. *Annu. Rev. Entomol.* 32:361–80

82. Ruben JA, Boucot AJ. 1989. The origin of the lungless salamanders (Amphibia: Plethodontidae). *Am. Nat.* 134:161–69

83. Ruben JA, Reagan NL, Verrell PA, Boucot AJ. 1993. Plethodontid salamander origins: a response to Beachy and Bruce. *Am. Nat.* 142:1038–51

84. Rydell J, Arlettaz R. 1994. Low-frequency echolocation enables the bat *Tadarida teniotis* to feed on tympanate insects. *Proc. R. Soc. Lond. B.* 257:175–78

85. Schemmel C. 1980. Studies on the genetics of feeding behaviour in the cave fish *Astyanax mexicanus* f. *Anoptichthys*. An example of apparent monofac-

torial inheritance by polygenes. *Zeit. Tierpsychol.* 53:9–22

86. Service PM, Rose MR. 1985. Genetic covariation among life-history components: the effect of novel environments. *Evolution* 39:943–45

87. Shubin NH, Alberch P. 1986. A morphogenetic approach to the origin and basic organization of the tetrapod limb. *Evol. Biol.* 20:319–87

88. Sket B. 1985. Why all cave animals do not look alike—a discussion on adaptive value of reduction processes. *NSS Bull.* 47:78–85

89. Spangler HG. 1988. Moth hearing, defense, and communication. *Annu. Rev. Entomol.* 33:59–81

90. Speakman JR. 1993. Flight capabilities in *Archaeopteryx. Evolution* 47:336–40

91. Speakman JR, Thomson SC. 1994. Flight capabilities of *Archaeopteryx. Nature* 370:514

92. Spence JR. 1989. The habitat templet and life history strategies of pond skaters (Heteroptera: Gerridae): reproductive potential, phenology, and wing dimorphism. *Can. J. Zool.* 67:2432–47

93. Spence JR, Andersen NM. 1994. Biology of water striders: interactions between systematics and ecology. *Annu. Rev. Entomol.* 39:101–28

94. Stoltzfus A, Spencer DF, Zucker M, Logsdon JM Jr, Doolittle WF. 1994. Testing the exon theory of genes: the evidence from protein structure. *Science* 265:202–7

95. Sullivan DT, Starmer WT, Curtiss SW, Menotti-Raymond M, Yum J. 1994. Unusual molecular evolution of an *Adh* pseudogene in *Drosophila. Mol. Biol. Evol.* 11:443–58.

96. Tabashnik BE. 1994. Evolution of resistance to *Bacillus thuringiensis. Ann. Rev. Entomol.* 39:47–79

97. Tabashnik BE, Finson N, Groeters FR, Moar WJ, Johnson MW, et al. 1994. Reversal of resistance to *Bacillus thuringiensis* in *Plutella xylostella. Proc. Natl. Acad. Sci. USA* 91:4120–24

98. Vandel A. 1964. *Biospeologie: La Bi-*

ologie des Animaux Cavernicoles. Paris: Gautheir-Villars. 619 pp.

99. Van Valen L. 1960. Nonadaptive aspects of evolution. *Am. Nat.* 94:305–8

100. Voneida TJ, Fish SE. 1984. Central nervous system changes related to the reduction of visual input in a naturally blind fish (*Astyanax hubbsi*). *Am. Zool.* 24:775–82

101. Wake DB. 1991. Homoplasy: the result of natural selection, or evidence of design limitations. *Am. Nat.* 138:543–67

102. Whitford WG. 1973. The effects of temperature on respiration in the Amphibia. *Am. Zool.* 13:505–12

103. Whitford WG, Hutchison VH. 1965. Gas exchange in salamanders. *Physiol. Zool.* 38:228–42

104. Whitford WG, Hutchison VH. 1966. Cutaneous and pulmonary gas exchange in ambystomatid salamanders. *Copeia* 1966:573–77

105. Wilder IW, Dunn ER. 1920. The correlation of lunglessness in salamanders with a mountain brook habitat. *Copeia* 1920:63–68

106. Wilkens H. 1971. Genetic interpretation of regressive evolutionary processes: studies on hybrid eyes of two *Astyanax* cave populations (Characidae, Pisces). *Evolution* 25:530–44

107. Wilkens H. 1988. Evolution and genetics of epigean and cave *Astyanax fasciatus* (Characidae, Pisces): support of the neutral mutation theory. *Evol. Biol.* 23:271–67

108. Wilkens H. 1993. Neutrale Mutationen und evolutionäre Fortentwicklung. *Z. Zool. Syst. Evolut.-Forsch.* 31:98–109

109. Wright S. 1964. Pleiotropy in the evolution of structural reduction and of dominance. *Am. Nat.* 98:65–69

110. Yager DD. 1990. Sexual dimorphism of auditory function and structure in praying mantises (Mantodea; Dictyoptera). *J. Zool.* 221:517–37

111. Yokoyama R, Yokoyama S. 1990. Convergent evolution of the red- and green-like visual pigment genes in fish, *Astyanax fasciatus,* and human. *Proc. Natl. Acad. Sci. USA* 87:9315–18

Annu. Rev. Ecol. Syst. 1995. 26:269–99

THE QUALITY OF THE FOSSIL RECORD: Populations, Species, and Communities

Susan M. Kidwell

Department of Geophysical Sciences, University of Chicago, 5734 South Ellis Avenue, Chicago, Illinois 60637

Karl W. Flessa

Department of Geosciences, University of Arizona, Tucson, Arizona 85721

KEY WORDS: ecology, paleoecology, taphonomy

ABSTRACT

Paleontologists have always been concerned about the documentary quality of the fossil record, and this has also become an important issue for biologists, who increasingly look to accumulations of bones, shells, and plant material as possible ways to extend the time-frame of observation on species and community behaviors. Quantitative data on the postmortem behavior of organic remains in modern environments are providing new insights into death and fossil assemblages as sources of biological information. Important findings include: 1. With the exception of a few circumstances, usually recognizable by independent criteria, transport out of the original life habitat affects few individuals. 2. Most species with preservable hardparts are in fact represented in the local death assemblage, commonly in correct rank importance. Molluscs are the most durable of modern aquatic groups studied so far, and they show highest fidelity to the original community. 3. Time-averaging of remains from successive generations and communities often prevents the detection of short-term (seasons, years) variability but provides an excellent record of the *natural range* of community composition and structure over longer periods. Thus, although a complex array of processes and circumstances influences preservation, death assemblages of resistant skeletal elements are for many major

0066-4162/95/1120-0269$05.00

groups good to excellent records of community composition, morphological variation, and environmental and geographic distribution of species, and such assemblages can record temporal dynamics at ecologically and evolutionarily meaningful scales.

INTRODUCTION

Every paleontologist can tell horror stories of long-distance transport of biological remains, of bones worn or comminuted beyond recognition, and of fossils found mixed with Recent shells along modern beaches. Every paleontologist also can recite some of the glories of the fossil record—for example, feathered *Archaeopteryx* from the Jurassic Solnhofen Limestone, the bizarre soft-bodied taxa of the late Neoproterozoic Ediacara Formation and Middle Cambrian Burgess Shale, exquisite plants from Carboniferous coal, and the spectacular assortment of bats, insects, frogs, and horses with undigested stomach contents in the Eocene Messel oil shale (2, 139). There is an unexceptional middle ground, however, comprising the vast majority of fossil and modern death assemblages, where the biological signal is expected to be damped and biased to some intermediate degree. One of the challenges of taphonomic research is to evaluate postmortem modification in these ordinary assemblages and derive practical guidelines for their use as sources of biological information. As Paul (91) has stressed, what matters is not that the fossil record is incomplete—all science is based on incomplete knowledge—but whether data incorporated in the fossil record are adequate to test particular hypotheses. Here we review the quality of the record with respect to reconstructing biological phenomena at the scale of populations, species, and communities, and we argue that even ordinary data in the fossil record are adequate to address many important biological questions.

There is a growing body of comparative and quantitative data on taphonomic (postmortem) phenomena and their consequences. Much of this work has focused on processes operating in modern environments as keys to understanding the past; this work has had two aims. One has been to use the quality of fossil preservation as evidence of (paleo)environmental conditions, such as wave/current energy, porewater chemistry, climate, directions of sediment transport, and duration of sedimentary hiatuses (3, 41, 77).

The second aim, and the focus here, is the effect of taphonomic processes on the biological signal. What is the quality of the record as an archive of ancient biotic interactions, species distributions, morphological variability in time and space, and community structure and dynamics? This, the original impetus of the field (44; Darwin devoted a chapter of the *Origin* to this), is critical in placing empirical constraints on interpretations of the fossil record. Research approaches include deductive analyses of fossil assemblages, in which preservational quality of individual specimens and sedimentary context are used to infer likely postmortem modification, and live:dead studies in modern environments, whereby the taxonomic or age-class composition of a death assemblage (shells, bones, leaf litter) is compared with that of the local living community. These empirical approaches have been complemented by probabilistic models and computer simulations aimed at testing both taphonomic and ecological (supply-side) controls on the nature of the record (11, 37, 38, 85, 86; several other papers in 71).

To the uninitiated, the major sources of postmortem bias are assumed to be out-of-habitat transport of carcasses, pollen, leaves, and shells, selective destruction of species and age-classes at the accumulation site, and especially the loss of soft, nonmineralized tissues. An equally important limit on interpretation, however, is time-averaging, that is, mixing of noncontemporaneous material so that specimens from successive generations or ecologically unrelated communities occur within a single sedimentary stratum (56, 57, 128). The interval between death and the time when remains are buried beyond the reach of everyday erosional exhumation and bioturbation determines, along with environmental conditions, the extent to which processes of selective destruction can act upon the raw input from mortality and shedding. The most complete records (in which soft tissues are preserved) of ancient communities are created almost exclusively by permanent, catastrophic burial upon death. In such instances, time-averaging is zero. Even environments prone to episodes of catastrophic burial, however, are characterized by intervening periods of slow or no appreciable net accumulation of sediment. Multiple cohorts of remains may accumulate in surficial sediments during these times, so that the final death assemblage is the product of time-integrated input and its progressive, generally selective, destruction (Figure 1). This is one of the most important adjustments for a biologist to make in dealing with the fossil record or even with modern death assemblages: Sediment accumulation rates are generally much slower than biological generation rates, and so most death assemblages composed of mineralized hardparts are time-averaged to some degree. Resolution in these instances is more analogous to a dataset built by pooling successive years of census data. Whether time-averaging is perceived as an insurmountable problem or a happy advantage, smoothing the noise of short-term variability (92), depends on the question being asked.

In this paper, we discuss the quality of the fossil record with respect to such attributes as spatial resolution, fidelity of species composition, and temporal resolution. For each of these topics, we address 1) the major processes that can modify the original biological signal, 2) estimates on the magnitudes and net effects of these processes, including the methods used to make such estimates, and 3) criteria for recognizing or anticipating bias in assemblages of unknown nature. We conclude with the implications of taphonomic bias for

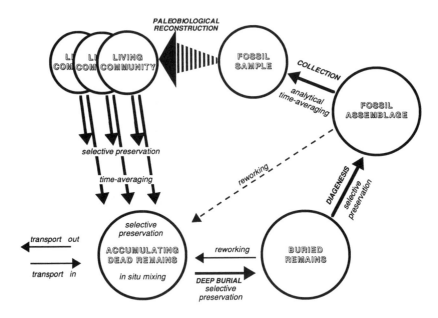

Figure 1 Taphonomic processes and circumstances that, during the fossilization of organic remains, have potential to modify the original biological signal at different postmortem phases.

current issues in (paleo)biology, such as the ability to detect events of speciation, extinction, and ecological perturbation. As invertebrate paleontologists, we stress marine systems, where most of the work has been done. We conclude with implications of these quantitative estimates for current issues in ecology, evolution, and paleobiology.

THE LOSS OF SOFT TISSUES

The most dramatic contrast in fossil preservation is between organisms having mineralized or highly refractory tissues (composed of calcium carbonate, calcium phosphate, silica, sporopollenin, or lignin; that is, "preservable taxa") and those that largely or entirely lack such materials ("soft-bodied taxa").

Soft-bodied taxa can be preserved under unusual conditions, such as rapid sedimentation that catastrophically buries part or all of a community and isolates the remains from scavengers and other taphonomic agents (20, 141). Individual organisms can be preserved with soft tissues intact in a number of small-scale circumstances, including sealing by amber, tar, and encrusting organisms; "pickling" in salt and humic acids; freezing; and mineralization under anoxic conditions (3, 139). Paleoecological, evolutionary, and biogeographic analysis of soft-bodied groups relies on the rare "preservational windows" created by such special circumstances. For some purposes, widely spaced windows (Fossil-Lagerstätten) are sufficient to track major morphological or ecological trends (21).

Under more ordinary environmental conditions such as relatively slow sediment accumulation on oxygenated seafloors and lakebeds, and on land surfaces characterized by moist and/or warm conditions, soft-bodied taxa have very low preservation potential (2, 3). These taphonomically unfavorable conditions commonly persist into the sediment for some depth because of sediment stirring by burrowing organisms, freeze-thaw cycles, and wave and tide reworking. Burial thus does not ensure preservation until it removes the remains to some threshold depth in the sedimentary column. The potential for destruction does not cease with permanent burial—there are still the perils of rock diagenesis (that is, compaction and chemical changes), tectonism, and erosion—but the many modifications that occur in the initial postmortem stages are certainly most tractable for study (Figure 1).

The destruction of soft-bodied organisms can represent a substantial loss in biological information. Most "reconstructions" of aquatic food webs and energy flows by paleoecologists thus differ fundamentally from those based on living communities both in intent and reality, and such reconstructions are useful only for broad comparisons among similarly preserved assemblages (105). Paleontologists consequently devote most effort to segments of communities with highest potential for preservation—the durable hardparts of molluscs, echinoderms, corals, arthropods, bryozoans, and vertebrates in marine communities, and vertebrates and woody or heavily cuticled plants and pollen in terrestrial systems. These "preservable taxa" are the primary focus of this review.

SPATIAL RESOLUTION OF HARDPARTS

Except for instances of catastrophic burial, most fossil remains consist of disarticulated elements separated some distance from other parts of the same skeleton or body. Such dispersed elements can only rarely be reassembled into their individual organisms. This difficulty also precludes determining the precise spatial distribution of individuals within their habitats. However, studies in modern environments have repeatedly demonstrated that most postmortem movement of organic remains is within the range of the original life habitat for the source population, or at least within the time-averaged range of the species. Significant, out-of-habitat transport appears limited to a few settings recognizable by independent geological criteria, and to particular sets of organisms.

Marine Systems

In level-bottom marine habitats, cluster analysis reveals strong agreement in live and dead molluscs, with most (80–95%) dead individuals belonging to species documented alive in the same habitat (Tables II-IV in 70; also see 99, 104). The same pattern is exhibited by crabs (93), echinoids (e.g. 89% fragment weight—89; 63), and freshwater molluscs (79–82% fidelity, based on 22, 36, 133). Exotic species thus generally account for few individuals in the death assemblage, even when they constitute a large part of the total species list. Most exotics are derived from immediately adjacent habitats (and see data on crinoids in 84, and data on benthic Foraminifera in 81). Small thin shells, especially of noncementing epifauna and shallow-burrowing infauna, are most likely to be moved from their life habitat (Table IX in 70). These exotics can often be recognized by their incongruence with the embedding sediment or with dominant associated species (e.g. species found in sand whose functional morphology or living relatives indicate ecologic preference for rocky shores).

It was originally assumed that damage to carcasses and individual hardparts could be used to rank transport distances. More recent work indicates, however, that fresh carcasses are highly durable even to energetic tumbling (1, 3) and that most long-distance transport is by relatively nondestructive means (e.g. floating, rafting, bulk movement; Table IX in Ref. 70). Most damage to hardparts results instead from processes operating within the original life habitat. In instances when already damaged hardparts are moved into another habitat, their taphonomic condition may well distinguish them from hardparts in the recipient assemblage, thereby allowing exotics to be identified, but this damage is not a function of distance traveled (e.g. 87).

Among macroinvertebrates, the record for long-distance transport must be held by nektonic cephalopods, with modern representatives showing 10^2 to 10^3 km of "necro-planktonic drift" outside their known oceanic ranges (101). Individuals may bear evidence of prolonged drift from epizoan overgrowths, such as has been recognized among extinct ammonoids and nautiloids (17, 106). Therefore, the entire group should be treated with caution if the objective is the determination of the original habitat (see also 102 on similar problems with fish and marine mammals). Diving birds can transport the shells of deep-water benthos to cliff-top middens (which may be composed of 100% exotic individuals; e.g. 76), creating bathymetrically anomalous assemblages; here the physical context of the assemblage should be the cause for suspicion. Along with offshore winds, birds can also be important agents of moving intertidal and terrestrial species into fully aquatic settings, but these exotics tend not to be numerically dominant in the recipient assemblage (22, 30, 36, 133; and see hermit crabs, 129). The most impressive out-of-habitat transport of benthic invertebrates is associated with steep depositional slopes or with settings having episodically very high pulse-type energy, where large numbers of shells can be transported via bulk flow of the seafloor (Table IX in 70). Sedimentary structures associated with the assemblage and its larger geological context are the most valuable clues that an assemblage might be strongly biased by or consist entirely of exotic material.

Terrestrial Systems

Live:dead comparisons for terrestrial vertebrates also indicate high fidelity of death assemblages to source communities at a habitat scale, despite opportunities for transport by predators, scavengers, and streams. Mammalian bone assemblages in eastern Africa have been studied in greatest detail because of the availability of data on the abundance and mortality rates of living megafauna in national parks. In the Amboseli Basin (11), Serengeti plains (16), and eastern (107) and central (122) Zaire, the relative bone frequencies of major species reflect their original habitat preferences, especially for the habitats in which mortality is concentrated (e.g. waterholes, ecotones between forest and grassland). Bones were not sampled from river channels in any of these studies, but field experiments on implanted bones (10; AK Behrensmeyer, personal communication) indicate relatively little downstream transport: The primary taphonomic effect of river channels seems to be in mixing material of different ages in the channel lag by erosional reworking of overbank deposits (Figure 2). Downstream floating of bloated carcasses may be the primary, although still rare, means of transport by water (rivers, coasts, open sea). Transport by scavengers and predators apparently does not exceed the natural range of movements of the prey when alive, especially for large-bodied prey (13, 77). Raptors also tend to sample relatively small areas (~ few km^2), but their concentrations may nonetheless include remains from several prey habitats, and so taphonomic evidence for this mode of concentration (distinctive fragmentation, size distribution, and selective partial digestion of skeletal elements) is an important clue to possible bias in the assemblage (4, 13). Sedimentary context (e.g. within a channel or in overbank deposits; Table 2.8 in Ref. 13) is also important in anticipating spatial resolution in fossil assemblages.

The spatial fidelity of plant death and fossil assemblages is more variable. At the high end, litter rain onto the forest floor is closely linked to source trees. In dense forests, leaves, seeds, fruits, and flowers generally do not fall further from the trunk than the height of the tree (~20° cone projected down from the treetop; 43, 50). Thus there is great potential for detailed reconstructions of spatial heterogeneity using terrestrial litter or its fossilized equivalent (26). Wind transport is a minor source of bias for comparatively heavy macrofloral

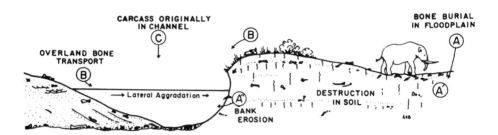

Figure 2 Model for postmortem input of bones to the sedimentary record of a river channel–floodplain system, showing potential for time-averaging, selective destruction, and reworking. (From Behrensmeyer 1982).

elements (43, 66), but water transport can be significant for species living along the water's edge (25, 62, 97, 113). Fortunately, variation in postmortem transport is fairly systematic. For example, riparian and coastal species are most prone to transport because they are most likely to fall directly into water; evergreen leaves float and resist decay longer than deciduous leaves (see review in 113); leaf assemblages from large, relatively low-energy rivers contain more long-distance exotics than do those from small, high-energy streams (e.g. 90% of species are exotic—58; 113); and lakes that are broader than the height of surrounding vegetation capture more wind-transported exotics than do lakes of small surface area (113). Attention to the sedimentary/environmental context of the assemblage is thus essential in anticipating the probable spatial resolution and in identifying those species most likely to be exotic. Unfortunately, the quiet forest-floor leaf assemblages that contain fewest exotics also have low preservation potential due to oxidation, root-bioturbation, and rapid decomposition by fungi and invertebrates during the initial seasons or years of exposure, depending on moisture and soil acidity (18a, 26, 62, 103). This does not diminish the usefulness of modern death assemblages for botanists, but it does diminish the likelihood of their being preserved in the older fossil record. Leaf assemblages that accumulate on the floors of deep (and thus more permanently anoxic) lakes and on lake-head deltas and stream point-bars of rapid sedimentation have the highest likelihood of preservation (103, 113). The larger the lake, the greater the inclusion of exotic material. This coarser spatial resolution may be an advantage for reconstructing regional vegetation in some situations (e.g. 123).

The spatial resolution of pollen assemblages varies depending upon the natural "trap" (lake, peat bog, moss polster, soil) (48, 49, 120) and on how the palynologist measures compositional changes. Accumulation–rate data from assemblages in lakes < 100 ha reveal only the broadest, regional-scale differences in pollen production (e.g. between tundra and forest) because of the effect of pollen resuspension and mixing within lakes (65), whereas pollen percentage data from the same assemblages typically allow the resolution of 10-km scale variation in forest composition (65, 134). In all studies of lake and peatland traps, the size of the trap and its distance from vegetation are the strongest factors in spatial resolution, as this determines the relative influence of wind-blown as opposed to gravity- and water-transported pollen (65, 120). Vegetational variation at the scale of hundreds of meters can commonly be resolved using small traps (15, 65). In contrast, pollen assemblages from alluvial fans, streams, and marine settings tend to have spatial resolutions at the scale of the drainage basin at best, since they are prone to selective water-transport, recycling of extant pollen from temporary reservoirs, and reworking of material from older rocks in the drainage basin (48, 125). Insect faunas have relatively similar postmortem dispersal behaviors (34, 46).

Common Patterns

Possibly the most important lesson on spatial resolution from these studies is that postmortem transport generally does not homogenize death assemblages across entire landscapes. In most settings, exotic species are not numerically dominant, and thus habitat-scale patchiness in species composition is commonly resolvable, even in time-averaged assemblages. Apparent discrepancies in the distribution of living organisms and dead remains are due primarily to habitat-shifting over time (and thus shifting in areas of input of various species) and not to the signal being smeared by postmortem transport of remains (and see numerical models 12, 85, 86). Settings in which exotics are significant should generally be identifiable by independent sedimentologic evidence. Moreover, the exotic species themselves tend to be smaller and less massive than indigenous remains, ecologically discordant with indigenous material or with the host sediment, characterized by different degrees of postmortem damage, or all of the above.

FIDELITY OF SPECIES COMPOSITION

Selective preservation, which operates even among "preservable" mineralized taxa, has been the focus of most taphonomic research. Anatomical attributes such as size, shape, density, mineralogy, and ultrastructure can influence the physical and chemical behavior of hardparts and also determine their attractiveness to bioeroders (raspers, borers) and other organisms that influence the fates of dead hardparts (3, 41, 77). These attributes vary among taxa, among age- or size-classes within species, and among skeletal elements of a single carcass. Staff et al (115), for example, documented postmortem "half-lives" of only ~2 months for 1 mm shells but immeasurably long half-lives for 10 mm adult specimens of the same molluscan species in Texas lagoons. Environments also differ in the likelihood of hardpart preservation: Bones, for example, appear to be destroyed more quickly in tropical rainforests than on dry savannas, and they last longest under temperate and subarctic conditions (5, 9, 67, 121; RW Graham, personal communication). The "law of numbers" also applies: taxa that are rare in life and/or have slow generation times (and thus produce few dead per unit time) have a lower likelihood of being fossilized in sufficient abundance to be discovered. Finally, the circumstances of death influence postmortem preservation: Catastrophic burial generally favors hardpart preservation, for example, whereas shell-crushing and prey-swallowing predators do not.

The net effects of these processes on the composition of the final assemblage is difficult to predict: Quantitative rate data are not available for all groups in all environments; we still have little insight on the constancy of rates over the decades to thousands of years that typify time-averaging; and we do not understand how the various processes reinforce or cancel out one another. Consequently, paleontologists have relied primarily upon empirical live:dead studies, whereby the composition of a modern death assemblage or Pleistocene fossil assemblage is compared with the local living community.

Molluscan Assemblages

Most live:dead studies have been conducted for shelled molluscs. At first reading of this literature, fidelity appears to vary widely, but metaanalysis suggests that much of the variance arises from differences in the quality of live data (especially single census versus replicates over time), in sampling density per habitat, and in differences in fidelity metrics (70; SM Kidwell, in preparation). When these artefacts are removed, live:dead agreement is high for many measures of community structure, despite the fact that postmortem processes probably destroy a huge proportion of the total shells produced.

For example, for 16 studies in temperate and subtropical latitudes, 90% of species living in intertidal settings are also recorded dead there; relevant numbers in other habitats are 98% in coastal subtidal settings and 75% in marine bights and on the continental shelf (Tables II-V in 70). In streams and lakes the value is 95% (22, 36, 133). Values are lower when live communities are compared with local Pleistocene fossil assemblages (e.g. 61% of molluscs living along San

Nicolas Island, California, are present in uplifted Pleistocene marine terraces; 99), but, as is also seen among modern death assemblages (70), values rise as the geographic scale becomes coarser (e.g. 77% of species found alive today within the Californian Province are also present in Pleistocene assemblages somewhere within that Province; 126). Molluscan species that fail to be recorded in death assemblages tend to be small and thin-shelled, numerically rare, composed of organic-rich microstructures (e.g. nacre, prismatic calcite), or all of the above (70, 126; SM Kidwell, in preparation). This selectivity is consistent with the results of short-term experimental studies on shell durability. Palmqvist (90) also found slight systematic differences in preservation potential of trophic groups among marine molluscs: 93% of primary consumers (filter-feeders, browsers, detritivores) leave dead remains versus 87% of predators, scavengers, and parasites. If live communities are compared with local Pleistocene assemblages, the numbers are 81% and 72% respectively (90).

Molluscan death assemblages typically have diversities (numbers of species) twice that of shelled molluscs sampled alive in the same habitat at any single time. Consequently, low estimates of fidelity result when calculated as the percentage of dead species also found alive (54% in intertidal habitats, 33% in coastal subtidal habitats, and 45% on the open shelf; Tables II-V in 70). Similar values are found in freshwater habitats (55%; 22, 36, 133). In most instances, however, low estimates of live:dead fidelity arise from an inadequately sampled live community rather than from taphonomic processes such as selective destruction and import of exotic species: When the death assemblage is compared to a species list compiled from replicate censusing of the live community over time, thereby improving the sampling of ephemeral and otherwise sparse species, agreement rises significantly to ~75%, demonstrating the importance of time-averaging in building dead diversity (70; similar pattern in crabs—93). Several lines of evidence (rarefaction of replicate live samples; maximum longevity of individuals in community; long-term ecological and hydrographic data) suggest that several years of replicate sampling are required to build an adequate species list for live communities in intertidal settings, several decades in coastal subtidal settings, and close to a century on the continental shelf, presumably because these are the time scales over which natural cycles in species recruitment and "rare events" attain a steady state condition (temperate-latitude data only; numbers are probably higher in the tropics) (70). By implication, these are minimum estimates for the duration of time-averaging for these death assemblages. The remaining 25% of the dead species list could be interpreted as the sum of additional ecological noise (e.g. as yet unsampled live species; species that have become extinct locally because of environmental change over the period recorded by the death assemblage) and true taphonomic bias (i.e. species that do not live and have never lived in the area, but which have been transported in from exotic habitats or exhumed from significantly older strata).

Intuitively, one expects poor agreement between live communities and death assemblages in terms of species' relative abundances, due to differences in population turnover rates (input) and preservation potential (destruction): The death assemblage should be biased numerically toward species that are short-lived and/or taphonomically robust. However, molluscan death assemblages consistently show strong fidelity to relative abundances in the live community. For example, in 12 out of 14 marine habitats in which data were based on at least five sample stations (but only a single census), rank orders are not significantly different between live and dead when the full species lists are compared (SM Kidwell, in preparation). Even higher live:dead agreements are attained if abundance is measured in terms of shell "biomass" (e.g. two habitats studied by 114). Numerical relative abundance relationships are also well preserved in freshwater molluscan assemblages (22, 36, 133), in many in-stances with precise agreement in rank order (total numbers of species and specimens here, however, are much smaller than in marine assemblages). Taphonomists have noted this agreement in individual datasets from the earliest live:dead studies (28, 131, 132), but paleontologists have been reluctant to accept it as a general pattern because it is counter-intuitive to the alarmingly high rates of postmortem destruction reported from some environments (e.g. the 60-day half-lives for bivalves mentioned above; 115).

There are at least two possible explanations for good agreement in molluscan relative abundance. The first is that species relative abundances in death assemblages are dominated by the most recent cohorts of dead shells added from the living community, rather than all cohorts being integrated over the entire period of time-averaging, a hypothesis consistent with dead-only species tending to be numerically rare. The second possible explanation is that although species vary in their rates of dead-shell production due to differences in population dynamics (opportunistic boom-and-bust behavior versus slow-growing and larger-bodied stress-tolerant and equilibrium species), destruction may vary in such a way as to take a proportionately higher toll on species having the highest dead-shell production rates, so that the net death assemblage more nearly resembles standing relative abundances. Such a systematic pattern of selective destruction would not be unreasonable, given that short-lived species tend to be smaller (75) and that juveniles are underrepresented in most assemblages (35). Distinguishing between these and other possible explana-tions requires more fieldwork, especially on shell carbonate budgets, because it is clear that much (most?) biogenic carbonate is recycled through dissolution (e.g. 130). However, the high correlation between molluscan death assemblage composition and community composition is useful even if the underlying causes are not yet known.

Other Benthos

Lower levels of live:dead fidelity are found for marine organisms with more fragile tests, as one might predict. In back- and peri-reef habitats along the Egyptian shore of the Red Sea, for example, only 76% of regular and irregular echinoid species found alive ($n = 17$) left recognizable fragments in the local death assemblage; 74% of species in the death assemblage ($n = 19$) were also documented alive (89). Species found both dead and alive comprised 89% of total fragment weight, and so, as for the molluscs, species suspected of being exotic are not dominant in the death assemblage. The only live:dead study of scleractinian corals indicates that death assemblages have lower diversities than does the source living community, suggesting that the aragonitic skeletons of at least some growth forms are more prone to destruction than echinoderm and mollusc skeletons (J. Pandolfi, personal communication–90a). The only live:dead study of crabs that we know (93) indicates that half the species found alive in a single census were present dead; the most abundant species in the death assemblage was not found alive but does occupy the area during other seasons of the year. Fidelities are thus strongly group dependent.

Land Animals

Live:dead data on vertebrate assemblage fidelity are limited to mammals on African savannas. In Amboseli basin, Kenya, bones surveyed for six major habitats were compared with six years of data on live species richness, relative abundances, and population turnover rates (12). Of the 47 species of wild mammals with body weights ≥ 1 kg, 72% were found in the bone assemblage, but as in molluscan assemblages, this varied with trophic group and body size. From 95 to 100% of large herbivores and carnivores (>15 kg body weight) were present, compared with only 60% and 21% of small herbivores and carnivores, respectively. Rank ordering of species in the bone assemblage showed strong agreement with that expected from turnover rates and standing abundances. Bias related to body size was still evident, however. Among ungulates, the largest herbivores (>200 kg) had higher than expected numbers in the bone assemblage (e.g. elephants and rhinos 2–10 times higher, although still rare), and small herbivores (<100 kg) were less abundant than expected. Bone assemblages were dominated by seasonal migrants, especially those of the dry season when mortality is highest. A study of four riverside habitats of eastern Zaire found a similar underrepresentation of small-bodied species in the bone assemblage (107, and see 111 for similar size bias among freshwater fishes). Rapid burial of small bones by trampling and by blowing sand are possible explanations for bias against small species, as are preferential fragmentation by trampling, more rapid rates of weathering, and more complete destruction by carnivores and scavengers (12, 107). In both mammalian studies, however, death assemblage diversity is not enriched relative to the living community the way it typically is among molluscs; the implication is that bones have significantly lower preservation potential in nonmarine records than do molluscs in shallow marine deposits.

A similar bias against small species is found in comparing the living mammal community with Pleistocene and older fossil assemblages (39, 142). Although small mammals may be preserved in correct rank order relative to one another, they are typically underrepresented relative to large-bodied species; this discrepancy is exacerbated when fossil assemblages are surface collected rather than bulk sampled (142).

Little taphonomic work has been published for insects. There is, however, a preservational bias against less sclerotized taxa and, based on a comparison between Quaternary packrat midden assemblages and modern pitfall traps in the same rockshelters, herbivores are underrepresented in the fossil faunas and scavengers are overrepresented (46).

Land Plants

The extent to which leaf litter records species diversity varies greatly with climatic regime, which produces differences in forest heterogeneity (24, 26). For example, a single sample of 350–450 naturally shed leaves captures ~80% or more of forest species in the surrounding hectare in temperate settings, but much less in tropical settings. Litter production is also closely correlated with plant relative abundance as measured by trunk basal area, at least for temperate forests (27, 43).

Pollen assemblages can provide high-fidelity records with respect to species presence/absence when averaged over a very large area (see discussion under spatial resolution), but the fidelity of relative abundance data varies greatly depending upon what part of the total pollen spectrum is examined. Assemblages from Quaternary lakes are dominated by wind-pollinated plants because their pollen is produced in larger quantities, is released more easily from reproductive organs, and is more aerodynamic; this dominance pertains even in wet tropical settings where the forest canopy is dominated by insect-pollinated trees (65). For this wind-pollinated part of the community, pollen assemblages are faithful recorders of plant relative abundances in the source area (measured as total number of trees or trunk basal area in the forest, or using the abundance of co-occurring macrofloral remains; 65, 66) within the effective pollen source area, especially when the forest is relatively homogeneous. By contrast, animal-pollinated plants are almost always grossly underrepresented because their larger, heavier pollen falls much nearer the source tree and in smaller amounts per tree. Spatial resolution is thus higher than for wind-pollinated taxa, but relative abundance in an assemblage is generally a poor proxy of the importance of these animal-pollinated plants in the surrounding vegetation (65).

TEMPORAL RESOLUTION

Time-averaging refers to the pooling of successive populations and communities into a single assemblage of remains, presumably because the rate of sediment accumulation and burial is slow relative to population turnover (128) (Figure 2). When sediment accumulation rates are low, noncontemporaneous remains can accumulate on the same surface. In addition, remains of one age can be mixed with those of earlier or later deposits by physical and biotic processes. Physical processes include decimeter-scale erosion and redeposition by storms and tides, and meter-scale erosion by stream cutbanks, all of which rework remains into younger deposits. Burrowing organisms displace skeletal remains over tens of centimeters, with occasional instances of meter-scale bioturbation. Mixing within the sedimentary column may alone produce a time-averaged fossil assemblage, but most time-averaged deposits are the result of a combination of both near-contemporaneous mixing and low sediment accumulation rates, which place sequential deposits in relatively close vertical succession.

Hypothetically, time-averaging can produce a fossil assemblage that differs significantly from an instantaneous census of the live community in several distinctive ways:

Summing of input. Preserved diversity may sum the biota from both normal conditions and short-term, chance recruitment events. Normally rare species that produce an abundance of skeletal remains during brief pulses may thus come, in the death assemblage, to outnumber the normally dominant species. Age or size-class distributions may also be modified in this way.

Cumulative destruction. Cohorts of organic remains deposited early in the period of time-averaging will experience a longer interval of mixing and postmortem destruction than will younger cohorts, and thus the earlier cohorts should suffer greater modification, loss, and repositioning within the sediment. Moreover, within any given cohort, less durable elements will be modified or lost before more durable ones. In the final assemblage, the most durable portions of the youngest cohorts thus might contribute the strongest biological signal.

Environmental change. The effects of time-averaging can be even more insidious in that local environmental conditions may change over the period of time-averaging, so that a single fossil assemblage telescopes the record of different conditions into a single deposit. Successive cohorts thus may be from different communities and subject to different styles of taphonomic modification. The longer the period of time-averaging, the more likely environmental change becomes.

Temporal mixing may also occur when fossils of significantly greater age are eroded from nearby deposits and deposited in an accumulation of recently dead remains. The disparity in ages can be great. For example, Miocene (~15 million year old) shark teeth and mollusc shells are currently eroding from the cliffs of Chesapeake Bay, where they mix with the shells of living molluscs. In the fossil record, such extreme age-mixing is recognized by the geological context (erosion surfaces, adjacent unmixed fossils) and the co-occurrence of species that are normally separated by large stratigraphic intervals. Fortunately, this magnitude of fossil mixing is rare. In a compilation of more than 3000 fossil mammal assemblages, only 10 contained species mixed from deposits more than one million years older than the principal fauna (J. Alroy, personal communication).

Determining the Duration of Time-Averaging in Modern Death Assemblages

The scale of time-averaging is the difference between the time of death of the oldest individual in the assemblage and the time of death of the youngest. This is not necessarily equal to the duration represented by the bed of enclosing sediments. Single volcanic ash falls and submarine turbidity flows, for example, occur over only a few hours or days, but nonetheless might entomb the remains of earlier generations as well as animals or plants killed during the sedimentary event. On the other hand, the duration represented by the fossil assemblage could be much less than that of the enclosing bed. Where rates of hardpart destruction are high, for example, fossils may represent only the youngest cohorts contributed to the sedimentary unit.

The age of the oldest shell (or bone) in a currently forming death assemblage is a simple measure of the time represented by the accumulation; in fossil or subfossil deposits, two or more dates are necessary to establish both the older and younger age-limits for the assemblage. The most obvious way to determine the duration of time-averaging is to date fossils directly, an approach that has been very useful in deposits younger than 40,000 years in which the radiocarbon method can be applied to shell carbonate, vertebrate bone, and plant material (52).

Amino acid racemization geochronology is another technique that can provide estimates of the age of bone and shells; it is based on the progressive postmortem increase in the ratio of D-alloisoleucine and L-isoleucine (A/I) (7, 136). Significant variation in A/I within an assemblage can indicate time-averaging (59). Although less precise than radiocarbon dating and subject to a variety of local effects (7, 88), the technique is less expensive and has a greater time range (up to 1.5 million years in mid-latitudes) (137).

Finally, historical records can also provide some absolute-age estimates of time-averaging. Bones of the American bison—locally extinct for more than 100 years—can be found among modern bones in the channel of the East Fork River, Wyoming (10), and shells of the American oyster *Crassostrea virginica* are still found on British beaches despite its apparent failure to become naturalized following commercial introductions that lasted until 1939 (18). The presence of such remains provides a minimum estimate for the duration of time-averaging in surficial sediments.

Enough evidence has now accumulated to estimate the duration of time-averaging for many of the depositional environments represented in the fossil record (71). Some evidence is from direct-dated fossils, while extrapolations from sedimentation rates, the likely persistence of hardparts, and resampling studies provide other estimates.

At the shortest time scales of days to seasons are the rare instances in which fragile tissues and behavioral ephemera are preserved in spectacular detail: the tracks of a passing herd, the digested remains of a meal preserved in dung and owl pellets, and individuals and in some instances entire communities trapped by tree sap, floods, or thick ash falls. Although some already-dead material can be swept up along with living organisms, these event-deposits nonetheless offer the finest temporal resolution available within the time-averaging spectrum.

Remains that accumulated over only a few seasons or decades are also rare. Notable examples are pollen, insect, and some fish assemblages from delicately laminated deposits in some lakes (14, 47, 135) and some marine predator middens (octopus, crab).

Vertebrate assemblages from terrestrial soils and land surfaces and plant material from unlaminated lake sediments and peats probably have temporal resolutions of decades to thousands of years (59, 61, 96). Vertebrate predator middens probably also mostly lie within this time span, as illustrated by boreal and temperate wolf dens (61) and accumulations in caves (19, 118).

Direct age data indicate that assemblages from many habitats are time-averaged over a hundred to several thousand years. These includes molluscan and benthic foraminiferal assemblages from nearshore marine settings (53, 80), and molluscan assemblages from large lakes (33). The record of benthic and planktonic microfossils in deep-sea sediments is also commonly resolvable to this scale (23, 42). Pleistocene and Holocene plant material in arid environments is sometimes preserved in packrat middens. These are rich sources of information on past changes in vegetation (15), but the typical time-span recorded by a midden sample is approximately 2000 years (KW Flessa, DM Smith, unpublished). Most vertebrate assemblages hosted by river channels also reflect accumulation over this time scale (10). Data from deltas, aggrading clastic shelves, carbonate platforms, and reefs are too sparse for confidence, but the high net sedimentation rates in these environments suggest that many of their fossil assemblages probably also fall within the hundreds-to-thousands-of-years range (this excludes assemblages concentrated during marine reworking of these deposits).

The longest durations of time-averaging commonly encountered are on the order of thousands to tens of thousands of years. These characterize marine and terrestrial settings with very low net rates of sediment accumulation. However, with the exception of modern sediment-starved and sediment-by-passed temperate shelves (53), relatively few directly dated assemblages constrain our estimates. The erosion and redeposition of major river deposits as the channel migrates across a floodplain suggest extensive time-averaging (10; Figure 2), and large caves are famous for their low sediment input and temporal persistence. Cave deposits can commonly be subdivided to produce higher resolution samples, but the presence on some modern cave floors of bones and dung from late Pleistocene vertebrates (e.g. 83) indicates significant time-averaging. Death assemblages from modern reefs have not been surveyed systematically. We expect that reef deposits are extensively time-averaged because their highly porous and invaginated structure permits colonization at many levels within the framework, especially during periods when construction is primarily outward rather than upward.

Recognizing Scales of Time-Averaging in Fossil Assemblages

Funds, facilities, and historical records are not always available for direct determination of time-averaging, and even though they may preserve organic carbon, most deposits are well beyond the range of radiocarbon and amino-acid methods. In such instances, indirect approaches must be used. These methods can also be used to rank the relative temporal resolution of modern death assemblages. Corroboration by direct-dating in Quaternary records increases our confidence in the use of indirect methods for the older fossil record (57, 70, 71).

ARGUMENT BY ANALOGY   Paleontologists commonly use analogous Recent deposits to guide interpretations of fossil assemblages. If the sedimentary deposit that contains the fossils is known in modern settings to be characterized by low time-averaging, then the fossil assemblage is inferred to have comparable temporal resolution. There are limitations, of course, to a strict uniformitarian approach. Some ancient depositional environments—for example, the vast shallow inland seas of the Mesozoic and Paleozoic—have few analogs in the modern world. Also, organisms producing hardparts have evolved substantially through geologic time, so that the durable aragonitic shells of Recent molluscs, for example, may not provide good taphonomic analogs for the thin calcitic brachiopod shells that characterize many Paleozoic assemblages. Kidwell & Brenchley (72) have argued that the greater durability of post-Paleozoic skeletonized fauna has in fact resulted in greater time-averaging in younger shallow marine deposits. Hardpart destroyers and modifiers have also evolved, introducing additional opportunities for rates of destruction to change.

STATE OF HARDPART PRESERVATION   Many indirect approaches to estimating time-averaging have focused on the condition of the fossils themselves, since a specimen's condition should only worsen with increasing exposure to taphonomic processes after death. However, attempts to calibrate such a "taphonomic clock" (69) in absolute years have yielded mixed results. Behrensmeyer (9) documented the progressive disarticulation and weathering of vertebrate skeletons on African savannas and found that the transformation from pristine to highly degraded elements occurred within ~15 years. Owing to differences in ultraviolet radiation and colonization by boring organisms, the disintegration of bones appears to be slower in temperate and subarctic conditions and faster in rainforests (5, 67, 121; RN Graham, personal communication). Among marine molluscs, shell color and luster are progressively altered in older shells, but other modifications such as abrasive rounding, fragmentation, and encrustation show little correlation with the specimen's age-since-death over the 10,000 years investigated so far (54, 74, 94, 138). These varying results may be explained by the frequent but unpredictable episodes of burial and exhumation that most hardparts experience, especially in shallow marine settings. Temporary burial clearly slows or stops the taphonomic clock for many kinds of shell damage, especially those that proceed most rapidly at the sediment-water interface. Outside of anoxic lake sediments, pollen undergoes an analogous series of decompositional stages caused by oxidation and corrosive porewaters (36a, 64), and the tests of Foraminifera and other microfossils reveal they are subject to partial dissolution by porewaters, bioeroders, and ingestion by predators (73, 124). Clearly, the resolving power of the taphonomic clock will vary with different kinds of damage and also among depositional environments.

Although the taphonomic condition of molluscs is a poor timekeeper over 10^2–10^4 years, it does allow discrimination of shells that have been reworked into Recent assemblages from 100,000-year-old deposits (52, 55, 138). Reworked fossils may also be distinguished by the presence of exotic sediment filling shell interiors, and such fossils are commonly composed of highly resistant material (56). The presence of reworked fossils in a deposit is an indication of extensive time-averaging, and when the disparity in age or in original environment is substantial, such potentially misleading fossils may be easily recognized and thereby eliminated from subsequent paleobiological analysis.

STRATIGRAPHIC CONTEXT   Insights into temporal resolution can also be drawn from surrounding rocks; this may be one of the strongest lines of evidence for the older stratigraphic record (69). Reworked fossils, for example, are commonly associated with erosional surfaces, and the low net sedimentation rates that permit prolonged time-averaging are indicated by diagnostic minerals, submarine crusts, and deep and well-developed soils. Fossil assemblages from beds containing evidence of extensive biological reworking are also likely to be time-averaged to some degree.

Relative Degrees of Time-Averaging

A direct quantitative determination of time-averaging is not always possible, and, in such instances, relative classification schemes are useful. The following scheme (70) divides the spectrum of time-averaging into four categories based on phenomena that can be recognized from the ecological and geological context of individual assemblages. These relative degrees can be quantified by analogy with the time scales at which these phenomena operate in the modern world.

ECOLOGICAL SNAPSHOTS   These reflect zero or minimal time-averaging. They are recognized by taphonomic features such as a high proportion of articulated specimens and/or soft-tissue preservation, and by ecological and sedimentological features indicating sudden death (escape structures, anoxic minerals). While selective mortality or preservation may not result in the participation of all species or age classes in the original community, snapshot assemblages approximate the samples an ecologist might take during a collecting trip. Such records are rare but provide an extraordinarily detailed, high-resolution picture of the life of the past. Quantitatively, snapshot assemblages are inferred to provide temporal resolution of minutes to years.

WITHIN-HABITAT TIME-AVERAGED ASSEMBLAGES   These are assemblages time-averaged from a single, temporally persistent community over a period of relative environmental stability. Many generations are mixed within the assemblage, including both transient and ecologically persistent species. Samples of pollen from lake laminae, bone assemblages within overbank deposits, and many shallow-marine shell accumulations fall within this category. Quantitatively, these assemblages probably mostly reflect accumulation over years to thousands of years.

ENVIRONMENTALLY CONDENSED ASSEMBLAGES   These are assemblages time-averaged over periods of significant environmental change (climate, substrate, water depth), and so species with environmental tolerances that did not overlap may be preserved together. Evidence for environmental change typically comes from the associated sediments but may be corroborated by the mixture of species that are ecologically incompatible and yet show no evidence of lateral transport. As an example, shells of intertidal and shallow subtidal molluscs found in surficial seafloor sediments at 50 m depth in the North Sea today are relics of communities indigenous to that site 8000 to 9000 years ago when the sea level was lower (29, 45). Vast areas of modern continental shelves are characterized by shelly sands produced by environmental condensation over the most recent, postglacial rise in sea level. Bone accumulations in caves are also good examples of environmental condensation through changes in terrestrial climate. Quantitatively, these assemblages form over time spans of centuries to tens of thousands of years.

BIOSTRATIGRAPHICALLY CONDENSED ASSEMBLAGES   These assemblages incorporate species with evolutionary ranges that do not overlap, indicating time-averaging over very long periods during which one or more local species become extinct or make their first appearance in the stratigraphic record. The period of accumulation, in many instances, spans major environmental changes as well as evolutionary time. Although noteworthy because of the potential to confound evolutionary patterns, examples are very rare. Quantitatively, this is time-averaging on an evolutionary time scale, reflecting accumulation and mixing at a site over hundreds of thousands to millions of years.

IMPLICATIONS: PUTTING THE DEAD TO WORK

Using Death Assemblages for Rapid Assessment of Modern Community Composition

The rain of dead hardparts and their accumulations in surficial sediments are, for many groups, good proxies for species' presence and relative abundances in the source community; paleontologists have established the spatial resolution of such samples and correction factors for biases.

Relative abundances of pollen in lake sediments, for example, reflect the distance-weighted "pollen source strengths" of wind-pollinated trees within a 10^1–10^2-km radius; these strengths are, in turn, closely correlated with the standard diversity measure of trunk basal area (65). Pollen and insect assemblages from packrat middens have higher spatial resolution but less relative abundance fidelity, and thus they are appropriate for a different set of questions (or provide answers with a different level of confidence). Leaf litter rain is an excellent means of approximating forest diversity and basal trunk area from 0.1–1.0 hectare areas, depending upon forest heterogeneity (linked to latitude); litter on the forest floor can be gathered and used as a substitute for "rain" because destruction on the ground is so rapid (24, 26). Bone assemblages also appear to have high fidelity to source community composition at the habitat scale (12, 122), but data as yet are limited to semiarid tropical African areas. Molluscan death assemblages, based on data largely limited to clastic settings in north temperate latitudes, show similar high fidelity at the habitat scale despite more prolonged time-averaging (70; SM Kidwell, in preparation).

Death Assemblages as Baselines of Modern Community Change

One way to assess the environmental impact of human activities is to compare the compositions of local biota before and after the event/change. In many instances, however, historical records on community composition do not extend far enough back in time, are based on anecdotal information, or must be extrapolated from other sites. Moreover, preimpact communities are often known only from a single census, rather than from time series that would reveal the full range of natural variation in the community. In such instances, the composition of the local death assemblage can serve as an excellent proxy for pooled, ecological data, especially if attention is focused on relatively durable groups (36, 40, 93, 95, 100). In some instances, time-resolution even within time-averaged assemblages is sufficient to reconstruct a fine-scale (decadal) history of biotic response, for example, pollen records of forest disease and land clearance by humans (65, 135) and diatom records of recent lake acidification (30a).

The widespread outbreak in the past 30 years by the coral predator *Acanthaster planci* (crown-of-thorns sea star) on Australia's Great Barrier Reef provides a revealing example of the strengths and weaknesses of death assemblage data in marine settings (140). Although it does identify *Acanthaster* as a long-standing member of the reef community, the sedimentary record has little promise for resolving questions such as whether past outbreaks have been idenitical in intensity, duration, or frequency over time-scales of concern to reef-management, because of mixing by bioturbation and uncertainties associated with radiocarbon dates (which can resolve ages hundreds of years apart, but not decades).

The Long-Term Stability of Communities

The high-resolution fossil records of pollen and insects preserved in undisturbed lake sediments have provided strong evidence for the individualistic behavior of species over the past 18,000 years: Community associations appear to have been transient in the face of climatic change (47, 112). The taphonomic reliability of these patterns lends credence to similar patterns reported for terrestrial vertebrates (60) and marine invertebrates (127), the individual assemblages of which are subject to more prolonged time-averaging. In such settings, mixing of noncontemporaneous faunas has been an alternative interpretation for ecologically anomalous associations of species. By focusing on those depositional settings and taxa that are not prone to time-averaging, the long-term cohesion of biotic communities can be tested with paleontological data.

The Significance of Morphological Variation

Morphology is the basis for recognizing fossil species, and morphological change is the evidence for evolution in the fossil record. Morphological variation may have taphonomic as well as biological causes, however, including the selective destruction of morphs, the production of "new" deformed morphs during post-burial compaction, and the mixing of multiple generations or populations through time-averaging (14, 68). These taphonomic components of variation complicate comparisons between fossil and living populations and comparisons among fossil samples. For example, because variation in a single population of hardparts is likely to be less than among the hardparts derived from many populations, an increase in variation between two successive fossil assemblages could simply reflect an increase in time-averaging. In addition, clinal variation in space or time can be pooled through range shifts during the period of time-averaging, obscuring microevolutionary trends (14). Samples for analysis of microevolutionary change in fossil populations should thus come from taphonomically comparable assemblages.

On the other hand, long-term evolutionary stasis can be assessed by spacing samples so widely that time-averaging is unlikely to produce artificial overlap between successive samples. This strategy has been effective in a growing number of analyses (e.g. late Cenozoic bryozoans and bivalves; 31, 116). However, any significant degree of time-averaging will complicate the interpretation of rapid evolutionary changes in morphology, whether in branching or phyletic mode. These effects can blur such changes into apparent gradualism or, if time is unrecorded by the rocks, sharpen them into apparent instantaneousness.

An additional source of uncertainty arises in some studies from "analytical time-averaging," that is, the pooling of specimens from overly thick stratigraphic intervals or the deliberate pooling of samples from different localities for which precise age-equivalence is not clear (13). This may artificially inflate levels of morphological variation as effectively as true, taphonomic time-averaging (109). For a complete discussion of the effects of "observational" as well as preservational completeness on evolutionary patterns, see McKinney (82).

Geographic and Environmental Distributions

Taphonomic research strongly suggests that the presence of a fossil at a given site is unlikely to mislead in terms of its original broad geographic distribution.

For example, lateral transport is rarely sufficient to shift fossils outside their original biogeographic province. For many taxa, including most benthic invertebrates, the presence of fossils in particular environments is also unlikely to be misleading, as burial is usually within the life habitat. This means that, with a few general caveats (e.g. be wary of data from environmentally condensed assemblages and of reliance on rare species, which are most likely to be exotic), the reconstruction of biogeographic and environmental histories for various taxonomic groups is justified.

Are Extinctions Sudden or Gradual?

There is increasing appreciation for the importance of taphonomic processes in shaping the fossil record of extinction. For example, many individuals with potential for fossilization can nonetheless be destroyed, with the result that the geographic and stratigraphic (evolutionary) ranges of species will tend to be less than those of the original living populations. Moreover, individuals that evade destruction can still be mixed with fossils of slightly different age; downward piping into older sediments will make a species appear to have arisen earlier than it really did, and reworking into younger sediments will delay its apparent extinction (37). The net result of these antagonistic trends is uncertainty in the actual timing of local and global origination, immigration, and extinction. Probabilistic models have been proposed to quantify error bars for species' ranges, using the actual patchiness of a fossil species' occurrence as a guide (78, 119).

These issues rise to the fore in debates over the nature and timing of mass extinctions, those geologically brief times in earth history when large and taxonomically diverse arrays of organisms disappeared from the record (51). For example, a strict literalist's reading of the fossil record suggests that the demise of many species preceded the Cretaceous-Tertiary boundary (32) and that a few dinosaurs even survived past it (98). When the effects of taphonomy and sampling are taken into account, however, the fossil record is more compatible with sudden extinction (108, 110). Post-Cretaceous dinosaur remains, which are restricted to river-channel deposits, are almost certainly reworked from older strata (6).

The timing of extinction is also important in the debate over the role of humans in animal extinctions during the late Pleistocene and Holocene. The hypothesis of Pleistocene overkill is that the demise of many North American mammals coincided with the arrival of humans on the continent (79). If the arrival of humans on Pacific islands several thousand years ago coincided with the extinction of many of the resident birds (117), human activity can again be implicated in many local extinctions. Establishing the relative and absolute temporal resolution of fossil assemblages is thus critical to establishing cause and effect relationships in extinction.

CONCLUSIONS

In a recent review of pollen taphonomy, ST Jackson (65) concluded:

> The relationship between pollen assemblages and their source vegetation is complex but comprehensible. The complexity derives from the numerous physical and biological processes intervening between the vegetation and the pollen assemblages. The comprehensibility derives from the fact that the effects of the processes can be predicted from theory and are supported by empirical correspondences. Distortions and biases, once understood, can be corrected for qualitatively or quantitatively.

Comparable statements can now be made for other major groups in the fossil record, most notably the comparatively well-studied macroflora, mammals, and benthic molluscs. We still have much to learn about environmental (including latitudinal) variation in fidelity and its underlying causes, and we need more work on developing criteria to recognize various degrees of time-averaging in the older fossil record. Moreover, work has barely begun on the question of long-term evolutionary changes in the production of preservable remains and the taphonomic processes that affect them. Studies of modern death assemblages have nonetheless established the basic dimensions of the problem, and they are giving us essential quantitative insights into rates of postmortem modification and possible correction factors for the Cenozoic at least. Also, despite clear limits on what can be inferred from hardpart assemblages due to the pervasiveness of time-averaging and selective destruction, these are still hugely informative records of past life: At biologically meaningful spatial and temporal scales, the fossil record is a robust archive of ecological and evolutionary information. Paleoecologists and evolutionary paleobiologists have become able to put rigorous confidence limits on their data and to gauge a system's suitability for answering a given question. As biologists increasingly appreciate the importance of processes operating on broad temporal and spatial scales, the fossil record will become an essential means of extending the time-frame of observation on population and community behaviors.

ACKNOWLEDGMENTS

We thank AH Cutler, D Jablonski, ST Jackson, and M Kowalewski for helpful reviews. We are grateful for support from NSF EAR 89-13442 and 94-05311 (to KW Flessa). This is Centro de Estudios de Almejas Muertas Contribution 12.

Literature Cited

1. Allison PA. 1986. Soft-bodied animals in the fossil record: the role of decay in fragmentation during transport. *Geology* 14:979–81
2. Allison PA. 1988. Konservat-Lagerstätten: cause and classification. *Paleobiology* 14:331–43
3. Allison PA, Briggs DEG, eds. 1991. *Taphonomy: Releasing the Data Locked in the Fossil Record.* New York: Plenum. 560 pp.
4. Andrews P. 1990. *Owls, Caves and Fossils.* Chicago: Univ. Chicago Press
5. Andrews P, Cook J. 1985. Natural modifications to bones in a temperate setting. *Man* 20:675–91
6. Argast S, Farlow JO, Gabet RM, Brinkman DL. 1987. Transport-induced abrasion of fossil reptilian teeth: implications for the existence of Tertiary dinosaurs in Hell Creek Formation, Montana. *Geology* 15:927–30
7. Bada JL. 1985. Amino acid racemization of fossil bones. *Annu. Rev. Earth Planet. Sci.* 13:241–68
8. Deleted in proof
9. Behrensmeyer AK. 1978. Taphonomic and ecologic information from bone weathering. *Paleobiology* 4:150–62
10. Behrensmeyer AK. 1982. Time resolution in fluvial vertebrate assemblages. *Paleobiology* 8:211–27
11. Behrensmeyer AK, Chapman RE. 1993. Models and simulations of time-averaging in terrestrial vertebrate accumulations. *Short Courses Paleontol.* 6: 125–49
12. Behrensmeyer AK, Dechant Boaz DE. 1980. The Recent bones of Amboseli Park, Kenya, in relation to East African paleoecology. In *Fossils in the Making,* ed. AK Behrensmeyer, AP Hill, pp. 72–92. Chicago: Univ. Chicago Press
13. Behrensmeyer AK, Hook RW. 1992. Paleoenvironmental contexts and taphonomic modes in the fossil record. In *The Evolutionary Paleoecology of Terrestrial Plants and Animals,* ed. AK Behrensmeyer, JD Damuth, WA DiMichele, R Potts, H-D Sues, SL Wing, pp. 15–136. Chicago: Univ. Chicago Press
14. Bell MA, Sadagursky MS, Baumgartner JV. 1987. Utility of lacustrine deposits for the study of variation within fossil samples. *Palaios* 2:455–66
15. Betancourt JL, Van Devener TR, Martin PS. 1990. Packrat middens. In *The Last 40,000 Years of Biotic Change.* Tucson: Univ. Ariz. Press. 467 pp.
16. Blumenschine RJ. 1989. A landscape taphonomic model of the scale of prehistoric scavenging opportunities. *J. Hum. Evol.* 18:345–71
17. Boston WB, Mapes RH. 1991. Ectocochleate cephalopod taphonomy. In *The Processes of Fossilization,* ed. SK Donovan, pp. 220–40. New York: Columbia Univ. Press
18. Bowden J, Heppel D. 1966. Revised list of British Mollusca 1. Introduction; Nuculacea - Ostreacea. *J. Conch.* 26:99–124
18a. Bradshaw RHW. 1981. Modern pollen-representation factors for woods in south-east England. *J. Ecol.* 69:45–70
19. Brain CK. 1981. *The Hunters or the Hunted? An Introduction to African Cave Taphonomy.* Chicago: Univ. Chicago Press. 365 pp.
20. Brett CE, Seilacher A. 1991 Fossil Lagerstätten: a taphonomic consequence of event stratification. In *Cycles and Events in Stratigraphy,* ed. G Einsele, W Ricken, A Seilacher, pp. 283–97. Berlin: Springer
21. Briggs DEG, Gall JC. 1990. The continuum in soft-bodied biotas from transitional environments: a quantitative comparison of Triassic and Carboniferous Konservat-Lagerstätten. *Paleobiology* 16:204–18
22. Briggs DJ, Gilbertson DD, Harris AL. 1990. Molluscan taphonomy in a braided river environment and its implications for studies of Quaternary cold-stage river deposits. *J. Biogeogr.* 17:623–37
23. Broecker WS, Klas M, Clark E, Bonani G, Ivy S, Wolfi W. 1991. The influence of $CaCO_3$ dissolution on core top radiocarbon ages for deep-sea sediments. *Paleoceanography* 6:593–608
24. Burnham RJ. 1989. Relationships between standing vegetation and leaf litter in a paratropical forest: implications for paleobotany. *Rev. Palaeobot. Palynol.* 58:5–32
25. Burnham RJ. 1990. Paleobotanical implications of drifted seeds and fruits from modern mangrove litter, Twin Cays, Belize. *Palaios* 5:364–70
26. Burnham RJ. 1993. Reconstructing richness in the plant fossil record. *Palaios* 8:376–84
27. Burnham RJ, Wing SL, Parker GG. 1992. The reflection of deciduous forest communities in leaf litter: implications for autochthonous litter assemblages from the fossil record. *Paleobiology* 18: 30–49

28. Cadée GC. 1968. Molluscan biocoenoses and thanatocoenoses in the Ria de Arosa, Galicia, Spain. *Rijksmus. Nat. Hist. Leiden Zool. Verhandl.* 95:1–121

29. Cadée GC. 1984. Macrobenthos and macrobenthic remains on the Oyster Ground, North Sea. *Neth. J. Sea Res.* 18:160–78

30. Cadée GC. 1989. Size-selective transport of shells by birds and its palaeoecological implications. *Palaeontology* 32:429–37

30a. Charles DF. 1987. Paleolimnological evidence for recent acidification of Big Moose Lake, Adirondack Mountains, New York. *Biogeochemistry* 3:267–96

31. Cheetham AH. 1986. Tempo of evolution in a Neogene bryozoan: rates of morphologic change within and across species boundaries. *Paleobiology* 12:190–202

32. Clemens WA. 1992. Dinosaur diversity and extinction. *Science* 256:159–60

33. Cohen AS. 1989. The taphonomy of gastropod shell accumulations in large lakes: an example from Lake Tanganyika, Africa. *Paleobiology* 15:26–45

34. Coope GR. 1970. Interpretations of Quaternary insect fossils. *Annu. Rev. Entomol.* 15:97–120

35. Cummins RH, Powell EN, Stanton RJ, Staff G. 1986. The size frequency distribution in palaeoecology: effects of taphonomic processes during formation of molluscan death assemblages in Texas bays. *Palaeontology* 29:495–518

36. Cummins RH. 1994. Taphonomic processes in modern freshwater molluscan death assemblages: implications of the freshwater fossil record. *Palaeogeogr. Palaeoclimatol. Palaeoecol.* 108:55–73

36a. Cushing EJ. 1967. Evidence for differential pollen preservation in Late Quaternary sediments in Minnesota. *Rev. Palaeobot. Palynol.* 4: 87–101

37. Cutler AH. 1993. Mathematical models of temporal mixing in the fossil record. *Short Courses Paleontol.* 6:169–87

38. Cutler AH, Flessa KW. 1990. Fossils out of sequence: computer simulations and strategies for dealing with stratigraphic disorder. *Palaios* 5:227–35

39. Damuth J. 1982. Analysis of the preservation of community structure in assemblages of fossil mammals. *Paleobiology* 8:434–46

40. Davies DJ. 1993. Taphonomic analysis as a tool for long-term community baseline delineation: taphonalysis in an environmental impact statement (EIS) for proposed human seafloor disturbances, Alabama continental shelf. *Geol. Soc. Am., Abstr. with Programs* 25:A459

41. Donovan SK. 1991. *The Processes of Fossilization.* New York: Columbia Univ. Press. 303 pp.

42. DuBois LG, Prell WL. 1988. Effects of carbonate dissolution on the radiocarbon age structure of sediment mixed layers. *Deep-Sea Res.* 35:1875–85

43. Dunwiddie PW. 1987. Macrofossil and pollen representation of coniferous trees in modern sediments from Washington. *Ecology* 68:1–11

44. Efremov JA. 1940. Taphonomy: new branch of paleontology. *Pan Am. Geol.* 74:81–93

45. Eisma D, Mook WG, Laban C. 1981. An early Holocene tidal flat in the Southern Bight. *Spec. Publs. Int. Assoc. Sediment* 5:229–37

46. Elias SA. 1990. Observations on the taphonomy of late Quaternary insect fossil remains in packrat middens of the Chihuahuan Desert. *Palaios* 5:356–63

47. Elias SA. 1994. *Quaternary Insects and Their Environments.* Washington, DC: Smithsonian Inst. Press. 284 pp.

48. Fall PL. 1987. Pollen taphonomy in a canyon stream. *Quat. Res.* 28:393–406

49. Fall PL. 1992. Pollen accumulation in a montane region of Colorado, USA: a comparison of moss polsters, atmospheric traps, and natural basins. *Rev. Palaeobot. Palynol.* 72:169–97

50. Ferguson DK. 1985. The origin of leaf-assemblages—new light on an old problem. *Rev. Palaeobot. Palynol.* 46: 117–88

51. Flessa KW. 1990. The "facts" of mass extinctions. *Geol. Soc. Am. Spec. Pap.* 247:1–7

52. Flessa KW. 1993. Time-averaging and temporal resolution in Recent shelly faunas. *Short Courses Paleontol.* 6:9–33

53. Flessa KW, Kowalewski M. 1994. Shell survival and time-averaging in nearshore and shelf environments: estimates from the radiocarbon literature. *Lethaia* 27:153–65

54. Flessa KW, Cutler AH, Meldahl KH. 1993: Time and taphonomy: quantitative estimates of time-averaging and stratigraphic disorder in a shallow marine habitat. *Paleobiology* 19:266–86

55. Frey RW, Howard JD. 1986. Taphonomic characteristics of offshore mollusk shells, Sapelo Island, Georgia. *Tulane Stud. Geol. Paleontol.* 19:51–61

56. Fürsich FT. 1978. The influence of faunal condensation and mixing on the preservation of fossil benthic communities. *Lethaia* 11:243–50

57. Fürsich FT, Aberhan M. 1990. Significance of time-averaging for paleocommunity analysis. *Lethaia* 23:143–52

58. Gastaldo RA, Douglass DP, McCarroll SM. 1987. Origin, characteristics, and provenance of plant macrodetritus in a Holocene crevasse splay, Mobile Delta, Alabama. *Palaios* 2:229–40
59. Goodfriend GA. 1989. Complementary use of amino-acid epimerization and radiocarbon analysis for dating of mixed-age fossil assemblages. *Radiocarbon* 31:1041–47
60. Graham RW. 1986. Plant-animal interactions and Pleistocene extinctions. In *Dynamics of Extinction,* ed. DK Elliott, pp. 131–154. New York: John Wiley
61. Graham RW. 1993. Processes of time-averaging in the terrestrial vertebrate record. *Short Courses Paleontol.* 6:102–24
62. Greenwood DR. 1991. The taphonomy of plant macrofossils. In *The Processes of Fossilization,* ed. SK Donovan, pp. 141–69. New York: Columbia Univ.
63. Greenstein BJ. 1993. Is the fossil record of regular echinoids really so poor? A comparison of living and subfossil assemblages. *Palaios* 8:587–601
64. Havinga AJ. 1967. Palynology and pollen preservation. *Rev. Palaeobot. Palynol.* 2:81–98
64a. Jackson ST. 1989. Postglacial vegetational change along an elevational gradient in the Adirondack Mountains (New York): a study of plant macrofossils. *N.Y. State Mus. Bull.* 465. 29pp.
65. Jackson ST. 1994. Pollen and spores in Quaternary lake sediments as sensors of vegetation composition: theoretical models and empirical evidence. In *Sedimentation of Organic Particles,* ed. A Traverse, pp. 253–86. Cambridge: Cambridge Univ. Pres
66. Jackson ST, Whitehead DR. 1991. Holocene vegetation patterns in the Adirondack Mountains. *Ecology* 72:641–53
67. Kerbis Peterhans JC, Wrangham RW, Carter ML, Hauser MD. 1993. A contribution to tropical rain forest taphonomy: retrieval and documentation of chimpanzee remains from Kibale Forest, Uganda. *J. Hum. Evol.* 25:485–514
68. Kidwell SK. 1986. Models for fossil concentrations: paleobiologic implications. *Paleobiology* 12:6–24
69. Kidwell SK. 1993. Patterns of time-averaging in the shallow marine fossil record. *Short Courses Paleontol.* 6:275–300
70. Kidwell SM, Bosence DWJ. 1991. Taphonomy and time-averaging of marine shelly faunas. In *Taphonomy: Releasing the Data Locked in the Fossil Record,* ed. PA Allison, DEG Briggs, pp. 115–209. New York: Plenum
71. Kidwell SK, Behrensmeyer AK, eds. 1993. Taphonomic approaches to time resolution in fossil assemblages. *Short Courses Paleontol.* 6. Knoxville: Paleontol. Soc. 302 pp.
72. Kidwell SK, Brenchley PJ. 1994. Patterns in bioclastic accumulation through the Phanerozoic: changes in input or in destruction? *Geology* 22:1139–43
73. Kotler E, Martin RE, Liddell WD. 1992. Experimental analysis of abrasion and dissolution resistance of modern reef-dwelling Foraminifera: implications for the preservation of biogenic carbonate. *Palaios* 7:244–76
74. Kowalewski M, Flessa KW, Aggen J. 1994. Taphofacies analysis of Recent shelly cheniers (beach ridges) northeastern Baja California, Mexico. *Facies* 31:209–42
75. Levinton JS. 1970. The paleoecological significance of opportunistic species. *Lethaia* 3:69–78
76. Lindberg DR, Kellogg MG. 1982. Bathymetric anomalies in the Neogene fossil record: the role of diving marine birds. *Paleobiology* 8:402–7
77. Lyman RL. 1994. *Vertebrate Taphonomy.* Cambridge: Cambridge Univ. Press. 524 pp.
78. Marshall CR. 1990. Confidence intervals on stratigraphic ranges. *Paleobiology* 16:1–10
79. Martin PS. 1984. Prehistoric overkill: the global model. In *Quaternary Extinctions: A Prehistoric Revolution,* ed. PS Martin, RG Klein, pp. 354–403. Tucson: Univ. Ariz. Press
80. Martin RE, Harris MS, Liddell WD. 1995. Taphonomy and time-averaging of foraminiferal assemblages in Holocene tidal flat sediments, Bahia la Choya, Sonora, Mexico (northern Gulf of California). *Mar. Micropalentol.* In press
81. Martin RE, Wright RC. 1988. Information loss in the transition from life to death assemblages of Foraminifera in back reef environments, Key Largo, Florida. *J. Paleontol.* 62:399–410
82. McKinney ML. 1991. Completeness of the fossil record: an overview. In *The Processes of Fossilization,* ed. SK Donovan, pp. 66–83. New York: Columbia Univ. Press
83. Mead JI, Agenbroad LD. 1992. Isotope dating of Pleistocene dung deposits from the Colorado Plateau, Arizona and Utah. *Radiocarbon* 34:1–19
84. Meyer DL, Meyer KB. 1986. Biostratinomy of Recent crinoids (Echinodermata) at Lizard Island, Great Barrier Reef, Australia. *Palaios* 1:294–302

85. Miller A, Cummins H. 1990. A numerical model for the formation of fossil assemblages: estimating the amount of post-mortem transport along environmental gradients. *Palaios* 5:303–16

86. Miller A, Cummins H. 1993. Using numerical models to evaluate the consequences to time-averaging in marine fossil assemblages. *Short Courses Paleontol.* 6:150–68

87. Miller A, Llewellyn G, Parsons KM, Cummins H, Boardman MR, Greenstein BJ, Jacobs DK. 1992. The effect of Hurricane Hugo on molluscan skeletal distributions, Salt River Bay, St. Croix, U.S. Virgin Islands. *Geology* 20:23–26

88. Miller GH, Brigham-Grette J. 1989. Amino acid geochronology: resolution and precision in carbonate fossils. *Quat. Int.* 1:111–28

89. Nebelsick JH. 1992. Echinoid distribution by fragment identification in the northern Bay of Safaga, Red Sea, Egypt. *Palaios* 7:316–28

90. Palmqvist P. 1993. Trophic levels and the observational completeness of the fossil record. *Rev. Espanola de Paleontol.* 8:33–36

90.a Pandolphi JM, Minchin PR. 1995. A comparison of taxonomic composition and diversity between reef coral life and death assemblages in Madang Lagoon, Papua, New Guinea. *Palaeogeogr. Palaeoclimatol. Palaeoecol.* 119: In press

91. Paul CRC. 1992. How complete does the fossil record have to be? *Rev. Espanola Paleontol.* 7:127–33

92. Peterson CH. 1977. The paleoecological significance of undetected short-term temporal variability. *J. Paleontol.* 51: 976–81

93. Plotnick RE, McCarroll S, Powell EN. 1990. Crab death assemblages from Laguna Madre and vicinity, Texas. *Palaios* 5:81–87

94. Powell EN, Davies DJ. 1990. When is an "old" shell really old? *J. Geol.* 98: 823–44

95. Powell EN, Staff G, Davies DJ, Callendar WR. 1989. Macrobenthic death assemblages in modern marine environments: formation, interpretation and application. *CRC Crit. Rev. Aquat. Sci.* 1:555–89

96. Retallack G. 1984. Completeness of the rock and fossil record: some estimates using fossil soils. *Paleobiology* 10:59–78

97. Rich FJ. 1989. A review of the taphonomy of plant remains in lacustrine sediments. *Rev. Palaeobot. Palynol.* 58: 33–46

98. Rigby JK, Jr, Newman KR, Smit J, Van der Kaars S, Sloan RE, Rigby JK. 1987. Dinosaurs from the Paleocene part of the Hell Creek Formation, McCone County, Montana. *Palaios* 2:296–302

99. Russell MP. 1991. Modern death assemblages and Pleistocene fossil assemblages in open coast high energy environments, San Nicolas Island, California. *Palaios* 6:179–91

100. Samtleben C. 1981. Die Muschelfauna der Schlei (Westliche Ostsee)—aktuopaläontolgische Untersuchungen an einem sterbenden Gewässer. *Meyniana* 33: 6–183

101. Saunders WB, Spinosa C. 1979. Nautilus movement and distribution in Palau, Western Caroline Islands. *Science* 204: 1199–1201

102. Schäfer W. 1972. *Ecology and Palaeoecology of Marine Environments.* Chicago: Univ. Chicago Press

103. Scheihing MH, Pfefferkorn HW. 1984. The taphonomy of land plants in the Orinoco delta: a model for the incorporation of plant parts in clastic sediments of Late Carboniferous age of Euramerica. *Rev. Palaeobot. Palynol.* 41: 205–40

104. Schneider DC, Haedrich RL. 1991. Post-mortem erosion of fine-scale spatial structure of epibenthic megafauna on the outer Grand Bank of Newfoundland. *Continental Shelf Res.* 11:1223–36

105. Scott RW. 1978. Approaches to trophic analysis of paleocommunities. *Lethaia* 11:1–14

106. Seilacher A. 1982. Posidonia shales (Toarcian, S. Germany)—stagnant basin model revalidated. In *Paleontology, Essential of Historical Geology,* ed. EM Gallitelli, pp. 25–55. Modena: STEM Mucchi

107. Sept JM. 1994. Bone distribution in a semi-arid riverine habitat in eastern Zaire: implications for the interpretation of faunal assemblages at early archaeological sties. *J. Archaeol. Sci.* 21:217–35

108. Sheehan PM, Fastovsky DE, Hoffmann RG, Berhaus CB, Gabriel DL. 1991. Sudden extinction of the dinosaurs: latest Cretaceous, upper Great Plains. *USA Sci.* 254:835–39

109. Sheldon PR. 1993. Making sense of microevolutionary patterns. In *Evolutionary Patterns and Processes,* ed. DR Lees, D Edwards, pp. 19–31. London: Academic

110. Signor PW III, Lipps JH. 1982. Sampling bias, gradual extinction patterns, and catastrophes in the fossil record. *Geol. Soc. Am. Spec. Pap.* 190:291–96

111. Smith GR, Stearley RF, Badgley CE.

1988. Taphonomic bias in fish diversity from Cenozoic floodplain environments. *Palaeogeogr. Paleoclimatol. Palaeoecol.* 63:263–73

112. Solomon AM, Webb T III. 1985. Computer-aided reconstruction of Late Quaternary landscape dynamics. *Annu. Rev. Ecol. Syst.* 16:63–84

113. Spicer RA. 1991. Plant taphonomic processes. In *Taphonomy: Releasing the Data Locked in the Fossil Record,* ed. PA Allison, DEG Briggs, pp. 71–113. New York: Plenum

114. Staff GM, Powell EN, Stanton RJ, Cummins H. 1985. Biomass: is it a useful tool in paleocommunity reconstruction? *Lethaia* 18:209–32

115. Staff GM, Stanton RJ Jr, Powell EN, Cummins H. 1986. Time averaging, taphonomy and their impact on paleocommunity reconstruction: death assemblages in Texas bays. *Geol. Soc. Am. Bull.* 97:428–43

116. Stanley SM, Yang X. 1987. Approximate evolutionary stasis for bivalve morphology over millions of years: a multivariate, multilineage study. *Paleobiology* 13:113–39

117. Steadman DW. 1995. Prehistoric extinctions of Pacific island birds: biodiversity meets zooarchaeology. *Science* 267:1123–31

118. Steiner MC. 1994. *Honor Among Thieves: A Zooarchaeological Study of Neandertal Ecology.* Princeton: Princeton Univ. Press 447 pp.

119. Strauss D, Sadler PM. 1989. Classical confidence intervals and Bayesian probability estimates for ends of local taxon ranges. *Math. Geol.* 21:411–27

120. Sugita S. 1993. A model of pollen source area for an entire lake surface. *Quat. Res.* 39:239–44

121. Tappan M. 1994a. Bone weathering in the tropical rain forest. *J. Archaeol. Sci.* 21:667–73

122. Tappan MJ. 1994b. Savanna ecology and natural bone deposition: implications for early hominid site formation, hunting and scavenging. *Curr. Anthropol.* 36:223–60

123. Thomasson JR. 1991. Sediment-borne "seeds" from Sand Creek, northwestern Kansas: taphonomic significance and paleoecological and paleoenvironmental implications. *Palaeogeogr. Palaeoclimatol. Palaeoecol.* 85:213–25

124. Thunell RC, Honjo S. 1981. Calcite dissolution and the modification of planktonic foraminiferal assemblages. *Mar. Micropaleo.* 6:169–82

125. Traverse A. 1990. Studies of pollen and spores in rivers and other bodies of water, in terms of source-vegetation and sedimentation, with special reference to Trinity River and Bay, Texas. *Rev. Palaeobot. Palynol.* 64:297–303

126. Valentine JW. 1989. How good was the fossil record? Clues from the Californian Pleistocene. *Paleobiology* 15:83–94

127. Valentine JW, Jablonski D. 1993. Fossil communities: compositional variation at many time scales. In *Species Diversity in Ecological Communities: Historical and Geographical Perspectives,* ed. RE Ricklefs, D Schluter, pp. 341–49. Chicago: Univ. Chicago Press

128. Walker KR, Bambach RK. 1971 The significance of fossil assemblages from fine-grained sediments: time-averaged communities. *Geol. Soc. Am., Abstr. Progr.* 3:783–84

129. Walker SE. 1989. Hermit crabs as taphonomic agents. *Palaios* 4:439–52

130. Walter LM, Burton E. 1990. Dissolution of Recent platform carbonate sediments in marine pore fluids. *Am. J. Sci.* 290:601–43

131. Warme JE. 1971. Paleoecological aspects of a modern coastal lagoon. *Univ. Calif. Publ. Geol. Sci.* 87:1–110

132. Warme JE, Ekdale AA, Ekdale SF, Peterson CH. 1976. Raw material of the fossil record. In *Structure and Classification of Paleocommunities,* ed. RW Scott, RR West, pp. 143–69. Stroudsberg, PA: Dowden, Hutchinson & Ross

133. Warren RE. 1991. Ozarkian fresh-water mussels (Unionoidea) in the upper Eleven Point River, Missouri. *Am. Malacol. Bull.* 8:131–37

134. Webb TW III. 1974. Corresponding distributions of modern pollen and vegetation in lower Michigan. *Ecology* 55:17–28

135. Webb TW III. 1993. Constructing the past from Late-Quaternary pollen data: temporal resolution and a zoom lens space-time perspective. *Short Courses Paleontol.* 6:79–101

136. Wehmiller JF. 1993. Applications of organic geochemistry for Quaternary research: aminostratigraphy and aminochronology. *In Organic Geochemistry,* ed. M Engel, S Macko, pp. 755–83. New York: Plenum

137. Wehmiller JF, Belknap DF, Boutin BS, Mirecki JE, Rahaim SD, York LL. 1988. A review of the aminostratigraphy of Quaternary mollusks from United States Atlantic Coastal Plain sites. *Geol. Soc. Am. Spec. Pap.* 227:69–110

138. Wehmiller JF, York LL, Bart ML. 1995. Amino acid racemization geochronol-

ogy of reworked Quaternary mollusks on US Atlantic coast beaches: Implications for chronostratigraphy, taphonomy, and coastal sediment transport. *Mar. Geol.*. 125: In press

139. Whittington HB, Conway Morris S, ed. 1985. Extraordinary fossil biotas: their ecological and evolutionary significance. *Philos. Trans. R. Soc. London Ser. B* 311:1–192

140. Wilkinson CR, Macintyre IG, eds. 1992. Special issue: the Acanthaster debate. *Coral Reefs* 11:51–122

141. Wing SL, Hickey LJ, Swisher CC. 1993. Implications of an exceptional fossil flora for Late Cretaceous vegetation. *Nature* 363:342–44

142. Wolff RG. 1975. Sampling and sample size in ecological analyses of fossil mammals. *Paleobiology* 1:195–204

Annu. Rev. Ecol. Syst. 1995. 26:301–21

HIERARCHICAL APPROACHES TO MACROEVOLUTION:
Recent Work on Species Selection and the "Effect Hypothesis"

Todd A. Grantham

Department of Philosophy, College of Charleston, Charleston, South Carolina 29424

KEY WORDS: macroevolution, hierarchy, species selection, effect hypothesis

ABSTRACT

After briefly introducing the hierarchical perspective, I discuss several theoretical issues concerning the nature of species selection and Vrba's effect hypothesis, and I review recent empirical evidence supporting hierarchical approaches to macroevolution. I argue that Vrba's influential definition of species selection is flawed in two ways. First, species selection does not require emergent traits because higher-level selection acting on aggregate traits can oppose lower-level selection. Second, clades do not play the same role in species selection that populations play in organismic selection. If explanatory and ontological reductionism are distinguished, then even though effect macroevolution does not involve a distinct macroevolutionary process, effect hypothesis explanations can be irreducible. A few well-documented cases of species selection and effect macroevolution suggest the need for a hierarchical expansion of neodarwinism.

INTRODUCTION

This paper reviews recent conceptual and empirical work on hierarchical theories of macroevolution, focusing primarily on species selection and effect macroevolution. Because the current debate is part of a much older controversy, I begin with some historical background and a brief introduction to hierarchical theories of macroevolution. In the three subsequent sections I

301

clarify the relationship between species selection and Vrba's "effect hypothesis," examine the widely accepted view that species selection occurs within clades, and review recent empirical work that demonstrates the value of hierarchical approaches to macroevolutionary phenomena.

Prior to the 1930s, many biologists thought that ordinary genetic and population-level processes would not be sufficient to explain macroevolution. With the rise of the "synthetic" theory of evolution, these doubts all but vanished. While the synthetic theory has undergone a great deal of refinement and articulation, many neodarwinians remain committed to the view that "the history of life at all levels—... embracing all macroevolutionary events—is fully accounted for by the processes that operate within populations and species" (30, p. 39; see also 3, 9, 40, 44). During the 1970s, paleontologists began to criticize the neodarwinian theory on the grounds that the processes identified by the neodarwinian theory are not sufficient to explain *all* macroevolutionary events. As Eldredge puts it, "My position ... is simply that the neo-Darwinian paradigm is indeed necessary—but not sufficient—to handle the totality of known phenomena" (15, p. 119). This claim—that the neodarwinian paradigm is insufficient to explain all macroevolutionary phenomena—is the central point at issue in the debate over hierarchical theories of macroevolution. I assess this claim by reviewing the evidence for and against (*a*) distinct higher-level processes (e.g. species selection) and (*b*) new modes of explanation that do not involve a distinct process but that nonetheless involve a hierarchical expansion of the synthetic theory (e.g. the effect hypothesis).

Although the term "macroevolution" has been used sporadically over the last 50 years, no single definition is widely accepted. For the purposes of this paper, I consider macroevolution to be the domain of evolutionary phenomena that require time spans long enough to be studied using paleontological techniques. This domain includes (but is not limited to) the emergence of higher taxa, the emergence of major "evolutionary novelties," adaptive radiations, long-term phyletic evolution, and mass extinctions. Notice that this definition is agnostic with respect to the causes of these phenomena. Because the correct explanation of these phenomena is precisely the point at issue, my definition should not presume that macroevolutionary events involve distinct processes.

Basic Hierarchical Concepts

Life on earth is hierarchically organized. The biotic world consists of many "levels," with the entities at each higher level composed of lower-level entities. Groups of cells form the tissues and organs out of which organisms are constructed, and organisms form various kinds of groups such as kin groups, populations, and species. While everyone recognizes the fact of hierarchical organization, advocates of hierarchical approaches make three more controversial claims: (*a*) Properties and processes at higher levels can affect the course of evolution, and (*b*) these higher-level properties and processes cannot be reduced to lower-level properties and processes. If we accept (*a*) and (*b*), then (according to advocates of hierarchy) we must accept a third claim: (*c*) A fully adequate theory of evolutionary dynamics must recognize these irreducible higher-level properties and processes (see e.g. 1, 15, 17, 18, 22, 63, 79, 81). Advocates of hierarchy often make several additional claims:

1. There is not just one hierarchy. Eldredge (15), for example, recognizes two primary hierarchies: the ecological hierarchy (molecules, cells, organisms, populations, communities) and the genealogical hierarchy (codons, genes, organisms, demes, species, monophyletic taxa). An entity is part of the genealogical hierarchy if it accurately reproduces itself and its informational content; units of interaction with the environment are considered part of the ecological hierarchy. Williams (88) draws a different distinction between the codical and material domains of selection. Hull (32) and Brandon (5) distinguish the hierarchy of interactors and the hierarchy of replicators.

2. Properties and processes at one level can affect the dynamics of evolution on adjacent levels. For example, gene mutations and population structure—phenomena at the gene and population levels—can affect selection at the organismic level. In the current jargon, these cross-level interactions are called "upward" causation when properties or processes at a lower level affect events at the focal level. "Downward" causation occurs when higher-level properties affect events at the focal level (82).

3. Once we recognize the possibility of upward and downward causation, we must then carefully distinguish selection and sorting (82). Sorting is the differential extinction or reproduction of entities, whereas selection is one possible cause of sorting. If, for example, groups of altruists tend to survive and send off propagules at greater rates than do groups of selfish individuals, then we have group-level sorting. But this sorting may not be caused by group selection (see 83, 87, 89 for discussions of group selection). Accepting this distinction implies that one should distinguish genuine species selection (cases in which the differential reproductive success of species is due to their differential adaptedness to a common environment—see 5) from effect macroevolution (a special case of upward causation in which selection at the lower-level causes higher-level sorting). (These definitions are discussed in greater detail below.)

4. Advocates of hierarchy generally focus on "taxic" or "cladogenetic" trends rather than anagenetic change within a single lineage (14, 23–25, 55, 57). Traditionally, evolutionists have focused on anagenetic or "transformational" trends—changes in a single, unbranching lineage. But when paleontologists began to frame their studies of macroevolution in terms of the theory of punctuated equilibria (16), it became clear that differential rates of extinction and speciation could drive trends. If, for example, the smaller-bodied species

of a clade tend to speciate faster than larger-bodied species, then the proportion of small-bodied species might increase, producing a clade-level trend toward small body size. A taxic trend toward small body size can, at least in principle, arise despite anagenetic trends toward large size within some of the lineages of that clade.

5. Although the differences between them are not always sharp, it is helpful to distinguish three different concepts of "reductionism."

Ontological reductionism is the view that there are no distinct higher-level processes or properties; all higher-level entities, properties, and processes can be fully expressed in terms of the entities, properties, and processes at lower levels. For example, an ontological reductionist might assert that all the properties and processes discussed in Mendelian genetics (e.g. dominance and epistasis) can be expressed in terms of molecular genetics. Those who deny ontological reductionism ("emergentists") assert that some processes or properties cannot be fully reduced. Emergentists avoid any appeal to vitalism by insisting that emergent properties arise from complex interactions of lower-level traits. While any particular instance of a higher-level property will involve a lower-level instantiation, it can be extremely difficult to express higher-level concepts in lower-level terms. Consider, for example, the concept of the "signal sequence" (a sequence of amino acids that directs protein transport within the cell). Although every particular signal sequence can be described biochemically, the concept "signal sequence" cannot be reduced. Whether a sequence of amino acids is a signal sequence depends on how the molecule functions in the cell, not on any particular biochemical structure. Even though every signal sequence has a specifiable biochemical structure, one cannot define the property signal sequences have in common (i.e. their function) in biochemical terms. Thus, "signal sequence" is an emergent property (41). Similar arguments show that the predicate terms of Mendelian genetics cannot be reduced (31, 43).

Explanatory reductionism is the view that a lower-level theory can explain all the facts explained by a higher-level theory. The only fully articulated accounts of explanatory reduction (e.g. 58, 65) require that the laws and properties of the higher-level be expressed in terms of the lower-level theory. That is to say, a full explanatory reduction seems to require an ontological reduction. *Methodological* reductionism is a research heuristic that advises scientists to try to understand a complex system by decomposing it into its component parts and studying these parts in isolation. Corresponding to each notion of reductionism is a sense in which a theory might be hierarchical. Thus, if lower-level theories cannot adequately explain phenomena explained by a higher-level theory, then the higher-level theory is explanatorily hierarchical.

This refined understanding of reductionism allows us to clarify the issues at stake in the current debate. Several of the most prominent critics of the hierarchical approach assert that without a distinct macroevolutionary *process*, no genuinely hierarchical theory can be produced (9, 30, 40). For example, Bock thinks the hierarchical approach will fail unless "at least one additional mechanism of evolutionary change unique to macroevolutionary events [is] postulated over and above those needed to explain the microevolutionary phenomena" (3, p. 21). Given the analysis of reductionism above, this view seems to be mistaken. Because ontological reductionism requires that we reduce both properties and processes, discovering emergent processes is not necessary for emergentism: The discovery of emergent properties is sufficient. Furthermore, if well-confirmed higher-level explanations refer to irreducible properties, then explanatory reductionism will also fail because a full explanatory reduction requires an ontological reduction. (This point will be developed in the following section.)

In sum, two questions are central in the current debate over hierarchical theories of macroevolution: (*a*) Are there distinct macroevolutionary properties or processes?, and (*b*) Is the neodarwinian theory sufficient to explain all macroevolutionary phenomena? While the first question is ontological, the second question is epistemological. Although several ways of modeling macroevolution deserve serious consideration [e.g. neodarwinian models (e.g. 3, 74), random models (60), and Cracraft's (11) "extrinsic control hypothesis"], this survey of recent work on species selection and the effect hypothesis will be sufficiently broad to answer these central questions.

SPECIES SELECTION AND THE EFFECT HYPOTHESIS

In the years since Stanley (72, 73) introduced the concept, "species selection" has been defined in a variety of ways (e.g. 5, 20, 48, 69, 77). Despite continuing debates over the meaning of "unit of selection" (see 70), there is significant agreement about a minimal conception of species selection. This minimal conception is based on Lewontin's (45) view that selection requires heritable variation in fitness and Hull's (32) claim that species selection requires that species be "interactors" (i.e. entities that interact as "cohesive wholes" with the environment so as to cause differential reproduction). Combining these elements, one arrives at the view that species selection occurs when the differential reproduction or extinction of species is caused by heritable differences in the fitness of species-level traits. Although these minimal conditions are widely accepted as necessary for species selection (e.g. 5, 47, 54, 79), some authors insist that additional requirements must be met (e.g. 30, 79; see 27 for a recent criticism of the "interactor approach"). Even though Williams (88) is correct in asserting that species selection is simply an instance of the broader class of "clade selection" phenomena, for the purposes of this paper, I focus on species selection.

In this section I adopt this minimal conception of species selection and address two further issues: First, I argue that emergent traits are not necessary for species selection. Second, I address the issue of whether the effect hypothesis is hierarchical or reductionistic.

The Lloyd-Vrba Debate

Although they agree on the minimal conception of species selection, Lloyd and Vrba disagree about whether the species-level traits must be emergent. The central point in dispute can be clarified with the aid of Figure 1. Vrba restricts the term species selection to region A, whereas Lloyd applies the term to both A and B. The issue is whether any of the class-B examples should count as species selection.

To ensure that species selection is an irreducible macroevolutionary process that differs from the effect hypothesis, Vrba defines species selection as "that interaction between heritable, emergent character variation and the environment which causes differences in speciation and/or extinction rates among the variant species within a monophyletic group" (77, p. 323). Thus, in addition to Lewontin's (45) necessary conditions, Vrba asserts that species selection

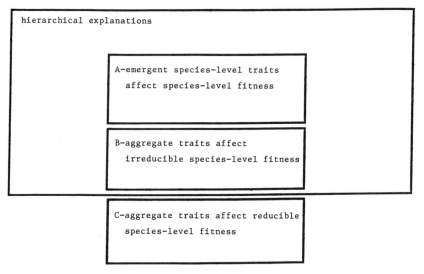

Figure 1 Three ways to explain species-level sorting. Vrba restricts species selection to class-A phenomena whereas Lloyd considers both A- and B-class phenomena to be species selection. This leads to a difference in the way they define effect macroevolution: For Vrba, both B and C count as effect macroevolution while Lloyd restricts this term to class-C phenomena.

requires emergent species-level traits. (Her claim that species selection occurs within monophyletic clades is discussed in *Clades and Populations*.) In order to assess Vrba's claim, we must determine what emergent traits are and why Vrba believes they are necessary for selection.

As discussed above, a property is called emergent if it cannot be reduced. Although this general account of "emergence" is widely accepted, participants in the units-of-selection debate use the term in two slightly different ways. The first approach (taken by Vrba, 79) asserts that a trait is emergent if and only if the trait cannot be attributed to any lower-level entity. According to Vrba, the frequency of brown eyes in a population is emergent because the frequency cannot be attributed to individual organisms. By contrast, the fact that all members of a species have white coat color is an aggregate ("sum of the parts") species-level character because "white coat" can be attributed to organisms.[1] Damuth & Heisler (13) offer a second approach. In their view, a population-level trait should be called emergent only if the population-level trait values cannot be obtained by combining measurements of organismic trait values (e.g. population density). Traits (such as relative frequencies) that can be obtained by combining measurements on individuals are considered aggregate. In this essay, I use Damuth & Heisler's stricter definition because it is the one generally used in the literature.

Vrba argues that "the acid test of a higher level selection process is whether it can in principle oppose selection at the next lower level" (80, p. 388). Species selection is supposed to be a distinct higher-level process, and if it is a genuinely distinct process, it must be capable of opposing organismic selection. Further, Vrba maintains that unless emergent traits are involved, higher-level sorting will just "proceed on directions from below" (80, p. 388). Thus, Vrba claims that higher-level sorting cannot oppose lower-level selection without emergent traits. [Vrba also argues that without emergent species-level traits, there can be no species-level adaptations.]

Based on her analysis of the meaning of "unit of selection," Lloyd rejects Vrba's approach as unnecessarily restrictive (47, 48; see 13 for a similar argument). Taking her inspiration from Wimsatt (91) and Hull (32), Lloyd defines "interactor" as a role in formal models of selection. Formal models represent evolution by natural selection in terms of correlations between trait values and additive variation in components of fitness. To ensure that the correlation between trait and fitness at a higher level is not simply an artifact of a correlation at lower levels, Lloyd stipulates that this correlation cannot be expressed at lower levels. Thus, Lloyd argues that selection is occurring at a

---

[1]As an anonymous reviewer pointed out, Vrba's definition has a peculiar consequence. Small changes in frequency (e.g. from 1.00 to 0.99) can produce an ontological leap from aggregate to emergent characters. Damuth & Heisler's approach avoids this peculiarity.

level when a correlation between a trait and the additive variation in a component of fitness at that level cannot be expressed by a correlation of traits and fitnesses at a lower level.

If Lloyd's approach to defining a unit of selection is correct,[2] then species selection does not require emergent traits. When a component of species-level fitness is correlated with an emergent trait, this correlation cannot be reduced because the trait cannot be represented at the lower level. Emergent traits are not, however, necessary for species selection. If an aggregate trait affects a component of species-level fitness (e.g. rate of speciation) and this component of fitness is irreducible, then the trait-fitness correlation will be irreducible. Thus, Lloyd claims that B-class phenomena are explanatorily irreducible. Furthermore, since selection on aggregate *can* oppose lower-level selection (Vrba's "acid test" for distinct processes), B-class phenomena constitute a distinct (irreducible) process.

Consider, for example, Wilson & Colwell's discussion of sex-ratio evolution (90). They analyze sex ratio evolution as the product of two opposing forces: selection within groups (favoring "Fisherian" ratios) and selection among groups (favoring female-biased "Hamiltonian" ratios). This model concerns species with a distinctive population structure. Groups are founded by one or more fertilized females that are randomly selected from the population. The offspring mature and reproduce (maybe for several generations) before synchronous dispersal back into the population for reproduction. Now, whenever the groups are founded by more than one female, within-group selection will favor Fisherian (0.5) sex ratios. Even though organismic (within group) selection favors Fisherian sex ratios, species with this population structure often have female-biased sex ratios. According to Wilson & Colwell, female-biased ratios prevail because groups with more "Hamiltonian" founders (i.e. females that produce female-biased broods) will contribute more genes to the global population than will groups with fewer Hamiltonian founders (at least under some parameter values). Thus, group-level selection on an aggregate trait (proportion of Hamiltonian founders) can oppose organismic (within group) selection. This is not to say that higher-level selection on aggregate traits will be a significant force in evolution. But even if selection on aggregate traits is generally swamped by selection at lower levels, the foregoing example shows that it is possible for selection on aggregate traits to oppose selection at a lower

---

[2]Several philosophers have raised doubts about the adequacy of Lloyd's approach (21, 64, 70). While I continue to find Lloyd's approach useful, the argument of this paper does not require a commitment to her approach. Rather than arguing for her position as a whole, I have tried to defend a single consequence of her view that can be accepted independently of a commitment to her definition of an interactor.

level. Thus, emergent traits are not necessary for species selection to be a distinct process.

Two final caveats are important. First, Vrba's definition of species selection has a crucial virtue: It accurately identifies the most interesting kind of species selection (class-A examples). Nonetheless, I believe that class-B examples can legitimately be viewed as species selection because selection on aggregate traits can oppose lower-level selection. Second, in arguing that species selection does not require emergent traits, I have not been criticizing Vrba. In Vrba's (79) sense of the word, emergence may be necessary for species selection. My claim is that the widely used (but more restrictive) notion introduced by Damuth & Heisler (14) is not necessary for selection at the species level.

Is The Effect Hypothesis Reductionist?

Kellogg (40) and Vrba (76) suggest that the effect hypothesis is reductionist. In her more recent work, however, Vrba (80) treats it as part of the hierarchical expansion of neodarwinism. To clarify the status of the effect hypothesis, I analyze two examples of effect macroevolution. My analysis leads to the conclusion that some of the cases usually considered instances of effect macroevolution are explanatorily irreducible.

The classic example of effect macroevolution is Vrba's (78) study of south African mammal clades. Vrba argues that species of specialized feeders (e.g. grazers) have become more common in several mammal clades because they have a higher speciation rate than generalist species. According to Vrba, this is an example of the effect hypothesis because no higher-level selection process is involved: The species-level sorting is merely an incidental effect of organismic selection. Although Vrba's explanation does not introduce a higher-level process, I would maintain that this explanation is not reducible. The explanation is irreducible because it appeals to differences in a component of species-level fitness (speciation rates) that cannot be reduced to the organismic level. Recall that explanatory reductions require a reduction of the properties mentioned in the explanation (See *Basic Hierarchical Concepts.*). Because no one has succeeded in expressing speciation rates as a function of organismic properties,[3] this explanation cannot be reduced. Furthermore, effect macroevolution is a significant break with standard neodarwinian explanations because

---

[3]The concept of "speciation rate" cannot be expressed at the organismic level because there is no simple set of organismic traits that determine speciation rate. Rather, a diverse set of organismic and population-level traits (including disperal ability, population structure, and behavioral compatibility between members of distant populations) affect gene flow and therefore affect speciation rates. Because of the large variety of factors affecting speciation rate, and because the way a single factor affects speciation rate is often context-sensitive, the higher-level property of "speciation rate" is, at best, extraordinarily difficult to express in organismic terms. The speciation rate of a taxon is irreducible in just the way signal hypothesis" is.

it treats aggregate species-level properties (e.g. the average degree of feeding specialization within a lineage) as causally relevant factors that affect speciation rate. Because some effect hypothesis explanations are hierarchical and quite different from standard neodarwinian accounts of macroevolution, I view effect macroevolution as part of a hierarchical theory of evolution.

While some of the cases Vrba treats as effect macroevolution are explanatorily hierarchical, others may not be. Consider, for example, Chatterton & Speyer's (10) account of the end-Ordovician extinction among the trilobites. They maintain that this extinction was caused by a global drop in ocean temperature. Glaciation led to a reduction in plankton populations, which led to food shortages and extinction for all species with pelagic adult lifestyles and nearly all species with planktotrophic larvae. Although Chatterton & Speyer do not say so explicitly, the differential extinction of trilobites seems to be a case of effect macroevolution: Differences in organismic traits (such as mode of larval development) led to differences in extinction rates without any distinct higher-level process of species selection. But in this case, the differences in extinction rates may be reducible. A species goes extinct if and only if every individual dies. Whereas differences in speciation rates cannot be expressed in organismic terms, differences in extinction rates will often be reducible (unless population-level traits such as variation matter; see 20). On these grounds, the end-Ordovician extinction among trilobite species seems to belong in category C of Figure 1.

CLADES AND POPULATIONS

In the paradigm cases of natural selection, changes in gene frequency are explained by differences in the adaptedness of the genes (or traits) within the population. Similarly, species selection explains changes in the frequency of species-level traits within a "population" of species. But what is the appropriate species-level analog of the population? Within what unit(s) should one study changes in the frequency of species-level traits?

One widely accepted view asserts that the population of species is the monophyletic clade. Vrba, for example, incorporates this view into her definition of species selection (see above). In this section I argue that if "population" is to have a univocal meaning across levels of the biological hierarchy, this view must be modified. As they are usually understood, populations have a hybrid status—they are neither purely ecological nor purely genealogical. For example, Futuyma defines a population as "a group of conspecific organisms that occupy a more or less well-defined geographic region and exhibit reproductive continuity from generation to generation. It is generally assumed that ecological and reproductive interactions are more frequent among these individuals than between them and members of other populations of the same species" (19, pp. 554–55). Thus, the term population has a genealogical component (populations consist of conspecific organisms) *and* an ecological component (these organisms are more likely to interact with one another than with members of other populations). Because clade is a purely genealogical notion, it seems ill-suited to function as the species-level population.

The standard view (expressed in 15, 79–81) is quite plausible. Populations consist of closely related organisms (organisms that share a recent common ancestor) and are maintained through the reproductive activity of organisms. Similarly, the species-level analog of a population should be a group of species that (*a*) shares a common ancestor and (*b*) is maintained by the survival and reproduction (i.e. speciation) of the component species. Thus, it would seem that the species-level population is the clade.

Although the standard account is appealing, Damuth (12) argues forcefully that it is not fully satisfactory because it ignores the "ecological" dimension of populations. Consider the following differences between clades and populations. Whereas organisms can migrate from one population to another, a species cannot migrate into a different clade. Populations are geographically restricted whereas clades are not. Furthermore—and this is the point I want to emphasize—the organisms within a population often compete for scarce resources. When one organism dies, other organisms in the population often exploit resources previously monopolized by the deceased individual. But members of a clade do not necessarily share this ecological relation. Imagine a clade consisting of two species that now inhabit different continents. Assuming the organisms can no longer travel between the continents, these species cannot compete with one another, nor could one of these species replace the other should it go extinct. Such a clade is a poor analog to the conventional notion of a population. To remedy this problem, I suggest that paleontologists focus on geographically constrained portions of monophyletic clades (e.g. all the species in a given taxon that inhabit a continent). Notice that I am not requiring that the entire clade be restricted to this region. Instead, all and only those members of the clade that inhabit the defined region should be considered members of the population.

Perhaps an example will help to clarify the situation. Consider a family of plant species that exist throughout North America. Because these plants cannot survive at high altitudes, no gene flow occurs across the Rockies. Thus, there is one pool of species[4] east of the Rockies and a second pool west of the Rockies. These two constrained "populations" of species are more cohesive

---

[4]For the purposes of this example, I adopt a restrictive use of the biological species concept. Even if the members of two populations on opposite sides of the Rockies are (in principle) capable of interbreeding, I treat them as distinct species. Nothing in my example hinges on this assumption, but it simplifies the discussion.

than the family as a whole because the species in these populations are demographically exchangeable. That is, if one of the eastern species goes extinct, its niche will be filled (if at all) by members of the eastern population. Furthermore, for a novel trait to come to fixation within this eastern sub-clade, it must originate within one of the eastern species. In my view, the eastern and western populations evolve independently. Even if they show the same species-level dynamics—e.g. if the same species-level traits increase species-level fitness—we ought to treat evolution within these two regions as distinct processes precisely because the species within these two regions are not demographically exchangeable. The same point can be made in a slightly different way: The species of these two populations do not share a common selective environment (see 5).

A word about competition. I agree with Williams's (88) claim that competition is not necessary for selection. I do, however, think that the potential for competition (i.e. demographic exchangeability) is important. Imagine that a competitively superior species has come to dominate the eastern sub-clade and would dominate the western sub-clade if it could cross the Rockies. In this case, the two sub-clades evolve independently. I require demographic exchangeability in order to avoid lumping together similar but independent selection processes (see 26 for further elaboration).

Although this way of articulating the concept of the species-level population differs from the standard view, it accords quite well with accepted methods for studying macroevolution. Vrba (78) focuses on the mammal clades in a limited geographic area. Similarly, Jablonski's (34) argument for species selection concerns just those mollusk species living along the Gulf and Atlantic coasts of North America. Finally, Brown & Maurer (6, 7) explicitly advocate studying geographically *and* taxonomically constrained biotas.

EVIDENCE FOR HIERARCHICAL THEORIES OF MACROEVOLUTION

The Macroevolutionary Consequences of Larval Ecology in Marine Invertebrates

Many studies have demonstrated that the mode of larval development can influence macroevolutionary patterns in gastropods and trilobites (e.g. 10, 28, 29, 34, 46, 66, 75). Jablonski (34, 35) argues that some of these patterns are the result of species selection. To make his case, he argues that species satisfy Lewontin's (45) conditions for units of selection. That is, he argues that species vary in *emergent* and heritable species-level traits that affect species-level fitnesses.

Jablonski begins by documenting a significant macroevolutionary pattern among the mollusks of the Gulf and Atlantic coasts of North America. He focuses on the distinction between planktotrophs (species with larvae that depend on plankton for nutrition during the free-swimming stage) and non-planktotrophs. In the late Cretaceous, planktotrophs have significantly larger geographic ranges and significantly longer temporal durations than do species with nonplanktotrophic larvae.

These correlations between planktotrophy, geographic range, and temporal duration should make ecological sense: Species with planktotrophic larvae will generally have widespread dispersal, leading to large geographic ranges. (This association has been documented in living species—see, e.g., 67.) Because species with large ranges are relatively immune to the local disruptions that may drive endemic species into extinction, species with large ranges have longer durations. Furthermore, planktotrophy tends to reduce speciation rates by promoting gene flow among populations. By contrast, nonplanktotrophs have a high per-taxon speciation rate precisely because they have low levels of gene flow and so have greater genetic differentiation between conspecific populations. Significantly, geographic range is not merely correlated with, but it actually causes, long duration. Because maximum geographic range is attained early in the duration of a species, and because range is so crucial to avoiding extinction, Jablonski (35) concludes that large ranges contribute to long duration. Finally, he argues that geographic ranges are heritable. In the case of organisms, we can determine heritability by comparing the traits of siblings or by studying parent-offspring regression. By performing analogous tests on species (e.g. regression of the trait values of mother and daughter species), he shows that geographic range is heritable (35). Thus, Jablonski has identified an emergent species-level trait (geographic range) that is variable and heritable, and that affects the fitness of species (i.e. the probability that a species will survive or reproduce).

Jablonski's case for species selection has been subject to two kinds of criticism. One objection comes from Russell & Lindberg (61, 62) who maintain that the correlation between range and duration might be an artifact—a result of sampling bias. As they say, "taxa having long geographic ranges are more likely to be preserved in the fossil record because they occur at a greater number of fossilization sites" (62). Russell & Lindberg have a legitimate concern: If sampling bias could produce the observed correlation, then we do not yet have a compelling reason to reject the null hypothesis of no correlation between duration and range. If the correlation is simply a result of sampling bias, then the pattern in well-preserved species should more closely approximate the null hypothesis of zero correlation than the pattern found among species with poorer preservation potential. But as Jablonski (37) points out, this prediction is not confirmed. Thus, the pattern does not seem to be a result merely of sampling bias. Marshall (51) confirms Jablonski's conclusion using mathematical modeling and computer simulation techniques. Although issues of sampling bias are always important in paleontological disputes, sampling bias does not seem sufficient to explain the patterns Jablonski has documented.

Second, one might object that the effect hypothesis provides a better explanation of these patterns. According to this objection, organismic traits (e.g. life history traits) are the crucial properties driving the trend, not species-level traits. Recall, for example, Chatterton & Speyer's (10) analysis of the mass extinction among trilobites. In this case, species-level properties do not seem to be significant (species with planktotrophic development went extinct because *organisms* of the species could not survive; see *Is the Effect Hypothesis Reductionist?*). While Chatterton & Speyer's model provides a viable interpretation of some trends, Jablonski's data show that the differential rates of speciation and extinction depend on *emergent* species level traits. Nonplanktotrophs speciate at a greater rate because of their population structure; planktotrophs have greater durations because they have larger geographic ranges. Even if mode of larval development (and consequently population structure and geographic range) are influenced by organismic selection, the explanations of these trends are not reducible because they appeal to emergent species-level properties.

The Evolution of Body Size

We now have a fairly detailed understanding of the microevolutionary forces that constrain body size evolution (see, e.g. 8, 49, 59, 68). In addition, paleontologists have identified robust macroevolutionary trends and patterns involving body size (e.g. Cope's Rule). Because both the microevolutionary processes and the macroevolutionary patterns are well understood, body size evolution provides an important opportunity to test the hierarchical perspective.

Some biologists (e.g. 4) have suggested that microevolutionary forces are *sufficient* to explain all aspects of body size evolution. Recently, Gould (23, 24) has cast doubt on this claim by reviving Stanley's (71) classic argument. According to Gould, trends toward increasing body size are fundamentally *cladogenetic* phenomena in which species diversify away from a boundary without any force consistently favoring size increase across all lineages. For example, Gould argues that the large size of modern horses (shown in Figure 2) arises through this kind of "passive diffusion process" (55–57 discuss the distinction between passive and driven trends). Presumably, the horse body plan is effective only above a certain minimum body size. If the earliest horses start out near this minimum size and the clade grows, variance in size can increase even in the absence of a directional force (such as selection) that favors size increase in all lineages. Thus, Gould rejects the standard view (i.e. that the trend arises because organismic selection consistently favors large size in most lineages) and suggests that the primary phenomenon is passive diffusion, increasing variance in size within the clade.

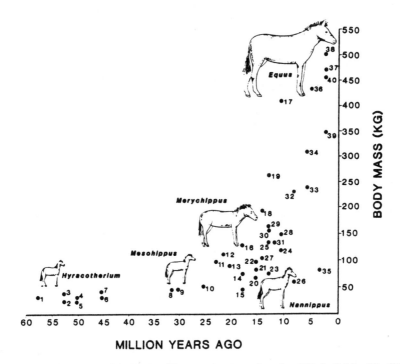

Figure 2   The Evolution of the Modern Horse: an instance of passive diffusion? After 30 million years of rough stasis in body size, horses in some lineages begin to increase in size, while those in other lineages remain quite small. Gould (24) argues that this trend is not driven by selection but arises through a passive diffusion process. (From 50; reprinted by permission of The Paleontological Society.)

Although Gould's conceptual point (that cladogenetic processes can cause trends) is valid, the particular case of horse evolution is problematic. MacFadden (50) and McShea (57) present evidence that size increase is driven (rather than passive diffusion). (Ref. 57 provides several tests for determining whether a trend is passive or driven.) While this particular example is problematic, studies of size increase in planktonic foraminifera (23) and late cretaceous mollusk clades (36) support the view that some trends in body size arise by "passive diffusion." These studies suggest the importance of the hierarchical (cladogenetic) perspective but do not provide evidence for species selection or the effect hypothesis.

Brown & Maurer argue that species selection has played a crucial role in shaping macroecological patterns (6, 7; see also 53). Figure 3 shows geographic range of North American terrestrial mammal species as a function of body size. This graph shows a clear upper bound on geographic range (the

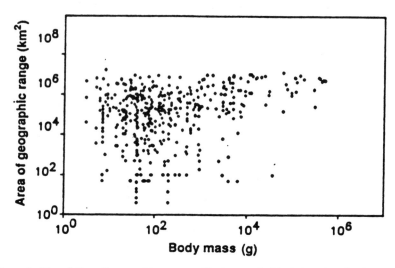

Figure 3 The relation of geographic range and body mass in North American land mammals. (Reprinted with permission from JH Maurer & BA Brown, "Macroecology", Science 243:1145-50. Copyright 1989 American Association for the Advancement of Science.)

size of the continent) and a constraint on minimum body size. For our purposes, the third ("fuzzy") side of the triangle-shaped figure is most important: Larger animals tend to have larger geographic ranges. According to Brown & Maurer, this correlation is a result of differential extinction. Large-bodied animals require more energy than do small-bodied organisms and are therefore constrained to live at lower population density. If a species lives at low density and has a small range, then it is at risk of extinction. Thus, species of large-bodied organisms with small ranges tend to go extinct.

Brown & Maurer argue that this pattern is the result of species selection: "The patterns we have documented appear to be in large part attributable to the dynamics of origination and extinction, and these processes depend greatly on properties, such as population density and area of geographical range, that cannot even be defined at the individual-organism level of the hierarchy" (6, p. 14). Within the set of large-bodied mammals, species with small ranges are more likely to go extinct *because* they have small ranges. The fitness of mammal species depends (in part) on the geographic range of the species. Assuming that range is heritable (a significant assumption), Brown & Maurer have a strong argument for species selection because the fitness of species depends on an emergent species-level trait. If range turns out not to be heritable, then this example cannot be considered species selection, but it would still demonstrate the need to consider how traits that emerge at different levels affect the course of evolution and would therefore support a hierarchical expansion of evolutionary theory.

Other Evidence

Several other lines of empirical evidence provide support for the hierarchical framework.

One important area of research concerns the relation between mass extinction and background extinction. Jablonski (33) argues that the end-Cretaceous mass extinction of marine bivalves and gastropods is qualitatively distinct from (i.e. not merely an intensification of) background extinction patterns. During background times, planktotrophic larval development and large geographic range enhance species survivorship. In contrast, neither property is strongly correlated with survivorship across the mass extinction. Nonetheless, surviving the mass extinction does not appear to be random: While the average range of component species does not affect the chances of clade survivorship, the geographic range of the clade does! This finding illustrates the importance of methodological nonreductionism. Because different patterns appear at different levels, we must carefully document evolutionary patterns at higher levels rather than assuming that they are simple extrapolations of lower-level trends. [Recent work on onshore-offshore patterns in the origin of higher taxa (e.g. 39) is another case in which different patterns are found at different levels of analysis.]

It remains unclear whether Jablonski's findings will apply to other taxa or to other mass extinctions. Chatterton & Speyer (10), for example, argue that the end-Ordovician mass extinction among trilobites was simply an intensification of ordinary background selection processes. By contrast, Westrop's (85, 86) account of the terminal Cambrian extinction among trilobites suggests distinct extinction regimes. (See 38 for a review of recent work on extinction.) While Jablonski and Westrop provide support for methodological nonreductionism, they do not prove that these patterns cannot be *explained* in lower-level terms. That is, these studies do not demonstrate the need for an explanatorily hierarchical theory.

Finally, several studies have documented the occurrence of effect macroevolution (42, 78, 84). Readers interested in these studies should consult the excellent review by Vrba (80). Bell (2) interprets patterns of pelvic reduction in sticklebacks as a case of effect macroevolution. More recently, Masters & Rayner (52) have argued that interspecific competition can be viewed as effect macroevolution.

CONCLUSIONS

Two questions lie at the center of the debate over hierarchical theories of macroevolution: (*a*) Are there distinct higher-level processes?, and (*b*) should we expand the neodarwinian paradigm to include new ways of explaining macroevolutionary phenomena? After two decades of theoretical and empirical work, both of these questions can be answered in the affirmative.

Even if one uses Vrba's restrictive definition of species selection, a few clear and forceful arguments for the reality of species selection can be found (6, 7, 34, 35). Furthermore, I argue that Vrba's definition is too restrictive because emergent species-level traits are not strictly necessary for species selection. If I am correct in this claim, then many of the cases that have previously been taken as evidence for the effect hypothesis (i.e. examples falling in class-B of Figure 1) may actually be instances of species selection. No matter how one resolves the Lloyd-Vrba dispute, species selection appears to be a real phenomenon because the best explanations of certain well-documented trends invoke species selection.

Furthermore, whether one treats them as cases of effect macroevolution or species selection, class-B examples are explanatorily hierarchical. Because an adequate explanation of these class-B phenomena must appeal to differences in species-level fitness that cannot be expressed at the organismic level, these phenomena are explanatorily irreducible. Thus, both class-A and class-B phenomena provide evidence that we need a hierarchical expansion of the neo-darwinian theory. At present, only a small handful of well-documented cases require hierarchical explanations. I suspect that the number will remain small. Given the problems of extracting information from the fossil record, providing convincing evidence for hierarchical explanations of macroevolution is difficult, though not impossible. Nonetheless, as paleobiologists begin actively to seek out evidence of hierarchy, more evidence supporting the expanded theory may be forthcoming.

The theory of species selection continues to undergo refinement and articulation. Recent work distinguishing selection and sorting (82) and emphasizing the significance of taxic trends (23–25, 55) has been particularly important. In this essay, I suggest two further refinements. First, I advocate a slight revision of the standard view of species-level "populations" (i.e. that species selection occurs within clades). In my view, species selection takes place among those members of a clade that occupy a restricted geographic area. This refinement makes species-level populations more closely analogous to populations of organisms and ensures that we have a single selection process rather than two independent but similar processes. Second, while the most interesting kind of species selection (class-A examples in Figure 1) requires emergent species-level traits, I argue that emergent traits are not strictly necessary for species selection.

ACKNOWLEDGMENTS

Comments by David Jablonski, David Hull, and an anonymous referee helped to improve this paper significantly. My research was supported by a Faculty Research and Development Grant from the College of Charleston.

Literature Cited

1. Arnold AJ, Fristrup K. 1982. The theory of evolution by natural selection: a hierarchical expansion. *Paleobiology* 8: 113–29
2. Bell MA. 1987. Interacting evolutionary constraints in pelvic reduction of threespine sticklebacks, *Gasterosteus aculeatus* (Pisces, Gasterosteidae). *Biol. J. Linn. Soc.* 31:347–82
3. Bock WJ. 1979. The synthetic explanation of macroevolutionary change: a reductionistic approach. *Bull. Carnegie Mus. Nat. Hist.* 13:20–69
4. Bonner JT. 1988. *The Evolution of Complexity.* Princeton, NJ: Princeton Univ. Press
5. Brandon R. 1990. *Adaptation and Environment.* Princeton, NJ: Princeton Univ. Press
6. Brown JH, Maurer BA. 1987. Evolution of species assemblages. *Am. Nat.* 130:1–17
7. Brown JH, Maurer BA. 1989. Macroecology: the division of food and space among species on continents. *Science* 243:1145–50
8. Calder WA III. 1984. *Size, Function, and Life History.* Cambridge, MA: Harvard Univ. Press
9. Charlesworth B, Lande R, Slatkin M. 1982. A neo-Darwinian commentary on macroevolution. *Evolution* 36(3):474–98
10. Chatterton BDE, Speyer SE. 1989. Larval ecology, life history strategies, and patterns of extinction and survivorship among Ordovician trilobites. *Paleobiology* 15:118–32
11. Cracraft J. 1982. A non-equilibrium theory for the rate-control of speciation and extinction and the origin of macroevolutionary patterns. *Syst. Zool.* 31: 348–65
12. Damuth J. 1985. Selection among 'species': a formulation in terms of natural functional units. *Evolution* 39:1132–46
13. Damuth J, Heisler IL. 1988. Alternative formulations of multilevel selection. *Biol. Phil.* 3:407–30
14. Eldredge N. 1979. Alternative approaches to evolutionary theory. *Bull. Carnegie Mus. Nat. Hist.* 13:7–19
15. Eldredge N. 1985. *The Unfinished Synthesis: Biological Hierarchies and Modern Evolutionary Thought.* New York: Oxford Univ. Press
16. Eldredge N, Gould SJ. 1972. Punctuated equilibrium: an alternative to phyletic gradualism. In *Models in Paleobiology*, ed. TJM Schopf, pp. 82–115. San Francisco, CA: Freeman
17. Eldredge N, Greene M. 1992. *Interactions.* New York, NY: Columbia Univ. Press
18. Eldredge N, Salthe S. 1984. Hierarchy and evolution. *Oxford Surveys in Evol. Bio.* 1:182–206
19. Futuyma DJ. 1986. *Evolutionary Biology.* Sunderland, MA: Sinauer. 2nd ed.
20. Gilinsky N. 1986. Species selection as a causal process. *Evol. Biol.* 20:249–73
21. Godfrey-Smith P. 1992. Additivity and the units of selection. *PSA 1992: Proc. 1992 Biennial Meet. Phil. Sci. Assoc.* 1:315–28
22. Gould SJ. 1982. Darwinism and the expansion of evolutionary theory. *Science* 216:380–87
23. Gould SJ. 1988. Trends as change in variance: a new slant on progress and directionality in evolution. *J. Paleontol.* 62:319–29
24. Gould SJ. 1990. Speciation and sorting as the source of evolutionary trends, or 'things are seldom what they seem.' In *Evolutionary Trends*, ed. KJ McNamara, pp. 3–27. Tucson, AZ: Univ. Ariz. Press
25. Gould SJ, Eldredge N. 1993. Punctuated equilibrium comes of age. *Nature* 366: 223–27
26. Grantham T. 1993. *Species selection and macroevolution.* PhD thesis. Northwestern Univ., Evanston, IL
27. Griffiths PE, Gray RD. 1994. Developmental systems and evolutionary explanation. *J. Phil.* 91:277–304
28. Hansen T. 1978. Larval dispersal and species longevity in lower Tertiary gastropods. *Science* 199:885–87
29. Hansen T. 1983. Modes of larval development and rates of speciation in early Tertiary neogastropods. *Science* 220:501–2
30. Hoffman A. 1989. *Arguments on Evolution: A Paleontologist's Perspective.* New York: Oxford Univ. Press
31. Hull D. 1974. *Philosophy of Biological*

Science. Englewood Cliffs, NJ: Prentice Hall

32. Hull D. 1980. Individuality and selection. *Annu. Rev. Ecol. Syst.* 11:311–32

33. Jablonski D. 1986. Background and mass extinctions: the alteration of macroevolutionary regimes. *Science* 231:129–33

34. Jablonski D. 1986. Larval ecology and macroevolution in marine invertebrates. *Bull. Mar. Sci.* 39:565–87

35. Jablonski D. 1987. Heritability at the species level: analysis of geographic ranges of Cretaceous mollusks. *Science* 238:360–63

36. Jablonski D. 1987. How pervasive is Cope's rule? *Geol. Soc. Am. Abstr.* 19(7):713–14 (Abstr.)

37. Jablonski D. 1988. Reply to letter by Russell and Lindberg. *Science* 240:969

38. Jablonski D. 1995. Extinctions in the fossil record. In *Extinction Rates*, ed. JH Lawton, RM May, pp. 25–44. Oxford: Oxford Univ. Press

39. Jablonski D, Bottjer D. 1991. Environmental patterns in the origins of higher taxa: the post-Paleozoic fossil record. *Science* 252:1831–33

40. Kellogg D. 1988. "And then a miracle occurs": weak links in the chain of argument from punctuation to hierarchy. *Biol. Phil.* 3:3–28

41. Kincaid H. 1990. Molecular biology and the unity of science. *Phil. Sci.* 57:575–93

42. Kitchell JA, Clark DL, Gombos AM. 1986. Biological selectivity of extinction: a link between background and mass extinction. *Palaios* 1:504–11

43. Kitcher P. 1984. 1953 and all that. *Phil. Rev.* 93:335–73

44. Levinton J. 1988. *Genetics, Paleontology, and Macroevolution*. Cambridge, UK: Cambridge Univ. Press

45. Lewontin RC. 1970. The units of selection. *Annu. Rev. Ecol. Syst.* 1:1–18

46. Lieberman BS, Allmon WD, Eldredge N. 1993. Levels of selection and macroevolutionary patterns in the turritellid gastropods. *Paleobiology* 19:205–15

47. Lloyd EA. 1988. *The Structure and Confirmation of Evolutionary Theory*. New York: Greenwood

48. Lloyd EA, Gould SJ. 1993. Species selection on variability. *Proc. Natl. Acad. Sci.* 90:595–99

49. Lomolino MV. 1985. Body size of mammals on islands: the island rule reexamined. *Am. Nat.* 125:310–16

50. MacFadden BJ. 1986. Fossil horses from "Eohippus"(*Hyracotherium*) to *Equus*: scaling, Cope's law, and the evolution of body size. *Paleobiology* 12:355–69

51. Marshall CR. 1991. Estimation of taxonomic ranges from the fossil record. In *Analytical Paleobiology*, ed. NL Gilinsky, PW Signor, pp. 4:19–38. Lawrence, KS: Paleontol. Soc.

52. Masters JC, Rayner RJ. 1993. Competition and macroevolution: the ghost of competition yet to come? *Biol. J. Linn. Soc.* 49:87–98

53. Maurer BA, Brown JH, Rusler RD. 1992. The micro and macro in body size evolution. *Evolution* 46:939–53

54. Mayo DG, Gilinsky NL. 1987. Models of group selection. *Phil. Sci.* 54:515–38

55. McKinney ML. 1990. Classifying and analyzing evolutionary trends. In *Evolutionary Trends*, ed. KJ McNamara, pp. 28–58. Tucson, AZ: Univ. Ariz. Press

56. McKinney ML. 1990. Trends in body size evolution. In *Evolutionary Trends*, ed. KJ McNamara, pp. 75–118. Tucson, AZ: Univ. Ariz. Press

57. McShea D. 1995. Mechanisms of large-scale evolutionary trends. *Evolution*. In press

58. Nagel E. 1961. *The Structure of Science*. New York: Harcourt

59. Peters RH. 1983. *The Ecological Implications of Size*. New York: Cambridge Univ. Press

60. Raup DM, Gould SJ, Schopf TJM, Simberloff DS. 1973. Stochastic models of phylogeny and the evolution of diversity. *J. Geo.* 81:525–42

61. Russell MP, Lindberg DR. 1988. Real and random patterns associated with molluscan spatial and temporal distributions. *Paleobiology* 14:322–30

62. Russell MP, Lindberg DR. 1988. Letter. *Science* 240:969

63. Salthe S. 1985. *Evolving Hierarchical Systems*. New York: Columbia Univ. Press

64. Sarkar S. 1994. The selection of alleles and the additivity of variance. *PSA 1994: Proc. 1994 Biennial Meet. Phil. Sci. Assoc.* 1:3–12

65. Schaffner K. 1976. Reduction in biology: prospects and problems. *PSA 1974: Proc. 1974 Biennial Meet. Phil. Sci. Assoc.* pp. 613–32

66. Scheltema RS. 1979. On the relationship between the dispersal of pelagic veliger larvae and the evolution of marine prosobranch gastropods. In *Marine Organisms*, ed. B Battaglia, J Beardmore, pp. 303–22. New York: Plenum

67. Scheltema RS. 1989. Planktonic and non-planktonic development among prosobranch gastropods and its relationship to the geographic range of species.

In *Reproduction, Genetics and Distributions of Marine Organisms*, ed. JS Ryland, PA Tyler, pp. 183–88. Fredensbord, Denmark: Olsen & Olsen

68. Schluter D, et al. 1985. Ecological character displacement in Darwin's finches. *Science* 227:1056–59

69. Sober E. 1984. *The Nature of Selection.* Cambridge, MA: MIT

70. Sober E, Wilson DS. 1994. A critical review of philosophical work on the units of selection problem. *Phil. Sci.* 61:534–55

71. Stanley S. 1973. An explanation of Cope's Rule. *Evolution* 27:1–26

72. Stanley S. 1975. A theory of evolution above the species level. *Proc. Natl. Acad. Sci. USA* 72:646–50

73. Stanley S. 1979. *Macroevolution: Pattern and Process.* San Francisco, CA: Freeman

74. Turner JRG. 1986. The genetics of adaptive radiation: a neo-darwinian theory of punctuational evolution. In *Patterns and Processes in the History of Life*, ed. DM Raup, D Jablonski, pp. 183–207. Berlin: Springer Verlag

75. Vermeij G. 1978. *Biogeography and Adaptation.* Cambridge, MA: Harvard Univ. Press

76. Vrba E. 1983. Macroevolutionary trends: new perspectives on the roles of adaptation and incidental effect. *Science* 221:387–89

77. Vrba E. 1984. What is species selection? *Syst. Zool.* 33:318–28

78. Vrba E. 1987. Ecology in relation to speciation rates: some case histories of Miocene-Recent mammal clades. *Evol. Ecol.* 1:283–300

79. Vrba E. 1989. Levels of selection and sorting with special reference to the species level. *Oxford Surveys in Evol. Biol.* 6:111–68

80. Vrba E. 1989. What are the biotic hierarchies of integration and linkage? In *Complex Organismal Functions*, ed. DB Wake, G Roth, pp. 379–401. Chichester: Wiley

81. Vrba E, Eldredge N. 1984. Individuals, hierarchies, and processes: towards a more complete evolutionary theory. *Paleobiology* 10:146–71

82. Vrba E, Gould SJ. 1986. The hierarchical expansion of sorting and selection: sorting and selection cannot be equated. *Paleobiology* 12:217–28

83. Wade M. 1978. A critical review of models of group selection. *Q. Rev. Biol.* 53:101–14

84. Werdelin L. 1987. Jaw geometry and molar morphology in marsupial carnivores: an analysis of a constraint and its macroevolutionary implications. *Paleobiology* 13:342–50

85. Westrop SR. 1989. Macroevolutionary implications of a mass extinction—evidence from an Upper Cambrian stage boundary. *Paleobiology* 15(1):46–52

86. Westrop SR. 1991. Intercontinental variation in mass extinction patterns: influence of biogeographic structure. *Paleobiology* 17(4):363–68

87. Williams GC. 1966. *Adaptation and Natural Selection.* Princeton, NJ: Princeton Univ. Press

88. Williams GC. 1992. *Natural Selection: Domains, Levels and Challenges.* Oxford: Oxford Univ. Press

89. Wilson DS. 1983. The group selection controversy: history and current status. *Annu. Rev. Ecol. Syst.* 14:159–87

90. Wilson DS, Colwell RK. 1981. Evolution of sex ratio instructured demes. *Evolution* 35:882–97

91. Wimsatt WC. 1980. Reductionistic research strategies and their biases in the units of selection controversy. In *Scientific Discovery: Case Studies*, ed. T Nickles, pp. 213–59. Dordrecht: Reidel

Annu. Rev. Ecol. Syst. 1995. 26:323–41

WOMEN IN SYSTEMATICS

Diana Lipscomb

Department of Biological Sciences, George Washington University, Washington, DC 20052

KEY WORDS: women in science, systematics, history of science

ABSTRACT

Although nothing specific has been written about women as systematic biologists, women have always been integral contributors to this scientific field. Like their male colleagues, they have contributed to systematics in a variety of ways and in roles compatible with their location in history. In the middle ages, women clerics kept the study of plant and animal diversity alive. During the zenith of natural history studies in the eighteenth and nineteenth centuries, women worked not only as assistants but also as independent professional naturalists collecting, describing, and classifying plants and animals. As science became professionalized at the end of the nineteenth century, women systematists joined the professional ranks first at women's colleges, then at coeducational universities, and at museums, herbaria, and government laboratories. What distinguishes women systematists from their male contemporaries are the added social obstacles they faced because of their gender. The ingenuity and perseverance they used to overcome these artificial barriers make their scientific achievements remarkable personal achievements as well.

INTRODUCTION

When we think of the past successes in the field of systematics, the role of women scientists is not immediately obvious. This is probably due either to the fact that the work has been disassociated from the names of the scientists that did it or to the tradition of referring to scientific work by the last names of scientists which hides the fact that many of these were women. A little detective work reveals an amazing number of women who have made valuable and lasting contributions to systematics. It is the intent of this paper to recog-

323

nize some of these earlier women and to look at their contributions. In these accounts of women systematists, two recurring themes emerge: their tremendous commitment and love for their work, and their pursuit of scientific problems because of their fascination with the organisms they were studying. Because space here is limited, the women profiled were chosen to represent a variety of historical settings and different career choices (see 34). I am a scientist, not a historian, and as a trespasser on the turf of these other scholars. I offer these brief accounts not as a definitive history but as a framework for further investigation. In the final section I provide a guide to the literature on women scientists.

THE MIDDLE AGES

In histories of biology, Aristotle is usually recognized as the first important scientific observer and given the title "father of natural history" (20). The first systematist is usually considered to be Carl Linnaeus (1707–1778), who constructed a scheme for classifying plants and developed a system of nomenclature (20). Nevertheless, from the time of Aristotle's first catalogs of nature until Linnaeus's natural system, some 2000 years elapsed. While it is true that during this time many of the descriptions and drawings of nature were either copies of ancient works or fabrications of fantastic lands with mythical flora and fauna, a facile interpretive leap from Aristotle to Linnaeus leaves out too much history—including history in which women played a far more prominent role in the culture of learning than they did in either Aristotle's ancient Greece or, arguably, in Linnaeus's seventeenth century Europe (30).

For much of this 2000-year time interval, the repositories of science, medicine, art, and other forms of intellectual activity were convents and monasteries. For part of this time, an androgynous Christian ideal was taken seriously in many of these convents, and aristocratic women gained significant control over both property and learning. Especially in Germany, women won respect as teachers, writers, and contemplatives (30). From this intellectual tradition came Hildegard of Bingen.

Hildegard of Bingen (1098–1179)

Hildegard spent most of her life within Benedictine convents. In 1147, she established a new Benedictine abbey, St. Rupert's on the Rhine. She remained abbess at St. Rupert's until her death at the age of 81. Hildegard was an unusual combination of administrator, priest, mystic, scientist, and physician. Hildegard's extraordinary influence during her life rested largely upon her reputation as a mystic and as a papally sanctioned prophetess. In this role, she corresponded with both kings and church leaders, and she published several books of visions (*Scivias*). Her studies, however, were versatile and encyclopedic;

she wrote at least 14 books (many in several volumes) on natural science, medicine, and religious philosophy.

Hildegard is considered by many to have been the first and foremost natural historian of her age (22). Despite her typically medieval mysticism, Hildegard's accounts of the natural history of western Europe come primarily from her own observations of nature itself rather than from her imagination, folklore, or the literature. This approach set her apart from her contemporaries, and her work has been considered to be the first rediscovery of Aristotle's scientific methodology in which observations are used to answer questions about the natural world (22). Although Hildegard has much to say about nature in several of her works, her greatest is a treatise entitled *Liber Subtilitatum Diversarum Naturarum Creaturarium* (usually known as *Physica*). It consists of nine books of descriptions of nearly 500 plants, fishes, birds, insects, mammals, amphibians, and reptiles, and it discusses their medicinal use. In these books, Hildegard includes descriptions of some animals and plants not native to her country. These accounts include mythological beasts such as dragons, griffins, and unicorns, and unknown (but real) animals such as lions and tigers. Here she relies entirely on the literature and tells stories in a typically medieval fashion. However, she carefully gives accurate and firsthand accounts of the plants and animals she knows, and when her observations conflict with folklore about a particular organism, she relies completely on her observations (22). Her book on mammals, for example, begins with fables of elephants, camels, and unicorns, but then proceeds to accurate descriptions of endemic species.

In her 230-chapter book on plants, she describes many native plants as well as varieties of grain and medicinal plants. In Hildegard's book on fishes, more than 30 local fishes are identified and described, while in her 72-chapter book on birds, she categorizes, describes, and names a great number of the common European birds. With pre-Darwinian logic, her book of birds includes the flying insects, and lists and describes bees, bumble bees, wasps, flies, mosquitoes, and locusts. Although they have long since been replaced by Latin binomials in the scientific literature, many names used by Hildegard are still in common use by the people of Germany. In her lifetime, Hildegard won great acclaim, and, although never canonized, she is often referred to as St. Hildegard (9, 22).

THE NATURALISTS

Natural history experienced some of its greatest popularity during the period from the 1600s through the 1800s. It was a time of worldwide exploration when many naturalists were engaged in discovering, describing, and naming new plants and animals. The study of natural history can be as simple as bird watching, insect collecting, or accurately drawing flowers. It can, therefore, be done by anyone who takes the time to learn the necessary skills. Because this was so, natural history was pursued with enthusiasm by clever girls, intelligent women, and keen old ladies.

The Invisible Women

Many women naturalists left little or no written record of their work, and these invisible women have been largely lost to history. The importance of such women to the work of other naturalists is hinted at in their writings. For example, as a collector of marine algae in England, Mrs. AW Griffiths of Torquay was warmly praised by many phycologists, and one genus and several species of seaweed were named after her. She published nothing in her own name, however, and now survives only as an acknowledgment in other people's work (3, 16).

Other women, as sisters, mothers, and wives, often worked with men on scientific projects for which only the men are remembered today. Sarah Sophia Banks (1744–1818), for example, had an inquiring mind (and an apparently eccentric personality) and discussed questions of plant biology with her brother, the botanist Joseph Banks. She influenced him greatly, and many of her ideas made their way into his writings. She also provided valuable support by recopying and editing the entire manuscript of Banks' Newfoundland voyage (published 1766) (32).

Drawing and illustrating plants and animals in monographs and descriptions written by men was a frequent way for women to participate in natural history studies. Many women illustrators greatly augmented systematists' work with their precise and elegant drawings. Maria Martin (1796–?), for example, was an associate and collaborator of John James Audubon. Working from nature, she painted numerous flower and plant backgrounds for Audubon's *Birds of North America* and *Viviparous Quadrupeds of North America,* and she contributed entire illustrations to John Edward Holbrook's *North American Herpetology* (4, 16).

Some historians tend to take the position that women's participation in science in the eighteenth and nineteenth century was primarily as authors of and audience for popular science, and as illustrative assistants to male scientists. Without downplaying the significance of the work of such women, it is important to recognize that some women played a central role in science in general (35) and systematics in particular (12).

The Early Naturalists

MARIA SIBYLLA MERIAN (1647–1717)   Merian became one of the most masterful insect naturalist-illustrators of her day. Born in Frankfurt, Germany, Maria received her artistic training from her painter stepfather and his apprentices at the family workshop. She married one of her stepfather's apprentices, Johanne Graff, and eventually took on female apprentices and developed her own trade, selling flower-painted textiles. In conjunction with her business, which depended on silk, she studied the life cycle of silkworms and undertook an (unsuccessful) search for worms that could produce a finer thread. As a result of her studies, she published an illustrated volume on the metamorphosis of caterpillars. After 17 years of marriage, Merian left her husband and resumed her maiden name. With her two daughters, she joined the Labadist, an ascetic religious community sympathetic to independent and accomplished women. At the age of 52, Maria, with her two daughters, left Europe for the South American Labadist mission in Suriname. There she eagerly proceeded with her studies and illustrations of insect life.

Upon her return to Holland, she published her major scientific work, a large illustrated volume on the life cycle of the caterpillars, worms, moths, butterflies, beetles, bees, and flies found in Suriname (*De Generatione et Metamorphose Insectorum Surinamensium*, Amsterdam, 1705). Besides descriptions of insects, there were also careful drawings of lizards, batrachians, and plants, as well as sympathetic explanations of the folk medicines and popular magic of indigenous peoples. This remarkable volume made Merian famous among leading naturalists of her day. Her descriptions reflect exact observations, and her meticulous illustrations are so striking that these drawings and paintings are still being exhibited today. In honor of her contributions, six plants, nine butterflies, and two beetles are named after her (35).

JANE COLDEN (1724–1766)   Colden is the first American woman scientist to gain recognition in any field of science. Her father, Cadwallader Colden, was an amateur but significant figure in eighteenth century botany. He educated his daughter in botany, including the Linnaean system of classifying plants. Because Colden helped her father with his large botanical correspondence, she became known to several American and European botanists, including Linnaeus. In her correspondence with Linnaeus, Colden did not mind pointing out Linnaeus's errors in applying his own "sexual" system (he used the number and arrangement of stamens and pistils to classify flowering plants). About *Clematis virginiana* she wrote "Neither [does] Linneaus take notice that there are some plants of the Clematis that bear only male flowers, but this I have observed with such care, that there can be no doubt about it." These correspondences between Colden and other botanists also indicate that she exchanged seeds and plants with them.

It is only certain that Colden wrote one published work, but there may have been others. The original manuscript (*Plantae Coldenghamae*) is in the British Museum of Natural History, and a portion of it was published in 1963. In it, she drew, described, and cataloged nearly 340 plants from the lower Hudson

River Valley. Many of the drawings were accompanied by detailed descriptions and local folklore and medicinal uses. Modern evaluations of this work vary, but in general her descriptions are considered excellent. Her drawings, on the other hand, are regarded as poor. It is possible that the drawings praised during her lifetime have been lost (13, 36, 39). The description and naming of the *Gardenia* is sometimes attributed to Colden, but, although she did propose that name (in honor of fellow botanist Alexander Garden) in her description of marsh St. Johnswort, it was adopted instead for the white-flowered hedge plant of the southern United States. Her plant was renamed *Hypericum virginicum* (4, 11).

Applied Descriptions

Today, systematic entomology provides an invaluable service in the identification and description of insect pests. This tradition was already alive and well in the nineteenth century and embodied in the work of Margaretta Morris and Charlotte Taylor.

MARGARETTA HARE MORRIS (1797–1867)   Morris was acquainted with a number of scientists and prepared illustrations for some botanical papers. Nothing is known about Morris's education and very little about her personal life, except that she lived her entire life unmarried in Germantown, Pennsylvania. Morris's important contribution to systematics is her published description of two devastating agricultural pests—the Hessian fly (*Cecidomyia destructor*) and the 17-year locust (*Cicada septemdecium*). Morris' work resulted in many awards in her life. In 1850, she and astronomer Maria Mitchell were the first women to be accepted into the American Association for the Advancement of Science, and, in 1859, she was elected to the Philadelphia Academy of Science (4, 13, 28).

CHARLOTTE DE BERNIER SCARBROUGH TAYLOR (1806–1861)   Taylor graduated from a private school in New York City and appears to have been self-taught in natural history. She became the wife of a well-to-do merchant of Savannah, Georgia, and mother of two daughters and a son. Taylor spent 15 years observing and drawing insects, especially parasites of cotton and wheat. At least 19 papers on entomological subjects by Taylor are known. In 1859 and 1860, she published her findings in "Insects Belonging to the Cotton Plant" in an article for *Harper's New Monthly Magazine*. She followed that with articles on natural history of both silkworms and spiders. For her time, she had an unusual awareness of the ecological relationships of insects and agriculture and, through her publications, she urged the informed control of pests. At the onset of the Civil War she left the South for the Isle of Man in the Irish Sea, where she died of tuberculosis (4, 28, 36).

Collectors

MARY ANNING (1799-1847)   Anning got her interest in collecting fossils from her father, a cabinet maker in Lyme Regis, Britain. Over her life she made several important paleontological finds, including the first complete ichthyosaur skeleton (1811), a plesiosaur, and a pterodactyl (1828). Anning's fossil collecting eventually provided her with a small income, including a government grant. She did not publish herself, but her finds and observations were of great importance to scientists of her day (32).

YNES ENRIQUETTA JULIETTA MEXIA (1870–1938)   Mexia studied natural history at the University of California but never completed a degree. Her major contribution was the numerous botanical specimens she collected from Central and South America. For example, in 1927 she returned from western Mexico with 1600 items, in sets of 15 specimens, including a new genus and approximately 50 new species. The most distinguished botanists of the day were vocal in their praise of her skill as well as her energy. On each of her trips to remote regions of Mexico, Brazil, Peru, Bolivia, Argentina, and Chile she studied the people and animals as well as the flora. Still, it was the botanical results that were the most impressive record. Several plants were named for her including the genus *Mexianthus* (4, 14, 36).

ANNIE TRUMBULL SLOSSON (1838–1926)   Slossen began by studying botany, and even corresponded with Asa Gray about some of her finds. In 1882 she reported locating one specimen of a rare plant, *Subularia aquatica*, at Echo Lake in Franconia, New Hampshire. But she began collecting insects one early spring when she was 48 years old because she wanted to know what bugs were infesting her garden. From then on she devoted most of her energy to studying insects. She clearly enjoyed finding new species, commenting in 1890 about a successful collecting season in Florida, "I have great hopes concerning my unnamed specimens. One large, oddly-marked Sphinx [moth] fills me with visions of a new genus as well as species, and I have already selected its name" (4). Because she collected all types of insects, she sent some of her specimens to experts for identification, thus providing them with new species and specimens. For her own large collection, she retained at least one of each species, giving away or selling her duplicates. Over the years, 100 insects were named for her by other entomologists whom she supplied with her own finds, including *Zethus slossonae*, a wasp that she discovered at Lake Worth, Florida. The species named for her did not represent all the new insects she discovered; some she named herself with names that usually honored the place she found them. Her methods were so successful that years later, entomologists were still praising her unusual ability to discover strange insects (4). With the money she made selling insect specimens, she helped support the *Journal of the New York Entomological Society*. Not only was she one of the first elected members of the society when it was started in 1892, but the meetings were held at her home until she persuaded the authorities of the American Museum of Natural History to let them meet in the museum. More than a collector, she was also a contributor to most entomological journals of the day (4, 8, 26, 28).

ANNIE MONTAGUE ALEXANDER (1867–1950)   Born in Honolulu, Hawaii, Alexander's childhood encouraged her interest in nature. When her family moved to the continental United States, she was introduced to biology and geology by Martha Beckwith, later a teacher at Mount Holyoke College, and to paleontology by Dr. John Mirriam of the University of California at Berkeley. On University-sponsored field trips, she quickly gained a reputation for her ability to find fossils and collect mammal skulls. She did not complete a degree; instead she spent much of her young adult life as a traveling companion for her adventurous father. Following his death in an accident while they were exploring in Africa, Alexander returned to the United States and began avidly to pursue collecting vertebrate fossils and skeletons.

Her collections became immense, especially after she began collecting in Alaska. Dr. Joseph Grinnell, a Pasadena naturalist working at the Troop Institute, convinced her that the west coast of the United States needed a museum to display the fauna of the western States from collections such as hers. Alexander insisted that the museum be housed at the University of California in Berkeley (where she had taken her first paleontology courses). She put up all the money for the Museum of Vertebrate Zoology and chose Grinnell as its first director. In 1909, to fill the newly opened museum, Alexander organized and participated in an expedition to the Quinn River in Nevada where she collected, among other things, fossil woolly rhinoceros, camels, and mastodons. In 1908 Alexander also helped found and fund the University of California's Department of Paleontology. Over the next 13 years, she began collecting small mammals and birds, and occasionally plants.

In the 1920s, Alexander founded and financed the Museum of Paleontology at Berkeley in much the same way that she continued to finance the Museum of Vertebrate Zoology. Over her lifetime, she and her collaborators collected 6,744 mammal, bird, amphibian, and reptile specimens for the Museum of Vertebrate Zoology and 17,851 plant specimens for the herbarium, as well as thousands of fossil specimens for the Museum of Paleontology. On one field trip she celebrated her eightieth birthday and collected over 4600 specimens of plants (including some new species) (4, 42).

AMALIE NELL DIETRICH (1821–1891)   Dietrich's interest in natural history began with her husband, Wilhelm, a "gentleman naturalist." Their marriage, however, was not a happy one. After he trained her as a collector of animal and plant specimens, he stayed home and sent her alone on trips throughout Europe to collect what he wanted. Upon returning from one long trip, she found that he had sent their daughter away as a household servant. Dietrich left her husband and decided to earn her living by collecting and selling specimens. Through RA Meyer, she met C Godeffroy, who was establishing, in Hamburg, a museum of the natural history of the South Pacific. Leaving her daughter to be educated by the Meyers, Dietrich was sent by Godeffroy to Australia and New Guinea where, from 1863 to 1873, she collected and identified birds, mammals, and plants. These collections added significantly to our understanding of the flora and fauna of Australia (29, 32).

Beginnings Of Professionalism

While some women were content to participate as collectors and illustrators, others were beginning to push to have their systematic work taken seriously as science. Increasingly in the 1870s and thereafter, women joined scientific organizations and sought work in museums and universities. In an article written in 1895 (12), Rosa Smith Eigenmann expressed dissatisfaction with the patronizing attitude toward women, assuming them to be popularizers of science rather than participants in the process.

ROSA SMITH EIGENMANN (1858–1947)   Eigenmann had no formal training as a scientist and very little college education. She was a newspaper reporter in 1880 when her first scientific paper on a new species of fish from the San Diego area was published. When ichthyologist David Starr Jordan heard her paper at a meeting of the San Diego Society of Natural History describing a new species she had discovered, he invited her to study with him at Indiana University. She worked with Jordan on his fish survey and in 1886 met and married one of his students, Carl H. Eigenmann. In the six years following their marriage in 1887, the Eigenmanns worked together and became widely known as authorities on the freshwater fishes of South America and western North America. In 1891, Carl Eigenmann returned to Indiana University where he replaced Jordan, who had moved to Stanford. This return to Indiana coincided with the end of Rosa Eigenmann's research. The challenge of rearing five children, including two with disabilities, restricted Rosa Eigenmann's systematics work to editing her husband's manuscripts. In her short scientific career, Rosa Eigenmann published 20 papers on the taxonomy of fishes on her own, 15 papers with her husband, and a monograph with Joseph Swain on the fishes of Johnson Island in the central Pacific Ocean. Rosa Eigenmann is often regarded as the first American woman to achieve prominence in ichthyology (19, 32, 36).

THE PROFESSIONALS

Role Of Women's Colleges

One of the major developments of the nineteenth century was the rise and transformation of higher education for women, from informal teaching to women's academies, to women's colleges. In the United States, this transformation took place because of the popularity of the idea that women must be educated so they could devote their lives to raising moral and patriotic sons (23, 34). Once established, women's colleges provided both a commitment to excellence in women's science education and, at the same time, employment as faculty members to many women scientists. It is difficult to learn about the work of these early professional women scientists. Possibly few of them did much research because few women's colleges had the financial means to support it. Furthermore, there was little incentive to do so. After all, these faculty members already held the best jobs available to women, so even with brilliant research accomplishments, they were not going to be offered jobs at universities with research laboratories and graduate students (34). Still, some of these women made lasting contributions to systematics.

CLARA EATON CUMMINGS (1855–1906)   After attending Wellesley College, Cummings worked there, beginning as a curator in the museum in 1878 and eventually rising to the rank of Hunnewell Professor of Cryptogamic Botany in 1906. In addition to many articles, Cummings published a catalog of North American mosses and liverworts (7). In her numerous extended field trips, Cummings made collections of mosses and lichens of New England, California, Jamaica, Alaska, and Europe, and she published several descriptions and classifications of them. Her works on the lichens of Alaska and Labrador represent especially important additions to the systematics of that group. She was a fellow of the American Association for the Advancement of Science, and member of the Society of Plant Morphology and Physiology (vice-president, 1904) and of the Torrey Botanical Club. A conservative nomenclaturist, Cummings made taxonomic changes only when the evidence was overwhelming. Consequently, she made few radical changes in the naming of lichens but left behind a body of solid descriptive material (32, 36).

JULIA WARNER SNOW (1863–1927)   Not only did many women's colleges provide employment to women scientists, but their graduates formed organizations to support and defend women scholars. The Association of Collegiate Alumnae (ACA) was one such organization. In 1890, the ACA established fellowships to support doctoral education of American women in German and Swiss universities. Julia Snow, a graduate of Cornell and later a professor of botany at Smith College, was one of the first recipients of an ACA fellowship; she earned her doctorate at Zurich in 1893 (34). Although some of her students remember her primarily for her extraordinary travels in Russia, Turkey, China, and India, it is as a systematic phycologist that she became known in science. Snow assisted with the United States Fish Commission's biological survey of the Great Lakes by studying algae (particularly chlorophytes) (37). The genus *Snowella* is named for her (36).

CARLOTTA JOAQUINA MAURY (1874–1938)   Although not always encouraged to do research themselves, an enthusiastic ésprit de corps among the science faculties of women's colleges resulted in their support of young women scientists. One such organization, the Naples Table Association for Promoting Laboratory Research by Women established several research fellowships and prizes including, in 1909, the Sara Berliner Research or Lecture Fellowship, which was probably the first movable postdoctoral fellowship for a scientist of either sex. One of its earliest recipients was Carlotta Maury (34). Maury was the younger sister of the Harvard astronomer Antonia Maury and, like her sister, was encouraged by her father to study science. She taught geology at a woman's college (Barnard) and the Huguenot College of South Africa. Maury was also an active researcher for most of her life. She was paleontologist for AC Veatch's geologic expedition to Venezuela (1910–1911), organized and carried out an expedition to the Dominican Republic (1916), and was consulting paleontologist for the Brazilian government and the Venezuelan division of the Royal Dutch Shell Petroleum Company. Specializing in Antillean, Venezuelan, and Brazilian stratigraphy and fossil faunas, Maury produced many publications, most in the *Bulletin of the American Museum of Natural History* (32, 36).

Women Who Worked At Coeducational Universities

In the early part of this century, women in the United States began taking up careers at coeducational institutions. It was often difficult to convince such universities to hire women (27, 40) and it was almost impossible for such women to rise above the rank of assistant professor (34). Nevertheless some women systematists succeeded in accomplishing remarkable research at these institutions.

KATHARINE JEANNETTE BUSH (1855–1937)   Many women working at coeducational institutions were research associates rather than faculty members. As the research associate of AE Verrill at Yale, Katharine Bush made significant contributions to the systematics of mollusks, annelids, and echinoderms. When she was awarded in 1901 the first doctorate in zoology given to a woman at Yale, she had already been publishing systematic descriptions of marine invertebrates for eight years. Although Bush worked at Yale from 1879 to 1913, she was paid for only 12 of these 34 years, and then by the United States Fish Commission, which supported the systematic classification of the large collections of sea animals obtained by government and private expeditions in the 1880s. Bush's thesis and published papers are excellent examples of careful descriptions and accurate drawings of many invertebrates, primarily the Mollusca (33, 34, 36).

JOSEPHINE ELIZABETH TILDEN (1869–1957)   Tilden was one of a few women who successfully rose through the academic ranks in the early half of this century. Josephine Tilden was appointed to an assistant professorship in botany at the University of Minnesota in 1895 and became a professor in 1910. As early as 1893 she began her work on the algae of Minnesota lakes, and in 1895 she started a bibliography of published work on algae. Her final book, *The Algae And Their Life Relations*, published in 1935 and 1937, represents the first American effort to summarize the known characteristics of these important freshwater and marine plants. Tilden did not confine her work to North America or to freshwater algae. For example, from 1934 to 1935, she led an algae-collecting expedition, which included 10 graduate students, from the Red Sea to Australia to California. She published descriptions, classifications, and evolutionary hypotheses on algae from thermal springs, in North and Central America, the West Indies, Hawaii, and China. In honor of her work in the Hudson Bay region, the genus *Tildenia* was named for her (32, 36).

Women Working At Museums, Herbaria, And Government Institutions

SOPHIA PEREYASLAWZEWA (18??–1904)   The Russian biologist Pereyaslawzewa, who received her doctorate from the University of Zurich, produced a long list of systematic articles and monographs. The most comprehensive is her *Monographic de Turbellaries de la Mer Noire,* published in 1892 in Odessa. This work won her a prize from the Congress of Naturalists in 1893. She went on to publish books and monographs, in German, French, and Russian, on the lower invertebrates of the Black and Mediterranean Seas. Pereyaslawzewa eventually became the director of the biological station at Sebastopol (21, 29).

MARY JANE RATHBUN (1860–1943)   Rathbun's interest and dedication to description of marine life were aided and encouraged her whole life by her brother Richard. As children they both became interested in the fish fossils they discovered in the family quarries near Buffalo, New York. When Richard became curator of marine invertebrates at the National Museum of the Smithsonian Institution in 1880, Rathbun accompanied him to the Woods Hole laboratory. Over the next few summers, they worked together sorting and studying the huge collections of marine fauna being brought in by the Fish Commission ships. Spencer F Baird, head of the commission and her brother's superior at the museum, was so impressed by Rathbun's interest and talent that in 1884 he offered her a full-time position at the National Museum. In the 53 years she worked at the Smithsonian, Rathbun worked on the systematic revision of crustaceans, in addition to her duties of identifying and cataloging the rapidly growing collection. Her scientific publications began in 1891 and include 158 titles (including four monographs averaging over 500 pages each) dealing primarily with brachyuran crabs. In 1917, George Washington University awarded her a PhD for her work on the Grapsoid Crabs (Grapsidae), and this resulted in a monograph the following year in the *Bulletin* of the National Museum. She became well-known for her efforts to establish sound zoological nomenclature for crustaceans and for her extensive collection. She obviously cared deeply about her colleagues as well, for in 1914, she returned her salary to the museum so that it could be reallocated to a younger male colleague with a large family. She then continued to work another 29 years without pay (32, 34, 36).

Geology attracted fewer women than other science fields, but the growth in the 1920s and 1930s of petroleum geology opened the doors for women to enter paleontology and add to our systematic knowledge of extinct fauna (34).

JULIA ANNA GARDNER (1882–1960)   Throughout her career, Gardner was interested in fossil mollusks. After completing her education at Bryn Mawr College and The Johns Hopkins University (PhD 1911), she was hired by the United States Geological Survey (USGS). During World War I, like many other graduates from women's colleges (34), she joined the Red Cross; she was wounded near Rheims. Following the war, she joined the United States Geological Survey as a staff member, and rose to associate geologist in 1924 and geologist in 1928. Most of her work between the World Wars focused on detailed studies of Tertiary mollusks of the Atlantic and Gulf coasts. In World War II, she again played an active role, this time working with the Military Geology Unit of the USGS. During this war she developed an interest in the geology of the western Pacific, and she studied its fossils while taking part in the geologic mapping of the area. Gardner was active in professional organizations and promoted the international exchange of data and specimens. At her retirement in 1952, she received the Distinguished Service Award from the Department of the Interior and was serving as president of the Paleontological Society. The following year she was elected a vice-president of the Geological Society of America (25, 36, 41).

WINIFRED GOLDRING (1888-1971)   Ranked first in her class at Wellesley, Goldring stayed to earn a master's degree in geology and to hold an instructorship for two years. Unlike many of her contemporaries, she did not remain at a women's college. Instead, in 1914, she began as a temporary "scientific expert" at the New York State Museum to prepare educational exhibits. Her handbooks and especially her exhibits were considered models for teaching and were often copied. She also wrote college textbooks and popular books on paleontology. Although she did not receive a permanent appointment at the museum until 1920, as early as 1916, Goldring had begun her own research on the description and classification of crinoids from the middle Paleozoic. (Apparently because these echinoderms are commonly called sea-lilies, Goldring is often incorrectly described as a paleobotanist in history of science books). Her numerous publications were highly acclaimed, especially the monograph on Devonian crinoids of New York, published in 1923. Goldring's working conditions were not pleasant. She was given the tasks that the male curators refused to do and was paid less than one half the salary of the museum's clerical staff. Exhaustion and stress finally caught up with her, and she had to take off a year in 1926 to recover her health. Despite it all, she accomplished a number of important studies and eventually became the first woman to be appointed state paleontologist. She was the first woman elected president of the Paleontological Society (1949), and she was vice-president of the Geological Society of America (1950) (24, 32, 34, 36).

MARY BRANDERGEE (1844-1920)   At the turn of the century, a number of botanists were working to name, describe, and classify the flora of the western United States. As might be expected, there were many controversies over the proper nomenclature for the region's diverse flora. One of the chief players in this debate was Mary Brandergee. At a time when most women employed by botanical gardens and herbaria were illustrators, Brandergee became curator of botany at the California Academy of Sciences (6, 34). Although an important systematist in her own right, Brandergee is often overshadowed by her assistant and successor Alice Eastwood.

ALICE EASTWOOD (1859-1953)   Eastwood spent her early childhood living on the grounds of the Toronto Asylum for the Insane (where her father was superintendent). After the death of her mother, Eastwood lived as a student in a convent in Colorado. One of her teachers, aware of her interest in plants, gave her a copy of Asa Gray's *Manual of Botany* and encouraged her. After she finished her education, Eastwood began her career inventorying and classifying plants of Colorado, the results of which form the nucleus of the University of Colorado herbarium (10). Eastwood replaced Brandergee as curator at the California Academy of Sciences in 1892 and instantly began building up the herbarium collection. She gained fame when, after the San Francisco earthquake, she had the courage and presence of mind to run back to the herbarium and save type specimens and catalogs before they could be destroyed by oncoming fires. Eastwood also deserves the bulk of the credit for rebuilding the herbarium's holdings (by adding over 340,000 specimens) and redeveloping the Academy's botanical library. Much of her systematic research focused on the plants of the California chaparral. She described one of the most characteristic shrubs, *Eastwood manzanita.* Several other plants of this habitat were named in her honor, including the daisy genus *Eastwoodia.* Eastwood was widely recognized as a scholar by her contemporaries (e.g. she was honorary president of the Seventh International Botanical Congress, Stockholm, 1950). Her papers, which number over 300, also include popular botany articles as well as scientific reports. These articles indicate that she was an early leader in the movement to promote public awareness of the importance of saving native plants (4, 32, 34, 36).

MARY AGNES CHASE (1869–1963)   In the 1920s and 1930s, the United States Department of Agriculture was the largest federal employer of women scientists and may, in fact, have been their largest employer of any kind in the United States (34). For more than 60 years Agnes Chase served in the Division of Forage Plants, where she began as an illustrator. In 1905 she became associated with AS Hitchcock (an expert in grasses), and her professional evolution began. She was first Hitchcock's assistant and then his collaborator in publication; by 1910 she was junior author on a revision of North American *Panicum.* She took charge of the whole grass project when Hitchcock died in 1936 and became the principal scientist in charge of systematic agrostology. Eventually she advanced to senior botanist. She also produced a three-volume checklist of grass species, for which the careful, critical evaluation of each species required her to make several trips to European herbaria to locate and verify type specimens (5). Not just a herbarium-based scientist, Chase also collected grasses in the United States, Mexico, Antilles and South America, until she was 70 years old. By the time she had made her last collecting trip in 1940 over 12,200 plants, mostly grasses, had been added to the National Herbarium. These specimens included many new grasses among the 4,500 she brought from Brazil (which increased the number of described grasses for Brazil by 10%). In doing so, she made the herbarium a unique research tool. Chase received many honors including a certificate of merit from the Botanical Society of America, a medal of service from Brazil, and an honorary DSc from the University of Illinois. The Smithsonian Institution made her its eighth honorary fellow, and she was unanimously elected a fellow of the Linnean Society (15). A militant suffragette, Chase demonstrated repeatedly for women's right to vote and went to jail twice for it (4, 15, 31, 32, 36).

EDITH PATCH (1904–1937)   In the United States a large increase in the number of women working for state and local governments occurred in the 1920s and 1930s. Edith Patch first sought a job at an experimental station in 1901, but she was rebuffed because the director thought that entomology was "unwomanly." Eventually she was hired (initially without pay) by the Maine Agricultural Experiment Station. In time Patch became head of entomology at the station. Patch's major interests were economic and ecological entomology, although she did some systematic revision as well (particularly on the Aphidae—one new genus and several species have been named for her). Overall she published 15 books and almost 100 articles. Patch received recognition for her contribution when in 1930 she was chosen both as the first woman president of the Entomological Association of America and as the president of the American Nature Study Society (2, 32).

Women Who Worked Without Professional Appointments

ELIZABETH GERTRUDE KNIGHT BRITTON (1858–1934)   Britton graduated from Hunter College (then Normal College) in 1875 and remained there as an assistant of natural science. Increasingly interested in botany, she joined the Torrey Botanical Club in 1879 and published the first of her many scientific papers in 1883. In 1885 she married Nathaniel Lord Britton, a geologist at Columbia College who had become so interested in botany that he moved to that department the next year. Following her marriage, E Britton left her job at Hunter College (see 34 for discussion of the requirement at many colleges that women resign upon marriage) and became the unofficial moss curator at Columbia University. Her efforts to develop this collection met with great success, and by various means, including the purchase of significant collections, it became a large and important research tool. During the 1890s, E Britton suggested the establishment of the New York Botanical Garden, and both she and her husband successfully promoted its foundation. Nathaniel became its first director, and together they carried out those responsibilities for more than 33 years. Britton was a very productive research worker who published 346 articles, mostly on mosses. The esteem in which she was held by the scientific community is reflected in the fact that 15 species and the genus *Bryobrittonia* were named for her. As editor of the *Bulletin of the Torrey Botanical Club,* she gained a reputation for her sharply critical reviews of what she considered careless scientific work. As she grew older, Britton became an important voice in the effort to educate the public on the need to preserve our natural wild flowers (4, 17, 18, 32).

LIBBIE HENRIETTA HYMAN (1888–1969)   It was said of Libbie Hyman, "As person and as a scholar, her career bears witness of devotion to the highest ideals" (38). Hyman grew up in Fort Dodge, Iowa, where she attended the public schools and graduated from high school in 1905 as valedictorian of her class. Following graduation, she took a job in a factory pasting labels on boxes, until a teacher helped secure a scholarship for her at the University of Chicago. She eventually entered the graduate school of the University as a student of CM Child. While a graduate teaching assistant in the introductory zoology and comparative vertebrate anatomy laboratories, she realized there was a need for better laboratory manuals. She wrote *A Laboratory Manual for Elementary Zoology* (1919) and *A Laboratory Manual for Comparative Vertebrate Anatomy* (1922). After receiving her PhD (1915), Hyman remained at Chicago as Child's research associate until his retirement. At that time, Hyman found she was financially independent because of royalties for her lab manuals.

The success of her manuals encouraged Hyman to write the first monographic study of invertebrates (her true interest) in English. She took an apartment in New York near the American Museum of Natural History in order to use its library. At first she worked at home, but in 1937 she was appointed a research associate at the Museum and was provided there with office and lab space, although never a salary. The first volume of *The Invertebrates, Protozoa Through Ctenophora,* appeared in 1940. For this and all subsequent volumes, she never had a secretary, an assistant, or a technician. Her histological preparations, her illustrations, and the text were entirely her own. To execute the drawings from living or prepared material, Hyman spent several summers at the Marine Biological Laboratory in Woods Hole, Massachusetts, and at other marine stations. Volumes II and III were published in 1951, volume IV in 1955, volume V in 1959, and volume VI in 1967. Progress on the final volume was retarded by Hyman's failing health; much of the work was done when she was unable to walk across the room without assistance. The coverage in these books is comprehensive, the treatment authoritative, and the illustrations clear and informative. Furthermore, *The Invertebrates* is more than a compilation; it includes incisive analysis, excellent evaluation, and masterly integration of information. In addition to the books, she published 136 papers, mostly on the morphology and systematics of the lower invertebrates.

Respected and admired, Hyman received many honors. She was president of Society of Systematic Zoology in 1959 and editor of the society's journal *Systematic Zoology* 1959–1963; she received honorary degrees from numerous institutions, including the University of Chicago. She was awarded the Daniel Giraud Elliot Medal of the National Academy of Sciences, 1951, and the Gold Medal of the Linnean Society of London, 1960. On April 9, 1969, at the Centennial Celebration of The American Museum of Natural History, she was awarded its Gold Medal for Distinguished Achievement in Science (36, 38).

CONCLUSION AND GUIDE TO THE LITERATURE

The women whose work I have discussed by no means represent all the major women contributors to systematics. Nonetheless, it must be apparent even from these few examples that women were indeed present in the practice of systematics from earliest times, and that their contributions were significant. Often, as with Hyman, their work changed the direction of their fields or propelled them forward. As with Britton, their work was often an original and personal achievement, even though a woman may have been allowed to pursue it only because of her husband's position.

What about the others? Who are they, and how does a scholar find information on these scientists? Several reference sources identify bibliographic information on women in science. Siegel & Finley (36) compiled an annotated bibliography, organized by field, for American women scientists. Oglivie (32) produced a useful bibliography of women in science of all nationalities through the nineteenth century. Several collections of biographies emphasize women scientists who have excelled in their respective fields, but few mention or discuss women in systematics or even natural history. Exceptions include Bonta's (4) and Norwood's (31) works on American field biologists, and Kass-Simon's (22) essay on women biologists in general. Other useful resources are the many obituaries scattered throughout the scientific literature (e.g. 38). A few historians have attempted to examine the lives and situations of women in science who were not famous. For insight into the lives of these women in systematics, Rossiter's book (34) examines the lives of American women, and Schiebinger's (35) work documents the role of women in the development of modern science in Europe. Also, Abir-Am & Outram (1) have edited a collection of essays documenting how ordinary botanists, ornithologists, and others try to balance their personal and professional lives.

Literature Cited

1. Abir-Am P, Outram D. 1987. *Uneasy Careers and Intimate Lives: Women In Science.* New Brunswick: Rutgers Univ. Press

2. Adams JB, Simpson GW. 1955. Edith Marion Patch. *Ann. Entomol. Soc. Am.* 48:313–14

3. Barber L. 1980. *The Hey-Day of Natural History, 1820–1870.* New York: Doubleday

4. Bonta MM. 1991. *Women in the Field: America's Pioneering Women Naturalists.* College Station, TX: Texas A&M Press. 299 pp.

5. Chase A, Niles CD. 1961. *Index to Grass Species.* Boston: CK Hall. 3 Vols.

6. Crosswhite FS, Crosswhite CD. 1985. The plant collecting Brandegees, with emphasis on Katharine Brandegee as a

liberated woman scientist of early California. *Desert Plants* 7:128–62

7. Cummings CE. 1885. *Catalogue of Musci and Hepaticae of North America, North of Mexico.* Natick: Howard & Stiles

8. Davis WT. 1926. Annie Trumbull Slosson. *J. New York Entomol. Soc.* 34:361–64

9. Dronke P. 1984. *Women Writers of the Middle Ages.* Cambridge: Cambridge Univ. Press

10. Eastwood A. 1893. *A Popular Flora of Denver, Colorado.* San Francisco: Zoe

11. Eifert VS. 1965. *Tall Trees and Far Horizons. Adventures and Discoveries of Early Botanists in America.* New York: Dodd, Mead

12. Eigenmann RS. 1895. Women in science. *Proc. Natl. Sci. Club* 1:13–7

13. Elliott CA. 1979. *Biographical Dictionary of American Science: The Seventeenth through the Nineteenth Centuries.* Westport, CT: Greenwood

14. Ewan J. 1971. Mexia, Ynes Enriquetta Julietta. *Notable Am. Women* 1:533–34

15. Fosburgh FR, Swallen JR. 1959. Agnes Chase. *Taxon* 8:145–51

16. Gould SJ. 1993. The invisible woman. *Nat. Hist.* 6/93:14–23

17. Grout AJ. 1935. Elizabeth Gertrude Knight Britton. *Bryologist* 38:1–3

18. Howe MA. 1934. Elizabeth Gertrude Knight Britton. *J. N. Y. Bot. Gard.* 35:97–103

19. Hubbs CL. 1971. Eigenmann, RS. *Notable Am. Women* 1:565–66

20. Hull DL. 1988. *Science as a Process.* Chicago: Univ. Chicago Press

21. Ireland NO. 1962. *Index to Scientists of the World from Ancient to Modern Times: Biographies and Portraits.* Boston: FW Faxton

22. Kass-Simon G. 1990. Biology is destiny. In *Women of Science: Righting the Record,* ed. G Kass-Simon, P Farnes, pp. 215–33. Bloomington: Ind. Univ. Press

23. Keber LK. 1976. The Republican Mother: Women and the Enlightenment—An American Perspective. *Am. Q.* 28:187–205

24. Kohlstedt SG. 1980. Goldring, Winifred. *Notable Am. Women* 4:282–83

25. Ladd HS. 1962. Memorial to Julia Anna Gardner (1882–1960). *Proc.* *Geol. Soc. Am. Annu. Rep. for 1960* pp. 87–92

26. Leng CW. 1918. History of the New York Entomological Society. *J. N. Y. Entomol. Soc.* 26:129–33

27. Lonn E. 1924. Academic status of women on university faculties. *J. Am. Assoc. Univ. Women* 17:5–11

28. Mallis A. 1971. *American Entomologists.* New Brunswick: Rutgers Univ. Press

29. Mozans HJ. 1913. *Women in Science.* New York: Appleton

30. Nobel DF. 1992. *A World Without Women. The Christian Clerical Culture of Western Science.* Oxford: Oxford Univ. Press

31. Norwood V. 1993. *Made From This Earth: American Women and Nature.* Raleigh: Univ. N. Carolina Press

32. Oglivie MB. 1986. *Women in Science: Antiquity Through the Nineteenth Century.* Cambridge: MIT Press

33. Remington JE. 1977. Katharine Jeanette Bush: Peabody's mysterious zoologist. *Discovery* 12:3–8

34. Rossiter MW. 1982. *Women Scientists in America: Struggles and Strategies to 1940.* Baltimore: Johns Hopkins Univ. Press

35. Schiebinger L. 1989. *The Mind Has No Sex? Women in the Origin of Modern Science.* Cambridge: Harvard Univ. Press

36. Siegel PJ, Finley KT. 1985. *Women in the Scientific Search: An American Biography, 1724–1979.* London: Scarecrow

37. Snow J.W. 1902. The plankton of Lake Erie. *US Fish Commision Bull.* 1902:369–94

38. Stunkard, HW. YEAR. In memorium, Libbie Henrietta Hyman, 1888–1969. In *Biology of the Turbellaria (Libbie H. Hyman Memorial Volume).* pp. ix-xii. New York: MacGraw-Hill

39. Vail AM. 1907. Jane Colden, an early New York botanist. *Toreya* 7:21–34

40. Welsch WH. 1922. Contribution of Bryn Mawr College to the Higher Education of Women. *Science* 56:1–8

41. Wilson D. 1961. Julia Anna Gardner. 1882–1960. *Nautilus* 75:1418–21

42. Zullo JL. 1969. Annie Montague Alexander: her work in paleontology. *J. West* 8:183–99

Annu. Rev. Ecol. Syst. 1995. 26:343–72

A GENETIC PERSPECTIVE ON THE ORIGIN AND HISTORY OF HUMANS[1]

N. Takahata

Coordination Center for Research and Education, The Graduate University for Advanced Studies, Hayama, Kanagawa 240-01, Japan

KEY WORDS: human evolution, hominids, *H. sapiens*, gene genealogy, coalescence

ABSTRACT

Recent topics in molecular anthropology are reviewed with special reference to hominoid DNA sequences and population genetics theory. To cover a wide range of possible demographic situations in the human lineage since the Miocene, a model is introduced that allows temporal changes in population structure and size. The coalescence process of neutral genes is formulated and used to make quantitative inferences on the origin and history of humans. Nuclear DNA sequence data support the theory that humans and chimpanzees diverged from each other 4.6 million years (mya) and the gorilla lineage branched off as early as 7.0–7.4 mya. The same data estimate the effective size of the Pliocene hominoid population as 10^5, a figure similar to that obtained independently from alleles that have persisted in the human population for more than 5 my. Hypotheses about the origin of *Homo sapiens*, genetic differentiation among human populations, and changes in population size are quantified. None of the hypotheses seems compatible with the observed DNA variation. The effective population size decreased to 10^4 in the Pleistocene, suggesting an important role of extinction/restoration in *H. sapiens* populations. Natural selection against protein variation might be relaxed in the Pleistocene. The

[1]Abbreviation and symbols: kb (kilo base pairs), bp (base pairs), yr (years), my (a) (million years (ago)), COII (cytochrome oxidase subunit II), rDNA (ribosomal DNA), mtDNA (mitochondrial DNA), MHC (major histocompatibility complex)

0066-4162/95/1120-0343$05.00

history of *H. sapiens* appears to have been more dynamical than is postulated by the hypotheses.

INTRODUCTION

"Much light will be thrown on the origin of man and his history" —Thus touches *The Origin of Species* simply upon the enigma of human history. Although Darwin believed strict continuity existed between ourselves and nature by means of natural selection, he dared not expose his then-unorthodox view to a public unwilling to accept it (21). All evidence for continuity was circumstantial, and no fossils had yet been found that linked us with close relatives in nature. During this century several important discoveries of fossil hominids (e.g. 1, 19, 59, 60, 96, 130, 131) have been made, including recently the Aramis fossil hominids dated to about 4.4 million years ago (mya) (125). In addition, during the last few decades an entirely new approach, molecular anthropology, has provided techniques to decipher the evolutionary change in genetic materials (26, 61). Here, I address several topics in the recent development of molecular anthropology fostered by DNA sequence data, recognizing that earlier data produced by immunological, DNA-DNA hybridization, and electrophoretic methods played important roles.

Even with DNA sequence data, we have no direct access to the processes of evolution, so objective reconstruction of the vanished past can be achieved only by creative imagination (9). Appropriate theory is necessary for this approach. This review therefore begins with a description of a demographic model of populations and presents several theoretical formulas that I believe are helpful in imagining the past.

THEORETICAL BACKGROUND

The following consideration is based on two genealogical processes. One is the coalescence of neutral genes sampled at random, and the other is the allelic genealogy—the family relationships among different alleles. Quite often, DNA sequences are for different alleles that are distinguished by electrophoresis or DNA typing. To such a set of DNA sequences, the coalescence theory is not directly applicable, and may even be inappropriate if alleles are subject to natural selection.

Gene Genealogy and Coalescence

Gene genealogy describes the ancestral relationships of genes at a locus, a simple consequence of random loss or multiplication of genes in the reproduction process. When a gene multiplied is transmitted to later generations, such a multiplication appears as a coalescence when looked at backward in time.

Coalescence theory provides a mathematical tool for inferring the genealogy of neutral genes that are sampled from a current population. For a randomly mating (panmictic) population with a constant number of breeding individuals (a constant effective population size), the theory is well established (46, 47, 53, 107, 116). In addition, several studies have been extended to the case of a geographically structured population (67, 68, 77, 78, 98, 109, 111, 113, 115) or a population of changing size (124). These extended theories are important because most, if not all, arguments about human evolution are concerned with changes in population structure and size. However, the mathematical formulation of coalescence processes becomes complicated when the sample size (j) is more than 2. For this reason, the sample size is restricted to 2. Despite this limitation, the formulation has wide applicability.

The demographic model used throughout this review assumes that there was an abrupt change in population structure x generations ago (Figure 1). The population was panmictic before this change, consisting of M breeding diploid individuals, but it subsequently subdivided into c descendant populations, each consisting of N breeding individuals. These descendant populations may or may not be different species, depending on time scales. Where descendant populations belong to a single species, they may be referred to as demes or local populations. In this case, demographic forces associated with population structure are taken into account. On the other hand, where descendant populations are different species, they are referred to simply as species. In this case, each descendant species is regarded as an independent evolutionary unit. Migration among demes strongly affects the coalescence process. The pattern of gene migration is assumed to be the same as that in the finite island model (51, 72, 132). The per-generation rate of gene migration is designated by m, and the rate from one deme to another particular one is $m^* = m/(c - 1)$. Wright (133) noted that if a species has a subdivided population structure, demes may be liable to frequent extinction with restoration by stray immigrants, and if this happens the effective size of the whole population can be substantially reduced. This possibility is also taken into account because it seems quite likely in the evolution of humans (113). The per-generation rate of extinction/restoration of a deme is designated by r.

The formulation of coalescence processes is based on two probabilities $P(t)$ and $Q(t)$, in which t is time measured in units of generations from the present to the past: $P(t)$ is the probability that two distinct ancestral lineages reside in the same deme at generation t, whereas $Q(t)$ is the probability that they reside in different demes. Migration affects both $P(t)$ and $Q(t)$ irrespective of whether two gene lineages exist in the same deme. However, this is not the case for extinction/restoration. When both lineages reside in the same deme, extinction/restoration does not change $P(t)$ and $Q(t)$, or it cannot separate the lineages into different demes. If a deme containing one lineage becomes extinct, one

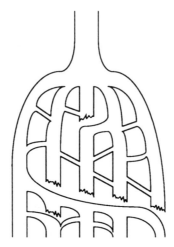

Figure 1   The demographic model of five descendant populations used throughout this review. The population structure is assumed to have changed x generations ago. The population was panmictic before this change and the five descendant populations experience migration and extinction/restoration.

of $c - 1$ demes restores it. And if the donor deme happens to have the other lineage, both lineages come into the same deme by restoration. Thus, $P(t)$ increases and $Q(t)$ decreases, but it is assumed that no immediate coalescence results from an extinction/restoration. In effect, the process resembles that of gene migration in that extinction/restoration can bring one lineage into the deme containing the other lineage.

Coalescence between two ancestral lineages occurs at a certain rate when they exist in the same deme. In each generation, coalescence reduces $P(t)$ by $1/(2N)$ when $t \le x$ and $1/(2M)$ when $t > x$. If all terms such as m^2, $m/(2N)$, $r/(2N)$ are ignored, $P(t)$ and $Q(t)$ satisfy

$$\dot{P}(t) = -(2m + \frac{1}{2N})P(t) + 2bQ(t), \quad \dot{Q}(t) = 2mP(t) - 2bQ(t) \quad \text{for } t \le x \qquad \text{1a.}$$

$$\dot{P}(t) = -\frac{1}{2M}P(t), \quad \dot{Q}(t) = 0 \quad \text{for } t > x, \qquad \text{1b.}$$

where the dot over $P(t)$ and $Q(t)$ indicates the differential with respect to t, $b = m^* + r^*$, $m^* = m/(c - 1)$, and $r^* = r/(c - 1)$. The solutions of Equation 1 lead to the probability $[f(t)]$ that coalescence occurs exactly at t: $f(t) = -\{\dot{P}(t) + \dot{Q}(t)\}$ which is equal to $P(t)/(2N)$ for $t \le x$ and $P(t)/(2M)$ for $t > x$. The

Laplace transform of $f(t)$, $f^*(p,x) = \int\limits_0^\infty f(t)e^{-pt}dt$, facilitates the computation, and

Equation 1 yields

$$f^*(p,x) = 1 - \frac{2Mp}{1+2Mp}\{P(x) + Q(x)\}e^{-px} - p\{P^*(p,x) + Q^*(p,x)\} \qquad 2.$$

In the above, $P^*(p,x) + Q^*(p,x) = \int\limits_0^x \{P(t) + Q(t)\}e^{-pt}dt$ is given by

$$P^*(p,x) + Q^*(p,x) = \frac{1}{F(p)}\{(2b+2m+p)G(p,x) + (2b+2m+\frac{1}{2N}+p)H(p,x)\}, \qquad 3.$$

where $F(p) = (2b + p)\{2m + 1/(2N) + p\} - 4mb$, $G(p,x) = P(0) - P(x)e^{-px}$, $H(p,x) = Q(0) - Q(x)e^{-px}$, and $P(0) + Q(0) = 1$. Since two genes necessarily coalesce to a common ancestor within a certain period of time, $f^*(0,x) = 1$ in Equation 2 holds true. The formula of $f^*(p,x)$ contains all information on coalescence time and related quantities. For instance, it can be used to compute the mean coalescence time (T) and the probability of identity (J): $T = -\partial f^*(p,x)/\partial p$ evaluated at $p = 0$ and $J = f^*(2v,x)$ where v is the mutation rate per locus per generation. The equation for J can be obtained by noting that the probability of no mutation for $2t$ generations equals e^{-2vt}. Of particular interest is mean coalescence time T, which is given by $T = 2M\{P(x) + Q(x)\} + \{P^*(0,x) + Q^*(0,x)\}$ or more explicitly:

$$T = 2N(1+\frac{m}{b}) + 2\{M-N(1+\frac{m}{b})\}\{P(x)+Q(x)\} + \frac{1}{2b}\{Q(0)-Q(x)\}. \qquad 4.$$

Three situations for the sample of genes are distinguished: $P(0) = 1$ and $Q(0) = 0$, $P(0) = 0$ and $Q(0) = 1$, and $P(0) = 1/c$ and $Q(0) = 1 - 1/c$. For the first or second case where two genes are sampled from the same deme or different demes, T is designated as T_0 or T_1, whereas for the third case of random sampling, T is retained. The effective size (N_e), the difference (T_d) between T_1 and T_0, and Wright's (132) F_{ST} are defined as

$$N_e = \frac{1}{2}T, \quad T_d = T_1 - T_0, \quad F_{ST} = \frac{(c-1)T_d}{cT} \qquad 5.$$

(77, 98, 113). The nucleotide differences (d_{XY}) between two genes sampled from different deme X and Y and the net nucleotide differences (d) are given as $d_{XY} = 2\mu T_1$ and $d = 2\mu T_d$ where μ is the nucleotide substitution rate per site per generation. The nucleotide differences $(d_X$ and $d_Y)$ within deme X and Y are identical and are given by $2\mu T_0$. Genetic distance (D) for gene frequency data is defined by $-ln\{J_{XY}/\sqrt{J_X J_Y}\}$. Here, J_X or J_Y is the identity probability within deme X or Y, which is computed as $f^*(2v,x)$ with $P(0) = 1$, whereas J_{XY} is the identity probability between X and Y, which is computed as $f^*(2v,x)$ with $Q(0) = 1$. Since $J_X = J_Y$ holds true as for d_X and d_Y, D reduces to $2vt$ if $m = r = 0$ and $N = M$. For these quantities and their statistical properties, see Nei (72) for review.

It remains to compute $P(x)$ and $Q(x)$ from Equation 1a, depending on how two genes are sampled from demes:

$$P(x) = \frac{1}{\sqrt{A}}[\{2bQ(0) - (2m+\frac{1}{2N}+\lambda_-)P(0)\}e^{\lambda_+^x} + \{(2m+\frac{1}{2N}+\lambda_+)P(0) - 2bQ(0)\}e^{\lambda_-^x}] \qquad \text{6a.}$$

$$Q(x) = \frac{1}{\sqrt{A}}[\{2mP(0) - (2b+\lambda_-)Q(0)\}e^{\lambda_+^x} + \{(2b+\lambda_+)Q(0) - 2mP(0)\}e^{\lambda_-^x}], \qquad \text{6b.}$$

where $A/4 = \{b + m + 1/(4N)\}^2 - b/N \geq 0$, $\lambda_\pm = -\{b + m + 1/(4N)\} \pm \sqrt{A}/2 \leq 0$, and in general $P(x) + Q(x) \leq 1$. The extreme case of $x = 0$ or $x = \infty$ corresponds to an equilibrium population. In the case of $x = 0$, Equation 2 becomes $1/(1 + 2Mp)$ [see also 123], while Equation 4 and 5 reduce to $T_0 = T_1 = T = 2M$, $N_e = M$, $T_d = 0$, and $F_{ST} = 0$, as they should be in a panmictic population of effective size M. In the case of $x = \infty$, both $P(x)$ and $Q(x)$ become 0. Equation 4 and 5 then become

$$T_0 = 2N(1+\frac{m}{b}),\ T_1 = T_0 + \frac{1}{2b},\ T = T_0 + \frac{c-1}{2cb} \qquad \text{7a.}$$

$$N_e = N(1 + \frac{m}{b}) + \frac{c-1}{4cb},\ T_d = \frac{1}{2b},\ F_{ST} = \frac{1}{1+4Nc(b+m)/(c-1)}, \qquad \text{7b.}$$

as they should be in the finite island model at equilibrium (77, 97, 98, 111, 112). In the absence of extinction/restoration, N_e is always larger than the total number (Nc) of breeding individuals in the whole population. The effect of extinction/restoration on T_d and F_{ST} is exactly the same as that of migration when m and r are appropriately interchanged. The situation is different for T_0. Unlike migration, extinction/restoration itself does not increase T_0 from $2N$. As a result, T and N_e can be kept small.

Allelic Genealogy

The ancestral relationship among different allelic lineages is called allelic genealogy. Obviously, an allelic divergence occurs only when a gene in an older lineage mutates to form a new one, a process that may or may not accompany a gene divergence. Allelic genealogy describes the ancestral history in a sample that contains different allelic lineages; any relationship among genes that belong to the same lineage is ignored. Allelic genealogy is applied to a random sample of different alleles.

Allelic genealogy under symmetric balancing selection is useful for understanding the evolution of genes such as those in the major histocompatibility complex (*MHC*) and for determining self-incompatibility in plants. "Symmetric" means that all alleles contribute equally to fitness of specified genotypes and "balancing" means that selection tends to maintain alleles for a long time. The "symmetric" attribute ensures that the topological relationships among alleles are the same as those of neutral genes, whereas the "balancing" attribute causes the timescale to be elongated greatly.

The allelic genealogy in a panmictic population of effective size N_e is briefly reviewed. The allelic genealogy (110) was constructed based on the distribution of time t_i at which i different alleles have been derived from $i - 1$ distinct ancestral alleles for the first time ($i \leq n$, the number of alleles in the population). By analogy with the coalescence process, the probability of $t_i > t$ is given by $\exp\{-i(i - 1)t/(4N_e f_s)\}$ for any $2 \leq i \leq n$ (46, 53, 107, 116). Here, f_s is the scaling factor as a function of v, s, and N_e, where s is the selection parameter and v is the mutation rate at which a new, non-preexisting allele appears at the locus per generation. The formula of f_s is approximately given by

$$f_s = \sqrt{\frac{s}{2N_e v^2}} \, [ln\left\{\frac{s}{8\pi N_e v^2}\right\}]^{-3/2}. \qquad 8.$$

The N_e for alleles under balancing selection is expanded as in a population of effective size $N_e f_s$. The mean values of the number (n) of alleles, homozygosity (J), and the nucleotide substitution rate (α) relative to v become

$$n^3 = 8\sqrt{2}N_e{}^2 svf_s, \quad J = \frac{1}{n}, \text{ and } \alpha = \frac{2\sqrt{2}N_e s}{n^2}. \qquad 9.$$

For $N_e v = 0.025$ and $N_e s = 125$, $f_s \approx 12$, $n = 7.5$, $J = 0.13$, and $\alpha = 6.4$. Formulas 9 were applied to *MHC* genes and used to estimate N_e over $N_e f_s$ generations (e.g. 55, 56).

HUMAN POPULATION IN THE MIOCENE AND PLIOCENE

Hominoid Trees

The time of the origin of a distinct hominid lineage has been a subject of controversy over the century (59, 60 for review). Based on the immunological dissimilarity between the serum albumins of hominoids and their clock-like progression (87), Sarich & Wilson (88) first suggested that if all the living Hominoidea derive from a late Miocene form, humans and the African apes share a Pliocene ancestor (about 5 mya). However, the phylogenetic relationships among humans and the African apes were left open (trichotomy problem). Using DNA-DNA hybridization, Sibley & Ahlquist (94, 95, see also 14) showed that humans are closer to chimpanzees than to gorillas, the gorilla lineage having diverged 9.3 mya and the human-chimpanzee split 5.9 mya. From G- and C-banding analysis of human and pongid chromosomes, Yunis & Prakash (135) also supported the human and chimpanzee clade. Based on standard electrophoresis, King & Wilson (52) demonstrated the close genetic relationship between humans and chimpanzees, and Goldman et al (35) reached the same conclusion based on two-dimensional electrophoresis. Genetic distances for 23 protein loci in nine hominoids indicated a significant resemblance between humans and pongids as well as their relatively recent divergences (13). However, these electrophoretic studies did not resolve the trichotomy problem.

DNA sequences have provided more definite conclusions about the branching order and the dating of hominoid gene divergences. Sequence analysis of the β-globin cluster (4, 58, 65, 79) suggested the close affiliation of humans, chimpanzees, and gorillas, with sequence divergence time in the range of 6.3–8.1 my. Many other phylogenetic studies among primates are based on DNA sequences (3, 11, 27, 29, 30, 33, 36, 39, 49, 84, 86, 90, 118, 119). Complete mtDNA sequences for 12 humans, one pygmy, and one common chimpanzee, one gorilla, and one orangutan have recently become available (44). The chimpanzee sequences are closest to human sequences, which is in accord with a comparison of COII sequences of 680 bp length (85, 85a), but out of accord with that of 12S rRNA sequences of 945 bp length (40). Under the assumption that the orangutan and African apes diverged 13 mya, the divergence time between human/chimpanzee and gorilla mtDNAs is estimated as 6.6 my and that between human and chimpanzee mtDNAs as 4.9 my (43, 44). Standard errors in these estimates are unprecedentedly small.

The phylogenetic tree of DNA sequences is not identical to the species relatedness, because orthologous genes in different species always diverged prior to the species splitting (72). At the face value, the divergence time of 4.9 my between the human and chimpanzee mtDNAs implies that these species became reproductively isolated after 4.9 mya. Likewise, splitting of the gorilla lineage must have occurred after 6.6 mya. Gene divergence times cannot delimit lower bounds of the species divergence times, a shortcoming that holds true for any gene tree. Therefore, the trichotomy problem cannot be resolved in principle by individual gene trees, even though estimated sequence divergence times are accurate.

When DNA sequences are available for unlinked, orthologous regions of the genome in related species, it is possible to develop a method for inferring a species tree (108, 114). It can generate a rooted tree without a reference species and also estimate the species divergence time (x) and the effective size (M) of the ancestral population simultaneously. The method is based on three premises: a molecular clock of silent substitutions (25, 64, 89, 93), no intragenic recombination, and neutrality of silent substitutions (50, 51). Here, the demographic model in Figure 1 is applied to different species, so that neither migration nor extinction/restoration is permitted. Equation 2 with $b = m = 0$ then reduces to

$$f^*(p,x) = \frac{P(0)-P(x)e^{-px}}{1 + 2Np} + \frac{\{P(x)+Q(x)\}e^{-px}}{1+2Mp},$$

10a.

and the formula for T becomes

$$T = 2NP(0) + 2(M-N)P(x) + (2M+x)Q(x).$$

10b.

In the above, $P(x) = P(0)e^{-x/(2N)}$ and $Q(x) = Q(0)$. For pairs of DNA sequences from two species, $Q(0) = Q(x) = 1$ and $P(0) = P(x) = 0$. Equation 10 then become as simple as $f^*(p,x) = e^{-px}/(1 + 2Mp)$ and $T_1 = 2M + x$. A pair of these DNA sequences would be different because of nucleotide substitutions accumulating since they were separated from a common ancestor. It is assumed that for separation time t, the number (k) of nucleotide substitutions between the two sequences follows the Poisson distribution with mean $2vt$. The probability generating function (pgf) of the Poisson distribution then becomes the exponential function of $e^{-2vt(1-z)}$ where z is a dummy variable for the pgf. However, since t is a random variable that follows the distribution of $f(t)$, we must take the average of $e^{-2vt(1-z)}$ over possible values of t. This average is immediately given by $f^*(2vt(1 - z),x)$, and the probability for k substitutions is given by the coefficient of z^k in $f^*(2vt(1 - z),x)$. For given pairs of DNA sequences at unlinked loci, the probability distribution of k is used to make the maximum likelihood (ML) estimates of $2vx$ and $4Mv$, population parameters scaled by the substitution rate. It is however necessary to take into account the possibility that v may differ from sequence to sequence because of different lengths (l). For this reason, $2vx$ and $4Mv$ are rewritten as $2l\mu x$ and $4Ml\mu$, respectively. Provided that l is given for each DNA sequence pair and that μ takes a constant value irrespective of sequences, the ML method estimates $2\mu x$ and $4M\mu$. For a known μ value, species divergence time x and ancestral population size M can be inferred.

The above ML method was applied to 13 DNA sequences of the human-chimpanzee pair, 7 of the human-gorilla pair, and 7 of the chimpanzee-gorilla pair (114 and references therein; Table 1). In this method, the nucleotide substitution rate is measured in units of generations, but it is often convenient to measure it in units of years. If generation time is g yr, the per-generation substitution rate μ becomes $g\mu_0$ where μ_0 is the per-year substitution rate. (In what follows, it is assumed that $g = 20$ yr in the Pleistocene, but $g = 10$–15 yr in the Pliocene and Miocene.) However, since the species divergence time (x) is also measured in units of generations, $\mu x = (\mu_0 g)(x_0/g) = \mu_0 x_0$ where x_0

Table 1 The ML estimates of species divergence time x and ancestral population size M, both of which are scaled by the synonymous substitution rate μ ($= 10^{-9}$; see 89).

	Human vs. chimp	Human vs. gorilla	Chimp vs. gorilla
$2\mu x$	0.0092	0.0140	0.0148
$4M\mu$	0.005	0.0046	0.0025

is the species divergence time measured in units of years. The ML estimate of $2\mu x$ between humans and chimpanzees is 0.0092. Hence, if $\mu_0 = 10^{-9}$ (89), the estimated divergence time of x_0 becomes 4.6 my. Likewise, the ML estimate of $2\mu x$ for the human-gorilla pair is 0.0140 ($x_0 = 7.0$ my), and that for the chimpanzee-gorilla pair is 0.0148 ($x_0 = 7.4$ my). The values of 4.6 my and 7.0–7.4 my are slightly different from those obtained from the hominoid mtDNA sequences, and they are smaller than those obtained from the DNA-DNA hybridization. In any case, current DNA sequence data favor the hypothesis that gorillas diverged prior to the divergence of humans from chimpanzees (114), a relationship denoted as ((H,C),G). The alternatives are as ((H,G),C) and ((C,G),H).

Rogers (81) recently enumerated published phylogenetic analyses of DNA sequences at seven nuclear loci and found that four support ((H,C),G), one is ambiguous, and two support ((C,G),H). These different gene trees may result from the ancestral polymorphism. Applying the formula for the probability with which gene and species (population) trees are concordant or discordant to each other (72), Rogers suggested that humans, chimpanzees, and gorillas diverged almost simultaneously. However, there seem to be some statistical problems. First, the formula (72) can be used only when the exact genealogy of three orthologous sequences is known. In reality, the genealogy must be inferred from nucleotide substitutions and so is subject to stochastic errors caused by factors other than random drift in the ancestral population. One such error is associated with nucleotide substitutions. A related second point is that information on each locus may not be equally informative about the branching order, because the number of sites examined and the number of substitutions both differ greatly from locus to locus. For example, the sequence used of the β-globin gene cluster is 11,855 bp long (4, 58), while the sequence of the *HOX2B* gene is only 238 bp long (84). Third, internal repeated sequences in involucrin (23, 24) used in the analysis might make their alignment difficult and unreliable (65). Thus, the proposed trifurcation of humans, chimpanzees, and gorillas (81) must be viewed with caution. Furthermore, if the three species indeed trifurcated, gene trees for ((H,C),G), ((H,G),C), and ((C,G),H) should occur with equal probability. The absence of DNA sequence data for ((H,G),C) implies that the three species did not diverge simultaneously. The reason for discordant gene trees is provided below, in light of the estimated effective size (*M*) of the ancestral population.

Early Demographic History of Human Lineage

The ML estimate of $4M\mu$ for the human-chimpanzee comparison is 0.005 (114). For $\mu_0 = 10^{-9}$ and $g = 10–15$ yr, $\mu = \mu_0 g = 1–1.5 \times 10^{-8}$ per generation. The estimated value of M is in the range of $0.8–1.3 \times 10^5$. A similar value $(0.8–1.2 \times 10^5)$ is obtained for the human-gorilla pair. On the other hand, the M value $(0.4–0.6 \times 10^5)$ estimated for the chimpanzee-gorilla pair is smaller than that for the human-chimpanzee pair, but the difference is not significant. Thus, it appears that M is of the order of 10^5 prior to the divergence between humans and chimpanzees. The application of the present method is not limited, at least in principle, to the comparison of humans and African apes. If DNA sequences of Old or New World monkeys are available, it is possible to estimate the effective size of the human lineage in the early Miocene and Oligocene or even in the Eocene (66). Although there are not enough such data to compare with those of humans, an alternative way to estimate the M is to use loci or regions at which alleles have been maintained in the population for a long time. An example is alleles at functional *MHC* loci. Major allelic lineages at the class II *DRB1* locus were established before the divergence between Hominoidea and Old World monkeys (54–55). Applying Equations 8 and 9 to such extraordinary *MHC* polymorphism allows us to infer the effective population size over some tens of my. The effective size is again of the order of 10^5 (Table 2), a conclusion that does not change even when the population is subdivided (the effect is weaker on long-persisting allelic lineages than on short-lived neutral lineages). Hence, both the ancestral neutral polymorphism and the *MHC* polymorphism under balancing selection provide a consistent estimate of M in the Miocene and Pliocene. It is about ten times

Table 2 Estimates of population parameters $2N_e s$ and $N_e v^a$ in the symmetric balancing selection model for five human *MHC* loci.

	A	*B*	*C*	*DRB1*	*DQB1*
$2N_e s$	2371	6301	674	2929	2320
$N_e v$	0.29	0.09	0.36	0.04	0.13
N_e	1.5×10^5	4.5×10^4	1.8×10^5	6.9×10^4	2.2×10^5
Mean N_e			1.3×10^5		

[a] The v is the per-generation rate of nonsynonymous substitutions in the peptide-binding region of MHC molecules (55): For $g = 15$, $v = 2 \times 10^{-6}$ at the class I loci (*A*, *B*, *C*) locus and 6×10^{-7} at the class II (*CRB1* and *DQB1*) locus.

larger than that in extant primates (73). The lineage leading to humans thus appears to have formed a fairly large population toward the end of the Miocene, although the global temperature dropped abruptly and there were widespread extinctions of primates throughout Eurasia (59, 60).

The nuclear DNA sequences indicate $y_0 = 7.2 - 4.6 = 2.6$ my differences between the divergences of gorillas and chimpanzees from the human lineage. For $g = 15$ yr, $y_0 = 2.6$ my amount to $y = 1.7 \times 10^5$ generations. We can now evaluate the discordance probability between species and gene trees. The probability (72) is given by $(2/3)\exp\{-y/(2N_e)\}$. The value is 0.28 for $y = 1.7 \times 10^5$ and $N_e = 10^5$. In other words, among seven nuclear loci examined for humans, chimpanzees, and gorillas (81), in about two cases gene trees are discordant to ((H,C),G). In fact, besides a dubious case of involucrin (23), the X-chromosomal pseudoautosomal boundary (27) supports ((C,G),H), and the immunoglobulin α locus (49) prefers none. If N_e were as small as 10^4, the discordance probability is almost nil, and all gene trees would have shown an identical phylogenetic relationship. In the case of mtDNA (44), $y_0 = 1.7$ my and $y = 1.1 \times 10^5$ generations, but since $2N_e$ must be replaced by the number of breeding females, the discordance probability becomes 0.074 for $2N_e = 5 \times 10^4$. If $y_0 = 2.6$ my as indicated by the nuclear DNA sequences, the probability is 0.022. In either case, the mtDNA tree for hominoids (43, 44) is likely identical to the species relatedness.

HUMAN POPULATION IN THE PLEISTOCENE

The first hominid species was Australopithecine, which lived from 4.4 mya (125) to 3 mya. If hominids existed as early as 4.4 mya, the gene divergence time in the human-chimpanzee pair must be older. All but one (39) dating of the oldest hominid inferred from gene genealogies meet this paleontological criterion (4, 14, 43, 44, 58, 65, 79, 88, 94, 95). From pelvic and lower limb material, some paleontologists believe *Australopithecus afarensis* to be an effective, well-adapted biped, but others say that it spent much of its time in the trees. However, the brain size of *A. afarensis* is 400 cm^3 to 500 cm^3, only a little larger than that of chimpanzees and nearly the same as that of gorillas. Other characters such as the rib cage are also ape-like.

Homo habilis emerged from one of the Australopithecines, some 2.5 mya (91, 130, 131). The genus *Homo* is traditionally associated with brain enlargement and acquisition of culture (130). The brain size of *H. habilis* ranges from 650 cm^3 to 800 cm^3. Whether *H. habilis* is the oldest species in the genus *Homo* is a matter of debate. Some paleontologists think that it differs in too few features from *Australopithecus*, whereas others think that some fossils of it are indistinguishable from those of a more advanced species, *H. erectus* (e.g. 96 for review). The debate concerning *H. habilis* thus centers on the definition of *Homo* and the morphological heterogeneity of *H. habilis*. Despite the lack of consensus about the lineage of *H. habilis*, the genus was at crucial stage in the development of human characters, especially the enlarged brain. Infants had to be born at an earlier stage of mental and physiological maturity. According to one popular view by Glynn Isaac (59 for review), females nurturing infants and children collected plant foods and males collected meat. These foods were brought to living sites, processed by stone tools, and shared in the group. This interdependence within groups acted as an agent of natural selection for social, intellectual, and communication skills. The period of *H. habilis* was short; it became extinct by 1.5 mya.

A more *Homo*-like species (*H. erectus*) emerged before 1.6 mya and migrated out of Africa more than 1 mya (103–105). The first migration might be as old as 1.8 mya, as suggested by the recent redating of fossils in Java, Indonesia (106). *Homo erectus* was the first hominid to use fire and live outside Africa. Brain size was about 80% of today's average of 1400 cm^3. Regional populations became established throughout Eurasia, from some of which Neanderthals differentiated in Europe and Western Asia. The rise of the Neanderthals was probably earlier than 0.15 mya, but they disappeared by 35,000 ya (70, 105). Because of this recent extinction, some anthropologists believe that Neanderthals cannot be the direct ancestor of *H. sapiens*. Although Neanderthals differed from *H. sapiens* in several striking ways, their brain size was as large as or slightly larger than 1400 cm^3. The skeleton was much more robust than that of modern humans, and both males and females are presumed to have been extremely muscled and strong. If Neanderthals were our ancestor, these characters must have been transformed in a very short period of time. But the term *H. sapiens* itself is ambiguous and controversial. One proposal defines *H. sapiens* based on shared derived characters, including relatively gracile skeleton, voluminous cranium, and orthognathous face (straight jaws). Novelties in ontogeny and behavioral and ecological adaptations are also suggested. *Homo sapiens* is divided into two evolutionary grades: basic or archaic *H. sapiens* and fully modern humans (*H. s. sapiens*). Archaic *H. sapiens* appears to have evolved more than 0.1 mya, and the fossil of the earliest *H. s. sapiens* is dated at 35,000 to 40,000 ya. However, there are not enough fossil records to answer when, where, and how *H. sapiens* emerged (45, 69, 92, 96, 100, 103, 104, 117, 127–129).

Hypotheses on the Origin of Homo sapiens

Among many hypotheses for the origin of *H. sapiens*, three are popular. One is the candelabra hypothesis, which envisages a single population of *H. erectus* that lived in Africa more than 1 mya, spread to Eurasia, and then subsequently local populations became completely isolated from each other (59 for review). The acquisition of modern humanity (e.g. as regards posture, range of move-

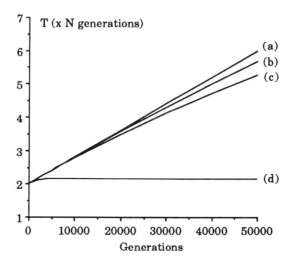

Figure 2   Mean coalescence time T (ordinate) against time x of population structure change (abscissa). (a) $m = r = 0$, (b) $Nm = 0.2$, $r = 0$, (c) $Nm = 2$, $r = 0$, and (d) $m = 0$, $Nr = 10$. $N = M = 10^4$. T can be as small as $2N$ only for (d).

ments, manipulation skill, and speech) must have occurred independently and in parallel, which means that *H. sapiens* had multiple origins, with each population having undergone the transition from *H. erectus* to *H. sapiens*. To examine this hypothesis quantitatively, we assume $x = 10^6/g = 5 \times 10^4$ for $g = 20$ yr as well as $r = m = 0$ in Equation 10 in the past 1 my. The situation is the same as that for different species except that the present value of x is one fifth of that for the human-chimpanzee pair. The mean coalescence time becomes $T_0 = 2N + 2(M-N)e^{-x/(2N)}$ for $P(0) = 1$ or $T_1 = x + 2M$ for $Q(0) = 1$. Since the ancestral population is assumed to have existed more than 5×10^4 generations ago, T_1 should always be longer than this time span. Because of this, T and F_{ST} values are expected also to be large (Figure 2 and 3). These are fairly strong predictions, and the candelabra hypothesis is rejected if there is a single case in which the observed T_1 is sufficiently shorter than $x = 5 \times 10^4$ generations.

In two regions is the estimated sequence-divergence time nearly or over 1 my: the intergenic region (58, 65) of the $\phi\eta$-β-globin cluster and the locus coding apolipoprotein C-II alleles (134). However, the sequence-divergence time estimated for the intergenic region may suffer from the enhanced substitution rate owing to rich CpG doublets (65, 79). In the case of apolipoprotein C-II alleles, the divergence time may not greatly exceed 1 my (112, 134). Although these divergence times are consistent with the candelabra hypothesis

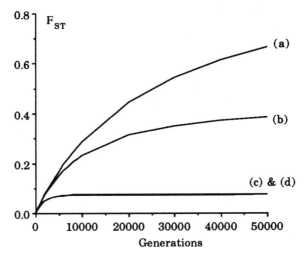

Figure 3 F_{ST} values (ordinate) against time x of population structure change (abscissa). (a) $m = r$ $= 0$, (b) $Nm = 0.2$, $r = 0$, (c) $Nm = 2$, $r = 0$, and (d) $m = 0$, $Nr = 10$. $N = M = 10^4$. Migration and extinction/restoration have similar effects on F_{ST}. The identity between (c) and (d) is expected from Equation 7b.

and its modified version (see below), the consistency cannot be taken as evidence for any hypothesis. By contrast, mtDNA is an excellent material to test the validity of the candelabra hypothesis. Because of its maternal inheritance and haploid nature, the effective population size of mtDNA is about one fourth of that for nuclear DNA, and so is the expected coalescence time. Nevertheless, the candelabra hypothesis predicts that T_1 cannot be shorter than divergence time x between local populations. Contrary to this prediction, all studies of human mtDNAs (e.g. 15, 38, 44) suggest that the coalescence time is shorter than 0.2 my (10^4 generations), definitely much shorter than 1 my (5×10^4 generations). In other words, the mtDNA diversity in the current human population was generated during the late Pleistocene. In a population of 5000 breeding females (73), the earliest sequence divergence time is expected to be 10,000 ± 5350 generations for a large sample (72, 107), and the probability that there was more than one distinct mtDNA lineage 5×10^4 generations ago is exceedingly small (116). Hence, the mtDNA genealogy clearly is contrary to the candelabra hypothesis.

A modified version of the candelabra hypothesis —the multiregional hypothesis (117, 128, 129)—holds the view that regional populations differentiated from each other, each subsequently evolving to *H. sapiens* more or less independently, but gene exchange occurred among local populations. The hypothesis is modeled by Figure 1 with $r = 0$ and $m > 0$ ($b = m^*$). With a limited amount of gene exchanges, the multiregional hypothesis is essentially the same as the candelabra hypothesis. No doubt, the human population has not been mating at random, and local populations have been mixed by migration to some extent. A problem in testing the multiregional hypothesis is that it does not specify the extent of gene exchanges. As a consequence, the argument for and against the hypothesis becomes inevitably qualitative (117, 127) and sometimes subjective. However, the expected N_e or $T/2$ must be larger than Nc or $x = 5 \times 10^4$, whichever is smaller (Equations 4 and 5). This prediction will be used later to show a difficulty in the multiregional hypothesis. The expected T and F_{ST} for various values of m, $c = 5$, and $N = M = 10^4$ are presented in Figure 2 and 3.

The Noah's Ark hypothesis (45) assumes that *H. sapiens* emerged at a certain time of the late Pleistocene (about 0.15 mya to 10,000 ya), replaced *H. erectus*, and spread over the world. Most controversial is the postulated rapid replacement of *H. erectus* by *H. sapiens*, without any mixing of their genes. Thus, the Noah's Ark hypothesis takes into account extinction of the entire population of *H. erectus*, but this process is not the same as that of extinction/restoration in Figure 1. For them to be similar it is necessary to allow extinction/restoration among local populations of *H. sapiens* as well. Only then is it possible that the coalescence time can be as short as $2N$ generations (Figure 2). Under the Noah's Ark hypothesis, it may also be claimed that Africans should be most variable (71) and genetic differentiation among major ethnic groups should be small, based on the assumption that the original population retained most of genetic diversity and newly founded populations inherited only part of it. While this assumption may be reasonable, the hypothesis itself does not specify source region of the original population, population sizes, or gene exchanges. Actually, human paleontologists have not specified the source region of the original population or placed alternatively in Europe, China, the Near East, Western Asia, sub-Saharan Africa, or Australia (e.g. 60). In this regard, the African origin hypothesis is more specific than the Noah's Ark hypothesis.

Late Demographic History of Human Lineage

The coalescence time of neutral genes within a species is on the order of the effective population size (N_e): the estimated N_e is automatically restricted to the demographic history of more or less the past N_e generations. This holds true irrespective of the population structure when N_e is defined from the mean coalescence time (T).

For multiple sequence data at a locus, Felsenstein (28) and Fu & Li (31) developed ML methods for estimating N_e, assuming randomly mating population and requiring a large sample of DNA sequences at a locus. At present, few such loci have been examined. In contrast, there are 49 loci at each of which a pair of nucleotide sequences is available (63, 114). In a panmictic population of effective size N_e, the probability for the number of substitutions between two sequences is geometric with mean $4N_e\mu$ (123). This probability distribution may be used to make the ML estimate from 49 pairs of DNA sequences (114). The ML estimate of $4N_e\mu$ is 0.00078. In a geographically structured population, the ML function can be derived from Equation 2, which is not geometric. Fortunately, however, the ML estimate is close to that obtained from the mean $4N_e\mu$ when the formula of N_e appropriate to the multiregional hypothesis is used (see also 77). This argues for $4N_e\mu = 0.00078$ even under the multiregional hypothesis. For $g = 20$ and $\mu_0 = 10^{-9}$, N_e is about 10^4, the 90% confidence limit of N_e ranging from 5,600 to 15,000. The value of $N_e = 10^4$ thus estimated and the expected coalescence time ($T = 2N_e$) both are significantly smaller than 10^5. The coalescence time amounts to 0.4 my. If more genes are sampled at a locus, the coalescence time may become twice as large, or 0.8 my. This expectation agrees with the β-globin sequence divergence within humans (32, 122).

As mentioned, the multiregional hypothesis predicts that N_e is always larger than the total number of breeding individuals (Nc) in the entire population. The estimated value of N_e imposes the boundary that Nc must have been less than 10^4 throughout the Pleistocene. This conclusion flies in face of the fact that the genus *Homo* has, for approximately 1 my, occupied three continents. If 5000 breeding couples were distributed uniformly in Africa (30,000,000 km^2), Asia (41,440,000 km^2), and Europe (9,723,000 km^2), each couple would occupy an area of 16,000 km^2. There would be only 22 couples in Germany and 35 in France. The distance between nearest neighboring couples would be 127 km. If people lived in bands, the distance would become larger. In hunters and gatherers of present-day Africa, the population density is about 0.2/km^2 (17). The mean distance between father-offspring and mother-offspring in northern Italy is estimated as 13 km and 8 km, respectively (17). If 10 km is the average migration distance per generation, the multiregional model requires that our ancestors could migrate over distances ten times longer than today's people do. The same difficulty is applied to the Noah's Ark hypothesis if *H. sapiens* populations are assumed to have been stable since their establishment and there was no population explosion in the late Pleistocene. Furthermore, if *H. erectus* populations died out not because of *H. sapiens* but due to other causes, why did the same thing not happen with then-contemporaneous *H. sapiens* populations? It seems likely that *H. sapiens* populations were also liable to extinction/restoration since their emergence.

Population Differentiation

The small effective size of the *H. sapiens* population due in part to the small extent of local differentiation is also suggested by other genetic studies. Elec-

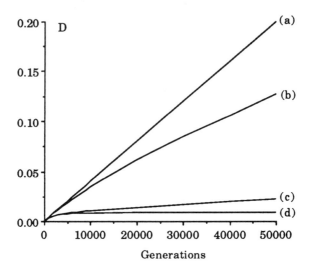

Figure 4   Genetic distance D (ordinate) against time x of population structure change (abscissa). (a) $m = r = 0$, (b) $Nm = 0.2$, $r = 0$, (c) $Nm = 2$, $r = 0$, and (d) $m = 0$, $Nr = 10$. $N = 10$. $N = M = 10^4$. D levels off even under a small extent of gene migration ($Nm = 0.2$).

trophoresis can be used to detect protein variation in local populations and to construct population trees (16–18, 74–76). Monomorphic loci do not contribute to discriminating local populations. This fact and others have resulted in an undesirable bias in the current data set of gene frequencies. The estimated heterozygosity $H = 1-J$ in the human population has increased from an earlier estimate of about 10–14% to a current one of 30%. The number of polymorphic loci now known is more than 200, or more than 60% of loci known (83). No other vertebrate species has the proportion of polymorphic loci over 60% and the heterozygosity more than 30% (73). Such biased data prohibit the estimation of population divergence times. To date population divergences, we must use a random set of loci, including monomorphic loci (76). The loci earlier examined appear to be more or less random, so 10% to 14% of heterozygosity observed (74) may be regarded as the unbiased extent in humans. Despite this relatively high heterozygosity, the estimated effective size is 10^4 (73, 75).

The number of loci in the unbiased data set is not large enough to estimate accurately the population relatedness as well as the divergence times (74, 76). Nevertheless, there is a suggestion that the earliest divergence among various ethnic groups occurred between African and Asian populations (75). The genetic distance (D) is 0.024 over 85 protein loci and blood groups. As expected in an ideal situation, if D is $2vx$ where x is the population divergence time, x can be estimated. For $D = 0.024$ and $v = 2 \times 10^{-6}$ per locus per

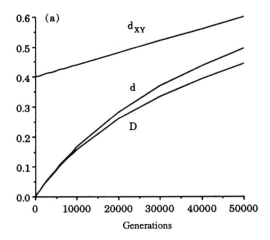

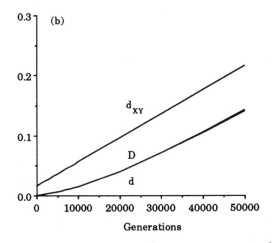

Figure 5   Net nucleotide differences d, d_{XY}, and D for $m = r = 0$, $v = 2 \times 10^{-6}$, and $N = 10^4$ as a function of population divergence time x (abscissa). To show how d and D depend on the size (M) of an ancestral population, M is set as $5N$ in (a) and as $N/5$ in (b). The substitution rate μ pertinent to d and d_{XY} is scaled by a factor of v/μ. If $\mu = 2 \times 10^{-8}$ per generation, the ratio of $v/\mu = 100$. This scaling demonstrates that d and D take very similar values.

generation, $x = 6000$ generations, or 0.12 my for $g = 20$ yr. This estimate is certainly consistent with the Noah's Ark hypothesis with $m = r = 0$. However, whether the same information can invalidate the multiregional hypothesis is a different matter. There is no guarantee for the validity of $D = 2vx$ under the effect of gene exchanges between local populations (Figure 4). In the presence of gene exchanges, D remains small even for long-term population subdivision. Accordingly, F_{ST} can be also small (e.g. 0.1 for proteins, blood groups, and DNA markers). Moreover, both D and d (or d_{XY}) for DNA sequences are sensitive to the assumption that genetic variation of the ancestral population can be inferred from that of current populations. When ancestral population size M is larger than current local population size N, the use of $D = 2vx$ and $d = 2\mu x$ overestimates x, whereas in the opposite situation, it underestimates x (Figure 5). The bias is hopelessly large, particularly when $M > N$, and makes it almost impossible to estimate population divergence time x.

In relation to the greatest diversity among Africans, as expected in the African origin hypothesis, the extent of protein polymorphism differs only slightly among major ethnic groups (75). The H value among Africans over 82 protein loci is about 13%, whereas among Caucasians over 121 protein loci and Asians over 73 loci it is 15% and 16%, respectively. The H for 20 to 70 blood groups is highest among Asians (20%) and lowest among Caucasians (11%). Most loci do not exhibit the significantly high diversity for the African population, including the latest data (76). It is likely that the diversity at nuclear loci is ancient, was passed on to local populations in the late Pleistocene, and therefore is not necessarily highest among Africans. In this respect, MHC loci provide more convincing evidence that the polymorphism was generated much earlier than the lifetime of hominids (54; The 11th International Histocompatibility Workshop 1991). A large number of alleles at MHC loci are detected by immunological methods even in small ethnic groups: In each ethnic group, the range in number is 6–24 at the A locus, 16–43 at the B locus, 6–13 at the C locus, 11–23 at the DRB1 locus, and 2–11 at the DQB1 locus. More than half the alleles are shared by the three major ethnic groups: 20 of 37 A alleles, 35 of 69 B alleles, 12 of 18 C alleles, 18 of 32 DRB1 alleles, and 10 of 12 DQB1 alleles. The H at each of all five loci is also higher than 80%–90%, and little differentiation is found among three major ethnic groups ($F_{ST} = 0.063$ at the A locus, 0.047 at the B locus, 0.053 at the C locus, 0.057 at the DRB1 locus, and 0.106 at the DQB1 locus).

There is a large body of mtDNA sequence data for the major ethnic groups (2, 5, 10, 12, 22, 80, 101, 102, 126). The per-site sequence divergence based on the restriction enzyme method is 0.47% among Africans, 0.35% among Asians, 0.25% among Australians and New Guineans, and 0.23% among Caucasians (15, 48, 71). The D-loop diversity is much higher (37, 41, 42, 57, 120, 121), with Africans (2.32%) more variable than Asians (1.25%) and

Europeans (0.95%). These observations for mtDNA appear to disagree with those for nuclear loci. One reason may be that the evolutionary timescale is different between the two genetic systems. Little mtDNA diversity was transmitted from the ancestral population that existed earlier than the late Pleistocene, and most has been generated in the past 0.1 my or so (38, 44). The highest mtDNA diversity among Africans may therefore reflect the relatively short demographic history of *H. sapiens.* However, the connection between the greatest diversity and the place of the original population is not straightforward. If some immigrant populations in Eurasia had expanded as early as some 0.1 mya, they could have been most variable in mtDNA.

The microsatellite DNAs scattered throughout chromosomes are more polymorphic than ordinary nuclear loci (7) and were recently used to construct a genealogical tree of human individuals (8). The polymorphism of microsatellites is produced by gain and loss of repeats owing to DNA slippage or unequal crossing over. Because of the high mutation rate, microsatellites may have a high resolution in phylogenetically discriminating closely related individuals. The heterozygosity averaged over some 30 microsatellite loci is 81% among Africans, 73% among Europeans, and 69% among Asians. Although the bulk of the variation occurs within groups, discrete clustering of ethnic groups in the tree indicates genetic differentiation among groups (8). The tree constructed for individuals is similar to a population tree because microsatellite loci dispersed throughout the genome are not linked and scrambled by recombination. This may explain why the genetic distances for pairs of individuals within a group are so similar to each other and at the same time branch lengths within groups are nearly as deep as those between groups. Slightly long branches between groups suggest that ethnic groups have been isolated from each other to some extent. A problem concerning the tree is the lack of time scale, without which it is difficult to choose among competing hypotheses. Although the rate of evolution of microsatellites is high, the present estimate ranges too widely from 10^{-2} to 10^{-5} per locus per generation. Moreover, even if a more accurate rate becomes available, the dating of population splitting based on distance measures will remain difficult in the presence of migration or extinction/restoration of local populations. This is the same difficulty mentioned in relation to genetic distance D for protein variation. Dating population divergences is still a challenging task to both experimentalists and theoreticians.

Changing Population Size

In relation to the Noah's Ark hypothesis, it is interesting to examine the possibility of population expansion in the late Pleistocene: It is unlikely that the population size of any species has remained the same for a long period of evolutionary time. If N in a panmictic population changes with time, we may express the probability of no coalescence until x generations ago as

$$P(x) = \exp\left\{-\frac{x}{2N_e(x)}\right\},$$ 11.

where $N_e(x)$ is the harmonic mean of time-dependent $N(t)$, or $1/N_e(x) = \frac{1}{x}\int_0^x \frac{dt}{N(t)}$. If the population size has been changing sufficiently rapidly relative to the coalescence time, $N_e(x)$ would be independent of x. Under this circumstance, the coalescence process becomes the same as that in a panmictic population of N_e. Of more interest may be the opposite situation in which $N(t)$ fluctuates in about the same time scale as the coalescence time. Here, we consider a simple model of changing N. We assume that there was a bottleneck between x and $x + y$ generations ago and that the $N(t)$ in the three phases is constant: N for $t < x$, M $(<N)$ for $x \le t < x + y$, and L for $t \ge x + y$. This is an extension of one stage bottleneck in Equation 10. It is straightforward to compute the Laplace transform of coalescence probability and the mean time as

$$f^*(p,x,y) = \frac{1-e^{(px+x*)}}{1+2Np} + \frac{\{1-e^{-(py+y*)}\}e^{-(px+x*)}}{1+2Mp} + \frac{e^{-\{x*+y*+p(x+y)\}}}{1+2Lp}$$ 12a.

$$T = 2N[1-e^{-x*} + M^*e^{-x*}(1-e^{-y*}) + L^*e^{-(x*+y*)}],$$ 12b.

where $x^* = x/(2N)$, $y^* = y/(2M)$, $M^* = M/N$, and $L^* = L/N$. Each term on the right hand side of Equation 12b corresponds to coalescence time, provided that coalescence occurred in the first, second, or third phase of changing $N(t)$. For the bottleneck effect to be manifest, it is required that $M < N(M^* < 1)$, $x < 2N$ $(x^* < 1)$, and $y > 2M(y^* > 1)$. The second condition implies that coalescence is likely to occur before x: If a bottleneck occurred too remotely, the current population has returned to equilibrium, and the coalescence process is unaffected. The third condition makes it likely that coalescence occurs during the bottleneck phase. For example, if $x^* = 0.5$ and $y^* = 2$, then $T = 2N(0.39 + 0.52M^* + 0.08L^*)$. For T to be much smaller than this, smaller x^* and larger y^* are required (Figure 6). In an extreme situation of population expansion after a severe bottleneck, y^* becomes indefinitely large and x^* becomes indefinitely small so that we have $T = x$. The coalescence time becomes nearly equal to the time period of population expansion, a necessary condition for a star-like genealogy (simultaneous divergences of multiple ancestral lineages). However, for i ancestral lineages of j genes $(j > 2)$ to remain distinct for x generations, it is also necessary that $i(i - 1)x^*/2$ is much smaller than 1 (99,

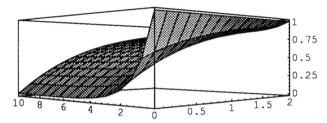

Figure 6   Mean coalescene time *T* relative to 2*N* (*z* axe ranging from 0 to 1). The value ranging from 0 to 2 stands for *x** = *x*/(2*N*) and the value ranging from 0 from 10 stands for *y** = *y*/(2*N*). The graph is based on Equation 6 with *L** = *L*/*N* = 1 and *M** = *M*/*N* = 0.01. The bottleneck effect (the small value of *T*/(2*N*) is manifest when *x** is small and *y** is large.

124). When *i* = 20, this requires that *x** is much smaller than 0.005 (*x* <<
0.01*N*). Also, for a star-like genealogy, the period of the bottleneck phase must
be shorter than that of the expansion phase, or *x* > *y*: Otherwise, it may take
time comparable to *x* for genes to coalesce during the bottleneck phase. Together with *y** >> 1 or *x* >> 2*M*, a star-like genealogy is expected when 2*N*
>> *x* > y >> 2*M*. For *x* = 5000 or 0.1 my with generation time being 20 yr, *y*
and *M* must be much smaller than 1000 and *N* must be much larger than 5 ×
10^5. In any case, only when there is a substantial discrepancy between *N* and
M, can we expect a star-like genealogy with 20 major branches.

Rogers & Harpending (82) re-derived the formula of Li (62) for the
so-called pairwise distribution of the number (*k*) of nucleotide substitutions
between two sequences. Strictly speaking, the formula can be applied to pairs
of sequences at unlinked loci, but it may approximate the actual genealogy
for a large sample of mtDNA sequences when the genealogy resembles a
star phylogeny (99). It should be noted, however, that even under an abrupt
population expansion, the pairwise distribution is not exactly a Poisson
distribution (simulation data not shown). Because the *k* values in all pairwise
comparisons are necessarily correlated by descent, the variance of *k* tends to
be smaller than the mean. While the mtDNA genealogy may signify the
population expansion, the sequence data from 49 nuclear loci did not show
such evidence (114). This may result from the low substitution rate, short
lengths, and small sample size of nuclear sequences. With the substitution
rate of 10^{-9} per site per year and 1-kb DNA sequences, one would not expect
to find even a single substitution between two sequences which diverged 0.1
mya. DNA sequences at nuclear loci are thus unsuited to demonstrate a recent
population expansion. On the other hand, it is possible that mtDNA sequences
can inscribe such a short period of expansion owing to the nearly 40 times
higher substitution rate (44).

DISCUSSION AND CONCLUSIONS

The phylogenetic relationship among humans, chimpanzees, and gorillas was analyzed based on various genetic data produced by immunological, electrophoretic, DNA-DNA hybridization, and DNA sequencing methods. The former two methods did not resolve the relationships, whereas the latter two generally suggested the human-chimpanzee clade. However, the clade is based entirely on individual gene trees, which are not identical to a species tree. Most tree-making methods based on DNA sequences reconstruct only individual gene trees and do not distinguish whether genes (or nucleotide sites) used are linked. Individual gene trees for unlinked loci may be discordant to each other, and this discordance can happen in principle for any species relatedness. The discordance among gene trees does not imply that the phylogenetic relationships of species are unresolved. A method was used to generate a rooted species tree without an out-group species (114). The application estimated the time difference between chimpanzee and gorilla divergences as 2.6 my and the effective size of the Pliocene hominoid as 10^5. These two figures show that the discordance probability among gene trees is 0.28, which is in remarkably good agreement with the observed proportion.

The effective population size in the human lineage does not appear to be constant over a long period of evolutionary time. It was rather large in the Pliocene and Miocene, but small in most of the Pleistocene. Analysis of mtDNA sequences (82) suggested that the population had begun to expand in the Late Pleistocene, although this conclusion is not supported by nuclear loci. A possible reason for the Pleistocene bottleneck may be that local human populations underwent frequent extinction/restoration because of increased dispersal and adverse environmental conditions in the Old World. However, since allelic lineages at some nuclear loci have been segregating in the population for more than 1 my, the bottleneck could not have been severe: either the reduced population size was not extremely small or the period the bottleneck lasted was short, or both.

As a caveat, the 14.3% heterozygosity of humans detected by electrophoresis is highest among primates—it is 2.1% in chimpanzees and 4.6% in gorillas—and higher than that in most vertebrates (13, 52, 73). This is puzzling in light of the fairly small extent of DNA polymorphism (63) and the small estimated value of N_e. If the protein polymorphism evolved as the neutral theory depicts (50), change in the degree of selective constraint or relaxation of negative selection pressure is the sole possibility. A question is whether or not the possibility of relaxation responsible for the protein polymorphism can be supported by paleo-anthropological data. It was emphasized that the human lineage has experienced dramatic changes in brain size, economic and social complexity, and culture over the past 2 my.

Although neither learning nor the mother-infant relationship is specific to human, the period of *H. habilis* was an incomparable stage of the development of humanity. In particular, the enlarged brain might have altered or even improved both internal and external environment. Therefore, it does not seem to be absurd to imagine that natural selection was relaxed and this relaxation permitted the incorporation of more amino acid replacement changes or electrophoretically detectable variants. In any case, further comparative study of DNA sequences between human and nonhuman primates will provide important information on the relationship between protein polymorphism and the history of humans in the Pleistocene.

Three hypotheses of the origin of *H. sapiens* were quantified to provide a common basis of arguments for and against them (6, 20, 26, 34, 45, 56, 61, 69, 86, 100, 103, 104, 117, 127, 134). One common feature about the hypotheses is that they are static with respect to population structure. This is a serious oversimplification to scrutinize the evolution of *H. sapiens*. The human population looks more unstable than hypotheses postulate. It is likely that local populations underwent repeated splits and reunions. Likewise, many local populations went to extinction, and the places they inhabited were reoccupied by entirely new comers. This can explain the rather small effective size or equivalently the rather short coalescence time estimated from genetic data (68, 113). Without extinction/restoration of local populations, the actual number of breeding individuals in the whole population is required to be no greater than 10^4.

Extinction at the population level is a general rule throughout the history of all organisms on the earth. Even *H. s. sapiens* populations could not escape this rule. Our ancestors tried to explore new habitats, but such explorations were unsuccessful in most cases. However, several ancestral populations were so successful that their descendants eventually conquered the globe. The close genealogical relationships among contemporary human individuals probably reflect such a dynamical process occurring until recently.

ACKNOWLEDGMENTS

This paper is dedicated to the memory of the great population geneticist Motoo Kimura, who died on 13 November 1994. Most work presented here has resulted from collaborations with Masatoshi Nei, Yoko Satta, Satoshi Horai, and Jan Klein, to all of whom I am grateful. I also thank Chris B. Stringer and John Wakeley for their comments on the early version of this paper.

Literature Cited

1. Andrews P. 1992. An ape from the south. *Nature* 356:106
2. Aquadro CF, Greenberg BD. 1983. Human mitochondrial DNA variation and evolution: analysis of nucleotide sequences of seven individuals. *Genetics* 103:287–312
3. Argaut C, Rigolet M, Eladari ME, Galibert F. 1991. Cloning and nucleotide sequence of the chimpanzee *c-myc* gene. *Gene* 97:231–37
4. Bailey WJ, Hayasaka K, Skinner CG, Kehoe S, Sieu LC, et al. 1992. Reexamination of the African hominoid trichotomy with additional sequences from the primate β-globin gene cluster. *Mol. Phylogenet. Evol.* 1:97–135
5. Ballinger SW, Schurr TG, Torroni A, Gan YY, Hodge JA, et al. 1992. Southern Asian mitochondrial DNA analysis reveals genetic continuity of ancient Mongoloid migrations. *Genetics* 130:139–52
6. Barinaga M. 1992. "African Eve" backers beat a retreat. *Science* 255:686–87
7. Bowcock AM, Osborne-Lawrence S, Barnes R, Chakravarti A, Washington A, Dunn C. 1993. Microsatellite polymorphism linkage map of human chromosome 13q. *Genomics* 15:376–86
8. Bowcock AM, Ruiz-Linares A, Tomfohrde J, Minch E, Kidd JR, Cavalli-Sforza LL. 1994. High resolution of human evolutionary trees with polymorphic microsatellites. *Nature* 368:455–57
9. Brenner S. 1991. Summary and concluding remarks. In *Evolution of Life: Fossils, Molecules, and Culture*, ed. S Osawa, T Honjo, pp. 391–413. Tokyo: Springer-Verlag
10. Brown WM. 1980. Polymorphism in mitochondrial DNA of humans as revealed by restriction endonuclease analysis. *Proc. Natl. Acad. Sci. USA* 77:3605–9
11. Brown WM, George M Jr, Wilson AC. 1979. Rapid evolution of animal mitochondrial DNA. *Proc. Natl. Acad. Sci. USA* 76:1967–71
12. Brown WM, Prager EM, Wang A, Wilson AC. 1982. Mitochondrial DNA sequences of primates: tempo and mode of evolution. *J. Mol. Evol.* 18:225–39
13. Bruce EJ, Ayala FJ. 1979. Phylogenetic relationships between man and the apes: electrophoretic evidence. *Evolution* 33:1040–56
14. Caccone A, Powell JR. 1989. DNA divergence among hominoids. *Evolution* 43:925–42
15. Cann RL, Stoneking M, Wilson AC. 1987. Mitochondrial DNA and human evolution. *Nature* 325:31–36
16. Cavalli-Sforza LL. 1991. Genes, peoples and languages. *Sci. Am.* 265:72–78
17. Cavalli-Sforza LL, Bodmer WF. 1971. *The Genetics of Human Populations.* San Francisco: WH Freeman
18. Cavalli-Sforza LL, Piazza A, Menozzi P, Mountain J. 1988. Reconstruction of human evolution: bringing together genetic, archaeological, and linguistic data. *Proc. Natl. Acad. Sci. USA* 85:6002–6
19. Conroy GC, Pickford M, Senut B, Couvering JV, Mein P. 1992. *Otavipithecus namibiensis*, first Miocene hominoid from southern Africa. *Nature* 356:144–48
20. Darlu P, Tassy P. 1987. Disputed African origin of human populations. *Nature* 329:111
21. Darwin C. 1872. *The Origin of Species by Means of Natural Selection.* London: John Murray
22. Denaro M, Blanc H, Johnson MJ, Chen KH, Wilmsen E, et al. 1981. Ethnic variation in *Hpa* I endonuclease cleavage patterns of human mitochondrial DNA. *Proc. Natl. Acad. Sci. USA* 78:5768–72
23. Djian P, Green H. 1989. Vectorial expansion of the involucrin gene and the relatedness of the hominoids. *Proc. Natl. Acad. Sci. USA* 86:8447–51
24. Djian P, Green H. 1989. The involucrin gene of the orangutan: generation of the large region as an evolutionary trend in the hominoids. *Mol. Biol. Evol.* 6:469–77
25. Easteal S. 1991. The relative rate of DNA evolution in primates. *Mol. Biol. Evol.* 8:115–27
26. Edelson E. 1991. Tracing human lineages. *MOSAIC* 22:56–63
27. Ellis N, Yen P, Neiswanger K, Shapiro LJ, Goodfellow PN. 1990. Evolution of the pseudoautosomal boundary in Old World monkeys and great apes. *Cell* 63:977–86
28. Felsenstein J. 1992. Estimating effective population size from samples of sequences: inefficiency of pairwise and segregation sites as compared to phylogenetic estimates. *Genet. Res.* 59:139–47
29. Ferris SD, Brown WM, Davidson WS, Wilson AC. 1981. Extensive polymorphism in the mitochondrial DNA of

apes. *Proc. Natl. Acad. Sci. USA* 78: 6319–23

30. Foran DR, Hixson JE, Brown WM. 1988. Comparison of ape and human sequences that regulate mitochondrial DNA transcription and D-loop DNA synthesis. *Nucleic Acids Res.* 16:5841–61

31. Fu YX, Li WH. 1993. Maximum likelihood estimation of population parameters. *Genetics* 134:1261–70

32. Fullerton SM, Harding RM, Boyce AJ, Clegg JB. 1994. Molecular and population genetic analysis of allelic sequence diversity at the human β-globin locus. *Proc. Natl. Acad. Sci. USA* 91:1805–9

33. Galili U, Swanson K. 1991. Gene sequences suggest inactivation of α-1,3-galactosyltransferase in catarrhines after the divergence of apes from monkeys. *Proc. Natl. Acad. Sci. USA* 88:7401–4

34. Gibbons A. 1992. Mitochondrial eve: wounded, but not dead yet. *Science* 257: 873–75

35. Goldman D, Giri PR, O'Brien SJ. 1987. A molecular phylogeny of the hominoid primates as indicated by two-dimensional protein electrophoresis. *Proc. Natl. Acad. Sci. USA* 84:3307–11

36. Gonzalez IL, Sylvester JE, Smith TF, Stambolian D, Schmickel RD. 1990. Ribosomal RNA gene sequences and hominoid phylogeny. *Mol. Biol. Evol.* 7:203–19

37. Greenberg BD, Newbold JE, Sugino A. 1983. Intraspecific nucleotide sequence variability surrounding the origin of replication in human mitochondrial DNA. *Gene* 21:33–49

38. Hasegawa M, Horai S. 1991. Time of the deepest root for polymorphism in human mitochondrial DNA. *J. Mol. Evol.* 32:37–42

39. Hasegawa M, Kishino H, Yono T. 1985. Dating of the human-ape splitting by a molecular clock of mitochondrial DNA. *J. Mol. Evol.* 22:160–74

40. Hixson JE, Brown WM. 1986. A comparison of the small ribosomal RNA genes from the mitochondrial DNA of the great apes and humans: sequence, structure, evolution, and phylogenetic implications. *Mol. Biol. Evol.* 3:1–8

41. Horai S, Hayasaka K. 1990. Intraspecific nucleotide sequence differences in the major noncoding region of human mitochondrial DNA. *Am. J. Hum. Genet.* 46:828–42

42. Horai S, Kondo R, Nakagawa-Hattori Y, Hayashi S, Sonoda S, Tajima K. 1993. Peopling of the Americas, founded by four major lineages of mi-

tochondrial DNA. *Mol. Biol. Evol.* 10: 23–47

43. Horai S, Satta Y, Hayasaka K, Kondo R, Inoue T, et al. 1992. Man's place in *Hominoidea* revealed by mitochondrial DNA genealogy. *J. Mol. Evol.* 35:32–43

44. Horai S, Hayasaka K, Kondo R, Tsugane K, Takahata N. 1995. The recent African origin of modern humans revealed by complete sequences of hominoid mitochondrial DNAs. *Proc. Natl. Acad. Sci. USA* 92:532–36

45. Howells WW. 1976. Explaining modern man: evolutionists versus migrationists. *J. Hum. Evol.* 5:477–95

46. Hudson RR. 1983. Testing the constant rate neutral allele model with protein sequence data. *Evolution* 37:203–17

47. Hudson RR. 1990. Gene genealogies and the coalescent process. In *Oxford Surveys in Evolutionary Biology,* ed. D Futuyma, J Antonovics, pp. 1–44. Oxford: Oxford Univ. Press

48. Johnson MJ, Wallace DC, Ferris SD, Rattazzi MC, Cavalli-Sforza LL. 1983. Radiation of human mitochondrial DNA types analyzed by restriction endonuclease cleavage patterns. *J. Mol. Evol.* 19:255–71

49. Kawamura S, Tanabe H, Watanabe Y, Kurosaki K, Saitou N, Ueda S. 1991. Evolutionary rate of immunoglobulin alpha noncoding region is greater in hominoids than in Old World monkeys. *Mol. Biol. Evol.* 8:743–52

50. Kimura M. 1968. Evolutionary rate at the molecular level. *Nature* 217:624–26

51. Kimura M. 1983. *The Neutral Theory of Molecular Evolution.* Cambridge: Cambridge Univ. Press

52. King MC, Wilson AC. 1975. Evolution at two levels in humans and chimpanzees. *Science* 188:107–16

53. Kingmann JFC. 1982. On the genealogy of large populations. *J. Appl. Prob.* 19A: 27–43

54. Klein J. 1986. *Natural History of the Major Histocompatibility Complex.* New York: John Wiley

55. Klein J, Satta Y, O'hUigin C, Takahata N. 1993. The molecular descent of the major histocompatibility complex. *Annu. Rev. Immunol.* 11:269–95

56. Klein J, Takahata N, Ayala FJ. 1993. *MHC* polymorphism and human origins. *Sci. Am.* 269:78–83

57. Kocher TD, Wilson AC. 1991. Sequence evolution of mitochondrial DNA in humans and chimpanzees: control region and a protein-coding region. In *Evolution of Life: Fossils, Molecules, and Culture*, ed. S Osawa, T Honjo, pp. 391–413. Tokyo: Springer-Verlag

58. Koop BF, Goodman M, Xu P, Chan K, Slightom JL. 1986. Primate η-globin DNA sequences and man's place among the great apes. *Nature* 319:234–38

59. Lewin R. 1984. *Human Evolution: An Illustrated Introduction.* New York: Freeman

60. Lewin R. 1988. *In the Age of Mankind.* Washington, DC: Smithsonian

61. Lewin R. 1991. *La naissance de l'anthropologie moléculaire. Recherche* 236:1242–51

62. Li WH. 1977. Distribution of nucleotide differences between two randomly chosen cistrons in a finite population. *Genetics* 85:331–37

63. Li WH, Sadler LA. 1991. Low nucleotide diversity in man. *Genetics* 129:513–23

64. Li WH, Tanimura M. 1987. The molecular clock runs more slowly in man than in apes and monkeys. *Nature* 326:93–96

65. Maeda N, Wu CI, Bliska J, Reneke J. 1988. Molecular evolution of intergenic DNA in higher primates: pattern of DNA changes, molecular clock, and evolution of repetitive sequences. *Mol. Biol. Evol.* 5:1–20

66. Martin RD. 1993. Primate origins: plugging the gaps. *Nature* 363:223–34

67. Marjoram P, Donnelly P. 1994. Pairwise comparisons of mitochondrial DNA sequences in subdivided populations and implications for early human evolution. *Genetics* 136:673–83

68. Maruyama T, Kimura M. 1980. Genetic variability and effective population size when local extinction and recolonization of subpopulations are frequent. *Proc. Natl. Acad. Sci. USA* 77:6710–14

69. Mellars P, Stringer CB. 1989. *The Human Revolution: Behavioral and Biological Perspectives on the Origin of Modern Humans.* Princeton: Princeton Univ. Press

70. Mercier N, Valladas H, Joron JL, Reyss JL, Léveque F, Vandermeersch B. 1991. Thermoluminescence dating of the late Neanderthal remains from Saint-Césaire. *Nature* 351:737–39

71. Merriwether DA., Clark AG, Ballinger SW, Schurr TG, Soodyall H, et al. 1991. The structure of human mitochondrial DNA variation. *J. Mol. Evol.* 33:543–55

72. Nei M. 1987. *Molecular Evolutionary Genetics.* New York: Columbia Univ. Press

73. Nei M, Graur D. 1984. Extent of protein polymorphism and the neutral mutation theory. *Evol. Biol.* 17:73–118

74. Nei M, Livshits G. 1989. Genetic relationships of Europeans, Asians and Africans and the origin of modern *Homo sapiens. Hum. Hered.* 39:276–81

75. Nei M, Roychoudhury AK. 1982. Genetic relationship and evolution of human races. *Evol. Biol.* 14:1–59

76. Nei M, Roychoudhury AK. 1993. Evolutionary relationships of human populations on a global scale. *Mol. Biol. Evol.* 10:927–43

77. Nei M, Takahata N. 1993. Effective population size, genetic diversity, and coalescence time in subdivided population. *J. Mol. Evol.* 37:240–44

78. Notohara M. 1990. The coalescent and the genealogical process in geographically structured population. *J. Math. Biol.* 29:59–75

79. Perrin-Pecontal P, Gouy M, Nigon VM, Trabuchet G. 1992. Evolution of the primate β-globin gene region: nucleotide sequence of the δ-β-globin intergenic region of gorilla and phylogenetic relationships between African apes and man. *J. Mol. Evol.* 34:17–30

80. Rienzo AD, Wilson AC. 1991. Branching pattern in the evolutionary tree for human mitochondrial DNA. *Proc. Natl. Acad. Sci. USA* 88:1597–601

81. Rogers J. 1993. The phylogenetic relationships among *Homo, Pan* and *Gorilla*: a population genetics perspective. *J. Hum. Evol.* 25:201–15

82. Rogers AR, Harpending HC. 1992. Population growth makes waves in the distribution of pairwise genetic differences. *Mol. Biol. Evol.* 9:552–69

83. Roychoudhury AK, Nei M. 1988. *Human Polymorphic Genes: World Distribution.* New York: Oxford Univ. Press

84. Ruano G, Rogers J, Ferguson-Smith AC, Kidd KK. 1992. DNA sequence polymorphism within hominoid species exceeds the number of phylogenetically informative characters for a *HOX2* locus. *Mol. Biol. Evol.* 9:575–86

85. Ruvolo M, Disotell TR, Allard MW, Brown WM, Honeycutt RL. 1991. Resolution of the African hominoid trichotomy by use of a mitochondrial gene sequence. *Proc. Natl. Acad. Sci. USA* 88:1570–74

85a. Ruvolo M, Pan D, Zehr T, Doldberg T, Disotell TR, von Dornum M. 1994. Gene trees and hominoid phylogeny. *Proc. Natl. Acad. Sci. USA* 91:8900–4

86. Saitou N, Omoto K. 1987. Time and place of human origins from mtDNA data. *Nature* 327:288

87. Sarich VM, Wilson AC. 1973. Generation time and genomic evolution in primates. *Science* 179:1144–47

88. Sarich VM, Wilson AC. 1967. Immu-

nological time scale for hominid evolution. *Science* 158:1200–3

89. Satta Y, O'hUigin C, Takahata N, Klein J. 1993. The synonymous substitution rate of the major histocompatibility complex loci in primates. *Proc. Natl. Acad. Sci. USA* 90:7480–84

90. Sawada I, Beal MP, Shen CJ, Chapman B, Wilson AC, Schmid C. 1983. Intergenic DNA sequences flanking the pseudo alpha globin genes of human and chimpanzee. *Nucleic Acids Res.* 11: 8087–101

91. Schrenk F, Bromage TG, Betzler CG, Ring U, Juwayeyi YM. 1993. Oldest *Homo* and Pliocene biogeography of the Malawi rift. *Nature* 365:833–36

92. Schwartz JH. 1984. Hominoid evolution: a review and a reassessment. *Curr. Anthropol.* 25:655–72

93. Seino S, Bell GI, Li WH. 1992. Sequences of primate insulin genes support the hypothesis of a slower rate of molecular evolution in humans and apes than in monkeys. *Mol. Biol. Evol.* 9: 193–203

94. Sibley CG, Ahlquist JE. 1984. The phylogeny of the hominoid primates, as indicated by DNA-DNA hybridization. *J. Mol. Evol.* 20:2–15

95. Sibley CG, Ahlquist JE. 1987. DNA hybridization evidence of hominoid phylogeny: results from an expanded data set. *J. Mol. Evol.* 26:99–121

96. Simons EL. 1989. Human origins. *Science* 245:1343–50

97. Slatkin M. 1990. Gene flow and the genetic structure of natural populations. In *Population Biology of Genes and Molecules*, ed. N Takahata, JF Crow, pp. 105–22. Tokyo:Baifukan

98. Slatkin M. 1991. Inbreeding coefficients and coalescence times. *Genet. Res.* 58: 167–75

99. Slatkin M, Hudson RR. 1991. Pairwise comparisons of mitochondrial DNA sequences in stable and exponentially growing populations. *Genetics* 129: 555–62

100. Smith FH, Spencer F. 1987. *The Origins of Modern Humans*. New York: Liss

101. Stoneking M, Bhatia K, Wilson AC. 1986. Rate of sequence divergence estimated from restriction maps of mitochondrial DNAs from Papua New Guinea. *Cold Spring Harbor Symp. Quant. Biol.*. LI:433–39

102. Stoneking M, Jorde LB, Bhatia K, Wilson AC. 1990. Geographic variation in human mitochondrial DNA from Papua New Guinea. *Genetics* 124:717–33

103. Stringer CB. 1990. The emergence of modern humans. *Sci. Am.* 263:68–75

104. Stringer CB, Andrews P. 1988. Genetic and fossil evidence for the origin of modern humans. *Science* 239:1263–68

105. Stringer CB, Grün R. 1991. Time for the last neanderthals. *Nature* 351:701–2

106. Swisher CC III, Curtis GH, Jacob T, Getty AG, Suprijo A, Widiasmoro. 1994. Age of the earliest known hominids in Java, Indonesia. *Science* 265: 1118–21

107. Tajima F. 1983. Evolutionary relationship of DNA sequences in finite populations. *Genetics* 105:437–60

108. Takahata N. 1986. An attempt to estimate the effective size of the ancestral species common to two extant species from which homologous genes are sequenced. *Genet. Res.* 48:187–90

109. Takahata N. 1988. The coalescent in two partially isolated diffusion populations. *Genet. Res.* 52:213–22

110. Takahata N. 1990. A simple genealogical structure of strongly balanced allelic lines and trans-species evolution of polymorphism. *Proc. Natl. Acad. Sci. USA* 87:2419–23

111. Takahata N. 1991. Genealogy of neutral genes and spreading of selected mutations in a geographically structured population. *Genetics* 129:585–95

112. Takahata N. 1993. Allelic genealogy and human evolution. *Mol. Biol. Evol.* 10:2–22

113. Takahata N. 1994. Repeated failures that led to the eventual success in human evolution. *Mol. Biol. Evol.* 11:803–5

114. Takahata N, Satta Y, Klein J. 1995. Divergence time and population size in the lineage leading to modern humans. *Theor. Pop. Biol.* In press

115. Takahata N, Slatkin M. 1990. Genealogy of neutral genes in two partially isolated populations. *Theor. Pop. Biol.* 38:331–50

116. Tavaré S. 1984. Lines-of-descent and genealogical processes, and their applications in population genetics models. *Theor. Pop. Biol.* 26:119–64

117. Thorne AG, Wolpoff MH. 1992. The multiregional evolution of humans. *Sci. Am.* 266:28–33

118. Ueda S, Watanabe Y, Saitou N, Omoto K, Hayashida H, et al. 1989. Nucleotide sequences of immunoglobulin-epsilon pseudogenes in man and apes and their phylogenetic relationships. *J. Mol. Biol.* 205:85–90

119. Ullrich A, Dull TJ, Gray A, Brosius J, Sures I. 1980. Genetic variation in the human insulin gene. *Science* 209:612–15

120. Vigilant L, Pennington R, Harpending H, Kocher TD, Wilson AC. 1989. Mi-

tochondrial DNA sequences in single hairs from a southern African population. *Proc. Natl. Acad. Sci. USA* 86: 9350–54

121. Vigilant L, Stoneking M, Harpending H, Hawkes K, Wilson AC. 1991. African populations and the evolution of human mitochondrial DNA. *Science* 253:1503–7

122. Wainscoat JS, Hill AVS, Boyce AL, Flint J, Hernandez M, et al. 1986. Evolutionary relationships of human populations from an analysis of nuclear DNA polymorphisms. *Nature* 319:491–93

123. Watterson GA. 1975. On the number of segregating sites in genetical models without recombination. *Theor. Pop. Biol.* 7:256–76

124. Watterson GA. 1984. Allele frequencies after a bottleneck. *Theor. Pop. Biol.* 26:387–407

125. White TD, Suwa G, Asfaw B. 1994. *Australopithecus ramidus,* a new species of early hominid from Aramis, Ethiopia. *Nature* 371:306–12

126. Whittam TS, Clark AG, Stoneking M, Cann RL, Wilson AC. 1986. Allelic variation in human mitochondrial genes based on patterns of restriction site polymorphism. *Proc. Natl. Acad. Sci. USA* 83:9611–15

127. Wilson AC, Cann RL. 1992. The recent African genesis of humans. *Sci. Am.* 266:22–27

128. Wolpoff MH, Zhi WX, Thorne AG. 1987. Modern *Homo sapiens* origins: a general theory of hominid evolution involving the fossil evidence from East Asia. In *The Origin of Modern Humans,* ed. FH Smith, F Spencer, pp. 411–83. New York: Liss

129. Wolpoff MH. 1989. Multiregional evolution: the fossil alternative to Eden. In *The Human Revolution—Behavioural and Biological Perspective on the Origin of Modern Humans,* ed. P Mellars, CB Stringer, pp. 62–108. New Jersey: Princeton Univ. Press

130. Wood B. 1992. Origin and evolution of the genus *Homo. Nature* 355:783–90

131. Wood B. 1993. Four legs good, two legs better. *Nature* 363:587–88

132. Wright S. 1931. Evolution in Mendelian populations. *Genetics* 16:97–159

133. Wright S. 1969. *Evolution and the Genetics of Populations.* Vol. 2. Chicago: Univ. Chicago Press

134. Xiong W, Li WH, Poster I, Yamamura T, Yamamoto A, et al. 1991. No severe bottleneck during human evolution: evidence from two apolipoprotein C-II deficiency alleles, *Am. J. Hum. Genet.* 48:383–89

135. Yunis JJ, Prakash O. 1982. The origin of man: a chromosomal pictorial legacy. *Science* 215:1525–30

Annu. Rev. Ecol. Syst. 1995. 26:373–401

HISTORICAL BIOGEOGRAPHY: Introduction to Methods

Juan J. Morrone and Jorge V. Crisci

Museo de La Plata, Paseo del Bosque, 1900 La Plata, Argentina

KEY WORDS: comparative biology, biogeography, cladistics, dispersalism, panbiogeography, cladistic biogeography

ABSTRACT

The five basic historical biogeographic methods are: dispersalism, phylogenetic biogeography, panbiogeography, cladistic biogeography, and parsimony analysis of endemicity. Dispersalism derives from the traditional concepts of center of origin and dispersal. Bremer's recent cladistic implementation of dispersalism estimates the relative probability that different areas were part of the ancestral distribution of a group. Phylogenetic biogeography applies the rules of progression and deviation to elucidate the history of the geographical distribution of a group. Panbiogeography consists of plotting distributions of different taxa on maps, connecting their distribution areas together with lines called individual tracks, and looking for coincidence among individual tracks to determine generalized tracks. Generalized tracks indicate the preexistence of widespread ancestral biotas, subsequently fragmented by geological or climatic changes. Cladistic biogeography assumes a correspondence between taxonomic relationships and area relationships, where comparisons between area cladograms derived from different taxa allow one to obtain general area cladograms. The most important cladistic biogeographic procedures are: component analysis, Brooks parsimony analysis, three-area statements, and reconciled trees. Parsimony analysis of endemicity (PAE) classifies areas by their shared taxa, analogous to characters, according to the most parsimonious solution. We think the various methods are not mutually-exclusive alternatives, but some of them can be integrated in a single biogeographic approach, with the capability of resolving different problems, such as the recognition of spatial

373

homology (panbiogeography), the identification of areas of endemism (PAE), and the formulation of hypotheses about area relationships (cladistic biogeography).

> O God, I could be bounded
> in a nutshell and count myself
> a King of infinite space
>
> *Hamlet,* II, 2

INTRODUCTION

Historical biogeography is going through an extraordinary revolution concerning its foundations, basic concepts, methods, and relationships to other disciplines of comparative biology (35, 36). In the last two decades considerable progress has been due especially to the development of cladistic biogeography (48, 71, 73, 120, 128, 133). Several quantitative methods have been proposed, and software is now available for applying most of them; however, confusion about methods has largely inhibited their application (100). On the other hand, most of the theoretical papers recently published on this subject are partisans of a particular method. For these reasons, we believe a critical review of the historical biogeographic methods available would be useful.

This paper is an introduction to historical biogeographic methods. We explain and illustrate the most frequently used procedures, briefly discuss the theoretical background of each, enumerate representative empirical studies, and provide information about relevant software. We also discuss an approach to integrate most of the methods as part of a single comprehensive analysis.

COMPARATIVE BIOLOGY AND BIOGEOGRAPHY

Metaphors are important components of any scientific paradigm, not only performing an explanatory function by bridging the gap between an abstract system and the real world, but also serving as the basic organizing relation of the paradigm (46). A metaphor created in 1964 by the Italian botanist Léon Croizat (39) suggests the central theme of comparative biology (120). This metaphor views the diversity of life as a historical phenomenon with three dimensions: form, space, and time. (Form in this context refers not only to the structure of organisms, but to all their attributes, be they structural, functional, molecular, or behavioral.)

If it is to allow us to understand the diversity of life, comparative biology must deal with three distinguishable elements: (*a*) similarities and differences in the attributes of organisms, (*b*) the history of organisms in space, and (*c*) the history of organisms in time (120). Biogeography is the discipline of comparative biology primarily concerned with the history of organisms in space.

Ecological and Historical Biogeography

Candolle (46a) was the first author to distinguish between ecological and historical biogeography. According to him, explanations for the former depend upon "physical causes operating at the present time," and for the latter, upon "causes that no longer exist today." Ecological explanations were early recognized to be insufficient, because areas on different continents with the same ecological conditions can be inhabited by totally different taxa (120).

Myers & Giller (107) view biogeography as distributed along a spatiotemporal gradient. At one end, ecological biogeography is concerned with ecological processes occurring over short temporal and small spatial scales. At the other end, historical biogeography deals with evolutionary processes occurring over millions of years on a large scale. Between the two extremes of ecological and historical biogeography is a compartment concerned with the effects of Pleistocene glaciations. Within each approach, various theories, hypotheses, and models have been proposed, but due to the different interests of the various biogeographic traditions, they have been largely noninteractive.

The division between ecological and historical biogeography reflects the past predominance of narrative rather than analytical methods. Narratives allow authors to cast their explanations in terms of rival beliefs rather than rigorous inferences. When analytical methods are used in biogeography, patterns may prove to be neither wholly historical nor wholly ecological, and testing and reasoning are needed if the effects of the processes causing these patterns are to be distinguished (148). We believe that disciplinary boundaries between ecological and historical biogeography are circumstantial and that they can be fruitfully unified into a single research program (94). Reviewing historical biogeographic methods is a step toward that needed synthesis.

Historical Explanations in Biogeography

Disjunct distribution patterns are the most intriguing problem for biogeographers. Related taxa may show such a pattern: Either their common ancestor originally occurred in one of the areas and later dispersed into the other one, where descendants survive to present day, or their ancestor was originally widespread in greater areas, which became fragmented, and its descendants have survived in the fragments until now. These historical explanations are named, respectively, dispersal and vicariance (119, 120).

In the dispersal explanation, the range of the ancestral population was limited by a barrier, which was crossed by some of its members. If they colonize the new area and remain isolated from the original population, they may eventually differentiate into a new taxon. In the vicariance explanation, the ancestral population was divided into subpopulations by the development of barriers they cannot cross. In time, the isolated subpopulations may differentiate into different taxa. In the vicariance explanation the appearance of the barrier causes the disjunction, so the barrier cannot be older than the disjunction. In the dispersal explanation the barrier is older than the disjunction.

Any particular distributional pattern, however, may be explained by either a dispersal or a vicariance explanation. Consider for example a taxon with three species (A, B, and C), one each in South America, New Zealand, and Australia (Figure 1a). According to their cladogram, the species from New Zealand is most closely related to the Australian species, and both constitute the sister taxon to the species from South America. Assuming a dispersal explanation (Figure 1b), the ancestor of B + C dispersed from South America to New Zealand, and the ancestor of C (or C itself) migrated from the latter

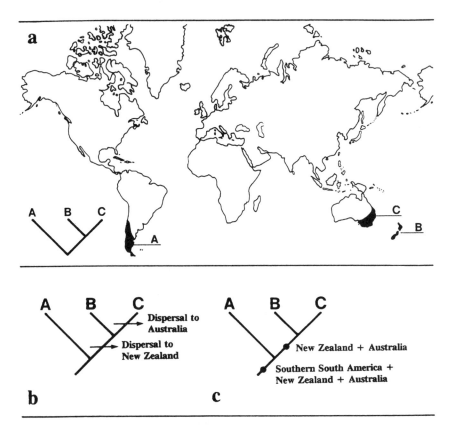

Figure 1 Historical explanations in biogeography. (*a*) geographical distribution and cladogram of three species (A, B, and C); (*b*) dispersal explanation; (*c*) vicariance explanation.

to Australia. (An alternative hypothesis is that the ancestor of B + C dispersed to Australia and the ancestor of B dispersed into New Zealand.) In the vicariance explanation (Figure 1c), the ancestor of the group occurred in South America, Australia, and New Zealand. When South America separated from Australia + New Zealand, A was separated from the ancestor of B + C, and the separation of New Zealand from Australia later caused the disjunction between B and C. Both dispersal and vicariance are natural processes, so neither dispersal nor vicariant explanations can be discounted a priori.

Dispersal was the dominant explanation for centuries, based on strict adherence to the geological concept of earth stability. In the 1950s, Hennig & Brundin (66) proposed phylogenetic biogeography, mainly based on dispersal, but accepting vicariance in some cases. Croizat (38) was one of the first scientists to challenge vocally the dispersal explanation and to promote vicariance as the most important process, in an approach called panbiogeography. In the last two decades, Croizat's and Hennig's ideas were combined, creating cladistic biogeography, which emphasizes the search for congruent biogeographic patterns using cladograms, disregarding both dispersal and vicariance explanations a priori. More recently, BR Rosen (149) proposed another pattern-oriented method—parsimony analysis of endemicity—which uses a cladistic algorithm to analyze geographical patterns of distribution. The taxonomy and a list of representative empirical studies of the current methods available are detailed in Table 1.

DISPERSALISM

In accordance with the biblical account of the Garden of Eden, Linnaeus proposed that species originated through creation in one small area, then dispersed to other areas available for colonization. Since Linnaeus's time, both centers of origin and dispersal have been the prevailing explanations in historical biogeography (120). Darwin (45) and Wallace (163, 164) considered that species originate in one center of origin, from which some individuals subsequently disperse by chance, and then change through natural selection. The Darwin-Wallace tradition has continued until this century; among its most prominent exponents have been Cain (14), Darlington (43, 44), Matthew (85), Mayr (88), Raven & Axelrod (144), and Simpson (160).

Dispersalism is based on five basic principles (170):

1. Higher taxa arise in centers of origin, where subsequent speciation occurs.
2. The center of origin of a taxon may be estimated by specific criteria.
3. The distribution of fossils is essential, because the oldest fossils are probably located near the center of origin.
4. New species evolve and disperse, displacing more primitive species toward

Table 1 A taxonomy of historical biogeographic methods.

Methods	Representative empirical studies
1.0 Dispersalism	14, 43, 44, 88, 144, 160
1.1 Ancestral areas	1, 6
2.0 Phylogenetic biogeography	11, 12, 67, 154
3.0 Panbiogeography	5, 18, 19, 38, 39, 47, 52, 59, 161, 162
3.2 Spanning graphs	—
3.3 Track compatibility	27, 93, 96, 99, 104
4.0 Cladistic biogeography	155
4.1 Reduced area cladogram	68, 69, 136, 152, 153
4.2 Quantitative phylogenetic biogeography	81, 90
4.3 Ancestral species map	2, 84, 170, 174
4.4 Component analysis	3, 6, 7, 17, 20, 21, 34, 72, 78, 81, 82, 96, 103, 120, 126, 157, 165
4.4.1 Component compatibility	46b, 147, 167, 176
4.4.2 Quantification of component analysis	34
4.5 Brooks parsimony analysis (BPA)	9, 21, 29, 34, 62, 76, 83, 86, 103, 172, 173
4.6 Three-area statements (TAS)	79, 95, 103, 105
4.7 Reconciled trees	133
5.0 Parsimony analysis of endemicity (PAE)	22, 29, 99, 103, 106, 149

the peripheral areas, away from the center of origin, where most apomorphic species will be found.

5. Organisms disperse as widely as their abilities and physical conditions of the environment permit, so derived taxa "push" primitive taxa toward the edges of the group's range.

There have been many criticisms of the dispersalist approach (32, 38, 39, 42, 73, 111, 120). Cain (14) evaluated the criteria for determining centers of origin, concluding that none of the criteria could be trusted independently and that some were even contradictory, e.g. the location of the most primitive forms vs. the location of the most advanced ones. Dispersal explanations reside in narrative frameworks, constituting irrefutable hypotheses that do not provide a general theory to explain distributional patterns, but rather individual case stories for each taxon. Panbiogeographers and cladistic biogeographers consider that dispersalism is an ad hoc discipline that requires external causes to explain the patterns analyzed (42, 55, 73, 120, 141). As Nelson (111) stated, concentrating on improbable dispersals as explanations for distributions results in the "science of the rare, the mysterious and the miraculous." In addition, the acceptance of dispersal as the primary causal factor of geographical distribution creates a methodological problem: If every disjunction is explained in terms of dispersal, biogeographic patterns that result from vicariance will never be discovered. Craw & Weston (32) applied the methodology of scientific research programs, developed by Lakatos (80), to discuss biogeographic approaches, concluding that dispersal biogeography was not a scientific program in Lakatos's sense.

Ancestral Areas

Bremer (6) recently formalized a cladistic procedure based on the dispersalist approach. This author considered that understanding ancestral areas for an individual group is a valid part of the study of the natural history of that group, and that it was the previous approach to search for centers of origin, not the search per se, which was spurious. Bremer's (6) procedure allows one to identify the ancestral area of a group from the topological information of its cladogram. Each area can be considered a binary character with two states (present or absent) and optimized on the cladogram, using Camin-Sokal parsimony. By comparing the numbers of gains and losses, it is possible to estimate areas most likely to have been part of the ancestral areas.

As an example, Bremer (6) considered a cladogram with four hypothetical species distributed in Malaya, Sumatra, Java, and Borneo (Figure 2a). Species *a, c,* and *d* are restricted to single areas, whereas *b* is widespread in both Malaya and Java. The simplest assumption implies that the ancestral area is identical to the area being considered, so all absences (equivalent to extinction or fragmentation due to vicariance) are plotted as losses (indicated by crosses in Figure 2b-e). Assuming that there were no losses and that all area presences are the result of gains, the ancestral area is empty, and the individual areas are

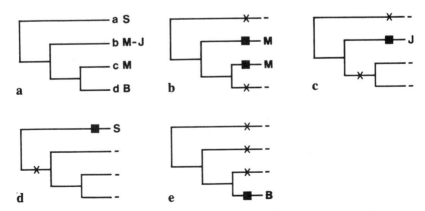

Figure 2 Bremer's ancestral areas approach. *a,* cladogram of four species (*a, b, c,* and *d*) inhabiting Sumatra (S), Malaya (M), Java (J), and Borneo (B); *b-e,* optimizations of the different areas.

plotted as gains (indicated by bars in Figure 2b-e). If there are more losses than gains for any individual area, it is excluded from the ancestral area. If there are more gains than losses, the individual area is identified as the ancestral area. The number of gains for Malaya (Figure 2b) and Sumatra (Figure 2d) equals the number of losses, and the number of losses for Java (Figure 2c) and Borneo (Figure 2e) exceeds the number of gains. Thus the ancestral area may have been limited to Malaya, Sumatra, or both, whereas Java and Borneo are less likely to have been part of the ancestral area for the group.

Ronquist (145) criticized Bremer's preference for Camin-Sokal parsimony instead of Wagner parsimony.

PHYLOGENETIC BIOGEOGRAPHY

Phylogenetic biogeography (11, 12, 66, 67) was the first approach to consider a phylogenetic hypothesis for a given group of organisms as the basis for inferring its biogeographic history. It was defined as the study of the history of monophyletic groups in time and space, taking into account cladogenesis, anagenesis, allopatry (evidence of vicariance), sympatry (evidence of dispersal), and paleogeographical events (12). Phylogenetic biogeography is based on two principles:

1. Closely related species tend to replace each other in space. Higher taxa also can be vicariant but usually show a certain degree of sympatry.
2. If different monophyletic groups show the same biogeographic pattern, they probably share the same biogeographic history. This principle had not been used, since phylogenetic biogeographers concentrated on the history of single groups rather than on congruent distributions shared by different groups inhabiting the same areas (170).

Phylogenetic biogeography applies two basic rules:

1. *Progression rule.* The primitive members of a taxon are found closer to its center of origin than more apomorphic ones, which are found on the periphery. Hennig (67) conceived that speciation was allopatric, involving peripheral isolates, and causally connected to dispersal. Within a continuous range of different species of a monophyletic group, the transformation series of characters run parallel with their progression in space.
2. *Deviation rule.* In any speciation event, an unequal cleavage of the original population is produced, where the species that originates near the margin is apomorphic in relation to its conservative sister species (12).

Although phylogenetic and dispersal biogeography may be lumped into the same approach, because both emphasize centers of origin and dispersal, some authors (73, 170) regard phylogenetic biogeography as an advance over dispersalism because of the explicit use of cladistic hypotheses instead of descriptive enumerations and scenarios. The progression rule is based on the peripheral isolation allopatric mode of speciation, so it cannot be applied when other modes of speciation are considered, because it is rejectable a priori (110). In addition, interpreting cladograms as phylogenetic trees rather than synapomorphy schemes requires ad hoc assumptions not fully justified by the information on which they are based (73).

PANBIOGEOGRAPHY

In contrast to the two previous methods, which focus on dispersal, Léon Croizat postulated that "earth and life evolve together," meaning that geographic barriers evolve together with biotas—essentially vicariance. From this metaphor grew up the concept of panbiogeography (37–40, 55, 57, 60, 127). Croizat's method was basically to plot distributions of organisms on maps and connect the disjunct distribution areas or collection localities together with lines called tracks. Individual tracks for unrelated groups of organisms were then superimposed, and if they coincided, the resulting summary lines were considered generalized tracks. Generalized tracks indicate the preexistence of ancestral biotas, which subsequently become fragmented by tectonic and/or climatic change.

There are three basic panbiogeographic concepts:

INDIVIDUAL TRACK   A track represents the spatial coordinates of a species or group of related species, and operationally is a line graph drawn on a map of their localities or distribution areas, connected according to their geographical proximity (23, 25–27, 38, 42, 60). In graph theory, a track is equivalent to a minimal spanning tree, which connects all localities to obtain the smallest possible link length (123). After the track is constructed, its orientation (i.e. rooting) can be determined using one or more of the following three criteria:

1. *Baseline* Features such as the crossing of an ocean or sea basin, or a major tectonic structure (25, 27, 28, 31).
2. *Main massing* A concentration of numerical, genetical or morphological diversity within a taxon in a given area (25–28, 30, 123).
3. *Phylogeny* If cladistic information is available, it can be used to direct the track from the most primitive to the most derived taxa (123).

GENERALIZED TRACK   Coinciding individual tracks for unrelated taxa or groups constitute a generalized or standard track (23, 28, 123), which provides a spatial criterion for biogeographic homology (56).

NODE   The area where two or more generalized tracks intersect (24, 25, 27, 60, 112, 123). It means that different ancestral biotic and geological fragments interrelate in space/time, as a consequence of terrain collision, docking, or suturing, thus constituting a composite area.

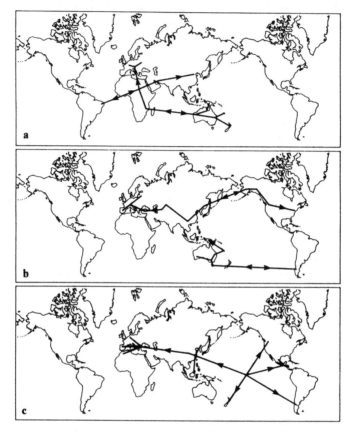

Figure 3 Individual tracks. (*a*) ratite birds; (*b*) *Nothofagus* (southern Hemisphere) and *Fagus* (northern Hemisphere); (*c*) *Leiopelma* and related taxa.

The panbiogeographic approach may be exemplified by analyzing three Austral taxa (26, 38, 56): the ratite birds (Figure 3a), the southern beeches (*Nothofagus*; Figure 3b), and the frog *Leiopelma* (Figure 3c). Their individual tracks show that these taxa do not share spatial homology. Only the Ratites are clearly Gondwanic, having their distribution oriented by the Atlantic and Indian Ocean basins. In spite of partial sympatry in Australia and southern South America, only *Leiopelma* and *Nothofagus* are geographically homologous, belonging to the same ancestral biota, which is different from that of the Ratites. This result contrasts with biogeographic studies in which Ratites and *Nothofagus* have been assumed a priori to belong to the same ancestral biota (68, 69, 136).

The last two decades have shown an intensification of the debate between proponents of panbiogeography and those of cladistic biogeography (24, 25,

27, 32, 35, 41, 74, 87, 101, 123, 130, 143, 156). The panbiogeographic approach has been subject to several criticisms. In many instances, panbiogeographers use systematic treatments in an uncritical way (136, 156). Main massings have been considered similar to centers of origin (74, 143). Platnick & Nelson (143) rejected the use of geographical proximity for drawing tracks, because they considered cladistic information a prerequisite to any historical biogeographic analysis.

Track Compatibility

Craw (27, 28, 29) developed a quantitative panbiogeographic procedure, which treats tracks as characters of the areas analyzed. Matrices of areas × tracks are then analyzed for track compatibility in a way analogous to character compatibility (89). Two or more individual tracks are regarded as compatible with each other if they are either included within, or replicated by, one another. (Panbiogeography uses the concept of compatibility in a restricted way, because nonoverlapping tracks are incompatible, although they would be compatible under the original concept.)

In the example of Figure 4, there are four individual tracks (A, B, C, and D; Figure 4a–d). The matrix of areas × tracks (Figure 4e), analyzed with a compatibility algorithm, produces a generalized track (Figure 4f) based on tracks A, B, and D, with C incompatible with them. For a track compatibility analysis, the *CLIQUE* computer program of *PHYLIP* package (all types of PCs; 51) can be used.

An alternative quantitative panbiogeographic procedure was proposed by Page (123), based on graph theory; however, it has not been yet applied to real data.

CLADISTIC BIOGEOGRAPHY

Cladistic biogeography was originally developed by DE Rosen, G Nelson, and N Platnick, (108–112, 119–121, 143, 151). Cladistic biogeography assumes that the correspondence between taxonomic relationships and area relationships is biogeographically informative. Comparisons between area cladograms derived from different plant and animal taxa that occur in a certain region allow general patterns to be elucidated (73, 124). A cladistic biogeographic analysis comprises two steps (Figure 5): the construction of area cladograms from different taxon cladograms, and the derivation of general area cladogram(s).

Construction of Area Cladograms

Area cladograms are constructed by replacing the names of terminal taxa with the names of the areas in which they occur. The construction of area

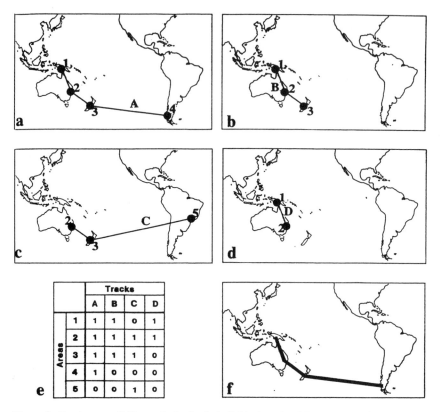

Figure 4 Track compatibility analysis. (*a–d*), individual tracks; (*e*), areas × species matrix; (*f*), generalized track.

cladograms is trivial if every taxon is endemic to a unique area and every area harbors one taxon, but it is complicated for taxon cladograms including widespread taxa (taxa present in more than one area), missing areas (areas absent in a cladogram), and redundant distributions (areas with more than one taxon). In these cases, area cladograms must be converted into resolved area cladograms, by applying assumptions 1 and 2 (120) and assumption 0 (176).

Figure 6 shows the treatment of a widespread taxon under the three assumptions. Under *Assumption 0*, widespread taxa become synapomorphies of the areas inhabited by them, so that the area relationships are considered to be monophyletic (sister areas). *Assumption 1* allows the area relationships to be mono- or paraphyletic in terms of the widespread taxon inhabiting them. Under *assumption 2* each occurrence is treated separately and can "float" on the

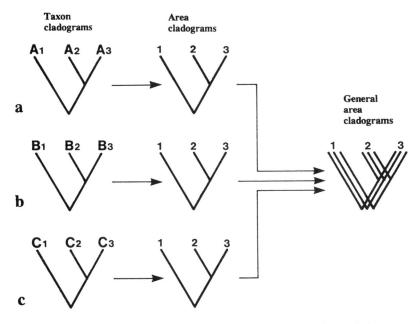

Figure 5 Steps of a cladistic biogeographic analysis: construction of area cladograms and derivation of a general area cladogram.

resolved area cladograms, so the area relationships can be mono-, para-, or polyphyletic. Missing areas are treated as uninformative under assumptions 1 and 2, and as primitively absent under assumption 0. Regarding redundant distributions, assumptions 0 and 1 consider that if two taxa are present in the same area, their occurrences are both valid, whereas under assumption 2, each occurrence of a redundant distribution is considered separately (e.g. in different resolved area cladograms). Assumptions are not mutually exclusive, so different assumptions can be combined to treat the different problems, such as treating widespread taxa under assumption 2 but redundant distributions under assumption 0 (128).

Authors generally prefer assumption 2 (70, 73, 100, 120, 128). Its implementation, however, can produce many resolved area cladograms in complex data sets (17, 36). Nelson & Ladiges (116) considered current implementations of assumption 2 to be deficient because it can obscure possibly real complexity. They suggested that the set of assumption 2 area cladograms could be further resolved by evaluating nodes in terms of three-area statements analysis, reducing widespread ranges in favor of endemics. A possible approach to minimize the impact of both widespread and redundant ranges might be to remove redundant, widespread distributions before analysis (98, 100).

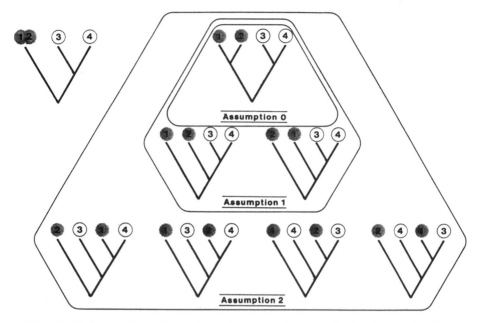

Figure 6 Cladogram with a widespread taxon in areas 1 and 2, and application of assumptions 0, 1, and 2 to produce resolved area cladograms.

Derivation of General Area Cladogram(s)

Based on the information in either the area cladograms or the resolved area cladograms, general area cladograms are derived. The main procedures for deriving general area cladograms are: component analysis (113, 120, 124), Brooks parsimony analysis (9, 171–173), three-area statements (114–116), and reconciled trees (133). Three other procedures are currently not applied: the reduced area cladogram (152, 153), quantitative phylogenetic biogeography (90), and the ancestral species map (169, 170).

Component Analysis

Component analysis (71–73, 113, 120, 124, 126, 128, 137, 176) derives sets of fully resolved area cladograms from the taxon cladograms under analysis, applying assumptions 0, 1, and 2. The general area cladogram is derived by the intersection of the sets of area cladograms for the taxa analyzed (113, 126). If no general area cladogram is found through intersection, or the intersect contains multiple cladograms, a consensus tree can be constructed (124). In Figure 7, application of assumption 2 produces 11 area cladograms for a taxon cladogram with a widespread taxon (Figure 7a), two area cladograms for a

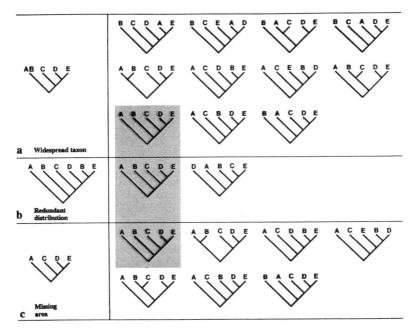

Figure 7 Application of component analysis under assumption 2: (*a*) area cladogram with a widespread taxon; (*b*) area cladogram with a redundant distribution; (*c*) area cladogram with a missing area. Intersection of the three sets of resolved area cladograms (indicated by shading) includes the general area cladogram.

taxon cladogram with a redundant distribution (Figure 7b), and seven area cladograms for a taxon cladogram with a missing area (Figure 7c). Their intersection leads to a single general area cladogram (indicated by shading). There is one software package available for applying component analysis: *COMPONENT,* version 1.5 (MS-DOS, IBM compatible; 125).

Further variations of component analysis consist of constructing a data matrix of components by areas, based on the area cladograms, and analyzing it with a compatibility algorithm (176) or with a Wagner parsimony algorithm (72). The former procedure is implemented in software CAFCA (all types of PCs; 175).

Wiley (171–173) criticized component analysis because of the preference for assumptions 1 and 2 instead of assumption 0, which he considered most parsimonious. Some authors (162b, 171, 173, 176) criticized the use of consensus techniques to obtain a general area cladogram. Page (126) argued that linking component analysis and consensus techniques is misleading, because it confounds the construction of area cladograms with the comparison of area

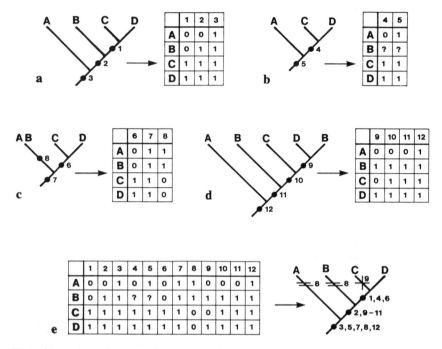

Figure 8 Brooks parsimony analysis. (*a*) area cladogram with complete data; (*b*) area cladogram with a missing area; (*c*) area cladogram with a widespread taxon; (*d*) area cladogram with a redundant distribution; (*e*) resulting data matrix (areas × components) and general area cladogram obtained after Wagner parsimony analysis.

cladograms. Clear justification of component analysis is provided by Page (124, 126, 128), Platnick (138), and Platnick & Nelson (142).

Brooks Parsimony Analysis

Brooks parsimony analysis (BPA) was proposed by Wiley (171, 172, 173), based on the ideas originally developed by Brooks (8, 9) for historical ecology. It is a Wagner parsimony analysis of area cladograms, which are coded and analyzed as characters. BPA is based on assumption 0, differing only in treating missing areas as uninformative rather than as primitively absent.

BPA uses an area × taxon matrix, produced by binary coding of the terminal taxa and their hypothetical ancestors. Four examples of data coding are shown in Figure 8: 1) a group with complete data (Figure 8a); 2) an example in which a member of the group is missing in one area (B in Figure 8b); 3) an example with a widespread taxon (present in areas A and B; Figure 8c); and 4) a redundant distribution (two taxa in area B; Figure 8d). In each case, the corresponding data matrix is also figured. All the information is combined in a single data matrix (Figure 8e), which, after applying Wagner parsimony analysis, results in a general area cladogram.

For applying BPA, an appropriate Wagner parsimony program like Hennig86 (MS-DOS, IBM compatible; 50) may be used. Kluge (76) presented a modification of Brooks parsimony analysis, which differs in the treatment of widespread taxa, which are considered irrelevant and so are coded as missing data, and in a weighting procedure for redundant distributions.

There has been extensive criticism of BPA (16, 115, 128, 133, 138, 146, 162b). According to Carpenter (16), the codings used in BPA to represent the taxon cladograms are not independent, and this can lead to bizarre results. The application of parsimony in biogeography has yet to be precisely defined and convincingly justified (77).

Three-Area Statements

Three-area statements (TAS) (114–116) code distributional data for area cladograms as a suite of three-item statements (117, 122), and the output is a data matrix for Wagner parsimony analysis. The data matrix can be obtained with the TAS program (MS-DOS, IBM compatible; 116), implemented for assumptions 0 and 1. Assumption 2 can be applied by prior manipulation of the data set (79, 116) or with the TASS program (MS-DOS, IBM compatible; 118). The matrix produced with TAS may then be analyzed with Hennig86 (50). Figure 9 shows the application of TAS to the same example used for BPA, with the corresponding three-area statements matrices, and the resulting general area cladogram.

The three-item statements approach has been criticized (77) mainly for its taxonomic applications. Some of these criticisms, e.g. the addition of missing data where none existed, which added ambiguity, may be also applied to TAS.

Reconciled Trees

The concept of reconciled trees arose independently in molecular systematics, parasitology, and biogeography as a means of describing historical associations between genes and organisms (53), hosts and parasites (91), and organisms and areas (128, 129, 132, 133). Page (133) proposed a cladistic biogeographic procedure that maximizes the amount of codivergence (shared history) among different area cladograms, which implies minimizing losses (i.e. extinctions or unsampled taxa) and duplications (i.e. speciation events independent of the vicariance of the areas) when combining different area cladograms into a single general area cladogram. Horizontal transfer (i.e. dispersal) should be also minimized, but that is not considered in the procedure. Page (134) describes a procedure to incorporate dispersal.

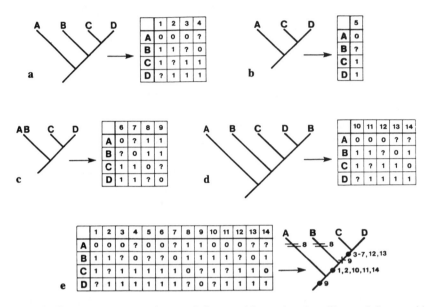

Figure 9 Three-area statements. *(a)* area cladogram with complete data; *(b)* area cladogram with a missing area; *(c)* area cladogram with a widespread taxon; *(d)* area cladogram with a redundant distribution; *(e)* resulting data matrix (area × three-area statements) and general area cladogram obtained after Wagner parsimony analysis.

Figure 10a shows a simple example of a reconciled tree between an area cladogram and its general area cladogram, where there is a maximum of codivergence. Figure 10b shows a more complex example of a similar situation, where a duplication (node f in the area cladogram) is needed to reconcile both trees. In biogeography there is often no host tree (general area cladogram) to reconcile with the associate (area cladogram). In that case we must search for the general area cladogram with maximal codivergence to the area cladograms.

Algorithms for obtaining reconciled trees are implemented in *COMPONENT* version 2.0 (Microsoft-Windows, IBM-compatible; 131). In order to identify the taxa that may have dispersed, each taxon can be deleted in turn and a reconciled tree computed for the remaining taxa. Those taxa whose deletion greatly increases congruence between area cladograms and taxon cladograms are likely to have dispersed (133).

Page (133) considered assumptions 0, 1, and 2 to suffer from the limitation that they simply follow an algorithm rather than optimizing an optimality criterion, which makes it impossible to find the general area cladogram that is optimal for two or more area cladograms. According to him the reconciled

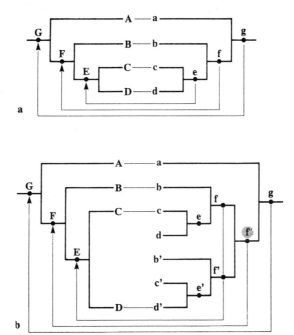

Figure 10 Reconciled trees. (*a*) between an area cladogram and its general area cladogram; (*b*) more complex example where a duplication is needed to reconcile both trees. Left, general area cladograms; right, area cladograms.

trees approach has an optimality criterion (i.e. maximize codivergence for all the area cladograms).

PARSIMONY ANALYSIS OF ENDEMICITY

Parsimony analysis of endemicity (PAE; 149, 150) classifies areas or localities (analogous to taxa) by their shared taxa (analogous to characters) according to the most parsimonious solution. PAE data consist of area × taxa matrices, and PAE cladograms represent nested sets of areas, in which terminal dichotomies represent two areas between which the most recent biotic interchange has occurred.

This method was originally proposed in a paleontological context, with cladograms based upon data collected from successively older geological horizons, and older interchange events in one horizon were assumed to be corroborated by the younger events in the next. With a poor fossil record or when treating only extant distributions, PAE is carried out on the data from a single time plane, using different taxonomic levels. This allows an interpretation of the history of space occupancy by taxa through time, assuming that subsequent dispersal has not obliterated the vicariant pattern, and that extinctions are random.

The main criticism of PAE is that it ignores cladistic relationships among taxa, considering only their distributions (70). Some authors (22, 99, 106) incorporate cladistic information to PAE, by adding supraspecific natural groups (containing two or more species) to the matrix.

DIFFERENT METHODS OR DIFFERENT PROBLEMS?

All the historical biogeographic methods discussed were originally proposed as alternatives. We believe, however, that most can be integrated into a single approach. Dispersalism and phylogenetic biogeography are excluded from this discussion because they mainly explain histories of single taxa instead of seeking replicated patterns. This integrative approach consists of using each method in a different step of one analysis, restricting its use to a specific problem. A historical biogeographic analysis should include at least three steps: recognition of spatial homology, identification of areas of endemism, and formulation of hypotheses about area relationships.

1. Recognition of spatial homology   The first step should consist of determining if the plant and animal taxa analyzed belong to the same biota. A panbiogeographic procedure could be employed (36, 61, 96, 98, 101) to find generalized tracks, which represent ancestral biotas and spatial homologies (56). Each generalized track then should be analyzed separately, thus avoiding the extreme incongruent patterns that result from mixing different ancestral biotas in the same analysis.

2. Identification of areas of endemism   Once biogeographic homologies have been recognized, we must identify the units of study. An area of endemism is defined by the congruent distributional boundaries of two or more species, where "congruent" does not demand complete agreement on those limits at all possible scales of mapping but does require relatively extensive sympatry (139).

Several authors have recently discussed the determination of areas of endemism (4, 33, 63, 65, 139). Morrone (97) proposed the use of PAE to identify areas of endemism, by using quadrats as operational units, and employing the sets of quadrats as a basis for choosing the species to be mapped. After drawing quadrats on a map of the region to be analyzed (Figure 11a), a data matrix $r \times c$ is constructed, where r (rows) represent the quadrats and c (columns) the species. An entry is 1 if a species is present and 0 if it is absent (Figure 11b).

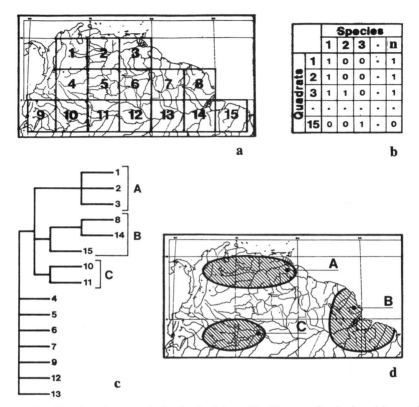

Figure 11 Use of parsimony analysis of endemicity to identify areas of endemism. (*a*) map of northern South America showing 15 quadrats; (*b*) *data matrix of species by quadrats;* (*c*) cladogram of the quadrats obtained applying Wagner parsimony; (*d*), areas of endemism based on the groups of the cladogram.

Application of Wagner parsimony analysis to the data matrix produces a cladogram (Figure 11c). Species endemic to each group of quadrats (defined by at least two species) are mapped, to delineate the boundaries of each area (Figure 11d).

3. Formulation of hypotheses about area relationships   Page & Lydeard (135) have suggested three criteria to choose taxa for this step: maximizing endemicity, sampling exhaustively within each clade, and including as many relevant areas as possible. The cladistic analysis of these taxa should then be carried out. Once we have the taxon cladograms, the four reviewed procedures (component analysis, Brooks parsimony analysis, three-area statements, and reconciled trees) can be applied to obtain the area cladograms and general area cladograms.

A GLIMPSE INTO THE FUTURE

We have reviewed the current analytical methods in historical biogeography. But the most important question has as yet gone unanswered: Which is the best method? The answer is not clear. In fact, despite their various shortcomings, each method makes a contribution in addressing a different type of biogeographical question. An integrative approach, like that proposed here, can take advantage of the merits of each method. One clear conclusion, however, is that the basic language of historical biogeography will be area cladograms, which allow the testing of hypotheses of general patterns (90, 119, 135).

To search for an optimal cladistic biogeographic procedure, Morrone & Carpenter (100) compared the application of component analysis, Brooks parsimony analysis, three-area statements, and reconciled trees to different data sets, mapping the area cladograms onto the general area cladograms (produced by the four procedures) and calculating the items of error (i.e. number of nodes added to the general area cladograms in order to explain the area cladograms). They also applied two accessory criteria: the number of cladograms produced and their degree of resolution. They found that none of the procedures was consistently superior. The lack of a consistent superiority of one of the procedures is caused by the influence of different sources of ambiguity (differentially present in the data sets), which seem to affect distinctively the alternative procedures. Primary sources of ambiguity are dispersal (100) and speciation events independent of the vicariance of the areas (that lead to multiple lineages), combined with extinction and unsampled taxa (129). For example, BPA is more affected by dispersal than is component analysis, whereas the latter is more affected by multiple lineages. Nelson & Ladiges (116) recommend treating clades having the same areas separately to avoid the ambiguity due to multiple lineages, but this might not generally resolve all area relationships, particularly when multiple lineages are combined with many widespread taxa.

Some statistical tests of biogeographical hypotheses have been proposed. Craw (27) formalized a test for assessing the significance of generalized tracks obtained by a track compatibility analysis. In cladistic biogeography, several tests have been proposed to determine if the agreement between area cladograms is greater than expected due to chance alone (10, 128, 129, 158, 159). The use of these tests has been criticized, based on problems with the definition of "chance" (see Farris [49]). Testing of the timing of biogeographic events using molecular divergence, based on molecular clocks, has been proposed by

RDM Page (129). Although most biologists now accept a broad correlation between the amount of molecular divergence (at least for proteins and DNA) and time, it is far from established that rates are constant (92). Therefore, the utility of molecular clocks in biogeography is, at this point, debatable.

The following represents a nonexhaustive list of critical issues in biogeography that need to be tackled:

1. Areas can have more than one history, in contrast with systematics where organisms have a single history (21, 34). This situation leads to complex and conflicting patterns of area relationships that represent obstacles to their discovery.
2. The origin of biogeographic patterns is never wholly historical nor wholly ecological (54, 94), but a combination of both, which is an obstacle for the progress of biogeography. Because biogeographic patterns reflect this complex origin, analysis of those patterns requires a combination of approaches.
3. The scarcity of high quality data hampers the development of historical biogeography (138; RDM Page, personal communication).
4. The progress of cladistic biogeography will depend heavily on the development of a procedure (or the improvement of one already proposed) that takes fully into account all the complexities of real data, like dispersal, multiple lineages, and extinction (100).
5. The molecular revolution is starting to influence biogeography (13, 64). More empirical studies are strongly needed to establish the promising connection between molecular systematics and biogeography (e.g. molecular clocks).
6. A critical evaluation is needed of the tests for assessing the significance of biogeographic hypotheses or the development of new ones, like those proposed in systematics (75).
7. Geological area cladograms derived from specific analyses of geological characters, in the same way that characters are analyzed in systematics (27), would be useful to allow comparisons with general area cladograms (100).
8. It has been recently suggested that the biodiversity question is really a biogeographic one, since it is a question of where the limited financial and human resources should be applied (140). Historical biogeographic analyses, however, are not playing the significant role in biodiversity conservation that they should (58, 61, 102, 105).

Prospects for research in biogeography are by no means hopeless, and the field for developing a new biogeographic synthesis is wide open. Anyone entering this field should be able to combine a feeling of intellectual adventure with imagination and a knowledge of tradition. New challenges will have to be faced, some biogeographic problems will have to be rethought, and new methods will have to be developed. In the next few years, biogeographers will witness, not without *horror vacui,* this fascinating perspective.

ACKNOWLEDGMENTS

We would like to thank Paul Berry, Jim Carpenter, John Grehan, Peter Hoch, Liliana Katinas, Rod Page, Norman Platnick, and Rino Zandee for their useful comments on the manuscript; Hugo Calvetti for the drawings; and Piero Marchionni for technical assistance. Our work was supported by National Geographic Society Grant 4662-91. We belong to the Consejo Nacional de Investigaciones Científicas y Técnicas (CONICET), Argentina, which continuous support we acknowledge.

Literature Cited

1. Albert VA, Bremer K. 1993. Flying kiwis and pattern information in biogeographic hypotheses. *Curr. Biol.* 3: 324–25
2. Anderberg AA, Freire SE. 1991. A cladistic and biogeographic analysis of the *Lucilia* group (Asteraceae, Gnaphaliae). *Bot. J. Linn. Soc.* 106:173–98
3. Andersen NM. 1991. Cladistic biogeography of marine water striders (Insecta, Hemiptera) in the Indo-Pacific. *Aust. Syst. Bot.* 4:151–63
4. Axelius B. 1991. Areas of distribution and areas of endemism. *Cladistics* 7: 197–99
5. Beauchamp AJ. 1989. Panbiogeography and rails of the genus *Gallirallus. NZ J. Zool.* 16:763–72
6. Bremer K. 1992. Ancestral areas: a cladistic reinterpretation of the center of origin concept. *Syst. Biol.* 4:436–45
7. Bremer K. 1993. Intercontinental relationships of African and South American Asteraceae: a cladistic biogeographic analysis. In *Biological Relationships Between Africa and South America,* ed. P Goldblatt, pp. 105–135. New Haven, CT: Yale Univ. Press
8. Brooks DR. 1985. Historical ecology: a new approach to studying the evolution of ecological associations. *Ann. Mo. Bot. Gard.* 72:660–80
9. Brooks DR. 1990. Parsimony analysis in historical biogeography and coevolution: methodological and theoretical update. *Syst. Zool.* 39:14–30
10. Brown JKM. 1993. Probabilities of evolutionary trees. *Syst. Biol.* 43:78–91
11. Brundin L. 1966. Transantarctic relationships and their significance. *Kungl. Svens. Vetenskapakad. Handl.* 11:1–472
12. Brundin L. 1981. Croizat's panbiogeography versus phylogenetic biogeography. See Ref. 122a, pp. 94–158
13. Caccone A, Milinkovitch MC, Sbordoni V, Powell JR. 1994. Molecular biogeography: using the Corsica-Sardinia microplate disjunction to calibrate mitochondrial rDNA evolutionary rates in mountain newts (*Euproctus*). *J. Evol. Biol.* 7:227–45
14. Cain SA. 1944. *Foundations of Plant Geography.* New York: Harper
15. Deleted in proof
16. Carpenter JM. 1992. Incidit in Scyllam qui vult vitare Charybdim. *Cladistics* 8:100–2
17. Carpenter JM. 1993. Biogeographic patterns in the Vespidae (Hymenoptera): two views of Africa and South America. In *Biological Relationships between Africa and South America,* ed. P Goldblatt, pp. 139–55. New Haven, CT: Yale Univ. Press
18. Climo F. 1988. Punctid snails: a path to panbiogeography. *Riv. Biol.—Biol. Forum* 81:533–51

19. Climo F. 1989. The panbiogeography of New Zealand as illuminated by the genus *Fectola* Iredale, 1915 and subfamily Rotadiscinae Pilsbry, 1927 (Mollusca: Pulmonata: Punctoidea: Charopidae). *NZ J. Zool.* 16:587–649

20. Cracraft J. 1982. Geographic differentiation, cladistics, and vicariance biogeography: reconstructing the tempo and mode of evolution. *Am. Zool.* 22: 411–24

21. Cracraft J. 1988. Deep-history biogeography: retrieving the historical pattern of evolving continental biotas. *Syst. Zool.* 37:221–36

22. Cracraft J. 1991. Patterns of diversification within continental biotas: hierarchical congruence among the areas of endemism of Australian vertebrates. *Aust. Syst. Bot.* 4:211–27

23. Craw RC. 1979. Generalized tracks and dispersal in biogeography: a response to R. M. McDowall. *Syst. Zool.* 28:99–107

24. Craw RC. 1982. Phylogenetics, areas, geology and the biogeography of Croizat: a radical view. *Syst. Zool.* 31:304–16

25. Craw RC. 1983. Panbiogeography and vicariance cladistics: Are they truly different? *Syst. Zool.* 32:431–38

26. Craw RC. 1985. Classic problems of southern hemisphere biogeography reexamined: panbiogeographic analysis of the New Zealand frog *Leiopelma,* the ratite birds and *Nothofagus.* *Z. Zool. Syst. Evolutionsforsch.* 23:1–10

27. Craw R. 1988. Continuing the synthesis between panbiogeography, phylogenetic systematics and geology as illustrated by empirical studies on the biogeography of New Zealand and the Chatham Islands. *Syst. Zool.* 37:291–310

28. Craw RC. 1989. Quantitative panbiogeography: introduction to methods. *NZ J. Zool.* 16:485–94

29. Craw RC. 1989. NZ biogeography: a panbiogeographic approach. *NZ J. Zool.* 16:527–47

30. Craw RC, Heads M. 1988. Reading Croizat: on the edge of biology. *Riv. Biol.—Biol. Forum* 81:499–532

31. Craw RC, Page RDM. 1988. Panbiogeography: method and metaphor in the new biogeography. In *Evolutionary Processes and Metaphors,* ed. M-W Ho, SW Fox, pp. 163–89. New York: John Wiley

32. Craw RC, Weston P. 1984. Panbiogeography: a progressive research program? *Syst. Zool.* 33:1–33

33. Crisci JV, Cigliano MM, Morrone JJ, Roig Juñent S. 1991. A comparative review of cladistic biogeography approaches to historical biogeography of southern South America. *Austr. Syst. Bot.* 4:117–26

34. Crisci JV, Cigliano MM, Morrone JJ, Roig Juñent S. 1991. Historical biogeography of southern South America. *Syst. Zool.* 40:152–71

35. Crisci JV, Morrone JJ. 1992. A comparison of biogeographic models: a response to Bastow Wilson. *Global Ecol. Biogeogr. Lett.* 2:174–76

36. Crisci JV, Morrone JJ. 1992. Panbiogeografía y biogeografía cladística: paradigmas actuales de la biogeografía histórica. *Ciencias (México),* nro. especial 6:87–97

37. Croizat L. 1952. *Manual of Phytogeography.* The Hague: Junk

38. Croizat L. 1958. *Panbiogeography.* Caracas, published by the author

39. Croizat L. 1964. *Space, Time, Form: The Biological Synthesis.* Caracas, published by the author

40. Croizat L. 1981. Biogeography: past, present, and future. See Ref. 122a, pp. 501–23

41. Croizat L. 1982. Vicariance/vicariism, panbiogeography, "vicariance biogeography", etc.: a clarification. *Syst. Zool.* 31:291–304

42. Croizat L, Nelson G, Rosen DE. 1974. Centers of origin and related concepts. *Syst. Zool.* 23:265–87

43. Darlington PJ Jr. 1957. *Zoogeography: The Geographical Distribution of Animals.* New York: Wiley

44. Darlington PJ Jr. 1965. *Biogeography of the Southern End of the World: Distribution and History of Far-Southern Life and Land, with an Assessment of Continental Drift.* Cambridge: Harvard Univ. Press

45. Darwin C. 1859. *On the Origin of Species by Means of Natural Selection or the Preservation of Favoured Races on the Struggle for Life.* London: John Murray

46. Derrida J. 1978. *Writing and Difference.* Chicago: Univ. Chicago Press

46a. de Candolle AP. 1820. Géographie botanique. In *Dictionnaire des Sciences Naturelles,* 18:359–422. Strasbourg: FG Levrault

46b. de Weerdt WH. 1989. Phylogeny and vicariance biogeography of North Atlantic Chalinidae (Haplosclerida, Demospongiae). *Beaufortia* 39:55–88

47. Dugdale JS. 1989. New Zealand Lepidoptera: a basic biogeography. *NZ J. Zool.* 16:679–87

48. Espinosa D, Llorente J. 1993. *Fundamentos de biogeografías filogenéticas.*

geographic analysis of some circum-Caribbean *Platynus* beetles (Carabidae). *Syst. Zool.* 37:385–409

82. Liebherr JK. 1991. Phylogeny and revision of the *Anchomenus* clade: the genera *Tetraleucus, Anchomenus, Sericoda,* and *Elliptoleus* (Coleoptera: Carabidae: Platynini). *Bull. Am. Mus. Nat. Hist.* 202:1–163

83. Liebherr JK. 1994. Biogeographic patterns of montane Mexican and Central American Carabidae (Coleoptera). *Can. Entomol.* 126:841–60

84. Livezey BC. 1986. Phylogeny and historical biogeography of steamer-ducks (Anatidae: *Tachyeres*). *Syst. Zool.* 35:458–69

85. Matthew WD. 1915. Climate and evolution. *Ann. New York Acad. Sci.* 24:171–318

86. Mayden RL. 1988. Vicariance biogeography, parsimony, and evolution in North American freshwater fishes. *Syst. Zool.* 37:329–55

87. Mayden RL. 1991. The wilderness of panbiogeography: a synthesis of space, time, and form? *Syst. Zool.* 40:503–19

88. Mayr E. 1946. History of the North American bird fauna. *Wilson Bull.* 58:3–41

89. Meacham C. 1984. Evaluating characters by character compatibility analysis. In *Cladistics: Perspectives on the Reconstruction of Evolutionary History,* ed. T Duncan, TF Stuessy, pp. 152–65. New York: Columbia Univ. Press.

90. Mickevich MF. 1981. Quantitative phylogenetic biogeography. See Ref. 51a, pp. 202–22

91. Mitter C, Brooks DR. 1983. Phylogenetic aspects of coevolution. In *Coevolution,* ed. DJ Futuyma, M Slatkin, pp. 65–98. Sunderland, MA: Sinauer

92. Moritz C, Hillis DM. 1990. Molecular systematics: context and controversies. In *Molecular Systematics,* ed. DM Hillis, C Moritz, pp. 1–10. Sunderland, MA: Sinauer

93. Morrone JJ. 1992. Revisión sistemática, análisis cladístico y biogeografía histórica de los géneros *Falklandius* Enderlein y *Lanteriella* gen. nov. (Coleoptera: Curculionidae). *Acta Entomol. Chil.* 17:157–74

94. Morrone JJ. 1993. Beyond binary oppositions. *Cladistics* 9:437–38

95. Morrone JJ. 1993. Cladistic and biogeographic analyses of the weevil genus *Listroderes* Schoenherr (Coleoptera: Curculionidae). *Cladistics* 9:397–411

96. Morrone JJ. 1993. Revisión sistemática de un nuevo género de Rhytirrhinini (Coleoptera: Curculionidae), con un análisis biogeográfico del dominio subantártico. *Bol. Soc. Biol. Concepción.* 64:121–45

97. Morrone JJ. 1994. On the identification of areas of endemism. *Syst. Biol.* 43:438–41

98. Morrone JJ. 1994. Systematics, cladistics, and biogeography of the Andean weevil genera *Macrostyphlus, Adioristidius, Puranius,* and *Amathynetoides,* new genus (Coleoptera: Curculionidae). *Am. Mus. Novit.* 3104:1–63

99. Morrone JJ. 1995. Distributional patterns of species of Rhytirrhinini (Coleoptera: Curculionidae) and the historical relationships of the Andean provinces. *Global Ecol. Biogeogr. Lett.* In press

100. Morrone JJ, Carpenter JM. 1994. In search of a method for cladistic biogeography: an empirical comparison of component analysis, Brooks parsimony analysis, and three-area statements. *Cladistics* 10. (2):99–153

101. Morrone JJ, Crisci JV. 1990. Panbiogeografía: fundamentos y métodos. *Evol. Biol. (Bogotá)* 4:119–40

102. Morrone JJ, Crisci JV. 1992. Aplicación de métodos cladísticos y panbiogeográficos en la conservación de la diversidad biológica. *Evol. Biol. (Bogotá)* 6:53–66

103. Morrone JJ, Katinas L, Crisci JV. 1995. Cladistic biogeography of Central Chile. *J. Biogeogr.* In press

104. Morrone JJ, Lopretto EC. 1994. Distributional patterns of freshwater Decapoda (Crustacea: Malacostraca) in southern South America: a panbiogeographic approach. *J. Biogeogr.* 21:97–109

105. Morrone JJ, Roig Juñent S, Crisci JV. 1994. Cladistic biogeography of terrestrial subantarctic beetles (Insecta: Coleoptera) from South America. *Natl. Geog. Res. Expl.* 10:104–15

106. Myers AA. 1991. How did Hawaii accumulate its biota? A test from the Amphipoda. *Global Ecol. Biogeogr. Lett.* 1:24–29

106a. Myers AA, Giller PS. 1988. *Analytical Biogeography: An Integrated Approach to the Study of Animal and Plant Distributions.* London & New York: Chapman & Hall

107. Myers AA, Giller PS. 1988. Process, pattern and scale in biogeography. See Ref. 106a, pp. 3–12

108. Nelson G. 1969. The problem of historical biogeography. *Syst. Zool.* 18:243–46

109. Nelson G. 1973. Comments on Leon

Croizat's biogeography. *Syst. Zool.* 22: 312–20

110. Nelson G. 1974. Historical biogeography: an alternative formalization. *Syst. Zool.* 23:555–58

111. Nelson G. 1978. From Candolle to Croizat: comments on the history of biogeography. *J. Hist. Biol.* 11:269–305

112. Nelson G. 1983. Vicariance and cladistics: historical perspectives with implications for the future. In *Evolution, Time and Space: The Emergence of the Biosphere,* ed. RW Sims et al, pp. 469–92. London & New York: Academic

113. Nelson G. 1984. Cladistics and biogeography. In *Cladistics: Perspectives on the Reconstruction of Evolutionary History,* ed. T Duncan, TF Stuessy, pp. 273–93. New York: Columbia Univ. Press.

114. Nelson G, Ladiges PY. 1991. Standard assumptions for biogeographic analyses. *Aust. Syst. Bot.* 4:41–58

115. Nelson G, Ladiges PY. 1991. Three-area statements: standard assumptions for biogeographic analysis. *Syst. Zool.* 40: 470–85

116. Nelson G, Ladiges PY. 1992. *TAS and TAX: MSDos computer programs for Cladistics.* Published by the authors, New York and Melbourne

117. Nelson G, Ladiges PY. 1993. Missing data and three-item analysis. *Cladistics* 9:111–13

118. Nelson G, Ladiges PY. 1994. *TASS. Three Area Subtrees.* Published by the authors, New York and Melbourne.

119. Nelson G, Platnick NI. 1980. A vicariance approach to historical biogeography. *Bioscience* 30:339–43

120. Nelson G, Platnick NI. 1981. *Systematics and Biogeography: Cladistics and Vicariance.* New York: Columbia Univ. Press

121. Nelson G, Platnick NI. 1988. Quantitative cladistic biogeography: constructing and comparing area cladograms. *Syst. Zool.* 37:254–70

122. Nelson G, Platnick NI. 1991. Three-taxon statements: a more precise use of parsimony? *Cladistics* 7:351–66

122a. Nelson G, Rosen DE. 1981. *Vicariance Biogeography: A Critique.* New York: Columbia Univ. Press

123. Page RDM. 1987. Graphs and generalized tracks: quantifying Croizat's panbiogeography. *Syst. Zool.* 36:1–17

124. Page RDM. 1988. Quantitative cladistic biogeography: constructing and comparing area cladograms. *Syst. Zool.* 37:254–70

125. Page RDM. 1989. *COMPONENT*

User's Manual. Release 1.5. Published by the author, Auckland, NZ

126. Page RDM. 1989. Comments on component-compatibility in historical biogeography. *Cladistics* 5:167–82

127. Page RDM. 1989. New Zealand and the new biogeography. *NZ J. Zool.* 16:471–83

128. Page RDM. 1990. Component analysis: a valiant failure? *Cladistics* 6:119–36

129. Page RDM. 1990. Temporal congruence in biogeography and cospeciation. *Syst. Zool.* 39:205–26

130. Page RDM. 1990. Tracks and trees in the Antipodes: a reply to Humphries and Seberg. *Syst. Zool.* 39:288–99

131. Page RDM. 1993. *COMPONENT user's manual.* Release 2.0. London: Nat. Hist. Mus.

132. Page RDM. 1993. Genes, organisms, and areas: the problem of multiple lineages. *Syst. Biol.* 42:77–84

133. Page RDM. 1994. Maps between trees and cladistic analysis of historical associations among genes, organisms, and areas. *Syst. Biol.* 43:58–77

134. Page RDM. 1994. Parallel phylogenies: reconstructing the history of host-parasite assemblages. *Cladistics.* In press

135. Page RDM, Lydeard C. 1994. Towards a cladistic biogeography of the Caribbean. *Cladistics* 10:21–41

136. Patterson C. 1981. Methods of paleobiogeography. See Ref. 122a, pp. 446–500

137. Platnick NI. 1981. Widespread taxa and biogeographic congruence. See Ref. 51a, pp. 223–27

138. Platnick NI. 1988. Systematics, evolution and biogeography: a Dutch treat. *Cladistics* 4:308–13

139. Platnick NI. 1991. On areas of endemism. *Aust. Syst. Bot.* 4:xi–xii

140. Platnick NI. 1992. Patterns of biodiversity. In *Systematics, Ecology, and the Biodiversity Crisis,* ed. N Eldredge, pp. 15–24. New York: Columbia Univ. Press.

141. Platnick NI, Nelson G. 1978. A method of analysis for historical biogeography. *Syst. Zool.* 27:1–16

142. Platnick NI, Nelson G. 1984. Composite areas in vicariance biogeography. *Syst. Zool.* 33:328–35

143. Platnick NI, Nelson G. 1988. Spanning-tree biogeography: shortcut, detour, or dead-end? *Syst. Zool.* 37:410–19

144. Raven PH, Axelrod DI. 1974. Angiosperm biogeography and past continental movements. *Ann. Missouri Bot. Gard.* 61:539–673

145. Ronquist F. 1994. Ancestral areas and parsimony. *Syst. Biol.* 43:267–74

146. Ronquist F, Nylin S. 1990. Process and pattern in the evolution of species associations. *Syst. Zool.* 39:323–44
147. Roos MC. 1990. Phylogenetic systematics of the Drynaroideae (Polypodiaceae). *Verh. Kon. Akad. Wetenschappen, Afd. Natuur. Tweede Reeks* 85:1–318
148. Rosen BR. 1988. Biogeographic patterns: a perceptual overview. See Ref. 106a, pp. 23–55
149. Rosen BR. 1988. From fossils to earth history: applied historical biogeography. See Ref. 106a, pp. 437–81
150. Rosen BR, Smith AB. 1988. Tectonics from fossils? Analysis of reef-coral and sea-urchin distributions from late Cretaceous to Recent, using a new method. In *Gondwana and Tethys Geol. Soc. Special Publ. No. 37*, ed. MG Audley-Charles, A Hallam, pp. 275–306. Oxford: Oxford Univ. Press
151. Rosen DE. 1976. A vicariance model of Caribbean biogeography. *Syst. Zool.* 24:431–64
152. Rosen DE. 1978. Vicariant patterns and historical explanation in biogeography. *Syst. Zool.* 27:159–88
153. Rosen DE. 1979. Fishes from the uplands and intermontane basins of Guatemala: revisionary studies and comparative geography. *Bull. Am. Mus. Nat. Hist.* 162:269–375
154. Ross HH. 1974. *Biological Systematics.* Reading: Addison-Wesley
155. Schuh RT, Stonedahl GM. 1986. Historical biogeography in the Indo-Pacific: a cladistic approach. *Cladistics* 2:337–55
156. Seberg O. 1986. A critique of the theory and methods of panbiogeography. *Syst. Zool.* 35:369–80
157. Seberg O. 1991. Biogeographic congruence in the South Pacific. *Aust. Syst. Bot.* 4:127–36
158. Simberloff D. 1987. Calculating the probabilities that cladograms match: a method of biogeographic inference. *Syst. Zool.* 36:175–95
159. Simberloff D, Heck KL, McCoy ED, Connor EF. 1981. There have been no statistical tests of cladistic biogeographic hypotheses. See Ref. 122a, pp. 40–63
160. Simpson GG. 1965. *The Geography of*
   Evolution. Philadelphia & New York: Chilton
161. Southey IC. 1989. The biogeography of New Zealand's terrestrial vertebrates. *NZ J. Zool.* 16:651–53
162. Tangney RS. 1989. Moss biogeography in the Tasman Sea region. *NZ J. Zool.* 16:665–78
162a. van Welzen PC. 1989. *Guioa* Cav. (Sapindaceae): Taxonomy, phylogeny, and historical biogeography. *Leiden Bot. Gard. Ser.* 12:1–315
162b. van Welzen PC. 1992. Interpretation of historical biogeographic results. *Acta Bot. Neerl.* 41:75–87
163. Wallace AR. 1876. *The Geographical Distribution of Animals.* New York: Hafner
164. Wallace AR. 1892. *Island Life.* London: Macmillan
165. Wallace CC, Pandolfi JM, Young A, Wolstenholme J. 1991. Indo-Pacific coral biogeography: a case study from the *Acropora selago* group. *Aust. Syst. Bot.* 4:199–210
166. Deleted in proof
167. Deleted in proof
168. Deleted in proof
169. Wiley EO. 1980. Phylogenetic systematics and vicariance biogeography. *Syst. Bot.* 5:194–220
170. Wiley EO. 1981. *Phylogenetics: The Theory and Practice of Phylogenetic Systematics.* New York: Wiley-Intersci.
171. Wiley EO. 1987. Methods in vicariance biogeography. In *Systematics and Evolution: A Matter of Diversity*, ed. P Hovenkamp et al, pp. 283–306. Utrecht: Inst. Syst. Bot., Utrecht Univ.
172. Wiley EO. 1988. Parsimony analysis and vicariance biogeography. *Syst. Zool.* 37:271–90
173. Wiley EO. 1988. Vicariance biogeography. *Annu. Rev. Ecol. Syst.* 19:513–42
174. Wiley EO, Mayden RL. 1985. Species and speciation in phylogenetic systematics, with examples from the North American fish fauna. *Ann. Missouri Bot. Gard.* 72:596–635
175. Zandee M. 1991. *CAFCA. A Collection of APL Functions for Cladistic Analysis.* Ver. 1.9.8. Leiden, The Netherlands
176. Zandee M, Roos MC. 1987. Component-compatibility in historical biogeography. *Cladistics* 3:305–32

Annu. Rev. Ecol. Syst. 1995. 26:403–22

MOLECULAR EVIDENCE FOR NATURAL SELECTION

Martin Kreitman and Hiroshi Akashi

Department of Ecology and Evolutionary Biology, University of Chicago, 1101 East 57th Street, Chicago, Illinois 60637

KEY WORDS: molecular evolution, protein evolution, neutral theory, nearly neutral theory, population genetics

---

ABSTRACT

Our understanding of the causes of molecular evolution is not as certain as it was a decade ago when Kimura's neutral theory appeared to explain major features of DNA conservation and change. The last ten years have seen the development of empirical approaches and statistical tests for detecting selection in DNA and a proliferation of data that challenge our current understanding of the molecular evolutionary process. We begin this review with a discussion of protein polymorphism and divergence: two major areas of research where the strictly neutral model cannot explain general patterns in the data. We then present a survey of statistical methods for detecting positive selection, which includes tests for balancing selection, for sequence convergence, and for unusually high rates of evolution that cannot be accounted for by neutral models. Finally, we present findings of a number of groups working on within- and between-species variation in *Drosophila*: These highlight the importance of adaptive evolution, purifying selection, and recombination in understanding levels and patterns of nucleotide variation.

INTRODUCTION

The success of the neutral theory of molecular evolution in explaining many patterns of DNA and protein variation in natural populations has created a serious challenge for evolutionary biologists. Under the neutral theory, genetic drift is the predominant force governing change at the molecular level. If adaptive evolution is a regular feature of phenotypic evolution, as it certainly

403

0066-4162/95/1120-0403$05.00

must be, then the inability to find evidence for natural selection in the genetic material itself (or in the case of allozymes, its proxy) raises far-reaching concerns about our understanding of evolution at the molecular level. Darwinian evolution requires adaptive substitutions in DNA.

To detect natural selection at the DNA level, the statistical analysis of variation within or between species must overcome two problems: identifying features of the process distinguishing genetic drift from natural selection, and detecting that signal when only a subset of mutational changes are under natural selection. Despite these obstacles, the development of methodologies for sequencing DNA and the advance of predictive theories to explain molecular evolutionary data have renewed interest in the statistical analysis of sequence variation. On the theoretical front, Kimura's neutral theory of molecular evolution (52) provides quantitative predictions for levels of variation both within and between species; this theory has laid the modern foundation for most thinking about molecular evolution (see, for example, 63). More recent theoretical advances include weak selection models (reviewed in 79), the reformulation of neutral theory using coalescence theory and gene genealogies (reviewed in 22, 41, 94), and the study of neutral variation linked to sites under directional selection (15, 49) and balancing selection (42, 48, 96). In addition, Gillespie (30) has developed models of selection in fluctuating environments.

The major theories of molecular evolution—Kimura's neutral model, Ohta's slightly deleterious model, and Gillespie's balancing and episodic selection models—are each consistent with at least some aspects of allozyme and DNA data. None of these models, however, can account for all available empirical observations. An understanding of the evidence, we believe, will require a comprehensive theory that emphasizes strong and weak forces acting simultaneously under constraints of genetic linkage and population size.

NEUTRAL THEORY AND PATTERNS OF PROTEIN VARIATION AND EVOLUTION

Allozyme Variation

The modern study of variation at the level of genes began with the development of a methodology for quantifying protein variation at single gene loci, by Hubby & Lewontin (40) and Harris (37), and the subsequent discovery of large amounts of polymorphism in *Drosophila pseudoobscura* (61, 84). The dilemma presented by the findings was quickly established. By 1970, Dobzhansky was satisfied that the large amounts of allozyme polymorphism in *Drosophila* and humans "are clearly in accord with the balance rather than the classical model of genetic population structure" (17, p. 224). However, recognizing arguments put forth by Kimura, Nei, and others, he went on to state that "The maintenance of abundant polymorphism and heterozygosity in populations demands, however, an explanation. ... The easiest way to cut the Gordian knot is, of course, to assume that a great majority of the polymorphisms observed involve gene variants that are selectively neutral, that is, have no appreciable effects on the fitness of their carriers."

A combination of biochemical studies and natural history data supports a role for natural selection in the maintenance of a number of enzyme polymorphisms (reviewed in 30, 95). Although case studies provide fascinating examples of biochemical adaptation, evidence for the selective maintenance of a few well-characterized enzyme polymorphisms does not help settle the more general issue of the relative contributions of selection and drift in determining observed levels of allozyme variation.

For allozymes, the lack of concrete statistical evidence for positive selection (be it balancing or directional) has led to the widespread belief that adaptive processes are infrequent in protein evolution. Many features of protein polymorphism and divergence are consistent with the neutral theory (52, 63, 74; see 30 for a rebuttal). Estimates of overall heterozygosity (73), the distribution of single locus heterozygosity (26, 75), variance in heterozygosity (32), number of alleles per locus (14), and correlation of single-locus heterozygosity between related species (13) can all be explained by the action of genetic drift on neutral variants. A positive correlation between the amount of allozyme polymorphism and the evolutionary rate is also predicted by neutral theory (87). However, Gillespie (31) has recently shown that such a correlation is also expected to occur at loci under balancing selection. Other authors have expressed caution about the overlap of predictions made by neutral and selection theories, especially as they relate to gene frequency data (21, 53).

ALLOZYME VS. DNA POLYMORPHISM   Under the strictly neutral theory, levels of polymorphism (measured by heterozygosity) will be proportional to the product of effective population size and the neutral mutation rate (52). Assuming similar mutation rates across species and populations at equilibrium, neutral theory predicts a positive correlation between genetic variation and population size. However, allozyme heterozygosity does not differ substantially between species. Nei & Graur (76) summarize data for hundreds of organisms in which 20 or more allozyme loci were examined: On average, invertebrates have at most only twice as much polymorphism as vertebrates. Why do polymorphism levels and population size fail to show the correlation predicted by the strictly neutral theory of molecular evolution? One possibility, proposed by Nei & Graur, is the occurrence of population bottlenecks in the recent history of many species.

However, comparison of DNA and protein polymorphism suggests that population bottlenecks do not explain the similarity of allozyme heterozygosity across species. In *Drosophila melanogaster* and *D. simulans,* the levels of allozyme variation are approximately the same (they may even be lower in *D. simulans*) (16). However, RFLP and DNA sequence studies indicate at least three- to six-fold higher levels of nucleotide polymorphism in *simulans* (reviewed in 4). If nucleotide heterozygosities are a measure of neutrally evolving mutations, and if allozyme variants are also maintained by genetic drift, then protein heterozygosity should show a similar difference between *D. melanogaster* and *D. simulans,* regardless of whether the populations are at equilibrium. The lack of such a pattern for allozyme variation cannot be explained by the strictly neutral theory. Aquadro (4) argues for weak selection against amino acid replacement mutations. Under a slightly deleterious model of protein evolution, selection will be more effective in removing weakly deleterious amino acid replacement mutations in the larger population size species.

Humans and *Drosophila*, for which there are abundant data, also show little difference in levels of allozyme polymorphism. Li & Sadler (64) addressed this issue by comparing human allozyme and DNA data. They restricted their analysis to alleles that were sequenced in the same lab, some 49 loci in total, to ensure that any differences between two sequences would have been checked. Note that a sequencing error rate of 1% may be 10 or 100 times higher than the naturally occurring polymorphic differences between two alleles, possibly as small as 0.1%–0.01%. What the study revealed was most surprising: The highest level of nucleotide polymorphism was only 0.11% for four-fold degenerate sites, some six-fold lower than estimates in *D. melanogaster* and more than ten-fold lower than in *D. simulans* and *D. pseudoobscura*.

The DNA data support a general intuition: There are (and have been) more flies than humans. But what about the roughly similar protein polymorphism levels in the two species? If the data are correct, and effective population size can be inferred from nucleotide data, the neutral model can again be rejected. Li & Sadler, like Aquadro, propose that slightly deleterious fitness effects for protein variants could account for these patterns. However, the authors do not attempt to evaluate the model quantitatively. Alternatively, pervasive balancing selection can also account for nearly equal levels of protein polymorphism (discussed below).

Protein Divergence

OVERDISPERSION OF THE MOLECULAR CLOCK   The clock-like divergence of many proteins was initially considered strong support for the neutral theory of molecular evolution (52). For strictly neutral mutations (with selection coefficients of zero), the rate of divergence will equal the mutation rate to neutral alleles, independent of population size. Variability of protein divergence, however, has been and remains a vexing problem for the neutral theory. The strictly neutral theory predicts, under a constant mutation rate to neutral alleles, that the expected variance in evolutionary rate will equal the mean rate (52). This is a simple consequence of modeling mutations as a Poisson point process. As early as 1971, Ohta and Kimura recognized that the rate constancy for amino acid replacement changes was violated for β-hemoglobin and cytochrome c (82). But the ratios, R, of the variance: mean ratio were not severely out of line with theoretical predictions (Rβ-globin = 2.05; Rα-globin = 1.37; $R_{cytochrome\ c}$ = 1.82). Langley & Fitch (56, 57), in more sophisticated analyses, found that rates of amino acid substitution for four proteins varied among mammalian lineages three-fold more than was predicted by neutral theory. Kimura admitted this apparent discrepancy but criticized detractors for not "seeing the forest for the trees" (52). Several follow-up studies by Gillespie (29, references cited therein) and the identification of an estimation bias (28) yielded a new average estimate of the overdispersion of the molecular clock, $R(t)$ = 7.75, and a renewed criticism of neutrality. Takahata (92, see also 24) modeled protein evolution with a changing neutral mutation rate to allow for an increased variance in the rate of substitutions. Gillespie, however, remained adamant, "That replacement substitutions are selected seems almost inescapable" (29, but see also 31).

LINEAGE AND GENERATION-TIME EFFECTS   The neutral theory predicts a constant rate of evolution equal to the mutation rate per generation. Given large differences in generation times (even within mammals) the clock-like behavior of protein divergence with absolute time is surprising (52). A generation-time effect observed in DNA evolution was therefore welcomed by neutralists. In a series of papers on DNA evolutionary rate, Li and coworkers (65–67) documented a two-fold slowdown in neutral divergence in humans compared to that of rhesus monkeys, approximately a ten-fold slowdown between humans and mice, and a two- to four-fold difference between humans and artiodactyls. The data are based on synonymous substitutions, noncoding sites (primarily introns) and pseudogenes (α and β-globin). There is also a suggestion of a rate slowdown for protein evolution in primates (humans in particular) compared to rodents (33, 35, 66, 67), but there are not yet enough data available to warrant any conclusion.

The generation-time effect seen for DNA may not hold for proteins. As presented in the now classic 1977 review of the molecular clock hypothesis, Wilson et al (97) rejected a generation-time effect in favor of a rate-constant dependency of protein substitution with absolute (geological) time. As with the allozyme and DNA heterozygosity, there may be an inconsistency between protein evolution and noncoding DNA evolution. Interestingly, Li et al (66) noted the same possible discrepancy in their DNA sequence analysis, but they offered no explanation.

Easteal & Collet (20) recently compared rates of replacement and silent substitution between rodent and primate lineages, using marsupials as an outgroup. At replacement sites, of 14 genes, 12 show greater divergence in the rodent lineage. Many of these loci show variable rates of silent evolution (some appear to be near saturation), but there is no overall lineage effect. Contrary to the findings of Li and coworkers, these results suggest a faster rate of protein evolution in rodents than in primates. Easteal & Collet invoke slightly deleterious mutations going to fixation in the presumably smaller populations of the rodent lineage to explain the higher rates of protein evolution. There is no evidence, however, to support the idea that historical population sizes have been smaller in rodents than in primates. Interestingly, weakly deleterious protein variants are also invoked (77) to explain the lack of a lineage effect in protein evolution found by Li and coworkers; the rate of protein evolution in the presumably smaller populations of primates increases by allowing more slightly deleterious mutations to go to fixation by genetic drift, thereby compensating for the slowdown caused by the intrinsically lower mutation rate per unit of absolute time. Again, models of adaptive evolution cannot be excluded as explanations for the protein data (31).

Slightly Deleterious Protein Evolution

One of the most appealing aspects of the neutral theory is its ability to make quantitative predictions both for expected levels of variation within populations and for divergence between species. Strictly neutral theory fails to explain either the lack of variation in levels of protein polymorphism in different species or the unexpectedly high levels of variation in rates of protein divergence (observed both as overdispersion of the clock and as lineage effects).

Ohta, Kimura, and others (reviewed in 52, 79) have developed the slightly deleterious model of molecular evolution to explain some of these discrepancies. The relative contributions of stochastic and deterministic forces to the evolutionary dynamics of slightly deleterious mutations depends critically on population size. This model posits the existence of a large class of protein variants with selection coefficients in the range $1/N_e$ (the reciprocal of the species effective population size). This allows nonneutral patterns to be explained by the dependence of deleterious selection on population sizes. Unfortunately, the lack of independent estimates of N_e in most natural populations allows great freedom to invoke "near neutrality" to explain many non-neutral patterns of protein evolution. More importantly, although many aspects of the evolutionary dynamics of weakly selected mutations have been investigated theoretically (79), there is little direct evidence that a proportion of amino acid mutations falls within this class of fitness effects. In *Drosophila*, where high levels of silent polymorphism suggest very large evolutionary effective population sizes, the region encompassing:

$|s| < 1/N_e$ may be smaller than $|s| < 10^{-6}$. Such a region could not even be represented by a thin line in the classic chromosome viability histograms (71, 72). We know virtually nothing about the distribution of fitness effects of new mutations around zero.

Perhaps the strongest available evidence for weak selection on protein variants comes from Hartl et al's (38) study of replacement and silent DNA polymorphism at the *Escherichia coli* 6-phosphogluconate dehydrogenase (gnd) locus. When compared to that of silent mutations, the frequency distribution of amino acid mutations is significantly skewed toward rare variants, implying the action of purifying selection on protein changes. However, maximum likelihood estimates show that the selection intensity for replacement mutations is probably not more than an order of magnitude greater than for silent changes. It would be very interesting to see whether rates of *gnd* protein evolution vary considerably between lineages with different effective population sizes.

Ohta (80) analyzed data on alcohol dehydrogenase genes in Hawaiian *Drosophila*, the *D. melanogaster* species subgroup, and the *D. obscura* species subgroup, showing that the replacement substitution rate relative to the synonymous rate is 40–50% higher in the (presumably) smaller populations of the Hawaiian species. The data are consistent with a slightly deleterious model, but adaptive alternatives such as the episodic selection model of Gillespie (30) cannot be rejected.

STATISTICAL EVIDENCE FOR ADAPTIVE PROTEIN EVOLUTION

Large-scale patterns of protein polymorphism and divergence allow us to reject the strictly neutral model of molecular evolution. Unfortunately, we cannot distinguish whether nearly neutral or adaptive models better account for these data. Locus-specific molecular evidence for adaptive protein evolution, however, has become increasingly abundant. We briefly review some of the best documented cases; all are based on recent work using DNA comparisons.

Balancing Selection

The theory that balancing selection maintains allozyme polymorphism is an obvious alternative to selection against slightly deleterious mutations (with shifting population size) to explain the lack of variation in allozyme heterozygosities among species. Under this model of selection, levels of polymorphism can be nearly independent of population size, instead being governed by environmental conditions and the rate at which balanced polymorphisms arise. Very few studies have convincingly demonstrated balancing selection in nature, and there are a number of theoretical arguments against a prominent role for it (52, 60, but see also 88). Single-locus examples, however, exist in both humans and *Drosophila*. Recent work by Berry & Kreitman (9) provides strong evidence for balancing selection maintaining the alcohol dehydrogenase (*Adh*) gene frequency cline along the east coast of the United States in *D. melanogaster*. Of some 20 polymorphic nucleotide sites in the *Adh* locus, only the Fast-Slow amino acid replacement polymorphism and an insertion in an intron (which increases gene expression levels—58) are significantly clinal.

The hypothesis that balancing selection maintains the Fast-Slow polymorphism is also supported by an analysis of silent DNA variation surrounding the single amino acid change. Under balancing selection, alleles can continue segregating in the population longer than would be expected for neutral variants. Neutral mutations will accumulate at sites closely linked to selected alleles, leading to unusually high levels of silent variation between them (42, 48). A conservative statistical test shows an excess of silent variation around the Fast-Slow replacement polymorphism that cannot be explained by genetic drift (43). The HKA test will detect only balanced polymorphisms that have been segregating in a population for a long period of time and only where the recombination rate surrounding the selected site is sufficiently low. The test has been applied to data for several other loci in *D. melanogaster* that are thought to have polymorphism maintained by selection; only *Adh*, and possibly alpha-glycerophosphate dehydrogenase (93) provide evidence for balancing selection. Balancing selection is either rare or short-lived, or else recombination often erases its footprint. In addition, interpretation of the HKA test for signature *Drosophila* data has become complicated by the presence of codon selection (2) as well as selective sweeps and background selection (discussed in later sections).

Hudson et al (44) have recently proposed a test to detect the action of balancing selection on more recently derived alleles. In a DNA sequence study of *D. melanogaster* superoxide dismutase (*Sod*) genes, 19 alleles of the slow allozyme variant share an identical haplotype, whereas 22 alleles of the (presumably older) fast allozyme have far more sequence variation. Hudson et al show that the high frequencies (around 50% in both California and Spain) of the recently derived Sod^s variant are unlikely to have occurred solely through the action of genetic drift. The action of natural selection may be necessary to explain the rapid spread of Sod^s.

A more general claim of overdominant selection maintaining allozyme polymorphism is made by Karl & Avise (51) for the American oyster, *Crassostrea virginica*. Populations of this oyster, which are found along the East coast of the United States, including the Gulf of Mexico, are highly polymorphic for allozymes. There is little genetic differentiation among populations—not surprising for a species with planktonic larvae. However, mitochondrial DNA analysis reveals substantial differentiation between Gulf and East Coast populations. Subsequent analyses of two random single-copy nuclear DNA polymorphisms appear to confirm differentiation between the two populations. Similar clines for a number of allozymes in the face of strong population subdivision suggests that balancing selection is maintaining the allozymes at relatively constant frequencies across the species's range. But with only two nuclear markers to confirm the mtDNA data, we cannot draw any strong conclusions about a role for selection. Indeed, divergent selection on mtDNA may provide a more parsimonious explanation for the data. Additional nuclear DNA data should be collected for this species.

Accelerated Protein Evolution

The emergence of DNA sequence data for many genes and species reveals a general principle about protein evolution: The vast majority of amino acid replacement mutations are disadvantageous and are eliminated by selection. The rate of amino acid replacement evolution can be as low as zero but is almost always less than the rate of silent evolution. Thus, rates of protein evolution can be explained by a combination of the elimination of deleterious replacement mutations by purifying selection and the fixation of selectively neutral mutations by genetic drift. Natural selection is undoubtedly vigilant in removing deleterious mutations. The question is whether all other (nondeleterious) amino acid changes observed within and between natural populations are neutral, weakly selected, or adaptive.

Gillespie (27, 31) has proposed a model of episodic selection to allow for sporadic bursts of change to account for the larger than expected rate variation in protein evolution. However, there are few, if any, specific examples of episodic selection. He cites several examples of accelerated evolution—baboon hemoglobin, visual pigments and human cytochromes—along with the caveat, "... the causes of most of the accelerations described ... are unknown."

GENE DUPLICATIONS   The strongest evidence for accelerated protein evolution is that which follows gene duplication (62, 78, 81). The phenomenon was first described for hemoglobins following the split of the α and β families (34, 36). Ohta cites a number of additional examples, including hemoglobin γ (Anthropoidia) and β (goat vs. sheep), stomach lysozyme ruminants, visual pigment and adrenergic receptor (human), histocompatibility antigen (antigen recognition site, human and mouse), immunoglobulins (mouse heavy chain and rat appa chain), and protease inhibitor (inhibitory site, many species). Other examples of accelerated evolution associated with gene duplication are provided by Li (62), including somatostatin (anglerfish and catfish), cytochrome c (*Drosophila*), and growth hormone genes (human and bovine). Rate accelerations have also been convincingly demonstrated for insulin in hystricomorph rodents (11, 30). In this case, however, the acceleration is attributed not to duplication but to a change in the active protein from a hexamer to a monomer in the guinea pig lineage.

In all the above examples, the evidence for rate acceleration is high value for protein divergence, k_a, relative to silent divergence, k_s, or relative to average k_a values in other lineages. In only one case does k_a exceed k_s (histocompatibility antigen). Given the large expected variance in evolutionary rates, the absence of adequate statistical testing, and a lack of functional information about the consequences of the substitutions, it is unlikely that this approach will be useful for distinguishing between adaptive evolution and relaxed constraints. Perhaps it is not surprising that Li's (62) conclusion about the likely cause of accelerated protein evolution following gene duplication, "... relaxation of selective constraints seems to be a more plausible explanation than advantageous mutation," contrasts sharply with Ohta's (78) conclusion about the same data: "Although it is difficult again to judge which of the two hypotheses is correct, it is likely that natural selection favored those individuals that possessed desired mutations in the duplicated gene copies."

A convincing case of adaptive evolution following gene duplication is the rapidly evolving *jingwei* gene in *D. yakuba* and *D. teissieri* (68). The two species form a monophyletic clade within the *D. melanogaster* species subgroup. A gene duplication arose from the *Adh* gene by retrotransposition in the *yakuba-teissiere* clade, landing on a different chromosome. Not only does k_a exceed k_s between paralogous genes within each species as well as between the homologous *jingwei* genes between species, there is also a significantly higher ratio of amino acid replacement to synonymous changes between species than to those within species. Such comparisons, if synonymous changes are neutral, reveal the action of positive selection for amino acid changes (70, 85, discussed below).

REPLACEMENT VS. SILENT DIVERGENCE   As discussed above, some proteins show faster rates of replacement than of synonymous DNA evolution. First, and possibly most dramatic, is the rapid evolution of the antigen recognition sites (ARS) of class I and II genes of the major histocompatibility loci of humans and mice. As Hughes & Nei (46) were able to show, replacement divergence in the ARS region is higher than at silent and noncoding sites in the same species. If silent sites are evolving neutrally, as is generally believed, then the higher rate for the antigenic sites must reflect the contribution of positive natural selection. In addition, nonsynonymous substitutions resulting in side-chain charge changes occur in the binding cleft of the ARS more frequently than predicted by chance (47). These same loci, it is worth noting, offer dramatic evidence for persistent balancing selection: Two highly diverged alleles are shared in mice and rats, suggesting a most recent common ancestry of at least 13 million years ago (23).

Several points should be made about this form of evidence for positive selection. First, requiring amino acid replacement changes to be more frequent than silent changes is an extremely stringent criterion for detecting selection. Because purifying selection is the most prominent form of selection on proteins (52), the rate of amino acid replacement substitution will tend to be much lower than that of synonymous divergence. Probably only in the rarest of instances will positive selection raise the rate to a level exceeding the neutral rate; hence, many instances of adaptive evolution may be missed. Second, in species exhibiting biased codon usage, purifying selection appears to constrain synonymous divergence (reviewed in 86). The validity of the $k_a > k_s$ comparison depends critically on the assumption of neutrality at silent sites, an assumption that must be examined carefully (2, 86). Third, the elevated rate of replacement substitution will likely be restricted to a single, and possibly small, domain of the protein. In the case of MHC, circularity in defining the selected region is avoided because the antigen recognition site has been identified independently of the pattern of replacement divergence.

Elevated protein divergence compared to synonymous divergence is observed for sperm lysins in 20 California abalone species, suggesting positive selection (58, 59). Amazingly, a number of pairwise comparisons show an excess of nonsynonymous divergence in the whole protein. In addition, Lee et al (58) find little bias in codon usage in these genes, suggesting that purifying selection at silent sites does not explain the excess of replacement divergence between species. The selection pressure driving adaptive evolution of abalone sperm lysins, however, remains to be established.

LINEAGE-BASED COMPARISONS   The very rapid evolution of antigenic sites in the hemaglutinins of the influenza A virus affecting humans has long been thought to be driven by selection to escape the immune response of their host. Fitch et al (25) provide statistical evidence supporting this contention. The protein can be divided into antigenic and nonantigenic sites (much about the structure and function of the protein is known, largely due to the efforts of Wiley and colleagues, cited in 25). In addition, this virus provides a unique opportunity to study molecular evolution. Because many (now extinct) strains were collected and kept in the laboratory, evolution in the surviving lineages can be contrasted with that of the extinct branches. Fitch et al show a significant excess of antigenic changes on the (surviving) trunk of the phylogenetic tree compared to the dead-end (extinct) branches. Positive selection appears to be driving the rapid evolution of the antigenic site in the influenza A virus.

Hughes (45) has tested the neutral theory prediction that the number of amino acid replacements in a given gene region should be a linear function of that in another region of the same gene. Applying this idea to members of the heat-shock protein 70 gene family, Hughes finds nonlinear relationships between three functional domains of the protein. The results are consistent with adaptive divergence among subfamily members of the gene, but the possibility of changes in functional constraint in one or more domains in some of the lineages cannot be rejected.

Convergent Evolution

Functional convergence through parallel substitutions in different evolutionary lineages demonstrates the action of Darwinian protein evolution. Although protein functional convergence can occur with or without structural or sequence convergence, rigorous evidence for adaptive protein evolution is limited to cases of convergence in both function and amino acid sequence.

To demonstrate sequence convergence, shared changes must be shown to be derived characters rather than conserved ancestral states or chance events; convergence can be discerned only in a phylogenetic context. Gut lysozyme evolution in the cow and langur monkey provides the strongest available evidence for adaptive protein convergence (89, 90). Lysozymes, expressed in macrophages, function as antibacterial enzymes. Two mammalian orders (primates and artiodactyls) have independently recruited lysozyme to digest cellulose in a fermentative foregut. The authors developed a phylogeny-based test for convergence by contrasting silent and replacement changes in four lineages: cow (artiodactyl), mouse (rodent), and rhesus and langur monkeys (primates). The test involves constructing the three possible relationships among the sequences of the lysozymes of the four species. Of 14 silent DNA changes, 13 support the well-established tree (primates together, cow and mouse together) while 6 of the 15 replacement sites support a genealogy linking the langur monkey with the cow. The distributions of silent and replacement changes on the standard tree are significantly different, suggesting that at least some of the 6 replacement changes have occurred in parallel.

There are other potential examples of convergent evolution. Yokoyama & Yokoyama (98) compared visual pigments in blind cave fish and humans. The authors argue for the independent evolution of the red visual pigments in both fish and humans from ancestral green pigments. However, the postulated convergence involves only two or three possible amino acid changes out of 15 variable residues, and it remains to be shown that the parallel changes are not merely coincidental events. Functional significance of the putative convergent changes should be investigated.

Other than the lysozymes, few claims of convergence are supported by phylogenetic analyses. An example lacking such analysis is rattlesnake cytochrome c, hypothesized to be convergent with the human homologue (3). A comparison of the amino acid sequence of rattlesnake cytochrome c to that of eight other vertebrates using a difference matrix indicates that rattlesnake cytochrome c is most similar to human cytochrome c. Human and rattlesnake cytochrome c differ by 14 of 104 amino acids. However, rattlesnake and monitor lizard cytochrome c differ by only 16 amino acids. Without any evidence of functional convergence between human and rattlesnake cytochrome c and in the absence of a phylogenetic analysis, the similarity between these two molecules could be explained as homoplasy resulting from accelerated substitution rather than convergent evolution.

Examples of convergent evolution supported by phylogenetic tests are few in number. Unfortunately, a phylogenetic analysis does not guarantee detection of convergence because the power of the statistical analysis depends on the number of parallel substitutions. Like the comparison of rates of replacement and silent divergence, this test will fail to detect instances where the number of adaptive changes is small.

Ratios of Polymorphism and Divergence

McDonald & Kreitman (70) tested a simple prediction of the strictly neutral theory for polymorphism within species and substitutions between species. According to theory, levels of polymorphism and rates of change are positively correlated, both being governed by the neutral mutation rate (52). A region of a gene with many possible neutral mutations should be more polymorphic and should evolve faster than a similar-sized region under more severe selective constraints. The prediction was tested for amino acid replacement changes and for synonymous changes at the *Adh* locus in three species, *D. melanogaster*, *D. simulans,* and *D. yakuba.* A statistically significant excess of amino acid replacement changes between species compared to synonymous changes was observed, suggesting that a significant fraction of the amino acid replacement changes between species was driven by natural selection.

Several other proteins appear to violate the neutral theory prediction for polymorphism and divergence. In a within- and between-species comparison of the *glucose-6-phosphate dehydrogenase* locus of *D. melanogaster* and *D. simulans* (18), only two amino acid replacement polymorphisms were detected in a total of 44 *G6pd* alleles from the two species. One, like *Adh-F/S*, is likely to be a balanced polymorphism. In contrast, there are 21 replacement differences between the species. The McDonald-Kreitman test is highly significant. Karotam et al's study of esterase 6 (50) showed that, unlike *Adh* and *G6pd*, *Est 6* is highly polymorphic for amino acid replacements. Nevertheless, there is a statistically significant excess of replacement substitutions between species compared to synonymous changes.

There are two possible alternative explanations for a significant departure from the neutral prediction in the direction of "too many" amino acid replacements between species. The first is that synonymous changes, rather than being neutral, are subject to weak negative selection. This is likely to be true for *Adh*, which is highly codon-biased. Compared to the neutral case, polymorphism will be less severely reduced than will substitutions; a departure from neutrality in the direction observed is expected if synonymous changes are negatively selected (2). The second alternative, suggested by Ohta (80), is that the population sizes of the extant species have recently increased compared to their evolutionary sizes. If amino acid replacement mutations are slightly deleterious (selection coefficients of order $1/N_e$), then the increase in population size would allow selection to remove the deleterious mutations from current populations. Under this scenario, a higher level of replacement divergence would be expected compared to polymorphism.

NATURAL SELECTION, RECOMBINATION, AND GENETIC VARIATION IN DROSOPHILA

Genetic Hitchhiking and Selective Sweeps

The final kind of evidence supporting the adaptive evolution hypothesis (but not the balancing selection hypothesis) is genetic hitchhiking in *Drosophila*. First studied theoretically by Maynard Smith & Haigh (69) and by Ohta & Kimura (83), and more recently by Kaplan et al (49), hitchhiking occurs when a neutral mutation changes frequency through genetic linkage to a mutation that is selected. Of particular interest is the effect of a recent adaptive fixation on the level of neutral polymorphism in a region surrounding the beneficial mutation. Depending on a number of factors—neutral mutation rate, recombination rate, population size, strength of selection, and time since the selective substitution—a selective sweep of a favored mutation will not only homogenize the population (or species) for the favored mutation, but it will also homogenize the population for sufficiently tightly linked neutral mutations. Genetic hitchhiking can reduce variation surrounding the site under selection.

Hitchhiking events (i.e. selective substitutions) can be inferred from a relative lack of synonymous or noncoding polymorphism at a locus in a species otherwise known to have relatively high levels of silent polymorphism. In addition, it must be shown that the lack of polymorphism, even if it is noncoding, cannot be attributed to selective constraints on the sites under consideration. This is relatively easily accomplished by comparing evolutionary divergence for the affected sites or region among species; selective constraint, but not hitchhiking, will result in a relatively smaller divergence between species.

In contrast to stringent tests for adaptive evolution described in the previous sections, which can require many changes concentrated in a small region of a gene, reduced variation via hitchhiking can result from only a single selection event. Unfortunately, a loss of precision is the penalty for the higher sensitivity of this test; the selected mutation cannot be localized within the region of reduced variation, nor can it be classified as a replacement or noncoding change.

The first evidence for a hitchhiking effect in *Drosophila* was a report of low DNA polymorphism levels in the yellow-achaete-scute region of the X chromosome in *D. melanogaster* (1, 6, 7, 19). This is especially so for *D. simulans,* where there is a dramatic lack of polymorphism. Yellow-achaete-scute is at the distal tip of the X chromosome, a region known to have severely reduced recombination, rendering its polymorphism level sensitive to selective sweeps at a distance.

Following the same reasoning, Berry et al (10) sequenced 19 *cubitus-interruptus-Dominant, ci^D,* genes in *D. melanogaster* and *D. simulans.* Only a single polymorphism was found, whereas the two species differed at 10% of sites. Clearly, selective sweeps have occurred in the relatively recent past of both species (they are too distantly related for a single sweep to have occurred in the common ancestor). ci^D is located on the fourth chromosome, which does not undergo recombination. These data also suggest that adaptive sweeps may be a regular feature of molecular evolution. There are only approximately 50 known complementation groups on the fourth chromosome (39), and independent sweeps have occurred in both *melanogaster* and *simulans.*

It is now becoming apparent that many if not all regions of (severely) reduced recombination in *Drosophila* exhibit reduced levels of polymorphism (5). Begun & Aquadro (8) have elevated the selective sweep process to prominence as a force in molecular evolution, suggesting the possibility of a high density of sweeps throughout the genome. Comparing recombination rates and polymorphism levels for 17 loci in *D. melanogaster*, they find a surprisingly strong correlation ($r^2 = 0.42$). Correctly noting the large expected variances for RFLP-based polymorphism estimates, the correlation suggests that many of the regions have sustained relatively recent selective sweeps. For this kind of comparison to be meaningful, the analysis must take into account variation in levels of selective constraint among loci, which can be estimated by quantifying between-species divergence levels. Unfortunately, such a correction complicates the development of an appropriate sampling theory and statistical test.

Background Selection

The recent development of an alternative model to explain the correlation between recombination rate and polymorphism level further complicates this picture. Charlesworth et al (15) show that selection against newly arising deleterious mutations, "background selection," can substantially reduce the level of linked neutral variation if deleterious mutations arise at sufficiently high rates and in tightly linked blocks. The mechanism for the reduction in polymorphism can be attributed to the reduction in the number of nondeleterious chromosomes in the population. According to the authors' calculations, this mechanism will reduce polymorphism levels by as much as 78% at the base of chromosomes two and three. It cannot account for the severe reduction in polymorphism observed for the fourth chromosome and for the tip of the X. Thus, at least for these regions, positive selective sweeps are likely to have occurred.

The frequency distribution of segregating mutations may provide information to distinguish between selective sweeps and background selection as explanations for reduced variation. A recent adaptive fixation will cause an excess of rare variants as well as a reduction in the nucleotide heterozygosity in the affected region. The effect is equivalent to the recovery of variation following a population bottleneck (12). Background selection, however, appears to have little effect on the expected frequency spectra of mutations (15). Tajima (91) has developed a statistical test to determine if frequency spectra deviate from that expected for neutral mutations at equilibrium. Braverman et al (12) suggest that this test should have sufficient power, in available data sets, to detect the excess of rare variants predicted by selective sweeps. They conclude that other forces (in addition to selective sweeps) must be invoked to explain the lack of evidence for skewed frequency distributions in regions of reduced variation in *Drosophila*. Aquadro & Begun's hypothesis remains highly contentious—it will certainly remain a major focus of attention in *Drosophila* population genetics.

As a final comment, we note that the reduction in variation observed for all regions of reduced recombination in *D. melanogaster* violates the balanced polymorphism hypothesis for the maintenance of variation. If a single balanced polymorphism was maintained on the fourth chromosome, for example, linked neutral polymorphism would be expected to accumulate throughout the two selected chromosomes (96). Recall, this was the explanation for the two highly diverged MHC alleles shared between mouse and rat. The lack of high levels of polymorphism in regions of reduced recombination allows us to reject balancing selection as a general explanation for the maintenance of genetic variation. Balancing selection, if it occurs with appreciable frequency, may not last long enough for the accumulation of linked neutral mutations.

CONCLUSIONS

A detailed picture of DNA polymorphism in *Drosophila* is emerging. Unfortunately there is no simple explanation for the complexity of the observed patterns. Levels of variation at a locus may depend on selection at the locus, selection (both positive and negative) in the chromosomal region of the locus, and the population dynamics of the species. Although none of the "standard" models of population genetics adequately explains all the molecular data, individual features of the data can be explained in terms of simple processes such as genetic drift or background selection against deleterious mutations.

Neither the strictly neutral model nor any model of molecular evolution can account for major features of protein evolution. Although genetic drift may play a major role in DNA evolution, the new data resurrect the question of what causes protein polymorphism and divergence. Curiously, we cannot distinguish between deleterious and adaptive models for much of the data. Whether adaptive evolution at the molecular level achieves the hegemony it enjoys in phenotypic evolution is debatable, but recent evidence suggests it deserves a new level of appreciation.

ACKNOWLEDGMENTS

Special thanks to Jennifer Hess for literature research and discussion of convergent evolution. We are also grateful to Peter Andolfatto, Eli Stahl, and Ling-Wen Zeng for their help in improving this manuscript. H Akashi is a Howard Hughes Medical Institute Predoctoral Fellow.

Literature Cited

1. Aguadé M, Miyashita N, Langley CH. 1989. Reduced variation in the yellow-achaete-scute region in natural populations of *Drosophila melanogaster*. *Genetics* 122:607–15

2. Akashi H. 1995. Inferring weak selection from patterns of polymorphism and divergence at 'silent' sites in *Drosophila* DNA. *Genetics* 139:1067–76

3. Ambler RP, Daniel M. 1991. Rattle-snake cytochrome c: a reappraisal of the reported amino acid sequence. *Biochem. J.* 274:825–31

4. Aquadro CF. 1991. Molecular population genetics of *Drosophila*. In *Molecular Approaches to Fundamental and Applied Entomology*, ed. J Oakeshott, MJ Whitten, pp. 222–66. New York: Springer

5. Aquadro CF, Begun DJ. 1993. Evidence for and implications of genetic hitchhiking in the *Drosophila* genome. In *Mechanisms of Molecular Evolution*, ed. N Takahata, AG Clark, pp. 159–78. MA: Sinauer

6. Beech RN, Leigh Brown AJ. 1989. Insertion-deletion variation at the yellow-achaete-scute region in two natural populations of *Drosophila melanogaster*. *Genet. Res.* 53:7–15

7. Begun DJ, Aquadro CF. 1991. Molecu-lar population genetics of the distal portion of the X chromosome in *Drosophila*: evidence for genetic hitchhiking of the yellow-achaete region. *Genetics* 129:1147–58

8. Begun DJ, Aquadro CF. 1992. Levels of naturally occurring DNA polymorphism correlate with recombination rates in *D. melanogaster*. *Nature* 356: 519–20

9. Berry AJ, Kreitman M. 1993. Molecular analysis of an allozyme cline: alcohol dehydrogenase in *Drosophila melanogaster* on the East coast of North America. *Genetics* 134:869–93

10. Berry AJ, Ajioka JW, Kreitman M. 1991. Lack of polymorphism on the *Drosophila* fourth chromosome resulting from selection. *Genetics* 129:1111–17

11. Blundell TL, Wood SP. 1975. Is the evolution of insulin Darwinian or due to selectively neutral mutations? *Nature* 257:197–203

12. Braverman JM, Hudson RR, Kaplan NL, Langley CH, Stephan W. 1995. The hitchhiking effect on the site frequency spectrum of DNA polymorphism. *Genetics.* 140:783–95

13. Chakraborty R, Fuerst PA, Nei M. 1978. Statistical studies on protein polymor-

phism in natural populations. *Genetics* 88:367–90

14. Chakraborty R, Fuerst PA, Nei M. 1980. Statistical studies on protein polymorphism in natural populations. III. Distribution of allele frequencies and the number of alleles per locus. *Genetics* 94:1039–63

15. Charlesworth B, Morgan MT, Charlesworth D. 1993. The effect of deleterious mutations on neutral molecular variation. *Genetics* 134:1289–303

16. Choudhary M, Singh R. 1987. A comprehensive study of genetic variation in natural populations of *Drosophila melanogaster*. III. Variations in genetic structure and their causes between *Drosophila melanogaster* and its sibling species *Drosophila simulans*. *Genetics* 117:697–710

17. Dobzhansky T. 1970. *Genetics of the Evolutionary Process*. New York: Columbia Univ. Press

18. Eanes WF, Kirchner M, Yoon J. 1993. Evidence for adaptive evolution of the G6PD gene in the *Drosophila melanogaster* and *D. simulans* lineages. *Proc. Natl. Acad. Sci. USA* 90:7475–79

19. Eanes WF, Labate J, Ajioka JW. 1989. Restriction-map variation in the yellow-achaete-scute region in five populations of *Drosophila melanogaster*. *Mol. Biol. Evol.* 6:492–502

20. Easteal S, Collet C. 1994. Consistent variation in amino-acid substitution rate, despite uniformity of mutation rate: protein evolution in mammals is not neutral. *Mol. Biol. Evol.* 11:643–47

21. Ewens WJ. 1979. *Mathematical Population Genetics. Biomathematics,* Vol. 9. New York: Springer-Verlag

22. Ewens WJ. 1989. Population genetics theory—the past and the future. In *Mathematical and Statistical Problems of Evolutionary Theory*, ed. S Lessard, pp. 177–227. Dordrecht: Kluwer Acad.

23. Figueroa F, Günther E, Klein J. 1988. MHC polymorphism pre-dating speciation. *Nature* 355:265–67

24. Fitch WM. 1971. Rate of change of concomitantly variable codons. *J. Mol. Evol.* 1:84–96

25. Fitch WF, Leiter JME, Li X, Palese P. 1991. Positive Darwinian evolution in human influenza A viruses. *Proc. Natl. Acad. Sci. USA* 88:4270–74

26. Fuerst PA, Chakraborty R, Nei M. 1977. Statistical studies on protein polymorphism in natural populations. I. Distribution of single locus heterozygosity. *Genetics* 86:455–83

27. Gillespie JH. 1984. The molecular clock may be an episodic clock. *Proc. Natl. Acad. Sci. USA* 81:8009–13

28. Gillespie JH. 1986. Variability of evolutionary rates of DNA. *Genetics* 113:1077–91

29. Gillespie JH. 1989. Lineage effects and the index of dispersion of molecular evolution. *Mol. Biol. Evol.* 6:636–47

30. Gillespie JH. 1991. *The Causes of Molecular Evolution*. New York: Oxford Univ. Press

31. Gillespie JH. 1994. Substitution processes in molecular evolution. II. Exchangeable models from population genetics. *Evolution* 48:1101–13

32. Gojobori T. 1982. Means and variances of heterozygosity and protein function. In *Molecular Evolution, Protein Polymorphism and the Neutral Theory*, ed. M Kimura, pp. 137–50. New York: Springer Verlag

33. Goodman M. 1961. The role of immunochemical differences in the phyletic development of human behavior. *Human Biol.* 33:131–62

34. Goodman M. 1976. Protein sequences in phylogeny. In *Molecular Evolution*, ed. FJ Ayala, pp. 141–59. Sunderland, MA: Sinauer

35. Goodman M, Barnabas J, Matsuda G, Moore GW. 1971. Molecular evolution in the descent of man. *Nature* 233:604–13

36. Goodman M, Czelusniak J, Koop BF, Tagle DA, Slightom JL. 1987. Globins: a case study in molecular phylogeny. *Proc. Cold Spring Harbor Symp. Quant. Biol.* 52:875–90

37. Harris H. 1966. Enzyme polymorphisms in man. *Proc. Roy. Soc. B.* 164:298–310

38. Hartl DL, Moriyama EN, Sawyer S. 1994. Selection intensity for codon bias. *Genetics* 138:227–34

39. Hochman B. 1976. The fourth chromosome of *Drosophila melanogaster*. In *The Genetics and Biology of Drosophila*, Vol. 1b, ed. M Ashburner, E Novitski, pp. 903–28. New York: Academic

40. Hubby JL, Lewontin RC. 1966. A molecular approach to the study of genic heterozygosity in natural populations. I. The number of alleles at different loci in *Drosophila pseudoobscura*. *Genetics* 54:577–94

41. Hudson RR. 1992. Gene genealogies and the coalescent process. In *Oxford Series in Ecology and Evolution*, vol. 7, ed. D Futuyma, J Antonovics, pp. 1–44. Oxford: Oxford Univ. Press

42. Hudson RR, Kaplan NL. 1988. The coalescent process in models with selection and recombination. *Genetics* 120:831–40

43. Hudson RR, Kreitman M, Aguadé M. 1987. A test of neutral molecular evo-

lution based on nucleotide data. *Genetics* 116:153–59

44. Hudson RR, Bailey K, Skarecky D, Kwiatowsky J, Ayala FJ. 1994. Evidence for positive selection in the superoxide dismutase (*Sod*) region of *Drosophila melanogaster*. *Genetics* 136:1329–40

45. Hughes AL. 1993. Nonlinear relationship among evolutionary rates identify regions of functional divergence in heatshock protein 70 genes. *Mol. Biol. Evol.* 10:243–55

46. Hughes AL, Nei M. 1989. Nucleotide substitution at major histocompatibility complex class II loci: evidence for overdominant selection. *Proc. Natl. Acad. Sci. USA* 86:958–62

47. Hughes AL, Ota T, Nei M. 1990. Positive Darwinian selection promotes diversity in the antigen-binding cleft of class I major-histocompatibility-complex molecules. *Mol. Biol. Evol.* 7:515–24

48. Kaplan NL, Darden T, Hudson RR. 1988. The coalescent process in models with selection. *Genetics* 120:819–29

49. Kaplan NL, Hudson RR, Langley CH. 1989. The "hitchhiking effect" revisited. *Genetics* 123:887–99

50. Karotam J, Boyce TM, Oakeshott J. 1993. Nucleotide variation at the hypervariable Esterase 6 isozyme locus of *Drosophila simulans*. *Mol. Biol. Evol.* 12:113–22

51. Karl SA, Avise JC. 1992. Balancing selection at allozyme loci in oysters: implications from nuclear RFLPs. *Science* 256:100–2

52. Kimura M. 1983. *The Neutral Theory of Molecular Evolution*. Cambridge: Cambridge Univ. Press

53. Kingman JFC. 1980. *Mathematics of Genetic Diversity. CBMS-NSF Regional Conf. Ser. in Appl. Math.*, vol. 34. Philadelphia: Soc. Indust. Appl. Math.

54. Kreitman M, Aguadé M. 1986. Genetic uniformity in two populations of *Drosophila melanogaster* as revealed by filter hybridization of four-nucleotide-recognizing restriction enzyme digests. *Proc. Natl. Acad. Sci. USA* 83:3562–66

55. Kreitman M, Hudson RR. 1991. Inferring the evolutionary histories of the *Adh* and *Adh-dup* loci in *Drosophila melanogaster* from patterns of polymorphism and divergence. *Genetics* 127:565–82

56. Langley CH, Fitch WM. 1973. The constancy of evolution: a statistical analysis of the a and b hemoglobins, cytochrome c, and fibrinopeptide A. In *Genetic Structure of Populations*, ed. NE Mor-

ton, pp. 246–62. Honolulu: Univ. Hawaii Press

57. Langley CH, Fitch WM. 1974. An estimation of the constancy of the rate of molecular evolution. *J. Mol. Evol.* 3:161–77

57a. Laurie CC, Stam LF. 1994. The effect of an intronic polymorphism on alcohol dehydrogenase expression in *Drosophila melanogaster*. *Genetics.* 138:379–85

58. Lee Y-H, Ota T, Vacquier VD. 1995. Positive selection is a general phenomenon in the evolution of abalone sperm lysin. *Mol. Biol. Evol.* 12:213–38

59. Lee Y-H, Vacquier VD. 1992. The divergence of species-specific abalone sperm lysins is promoted by positive Darwinian selection. *Biol. Bull.* 182:97–104

60. Lewontin RC, Ginzburg L, Tuljapurkar S. 1978. Heterosis as an explanation for large amounts of genic polymorphism. *Genetics* 88:149–70

61. Lewontin RC, Hubby JL. 1966. A molecular approach to the study of genic heterozygosity in natural populations. II. Amount of variation and degree of heterozygosity in natural populations of *Drosophila pseudoobscura*. *Genetics* 54:595–609

62. Li W-H. 1985. Accelerated evolution following gene duplication and its implication for the neutralist-selectionist controversy. In *Population Genetics and Molecular Evolution*, ed. T Ohta, K Aoki, pp. 333–52. Tokyo: Japan Sci. Soc. Press

63. Li W-H, Graur D. 1991. *Fundamentals of Molecular Evolution*. Sunderland, MA: Sinauer

64. Li W-H, Sadler LA. 1991. Low nucleotide diversity in man. *Genetics* 129:513–23

65. Li W-H, Tanimura M. 1987. The molecular clock runs more slowly in man than in apes and monkeys. *Nature* 326:93–96

66. Li W-H, Tanimura M, Sharp PM. 1987. An evaluation of the molecular clock hypothesis using mammalian DNA sequences. *J. Mol. Evol.* 25:330–42

67. Li W-H, Wu C-I. 1987. Rates of nucleotide substitution are evidently higher in rodents than in man. *Mol. Biol. Evol.* 4:74–77

68. Long M, Langley CH. 1993. Natural selection and the origin of *jingwei*, a processed functional gene in *Drosophila. Science* 260:91–95

69. Maynard-Smith J, Haigh J. 1974. The hitchhiking effect of a favorable gene. *Genet. Res.* 23:23–35

70. McDonald JH, Kreitman M. 1991.

Adaptive protein evolution at the *Adh* locus in *Drosophila. Nature* 351:652–54

71. Mukai T. 1964. The genetic structure of natural populations of *Drosophila melanogaster*. I. Spontaneous mutation rate of polygenes controlling viability. *Genetics* 50:1–19

72. Mukai T, Chigusa SI, Mettler LE, Crow JF. 1972. Mutation rate and dominance of genes affecting viability in *Drosophila melanogaster. Genetics* 72:335–55

73. Nei M. 1983. Genetic polymorphism and the role of mutation in evolution. In *Evolution of Genes and Proteins,* ed. M Nei, RK Koehn, pp. 165–90. Sunderland, MA: Sinauer

74. Nei M. 1987. *Molecular Evolutionary Genetics.* New York: Columbia Univ. Press

75. Nei M, Fuerst PA, Chakraborty R. 1976. Testing the neutral mutation hypothesis by distribution of single locus heterozygosity. *Nature* 262:491–93

76. Nei M, Graur D. 1984. Extent of protein polymorphism and the neutral mutation theory. *Evol. Biol.* 17:73–118

77. Ohta T. 1987. Very slightly deleterious mutations and the molecular clock. *J. Mol. Evol.* 26:1–6

78. Ohta T. 1991. Multigene families and the evolution of complexity. *J. Mol. Evol.* 33:34–41

79. Ohta T. 1992. The nearly neutral theory of molecular evolution. *Annu. Rev. Ecol. Sys.* 23:263–86

80. Ohta T. 1993. Amino acid substitution at the *Adh* locus of *Drosophila* is facilitated by small population size. *Proc. Natl Acad. Sci. USA* 90:4548–51

81. Ohta T. 1994. Further examples of evolution by gene duplication revealed through DNA sequence comparisons. *Genetics* 138:1331–37

82. Ohta T, Kimura M. 1971. On the constancy of the evolutionary rate of cistrons. *J. Mol. Evol.* 1:18–25

83. Ohta T, Kimura M. 1975. The effect of a selected linked locus on heterozygosity of neutral alleles (the hitchhiking effect). *Genet. Res.* 28:307–8

84. Prakash S, Lewontin RC, Hubby JL. 1969. A molecular approach to the study of genic heterozygosity in natural populations. IV. Patterns of genic variation in central, marginal and isolated populations of *Drosophila pseudoobscura. Genetics* 61:841–58

85. Sawyer SA, Hartl DL. 1992. Population genetics of polymorphism and divergence. *Genetics* 132:1161–76

86. Sharp PM. 1989. Evolution at 'silent' sites in DNA. In *Evolution and Animal Breeding: Reviews in Molecular and Quantitative Approaches in Honour of Alan Robertson,* ed. WG Hill, TFC Mackay, pp. 23–32. Wallingford, UK: CAB Int.

87. Skibinski DO, Woodwark M, Ward RD. 1993. A quantitative test of the neutral theory using pooled allozyme data. *Genetics* 135:233–48

88. Spencer HG, Marks RW. 1992. The maintenance of single-locus polymorphism. IV. Models with mutation from existing alleles. *Genetics* 130:211–21

89. Stewart C-B, Wilson AC. 1987. Sequence convergence and functional adaptation of stomach lysozymes from foregut fermenters. *Cold Spring Harbor Symp. Quant. Biol.* 52:891–99

90. Swanson KW, Irwin DM, Wilson AC. 1991. Stomach lysozyme gene of the langur monkey: tests for convergence and positive selection. *J. Mol. Evol.* 33:418–25

91. Tajima F. 1989. Statistical method for testing the neutral mutation hypothesis by DNA polymorphism. *Genetics* 123:585–95

92. Takahata N. 1987. On the overdispersed molecular clock. *Genetics* 116:169–79

93. Takano TS, Kusakabe S, Mukai T. 1993. DNA polymorphism and the origin of protein polymorphism at the *Gpdh* locus of *Drosophila melanogaster*. In *Mechanisms of Molecular Evolution,* ed. N Takahata, AG Clark, pp. 179–90. Sunderland, MA: Sinauer

94. Tavaré S. 1984. Line-of-descent and genealogical processes, and their applications in population genetic models. *Theor. Pop. Biol.* 26:119–64

95. Watt WB. 1994. Allozymes in evolutionary genetics: self-imposed burden or extraordinary tool? *Genetics* 136:11–16

96. Watterson GA. 1982. Mutant substitutions at linked nucleotide sites. *Adv. Appl. Prob.* 14:206–24

97. Wilson AC, Carlson SS, White TJ. 1977. Biochemical evolution. *Annu. Rev. Biochem.* 46:573–639

98. Yokoyama R, Yokoyama S. 1990. Convergent evolution of the red- and greenlike visual pigment genes in fish, *Astyanax fasciatus,* and human. *Proc. Natl. Acad. Sci USA* 87:9315–18

Annu. Rev. Ecol. Syst. 1995. 26:423–44

GENETIC MOSAICISM IN PLANTS AND CLONAL ANIMALS

Douglas E. Gill, Lin Chao, Susan L. Perkins, and Jason B. Wolf[1]

Department of Zoology, University of Maryland, College Park, Maryland 20742

KEY WORDS: somatic mutation, intraplant variation, modular growth, plant evolution, herbivorous pests

ABSTRACT

The genetic mosaicism hypothesis (GMH) proposed that arborescent plants accumulate spontaneous mutations and become genetically mosaic as they grow. GMH predicted that the intraplant heterogeneity influences plant-pest interactions ecologically and provided a partial solution to the problem of how long-lived trees evolve resistance to short-lived pests. Theoretical models predict that genetic mosaics should be rare (about 5%) and that genetic variation within a clonal unit should be difficult to detect. Somatic mutations can contribute more to standing genetic variation in populations than do gametic mutations and thereby can increase plant evolutionary rates. If population size is small, somatic mutations can increase heterozygosity by two or more orders of magnitude. Reported frequencies of somatic mutants match the values expected in theory: The average value of mutant frequencies per locus is 10^{-6}; the observed frequency for polygenic traits (such as chlorophyll-less tissues) is 6.3×10^{-4} per genome; and spontaneous mutants occur 0.1–19% in asexual plants. Like plants, many clonal animals violate Weismann's doctrine (separation of germlines from soma), and GMH should apply, but no estimates of mutant frequencies or mutation rates within colonies of clonal invertebrate animals are yet available. Pests respond to intraplant heterogeneity and can impose selective differentials on modules, but the significance of clumped patterns of galling-aphids on witch-hazels previously reported by Gill (40) as supporting GMH is refuted here.

[1]Present address: Center for Ecology, Evolution and Behavior, TH Morgan School of Biological Sciences, University of Kentucky, Lexington, KY 40506-0225

423

0066-4162/95/1120-0423$05.00

INTRODUCTION

Fifteen years ago, Whitham (138) proposed a hypothesis that the foliage heterogeneity of an arborescent plant strongly influences the interaction of the plant with its pests: Variation in leaf quality makes the plant less apparent to its pests, increases the level of competitive interactions among pests for superior forage sites, and concentrates pests into clumps that expose them to greater vulnerability to their enemies. The idea of phenotypic heterogeneity governing ecological interactions was expanded to the concept of genotypic variability shaping the coevolution of clonal plants and their pests (139–141). The general proposition has become known as the genetic mosaicism hypothesis (GMH).

Summarized, the GMH asserts:

1. Spontaneous mutations doubtless occur among the proliferating meristems of a growing perennial.
2. The meristematic and modular basis of plant development assures that many novel mutations are preserved and expanded hierarchically among modules as the plant grows.
3. Some of the easily observed phenotypic variations must have a genetic basis (broad-sense heritability).
4. The differential growth and survival of modules (ramets, branches, shoots, etc) should alter the genotypic configuration of the plant as it grows.
5. The phenotypic heterogeneity (both heritable and induced) affects the response of pests, especially herbivores, in two ways: *a*) rather than being offered a uniform food resource, herbivorous pests (insects) encounter a diverse array of leaves that imposes sharp penalties on short-term, demographic variables of growth, survival, and reproduction, and *b*) coevolutionary responses of the pests are sharply constrained because each population of pest on a plant is itself subdivided and subject to myriad directional selective pressures from the intraplant phenotypic heterogeneity.

Developing the genetic mosaic hypothesis independently, Gill (40) summarized the evidence that: 1. Plants have modular construction; 2. the GMH is consistent with current views of meristem anatomy and plant development; 3. phenotypic variation is prevalent among plant parts; 4. the propagation of bud sports in horticulture proves the existence and importance of genetic variation within clonal plants; 5. fitness differentials among modules are documentable, often driven by localized herbivory; 6. the genotype of single plants can change with age by the differential proliferation of modules; 7. at least theoretically, the GMH helps solve the dilemma of disparate generation times in the coevolution of long-lived plants and their short-lived enemies, and 8. the GMH has important bearing on the evolution of breeding systems of mass-flowering plants. The GMH was apparently provocative because numerous papers and symposia on the ecological and evolutionary implications of the GMH appeared in the early 1980s, including the symposia volumes *Population Biology and Evolution of Clonal Organisms* (64), *Developmental Mutants in Higher Plants* (120), and *Plasticity in Plants* (67).

In this review we ask what assumptions and conclusions of the GMH as stated by Whitham & Gill need to be revised in light of new evidence. Although the evolutionary importance of genetic mosaicism was developed for plants with modular architecture and meristematic development, the concept seems to apply more broadly to many other organisms that are structured as clones or colonies. These include nonwoody, nonarborescent but herbaceous, planar, rhizomatous plants like many grasses and forbs (9, 44, 142), all viruses, plasmids, bacteria, and unicellular protists (86), and multicellular, clonal invertebrate animals such as some cnidarians, bryozoans, cladocerans, and aphids (2, 14, 63, 65, 116). We avoid repeating here the meticulous arguments that Whitham & Gill have made and the literature reviewed in the original papers.

The Concept of Individuality

The GMH redefines both the unit of plant evolution and the concept of individual. The problem of defining Darwinian fitness in clonal plants has troubled evolutionary biologists for a long time and has yet to reach resolution (15, 34, 36, 46, 47, 62, 101, 106, 121–124, 131). Argument about the difference between an individual and a colony has also been debated by invertebrate zoologists interested in evolutionary processes: Urbanek (125) discussed the problem, while Mackie (87) dismissed the issue as a *Scheinproblem*, a non-issue. Points of consensus are that development in plants and other clonal organisms is often hierarchical, fractal, or modular, that changes in one level will influence levels above and below it, and that natural selection can be a potent force of evolution at any and all levels. Lewontin (84) had already set the stage for this consensus.

Words like clones, genets, ramets, sectors, modules, and metamers have come into common use to identify the level of structural organization of the plant (121, 126, 137). The GMH addresses the problem of genetic variants arising in the various forms of clones, when new genetic lineages spontaneously emerge in an expanding collection of cells, tissues, individuals, or demes that descend from a progenitive cell by mitosis or asexual reproduction. Spatial connection, as in ramets, rhizotamous stands, and invertebrate colonies, is often helpful in identifying the clone's limits and ancestry. Obversely, it is difficult to identify a clone when its members are physically separate, for example, bacteria, unicellular algae and protists, aphids and cladocera swarms, and duckweed and water hyacinth mats in ponds. Thus, we treat as clones an oak tree ascending from an acorn, a 1000-yr-old genet of goldenrods with 10,000 herbaceous ramets, and a stand of aspens that covers 100 ha with 47,000 trees all produced through asexual reproduction (42, 66, 141). The GMH poses the question "Does the oak tree have genetically mosaic branches?" or "Do goldenrod genets have genetically different stems?" An aspen stand illustrates the problem of multiple levels of intraclonal variation, i.e. how many genetically distinct trees are likely to be found in the stand (analog to the goldenrods) or among branches within a single tree (like the oak)?

Modules: Independent or Integrated

The fact that much of plant development proceeds as the repetitious multiplication of a basic subunit, the metamer (an internode and node bearing one or more leaves with associated axillary meristems) into structurally equivalent modules is now widely accepted by biologists (133, 136, 137, e.g. 97, p. 669). Simple computer algorithms assemble modules with surprisingly few variables and now generate a large diversity of plant architectures (48, 132). Generated by apical meristems, each module develops independently and divergently from the others, and older modules give rise to new ones by lateral budding. In a very real sense, each module undergoes a complete life cycle of birth, growth, maturation, senescence, and death. As a consequence, a large mature tree with its myriad branches is really a population of modules with a distinct age structure that lends itself to rigorous demographic analysis (50, 132, 136).

Experimental defoliation of branches routinely demonstrates functional independence of modules, producing effects that are invariably highly localized within the plant (40). The experiments reviewed in (40) all dealt with the growth of the module, both short term within the growing season and longer term over several seasons, and with the quality and quantity of fruit production at the end of the growing season in which the experiment was performed. Only one recent paper (93) reported no effects of 12%, 25%, 50%, and 75% defoliations on the percentage of flowers producing fruit or the phenology of fruit abortion and fruit abortion rate of *Ligustrum vulgare* (Oleaceae).

Modules also have a corresponding hierarchical physiological independence. Watson & Casper (134) reviewed the morphogenetic constraints on carbon distributions in plants and concluded that there are integrated physiological units (IPUs) between which the production, distribution, and utilization of products is sharply restricted. For example, the subtending leaf of chaparral, *Diplacus aurantiacus*, provides one third of the assimilates required by its primary flower, but no assimilate is obtained from the opposite leaf at the same node. There is compelling evidence that arborescent woody flowering plants are often structured in sectors that may comprise several modules, so that photosynthates are transported to and stored in sites located directly under the modules responsible for production. While Watson (133) expanded her concept of IPU to include integrated units for mineral and hormonal transport as well as carbon, Sprugel et al (114) argued that branches are essentially autonomous subunits with respect to water flow. Hardwick (45) argued philosophically that modular plants must have mechanisms to detect and eliminate somatic perturbations and mutations that occur during development in order to maintain their structural stability. He asserted that the principal mechanisms are genetic repair and diplontic selection (growth of healthy cells over genetically defective cells—see 38) and concluded that the physiology of modular plants is better described in terms of cooperation than competition.

The current view of plant development and structure is: Reiteration of metamers produces modules that can retain morphological and genetic integrity and autonomy, and that production, transport, and use of nutrients and some plant compounds are integrated into rather well-defined physiological units (IPUs) that range in size from subelements of metamers to vertical sectors of full-grown trees. IPUs often integrate nonadjacent flowers and fruits as a consequence of the phyllotaxy during development (133).

GENETIC VARIATION WITHIN PLANTS

Theoretical Expectations

Compared to the number of verbal discussions it stimulated, the GMH has generated surprisingly few theoretical models that predict the amount of genetic variation to be expected within clones and the parameter values that define evolutionary importance. The role of somatic mutations in plant evolution has been examined from two perspectives: whether a clonal unit is statistically likely to be a chimera or mosaic, and whether somatic mutations contribute significant genetic variation for evolution when compared to gametic mutations.

PROBABILITY OF FINDING MOSAICISM   The simplest process of clonal reproduction is by fission; the classic model is a growing bacterial colony. The probability of finding a mosaic unit is given by the frequency distribution of colonies with >1 genotypes. It is important to remember the distinction between mutants (the outcome) and mutation (the process); a growing population experiences mutations and accumulates mutants. While mutation is a Poisson process on the mutable units available at an instant in time, Luria & Delbruck (86) demonstrated that the final distribution of mutants across colonies deviates from the Poisson because growth of mutant and nonmutant lineages is geometric.

Various approximations exist for the Luria-Delbruck distribution (82, 104, 115). In general, the distribution has the property of having the variance-to-mean ratio much greater than one because a rare mutation early in development produces a clone with an extremely high proportion of mutants (analogous to a "jackpot" in a slot-machine—86); the maximum is 50% mutant when the mutation occurs in one of the first two daughter cells. Although jackpots are theoretically possible, the Luria-Delbruck probability of finding a jackpot is discouragingly small in practice. If colony size N is 10^6 cells and mutation rate v is 10^{-7} per locus per generation, most colonies (95.1%) will have no mutants; conversely the probability of finding a colony with at least 1 mutation is 4.9%, and the most common mosaic is expected to have only one mutant. The probability of finding a (large) jackpot colony with 64 or more mutants is approximately 1.5×10^{-3} (82). If colony size is increased to 10^8 cells, almost all colonies will be mosaics, but the most common mosaic still has only 17–31 mutants among 100 million cells. Thus, unless clonal units and mutation rates are much larger than those stated above, mosaic clones are extremely rare and difficult to find.

The Luria-Delbruck model is, however, inadequate for organisms that may have several rounds of replication before fission. Unlike bacterial colonies, the meristems of most vascular plants are complicated tissues with multiple proliferating developmental layers. Pteridophyte meristems have single apical initials and are, perhaps, the easiest group in which to relate mutations to modules, but the nonstratified meristems of most gymnosperms and the tunicacorpus meristems in all angiosperms and some gymnosperms impose difficulties in modelling the expected distribution of mutants in the mature plant. Moreover, the system of L1-L2-L3 layers of angiosperm meristems (which produce epidermis, parenchyma and gametophytes, and vascular tissues, respectively) further constrains the biological impact of random developmental mutations, because only those mutations that occur in L2 will have inheritance through gametophytes (10, 40, 79, 81). Thus, in a branching tree, there may be many cell divisions before an apical meristem gives rise to multiple bud meristems.

Klekowski & Karinova-Fukshansky (76, 77) discussed the fixation of selectively neutral and selective loss of deleterious cell genotypes in the meristems of plants. Antolin & Strobeck (6) determined that the probability that two buds randomly chosen from the same tree would have different alleles is $2(n-1)mv$ [Equation 1], where m is the number of cell divisions between branching (which can be a function of the pattern of ontogeny, see 78), n is the exponential rate of bud production, and v is the somatic mutation rate. Assuming that $m = 5$–20, $n = 20$, and $v = 10^{-6}$ per locus per generation, the probability ranges from 2–7.6×10^{-4}. Thus, unless the mutation rate, number of replications, or both are much higher, the probability of finding a mosaic is as small as the Luria & Delbruck expectation.

In the above example, the mutant genotype was assumed to be selectively neutral. Obviously, selection for a favorable mutant genotype within a clonal unit can increase the probability of finding mosaics. This possibility was also examined by Antolin & Strobeck (6). One difficulty in modeling this mutation-selection process is that the growth of most trees is not simple dichotomous branching. For example, bud production is greatest in apical meristems, and it decreases in meristems on the side shoots. Using probabilities for bud production estimated for silver birch, Antolin & Strobeck (6) considered the effect of neutral and favorable mutants during the growth of the tree. The results were again not in favor of finding mosaicism in clonal units. For a neutral mutation with a rapid mutation rate of 10^{-4}, the expected proportion of mutant buds in a tree is less than 1%. For a favorable mutant with a selective advantage as great as 25%, the expected proportion increases only to about 5%. Thus, two bud meristems plucked at random from a tree will be identical most of the time.

Having specified that apical meristems have the highest budding rates (compared to laterals), Antolin & Strobeck's model predicts that most mutants will occur as singletons near the tips of branches. That vanishingly rare distribution makes such mutants difficult to detect by researcher and herbivore alike, and such mutants, even if intrinsically highly favorable, will manifest negligible selective impact.

The general conclusion from these theoretical considerations is that modules in arborescent plants are unlikely to be mutant; trees are unlikely to be mosaics. Moreover, if any unit is a mosaic, it will be difficult to identify because the proportion of mutants will be very small within the unit. These conclusions apply equally to both nuclear genomes and replicating organelles in the cytoplasm (such as mitochondria, chloroplasts, etc) because the growth process is straightforward asexual cellular division generating cellular lineages. By the same argument, the probability of gametophytes acquiring cytoplastidic mutants through developmental ("somatic") mutation is the same as that of acquiring a nuclear mutant. These same results, however, also highlight the conditions in which mosaics may be easier to find: The older the clonal unit is, the greater is the probability of finding mosaics because n and m are large, and even the mutation rate v may be greater because of senescence. Thus, 1. the chance of finding mosaicism among branches in an old oak tree or a large grove of aspens (42) is appreciably greater than in saplings; 2. mosaicism is most likely among those loci with the highest mutation rates; and 3. clonal units that are subject to higher rates of mutagenesis, or generate a larger number of new buds (or equivalent) from the lower order branches (or equivalent), have a better chance of becoming a mosaic with a higher proportion of mutants.

Phenotypic mosaicism is expected much more frequently than mosaicism at marker loci. When considering polygenic traits, the likelihood of finding a mosaic is increased because mutation at any locus may produce the same mutant phenotype. Consider chlorophyll-deficient variegations (10). There are about 300–500 nuclear genes controlling chlorophyll structure and function (74); defective mutants in any of them result in chlorophyll deficiency, often visible as variegation. Jorgensen & Jensen (69) estimated that the spontaneous frequency of chlorophyll-deficient mutants among 1.43 million barley seedlings was 1.6×10^{-4} (This estimate is generated by selfing from 6.3×10^{-4} mutations per diploid genome per generation). Applying this value to Equation 1 for a full-grown tree, the probability that two buds from the same tree are phenotypically different increases to 0.06–0.23. Phenotypic mosaics should be easily found.

These expectations, however, run quickly into modelling trouble and into the conceptual ambiguities of individuals, clones, and populations in organisms with physically separated members. The difference between a clone and an asexual population vanishes as the progenitive ancestral cell disappears into ancient history. How does one distinguish between clonal variation and intrapopulational genetic polymorphism? If mosaics are to be found only in old clonal units, one must account for the history of selection and genetic drift that has already transpired within and among genets. Part of the appeal of mosaicism is the origin of a clonal unit from a single cell, which we can recognize only in young phases of the clonal growth process. Calling variation in a stand of ferns "mosaicism" and variation in a *Daphnia* population "genetic polymorphism" may be making much ado about nothing.

SOMATIC VS GAMETIC MUTATIONS   If somatic mutations can be acquired by the gametophytes, they can be passed to future generations, and they have the potential to influence the rates of evolution of plants. However, whether they will accelerate plant evolution significantly depends on their effect relative to gametic mutations. To determine the relative effect of gametic and somatic mutations, Slatkin (113) and Antolin & Strobeck (6) estimated their relative contributions to genetic variation, as measured by the average heterozygosity, in a population. In general, their result is that $H = 4N(u + nmv) / [1 + 4N(u + nmv)]$ [Equation 2] where H is heterozygosity, N is population size, m is the number of cell divisions between branching, n is the exponential rate of bud production, v is the somatic mutation rate, and u is the gametic mutation rate. Thus, if $u < nmv$, the contribution of somatic mutation will be greater than that of gametic mutation to the variation within a population. Most estimates of u and v find the two parameters to be approximately the same (140, 141). Therefore, the effect of somatic mutation outweighs that of gametic mutation in modular plant evolution.

Depending on values of the parameters, the contribution of somatic mutations can be dramatic. For example, using Antolin & Strobeck's estimate of $n = 20$, $m = 10$, and $u = v = 10^{-6}$ per locus per generation, in a population of size $N = 500$, the expected heterozygosity is only $H = 0.002$ in the absence of somatic mutations. Including somatic mutations increases the heterozygosity to $H = 0.29$, a change of more than two orders of magnitude. In general, the effect will be greatest when Nu is small and nm is large. This leads to the prediction that the relative contribution of somatic mutations can be especially important in small populations.

Measured Amounts of Intraplant Genetic Variation

The theoretical models provide a context, both qualitative guidelines and coarse quantitative predictions, to evaluate the significance of observations pertinent to the GMH. The predicted frequency of mosaics depends on the critical parameters of the models [Equations 1 & 2], especially n, m, and v. This section reviews published reports of genetic mosaicism as they relate to the theoretical expectations and as they provide realistic values for the model parameters. Although the GMH was originally proposed as a model for long-lived trees, we considered evidence reports for any kind of clonal plants and animals.

Qualitative evidence of mosaicism is abundant. In a review of vegetative propagation of bud and branch sports, histogenic origins of cytogenetic and chloroplastidic variegations, and inheritance of somatic mutations through the gametophytes, Gill (40) concluded that all three areas of research supported the GMH. Commercial horticulture has capitalized on desirable spontaneous genetic sports being preserved through vegetative propagation of cuttings. Probably the most famous example is the pink grapefruit being found as a single branch bud sport in 1906 (51). Shaw (111) documented intraclonal variation in many polygenic traits in morphology, germination, gametophytic growth, and copper tolerance in the moss *Funaria*. Cytogenetic variation within plants is known, including the famous cytogenetic variation in *Claytonia virginica* (83). In *Tradescantia*, mutations in the stamen hairs and petals result in a pink phenotype rather than the blue wild type (103). The average frequency of pink mutant stamen hairs events was 0.0013 in a sample of over 1.89×10^5 hairs.

The detection of somatic novelties has been aided by molecular techniques. Apomictic species of *Taraxacum* have been particularly informative. King & Schaal (70) discovered two parents (themselves siblings) from 31 lineages (26 genotypes, 714 offspring examined) of *Taraxacum officinale* that independently gave rise to progeny with a nonparental genotype (differing in restriction site maps of *Eco*RI). Ford & Richards (37) found no variation in tyrosinase and phosphatase in ten agamospecies of *Taraxacum*, but they did find a mean variability of 19% in esterase in half of the agamospecies. Mogie (90) reported extensive variation in morphological and electrophoretic characters in several obligate apomictic *Taraxacum* species. Mogie & Richards (91) found that 2 of 232 individuals had nonparental esterase zymograms in part of their tissues.

These studies indicate detectable rates of mosaics of 6.1%, 19%, and 0.86% spontaneous mutations per locus, values that agree favorably with the expected detection rate of 4.9%.

Vasseur et al (127) found an average of 19.6 allozymic genotypes per population (18 loci from 13 enzyme systems) in eight apomictic populations of *Lemna minor*, and they attribute the genetic diversity to spontaneous mutation. That abundant genetic variation is routinely found in asexual/apomictic species suggests that high rates of somatic mutation are compensatory to the absence of recombination from sexual reproduction and may allow these primarily asexual species to adapt to changing environmental conditions (90, 99). These investigations clearly refute the assumption that clonal plants have less genetic variability than do sexual ones (33, 112).

Natural levels of genetic variability have been estimated among ramets within genets within a few populations. Using restriction site and restriction fragment polymorphisms in the ribosomal DNA (rDNA) in white clover (*Trifolium repens*), Capossela et al (21) found at least one somatic variant in the 16 samples surveyed. The only other known case at the time (105) of somatic rDNA variation showed restriction site variation in *Eco*RI and *Eco*RV restriction sites among ramets within *Solidago altissima* clones. There is some evidence that changes in the rDNA may be acted upon by natural selection. Saghai-Maroof et al (102) showed that directional selection in barley had taken place and had changed the frequency of an rDNA s1 phenotype: A phenotype that had started out in low frequency in the population gained dominance and became common after only 54 generations.

The frequency of somaclonal variants in karyotype, morphology, biochemical attributes is often 0.05–5% of micropropagated clones (109). The rates compared favorably with those theoretically expected but are viewed as unacceptably high rates of production of undesirable off-types by the horticultural industries geared to the rapid propagation of desirable cultivars. Gill (40) had hoped that micropropagational techniques would provide definitive measures of standing levels of somatic mutations, but nearly all somaclonal variation is explicable as unintended side effects of the culture technology. There remains little clarity about what portion of the variants resided in the somatic tissues of the original plant before fragmentation.

In the example of chlorophyll-deficiency (cd) in barley (10, 74), the rate of somatic mutation per locus per haploid genome per generation v is between 6×10^{-7} and 1.1×10^{-6}. The estimate of 0.002 cd mutations per gamete for *Zea mays* is comparable, assuming that a similar number (300–500) loci are involved in corn chlorophyll function. Estimated rates for other polygenic traits in *Zea mays*, e.g. endosperm characteristics, range from 5×10^{-4} to 1×10^{-6} (28). Assessing the rate of chlorophyll-deficiency in the woody perennial red mangroves *Rhizophora mangle,* Lowenfeld & Klekowski (85) found a mean rate of 5.8×10^{-3} recessive mutations per nuclear genome per generation in three primarily self-pollinated populations; in a fourth population the rate was one third lower (0.0021) and may be associated with a higher rate of outcrossing; heterozygotes doubtless went undetected (26, 80). These long-lived perennials have mutation rates about 25 times higher than those reported for most annuals (75). This rate may be attributed to the longer life span which leads to higher mutation rates per genome per generation, but it may be similar on a per-mitotic-event basis if we knew the values of n and m. Moreover, these estimates of rate of chlorophyll-deficient mutations are likely to be underestimates when we consider that chlorophyll-deficiency probably leads to lower fitness, and many deleterious mutants were doubtless shed by the time the investigators made their counts. Of course, as screening procedures become more refined, the estimated rates will necessarily increase. It would be interesting to learn whether high mutation rates exist in genes subject to frequency-dependent selection, e.g. the self-compatibility loci.

Studies that document the mutation rate per genome are particularly rare. Klekowski (71, 72) documented the gametophytic mutation frequencies per ramet generation (defined as the time necessary to double the number of ramets or shoot apices) for three species of ferns. All rates reported are on the 10^{-2} order of magnitude with a mean of 0.0286 gametophytic mutations per ramet generation (72). There is no good estimate of the actual number of genes that are being monitored for mutation in this case, but because pteridophyte meristems have single well-defined apical initials, these frequencies are also rates per cell. Klekowski (73) reported the frequencies of mutant ramets within a clone (genet) of two of these fern species. He found that an average of 0.22% of all clones contained chimeric individuals with average frequencies of mutant apices per mosaic clone of 0.21% in *Onoclea* and 0.4782 in *Matteuccia*. He attributes the difference in mutation buildup to the fact that *Onoclea* clones are shorter-lived because they are frequently invaded by new sexually produced competitors, while *Matteuccia* clones are longer-lived and rarely reproduce sexually (73).

Mitotic crossing-over can generate new *cis-trans* recombinants that may alter the expression of the genes and expose the mutant tissue to spontaneous diplontic selection. The frequency at which mitotic crossing-over occurs has been estimated as 10^{-5} to 10^{-4} per cell for several species of plants (35). Because this number is for a pair of homologous chromosomes, the actual frequency per genome may be as high as 10^{-4} to 10^{-3} mitotic crossovers per cell (78). Somatic chromosome rearrangement may play a role in *Taraxacum* agamospecies (99). High levels of somatic chromosomal rearrangement were detected in three out of five obligate agamosperms (13 of 86 chromosomes surveyed), while no rearrangements were found in four individuals of a sexual species (113 chromosomes surveyed).

The rate of somatic genome modification may be affected by environmental factors such as stress or shock (88). Best known are the environmentally induced, potentially adaptive changes in the flax genome (30, 107). Greatly exceeding the rate of gametic mutation (29), these pseudo-Lamarkian genetic changes are distinct from the DNA mutations induced by UV light, gamma rays, or chemical mutagens. Heritable phenotypic changes such as plant mass, peroxidase and acid phosphatase isozyme mobilities, and number of hairs on the seed-capsule septa occur in response to nutrient or temperature conditions. Associated with these changes are detectable changes in the genome such as total amount of nuclear DNA and number of genes coding for highly repeated sequences such as ribosomal RNA (30).

RESPONSES OF PESTS TO INTRAPLANT HETEROGENEITY

Several papers apparently were stimulated by Gill's (39) and Gill & Halverson's (41) use of nested ANOVA to determine the relative significance of pest infestation and fruit production among genets/trees, ramets/branches within genets, etc. Acosta et al (1) reported significant contributions from all levels (shoot modules, branches, and genets) in seed production in *Cistus ladanifer*. Multilevel analysis of variance of flower morphology of *Dalechampia scandens* yielded significant covariance terms within genets, among genets, and among populations that Armbruster (7) interpreted might signal genetic and selective importance. In a similar hierarchical analysis on trees, ramets, branches, short shoots, and leaves of *Betula pubescens* ssp. *tortuosa*, Suomela & Ayres (117) found that all levels contributed significant variation. For water content and leaf toughness, the variation among ramets was greater than among trees, but leaf nitrogen content was invariant across ramets. Thus, statistical partitioning of variance of phenotypic variation in the hierarchical structure of trees adds a satisfying quantitative support to the predictions of the GMH. There continues to be broad and abundant support for the conclusion that pests, especially insect herbivores, respond to the heterogeneity presented by the foliage of arborescent trees (31, 89). A wide variety of herbivores can distinguish leaves of different sizes and shapes, ages, biochemical characters, etc among branches *within* plants. The imagoes of 18 species of gall-makers avoid small leaves and are found clumped on large leaves of *Quercus cerris* (119). The density and grazing rate of insect herbivores respond to the variable hairy trichomes and nutrient condition of the leaves of *Wigandia urens* (Hydrophyllaceae); those leaf qualities are determined by weekly water availability (20). Foraging tactics of caterpillars lead to microhabitat selection of the patches within heterogeneous trees (108).

Suomela & Nilson (118) used multilevel ANOVAs to show that larval growth of the geometrid *Epirrita autumnata* significantly differed among ramets within trees, among branches within ramets, and among short shots within branches. The impact on the tree delayed effects on growth and morphology of the birches in the next year, and the feedback interactions, including temperature conditions in successive years (110), in part account for the 9–10 cycles in moth abundance in the birch forests of Fennoscandia (55). An isolated exception is Quiring's (98) report that the differences in herbivory, density, and survivorship of bud moths were not related to intra-tree heterogeneity.

One of the important factors generating intraclonal heterogeneity and influencing pest responses is the induction of plant defenses by the pests themselves (52). Inducible defenses play a major role in the interaction of herbivorous insects and mountain birch (57). The variation in the reproductive biology of the bronze birch borer (Coleoptera: Buprestidae) among tree species is in part explained by inducible defenses of the different species (3). On the other hand, the best explanation for the curious selective and partial defoliations by leaf-cutting ants is leaf quality among and within trees and not the rapid induction of plant resistance (58).

Supporting their earlier work known as the deme-formation hypothesis, Alstad & Edmunds (4) continue to provide empirical and experimental-transfer evidence that the interactions among clones of black pine scale, *Nuculaspis californica*, (Homoptera: Diaspididae) and individual ponderosa pine trees in the western United States are sufficiently intense to generate genetically distinct demes of scales on individual trees and within branches of trees. Clones typically fail to spread onto neighboring trees because of their genetic specializations to unique chemical properties of each tree. Systematic changes in the sex ratios of the scales within demes also were taken as evidence of intrademal evolution (5). In a contrasting report, Cobb & Whitham (27) were unable to find evidence for a similar selective process in the pinyon pine needle scale on pinyon pine, *Pinus edulis*, and they offered the guarded hope that deme formation in the pinyon pine needle scale might yet occur at a much larger geographic scale than Alstad & Edmunds envision.

Herbivory may change the evolutionary trajectories of plants in unexpected ways. Haukioja (56) argues that the influence of grazing on the evolution of modular plants is not so much to generate new traits in morphology and physiology as it is to modify the regulation of plant structures and function, and the rules governing interactions among modules. Contrary to the popular notion that herbivores attack stressed plants, Price (96) defended the plant vigor hypothesis and argues that herbivores feed preferentially on vigorous plants or plant modules because of the higher nutritional levels.

The Witch-Hazel Aphid

Based on preliminary empirical evidence, Gill (40) and Gill & Halverson (41) came to the erroneous conclusion that the within-plant distribution of the witch-hazel aphids, *Hormaphis hammamelidis*, on witch-hazel, *Hamamelis virginiana*, was consistent with the hypothesis of mosaic variation in resistance/susceptibility among ramets within genets. The claim was based on year-to-year autocorrelation of levels of infestation among branches and ramets within clones of witch-hazels, and it assumed that the aphids at the study site alternated hosts between witch-hazel (primary) and birches (secondary) at their mountain study site during the year (94). This further assumed that the pattern of infestation on the witch-hazels was reset each year from randomly settling propagules migrating from the secondary birch plant-hosts, and that the clumped distributions of galls seen in the Spring were the result of active choice by aphids, either the sexuaparae and/or sexuales in the fall, or the fundactrices in the spring, for favorable and against undesirable witch-hazel leaves.

Subsequent work by von Dohlen and colleagues (128, 129, 130) demonstrated that the populations of galling aphids at the mountain study site were, in fact, autoecious on witch-hazels, and had short (3-generational) life cycles that had deleted the birch phase. Hence, the highly localized clusters of galls that occurred annually on the same branches and not on others were, in fact, perenniating clone-demes. It is also possible that these aphids produce and respond to a self-generated aggregating pheromone the way *Rhopalosiphum padi* (L.) gynoparae do on their *Prunus padus* winter hosts (95). Thus for two compelling reasons, the correlational evidence previously reported by Gill (40) and Gill & Halverson (41) is not evidence of aphid-plant interactions that partitioned susceptible from resistant modules.

INTRACLONAL VARIATION IN ANIMALS

Because research interest in the biology of clonal animals blossomed at the same time the GMH was proposed, and the question of the evolutionary importance of intraclonal mutations was repeatedly discussed (22), it is important to include in this review evidence of genetic mosaicism from the work of invertebrate zoologists that relates to the GMH.

Perhaps the reason that animals were traditionally ignored in the context of somatic mosaicism is because those most familiar to many researchers, including insects and vertebrates, have preformistic development of primordial germ cells and so obey Weismann's doctrine of the separation of germ and somatic lines. Weismann (135) argued that the sequestering of precursors to the germ line very early in embryogenesis prevented somatic mutations from being inherited, no matter how frequently they occurred. This was considered adaptive for animals that are incapable of sloughing off body parts that have acquired mutations. Selection for low mutation rates and rapid disposal of variant cell types was proposed by Weismann to be strong in most animals

(14, 49). Whitham & Slobodchikoff (140) expanded this prediction: Natural selection should encourage somatic mutations in modular plants and suppress somatic mutation rates in animals. We now know that rates of mutation are themselves under genetic control and can be accelerated or slowed by natural selection (24). While a mechanism exists for evolution to fulfill Whitham & Slobodchikoff's prediction, the limited data considered in this review do not support their prediction: Somatic mutation rates in plants are not noticeably higher than those in animals.

The GMH rekindled interest in critically examining Weismann's doctrine: The literature we reviewed repeatedly emphasized how many systems were fundamentally violating Weismann's doctrine (126). In addition to plants with modular structure, animals in at least 19 phyla do not have preformistic development (13, 92). These organisms, which include cnidarians, platyhelminthes, bryozoans, annelids, and entoprocts, either do not sequester germ cells at all or do so much later in development (epigenesis). In such animals, it is possible for somatic mutations to find their way into the germ line and be inherited. Moreover, animals that alternate sexual and asexual generations have a much greater effective number of cell divisions between sexual stages, so that the opportunity for genetic variation both to arise and to be inherited is greater than in preformistic animals. For example, while *Drosophila*, a preformistic animal, undergoes only 13 nuclear divisions per sexual generation, *Hydra* will undergo many more, in fact, "an astronomical number of divisions" (15). Even some preformistic animals such as *Daphnia*, aphids, and bdelloid rotifers may allow somatic mutations to proliferate through a parthenogenetic clone (32, 61, 65). It is still necessary that the mutations occur at an appropriate time to be expressed in germ cells or asexually-reproducing ramets; whether this is a common phenomenon is not yet known. A detailed animal analog of the GMH has not yet been modeled (68).

Several somatic variants have been anecdotally reported in clonal animals. Hughes (59) noticed that supposedly genetically identical colonies of bryozoans exhibited very different growth rates and proposed that these might be the result of a somatic mutation. Similarly, Chaplin (25) found intraclonal variation in ostracods that she speculated may have been caused by a spontaneous mutation. Harvell (54) reported one variant bryozoan that possessed only zooids with double lophophores. Calling this aberration a "hopeful monster," she proposed that such variant morphologies arising spontaneously as mutations or budding mistakes may be precursors to polymorphism in colonial invertebrates.

We were unable to identify any systematic studies that attempted to quantify variation due to somatic mutations in animals, possibly because it may be technically difficult in most groups. Detecting clonal animal mosaics, especially with phenotypically neutral markers, is obviously a technically difficult task. Natural populations of clonal organisms such as *Daphnia* and aphids are especially problematic because they exist as separate individuals, thus complicating efforts to trace clonal lineages. Conversely, in colonial animals, intraclonal mutations may be confused with chimeras formed as a result of colony fusion (see below). With the molecular genetics techniques currently available, this problem may become more tractable. Several studies of clonal animals that incorporated molecular techniques such as polymerase chain reaction (PCR) and DNA fingerprinting have been successful at detecting intraclonal variation (8, 23).

Despite the absence of data, the evolutionary implications of somatic mutations in animals have been thoroughly discussed. Somatic mutations in clonal, and especially modular, animals may provide the variation necessary to deal with fluctuating environments (60). Life spans of clonal organisms can greatly exceed those of most of their parasites and predators (63), but the turnover of modules is much more rapid. Bryozoan zooids and individual aphids, for example, live only about two weeks before they are replaced (32, 53). Thus, there will be a unit of selection with generation times more to the scale of the predators or parasites available to the colony. Somatic mutations may have also played an important role in the evolution of allo-recognition or self-compatibility systems. Upon contact with another colony, several sessile, clonal animals will fuse with the same clonal lineage (or a close relative) or fight aggressively if unrelated (11, 12, 16, 18, 43, 100). In one described case, a hydroid ramet, upon meeting the original clone from which it came, attacked and killed it (19). It seems very likely that a mutation occurred during the asexual spread of the hydroid, rendering the original and derivative clones so genetically different as to be antagonists. It is also plausible that somatic mutations may play a role in producing highly variable regions such as self-recognition and histocompatibility loci (17).

CONCLUSIONS

The Genetic Mosaicism Hypothesis (GMH) was raised 15 years ago in two contexts: how pests respond to within-tree foliage heterogeneity, and how long-lived trees evolve resistance to short-lived (and presumably faster evolving) pests. It remains controversial how the principles of Darwinian natural selection apply to plants and animals with repetitious modular structure. Much evidence, including experimental defoliations, supports the concept of modules attaining developmental, genetic, reproductive, and ecological autonomy, but new work has revealed some physiological integration of modules.

Theoretical models predict that genetic mosaics should be rare (about 5%) and that genetic variation within a clonal unit should be difficult to detect. Only if somatic mutation rates are high ($\geq 10^{-3}$ per locus per meristem cycle) and selective benefits are large (= 25%) will mutants be frequent enough in a mosaic to be detectable. Somatic mutations can therefore significantly increase the genetic variation and evolutionary rates within a population. The rates of somatic and gametic mutations are reported equivalent. In theory this observation predicts that somatic mutations should contribute more to standing variation than do gametic mutations. If population size is small, somatic mutations can increase heterozygosity by two or more orders of magnitude.

Spontaneous mutants are found frequently (0.1–19%) within clones of primarily asexual plants such as dandelion, duckweed, and clover. An average value of mutant frequencies per locus is 10^{-6}. For polygenic traits the frequency of observable mutants (such as chlorophyll-less tissues) is 6.3×10^{-4} per genome. These estimates are probably too conservative but agree with theoretical expectations. The prediction that plants should have higher rates of somatic mutation than should animals is not yet supported by the available quantitative data.

Arborescent plants present heterogeneous foliage, and pests respond to the variation at all scales and in ways that affect their ecology and evolution as the GMH predicts. Few papers have attempted to determine the heritability of the within-plant phenotypic variation. The significance of clumped patterns of galling-aphids on witch-hazels previously reported by Gill (40) as supporting the GMH is refuted here.

The GMH seems to apply broadly to many organisms that are structured as clones or colonies that violate Weismann's Doctrine. No estimates of mutant frequencies or mutation rates within colonial invertebrates are yet available. "The most appropriate dividing line is not between plants and animals but between modular and unitary organisms" (68).

Literature Cited

1. Acosta FJ, Serrano JM, Pastor C, Lopez F. 1993. Significant potential levels of hierarchical phenotypic selection in a woody perennial plant, *Cistus ladanifer*. *Oikos* 68:267–72

2. Addicott JF. 1979. On the population biology of aphids. *Am. Nat.* 114:762–63

3. Akers R, Nielsen D. 1990. Reproductive biology of the bronze birch borer (Coleoptera:Buprestidae) on selected trees. *J. Entomol. Sci.* 25:196–203

4. Alstad DN, Edmunds GF Jr. 1983. Adaptation, host specificity, and gene flow in the Black Pineleaf Scale. See Ref. 31, pp 413–26

5. Alstad DN, Edmunds GF Jr. 1983. Selection, outbreeding depression, and the sex ratio of scale insects. *Science* 220: 93–95

6. Antolin MF, Strobeck C. 1985. The population genetics of somatic mutation in plants. *Am. Nat.* 126:52–62

7. Armbruster WS. 1991. Multilevel analysis of morphometric data from natural plant populations: insights into ontogenetic, genetic, and selective correlations

in *Dalechampia scandens. Evolution* 45: 1229–44

8. Black WC IV, DuTeau NM, Puterka GJ, Nichols JR, Petteroini JM. 1992. Use of the random amplified polymorphic DNA polymerase chain reaction (RAPD-PCR) to detect DNA polymorphism in aphids. *Bull. Entomol. Soc.* 82:151–60

9. Breese EL, Hayward MD, Thomas AC. 1965. Somatic selection in perennial ryegrass. *Heredity* 20:367–79

10. Burk LG, Stewart RN, Dermen H. 1964. Histogenesis and genetics of a plastid-controlled chlorophyll variegation in tobacco. *Am. J. Bot.* 51:713–24

11. Burnet FM. 1971. "Self-recognition" in colonial marine forms and flowering plants in relation to the evolution of immunity. *Nature* 232:230–35

12. Buss LW. 1982. Somatic cell parasitism and the evolution of somatic tissue compatibility. *Proc. Natl. Acad. Sci. USA* 79:5337–41

13. Buss LW. 1983. Evolution, development, and the units of selection. *Proc. Natl. Acad. Sci. USA* 80:1387–91

14. Buss LW. 1985. The uniqueness of the individual revisited. See Ref. 64, pp. 467–505

15. Buss LW. 1987. *The Evolution of Individuality.* Princeton: Princeton Univ. Press

16. Buss LW. 1990. Competition within and between encrusting colonial invertebrates. *Trends Ecol. Evol.* 5:352–56

17. Buss LW, Green DR. 1985. Histocompatibility in vertebrates: the relict hypothesis. *Dev. Comp. Immun.* 9:191–201

18. Buss LW, Grosberg RK. 1990. Morphogenetic basis for phenotypic differences in hydroid competitive behaviour. *Nature* 343:63–65

19. Buss LW, Moore JL, Green DR. 1985. Autoreactivity and self-tolerance in an invertebrate. *Nature* 313:400–2

20. Cano-Santana Z, Oyama K. 1992. Variation in leaf trichomes and nutrients in *Wigandia urens* (Hydrophyllaceae) and its implications for herbivory. *Oecologia* 92:405–9

21. Capossela A, Silander JA Jr., Jansen RK, Bergen B, Talbot DR. 1992. Nuclear ribosomal DNA variation among ramets and genets of white clover. *Evolution* 46:1240–47

22. Carvalho GR. 1994. Evolutionary genetics of aquatic clonal invertebrates: concepts, problems, and prospects. In *Genetics and Evolution of Aquatic Organisms,* ed. AR Beaumont, pp. 291–323. London: Chapman & Hall

23. Carvalho GR, MacLean N, Wratten SD,

Carter RE, Thurston JP. 1991. Differentiation of aphid clones using DNA fingerprints from individual aphids. *Proc. R. Soc. London B* 243:109–14

24. Chao L, Cox EC. 1981. Evolution of mutation rates in *E. coli. Evolution* 37: 125–34

25. Chaplin J. 1992. Variation in the mode of reproduction among individuals of the ostracod *Candonocypris novaezelandiae. Heredity* 68:411–24

26. Charlesworth D. 1989. A high mutation rate in a long-lived plant. *Nature* 340: 346–47

27. Cobb N, Whitham T. 1993. Herbivore deme formation on individual trees: a test case. *Oecologia* 94:496–502

28. Crumpacker DW. 1967. Genetic loads in maize (*Zea mays* L.) and other cross-fertilized plants and animals. *Evol. Biol.* 1:306–424

29. Cullis CA. 1986. Unstable genes in plants. See Ref. 67, pp. 77–84

30. Cullis CA. 1987. The generation of somatic and heritable variation in response to stress. *Am. Nat.* 130(S):62–73

31. Denno RF, McClure MS. 1983. *Variable Plants and Herbivores in Natural and Managed Ecosystems.* New York: Academic

32. Dixon AFG. 1977. Aphid ecology: life cycles, polymorphism, and population regulation. *Annu. Rev. Ecol. Syst.* 8:329–53

33. Ellstrand NC, Roose WR. 1987. Patterns of genotypic diversity in clonal plant species. *Am. J. Bot.* 74:123–31

34. Eriksson O, Jerling L. 1990. Hierarchical selection and risk spreading in clonal plants. In *Clonal Growth in Plants: Regulation and Function,* ed. J van Groenendael, H Kroon, pp. 79–94. The Hague: SPB Academic

35. Evans DA, Paddock EF. 1979. Mitotic crossing-over in higher plants. In *Plant Cell and Tissue Culture, Principles and Applications,* ed. WR Sharp, PO Lason, EF Paddock, V Raghavan, pp. 315–51. Columbus: Ohio State Univ. Press

36. Fagerstroem T. 1992. The meristem-meristem cycle as a basis for defining fitness in clonal plants. *Oikos* 63:449–53

37. Ford H, Richards AJ. 1985. Isozyme variation within and between *Taraxacum* agamospecies in a single locality. *Heredity* 55:289–91

38. Gaul H. 1965. Selection in M<d<1 generation after mutagenic treatment of barley seeds. In *Induction of Mutations and the Mutation Process,* ed. J Veleminsky, T Gichner, pp 62–71. Prague: Czechoslovak Acad. Sci.

39. Gill DE. 1983. Within tree variation in

fruit production and seed set in *Capparis odoratissima* Jacq. in Costa Rica. *Brenesia* 21:33–40

40. Gill DE. 1986. Individual plants as genetic mosaics: ecological organisms versus evolutionary individuals. In *Plant Ecology*, ed. MJ Crawley, pp. 321–43. Oxford: Blackwell

41. Gill DE, Halverson TG. 1984. Fitness variation branches within trees. In *Evolutionary Ecology*, ed. B Shorrocks, pp. 105–116. Oxford: Blackwell

42. Green KA, Zasada JE, Van Cleve K. 1971. An albino aspen sucker. *For. Sci.* 17:272

43. Grosberg RK. 1988. The evolution of allorecognition specificity in clonal invertebrates. *Q. Rev. Biol.* 63:377–412

44. Harberd DJ. 1967. Observations on natural clones in *Holcus mollis*. *New Phytol.* 66:401–08

45. Hardwick RC. 1986. Physiological consequences of modular growth in plants. *Phil. Trans. R. Soc. London B* 313:161–73

46. Harper JL. 1967. A Darwinian approach to plant ecology. *J. Ecol.* 55:247–70

47. Harper JL. 1977. *Population Biology of Plants*. London: Academic

48. Harper JL, Bell AD. 1979. The population dynamics of growth form in organisms with modular construction. In *Population Dynamics*, ed. RM Anderson, BD Turner, LR Taylor, pp 29–52. Oxford: Blackwell

49. Harper JL, Rosen BR, White J. 1986. Preface to *The Growth and Form of Modular Organisms*. *Phil. Trans. R. Soc. London B* 313:3–5

50. Harper JL, White J. 1974. The demography of plants. *Annu. Rev. Ecol. Syst.* 5:419–63

51. Hartmann HT, Kester DE. 1975. *Plant Propagation*. Englewood Cliffs: Prentice-Hall

52. Harvell CD. 1990. The ecology and evolution of inducible defenses. *Q. Rev. Biol.* 65:323–40

53. Harvell CD. 1991. Coloniality and inducible polymorphism. *Am. Nat.* 138:1–14

54. Harvell CD. 1994. The evolution of polymorphism in colonial invertebrates and social insects. *Q. Rev. Biol.* 69:155–85

55. Haukioja E. 1991. Cyclic fluctuations in density: interactions between a defoliator and its host tree. *Acta Oecologia* 12:77–88

56. Haukioja E. 1991. The influence of grazing on the evolution, morphology, and physiology of plants as modular organisms. *Phil. Trans. R. Soc. Lond.* 333:241–47

57. Haukioja E, Ruohomaki K, Senn J, Suomela J, Walls M. 1990. Consequences of herbivory in the mountain birch (*Betula pubescens* spp. *tortuosa*): importance of the functional organization of the tree. *Oecologia* 82:238–47

58. Howard J. 1990. Infidelity of leafcutting ants to host plants: resource heterogeneity or defense induction? *Oecologia* 82:394–401

59. Hughes DJ. 1992. Genotype-environment interactions and relative clonal fitness. *J. Anim. Ecol.* 61:291–306

60. Hughes RN. 1989. *A Functional Biology of Clonal Animals*. London: Chapman & Hall

61. Hughes RN, Cancino JM. 1985. An ecological overview of cloning in metazoa. See Ref. 64, pp. 153–186

62. Hull DL. 1990. Individuality and selection. *Annu. Rev. Ecol. Syst.* 11:311–32

63. Jackson JBC. 1985. Distribution and ecology of clonal and aclonal benthic invertebrates. See Ref. 64, pp. 297–355

64. Jackson JBC, Buss LW, Cook RE. 1985. *Population Biology and Evolution of Clonal Organisms*. New Haven: Yale Univ. Press

65. Janzen DH. 1977. What are dandelions and aphids? *Am. Nat.* 111:586–89

66. Jelinski D, Cheliak W. 1992. Genetic diversity and spatial subdivision of *Populus tremuloides* (Salicaceae) in a heterogeneous landscape. *Am. J. Bot.* 79:728–36

67. Jennings DH, Trewavas AJ. 1986. *Plasticity in Plants*. Cambridge: Co. Biol.

68. Jerling L. 1985. Are plants and animals alike? A note on evolutionary plant population ecology. *Oikos* 45:150–53

69. Jorgensen JH, Jensen HP. 1986. The spontaneous chlorophyll mutation frequency in barley. *Hereditas* 105:71–72

70. King LM, Schaal BA. 1990. Genotypic variation with asexual lineages of *Taraxacum officinale*. *Proc. Natl. Acad. Sci. USA* 87:998–1002

71. Klekowski EJ Jr. 1984. Mutational load in clonal plants: a study of two fern species. *Evolution* 38:417–26

72. Klekowski EJ Jr. 1988. *Mutation, Developmental Selection and Plant Evolution*. New York: Columbia Univ. Press

73. Klekowski EJ Jr. 1988. Progressive cross- and self-sterility associated with aging in fern colonies and perhaps other plants. *Heredity* 61:247–53

74. Klekowski EJ Jr. 1992. Mutation rates in diploid annuals: Are they immutable? *Int. J. Plant. Sci.* 153:462–65

75. Klekowski EJ Jr, Godfrey PJ. 1989.

Ageing and mutation in plants. *Nature* 340:389–90

76. Klekowski EJ Jr, Karinova-Fukshansky NK. 1984. Shoot apical meristems and mutation: fixation of selectively neutral cell genotypes. *Am. J. Bot.* 71:22–27

77. Klekowski EJ Jr, Karinova-Fukshansky NK. 1984. Shoot apical meristems and mutation: selective loss of advantageous cell genotypes. *Am. J. Bot.* 71:28–34

78. Klekowski EJ Jr, Kazarinova-Fukshansky NK, Fukshansky L. 1989. Patterns of plant ontogeny that may influence genomic stasis. *Am. J. Bot.* 76:185–95

79. Klekowski EJ Jr, Kazarinova-Fukshansky NK, Mohr H. 1985. Shoot apical meristems and mutation: stratified meristems and angiosperm evolution. *Am. J. Bot.* 72:1788–1800

80. Klekowski EJ Jr, Lowenfield R, Helper PK. 1994. Mangrove genetics. II. Outcrossing and lower spontaneous mutation rates in Puerto Rican *Rhizophora*. *Int. J. Plant Sci.* 155:373–81

81. Klekowski EJ Jr, Mohr H, Kazarinova-Fukshansky NK. 1986. Mutation, apical meristems and developmental selection in plants. In *Genetics, Molecules, and Evolution*, ed. JP Gustafson, GL Stebbins, FJ Ayala, pp. 79–113. New York: Plenum

82. Lea DE, Coulson AC. 1949. The distribution of the numbers of mutants in bacterial populations. *J. Genet.* 49:264–85

83. Lewis WH, Oliver RL, Luikart TK. 1971. Multiple genotypes in individuals of *Claytonia virginica*. *Science* 172:564–65

84. Lewontin RC. 1970. The units of selection. *Annu. Rev. Ecol. Syst.* 1:1–11

85. Lowenfield R, Klekowski EJ Jr. 1992. Mangrove genetics. I. Mating system and mutation rates of *Rhizophora mangle* in Florida and San Salvador Island, Bahamas. *Int. J. Plant Sci.* 153:394–99

86. Luria SE, Delbruck M. 1943. Mutations of bacteria from virus sensitivity to virus resistance. *Genetics* 28:491–511

87. Mackie GO. 1963. Siphonophores, bud colonies, and superorganisms. In *The Lower Metazoa*, ed. EC Dougherty et al, pp 329–37. Berkeley: Univ. Calif. Press

88. McClintock B. 1984. The significance of responses of the genome to challenge. *Science* 26:792–801

89. Mitter C, Futuyma DJ. 1983. An evolutionary–genetic view of host plant utilization by insects. See Ref. 31, pp 427–59

90. Mogie M. 1985. Morphological, developmental and electrophoretic variation within and between obligately apomictic *Taraxacum* species. *Biol. J. Linn. Soc.* 24:207–16

91. Mogie M, Richards J. 1983. Satellited chromosomes, systematics and phylogeny in *Taraxacum* (Asteraceae). *Plant Syst. Evol.* 141:219–29

92. Nieuwkoop PD, Sutasurya LA. 1981. *Primordial Germ Cells in the Invertebrates*. Cambridge: Cambridge Univ. Press

93. Obeso JR, Grubb PJ. 1993. Fruit maturation in the shrub *Ligustrum vulgare* (Oleaceae): lack of defoliation effects. *Oikos* 68:309–16

94. Pergande T. 1901. *The life history of two species of plant-lice. Tech. Series No. 9, 7–44*. Dep. Agric., Washington, DC: US Govt. Print. Off.

95. Pettersson J. 1993. Odour stimuli affecting autumn migration of *Rhopalosiphum padi* (L.) (Hemiptera:Homoptera). *Ann. Appl. Biol.* 122:417–25

96. Price P. 1991. The plant vigor hypothesis and herbivore attack. *Oikos* 62:244–51

97. Purves WK, Orians GH, Heller HC. 1995. *Life: The Science of Biology*, Sunderland, MA: Sinauer. 4th ed.

98. Quiring D. 1993. Influence of intra-tree variation in time of budburst of white spruce on herbivory and the behaviour and survivorship of *Zeiraphera canadensis*. *Ecol. Entomol.* 18:353–64

99. Richards AJ. 1989. A comparison of within-plant karyological heterogeneity between agamospermous and sexual *Taraxacum* (Compositae) as assessed by the nucleolar organiser chromosomes. *Plant Syst. Evol.* 153:177–85

100. Rinkevich B, Weissman IL. 1987. A long-term study on fused subclones in the ascidian *Botryllus schlosseri*: the resorption phenomenon (Protochordata:Tunicata). *J. Zool. London* 213:717–33

101. Sackville Hamilton NR, Schmid B, Harper JL. 1987. Life-history concepts and the population biology of clonal organisms. *Proc. R. Soc. London B.* 232:35–57

102. Saghai-Maroof MA, Soliman KM, Jorgensen RA, Allard RW. 1984. Ribosomal DNA spacer-length polymorphisms in barley: Mendelian inheritance, chromosomal location and population dynamics. *Proc. Natl. Acad. Sci. USA.* 81:8014–18

103. Sanda-Kamigawara M, Ichikawa S, Watanabe K. 1991. Spontaneous radiation- and EMS-induced somatic pink mutation frequencies in the stamen hairs and petals of a diploid clone of *Trades-*

cantia, KU 27. *Environ. Exp. Botany* 31:413–21.

104. Sarkar S. 1991. Haldane's solution of the Luria-DelBruck distribution. *Genetics* 127:257–61

105. Schaal BA, Learn GH Jr. 1988. Ribosomal DNA variation within and among plant populations. *Ann. Mo. Bot. Gard.* 75:1207–16

106. Schmid B. 1990. Some ecological and evolutionary consequences of modular organization and clonal growth in plants. *Evol. Trends Plants* 4:25–34

107. Schneeberger RG, Cullis CA. 1991. Specific DNA alterations associated with the environmental induction of heritable changes in flax. *Genetics* 128:619–30

108. Schultz JC. 1983. Habitat selection and foraging tactics of caterpillars in heterogeneous trees. See Ref. 31, pp 61–90

109. Scowcroft WR. 1985. Somaclonal variation: the myth of clonal uniformity. In *Genetic Flux in Plants,* ed. B Hohn, ES Dennis, pp 217–245. Wien: Springer-Verlag

110. Senn J, Hanhimaki S, Haukioja E. 1992. Among-tree variation in leaf phenology and morphology and its correlation with insect performance in mountain birch. *Oikos* 63:215–32

111. Shaw AJ. 1990. Intraclonal variation in morphology, growth, and copper tolerance in the moss, *Funaria hygrometrica. Evolution* 44:441–47

112. Silander JA Jr. 1985. Microevolution in clonal plants. See Ref. 64, pp. 107–152

113. Slatkin M. 1984. Somatic mutations as an evolutionary force. In *Evolution: Essays in Honour of John Maynard Smith,* ed. PJ Greenwood, PH Harvey, M Slatkin, pp. 19–30. Cambridge: Cambridge Univ. Press

114. Sprugel DG, Hinckley TM, Schaap W. 1991. The theory and practice of branch autonomy. *Annu. Rev. Ecol. Syst.* 22:309–34

115. Stewart FM, Gordon DM, Levin BR. 1990. Fluctuation analysis: the probability distribution of the number of mutants under different conditions. *Genetics* 124:175–85

116. Stirling DG. 1993. *Genetic and environmental determinants of variation in vertical distributions of* Daphnia galeata mendotae. PhD thesis. Univ. Md, College Station. 208 pp.

117. Suomela J, Ayres MP. 1994. Within-tree and among-tree variation in leaf characteristics of mountain birch and its implications for herbivory. *Oikos* 70:212–22

118. Suomela J, Nilson A. 1994. Within-tree and among-tree variation in growth of *Epirrita autumnata* on mountain birch leaves. *Ecol. Entomol.* 19:45–56

119. Szabo L. 1992. Interleaf and intraleaf distribution of gall formers on *Quercus cerris;* the organization of the leaf-galling community. *Oecologia* 13:269–77

120. Thomas H, Grierson D. 1987. *Developmental Mutants in Higher Plants.* Cambridge: Cambridge Univ. Press

121. Tuomi J, Vuorisalo T. 1989. Hierarchical selection in modular organisms. *Trends Ecol. Evol.* 4:209–13

122. Tuomi J, Vuorisalo T. 1989. What are the units of selection in modular organisms? *Oikos* 54:227–33

123. Tuomi J, Vuorisalo T. 1990. Modularity as an organizational constraint on selection. In *Organizational Constraints on the Dynamics of Evolution,* ed. J. Maynard Smith, G. Vida. Manchester: Manchester Univ. Press

124. Tuomi J, Salo J, Haukioja E, Niemela P, Hakala T, Mannila R. 1983. The existential game of individual self-maintaining units: selection and defence tactics of trees. *Oikos* 40:369–76

125. Urbanek A. 1973. Organization and evolution of graptolite colonies. In *Animal Colonies: Development and Function Through Time,* ed. RS Boardman, AH Cheetham, WA Oliver, Jr, pp. 441–514. Stroudsburg: Dowden, Hutchinson, & Ross

126. Van Valen LM. 1987. Non-Weismann evolution. *Evol. Theory* 8:101–7

127. Vasseur L, Aarssen LW, Bennett T. 1993. Allozyme variation in local populations *Lemna minor* (Lemnaceae). *Am. J. Bot.* 80:974–79

128. von Dohlen CD. 1990. Evolutionary loss of the secondary host in heteroecious aphids: phylogenetic evidence. *Acta Phyto. Entomol. Hung.* 25:243–52

129. von Dohlen CD, Gill DE. 1989. Geographic variation and evolution in the life cycle of the witch-hazel leaf gall aphid, *Hormaphis hamamelidis. Oecologia* 78:165–75

130. von Dohlen CD, Stoetzel MB. 1991. Separation and redescription of *Hormaphis hamamelidis* (Fitch, 1851) and *Hormaphis cornu* (Shimer, 1867) on witch-hazel in eastern North America. *Proc. Entomol. Soc. Wash.* 93:533–48

131. Vuorisalo T, Tuomi, J. 1986. Unitary and modular organisms: criteria for ecological division. *Oikos* 47:382–85

132. Waller DM. 1986. The dynamics of growth and form. In *Plant Ecology,* ed. MJ Crawley, pp. 291–320. Oxford: Blackwell

133. Watson MA. 1986. Integrated physi-

ological units in plants. *Trends Ecol. Evol.* 4:119–23

134. Watson MA, Casper BB. 1984. Morphogenetic constraints on patterns of carbon distribution in plants. *Annu. Rev. Ecol. Syst.* 15:233–58

135. Weismann W. 1893. *The Germ Plasm: A Theory of Heredity.* London: Scott

136. White J. 1979. The plant as a metapopulation. *Annu. Rev. Ecol. Syst.* 10: 109–46

137. White J. 1984. Plant metamerism. In *Perspectives in Plant Population Ecology,* ed. R Dirzo, J Sarukhan, pp. 15–47. Sunderland, MA: Sinauer

138. Whitham TG. 1981. Individual trees as heterogeneous environments: adaptation to herbivory or epigenetic noise? In *Insect Life History Patterns: Habitat and Geographic Variation,* ed. RF Denno, H Dingle, pp. 9–27. New York: Springer-Verlag

139. Whitham TG. 1983. Host manipulation of parasites: within-plant variation as a defense against rapidly evolving pests. See Ref. 31, pp. 15–41

140. Whitham TG, Slobodchikoff CN. 1981. Evolution by individuals, plant-herbivore interactions, and mosaics of genetic variability: the adaptive significance of somatic mutations in plants. *Oecologia* 49:287–92

141. Whitham TG, Williams AG, Robinson AM. 1984. The variation principle: individual plants as temporal and spatial mosaics of resistance to rapidly evolving pests. In *A New Ecology,* ed. PW Price, CN Slobodchikoff, WS Gaud, pp 15–51. New York: John Wiley

142. Wikberg S, Svenson BM, Carlsson BA. 1994. Fitness, population growth rate, and flowering in *Carex bigelowii,* a clonal sedge. *Oikos* 70:57–64

Annu. Rev. Ecol. Syst. 1995. 26:445–71

DINOSAUR BIOLOGY

James O. Farlow

Department of Geosciences, Indiana-Purdue University at Fort Wayne, Indiana
46805

Peter Dodson

School of Veterinary Medicine, University of Pennsylvania, 3800 Spruce Street,
Philadelphia, Pennsylvania 19104

Anusuya Chinsamy

South African Museum, Post Office Box 61, Cape Town 8000, South Africa

KEY WORDS: Mesozoic Era, terrestrial paleoecology, body size, functional morphology,
             paleobiology

ABSTRACT

Most aspects of dinosaur biology cannot be observed directly but must be
reconstructed by a variety of often speculative approaches. Overall body form
can be established if good skeletal material of a dinosaur species is available.
From a skeletal reconstruction, interpretations of the animal's soft parts, and
inferences about how the creature's skeleton functioned as a living machine,
can be made. Inferences about dinosaur habitat preferences and sociality are
made from observations of the preservational contexts of skeletons, nesting
sites, and trackways. Some aspects of dinosaur biology are interpreted on the
basis of relationships between body size and physiological and ecological
parameters in living animals, but this involves much uncertainty.

Primary tissues of dinosaur bone suggest that dinosaurs had rapid growth rates,
but calibrating dinosaur growth rates in terms of body mass gained per unit time
is difficult. It is uncertain whether dinosaurs needed metabolic rates comparable
to those of living birds and mammals in order to grow quickly enough to form the
primary bone tissues commonly found in dinosaur skeletons.

No evidence convincingly shows that dinosaurs were endotherms, and some
evidence suggests that they were not. Dinosaurs routinely achieved considerably larger body sizes than do terrestrial mammals, and they maintained viable

populations in smaller geographic areas than is possible for elephant-sized mammals. This suggests that dinosaurian food requirements were proportionately less than those of birds and mammals, thus permitting large population densities.

INTRODUCTION

Dinosaurs have become such prominent features of popular culture that it is easy to forget that they were once living animals. Countless movies and works of pulp fiction have made them symbols of immense, ravening destruction, our culture's answer to the dragons of earlier mythology—Rahab, Fafnir, and the Hydra displaced by *Tyrannosaurus*, *Velociraptor*, and Godzilla.

Perhaps contributing to this mythic "monster" image is that so many features that characterize the lives of real animals cannot be observed, but only inferred, for dinosaurs. We do not know exactly what the external appearance of any dinosaur was like. We cannot directly determine how much they weighed, how quickly they grew, how long they lived, or how often they reproduced. It is a real accomplishment when, for any dinosaur species, we can pin down natural history details of the kind that ecologists working on living animals can almost take for granted.

Unfortunately, the popular media often seem unable to distinguish speculation about dinosaur paleobiology from actual observations, and sometimes paleontologists themselves do not make this distinction sufficiently clear. Once published, hypotheses become "evidence" cited to support secondary hypotheses. Hypothetical, artistic restorations of living dinosaurs themselves become rhetorical ammunition in debating interpretations of dinosaur biology.

There is nothing wrong with responsible speculation, though, as long as one remembers that the plausibility of a hypothesis does not guarantee its truth. Informed speculation can focus the attention of paleontologists on questions that might not otherwise be asked, and in the process can suggest new observations, or ways of looking at previous discoveries, that can either disprove or corroborate hypotheses about dinosaur biology.

Here we review the kinds of evidence used to reconstruct dinosaurs as living animals. We then consider selected aspects of dinosaur biology in more detail.

THE "EPISTEMOLOGY" OF DINOSAUR BIOLOGY: WHAT DO WE KNOW, AND WHAT CAN WE KNOW?

The Fossil Record of Dinosaurs

Inferences about dinosaur biology range from the very robust to the highly speculative, depending on the kind of evidence adduced in their support.

Generally, the more directly an interpretation is based on the hard evidence of the fossils themselves, the greater the confidence that can be placed on it. Our primary database for understanding dinosaurs is thus the fossil record itself, with all its biases and imperfections (43). Of 285 genera of dinosaurs recognized in 1988, about 45% were based on only a single articulated specimen, and only about 20% on essentially complete skulls and skeletons. Rough estimates of the total number of dinosaur genera that lived over the course of the Mesozoic Era range from as few as 675 to as many as 3400 (43, 120).

Morphological/Comparative Anatomical Arguments

GROSS BODY FORM   The popular stereotype notwithstanding, paleontologists cannot reconstruct an entire skeleton from a single bone unless more complete remains of the same or a closely related taxon have previously been found. Given a reasonable skeletal reconstruction (or better still a complete skeleton), however, paleontologists can make a reasonable conjecture about the soft parts of the living dinosaur. Even without direct fossil evidence, we may confidently assert that dinosaurs had hearts, carotid arteries, lungs, and other organs that all amniotes possess—but we can do much better than that. Comparing the occurrence of soft tissue structures in cladistic crown-groups [the extant phylogenetic bracket (EPB); 15, 138] sharpens our ability to make phylogenetic inferences about their occurrence in extinct forms. Crocodilians and birds form the EPB for dinosaurs. Soft structures that are present (e.g. a four-chambered heart) or absent (e.g. a muscular diaphragm) in both crocodilians and birds probably were correspondingly present or absent in dinosaurs. When a structure occurs in only one of the crown-groups, its presence in dinosaurs is more speculative. A sacral glycogen body occurs in birds, but not in crocodilians. The same body has been thought to have been present in some dinosaurs (58); this is not unlikely but is more speculative on phylogenetic grounds. Even more speculative are structures found neither in crocodilians nor birds, such as the putative "carotid hearts" suggested to have elevated blood up the necks of sauropods (29).

In general, muscles with fleshy attachments leave little osteological evidence of their presence, but muscles with tendinous or aponeurotic attachments are likely to leave skeletal signatures in the form of crests, ridges, or scars. Based on the limited number of species studied so far (15, 88, 94), reptiles and birds seem much less likely to show muscle scars than do mammals—bad news from the standpoint of restoring dinosaur musculatures. More positively, however, body size may correlate positively with the prominence of muscle scars on bones; the large sizes of most dinosaurs may enhance our ability to find and interpret muscle scars on their bones. Use of the EPB (with crocodilians and birds again the crown-groups) extends our interpretation of dinosaurian musculature beyond the limited evidence from muscle scars, and numerous restorations of muscle arrangements in particular dinosaurs have been published (e.g. 94, 101).

The scant evidence from fossil integument impressions of nonavian dinosaurs shows a skin with a pebbly, reptilian texture (10, 84, 126). Impressions of soft, elevated, rectangular scutes have been reported in the tail of a hadrosaur (64), and conical tail scutes up to 18 cm high from an undescribed sauropod (36). Such finds offer a rare glimpse into species-specific patterns of ornamentation. Dinosaur color patterns and the use of color in behavioral displays are completely conjectural, but the EPB method suggests that it is reasonable to infer that crests, frills, or other features used as display structures were colorful.

FUNCTIONAL MORPHOLOGY   Analyses of the shapes of bones and teeth, tooth wear surfaces, skeletal proportions, and joint configurations have been used to reconstruct styles of locomotion, feeding, and even intraspecific courtship and agonistic behavior in dinosaurs (52, 56, 60, 75, 96, 98, 136). Inferences of this kind often rely, at least in part, on comparisons with the functional morphology and natural history of living animals. For example, speculations that the horns of ceratopsids were employed as much or more during intraspecific interactions than as anti-predator devices are based partly on observations of the way horns and hornlike organs are used by living mammals and reptiles (50, 90), and on what is plausibly interpreted as sexual dimorphism in the deployment of ceratopsid horns (81).

Sometimes interpretations based on functional interpretations of skeletal material can be corroborated by other kinds of evidence. The inference that dinosaurs had an erect limb carriage (at least for the hindlimb—75) is amply confirmed by the narrow "gauge" of dinosaur trackways (129). Identification of dinosaur species as carnivores or herbivores on the basis of tooth shapes and wear surfaces can sometimes be corroborated by stable isotope ratios of dinosaur bones (9, 99), the occurrence of bite marks in fossil bone, or the analysis of gut contents or fossilized droppings of dinosaurs (129). Even the occurrence of injuries in fossil bone, or the development of unusual ossifications, may shed light on stresses to which the bones were subjected and thus constrain functional interpretations of osteological structures (115).

BONE HISTOLOGY   Although an array of bone tissue types exists in vertebrates, the basic components of bone remain essentially the same (55). Differences in bone tissue types reflect differences in the rate of bone formation (106, 110). Although the organic components of bone decompose after death, the inorganic fraction is remarkably stable, permitting characterization of fossil bone microstructure.

Fibro-lamellar primary bone is formed during very rapid bone deposition;

it is typical of bird and mammal bones but relatively rare in bones of extant reptiles, although it has been reported in turtles and crocodilians (46, 108). More typical of living reptiles is lamellar-zonal bone, indicative of seasonal alternations between fast and slow rates of bone deposition. Zonal bone is uncommon in endotherms but does occur (27).

Because the histology of primary bone provides a record of the processes of growth, it is possible to interpret bone growth patterns in dinosaur species. However, the task is complicated by the fact that, histologically, dinosaur bone is not exactly like that of either modern ectotherms or endotherms (see below).

Taphonomic/Aktuopaleontological Arguments

DINOSAUR HABITATS   Dinosaur remains are preserved in environments ranging from shallow marine through arid, eolian deposits, but the most frequently sampled environments are moist lowland settings in the fluvio-deltaic complex (35, 53, 65, 70, 72, 80, 93, 114). Effective use of taphonomic data for interpreting dinosaur paleoecology depends on statistical patterns of occurrence of fossils in different sedimentary situations. It is unwise to draw conclusions from the occurrence of a single skeleton in a single depositional setting. The best bases for making such inferences are situations in which the same taxa are distributed across a suite of environments, or instances in which a transect across contemporaneous, environmentally different rock formations can be made.

In the Late Jurassic Morrison Formation of the western United States, the same dinosaur fauna occurs at sites distributed over a distance of 1000 km. The distribution of sauropod genera across a spectrum of preservational environments demonstrates that these huge herbivores were not confined to single environments, but rather roamed across all habitats, and may even have migrated on an annual basis in the face of periodic drought (44). The Morrison sedimentary record also suggests that stegosaurs may have preferred somewhat drier habitats than did their sauropod neighbors.

Dinosaur diversity of the moist, fluvio-deltaic, coastal lowland environments of the Late Cretaceous Judith River Formation of western North America is high in comparison with that of the drier, more inland Two Medicine Formation of the same age and region; there is little indication of faunal mixing between the two formations (7, 12, 65, 133). The ornithopods *Maiasaura* and *Orodromeus* are absent or uncommon in the lowland faunas, and the common hadrosaurs and ceratopsians from the lowlands do not occur in the upland deposits.

Although ceratopsids are common members of both lowland and upland faunas of the Late Cretaceous of western North America, protoceratopsids are not found in coastal lowland deposits. Protoceratopsids are rare in North

America, but along with psittacosaurs can be abundant in xeric inland deposits of China and Mongolia (40, 74, 80, 97).

In Mongolia, stratigraphically successive Late Cretaceous units demonstrate strong environmental control of faunas (40, 74, 97). In the older Djadochta and Barun Goyot Formations, oxidized red sediments record dry conditions, and the dinosaur fauna is dominated by small and medium-sized species (protoceratopsids, small theropods, a small ankylosaurid), with rare evidence of large hadrosaurs and theropods. The youngest unit, the Nemegt Formation, records the return of moist conditions and fluvial systems. Here large hadrosaurs and tyrannosaurs dominate the fauna, and even rare sauropods are found.

In many Mesozoic sedimentary units, the number of dinosaur skeletons known is greatly exceeded by the number of fossilized dinosaur trackways (82). Jurassic and Cretaceous footprint assemblages in carbonate rocks are often dominated by tracks of large theropods and sauropods, while track assemblages in siliciclastic rocks of comparable age are frequently dominated by ornithopod prints (83). Thus the footprint record, like the skeletal record, allows at least broad inferences about the habitat preferences of particular dinosaur groups.

Paleogeographic reconstructions of continental positions over the course of the Mesozoic Era (123), combined with paleoenvironmental reconstructions for given regions, allow "retrodictions" of ancient climatic regimes (5, 59, 124, 131, 139). The Mesozoic Era seems to have been a time of fairly warm conditions over much of the planet. At least seasonally dry climates were widespread in the Triassic Period, and they remained prevalent at low latitudes through the remainder of the era; moister conditions became established at midlatitudes in the later Mesozoic, however (6). As a group, dinosaurs coped with the entire range of Mesozoic climates; some species occupied high-latitude regions that experienced seasonally dark and cool conditions (30, 91a, 109).

DINOSAUR BEHAVIOR   Monotypic bonebeds suggest that certain dinosaurs lived in single-species groups. This interpretation becomes more compelling when multiple bonebeds are known for a particular species of dinosaur, thus demonstrating a repeated tendency rather than a unique event. The Ghost Ranch Quarry, New Mexico, contains remains of 1000 or more juvenile and adult specimens of the small Late Triassic theropod *Coelophysis* that may have perished in a severe drought (122). Because this deposit is unique, it does not form any basis for generalization about sociality in *Coelophysis*. However, the existence of a similar mass kill of a closely related species (105a) supports the idea that these small theropods were gregarious. Monotypic bonebeds are common for the prosauropod *Plateosaurus*, ceratopsids, and hadrosaurids (34,

121, 134), strongly suggesting that these dinosaurs lived (and perished) in herds.

It is common for dinosaur trackway sites to record preferred directions of travel of trackmakers, but in many instances those preferred directions parallel ancient shorelines (82). Although it is possible that such sites record movements of herds along a shore, a bimodal, mirror-image pattern of predominant trackway orientations could merely record movements of solitary animals back and forth along the shoreline. However, if one trackmaker type shows a unimodal pattern of trail orientations, while other kinds of trackmakers left the mirror-image pattern, this suggests that the first set of trails was made by a group of animals. Even more persuasive trackway evidence of dinosaur sociality comes from sites where parallel trails of a particular kind show abrupt, side-by-side changes in direction, suggesting that a group of animals simultaneously adjusted their movements to avoid colliding with one another (33).

Dinosaur nest sites have yielded a wealth of information about dinosaur nest building, egg laying, and clutch sizes (19). The spacing of individual nests, and their occurrence on successive bedding planes, indicate communal nesting behavior and repeated use of particular nest sites by certain ornithopods (63, 69).

However, identification of the egg-layers is only possible when identifiable embryos occur within eggs; the mere discovery of a range of growth stages of a particular dinosaur in the vicinity of a nesting ground is not enough. Mongolian eggs long ascribed to *Protoceratops* turned out, once identifiable embryos were found inside eggs (95), to have been laid by the peculiar theropod *Oviraptor*. This discovery suggests that the parent *Oviraptor* may have guarded its nest in the manner of many birds and crocodilians (32).

Modeling and Scaling Arguments

A fairly speculative approach to dinosaur biology is the use of mathematical or even three-dimensional, experimental models to constrain interpretations of the functions of specialized features of skeletal anatomy. The hollow crests of lambeosaurine hadrosaurs have been viewed as resonating devices, in part on the basis of acoustic modeling of the resonant frequencies of such features (135). The highly vascularized dorsal bony plates of *Stegosaurus* were suggested to have had a thermoregulatory function because wind-tunnel experiments with model stegosaurs indicated that the plates were deployed in an ideal arrangement for dumping excess body heat by forced convection (51).

Among the most interesting ideas about dinosaur biology are those based on scaling arguments, but these are at the same time the interpretations farthest removed from the fossils themselves. A burgeoning literature considers the manifold relationships between animal body size and a variety of physiological and ecological parameters (16, 37, 100, 104). As the largest land animals in earth history, dinosaurs become highly appropriate fodder for such rumination.

However, there is no way of directly weighing a dinosaur. Given a fairly complete skeleton, a scientific artist can sculpt a life restoration that should come reasonably close to the body proportions of the living animal. From the volume of the model and its scale one can then estimate the volume of the full-sized animal, and given reasonable values of the creature's specific gravity, an estimate of the animal's live mass can be made (1). Alternatively, one can do a regression of body mass on some skeletal measurement of living animals, and use the regression equation to predict dinosaur masses from their bones (2, 17).

Each of these approaches has obvious shortcomings. There is no way to be sure that an artist has correctly restored the musculature and other soft tissues on a dinosaur model. One cannot be certain that regression equations based on living reptiles, birds, or mammals are entirely appropriate for dinosaurs, and there is the additional problem that some dinosaurs were considerably larger than living land animals, such that measurements of their bones may lie well outside the range of data used to create the regression equations. These uncertainties mean that estimates of dinosaur masses are themselves hypotheses and not observations. Arguments that rely on such mass estimates can be considered robust only if they hold true for a range of body mass estimates above and below the estimate actually used.

Scaling arguments and estimates of dinosaur body masses have been used in discussions of dinosaur food consumption rates, predator/prey ratios, reproductive rates, locomotor capabilities, and body sizes (1, 21, 48, 49, 57, 62, 73). Their most sophisticated use has been in models of dinosaur thermoregulation and life history variation, based on data from living reptiles (45, 125).

As useful as such studies have been in focusing attention on plausible scenarios for various aspects of dinosaur biology, they are only as good as the assumptions and data on which they are based. Although their conclusions may be fairly reliable, they probably cannot be considered as well established as interpretations made more directly on fossil material.

ASPECTS OF DINOSAUR BIOLOGY

Reproduction and Growth

SEXUAL DIMORPHISM   Recognition of sexual dimorphism in dinosaurs is based on analyses of series of specimens that potentially represent single biological populations (as from a monotypic bonebed) or that putatively represent single species (specimens from a single geologic formation from a particular region).

All or nearly all dinosaurs were oviparous (19). Because dinosaur eggs are not especially large, compared to the sizes of presumed egg-layers, pelvic architecture is rarely informative as to the sex of individual dinosaurs. However, in male crocodilians the large, first chevron articulates with the second caudal centrum, and it is big in order to serve as an anchor for retractor muscles of an intromittent organ (78). In female crocodilians, in contrast, a relatively smaller first chevron articulates with the third caudal centrum. Consequently the female cloaca is larger than that of males, presumably to allow more room for eggs to pass. The same set of pelvic differences observed in crocodilians is said to occur in *Tyrannosaurus* (78), but unlike crocodilians, and like birds of prey, it is the putative female morph that is larger and more massively built in *Tyrannosaurus*.

Dimorphism of display structures is well known in extant, visually oriented vertebrates. As a working hypothesis, we assume that those individuals of a particular dinosaur species showing the most conspicuously developed potential display structures are males, based on what is commonly seen in living animals. Possible dimorphism of this kind has been inferred for the horns and frills of ceratopsians, the thickened and ornamented skull domes of pachycephalosaurs, the cranial crests of lambeosaurine hadrosaurs, and more subtly in the cranial rugosities of theropods and the body armor of ankylosaurs (18, 41, 42, 81, 91, 128).

Robust and gracile morphs occur in small and large species of theropods (18, 31, 105a). These are usually identified as sexual dimorphs and, as already noted for *Tyrannosaurus*, the female is usually suggested to have been larger and more massively built than the male.

GROWTH RATES   Some research has suggested that large ectothermic dinosaurs would have taken several decades to reach maturity (21). Models based on living reptiles, however, indicate that very slow growth rates would require impossibly high juvenile survivorship rates; it is more likely that even the largest dinosaurs reached maturity within 20 years (45).

Dinosaur primary bone commonly shows fibro-lamellar bone deposited in zones separated by lines of arrested growth (LAGs—107, 111). Episodes of fairly rapid growth were punctuated by intervals when growth slowed or even ceased. If growth spurts occurred during a single season of the year, the minimum ages of dinosaurs can be estimated from the number of growth lines preserved in bones. Such LAG counts have been used to suggest that the small theropod *Syntarsus* required 7–8 years to reach full size, another small theropod (*Troodon*) 3–5 years, and the prosauropod *Massospondylus* 15 years (22, 23, 132). However, deposition of azonal fibro-lamellar bone occurs in the small to medium-sized ornithopod *Dryosaurus*, suggesting continuous rapid growth with no pauses or decelerations (25).

Closely spaced, peripherally positioned lines of arrested growth occur in compacta formed during late growth in mammals, generated at a time when growth has slowed or ceased upon the attainment of sexual maturity. Similar peripheral LAGs occur in the bones of many dinosaurs (22, 107, 132). However, they are not found in the largest individuals of *Massospondylus* and *Dryosaurus*. Either these individuals do not represent fully grown animals, or these were species with an indeterminate pattern of growth, as opposed to the more determinate growth pattern seen in birds, mammals, and other dinosaurs (23, 25).

A scenario of rapid ontogenetic growth in the hadrosaurid *Maiasaura* has been formulated on studies of nesting sites, young individuals, and monotypic bone beds (63, 66–69, 71, 119, 134). Numerous skeletons of young individuals were found within nests. Wear surfaces on their teeth suggested that they had been feeding prior to death, but the unfinished appearance of bony tissues of femoral epiphyses suggested that the nestlings were altricial (in contrast to another member of the same fauna, the hypsilophodontid *Orodromeus*), and unable to leave the nest. The largest *Maiasaura* found within nests were a little more than a meter long, suggesting that hatchlings were nestbound until they left the nest upon growing to that size. The presence of broken eggshells within *Maiasaura* nests, rather than the more complete half-eggs found in abandoned nests of the precocial *Orodromeus*, was attributed to trampling of hatched eggs by the young hadrosaurs. Adult *Maiasaura* were thought to have fed their young regurgitated plant material during the time they were nestbound.

A catastrophic kill of perhaps thousands of individuals of *Maiasaura* was found in the same formation as the nest sites. There were said to be distinct size classes of individuals, with modes corresponding to body lengths of 3, 4, 5.2, and 7 meters (66, 119, 134). On the assumption that these size classes were year classes, the three-meter-long *Maiasaura* were interpreted as yearlings, and the largest size class as individuals four (or five) years old (and older). Consequently *Maiasaura* was thought to reach full size in roughly 4 or 5 years. If sexual maturity was reached at 67–75% of the final body length (20, 21), this would correspond to the 5.2-m size class, and an age of roughly three years. This is considerably shorter than the 10–12 years to sexual maturity predicted for *Maiasaura* by a model based on the physiological ecology of living reptiles (45).

However, this scenario is obviously based on rather circumstantial evidence. Although it is possible that young *Maiasaura* were nestbound, it would be somewhat surprising for the parents to have left eggshells in the nest where they would be subject to trampling, instead of removing them from the nest (the latter as commonly done by altricial birds—3). The interpretation that the unfinished articular surfaces of young *Maiasaura* indicate altriciality (71) was not documented by a comparison of limb articular surfaces in a variety of altricial as opposed to precocial bird species. In fact, the articular surfaces of

young *Maiasaura* show marked similarities to those of young—and precocial—domestic chickens (4).

The scenario requires the assumption that size classes in *Maiasaura* bonebeds correspond to year classes (not unreasonable), and that the age classes
have been correctly identified (less certain). Until independent criteria for
assessing ontogenetic ages are established, growth rates interpreted on the basis
of the *Maiasaura* scenario should be regarded as hypotheses rather than observations.

Discussions of dinosaur growth rates and primary bone texture figure prominently in interpretations of dinosaur physiology (26, 27, 106–108, 110, 132).
The argument usually involves two steps: The presence of fibro-lamellar tissue
and a high degree of vascularity are taken as indicative of rapid growth, and
rapid growth rates are then presumed to require rapid metabolic rates.

Although a link between fibro-lamellar primary bone and fast growth rates
seems likely, calibrating the relationship is difficult. Just how fast bone deposition has to be before an animal shifts from making lamellar-zonal to fibro-
lamellar bone, and how this bone deposition rate relates to overall animal
growth in units of body mass per time, are uncertain.

The overall vascularity of the femora of juvenile ostriches is greater than
for young Nile crocodiles—but not by much—and young crocodiles have more
highly vascularized femora than do young secretary birds (24). Although a
link between bone vascularity and growth rate is plausible, the correlation does
not seem very tight.

The relationship between growth and metabolic rates may be even looser.
The absolute mass gain (g/day) during the interval of fastest growth is positively correlated with adult body mass in vertebrates (20), but how the growth
rates of reptiles and endotherms of a given mass compare depends on how
adult body mass is defined. Birds and mammals first reproduce when they
have grown to nearly their final body mass, but reptiles first breed at roughly
30–40% of maximum body mass. If juvenile growth rate is regressed against
final adult mass, reptilian growth rates are about an order of magnitude less
than those of birds and mammals (20). However, if growth rates are standardized against body mass at sexual maturity, the discrepancy is much less;
alligators may equal and even exceed the absolute growth rate observed in
some mammals of comparable mass at first reproduction (117).

Were Dinosaurs Endotherms or Ectotherms?

Dinosaur bone histology seems not to provide unambiguous evidence about
dinosaur metabolic rates. It therefore joins numerous other lines of evidence
that have failed to settle the contentious question of whether dinosaurs were
endotherms, ectotherms, or transitional between these two metabolic states
(48, 117). Although we wonder if this matter will ever be settled definitively,

some hope is offered by the recent discovery that the presence of respiratory turbinates (which occur in living birds and mammals, but not reptiles) is tightly linked to the rapid rates of pulmonary ventilation associated with endothermy (61). Skeletal attachment sites for respiratory turbinates occur in both birds and mammals, and they are also found in therocephalian and cynodont therapsids (61). So far there is no evidence for such nonsensory, respiratory turbinates in any dinosaurs (J. Ruben, unpublished observations), which suggests that these reptiles were not endotherms.

Body-Size Distributions in Dinosaur Faunas: Ectothermic Giants in a Greenhouse World?

If dinosaurs were metabolically more like living ectothermic reptiles than like endothermic birds and mammals, this may help explain one of the most conspicuous features of these ancient reptiles: the large body sizes of most dinosaur species (103).

Recent studies (14, 87) have described the body-size frequency distributions of Holocene (modern) nonvolant, terrestrial mammalian species and genera on large and small continents. Log body mass distributions are positively skewed for the faunas of large continents, but more nearly normally distributed for the faunas of small continental masses, due to the absence of the right-hand tail of large-bodied species on the smaller continents. Mammalian faunas sampled at a progressively finer scale (continent to biome to "local patches of uniform habitat") show a shift from positively skewed to nearly uniform log body mass distributions.

The present-day North American mammalian fauna is depleted in very large species, reflecting the geologically recent extinction of the megafauna (77). Figure 1 (a—shaded) shows a body-size distribution for the late Pleistocene (25,000–10,000 BP) mammalian fauna of North America (actually the United States), based on species presently known from fossil evidence. Body-size data are also summarized for members of the Order Carnivora and for plant-eating mammals that attain a body size of roughly 1 kg or more (a taxonomically heterogeneous assemblage).

Like that of the modern mammalian fauna, the late Pleistocene body-size distribution is positively skewed. Most species occur in size classes 2 and 3, and the median is size class 3, but the relative importance of these smaller mammals is less marked than in the Holocene body-size distribution. Rather than indicating a real paucity of small mammals in the latest Pleistocene fauna, this probably reflects a taphonomic bias against preservation of the fossils of small-bodied animals. Because all (or nearly all) Holocene mammal species were probably in existence by the latest Pleistocene, a "theoretical" body-size distribution for the latest Pleistocene fauna can be created by adding present-day species not yet found as fossils to those known from fossil deposits (Figure

1 [a—unshaded]). If this is done, the Pleistocene body-size distribution does not look markedly different from that of the Holocene fauna (14).

What is different, of course, is that there are more large-bodied species in the Pleistocene fauna, particularly among the large herbivores. All of the really big (size class 7) latest Pleistocene mammals were plant-eaters, representing a size class not present in the modern fauna.

As one of the smaller continents, Holocene Australia has few large-bodied mammalian species (87). The mode for the fauna is size class 2, and the median is size class 3; the largest species occur in size class 5 (Figure 1 [b]). However, like many other parts of the world, Australia has experienced late Quaternary extinctions of megafaunal species (92). The mode and median for Australian Pleistocene megafaunal herbivores are size class 5, and only one species occurs in size class 7 (Figure 1 [c]). The few megafaunal carnivores all occur in size class 5. In fact, the largest Australian megafaunal predator was not even a mammal, but rather a huge varanid lizard (54, 92).

The scarcity of large mammalian carnivores in Australia has been attributed to the nutrient-poor nature of Australian soils, and the variability of rainfall patterns due to the El Niño-Southern Oscillation cycle, which are thought to result in low and unpredictable levels of plant productivity (54). This would in turn prevent the establishment of large populations of herbivorous mammals and thus preclude the existence of a sufficiently large resource base to support large carnivorous mammals. However, the small land area of the continent probably is also important in having prevented the evolution of very large Australian mammals (87).

The mammalian faunas of the Pliocene Laetolil Beds of Tanzania, the Miocene Dhok Pathan Formation of Pakistan, and the Pleistocene of the Ozark region of the United States (Figure 1 [d-f]) do not represent the overall mammalian faunas of entire continents, but rather samples of mammals from three geographically circumscribed sets of sites on large continents. The faunas from these three units probably reflect spatial scales intermediate between the biome and the "local patches of uniform habitat" (14) levels. This may partly account for the lesser strength of the smaller size classes in the body-size distributions of these three faunas, with the taphonomic bias against small-bodied animals constituting the rest (perhaps the greater part) of the deficit.

To generalize across the various mammalian faunas, when allowances are made for geographic scale and taphonomic biases, small-bodied species domi-

→

Figure 1   Body size distributions in mammalian faunas. Body masses are expressed in terms of logarithmic size classes because of the uncertainties in estimating body masses of extinct forms. Size class 1 = mass greater than/equal to (GE) 0.001 kg and less than (LT) 0.01 kg, size class 2 = mass GE 0.01 kg and LT 0.1 kg, and so on. (Data from 14, 39, 77, 79, 85, 92, 127; and C Badgley, J Damuth, R Graham, unpublished observations).

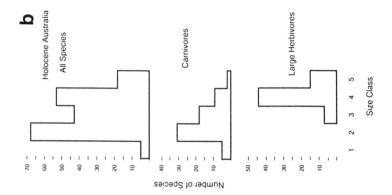

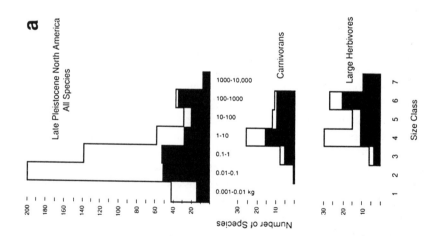

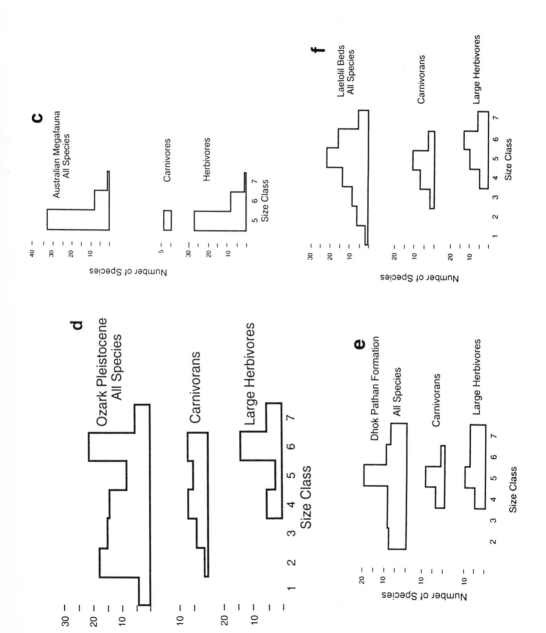

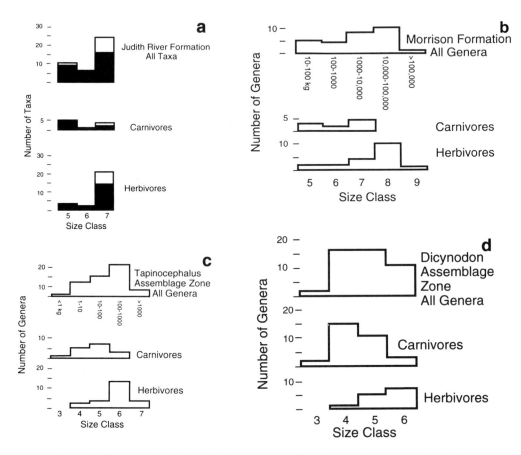

Figure 2 Body-size distributions in dinosaurian (a, b) and therapsid faunas (c, d). Ornitho-mimosaurs and oviraptorosaurs are included in the distribution for all Judith River dinosaurs, but not in the carnivore or herbivore distributions, due to uncertainty about their diets. Data from 1, 2, 76, 118, 137, and T Holtz, N Hotton, G King, B Rubidge, H-D Sues (unpublished observations).

nate faunas that sample entire continents. The greatest species richness of large carnivores occurs in the 1–10 or 10–100 kg size class. Continent-wide large herbivore faunas show their greatest richness in size class 4 but are also strong in bigger size classes, particularly on large continents. Where really big species occur in a mammalian fauna, these are always plant-eaters.

The situation is very different for dinosaurian faunas. Figure 2 (*a*—un-shaded) shows the body-size distribution of Judith River Formation dinosaurs [based on current species-level classifications (137)]. Figure 2 (*a*—shaded) shows the Judith River body-size distribution on the conservative assumption that there is only one valid species per genus.

The present species-level taxonomy of Morrison Formation dinosaurs is almost surely overly split, particularly for sauropods, although there may well be more than one valid species for some Morrison genera (137). We choose not to attempt to decide such systematic questions, and in Figure 2 (b) illustrate only a genus-level body-size distribution.

Like the mammalian faunas of the Laetolil Beds, the Dhok Pathan Formation, and the Ozark Pleistocene, the Judith River dinosaurian fauna probably represents a scale intermediate between the homogeneous local habitat and the biome levels, and so the total taxonomic richness of dinosaurs may reflect a mixture of within-habitat and between-habitat diversity (7, 12). The species-level and genus-level body-size distributions for Judith River dinosaurs show identical patterns. The smallest taxa occur in size class 5—a size class three orders of magnitude larger than the modal size class for North American late Pleistocene mammals (theoretical), and the same as the modal size class for mammals from the Dhok Pathan Formation, the Laetolil Beds (although, as previously noted, these assemblages are likely biased against very small-bodied mammal species), and the Australian Pleistocene megafauna.

The median and modal size class of Judith River dinosaurs (a size class dominated by herbivorous taxa) is size class 7. This is two orders of magnitude larger than the mode for the Dhok Pathan and Australian megafaunal herbivores, and an order of magnitude larger than the modes for Laetolil and Ozark large herbivores, as well as the second mode (size class 6) seen in the theoretical latest Pleistocene large herbivores of North America. Although there is overlap in body size between herbivorous dinosaurs and large mammalian herbivores, for mammals size class 7 represents a relatively minor category, while for plant-eating Judith River dinosaurs this is the strongest size class.

Judith River carnivorous dinosaurs show a mode in size class 5, as do carnivores from most of the mammalian faunas. In addition, though, the Judith River theropods have representatives in size class 7, a size greater than that reached by any known terrestrial mammalian carnivore, living or extinct (49).

Morrison Formation dinosaurs show much the same pattern as the Judith River dinosaurs (Figure 2 [b]), except that Morrison sauropods attained body masses even greater than those of Judith River herbivores, reaching 10,000–100,000 kg (or more), a size approached by only a few mammalian herbivores.

Although the same taphonomic bias against small-bodied animals undoubtedly occurs in Mesozoic as in Cenozoic terrestrial vertebrate faunas, this is probably not responsible for the scarcity of small-bodied dinosaur species. Dinosaurs with adult masses of a few kilograms are known (137), but the small vertebrates commonly found in well-sampled units like the Judith River and Morrison Formations represent non-dinosaurian groups (6). There is no fossil evidence that even hints at a diversity of non-avian dinosaurs in size classes 1–4. The shape of the Judith River and Morrison body-size distributions actually suggests that a continent-scale body-size distribution of dinosaur taxa would be negatively skewed (cf. 49, 103, and J Damuth, unpublished observations).

Certain features of the geographic occurrence of dinosaur taxa make the greater body-size of dinosaurs than of terrestrial mammals even more remarkable. During the Campanian Age of the Late Cretaceous (when sediments of the Judith River and Two Medicine Formations were deposited), the western part of North America was separated from the eastern part of the continent by a shallow seaway. Western North America was connected with eastern Asia above the Arctic Circle to form a single landmass, Asiamerica, that had a distinctive dinosaurian fauna as compared with the rest of the world.

Although the Asian and American portions of Asiamerica have several dinosaurian families and one or more genera in common (74, 137), they seem not to have shared any species, suggesting that the Arctic connection between the two segments of Asiamerica acted as an ecological filter preventing free exchange of species. If so, then the Campanian dinosaurs of the Judith River Formation and correlative western North American stratigraphic units were largely restricted to the eastern portion of Asiamerica, a landmass with an area approximately the same as that of present-day Australia (49, 120). Judith River dinosaurs not only attained larger sizes than did most terrestrial mammalian species, but also did so on a small landmass—in contrast to what occurs in mammalian faunas.

Among modern mammals, population density decreases with increasing body size, a relationship strongly influenced by diet (37, 38, 112). At any given body mass, carnivores have lower population densities than do herbivores. Population turnover rates likewise decrease with increasing body size (16, 104). Consequently, populations of very large mammals are thinly spread and recover only slowly from environmental perturbations. To maintain viable populations under these conditions, very large mammals should show relatively little habitat specialization or have large geographic ranges, or both (100).

Once again the Campanian dinosaurs of western North America seem at odds with expectations for dinosaur-sized mammals. There is habitat zonation of dinosaur taxa within the Judith River Formation, and between the Judith River and Two Medicine Formations (7, 12, 65, 133). Furthermore, there may be north-south differences in Campanian dinosaur faunas as well (53, 116). It would be surprising to find this much large-vertebrate habitat zonation or biogeographic subdivision in a regional or continental fauna dominated by a diversity of elephant-sized mammals. Not only did dinosaurs routinely reach body sizes much larger than mammals, and on a rather small landmass, but many large-bodied dinosaur species also may not even have occurred in all the available terrestrial habitats on that small landmass.

Dinosaurian oviparity, as opposed to the combination of viviparity and lactation employed by most mammals, has been invoked to explain the ability of dinosaurs to achieve larger body sizes than are typical of mammals (73, 102). As egg-layers, dinosaurs may not have shown the decrease in annual and lifetime reproductive output that in terrestrial mammals accompanies large body size. Dinosaur populations may therefore have been able to recover quickly from environmental catastrophes that would doom elephant-sized mammals.

Although differences in reproductive biology may well have been a contributing factor, we do not think this a sufficient explanation for differences in body-size distributions of dinosaurian as opposed to mammalian faunas. If reproductive differences between dinosaurs and mammals were the single factor that permitted differences in body size in the two groups, we would expect that ground-living, nonpasserine birds, which are thought to be fairly similar to dinosaurs in reproductive biology (73), would have evolved at least some species comparable in mass to large dinosaurs. Although large flightless birds are known from several Cenozoic faunas (17, 86, 92), the largest forms probably weighed only about 500 kg. This is not only much less than the masses routinely achieved by dinosaurs, but it is also smaller than the masses of mammals that were contemporaries of, and in some cases members of the same faunas as, the big ground birds.

Cenozoic birds, like mammals, are endotherms with significantly higher food requirements than ectothermic animals of comparable body mass (16, 104). Elephant-sized birds, like huge mammals, should therefore have very low population densities. To explain how dinosaurian giants maintained viable populations in relatively small areas, it is probably necessary to consider mechanisms that would permit the maintenance of higher dinosaurian population densities than possible for dinosaur-sized birds and mammals.

The Mesozoic Era was a time of warmer climates than at present over much of the earth, possibly due in part to higher concentrations of CO_2 in the atmosphere than in the modern world (5, 8, 131). The combination of warmer temperatures and higher atmospheric CO_2 levels conceivably might have stimulated terrestrial primary productivity, or at least extended growing seasons, and this in turn might have supported larger populations of very large animals than would be possible on the modern earth.

However, although Mesozoic climates were warm, they were also in many places rather dry (6, 133), including (at least at times) much of the area over which Late Jurassic and Late Cretaceous sediments of western North America were deposited. Dry conditions coupled with warm temperatures would probably not have permitted significantly higher levels of primary productivity than seen in the Cenozoic world. Furthermore, many of the plant taxa that dominated Mesozoic vegetations were probably of poor quality as forage (113) and also slowly growing and/or slow to recover from heavy browsing by herbivores (6, 11, 130).

This leaves us with what seems the most important factor that permitted dinosaurian gigantism: that dinosaurs had lower food consumption rates than expected for equally large birds or mammals. If large dinosaurs had metabolic rates more nearly like those of living reptiles than those of modern endotherms, they might have maintained population densities several times higher than possible for equally large mammals or birds (37, 49, 89). Although most herbivorous dinosaurs were unquestionably megaherbivores (100) in the sense of being large-bodied plant-eaters, the per-animal impact on Mesozoic vegetations may have been less than for elephant-sized mammals. Most ceratopsians and hadrosaurs may have been ecologically more like antelopes than elephants in this regard, and only the biggest herbivorous dinosaurs (particularly sauropods) would have been the energetic equivalents of mammalian megaherbivores. Higher population densities than expected for equally huge mammals may have been even more critical in permitting the evolution of gigantic predatory dinosaurs, given the tighter energetic constraints on large carnivores than on herbivores (49, 54, 112).

Even if low food consumption rates are a precondition for the existence of diverse faunas of very large land animals, this does not mean that tetrapods with low metabolic rates will inevitably evolve such giants. Some of the carnivorous species of Late Permian therapsid faunas may have been endotherms, but herbivorous therapsids probably were not (28, 61, 76). Therapsids were larger on average than terrestrial mammals (Figures 1, 2 [c, d]), but the biggest therapsid size classes are no bigger than those of mammals.

Why Did Plant-Eating Dinosaurs Produce More Very Large Species than Herbivorous Therapsids Did?

Conceivably this relates to changes in vegetation structure between late Paleozoic and early Mesozoic ecosystems, with dinosaurs evolving in association with higher-crowned plants than those typical of the late Permian (6, 130). However, it is intriguing that the first large dinosaurs appeared at a time when atmospheric CO_2 levels are thought to have been well above those of the modern world, and at the high point after recovery from low values of the later Paleozoic Era (8). If greenhouse conditions resulted in longer growing seasons and higher levels of plant productivity than prevailed in the late Paleozoic, this might have permitted larger body sizes in dinosaurs than in late Paleozoic therapsids.

We suggest, then, that dinosaurian gigantism was permitted by a concatenation of factors that made the Mesozoic world and its large-vertebrate inhabitants different from those of the preceding Paleozoic and the following Cenozoic Eras. Although in features of reproductive biology and metabolic rates dinosaurs were probably similar to herbivorous therapsids, the greenhouse conditions of the Mesozoic world may have been more suitable for the evolution of very large, low-food-requirement herbivores and their predators than were climatic conditions of the late Paleozoic. Even though greenhouse conditions gradually deteriorated as the Cenozoic world lurched toward the modern icehouse configuration (105), levels of plant productivity may have remained comparable to those of the Mesozoic, or even increased, due to the dominance of vegetations by fast-growing angiosperm species that recover quickly from heavy cropping (6, 11, 130). Cenozoic mammals may have been prevented from evolving gigantism to the extent dinosaurs did, not so much because of environmental conditions, but rather because of their higher food requirements and their generally viviparous mode of reproduction.

Even if our speculations about the constraints that prevented gigantism in mammals (and perhaps therapsids), but not dinosaurs, have merit, our hypotheses do not identify the positive selective factors that actually caused dinosaurs to evolve very large body sizes. These might have involved thermoregulation, greater vagility, access to higher plant crowns, digestive physiology, predator-prey interactions, or reproductive dynamics (6, 13, 47, 125, 130; J Damuth, unpublished observations), but evaluating these alternatives would necessitate our being even more speculative than we have have already been.

THE VALUE OF DINOSAURS

Because of small sample sizes of specimens of these generally large animals, dinosaurs are not the ideal subjects for studies of evolutionary rates and modes; nor are their remains the most useful fossils for biostratigraphic work. What, then, is the value of dinosaur paleontology?

Dinosaurs push the envelope of what it means to be a large, terrestrial vertebrate. Their very existence poses questions about how body size is related to locomotion, reproduction, growth, metabolism, and trophic ecology that are broader than would be asked if land animals of the modern world were the only terrestrial creatures we knew. Can theories about the factors affecting the structure and function of terrestrial biological communities based on studies of the Holocene biota account for the features of any terrestrial biota, regardless of its taxonomic composition? Or do such theories work only for floras and faunas of the kind we see today? Attempting to understand dinosaur biology confronts us with such questions in an unusually forceful way.

ACKNOWLEDGMENTS

We thank C Badgley, J Damuth, R Graham, T Holtz, N Hotton, G King, R Reid, J Ruben, B Rubidge, D Russell, and H-D Sues for discussions and access to unpublished information.

Literature Cited

1. Alexander R McN. 1985. Mechanics of posture and gait of some large dinosaurs. *Zool. J. Linn. Soc.* 83:1–25
2. Anderson JF, Hall-Martin A, Russell DA. 1985. Long-bone circumference and weight in mammals, birds and dinosaurs. *J. Zool. Lond. A* 207:53–61
3. Armstrong EA. 1964. Parental care. In *A New Dictionary of Birds*, ed. AL Thomson, pp. 597–600. New York: Mc-Graw-Hill
4. Barreto C, Albrecht RM, Bjorling DE, Horner JR, Wilsman NJ. 1993. Evidence of the growth plate and the growth of long bones in juvenile dinosaurs. *Science* 262:2020–23
5. Barron EJ, Fawcett PJ, Pollard D, Thompson S. 1993. Model simulations of Cretaceous climates: the role of geography and carbon dioxide. *Philos. Trans. R. Soc. Lond. B* 341:307–16
6. Behrensmeyer AK, Damuth JD, Di-Michele WA, Potts R, Sues H-D, Wing SL, eds. 1992. *Terrestrial Ecosystems through Time: Evolutionary Paleoecology of Terrestrial Plants and Animals*. Chicago: Univ. Chicago Press. 568 pp.
7. Béland P, Russell DA. 1978. Paleoecology of Dinosaur Provincial Park (Cretaceous), Alberta, interpreted from the distribution of articulated dinosaur remains. *Can. J. Earth Sci.* 15:1012–24
8. Berner RA. 1994. Geocarb II: a revised model of atmospheric CO_2 over Phanerozoic time. *Am. J. Sci.* 294:56–91
9. Bocherens H, Brinkman DB, Dauphin Y, Mariotti A. 1994. Microstructural and geochemical investigations on Late Cretaceous archosaur teeth from Alberta, Canada. *Can. J. Earth Sci.* 31:783–92
10. Bonaparte JF, Novas FE, Coria RA. 1990. *Carnotaurus sastrei*, the horned, lightly built carnosaur from the Middle Cretaceous of Patagonia. *Contrib. Sci. Nat. Hist. Mus. Los Angeles County* 416:1–42
11. Bond WJ. 1989. The tortoise and the hare: ecology of angiosperm dominance and gymnosperm persistence. *Biol. J. Linn. Soc.* 36:227–49
12. Brinkman DB. 1990. Paleoecology of the Judith River Formation (Campanian) of Dinosaur Provincial Park, Alberta, Canada: evidence from vertebrate microfossil localities. *Palaeogeogr. Palaeoclimatol. Palaeoecol.* 78:37–54
13. Brown JH, Marquet PA, Taper ML. 1993. Evolution of body size: consequences of an energetic definition of fitness. *Am. Nat.* 142:573–84
14. Brown JH, Nicoletto PF. 1991. Spatial scaling of species composition: body masses of North American land mammals. *Am. Nat.* 138:1478–512
15. Bryant HN, Seymour KL. 1990. Observations and comments on the reliability of muscle reconstruction in fossil vertebrates. *J. Morphol.* 206:109–17
16. Calder WA. 1984. *Size, Function, and Life History*. Cambridge (MA): Harvard Univ. Press. 431 pp.
17. Campbell KE Jr, Marcus L. 1992. The relationship of hindlimb bone dimensions to body weight in birds. In *Papers in Avian Paleontology*, ed. KE Campbell Jr, pp. 395–412. Los Angeles: Sci. Ser. 36, Nat. Hist. Mus. Los Angeles County
18. Carpenter K, Currie PJ, eds. 1990. *Dinosaur Systematics: Perspectives and Approaches*. Cambridge: Cambridge Univ. Press. 318 pp.
19. Carpenter K, Hirsch KF, Horner JR, eds. 1994. *Dinosaur Eggs and Babies*. Cambridge: Cambridge Univ. Press. 372 pp.
20. Case TJ. 1978. On the evolution and adaptive significance of postnatal growth rates in the terrestrial vertebrates. *Quart. Rev. Biol.* 53:243–82
21. Case TJ. 1978. Speculations on the growth rate and reproduction of some dinosaurs. *Paleobiology* 4:320–28
22. Chinsamy A. 1990. Physiological implications of the bone histology of *Syntarsus rhodesiensis* (Saurischia: Theropoda). *Palaeont. afr.* 27:77–82
23. Chinsamy A. 1993. Bone histology and growth trajectory of the prosauropod dinosaur *Massospondylus carinatus* (Owen). *Mod. Geol.* 18:319–29
24. Chinsamy A. 1993. Image analysis and the physiological implications of the vascularisation of femora in archosaurs. *Mod. Geol.* 19:101–8
25. Chinsamy A. 1995. Ontogenetic changes in the bone histology of the Late Jurassic ornithopod *Dryosaurus*

lettowvorbecki. J. Vert. Paleontol. 15: 96–104

26. Chinsamy A, Chiappe LM, Dodson P. 1994. Growth rings in Mesozoic birds. *Nature* 368:196–97
27. Chinsamy A, Dodson P. 1995. Inside a dinosaur bone. *Am. Sci.* 83:174–80
28. Chinsamy A, Rubidge BS. 1993. Dicynodont (Therapsida) bone histology: phylogenetic and physiological implications. *Palaeont. afr.* 30:97–102
29. Choy DSJ, Altman P. 1992. The cardiovascular system of *Barosaurus*: an educated guess. *Lancet* 340:534–36
30. Clemens WA, Nelms LG. 1993. Paleoecological implications of Alaskan terrestrial vertebrate fauna in latest Cretaceous time at high paleolatitudes. *Geology* 21:503–6
31. Colbert EH. 1989. The Triassic dinosaur *Coelophysis. Bull. Mus. N. Arizona* 57: 1–160
32. Coombs WP Jr. 1989. Modern analogs for dinosaur nesting and parental behavior. In *Paleobiology of the Dinosaurs*, ed. JO Farlow, pp. 21–53. Boulder: Geol. Soc. Am. Spec. Pap. 238
33. Currie PJ. 1983. Hadrosaur trackways from the Lower Cretaceous of Canada. *Acta Palaeontol. Polonica* 28:63–73
34. Currie PJ, Dodson P. 1984. Mass death of a herd of ceratopsian dinosaurs. In *Third Symp. Mesozoic Terrestrial Ecosystems, Short Papers*, ed. W-E Reif, F Westphal, pp. 61–66. Tübingen: Attempto Verlag
35. Currie PJ, Eberth DA. 1993. Palaeontology, sedimentology and palaeoecology of the Iren Dabasu Formation (Upper Cretaceous), Inner Mongolia, People's Republic of China. *Cretac. Res.* 14:127–44
36. Czerkas SA. 1992. Discovery of dermal spines reveals a new look for sauropod dinosaurs. *Geology* 20:1068–70
37. Damuth J. 1987. Interspecific allometry of population density in mammals and other animals: the independence of body mass and population energy-use. *Biol. J. Linn. Soc.* 31:192–246
38. Damuth J. 1993. Cope's rule, the island rule and the scaling of mammalian population density. *Nature* 365:748–50
39. Damuth J, MacFadden BJ, eds. 1990. *Body Size in Mammalian Paleobiology: Estimation and Biological Implications.* Cambridge : Cambridge Univ. Press. 397 pp.
40. Dashzeveg D, Novacek MJ, Norell MA, Clark JM, Chiappe LM, et al. 1995. Extraordinary perservation in a new vertebrate assemblage from the Late Cretaceous of Mongolia. *Nature* 374:446–49
41. Dodson P. 1975. Taxonomic implications of relative growth in lambeosaurine hadrosaurs. *Syst. Zool.* 24: 37–54
42. Dodson P. 1976. Quantitative aspects of relative growth and sexual dimorphism in *Protoceratops. J. Paleontol.* 50:929–40
43. Dodson P. 1990. Counting dinosaurs: How many kinds were there? *Proc. Nat. Acad. Sci. USA* 87:7608–12
44. Dodson P, Behrensmeyer AK, Bakker RT, McIntosh JS. 1980. Taphonomy and paleoecology of the dinosaur beds of the Jurassic Morrison Formation. *Paleobiology* 6:208–32
45. Dunham AE, Overall KL, Porter WP, Forster CA. 1989. Implications of ecological energetics and biophysical and developmental constraints for life-history variation in dinosaurs. In *Paleobiology of the Dinosaurs*, ed. JO Farlow, pp. 1–19. Boulder: Geol. Soc. Am. Spec. Pap. 238
46. Enlow DH, Brown SO. 1957. A comparative histological study of fossil and recent bone tissue. Part 2. *Texas J. Sci.* 9:186–214
47. Farlow JO. 1987. Speculations about the diet and digestive physiology of herbivorous dinosaurs. *Paleobiology* 13: 60–72
48. Farlow JO. 1990. Dinosaur energetics and thermal biology. In *The Dinosauria*, ed. D Weishampel, P Dodson, H Osmólska, pp. 43–55. Berkeley: Univ. Calif. Press
49. Farlow JO. 1993. On the rareness of big, fierce animals: speculations about the body sizes, population densities, and geographic ranges of predatory mammals and large carnivorous dinosaurs. *Am. J. Sci.* 293-A:167–199
50. Farlow JO, Dodson P. 1975. The behavioral significance of frill and horn morphology in ceratopsian dinosaurs. *Evolution* 29:353–61
51. Farlow JO, Thompson CV, Rosner DE. 1976. Plates of the dinosaur *Stegosaurus*: forced convection heat loss fins? *Science* 192:1123–25
52. Fiorillo AR. 1991. Dental microwear on the teeth of *Camarasaurus* and *Diplodocus*: implications for sauropod paleoecology. In *Fifth Symp. on Mesozoic Terrestrial Ecosystems and Biota*, ed. S Kielan-Jaworowska, N Heintz, HA Nakrem, pp. 23–24. Contrib. Paleontol. Mus. Univ. Oslo 364
53. Fiorillo AR, Currie PJ. 1994. Theropod teeth from the Judith River Formation

(Upper Cretaceous) of South-Central Montana. *J. Vert. Paleontol.* 14:74–80

54. Flannery T. 1991. The mystery of the Meganesian meat-eaters. *Australian Natural History* 23:722–29

55. Francillon-Vieillot H, de Buffrénil V, Castanet J, Géraudie J, Meunier FJ, et al. 1990. Microstructure and mineralization of vertebrate skeletal tissues. In *Skeletal Biomineralization: Patterns, Processes and Evolutionary Trends*, ed. JG Carter, 1:471–529. New York: Van Nostrand Reinhold

56. Gatesy SM. 1990. Caudofemoral enlargements and the evolution of theropod locomotion. *Paleobiology* 16: 170–86

57. Gatesy SM. 1991. Hind limb scaling in birds and other theropods: implications for terrestrial locomotion. *J. Morphol.* 209:83–96

58. Giffin EB. 1991. Endosacral enlargements in dinosaurs. *Mod. Geol.* 16:101–12

59. Hallam A. 1993. Jurassic climates as inferred from the sedimentary and fossil record. *Philos. Trans. R. Soc. Lond. B* 341:287–96

60. Heinrich RE, Ruff, CB, Weishampel DB. 1993. Femoral ontogeny and locomotor biomechanics of *Dryosaurus lettowvorbecki* (Dinosauria, Iguanodontia). *Zool. J. Linn. Soc.* 108:179–96

61. Hillenius WJ. 1994. Turbinates in therapsids: evidence for Late Permian origins of mammalian endothermy. *Evolution* 48:207–29

62. Holtz TR Jr. 1994. The arctometatarsalian pes, an unusual structure of the metatarsus of Cretaceous Theropoda (Dinosauria: Saurischia). *J. Vert. Paleontol.* 14:480–519

63. Horner JR. 1982. Evidence of colonial nesting and 'site fidelity' among ornithischian dinosaurs. *Nature* 297:675–76

64. Horner JR. 1984. A "segmented" epidermal tail frill in a species of hadrosaurian dinosaur. *J. Paleontol.* 58: 270–71

65. Horner JR. 1984. Three ecologically distinct vertebrate faunal communities from the Late Cretaceous Two Medicine Formation of Montana, with discussion of evolutionary pressures induced by interior seaway fluctuations. *Montana Geol. Soc. 1984 Field Conf.*, pp. 299–303

66. Horner JR. 1992. Dinosaur behavior and growth. In *Fifth North American Paleontological Convention, Abstracts and Program.* ed. RS Spencer. *Paleontol. Soc. Special Publ.* 6:135 (Abstr.)

67. Horner JR. 1994. Comparative taphonomy of some dinosaur and extant bird colonial nesting grounds. In *Dinosaur Eggs and Babies*, ed. K Carpenter, KF Hirsch, Horner JR, pp. 116–123. Cambridge : Cambridge Univ. Press

68. Horner JR, Gorman J. 1988. *Digging Dinosaurs*. New York: Workman. 210 pp.

69. Horner JR, Makela R. 1979. Nest of juveniles provides evidence of family structure among dinosaurs. *Nature* 282: 296–98

70. Horner JR, Varricchio DJ, Goodwin MB. 1992. Marine transgression and the evolution of Cretaceous dinosaurs. *Nature* 358:59–61

71. Horner JR, Weishampel DB. 1988. A comparative embryological study of two ornithischian dinosaurs. *Nature* 332: 256–57

72. Insole AN, Hutt S. 1994. The palaeoecology of the dinosaurs of the Wessex Formation (Wealden Group, Early Cretaceous), Isle of Wight, Southern England. *Zool. J. Linn. Soc.* 112:197–215

73. Janis CM, Carrano M. 1992. Scaling of reproductive turnover in archosaurs and mammals: why are large terrestrial mammals so rare? *Ann. Zool. Fennici* 28:201–16

74. Jerzykiewicz T, Russell DA. 1991. Late Mesozoic stratigraphy and vertebrates of the Gobi Basin. *Cretac. Res.* 12:345–77

75. Johnson R, Ostrom JH. The forelimb of *Torosaurus*, and an analysis of the posture and gait of ceratopsian dinosaurs. In *Functional Morphology in Vertebrate Paleontology*, ed. J Thomason. Cambridge : Cambridge Univ. Press. In press

76. King G. 1990. *The Dicynodonts: a Study in Palaeobiology*. London: Chapman and Hall. 233 pp.

77. Kurtén B, Anderson E. 1980. *Pleistocene Mammals of North America*. New York: Columbia Univ. Press. 442 pp.

78. Larson PL. 1994. *Tyrannosaurus* sex. In *Dino Fest*, ed. GD Rosenberg, DL Wolberg, pp. 139–155. *Paleontol. Soc. Spec. Pub.* 7

79. Leakey MD, Harris JM, eds. 1987. *Laetoli: a Pliocene Site in Northern Tanzania*. Oxford: Clarendon Press. 561 pp.

80. Lehman TM. 1987. Late Maastrichtian paleoenvironments and dinosaur biogeography in the western interior of North America. *Palaeogeogr. Palaeoclimatol. Palaeoecol.* 60:189–217

81. Lehman TM. 1990. The ceratopsian family Chasmosaurinae: sexual dimorphism and systematics. In *Dinosaur Systematics: Approaches and Perspectives*,

ed. K Carpenter, PJ Currie, pp. 211–29. Cambridge: Cambridge Univ. Press

82. Lockley MG. 1991. *Tracking Dinosaurs: A New Look at an Ancient World.* Cambridge : Cambridge Univ. Press. 238 pp.

83. Lockley MG, Hunt AP, Meyer CA. 1994. Vertebrate tracks and the ichnofacies concept: implications for palaeoecology and palichnostratigraphy. In *Paleobiology of Trace Fossils*, ed. S Donovan. New York: Belhaven

84. Lull RS, Wright NE. 1942. Hadrosaurian dinosaurs of North America. *Geol. Soc. Am. Spec. Pap.* 40:1–242

85. Lundelius EL Jr, Graham RW, Anderson E, Guilday J, Holman JA, et al. 1983. Terrestrial vertebrate faunas. In *Late-Quaternary Environments of the United States: The Late Pleistocene*, ed. SC Porter, 1:311–53. Minneapolis: Univ. Minn. Press

86. Marshall LG. 1994. The terror birds of South America. *Sci. Am.* 270(2):90–95

87. Maurer BA, Brown JH, Rusler RD. 1992. The micro and macro in body size evolution. *Evolution* 46:939–53

88. McGowan C. 1979. The hind limb musculature of the brown kiwi, *Apteryx australis mantelli. J. Morphol.* 160:33–74

89. McNab BK. 1994. Resource use and the survival of land and freshwater vertebrates on oceanic islands. *Am. Nat.* 144: 643–60

90. Molnar RE. 1977. Analogies in the evolution of combat and display structures in ornithopods and ungulates. *Evol. Theor.* 3:165–90

91. Molnar RE. 1991. The cranial morphology of *Tyrannosaurus rex. Palaeontographica A* 217:137–76

91a. Molnar RE, Wiffen J. 1994. A Late Cretaceous polar dinosaur fauna from New Zealand. *Cretac. Res.* 15:689–706

92. Murray P. 1991. The Pleistocene megafauna of Australia. In *Vertebrate Paleontology of Australasia*, ed. P Vickers-Rich, JM Monaghan, RF Baird, TH Rich, EM Thompson, C Williams, pp. 1071–1164. Melbourne: Pioneer Design Studio/Monash Univ.

93. Nadon GC. 1993. The association of anastomosed fluvial deposits and dinosaur tracks, eggs, and nests: implications for the interpretation of floodplain environments and a possible survival strategy for ornithopods. *Palaios* 8:31–45

94. Nicholls EL, Russell DA. 1985. Structure and function of the pectoral girdle and forelimb of *Struthiomimus altus* (Theropoda: Ornithomimidae). *Palaeontology* 28:643–77

95. Norell MA, Clark JM, Dashzeveg D, Barsbold R, Chiappe L, et al. 1994. A theropod dinosaur embryo and the affinities of the Flaming Cliffs dinosaur eggs. *Science* 266:779–82

96. Norman DB. 1980. On the ornithischian dinosaur *Iguanodon bernissartensis* of Bernissart (Belgium). *Mem. de l'Inst. Roy. Sci. Nat. Belg.* 178:1–103

97. Osmólska H. 1980. The Late Cretaceous vertebrate assemblages of the Gobi Desert, Mongolia. *Mem. Soc. Geol. France* 139:145–50

98. Ostrom JH. 1969. Osteology of *Deinonychus antirrhopus*, an unusual theropod from the Lower Cretaceous of Montana. *Bull. Peabody Mus. Nat. Hist.* 30: 1–165

99. Ostrom P, Macko SA, Engel MH, Russell DA. 1993. Assessment of trophic structure of Cretaceous communities based on stable nitrogen isotope analyses. *Geology* 21:491–94

100. Owen-Smith RN. 1988. *Megaherbivores: The Influence of Very Large Body Size on Ecology.* Cambridge: Cambridge Univ. Press. 369 pp.

101. Paul GS. 1987. The science and art of restoring the life appearance of dinosaurs and their relatives: a rigorous how-to guide. In *Dinosaurs Past and Present* Vol. II, ed. SJ Czerkas, Olson EC, pp. 4–49. Seattle: Univ. Wash. Press

102. Paul GS. 1994. Dinosaur reproduction in the fast lane: implications for size, success, and extinction. In *Dinosaur Eggs and Babies*, ed. K Carpenter, KF Hirsch, JR Horner, pp. 244–255. Cambridge : Cambridge Univ. Press

103. Peckzis J. 1994. Implications of body-mass estimates for dinosaurs. *J. Vert. Paleontol.* 14:520–33.

104. Peters RH. 1983. *The Ecological Implications of Body Size.* Cambridge : Cambridge Univ. Press. 329 pp.

105. Prothero DR. 1994. *The Eocene-Oligocene Transition: Paradise Lost.* New York: Columbia Univ. Press. 291 pp.

105a. Raath MA. 1990. Morphological variation in small theropods and its meaning in systematics: evidence from *Syntarsus rhodesiensis*. In *Dinosaur Systematics: Approaches and Perspectives*, ed. K Carpenter, PJ Currie, pp. 91–105. Cambridge: Cambridge Univ. Press

106. Reid REH. 1987. Bone and dinosaurian "endothermy." *Mod. Geol.* 11:133–54

107. Reid REH. 1990. Zonal "growth rings" in dinosaurs. *Mod. Geol.* 15:19–48

108. Reid REH. 1993. Apparent zonation and slowed growth in a small Cretaceous theropod. *Mod. Geol.* 18:391–406

109. Rich THV, Rich, PV. 1989. Polar di-

nosaurs and biotas of the Early Cretaceous of southeastern Australia. *Nat. Geogr. Res.* 5:15–53

110. Ricqlès A de. 1980. Tissue structure of dinosaur bone: functional significance and possible relation to dinosaur physiology. In *A Cold Look at the Warm Blooded Dinosaurs*, ed. RDK Thomas, EC Olson, pp. 103–139. Boulder: Westview

111. Ricqlès A de. 1983. Cyclical growth in the long limb bones of a sauropod dinosaur. *Acta Palaeontol. Polonica* 28: 225–32

112. Robinson JG, Redford KH. 1986. Body size, diet, and population density of Neotropical forest mammals. *Am. Nat.* 128: 665–680

113. Robinson JM. 1990. Lignin, land plants, and fungi: biological evolution affecting Phanerozoic oxygen balance. *Geology* 18:607–10

114. Rogers RR. 1990. Taphonomy of three dinosaur bone beds in the Upper Cretaceous Two Medicine Formation of northwestern Montana: evidence for drought-related mortality. *Palaios* 5: 394–413

115. Rothschild BM, Martin LD. 1993. *Paleopathology: Disease in the Fossil Record.* Boca Raton: CRC. 386 pp.

116. Rowe T, Cifelli RL, Lehman TM, Weil A. 1992. The Campanian Terlingua local fauna, with a summary of other vertebrates from the Aguja Formation, Trans-Pecos, Texas. *J. Vert. Paleontol.* 12:472–93

117. Ruben J. 1995. The evolution of endothermy in mammals and birds: from physiology to fossils. *Annu. Rev. Physiol.* 57:69–95

118. Rubidge B, ed. *Biostratigraphy of the Beaufort Group (Karoo Supergroup), South Africa.* South African Committee for Stratigraphy. In press

119. Russell DA. 1989. *An Odyssey in Time: The Dinosaurs of North America.* Toronto: Univ. Toronto Press. 240 pp.

120. Russell DA. China and the lost worlds of the dinosaurian era. *Hist. Biol.*. In press

121. Sander PM. The Norian *Plateosaurus* bonebeds of central Europe and their taphonomy. *Palaeogeogr. Palaeoclimatol. Palaeoecol.* 93:255–99

122. Schwartz HL, Gillette DD. 1994. Geology and taphonomy of the *Coelophysis* quarry, Upper Triassic Chinle Formation, Ghost Ranch, New Mexico. *J. Paleontol.* 68:1118–30

123. Scotese CR, Golonka J. 1992. *Paleogeographic Atlas*, PALEOMAP Project, Univ. Texas, Arlington. 38 pp.

124. Sellwood BW, Price GD, Valdes PJ. 1994. Cooler estimates of Cretaceous temperatures. *Nature* 370:453–55

125. Spotila JR, O'Connor MP, Dodson P, Paladino FV. 1991. Hot and cold running dinosaurs: body size, metabolism and migration. *Modern Geol.* 16:203–27

126. Sternberg CM. 1925. Integument of *Chasmosaurus belli. Can. Field-Nat.* 39: 108–10

127. Strahan R, ed. 1983. *The Australian Museum Complete Book of Australian Mammals: the National Photographic Index of Australian Wildlife.* London: Angus and Robertson. 530 pp.

128. Sues H-D, Galton PM. 1987. Anatomy and classification of the North American Pachycephalosauria (Dinosauria: Ornithischia). *Palaeontographica A* 198:1–40

129. Thulborn T. 1990. *Dinosaur Tracks.* London: Chapman and Hall. 410 pp.

130. Tiffney BH. 1992. The role of vertebrate herbivory in the evolution of land plants. In *Essays in Evolutionary Plant Biology*, ed. BS Venkatachala, DL Dilcher, HK Maheshwari, pp. 87–97. Lucknow: Birbal Sahni Inst. Palaeobotany

131. Valdes P. 1993. Atmospheric general circulation models of the Jurassic. *Philos. Trans. R. Soc. Lond. B* 341:317–26

132. Varricchio DJ. 1993. Bone microstructure of the Upper Cretaceous dinosaur *Troodon formosus. J. Vert. Paleontol.* 13:99–104

133. Varricchio D. 1993. Montana climatic changes associated with the Cretaceous Claggett and Bearpaw transgressions. In *Energy and Mineral Resources of Central Montana: 1993 Field Conference Guidebook*, ed. LDV Hunter, pp. 97–102. Billings: Montana Geol. Soc.

134. Varricchio DJ, Horner JR. 1993. Hadrosaurid and lambeosaurid bone beds from the Upper Cretaceous Two Medicine Formation of Montana: taphonomic and biologic implications. *Can. J. Earth Sci.* 30:997–1006

135. Weishampel DB. 1981. Acoustic analyses of potential vocalization in lambeosaurine dinosaurs (Reptilia: Ornithischia). *Paleobiology* 7:252–61

136. Weishampel DB. 1984. Evolution of jaw mechanisms in ornithopod dinosaurs. *Adv. Anat. Embry. Cell Biol.* 87:1–110

137. Weishampel DB, Dodson P, Osmólska H, eds. 1990. *The Dinosauria.* Berkeley: Univ. CA Press. 733 pp.

138. Witmer LM. 1994. The extant phylogenetic bracket and the importance of

reconstructing the soft parts of fossils. In *Functional Morphology in Vertebrate Paleontology*, ed. J Thomason, pp. 19–33. Cambridge: Cambridge Univ. Press

139. Ziegler AM, Parrish JM, Jiping Y, Gyllenhaal ED, Rowley DB, et al. 1993. Early Mesozoic phytogeography and climate. *Philos. Trans. R. Soc. Lond. B* 341:297–305

Annu. Rev. Ecol. Syst. 1995. 26:473–503

THE ROLE OF NITROGEN IN THE RESPONSE OF FOREST NET PRIMARY PRODUCTION TO ELEVATED ATMOSPHERIC CARBON DIOXIDE[1]

A. David McGuire and Jerry M. Melillo

The Ecosystems Center, Marine Biological Laboratory, Woods Hole, Massachusetts 02543

Linda A. Joyce

Rocky Mountain Forest and Range Experiment Station, US Department of Agriculture Forest Service, Fort Collins, Colorado 80526

KEY WORDS: carbon dioxide, net primary production, nitrogen cycle, photosynthesis, respiration

ABSTRACT

We review experimental studies to evaluate how the nitrogen cycle influences the response of forest net primary production (NPP) to elevated CO_2. The studies in our survey report that at the tissue level, elevated CO_2 reduces leaf nitrogen concentration an average 21%, but that it has a smaller effect on nitrogen concentrations in stems and fine roots. In contrast, higher soil nitrogen availability generally increases leaf nitrogen concentration. Among studies that manipulate both soil nitrogen availability and atmospheric CO_2, photosynthetic response depends on a linear relationship with the response of leaf nitrogen concentration and the amount of change in atmospheric CO_2 concentration. Although elevated CO_2 often results in reduced tissue respiration rate per unit biomass, the link to changes in tissue nitrogen concentration is not well studied.

[1]The US government has the right to retain a nonexclusive, royalty-free license in and to any copyright covering this paper.

At the plant level, soil nitrogen availability is an important factor that often constrains the response of woody plant growth to elevated CO_2. Also, increased nitrogen availability and elevated CO_2 have opposite effects on the relative allocation of carbon to aboveground and belowground biomass. At the ecosystem level, the effects of elevated CO_2 on tissue nitrogen concentration, plant growth, and biomass allocation have the potential to alter soil nitrogen availability indirectly by influencing decomposition, nitrogen mineralization, and nitrogen fixation. Our analyses in this review indicate that the nitrogen cycle plays an important role in the response of forest NPP to elevated CO_2. Because interactions between the nitrogen cycle and elevated CO_2 are complex and our understanding is incomplete, additional research is required to elucidate how such interactions affect forest NPP.

INTRODUCTION

Net primary production (NPP) is the net rate at which the vegetation in an ecosystem captures carbon from the atmosphere. Forests, which cover 43% of the terrestrial biosphere, are potentially responsible for 72% of annual global terrestrial NPP (69). Humans rely on a portion of this production for fiber, fuel, and food. During the past 250 years the combustion of fossil fuels and deforestation have increased atmospheric carbon dioxide from preindustrial levels of approximately 280 ppmv to 353 ppmv in 1990 (128). The projection is that CO_2 concentrations will reach 500 ppmv by the year 2040, and 800 ppmv by the year 2100, if no steps are taken to limit CO_2 emissions (128). This projection necessitates that the scientific community advance its understanding concerning the sensitivity of forest NPP to elevated CO_2.

The availability of inorganic nitrogen often limits production in terrestrial ecosystems, and increased forest production in response to nitrogen fertilization has been observed in numerous studies (63–65, 122). A number of studies have recently reviewed various aspects of NPP response to elevated CO_2 (3, 14, 16, 38, 42, 44, 76, 83, 93, 98, 102, 127, 134). Many of the reviews identify uncertainties that represent gaps in our knowledge about the role of nitrogen in the response of forest ecosystems to elevated CO_2. Knowledge about the influence of nitrogen on forest carbon dynamics is a major issue that limits, in part, the ability of ecologists to model the response of terrestrial ecosystems to global change (121a). In this study we discuss the potential role of nitrogen in the response of forest NPP to elevated CO_2.

MAJOR LINKAGES BETWEEN THE CARBON AND NITROGEN CYCLES

The carbon and nitrogen cycles are closely coupled in terrestrial ecosystems (Figure 1). Nitrogen exerts control over the rates of several carbon cycling

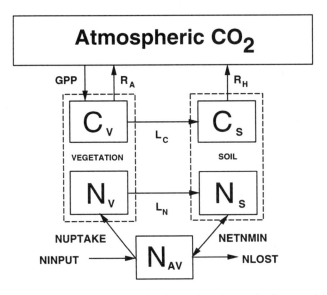

Figure 1 A generalized representation of carbon and nutrient cycles in terrestrial ecosystems. Carbon enters the vegetation pool (C_V) as gross primary production (GPP) and transfers either to the atmosphere as autotrophic (plant) respiration (R_A) or to the soil pool (C_S) as litter production (L_C); it leaves the soil pool as heterotrophic respiration (R_H). Nitrogen enters the vegetation pool (N_S) from the inorganic nitrogen pool of the soil (N_{AV}) as NUPTAKE. It transfers from the vegetation to the organic soil pool (N_S) in litter production as the flux L_N. Net nitrogen mineralization (NETNMIN) accounts for nitrogen exchanged between the organic and inorganic nitrogen pools of the soil. Nitrogen inputs from outside the ecosystem (NINPUT) enter the inorganic nitrogen pool; losses leave this pool as the flux NLOST.

processes including net primary production (NPP). Net primary production is the difference between gross primary production (GPP; i.e. gross assimilation of carbon captured through photosynthesis), and plant respiration (R_A; the energy cost of metabolic activity). Because both gross primary production and plant respiration represent biochemical processes that are catalyzed by nitrogen-rich enzymes, the rate of these processes depends, in part, on the nitrogen content of tissue. Also, because the construction of new tissue requires nitrogen in addition to carbon, gross primary production may depend on the nitrogen status of the plant. Nitrogen status is influenced by both the amount of nitrogen stored in vegetation (N_V) and the supply of nitrogen to vegetation (NUPTAKE). The supply to vegetation depends on effort expended by the plant to obtain nitrogen from the soil and the amount of nitrogen available in the soil solution (N_{AV}). Soil nitrogen availability is influenced by plant uptake (NUPTAKE), the net amount of nitrogen mineralized during the decomposition of soil organic matter (NETNMIN), inputs from the atmosphere (NINPUT) that include nitrogen fixation and deposition of atmospheric nitrogen, and nitrogen losses both to the atmosphere and to groundwater (NLOST). Thus, nitrogen may play a role in the response of forest NPP to elevated CO_2 by influencing tissue, plant, and ecosystem processes.

The effects of elevated CO_2 on NPP have been investigated at the tissue, plant, and ecosystem levels. Studies at the tissue level have focused primarily on the response of net photosynthesis and tissue respiration. Net photosynthesis is the net amount of carbon assimilated during photosynthesis and is the difference between gross assimilation and the leaf respiration that occurs simultaneously with photosynthesis (36). In contrast to studies at the tissue level, those at the level of the individual plant have focused primarily on the response of growth, which is NPP minus biomass losses such as herbivory and litter production (L_C in Figure 1). Because growth is essentially equivalent to NPP if biomass losses are negligible, growth is generally a better integrative measure of NPP than are net photosynthesis and respiration because of the difficulties in continually measuring both of these processes for entire plants. For practical reasons, studies at the plant level generally focus on the response of "potted" seedlings in growth chambers, greenhouses, and field chambers. Although these studies integrate the response of photosynthesis and respiration for individual organisms, they do not necessarily capture the feedback between plant and soil processes that operates in ecosystems. Studies at the ecosystem level focus primarily on how growth responds to elevated CO_2 in the context of plant and soil interactions.

TISSUE-LEVEL RESPONSES

Tissue-level processes that may be affected by elevated atmospheric CO_2 include photosynthesis and respiration. Net photosynthesis in plant leaves represents both carbon gain and loss during the process of photosynthesis; carbon loss is caused by aerobic respiration occurring simultaneously with gross assimilation. Aerobic respiration, which represents the oxidative energy cost of numerous enzyme-catalyzed biochemical pathways, results in carbon loss in the form of CO_2 from all plant tissues. One way that the nitrogen cycle potentially interacts with elevated atmospheric CO_2 to influence tissue metabolism is through effects on enzyme concentrations in tissue.

Nitrogen is a major constituent of enzymes, and changes in nitrogen concentration of tissue generally reflect changes in enzyme concentration. Although nitrogen concentration of woody plant tissues is commonly observed to decline in response to long-term exposure to elevated atmospheric CO_2, much more information is available for leaf tissue (77 reports in Table 1) than for stems (18 reports) and fine roots (26 reports). Among the reports in our survey, the mean decrease of leaf nitrogen concentration is 21% in response to elevated CO$_2$. In 10 reports no change in nitrogen concentration occurs, and in 2 it increases. Decreases in leaf nitrogen concentration are greater than decreases in other tissues (Kruskal-Wallis Test, $H = 24.1$, $P < 0.0001$, df = 2); decreases in stems (7%) and fine roots (7%) are not statistically distinguishable. It is not clear whether decreases in stem and fine root nitrogen concentration are different from no change; tests for differences are not significant but have low power to detect differences (0.22 for stems and 0.33 for roots vs. desired 0.80). Among 33 reports in our survey, the mean decrease in plant nitrogen concentration is 15%, which is statistically different from no change.

Although elevated CO$_2$ generally reduces leaf nitrogen concentration when the nitrogen fertilization regime is held constant, a different pattern emerges if changes in nitrogen concentration are examined across fertilization treatments. When compared to the nitrogen concentration at the lowest level of nitrogen availability, higher levels of nitrogen availability generally lessen the reduction or increase the nitrogen concentration of leaves in woody plants grown at elevated CO$_2$ (Table 2; Paired-sample t-test, $t = 4.31$, $P = 0.0003$, df = 23). Of the 24 comparisons in Table 2, a further reduction in leaf nitrogen concentration is observed under conditions of higher nitrogen availability only for *Eucalyptus grandis* and the nitrogen-fixing species *Alnus rubra*. Leaf nitrogen concentrations increase for *Pinus taeda, Populus tremuloides,* and *Salix* × *dasyclados* when elevated CO$_2$ is accompanied with nitrogen fertilization. Although increased nitrogen availability and elevated CO$_2$ have opposite effects on leaf nitrogen concentration, the extant data are too few to determine whether nitrogen concentrations in stems, fine roots, and whole plants of woody vegetation are similarly affected. Clearly, more information is needed on how elevated CO$_2$ interacts with nitrogen availability to affect nitrogen concentrations in stems, fine roots, and whole plants in woody vegetation.

Effects on Net Photosynthesis

For plants grown in elevated CO$_2$, three photosynthetic acclimation responses are observed: downregulation, upregulation, and depressed photosynthesis (58). Downregulation occurs when the photosynthetic capacity of plants grown in elevated CO$_2$ decreases in comparison to plants grown at baseline CO$_2$, but the rate of photosynthesis for plants grown and measured at elevated CO$_2$ is still higher than the rate for plants grown and measured at baseline CO$_2$. For plants grown at elevated CO$_2$ compared to those grown at baseline CO$_2$, higher photosynthesis measured at both baseline and elevated CO$_2$ is defined as upregulation, and lower photosynthesis measured at both baseline and elevated CO$_2$ is defined as depressed photosynthesis.

The long-term responses of net photosynthesis have been reviewed for

Table 1 Effects of elevated atmospheric carbon dioxide on the nitrogen concentration of leaf, stem, root, and whole plant tissue of woody vegetation.

Species	Baseline CO$_2$ (ppmv)	Elevated CO$_2$ (ppmv)	Growth apparatus[a]	Other details[b]	Percent change in nitrogen concentration (% gN gdm^{-1})[c]				Reference
					Leaf	Stem	Root	Plant	
Acer pseudo-platanus	390	+130	GH	—	—	—	—	-10%	86
Acer saccharum	390	+260	GH	—	—	—	—	-17%	56
Acer saccharum	350	+300	GC	—	-17%	—	NSD	—	95
Alnus glutinosa	350	+300	GC	—	NSD[d]	—	NSD	—	75
	350	+350	GC	—	-46%	—	—	—	
Alnus rubra	350	+300	GC	No nod; +N	+19%	+14%	-7%	+14%	
	350	+300	GC	Nod; No N	-11%	-5%	-2%	-6%	
	350	+300	GC	Nod; +N	-14%	-4%	+5%	-11%	4
Artemisia tridentata	350	+300	GC	—	-7%	—	—	—	48
Artemisia tridentata	350	+300	GC	low N	-17%	—	—	—	
	350	+300	GC	high N	-28%	—	—	—	49
Betula alleghaniensis	350	+350	GH	—	-30%	—	—	—	97
Betula lenta	350	+350	GH	—	-25%	—	—	—	97
Betula papyrifera	350	+350	GH	—	-33%	—	—	—	97
Betula papyrifera	350	+300	GC	—	-20%	—	—	—	99
Betula pendula	350	+350	GC	—	-14%	-4%	+1%	—	92
Betula pendula	350	+350	GC	low N	—	—	—	-24%	
	350	+350	GC	medium N	—	—	—	-20%	
	350	+350	GC	high N	—	—	—	-7%	94
Betula populifolia	350	+350	GC	high N	-33%	—	—	—	97
Betula populifolia	350	+150	GH	—	—	—	—	-18%	
Bottomland species	350	+250	GC	—	—	—	—	-36%	124
Castanea sativa	350	+350	GC	No fert	—	lower	lower	-42%	73
	350	+350	GH	—	—	-11%	-16%	-13%	

Species	CO₂	+CO₂	Method	N treatment					Ref
Castanea sativa	350	+350	GH	fert	—	-19%	-28%	-23%	32
Castanea sativa	350	+350	GH	18 months	-36%	-26%	NSD	—	100
Elaeagnus angustifolia	350	+350	GC	—	-31%	—	—	-29%	75
Eucalyptus camaldulensis	330	+330	GH	low N	-26%	—	—	-27%	126
	330	+330	GH	high N	-30%	—	—	-21%	126
Eucalyptus cypellocarpa	330	+330	GH	low N	-25%	—	—	-29%	126
	330	+330	GH	high N	-38%	—	—	-22%	
Eucalyptus grandis	340	+320	GC	low N	-60%	—	—	—	21
Eucalyptus miniata	340	+320	GC	highest N	NSD	—	—	—	29
	355	+345	GH	—	-22%	—	—	-21%	
Eucalyptus pauciflora	330	+330	GH	low N	-21%	—	—	-16%	126
	330	+330	GH	high N	-18%	—	—	-17%	
	330	+330	GH	low N	-17%	—	—	-15%	
Eucalyptus pulverulenta	330	+330	GH	high N	-33%	—	—	—	126
Eucalyptus tetrodonta	355	+345	GH	—	NSD	—	—	—	29
Fagus grandifolia	350	+300	GC	—	NSD	—	+13%	-9%	95
Fagus sylvatica	390	+130	GH	—	—	—	—	-10%	
	390	+260	GH	—	—	—	—	-11%	86
Gliricidia sepium	350	+300	GC	no N	-24%	NSD	NSD	—	
	350	+300	GC	+N	-14%	NSD	NSD	—	115
Lindera benzoin	350	+340	OTC	—	-11%	—	—	NSD	6
Liriodendron tulipifera	367	+325	GC	low N	—	—	—	-33%	84
	371	+122	GC	no fert	-14%	-7%	-9%	—	
	371	+416	GC	no fert	-28%	-10%	-14%	—	
	371	+122	GC	fert	-12%	-5%	-4%	—	
Liriodendron tulipifera	371	+416	GC	fert	-30%	-18%	-28%	—	
	355	+150	OTC	—	-24%	—	—	—	79

Table 1 (continued)

Species	Baseline CO$_2$ (ppmv)	Elevated CO$_2$ (ppmv)	Growth apparatus[a]	Other details[b]	Percent change in nitrogen concentration (% gN gdm^{-1})[c]				Reference
					Leaf	Stem	Root	Plant	
Liriodendron tulipifera	355	+300	OTC	—	-32%	—	—	—	77
	355	+150	OTC	—	-47%	—	—	—	
Liriodendron tulipifera	355	+150	OTC	—	-45%	—	—	—	131
Picea mariana	350	+350	GC	—	-25%	—	—	—	47
Pinus strobus	350	+300	GC	—	NSD	—	—	—	99
Pinus sylvestris	406	+348	GC	1 year old	NSD	—	—	—	
	406	+348	GC	current yr	-23%	—	—	—	89
	350	+150	GH	low N	-38%	—	—	—	
	350	+300	GH	low N	-29%	—	—	—	
	350	+150	GH	high N	NSD	—	—	—	
Pinus taeda	350	+300	GH	high N	NSD	—	—	—	40
	375	+335	GH	low N	-20%	-20%	NSD	NSD	
Pinus taeda	375	+335	GH	high N	NSD	NSD	NSD	NSD	54
	355	+355	GH	low P; -myc	—	—	NSD	—	
	355	+355	GH	high P; -myc	—	—	NSD	—	
	355	+355	GH	low P; +myc	—	—	NSD	—	
Pinus taeda	355	+355	GH	high P; +myc	—	—	NSD	—	55
	350	+350	GC	low water	—	—	-29%	—	
Pinus taeda	350	+350	GC	high water	—	—	-26%	—	118
	350	+150	OTC	—	-20%	—	—	—	
Pinus taeda	350	+300	OTC	—	-28%	—	—	—	123
Pinus virginiana	340	+600	OTC	—	—	—	—	+4%	60
	361	+346	OTC	45 days	-11%	—	—	—	
	361	+346	OTC	70 days	0%	—	—	—	
	350	+400	GC	low N	-44%	—	—	—	
Populus grandidentata	350	+400	GC	medium N	-29%	—	—	—	25

Species									
Populus tremuloides	350	+400	GC	high N	-21%	—	—	—	11
Populus tremuloides	350	+300	GC		-24%	—	—	—	56
Quercus alba	362	+328	GC		-19%	-27%	-17%	—	81
Quercus alba	355	+150 to +650	Natural		+9%	—	—	—	52
Quercus pubescens	355	+150 to +650	Natural		-15%	—	—	—	52
Quercus rubra	350	+300	GC		NSD	—	—	—	56
Robinia pseudoacacia	350	+350	GC		-32%	—	—	—	75
	300	+200	GH	lowest N	-29%	—	—	—	
	300	+400	GH	lowest N	-23%	—	—	—	
	300	+700	GH	lowest N	-14%	—	—	—	
	300	+200	GH	low N	-4%	—	—	—	
	300	+400	GH	low N	-17%	—	—	—	
	300	+700	GH	low N	-13%	—	—	—	
	300	+200	GH	high N	-40%	—	—	—	
	300	+400	GH	high N	-38%	—	—	—	
	300	+700	GH	high N	-35%	—	—	—	
	300	+200	GH	highest N	-4%	—	—	—	
	300	+400	GH	highest N	-31%	—	—	—	
	300	+700	GH	highest N	-11%	—	—	—	
Salix x dasyclados	300	+700	GH			—	—	-10%	109
Tropical vegetation	340	+270	GH			—	—	NSD	51
	350	+150	GC			—	—	NSD	
Upland species	350	+250	GC			NSD	NSD	—	124
Upland and Bottomland	350	+150	GC		-25%	NSD	NSD	—	124
species	350	+250	GC		-25%	NSD	-20%	—	124

[a] GC—growth chamber experiments, GH—greenhouse experiments, OTC—open-top chamber experiments.
[b] Nod—nodulated, N—nitrogen, fert—fertilized, P—phosphorus, myc—mycorrhizae.
[c] Change in nitrogen concentration relative to concentration for baseline CO₂ at same fertilization level.
[d] NSD—no significant difference from nitrogen concentration at baseline CO₂.

Table 2 Effects of elevated atmospheric carbon dioxide and nitrogen fertilization on the nitrogen concentration of leaf, stem, root, and whole plant tissue of woody vegetation.[a]

Species	Baseline CO_2 (ppmv)	Elevated CO_2 (ppmv)	Growth apparatus[b]	Other details[c]	Leaf	Stem	Root	Plant	Reference
Alnus rubra	350	+300	GC	Nod; No N	-11%	-5%	-2%	-6%	
	350	+300	GC	Nod; +N	-24%	-20%	+9%	-24%	4
Artemisia tridentata	350	+300	GC	low N	-17%	—	—	—	
	350	+300	GC	high N	-9%	—	—	-24%	49
Betula pendula	350	+350	GC	low N	—	—	—	+65%	
	350	+350	GC	medium N	—	—	—	+147%	
	350	+350	GC	high N	—	-11%	-16%	-13%	94
Castanea sativa	350	+350	GH	no fert	—	—	—	-34%	
	350	+350	GH	fert	—	-21%	-32%	-27%	32
Eucalyptus camaldulensis	330	+330	GH	low N	-31%	—	—	0%	
	330	+330	GH	high N	-5%	—	—	-29%	126
Eucalyptus cypellocarpa	330	+330	GH	low N	-30%	—	—	-7%	
	330	+330	GH	high N	-8%	—	—	—	126
Eucalyptus grandis	340	+320	GC	low N	-38%	—	—	—	
	340	+320	GC	highest N	-43%	—	—	-21%	21
Eucalyptus pauciflora	330	+330	GH	low N	-22%	—	—	-4%	
	330	+330	GH	high N	-4%	—	—	-17%	126
Eucalyptus pulverulenta	330	+330	GH	low N	-18%	—	—	—	
	330	+330	GH	high N	-8%	—	—	+9%	126
	350	+300	GC	no N fert	-24%	NSD[d]	NSD	-11%	126

Species									
Gliricidia sepium	350	+300	GC	N fert	-14%	NSD	NSD	NSD	115
Liriodendron tulipifera	371	+122	GC	no fert	-14%	-7%	-9%	—	79
	371	+122	GC	fert	-9%	-7%	-26%	—	
Liriodendron tulipifera	371	+416	GC	no fert	-28%	-10%	-14%	—	79
	371	+416	GC	fert	-28%	-19%	-45%	—	
Pinus taeda	350	+150	GH	low N	-38%	—	—	—	40
	350	+150	GH	high N	+49%	—	—	—	
Pinus taeda	350	+300	GH	low N	-29%	—	—	—	40
	350	+300	GH	high N	+49%	—	—	—	
Pinus taeda	375	+335	GH	low N	-20%	-20%	NSD	NSD	54
	375	+335	GH	high N	NSD	—	—	NSD	
Populus tremuloides	350	+400	GC	low N	-44%	—	—	—	
	350	+400	GC	medium N	-41%	—	—	—	11
	350	+400	GC	high N	+19%	—	—	—	
Salix x dasyclados	300	+200	GH	lowest N	-29%	—	—	—	
	300	+200	GH	low N	+31%	—	—	—	
	300	+200	GH	high N	+37%	—	—	—	
	300	+200	GH	highest N	+146%	—	—	—	109
Salix x dasyclados	300	+400	GH	lowest N	-23%	—	—	—	
	300	+400	GH	low N	+14%	—	—	—	
	300	+400	GH	high N	+43%	—	—	—	
Salix x dasyclados	300	+400	GH	highest N	+106%	—	—	—	109
Salix x dasyclados	300	+400	GH	highest N	+106%	—	—	—	109
	300	+700	GH	lowest N	-14%	—	—	—	
	300	+700	GH	low N	+20%	—	—	—	
	300	+700	GH	high N	+49%	—	—	—	
Salix x dasyclados	300	+700	GH	highest N	+186%	—	—	—	109

[a] relative to the nitrogen concentration for the treatment that uses baseline CO$_2$ and the lowest level of nitrogen fertilization.
[b] GC—growth chambers experiments, GH—greenhouse experiments.
[c] Nod—nodulated, N—nitrogen, fert—fertilized.
[d] NSD—no significant difference from nitrogen concentration at baseline CO$_2$.

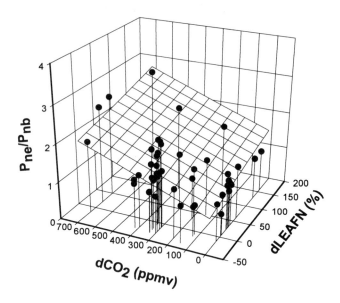

Figure 2 The relationship between photosynthetic response (P_{ne}/P_{nb}), percent change in nitrogen concentration of leaf tissue (dLEAFN), and amount of change in atmospheric CO_2 in ppmv (dCO2), where P_{nb} is the net photosynthetic rate per unit area for plants grown and measured at both baseline CO_2 and the lowest level of nitrogen fertilization in the experiment, P_{ne} is the net photosynthetic rate for plants grown and measured at elevated CO_2 and/or higher levels of nitrogen fertilization. The plane is described by $P_{ne}/P_{nb} = 0.95924 + 0.00298$ dLEAFN $+ 0.00178$ dCO2 ($F = 33.1$, $P < 0.0001$, df = 2,43). Data are from studies than manipulate both soil nitrogen availability and elevated CO_2 for woody species, and these are documented in Tables 2 and 3.

woody species grown in elevated CO_2 (30, 16, 42). In a review of the photosynthetic responses of 16 woody species described in studies published in the 1980s, Eamus & Jarvis (30) observed that, for most experiments, carbon assimilation of plants grown and measured at elevated CO_2 is greater than that of plants grown and measured at baseline CO_2. Similarly, in a review of studies published in the early 1990s, Ceulemans & Mousseau (16) observed that elevated CO_2 enhances photosynthesis by an average 40% among 12 conifer species and 61% among 53 broadleaf species. Among 69 reports in Gunderson & Wullschleger (42), net photosynthesis is 44% higher on average for plants grown at elevated CO_2. However, when measured at baseline CO_2, photosynthesis for plants grown at elevated CO_2 is an average 21% lower than for plants grown at baseline CO_2. The observations of Gunderson & Wullschleger (42) indicate downregulation; only 8 of 69 reports indicate upregulation, and only 4 of 20 reports indicate depressed photosynthesis. Downregulation appears to be the predominant photosynthetic acclimation response of woody plants to elevated CO_2.

Among studies that manipulate both CO$_2$ and nitrogen availability, the mean enhancement of photosynthesis to elevated CO$_2$ at the lowest level of nitrogen availability is 40%, while the mean enhancement at higher levels of nitrogen availability is 59% (Table 3; paired-sample t-test, $t = 2.41$, $P = 0.0239$, df = 24). Relative to photosynthesis and nitrogen concentrations of the lowest fertilization treatment in each experiment, a linear relationship exists between photosynthetic enhancement, change in leaf nitrogen concentration, and the amount of CO$_2$ change (Figure 2):

$$P_{ne}/P_{nb} = 0.95924 + 0.00298 \text{ dLEAFN} + 0.00178 \text{ dCO}_2 , \qquad 1.$$

where P_{nb} is net photosynthesis per unit leaf area for plants grown and measured at both baseline CO$_2$ and the lowest level of fertilization in the experiment, P_{ne} is net photosynthesis rate for plants grown and measured at elevated CO$_2$ and/or higher levels of nitrogen fertilization; dLEAFN is the percent change in nitrogen concentration between leaves corresponding to the measurement of P_{nb} and those of P_{ne}; and dCO$_2$ is the concentration difference in ppmv between elevated and baseline CO$_2$. Baseline CO$_2$ was defined for each experiment as the CO$_2$ concentration that best represents contemporary atmospheric CO$_2$. The relationship explains 61% of the variability in P_{ne}/P_{nb} ($F = 33.1$, $P < 0.0001$, df = 2,43). Both independent variables are significant (dLEAFN: $P = 0.0154$; dCO$_2$: $P < 0.0001$), and each variable contributes significantly to the variance explained by the relationship. The intercept, which is significantly different from 0 ($P < 0.0001$), does not significantly differ from the expected 1.0 for no changes in leaf nitrogen and CO$_2$. In the absence of changes in nitrogen availability, photosynthetic response depends on how leaf nitrogen concentration responds to elevated CO$_2$. Severe reductions in leaf nitrogen cause depressed photosynthesis, moderate to small reductions cause downregulation, and increases cause upregulation. Because nitrogen availability affects dLEAFN, the response of photosynthesis to elevated CO$_2$ also depends on changes in nitrogen availability. Although other factors such as water availability would be useful to include in a relationship of photosynthetic response to elevated CO$_2$, the relationship indicates that nitrogen availability and nitrogen allocation play important roles in the acclimation of photosynthesis to elevated CO$_2$.

Acclimation to elevated CO$_2$ can occur by affecting one or more of three aspects of leaf-level carbon assimilation (93, 102): carboxylation, light harvest, and carbohydrate synthesis. Under saturating light conditions at low levels of intercellular CO$_2$, assimilation is limited by the quantity and activity of ribulose bisphosphate carboxylase (rubisco), the enzyme that is primarily responsible for capturing atmospheric carbon in the production of sugars. Rubisco may accept either CO$_2$ (carboxylation) or O$_2$ (oxygenation) as a substrate; oxygenation is responsible for photorespiration. Because CO$_2$ competes with O$_2$ for rubisco binding sites, enhancement of photosynthesis by elevated CO_2 is possible through increased carboxylation and decreased oxygenation. Carboxylation increases with rising intercellular CO_2 to levels at which the regeneration of rubisco, and thus the ability to fix carbon, is limited by the light-harvesting machinery of photosynthesis. At high levels of intercellular CO_2, the enzymatically controlled rate of carbohydrate synthesis, which affects the phosphate regeneration that is necessary for harvesting light energy, may regulate the fixation of carbon.

Because rubisco represents a substantial proportion of leaf nitrogen (35), photosynthetic rate is generally correlated with the nitrogen content of leaves (35, 37). Reduced nitrogen availability has often been observed to decrease both leaf nitrogen content and photosynthesis (18, 34, 41, 53, 103, 104, 125). If lower nitrogen concentrations of leaves in response to elevated CO_2 primarily reflect lower rubisco concentrations, then lower assimilation over the carboxylation-limited range of intercellular CO_2 is expected in plants grown at elevated CO_2. It has been suggested that if elevated CO_2 causes intercellular CO_2 generally to rise above this region, then lower rubisco levels may be advantageous because they represent the allocation of nitrogen away from excess rubisco capacity (102) to other activities such as fine root function (33) and enzymes of the light-harvesting machinery and carbohydrate synthesis (38, 102, 105, 112, 117, 127).

One explanation for acclimation to elevated CO_2 is that restricted rooting volume in small pots causes photosynthesis to be regulated by sink activity (5). A mechanism proposed to explain this phenomenon is that the accumulation of carbohydrates in leaves induces feedback to reduce phosphate regeneration (10, 105), a phenomenon labeled "end-product inhibition." This explanation is consistent with the observed accumulation of photosynthate in leaves of some plants that have received long-term exposure to elevated CO_2 (15, 26, 28, 31, 43, 87, 111, 129). Most of the evidence concerning end-product inhibition is from studies of herbaceous plants, and the hypothesis does not explain acclimation in trees when rooting volume is not restricted (42). Because end-product inhibition may represent an artifact of inappropriate pot size (5, 10, 116; but see 8, 50, 61, 62), true photosynthetic acclimation in response to elevated CO_2 may be best understood in terms of the allocation of nitrogen and other components to leaf function, i.e. carbon capture, and root function, i.e. the acquisition of nutrients and water (see 58).

Effects on Respiration

The study of respiration responses to elevated CO_2, a rapidly expanding field, has recently been reviewed by Amthor (3), Bunce (14), and Wullschleger et al (134). Emerging evidence indicates that the long-term acclimation of woody plants to elevated CO_2 often results in reduced leaf respiration rates (6, 45, 46,

95, 130, 131, 133) and perhaps in reduced respiration rates for whole seedlings (13, 72, 95). When the leaf respiration response has been partitioned into growth and maintenance components, the reduction is dominated by maintenance respiration (131). Although growth respiration per unit leaf biomass may decrease (130, 131), larger leaves in elevated CO_2 may compensate for the reduced rate (131; see also 95). Similarly, Norby (76) reports that although the respiration rate per unit fine root biomass was decreased in *Quercus alba* grown at elevated CO_2, increased fine root density probably results in higher total fine root respiration. Reid & Strain (95) observed decreased respiration per belowground biomass for *Acer saccharum*, but not for *Fagus grandifolia;* total belowground respiration was not affected by elevated CO_2 in either species. In contrast, both maintenance respiration per stem volume and growth respiration per stem increment in *Q. alba* are unaffected by long-term acclimation to elevated CO_2 (132), but higher rates of stem growth cause higher total stem respiration.

Changes in tissue nitrogen concentrations may have effects on both growth and maintenance respiration. The energy required to construct tissues with reduced nitrogen/protein concentrations may be less at elevated CO_2 (131, 134; but see 57). Maintenance respiration involves the energy cost of many cell functions, which include numerous biochemical pathways, cell repair, membrane function, and protein synthesis and maintenance. Protein synthesis and maintenance represent a substantial proportion of maintenance respiration (88), and a linear relationship often exists between tissue maintenance respiration and tissue nitrogen concentrations (101). Wullschleger et al (131) documented a linear relationship between leaf respiration rates and leaf nitrogen concentrations among *Liriodendron tulipifera* plants grown at ambient and elevated levels of atmospheric CO_2, but Azcon-Bieto et al (6) observed that respiration per unit leaf nitrogen declined for *Lindera benzoin* plants grown in elevated CO_2. To our knowledge, these are the only reported relationships among tissue respiration, nitrogen concentration, and elevated CO_2 for woody species. Additional research is required to elucidate the role that changes in tissue nitrogen concentration play in the acclimation response of tissue respiration to elevated CO_2 (134).

PLANT-LEVEL RESPONSES

Studies of tissue-level responses to elevated CO_2 and nitrogen availability help us to understand how these two variables interact to affect the exchange of carbon between the plant and the atmosphere on a per unit biomass or per unit leaf area basis. For us to understand how whole-plant carbon exchange is affected, we need to know how growth and biomass allocation are influenced. The responses of growth and allocation are important to consider because

Table 3 Effects of elevated atmospheric carbon dioxide and nitrogen fertilization on net photosynthesis (P_n), growth, and root to shoot ration (R/S)

Species	Baseline CO$_2$ (ppmv)	Elevated CO$_2$ (ppmv)	Growth apparatus[a]	Other details[b]	Change[c] P_n	Growth	R/S	Reference
Acer pennsylvanicum	350	+350	GH	low PAR; low N	—	1.3/1.3	+20/+20	7
	350	+350	GH	low PAR; high N	—	1.6/6.4	+1/+4	
	350	+350	GH	high PAR; low N	—	1.2/1.2	+33/+33	7
Acer pennsylvanicum	350	+350	GH	high PAR; high N	—	1.2/11.3	+11/−5	7
	350	+350	GH	low PAR; low N	—	1.3/1.3	−6/−6	
Acer rubrum	350	+350	GH	low PAR; high N	—	1.6/8.9	−6/−22	7
	350	+350	GH	high PAR; low N	—	1.1/1.1	+3/+3	
Acer rubrum	350	+350	GH	high PAR; high N	—	1.4/28.0	+0/−22	7
Alnus rubra	350	+300	GC	Nod; No N	2.2/2.2	1.2/1.2	−19/−19	4
	350	+300	GC	Nod; +N	2.0/2.0	1.7/3.0	+16/+24	
	350	+300	GC	low N	1.5/1.5	1.0/1.0	−19/−19	
Artemisia tridentata	350	+300	GC	high N	1.1/1.5	1.7/2.3	−49/−56	49
Betula allegheniensis	350	+350	GH	low PAR; low N	—	1.2/1.2	+8/+8	7
	350	+350	GH	low PAR; high N	—	1.6/9.0	−13/−32	
Betula allegheniensis	350	+350	GH	high PAR; low N	—	1.2/1.2	−3/−3	7
	350	+350	GH	high PAR; high N	—	1.1/10.6	−1/−15	
Betula allegheniensis	350	+350	GC	low N	1.4/1.4	—	+4/+4	7
	350	+350	GC	medium N	1.0/1.5	—	+9/−48	
Betula pendula	350	+350	GC	high N	1.3/2.1	—	−5/−60	94
Betula populifolia	350	+350	GH	low PAR; low N	—	1.1/1.1	+5/+5	7
	350	+350	GH	low PAR; high N	—	1.4/6.8	+8/−29	
Betula populifolia	350	+350	GH	high PAR; low N	—	1.0/1.0	+9/+9	7
	350	+350	GH	high PAR; high N	—	1.2/11.4	+5/−9	
Betula populifolia	350	+350	GH	no fert	—	1.2/1.2	+8/+8	7
Castanea sativa	350	+350	GH	fert	—	1.2/2.9	−11/−34	32
	330	+330	GH	low N	1.1/1.1	1.8/1.8	+6/+6	

Species								
Eucalyptus camaldulensis	330	+330	GH	high N	1.2/1.3	2.7/6.9	+0/-32	126
	330	+330	GH	low N	1.1/1.1	1.9/1.9	+7/+7	
Eucalyptus cypellocarpa	330	+330	GH	high N	1.3/1.6	3.1/11.7	+0/-31	126
	340	+320	GC	lowest N	—	2.5/2.5	—	
	340	+320	GC	low N	—	3.1/3.3	—	
	340	+320	GC	high N	—	3.1/4.2	—	
Eucalyptus grandis	340	+320	GC	highest N	1.4/1.4	4.0/5.0	-9/-9	22
Eucalyptus pauciflora	330	+330	GH	low N	1.2/1.9	3.1/3.1	-10/-45	126
	330	+330	GH	high N	1.3/1.3	2.4/10.1	-25/-25	
Eucalyptus pulverulenta	330	+330	GH	low N	1.3/1.4	2.6/2.6	-40/-34	126
	330	+330	GH	high N	—	3.2/9.2	+2/+2	
Fraxinus americana	350	+350	GH	low PAR; low N	—	1.1/1.1	+20/-21	7
	350	+350	GH	low PAR; high N	—	1.6/18.5	-27/-27	
Fraxinus americana	350	+350	GH	high PAR; low N	—	0.7/0.7	+3/-33	7
	350	+350	GH	high PAR; high N	—	1.3/8.1	-35/-35	
Gliricidia sepium	350	+300	GC	no N	0.8/0.8	1.1/1.1	-1/-34	115
	350	+300	GC	+N	1.5/0.8	1.3/2.4	+25/+25	
Liriodendron tulipifera	371	+122	GC	no fert	1.2/1.2	1.2/1.2	-12/+11	79
	371	+122	GC	fert	2.2/1.3	1.1/11.6	+46/+46	
Liriodendron tulipifera	371	+416	GC	no fert	—	1.2/1.2	+16/+34	79
	371	+416	GC	fert	—	1.2/13.0	+8/+8	
	350	+400	GC	low N	—	1.3/1.3	+31/-32	
	350	+400	GC	medium N	—	1.2/9.1	-8/-61	
Picea glauca	350	+400	GC	high N	—	1.5/15.1	-13/-13	12
Picea rubens	362	+349	GH	low fert	1.9/1.9	1.8/1.8	-13/-13	106
	362	+349	GH	high fert	1.9/1.9	1.8/1.8	—	
Picea rubens	374	+339	GH	low fert	—	—	—	107
	374	+339	GH	high fert	—	—	—	
Pinus taeda	350	+150	OTC	high P; low N	—	1.3/1.3	—	113
	350	+150	OTC	high P; low N	—	1.2/5.1	—	
	350	+300	OTC	high P; low N	—	1.1/1.1	—	

Table 3 (*continued*)

Species	Baseline CO_2 (ppmv)	Elevated CO_2 (ppmv)	Growth apparatus[a]	Other details[b]	Change[c] P_n	Growth	R/S	Reference
Pinus taeda	350	+300	OTC	high P; low N	—	1.3/5.4	—	113
	350	+150	GH	low N	1.0/1.0	1.0/1.0	—	
Pinus taeda	350	+150	GH	high N	1.1/1.5	1.4/4.0	—	40
	350	+300	GH	low N	1.0/1.0	1.0/1.0	—	
Pinus taeda	350	+300	GH	high N	1.1/1.5	1.6/5.8	—	40
	375	+335	GH	low N	2.1/2.1	1.6/1.6	+6/+6	
Pinus taeda	375	+335	GH	high N	2.0/2.2	1.8/1.8	-3/+0	54
	350	+300	OTC	low N	0.9/09	—	—	
Pinus taeda	350	+300	OTC	high N	1.6/2.1	—	—	117
	350	+300	GH	high P; low N	1.0/1.0	—	—	
Pinus taeda	350	+300	GH	high P; high N	1.6/2.2	—	—	114
	350	+400	GC	low N	—	1.3/1.3	+3/+3	
Populus tremuloides	350	+400	GC	medium N	—	1.0/8.0	+1/+8	12
	350	+400	GC	high N	—	1.0/17.8	-17/+21	
	389	+107	GC	high P; no N	—	1.1/1.1	-16/-16	
Quercus alba	389	+107	GC	high P; +N	—	1.2/1.3	-12/-35	78
	389	+404	GC	high P; no N	—	1.1/1.1	-28/-28	
Quercus alba	389	+404	GC	high P; +N	—	1.2/1.3	-12/-35	78
	350	+350	GH	low PAR; low N	—	1.8/1.8	+23/+23	
Quercus rubra	350	+350	GH	low PAR; high N	—	3.0/8.4	-17/-32	7
	350	+350	GH	high PAR; low N	—	1.4/1.4	-9/-9	
Quercus rubra	350	+350	GH	high PAR; high N	—	1.2/9.0	+39/-19	7
	300	+200	GH	lowest N	1.3/1.3	1.1/1.1	—	
	300	+200	GH	low N	1.1/1.1	1.1/1.7	—	
	300	+200	GH	high N	1.2/1.3	1.5/3.2	—	

Species								
Salix x dasyclados	300	+200	GH	highest N	1.6/1.7	1.5/3.7	—	109
	300	+400	GH	lowest N	1.4/1.4	1.2/1.2	—	
	300	+400	GH	low N	1.3/1.4	1.1/1.7	—	
	300	+400	GH	high N	1.7/1.8	1.5/3.2	—	
Salix x dasyclados	300	+400	GH	highest N	2.1/2.3	1.5/3.6	—	109
	300	+700	GH	lowest N	1.9/1.9	1.0/1.0	—	
	300	+700	GH	low N	2.5/2.6	1.0/1.5	—	
	300	+700	GH	high N	2.5/2.7	1.8/3.9	—	
Salix x dasyclados	300	+700	GH	highest N	2.4/2.6	1.4/3.5	—	109
	300	+200	GH	lowest N	—	1.5/1.5	—	
	300	+200	GH	low N	—	0.9/1.1	—	
	300	+200	GH	high N	—	1.1/2.2	—	
Salix phylicifolia	300	+200	GH	highest N	—	3.8/6.8	—	110
	300	+400	GH	lowest N	—	1.0/1.0	—	
	300	+400	GH	low N	—	1.7/2.0	—	
	300	+400	GH	high N	—	1.8/3.5	—	
Salix phylicifolia	300	+400	GH	highest N	—	1.3/6.0	—	110
	300	+700	GH	lowest N	—	1.0/1.0	—	
	300	+700	GH	low N	—	1.7/2.0	—	
	300	+700	GH	high N	—	2.5/5.0	—	
Salix phylicifolia	300	+700	GH	highest N	—	2.7/4.8	—	110

[a] GC—growth chamber experiments, GH—greenhouse experiments.

[b] PAR—photosynthetically active radiation, N—nitrogen, Nod—nodulated, fert—fertilized, P—phosphorus.

[c] Change is indicated as the ratio of the quantities at elevated and baseline CO$_2$ for P$_n$ and growth, and as percent change from the quantity at baseline CO$_2$ for R/S. P$_n$ values are from measurements made at growth CO$_2$. Quantity on the left is relative to the treatment that uses baseline CO$_2$ and the same level of fertilization, and the quantity on the right is relative to the treatment that uses baseline CO$_2$ and the lowest level of nitrogen fertilization.

resources may be allocated so that growth becomes equally limited by all resources, i.e. carbon, nutrients, and water (2, 9, 17, 96). In this section we examine how changes in nitrogen availability and atmospheric CO_2 interact to influence growth and allocation.

Effects on Growth

Among the studies reviewed by Eamus & Jarvis (30), a doubling of CO_2 reportedly increased biomass approximately 40%. Ceulemans & Mousseau (16) observed that biomass increased 38% for conifer species and 63% for broadleaf species in response to elevated CO_2. In studies in which both CO_2 and nitrogen availability were manipulated, the mean increase in biomass to elevated CO_2 at the lowest level of availability is 35%, while at higher levels of nitrogen availability, the mean increase is 71% (Table 3; Wilcoxin Signed Rank Test, $W = 883.0$, $P < 0.0001$, $N = 55$). In comparison to the biomass accumulation at baseline CO_2 and the lowest level of nitrogen availability, elevated CO_2 accompanied by increased nitrogen availability enhanced biomass an average 6.5 times the baseline biomass among the 55 reports in our survey. These analyses indicate that low nitrogen availability constrains the response of growth to elevated CO_2; they contrast with the conclusion of Idso & Idso (44) that "the percentage increase in plant growth produced by raising the air's CO_2 content is generally not reduced by less than optimal levels of ... soil nutrients."

Effects on Allocation

Eamus & Jarvis (30) observed that, under conditions of low nutrient availability, trees increase the proportion of root biomass in response to elevated CO_2, but under conditions of high nutrient availability, root proportion may decrease, remain unaltered, or increase. Citing unpublished work (by SD Wullschleger), Norby (76) indicates that the mean response of root/shoot ratio to elevated CO_2 is an increase of 6% among 224 observations for woody species, and that there is no effect of nutrient status on the response. Reports in the review by Ceulemans & Mousseau (16) are dominated by increases in root/shoot ratio, which the authors interpreted as investment to ensure better acquisition of mineral nutrients in poor forest soils. They also observed that, at higher levels of nutrients, the change in root/shoot ratio is less. For studies that manipulate nitrogen availability, we observe trends for these patterns, but they are not statistically significant. At the lowest levels of nitrogen availability, the mean increase in root/shoot ratio is 1%, and at higher levels the mean change is –2% (Table 3). These observations are not statistically distinguishable from each other (Paired-sample t-test, $t = 0.764$, $P = 0.4508$, df $= 32$) and are not different from no change in root/shoot ratio. Because the power of the test for differences is low (0.05 vs. desired 0.80), the conclusion from Table 3 that elevated CO$_2$ has no effect on root/shoot ratio among studies that manipulate nitrogen availability is probably an artifact of inadequate sample size. Studies that manipulate nitrogen availability in woody species clearly identify an effect of increased nitrogen availability in reducing root/shoot ratio. In comparison to root/shoot ratios at baseline CO$_2$ and the lowest level of nitrogen availability, the mean decrease in root/shoot ratios for elevated CO$_2$ accompanied with higher nitrogen availability is 21% (Table 3), which is significantly lower than changes in root/shoot ratio at the lowest level of nitrogen availability (Paired-sample t-test, $t = 5.30$, $P < 0.0001$, df = 31). Thus, elevated CO$_2$ and increased soil nitrogen availability have opposite effects on relative allocation of aboveground and belowground biomass.

ECOSYSTEM-LEVEL RESPONSES

The CO$_2$ responses of tissue nitrogen concentration, growth, and relative allocation of biomass to root function appear to be functionally linked (58). Our analyses in this review indicate that tissue nitrogen concentration, growth, and root/shoot ratio in woody species are affected by changes in atmospheric CO$_2$ and soil nitrogen availability. Because most experiments with woody plants involve seedlings or saplings, the possibility exists that reduced tissue nitrogen concentration and root/shoot ratio may, in part, represent the indirect effect of elevated CO$_2$ in accelerating development (1, 20, 76, 20). If the responses of nitrogen concentration, growth, and biomass allocation persist throughout development, they have the potential to alter soil nitrogen availability indirectly. In this section we examine how elevated CO$_2$ may influence soil nitrogen availability through effects on plant nitrogen concentration and on plant growth and allocation.

Effects of Changes in Plant Nitrogen Concentration

The effect of elevated CO$_2$ in reducing nitrogen concentration of plant tissue may alter soil nitrogen availability by influencing decomposition. Rates of leaf decomposition are often correlated with several indices of nitrogen litter quality, which include nitrogen concentration, carbon/nitrogen ratio, and lignin/nitrogen ratio (67). Nitrogen concentration generally is positively correlated with decomposition, whereas the other two indices generally are negatively correlated. Compared to leaf litter of woody plants grown at baseline CO$_2$, decreased nitrogen concentration for leaf litter of plants grown at elevated CO$_2$ has been observed for *Liquidambar styraciflua* (sweetgum; 66), *Quercus alba* (white oak;

82), *Castanea sativa* (sweet chestnut; 24), *Fraxinus excelsior* (ash; 23), *Betula pubescens* (birch; 23), *Acer pseudoplantanus* (sycamore; 23), *Picea sitchensis* (sitka spruce; 23), *Liriodendron tulipifera* (yellow poplar; 83), *Acer rubrum* (red maple; 68), and *Acer pennsylvanicum* (striped maple; 68). Increased carbon/nitrogen ratio has been observed for all these species except sitka spruce and yellow poplar, where carbon/nitrogen ratio was the same. Increased lignin/nitrogen ratio was observed for all species except white oak, in which it decreased from 5.7 to 4.8, and yellow poplar, in which it was the same.

The predicted decay rates for white oak, which were determined from lignin/nitrogen and lignin/phosphorus ratios of leaf litter, suggest there would be no difference between litter derived from plants grown in baseline and those in elevated CO_2 (82). In contrast, rates of decay for maple species, also determined from lignin/nitrogen ratios, suggest decay rates per unit of litter would be slower for elevated-CO_2 material (68). For sweet chestnut leaf litter incubated with only microflora and protozoa, mass loss was 60% less for litter derived from elevated CO_2 plants than that from baseline CO_2 plants (24). However, mass loss was similar between the CO_2 treatments for litter incubated with nematodes and collembola in addition to microflora and protozoa; the addition of isopods increased mass loss by 30% in the elevated CO_2 treatment. The enhanced decomposition was attributed to a change in the microflora community, which became dominated by white-rot fungus. Among decomposition experiments with leaf litter of ash, birch, sycamore, and sitka spruce, cumulative respiration rates were lower for litter derived from elevated CO_2 plants among the three deciduous species, but rates were similar for spruce (23). Significantly lower mass loss was observed for both birch and spruce, but there was also a nonsignificant trend for lower mass loss in ash. Cumulative nitrogen mineralization did not differ between CO_2 treatments for any of the four species. No difference in mass loss rates were observed for yellow poplar after two years of decomposition in litter bags (83).

If decomposition and nitrogen mineralization are depressed because of CO_2-induced changes in litter quality, soil nitrogen availability may be reduced in ecosystems. Our earlier analyses suggest that reduced nitrogen availability has the potential to limit both photosynthetic and growth responses to elevated CO_2. Thus, reduced litter quality resulting from elevated CO_2 has the potential to cause long-term negative feedback to constrain the response of NPP. Reductions in leaf litter quality seem to be common among woody species, but these may not be universal. If nitrogen is not resorbed from fine roots prior to senescence, as suggested by Nambiar (74), then the response of fine root nitrogen concentration may be a good indicator of changes in fine root litter quality (76). Small reductions in the nitrogen concentrations of fine roots and stems may contribute to lower total litter quality, but to our knowledge no published studies examine how elevated

CO_2 affects decomposition of fine root and stem litter in woody plants. One study of fine root decomposition in a grass reports lower decomposition rates for root residue from plants grown in elevated CO_2; a lower decomposition rate is associated with an increase in carbon/nitrogen ratio from 18 in baseline plants to 32 in elevated-CO_2 plants (39). At present, the available data suggest that CO_2-induced reductions in litter quality may depress decomposition rates, but the data are ambiguous. Effects on nitrogen mineralization of CO_2-induced reductions in litter quality are less well documented. Also, there is no information on how nitrogen mineralization might be affected by the combination of CO_2-induced reductions in decay rate and enhancements in litter production. Clearly, more research is needed on the potential for CO_2-induced reductions in litter quality to decrease soil nitrogen availability and cause long-term feedback that constrains the response of forest NPP to elevated CO_2.

Effects of Changes in Growth and Allocation

Increased growth and root/shoot ratio in response to elevated CO_2 have the potential to increase production of belowground biomass in forest ecosystems. Belowground biomass represents both storage in coarse roots and investment for the acquisition of nutrients and water by fine roots. To the extent that increased belowground inputs are manifested in enhanced fine root growth, nitrogen uptake to plants might increase because of additional or more efficient exploration of soil volume by rooting systems. However, in mature ecosystems it is not clear whether the soil exploration by rooting systems is saturated under present CO_2 conditions. Increased production in response to elevated CO_2 may also enhance the transfer of carbon to microbes by means of increased fine root turnover or by the exudation of soluble organic carbon from roots into the soil. If elevated CO_2 causes greater inputs of root-derived carbon into the soil, it may increase nitrogen availability by enhancing nitrogen fixation (59) or nitrogen mineralization (135).

Symbiotic nitrogen fixers acquire carbon from their host plants and provide inorganic nitrogen to their hosts. Elevated CO_2 enhances nitrogen fixation per plant for symbiotic associations involving woody species by increasing nodule mass (4, 75, 115), nodule number (4), or nitrogenase activity per nodule (4). In addition to carbon availability, nitrogen fixation may be limited by phosphorus availability (27). Elevated CO_2 enhances colonization of ectomycorrhizae in woody plants (55, 83, 85), presumably because of enhanced root exudation of soluble carbon (80). Because mycorrhizae are important for supplying phosphorus to plants, increased inputs of root carbon into the soil may indirectly affect nitrogen availability by helping to supply phosphorus to symbiotic nitrogen fixers. Effects of elevated CO_2 on nitrogen fixation may have important consequences for NPP of tropical forests, where symbiotic nitrogen fixation is more important than in extratropical regions (83). The consequences may be especially relevant for tropical forests that occur on extremely weathered soils, which are often deficient in phosphorus (108). Whereas the effects of elevated CO_2 on symbiotic nitrogen fixation have received some attention, the effects on asymbiotic nitrogen fixation have not been addressed by the scientific community. Because asymbiotic nitrogen fixation may be important for most of the atmospheric nitrogen fixed in some ecosystems (83), progress is needed to understand how elevated CO_2 may affect this potentially important process.

It has also been hypothesized that if elevated CO_2 results in a greater flow of carbon from roots to soil, then nitrogen mineralization may be enhanced (135). This hypothesis makes the prediction that: 1. microbial growth in the vicinity of the root will be enhanced by increased root turnover or exudation, and 2. that nitrogen mineralization will be increased by higher rates of protozoan grazing on microbial populations (19) or by increased rates of organic matter decomposition (135). Enhanced decomposition of organic matter could increase nitrogen availability through greater mineralization of microbial nitrogen derived from either root residues or native soil organic matter (135). Zak et al (135) tested the hypothesis for *Populus grandidentata* grown in open top chambers. For elevated CO_2 treatments, they observed root and microbial biomass increased, net nitrogen mineralization increased in short-term laboratory incubations of the bulk soil, respiration rates were higher in the rhizosphere, and there were nonsignificant trends for higher rates of respiration and nitrogen mineralization in the bulk soil.

Although the results of the experiment are consistent with the hypothesis of Zak et al (135), an alternative explanation for the results may be related to the fact that they added inorganic nitrogen to all treatments (4.5 g N m^{-2} over a 47-day period). If microbial growth were simultaneously limited by both carbon and nitrogen availability, then this inorganic nitrogen addition may have stimulated microbial growth in the elevated-CO_2 treatments because of increased root-derived carbon inputs into the soil (see 121). The enhanced microbial growth, if it causes increased grazing by protozoa or increased microbial turnover, could result in higher mineralization rates. This interpretation of the Zak et al (135) results, if correct, has important implications for the effects of elevated CO_2 on forest growth in regions of the world where soils receive substantial inputs of anthropogenic nitrogen from the atmosphere (see 70). Clearly, the link between nitrogen availability and CO_2-induced inputs of carbon into the soil is complex, and additional research is required to elucidate how elevated CO_2 and nitrogen availability interact to influence nitrogen cycling in forest ecosystems.

Univ. Nacional Autónoma de México, México DF

49. Farris JS. 1981. Discussion. See Ref. 122a, pp. 73–84

50. Farris JS. 1988. *Hennig86 reference.* Version 1.5. Published by the author, Port Jefferson, New York.

51. Felsenstein J. 1993. *PHYLIP. Phylogeny Inference Package.* 3.5. Univ. Wash., Seattle

51a. Funk VA, Brooks DR, eds. 1981. *Advances in Cladistics: Proceedings of the First Meeting of the Willi Hennig Society.* Bronx, NY: New York Bot. Gard.

52. Gibbs GW. 1989. Local or global? Biogeography of some primitive Lepidoptera in New Zealand. *NZ J. Zool.* 16:689–98

53. Goodman M, Czelusniak J, Moore GW, Romero-Herrera AE, Matsuda G. 1979. Fitting the gene lineage into its species lineage: a parsimony strategy illustrated by cladograms constructed from globin sequences. *Syst. Zool.* 28:132–68

54. Gray R. 1989. Oppositions in panbiogeography: Can the conflicts between selection, constraint, ecology, and history be resolved? *NZ J. Zool.* 16:787–806

55. Grehan JR. 1988. *Panbiogeography: evolution in space and time. Riv. Biol.—Biol. Forum* 81:469–98

56. Grehan JR. 1988. Biogeographic homology: ratites and the southern beeches. *Riv. Biol.—Biol. Forum* 81:577–87

57. Grehan JR. 1989. New Zealand panbiogeography: past, present, and future. *NZ J. Zool.* 16:513–25

58. Grehan JR. 1989. Panbiogeography and conservation science in New Zealand. *NZ J. Zool.* 16:731–48

59. Grehan JR. 1991. A panbiogeographic perspective of pre-Cretaceous angiosperm-Lepidoptera coevolution. *Aust. Syst. Bot.* 4:91–110

60. Grehan JR. 1991. Panbiogeography 1981–91: development of an earth/life synthesis. *Progr. Phys. Geogr.* 15:331–63

61. Grehan JR. 1993. Conservation biogeography and the biodiversity crisis: a global problem in space time. *Biodiversity Lett.* 1:134–40

62. Griswold CE. 1991. Cladistic biogeography of afromontane spiders. *Aust. Syst. Bot.* 4:73–89

63. Harold AS, Mooi RD. 1994. Areas of endemism: definition and recognition criteria. *Syst. Biol.* 43:261–66

64. Hedges SB, Hass CA, Maxson LR. 1992. Caribbean biogeography: molecular evidence for dispersal in West Indian terrestrial vertebrates. *Proc. Natl. Acad. Sci. USA* 89:1909–13

65. Henderson IM. 1991. Biogeography without area? *Aust. Syst. Bot.* 4:59–71

66. Hennig W. 1950. *Grundzüge Einer Theorie der Phylogenetischen Systematik.* Berlin: Deutscher Zentralverlag.

67. Hennig W. 1966. *Phylogenetic Systematics.* Urbana: Univ. Ill. Press

68. Humphries CJ. 1981. Biogeographical methods and the southern beeches (Fagaceae: Nothofagus). See Ref. 51a, pp. 177–207

69. Humphries CJ. 1981. Biogeographical methods and the southern beeches. In *Chance, Change and Challenge.* Vol. 2. *The Evolving Biosphere,* ed. PL Forey, pp. 283–97. London: Br. Mus. (Nat. Hist.) and Cambridge Univ. Press

70. Humphries CJ. 1989. Any advance on assumption 2? *J. Biogeogr.* 16:101–2

71. Humphries CJ. 1992. Cladistic biogeography. In *Cladistics: A Practical Course in Systematics,* ed. PL Forey et al, pp. 137–59. *Oxford: Syst. Assoc. Publ. No. 10.* Clarendon: Oxford Univ. Press

72. Humphries CJ, Ladiges PY, Roos M, Zandee M. 1988. Cladistic biogeography. See Ref. 106a, pp. 371–404

73. Humphries CJ, Parenti LR. 1986. *Cladistic Biogeography.* Oxford: Oxford Univ. Press

74. Humphries CJ, Seberg O. 1989. Graphs and generalized tracks: some comments on method. *Syst. Zool.* 38:69–76

75. Källersjo M, Farris JS, Kluge AG, Bult C. 1992. Skewness and permutation. *Cladistics* 8:275–87

76. Kluge AG. 1988. Parsimony in vicariance biogeography: a quantitative method and a Greater Antillean example. *Syst. Zool.* 37:315–28

77. Kluge AG. 1993. Three-taxon transformation in phylogenetic inference: ambiguity and distortion as regards explanatory power. *Cladistics* 9:246–59

78. Ladiges PY, Newnham MR, Humphries CJ. 1989. Systematics and biogeography of the Australian "green ash" eucalypts (*Monocalyptus*). *Cladistics* 5:345–64

79. Ladiges PY, Prober SM, Nelson G. 1992. Cladistic and biogeographic analysis of the 'blue ash' eucalypts. *Cladistics* 8:103–24

80. Lakatos I. 1970. Falsification and the methodology of scientific research programmes. In *Criticism and the Growth of Knowledge,* ed. I Lakatos, A Musgrave, pp. 91–176. Cambridge: Cambridge Univ. Press

81. Liebherr JK. 1988. General patterns in West Indian insects, and graphical bio-

CONCLUSION

The interaction between soil nitrogen availability and elevated CO$_2$ is important to consider because: 1. nitrogen availability is spatially variable (65), and 2. elevated temperature, which might accompany elevated CO$_2$ (71), has the potential to affect soil nitrogen availability by influencing decomposition (63, 65, 69, 90, 91, 119, 120). To make progress in modeling the response of forest ecosystems to global change, the scientific community needs to improve its understanding of how nitrogen availability and elevated CO$_2$ interact to affect forest NPP. In this study we identified important influences of the nitrogen cycle in the potential response of forest NPP to elevated CO$_2$. At the tissue level, effects appear to be related to changes in tissue nitrogen concentration, and they may influence photosynthetic and respiration responses to elevated CO$_2$. However, it is important to recognize that increased nitrogen availability and elevated CO$_2$ have opposite effects on nitrogen concentration of leaf tissue. Although more research is needed to understand how changes in nitrogen concentration affect biochemical and physiological processes, an important challenge will be to understand the mechanisms responsible for changes in tissue nitrogen concentration. Research to address this issue requires attention at the levels of both tissue and plant. At the plant level, soil nitrogen availability is an important factor that often constrains the response of woody plant growth to elevated CO$_2$. Also, increased nitrogen availability and elevated CO$_2$ have opposite effects on the relative allocation of carbon to aboveground and belowground biomass. Thus, changes in nitrogen availability in response to climatic changes influence the ability of vegetation to incorporate elevated CO$_2$ into production. Effects of elevated CO$_2$ at the tissue and plant levels may have important consequences for nitrogen cycling at the ecosystem level, but our knowledge of how CO$_2$-induced changes in litter quality and in root-derived soil carbon influence nitrogen availability is based on a small number of studies. Additional research is required at the ecosystem level to understand how interactions of the nitrogen cycle and elevated CO$_2$ affect forest NPP. Factorial studies that manipulate both atmospheric CO$_2$ and soil nitrogen availability for whole ecosystems would advance understanding.

ACKNOWLEDGMENTS

We thank B. G. Drake, D. W. Kicklighter, H. A. Mooney, and M. G. Ryan for comments on an earlier draft of this paper. This study was funded by the Earth Observing System Program of the National Aeronautics and Space Administration (NAGW-2669) and the USDA Forest Service Resources Program and Assessment Staff.

Literature Cited

1. Agren GI. 1994. The interaction between CO_2 and plant nutrition: comments on a paper by Coleman, McConnaughay and Bazzaz. *Oecologia* 98: 239–40
2. Agren GI, Ingestad T. 1987. Root:shoot ratio as a balance between nitrogen productivity and photosynthesis. *Plant, Cell & Environ.* 10:579–86
3. Amthor JS. 1994. Plant respiratory responses to the environment and their effects on the carbon balance. In *Plant-Environment Interactions*, ed. RE Wilkerson, pp. 501–54. New York: Marcel Dekker
4. Arnone JA III, Gordon JC. 1990. Effect of nodulation, nitrogen fixation, and CO_2 enrichment on the physiology, growth and dry mass allocation of seedlings of *Alnus rubra* Bong. *New Phytol.* 116:55–66
5. Arp WJ. 1991. Effects of source-sink relations on photosynthetic acclimation to elevated CO_2. *Plant, Cell & Environ.* 14:869–75
6. Azcon-Bieto J, Gonzalez-Meler MA, Doherty W, Drake BG. 1994. Acclimation of respiratory O_2 uptake in green tissues of field-grown native species after long-term exposure to elevated atmospheric CO_2. *Plant Physiol.* 106: 1163–68
7. Bazzaz FA, Miao SL. 1993. Successional status, seed size, and responses of tree seedlings to CO_2, light, and nutrients. *Ecology* 74:104–12
8. Bernston GM, McConnaughay KDM, Bazzaz FA. 1993. Elevated CO_2 alters deployment of roots in "small" growth containers. *Oecologia* 93:558–64
9. Bloom AJ, Chapin FS III, Mooney HA. 1985. Resource limitation in plants—an economic analogy. *Annu. Rev. Ecol. Syst.* 16:363–92
10. Bowes G. 1991. Growth at elevated CO_2: photosynthetic responses mediated through Rubisco. *Plant, Cell & Environ.* 14:795–806
11. Brown KR. 1991. Carbon dioxide enrichment accelerates the decline in nutrient status and relative growth rate of *Populus tremuloides* Michx. seedlings. *Tree Physiol.* 8:161–73
12. Brown K, Higginbotham KO. 1986. Effects of carbon dioxide enrichment and nitrogen supply on growth of boreal tree seedlings. *Tree Physiol.* 2:223–32
13. Bunce JA. 1992. Stomatal conductance, photosynthesis and respiration of temperate deciduous tree seedlings grown outdoors at an elevated concentration of carbon dioxide. *Plant, Cell & Environ.* 15:541–49
14. Bunce JA. 1994. Response of respiration to increasing atmospheric carbon dioxide concentrations. *Physiol. Plant.* 90: 427–30
15. Cave G, Tolley LC, Strain BR. 1981. Effect of carbon dioxide enrichment on chlorophyll content, starch content and starch grain structure in *Trifolium subterraneum* leaves. *Physiol. Plant.* 51: 171–74
16. Ceulemans R, Mousseau M. 1994. Effects of elevated atmospheric CO_2 on woody plants. *New Phytol.* 127:425–46
17. Chapin FS III, Bloom AJ, Field CB, Waring RH. 1987. Plant responses to multiple environmental factors. *BioScience* 37:49–57
18. Chapin FS III, Walter CSH, Clarkson DT. 1988. Growth response of barley and tomato to nitrogen stress and its control by abscisic acid, water relations, and photosynthesis. *Planta* 173:352–66
19. Clarholm M. 1985. Interactions of bacteria, protozoa and plants leading to mineralization of soil nitrogen. *Soil Biol. Biochem.* 17:181–87
20. Coleman JS, McConnaughay, Bazzaz FA. 1993. Elevated CO_2 and plant nitrogen-use: Is reduced tissue nitrogen concentration size-dependent? *Oecologia* 97:195–200
21. Conroy JP. 1992. Influence of elevated CO_2 concentrations on plant nutrition. *Aust. J. Bot.* 40:445–56
22. Conroy JP, Milham PJ, Barlow EWR. 1992. Effect of nitrogen and phosphorus availability on the growth response of *Eucalyptus grandis*. *Plant, Cell & Environ.* 15:843–47
23. Cotrufo MF, Ineson P, Rowland AP. 1994. Decomposition of tree leaf litters grown under elevated CO_2: effect of litter quality. *Plant & Soil* 163:121–30
24. Couteaux MM, Mousseau M, Celerier ML, Bottner PP. 1990. Increased atmos-

pheric CO$_2$ and litter quality: decomposition of sweet chestnut leaf litter with animal food webs of different complexities. *Oikos* 61:54–64

25. Curtis PS, Teeri JA. 1992. Seasonal responses of leaf gas exchange to elevated carbon dioxide in *Populus grandidentata. Can. J. For. Res.* 22:1320–25

26. DeLucia EH, Sasek TW, Strain BR. 1985. Photosynthetic inhibition after long-term exposure to elevated levels of atmospheric carbon dioxide. *Photosynth. Res.* 7:175–84

27. Dixon ROD, Wheeler CT. 1983. Biochemical, physiological and environmental aspects of symbiotic nitrogen fixation. In *Biological Nitrogen Fixation in Forest Ecosystems: Foundations and Applications,* ed. JC Gordon, CT Wheeler, pp. 107–171. The Hague: Martinus Nijhoff/Dr. W. Junk

28. Du Cloux H, Andre M, Gerbaud A, Daguenet A. 1989. Wheat response to CO$_2$ enrichment: effect on photosynthetic and photorespiratory characteristics. *Photosynthetica* 23:145–53

29. Duff GA, Berryman CA, Eamus D. 1994. Growth, biomass allocation and foliar nutrient contents of two Eucalyptus species of the wet-dry tropics of Australia grown under CO$_2$ enrichment. *Funct. Ecol.* 8:502–8

30. Eamus D, Jarvis PG. 1989. The direct effects of increase in the global atmospheric CO$_2$ concentration on natural and commercial temperate trees and forests. *Adv. Ecol. Res.* 19:1–55.

31. Ehret DL, Jolliffe PA. 1985. Leaf injury to bean plants grown in carbon dioxide enriched atmospheres. *Can J. Bot.* 63: 2015–20

32. El Kohen A, Rouhier H, Mousseau M. 1992. Changes in dry weight and nitrogen partitioning induced by elevated CO$_2$ depend on soil nutrient availability in sweet chestnut (*Castanea sativa* Mill). *Ann. Sci. For.* 49:83–90

33. El Kohen A, Venet L, Mousseau M. 1993. Growth and photosynthesis of two deciduous forest species at elevated carbon dioxide. *Funct. Ecol.* 7:480–86

34. Evans JR. 1983. Nitrogen and photosynthesis in the flag leaf of wheat (*Triticum aestivum* L.). *Plant Physiol.* 72:297–302

35. Evans JR. 1989. Photosynthesis and nitrogen relationships in leaves of C$_3$ plants. *Oecologia* 78:9–19

36. Farquhar GD, von Caemmerer S, Berry JA. 1980. A biochemical model of photosynthetic CO$_2$ assimilation in leaves of C$_3$ species. *Planta* 149:79–90

37. Field CB. 1991. Ecological scaling of

carbon gain to stress and resource availability. In *Response of Plants to Multiple Stresses,* ed. HA Mooney, WE Winner, pp. 35–65. San Diego:Academic

38. Gifford RM. 1994. The global carbon cycle: a viewpoint on the missing sink. *Aust. J. Plant Physiol.* 21:1–15

39. Gorissen A, van Ginkel JH, Keurentjes JJB, van Veen JA. 1995. Grass root decomposition is retarded when grass has been grown under elevated CO$_2$. *Soil Biol. Biochem.* 27:117–20

40. Griffin KL, Thomas RB, Strain BR. 1993. Effects of nitrogen supply and elevated carbon dioxide on construction cost in leaves of *Pinus taeda* (L.) seedlings. *Oecologia* 95:575–80

41. Gulmon SL, Chu CC. 1981. The effects of light and nitrogen on photosynthesis, leaf characteristics, and dry matter allocation in the chaparral shrub, *Diplacus aurantiacus. Oecologia* 49:207–12

42. Gunderson CA, Wullschleger SD. 1994. Photosynthetic acclimation in trees to rising atmospheric CO$_2$: a broader perspective. *Photosynth. Res.* 39:369–88.

43. Huber SC, Rogers HH, Mowry FL. 1984. Effects of water stress on photosynthesis and carbon partitioning in soybean (*Glycine max* L. Merr.) plants grown in the field at different CO$_2$ levels. *Plant Physiol.* 76:244–49

44. Idso KE, Idso SB. 1994. Plant responses to atmospheric CO$_2$ enrichment in the face of environmental constraints: a review of the past 10 years research. *Agric. For. Meteorol.* 69:153–203

45. Idso SB, Kimball BA. 1992. Effects of atmospheric CO$_2$ enrichment on photosynthesis, respiration, and growth of sour orange trees. *Plant Physiol.* 99: 341–43

46. Idso SB, Kimball BA. 1993. Effects of atmospheric CO$_2$ enrichment on net photosynthesis and dark respiration rates of three Australian tree species. *J. Plant. Physiol.* 141:166–71

47. Johnsen KH. 1993. Growth and ecophysiological responses of black spruce seedlings to elevated CO$_2$ under varied water and nutrient additions. *Can. J. For. Res.* 23:1033–42

48. Johnson RH, Lincoln DE. 1990. Sagebrush and grasshopper responses to atmospheric carbon dioxide concentration. *Oecologia* 84:103–10

49. Johnson RH, Lincoln DE. 1991. Sagebrush carbon allocation patterns and grasshopper nutrition: the influence of CO$_2$ enrichment and soil mineral limitation. *Oecologia* 87:127–34

50. Kerstiens G, Hawes CV. 1994. Response of growth and carbon allocation

to elevated CO_2 in young cherry (*Prunus avium* L.) saplings in relation to root environment. *New Phytol.* 128: 607–14

51. Korner C, Arnone JA III. 1992. Responses to elevated carbon dioxide in artificial tropical ecosystems. *Science* 257:1672–75

52. Korner C, Miglietta F. 1994. Long term effects of naturally elevated CO_2 on mediterranean grassland and forest trees. *Oecologia* 99:343–51

53. Lajtha K, Whitford WG. 1989. The effect of water and nitrogen amendments on photosynthesis, leaf demography, and resource-use efficiency in *Larrea tridentata*, a desert evergreen shrub. *Oecologia* 80:341–48

54. Larigauderie A, Reynolds JF, Strain BR. 1994. Root response to CO_2 enrichment and nitrogen supply in loblolly pine. *Plant & Soil* 165:21–32

55. Lewis JD, Thomas RB, Strain BR. 1994. Effect of elevated CO_2 on mycorrhizal colonization of loblolly pine (*Pinus taeda* L.) seedlings. *Plant & Soil* 165: 81–88

56. Lindroth RL, Kinney KK, Platz CL. 1993. Responses of deciduous trees to elevated atmospheric CO_2: productivity, phytochemistry, and insect performance. *Ecology* 74:763–77

57. Loomis RS, Lafitte HR. 1987. The carbon economy of a maize crop exposed to elevated CO_2 concentrations and water stress, as determined by elemental analysis. *Field Crop Res.* 17:63–74

58. Luo Y, Field CB, Mooney HA. 1994. Predicting responses of photosynthesis and root fraction to elevated $[CO_2]$a: interactions among carbon, nitrogen, and growth. *Plant, Cell & Environ.* 17: 1195–204

59. Luxmoore RJ. 1981. CO_2 and phytomass. *BioScience* 31:626

60. Luxmoore RJ, O'Neill EG, Ellis JM, Rogers HH. 1986. Nutrient uptake and growth responses of Virginia pine to elevated atmospheric carbon dioxide. *J. Environ. Qual.* 15:244–51

61. McConnaughay KDM, Bernston GM, Bazzaz FA. 1993. Plant responses to carbon dioxide. *Nature* 361:24

62. McConnaughay KDM, Bernston GM, Bazzaz FA. 1993. Limitations to CO_2-induced growth enhancement in pot studies. *Oecologia* 94:550–57

63. McGuire AD, Joyce LA, Kicklighter DW, Melillo JM, Esser G, Vorosmarty CJ. 1993. Productivity response of climax temperate forests to elevated temperature and carbon dioxide: a North American comparison between two

global models. *Clim. Change.* 24:287–310

64. McGuire AD, Joyce LA. 1995. Responses of net primary production in temperate forests to potential changes in carbon dioxide and climate. In *Gen. Tech. Report for the 1993 RPA Assessment Update*, ed. LA Joyce. Fort Collins: USDA For. Serv. In press

65. McGuire AD, Melillo JM, Kicklighter DW, Grace AL, Moore B III, Vorosmarty CJ. 1992. Interactions between carbon and nitrogen dynamics in estimating net primary productivity for potential vegetation in North America. *Global Biogeochem. Cycles* 6:101–24

66. Melillo JM. 1983. Will increases in atmospheric CO_2 concentrations affect decay processes? In *The Ecosystems Center Annual Report*, pp. 10–11. Woods Hole: Mar. Biol. Lab.

67. Melillo JM, Aber JD, Muratore JF. 1982. The influence of substrate quality of leaf litter decay in a northern hardwood forest. *Ecology* 63:621–26

68. Melillo JM, Kicklighter DW, McGuire AD, Peterjohn WT, Newkirk KM. 1995. Global change and its effects on soil organic carbon stocks. In *Role of Non-living Organic Matter in the Earth's Carbon Cycle*, ed. RG Zepp, C Sonntag, pp. 175–189. New York: John Wiley & Sons

69. Melillo JM, McGuire AD, Kicklighter DW, Moore B III, Vorosmarty CJ, Schloss AL. 1993. Global climate change and terrestrial net primary production. *Nature* 363:234–40

70. Melillo JM, Steudler PA, Aber JD, Bowden RD. 1989. Atmospheric deposition and nutrient cycling. In *Exchange of Trace Gases between Terrestrial Ecosystems and the Atmosphere*, ed. MO Andreae, DS Schimel, pp. 263–80. New York: John Wiley & Sons

71. Mitchell JFB, Manabe S, Meleshko V, Tokioka T. 1990. Equilibrium climate change—and its implications for the future. In *Climate Change: The IPCC Scientific Assessment*, ed. JT Houghton et al, pp. 131–72. Cambridge: Cambridge Univ. Press

72. Mousseau M. 1993. Effects of elevated CO_2 on growth, photosynthesis and respiration of sweet chesnut (*Castanea sativa* Mill.). *Vegetatio* 104/105:413–19

73. Mousseau M, Enoch HZ. 1989. Carbon dioxide enrichment reduces shoot growth in sweet chestnut seedlings (*Castanea sativa* Mill.). *Plant, Cell & Environ.* 12:927–34

74. Nambiar EKS. 1987. Do nutrients re-

translocate from fine roots? *Can J. For. Res.* 17:913–18

75. Norby RJ. 1987. Nodulation and nitrogenase activity in nitrogen-fixing woody plants stimulated by CO$_2$ enrichment of the atmosphere. *Physiol. Plant.* 71:77–82

76. Norby RJ. 1994. Issues and perspectives for investigating root responses to elevated atmospheric carbon dioxide. *Plant & Soil* 165:9–20

77. Norby RJ, Gunderson CA, Wullschleger SD, O'Neill EG, McCracken MK. 1992. Productivity and compensatory responses of yellow-poplar trees in elevated CO$_2$. *Nature* 357:322–24

78. Norby RJ, O'Neill EG. 1989. Growth dynamics and water use of seedlings of *Quercus alba* L. in CO$_2$-enriched atmospheres. *New Phytol.* 111:491–500

79. Norby RJ, O'Neill EG. 1991. Leaf area compensation and nutrient interactions in CO$_2$-enriched seedlings of yellow-poplar (*Liriodendron tulipifera* L.). *New Phytol.* 117:515–28

80. Norby RJ, O'Neill EG, Hood WG, Luxmoore RJ. 1987. Carbon allocation, root exudation and mycorrhizal colonization of *Pinus echinata* seedlings grown under CO$_2$ enrichment. *Tree Physiol.* 3:203–10

81. Norby RJ, O'Neill EG, Luxmoore RJ. 1986. Effects of atmospheric CO$_2$ enrichment on the growth and mineral nutrition of *Quercus alba* seedlings in nutrient-poor soil. *Plant Physiol.* 82:83–89

82. Norby RJ, Pastor J, Melillo JM. 1986. Carbon-nitrogen interactions in CO$_2$-enriched white oak: physiological and long-term perspectives. *Tree Physiol.* 2:233–41

83. O'Neill EG. 1994. Responses of soil biota to elevated atmospheric carbon dioxide. *Plant & Soil* 165:55–65

84. O'Neill EG, Luxmoore RJ, Norby RJ. 1987. Elevated atmospheric CO$_2$ effects on seedling growth, nutrient uptake, and rhizosphere bacterial populations of *Liriodendron tulipifera* L. *Plant & Soil* 104:3–11

85. O'Neill EG, Luxmoore RJ, Norby RJ. 1987. Increases in mycorrhizal colonization and seedling growth in *Pinus echinata* and *Quercus alba* in an enriched CO$_2$ atmosphere. *Can. J. For. Res.* 17:878–83

86. Overdieck D. 1993. Elevated CO$_2$ and the mineral content of herbaceous and woody plants. *Vegetatio* 104/105:403–11

87. Peet MM, Huber SC, Patterson DT. 1986. Acclimation to high CO$_2$ in mono-ecious cucumbers. II. Carbon exchange rates, enzyme activities, and starch and nutrient concentrations. *Plant Physiol.* 80:63–67

88. Penning de Vries FWT. 1975. The cost of maintenance processes in plant cells. *Annals of Botany* 39:77–92

89. Perez-Soba M, Van der Eerden L, Stulen I, Kuiper PJC. 1994. Gaseous ammonia counteracts the response of Scots pine needles to elevated atmospheric carbon dioxide. *New Phytol.* 128:307–13

90. Peterjohn WT, Melillo JM, Bowles FP, Steudler PA. 1993. Soil warming and trace gas fluxes: experimental design and preliminary flux results. *Oecologia* 93:18–24

91. Peterjohn WT, Melillo JM, Steudler PA, Newkirk KM, Bowles FP, Aber JD. 1994. Responses of trace gas fluxes and N availability to experimentally elevated soil temperatures. *Ecol. Appl.* 4:617–25

92. Pettersson R, McDonald AJS. 1992. Effects of elevated carbon dioxide concentration on photosynthesis and growth of small birch plants (*Betula pendula* Roth.) at optimal nutrition. *Plant, Cell & Environ.* 15:911–19

93. Pettersson R, McDonald AJS. 1994. Effects of nitrogen supply on the acclimation of photosynthesis to elevated CO$_2$. *Photosynth. Res.* 39:389–400

94. Pettersson R, McDonald AJS, Stadenberg I. 1993. Response of small birch plants (*Betula pendula* Roth.) to elevated CO$_2$ and nitrogen supply. *Plant, Cell & Environ.* 16:1115–21

95. Reid CD, Strain BR. 1994. Effects of CO$_2$ enrichment on whole-plant carbon budget of seedlings of *Fagus grandifolia* and *Acer saccharum* in low irradiance. *Oecologia* 98:31–39

96. Reynolds JF, Thornley JHM. 1982. A shoot:root partitioning model. *Ann. Bot.* 49:585–97

97. Rochefort L, Bazzaz FA. 1992. Growth response to elevated CO$_2$ in seedlings of four co-occurring birch species. *Can. J. For. Res.* 22:1583–87

98. Rogers HH, Runion GB, Krupa SV. 1994. Plant responses to atmospheric CO$_2$ enrichment with emphasis on roots and the rhizosphere. *Environ. Pollut.* 83:155–89

99. Roth SK, Lindroth RL. 1994. Effects of CO$_2$-mediated changes in paper birch and white pine chemistry on gypsy moth performance. *Oecologia* 98:133–38

100. Rouhier H, Billes G, El Kohen A, Mousseau M, Bottner P. 1994. Effects of elevated CO$_2$ on carbon and nitrogen distribution within a tree (*Castanea sa-*

tiva Mill.)—soil system. *Plant Soil* 162: 281–92

101. Ryan MG. 1991. Effects of climate change on plant respiration. *Ecol. Appl.* 1:157–67

102. Sage RF. 1994. Acclimation of photosynthesis to increasing atmospheric CO_2: the gas exchange perspective. *Photosynth. Res.* 39:351–68

103. Sage RF, Pearcy RW. 1987. The nitrogen use efficiency of C_3 and C_4 plants. I. Leaf nitrogen, growth, and biomass partitioning in *Chenopodium albion* (L.) and *Amaranthus retroflexus* (L.). *Plant Physiol.* 84:954–58

104. Sage RF, Pearcy RW. 1987. The nitrogen use efficiency of C_3 and C_4 plants. II. Leaf nitrogen effects on the gas exchange characteristics of *Chenopodium album* (L.) and *Amaranthus retroflexus* (L.). *Plant Physiol.* 84:959–63

105. Sage RF, Sharkey TD, Seeman JR. 1989. The acclimation of photosynthesis to elevated CO_2 in five C_3 species. *Plant Physiol.* 89:590–96

106. Samuelson LJ, Seiler JR. 1993. Interactive role of elevated CO_2, nutrient limitations, and water stress in the growth responses of red spruce seedlings. *For. Sci.* 39:348–58.

107. Samuelson LJ, Seiler JR. 1994. Red spruce seedling gas exchange in response to elevated CO_2, water stress, and soil fertility treatments. *Can. J. For. Res.* 24:954–59

108. Sanchez PA, Bandy DE, Villachica JH, Nicholaides JJ. 1982. Amazon Basin soils: management for continuous crop production. *Science* 216:821–27

109. Silvola J, Ahlholm U. 1992. Photosynthesis in willows (*Salix × dasyclados*) grown at different CO_2 concentrations and fertilization levels. *Oecologia* 91: 208–13

110. Silvola J, Ahlholm U. 1993. Effects of CO_2 concentration and nutrient status on growth, growth rhythm and biomass partitioning in a willow, *Salix phylicifolia. Oikos* 67:227–34

111. Sionit N, Rogers HH, Bingham GE, Strain BR. 1984. Photosynthesis and stomatal conductance with CO_2-enrichment of container- and field-grown soybeans. *Agronomy J.* 76:447–51

112. Stitt M, Schulze D. 1994. Does *Rubisco* control the rate of photosynthesis and plant growth? An exercise in molecular ecophysiology. *Plant, Cell & Environ.* 17:465–87

113. Strain BR, Thomas RB. 1992. Field measurements of CO_2 enhancement and climate change in natural vegetation. *Water, Air, Soil Poll.* 64:26–60

114. Thomas RB, Lewis JD, Strain BR. 1994. Effects of leaf nutrient status on photosynthetic capacity in loblolly pine (*Pinus taeda* L.) seedlings grown in elevated atmospheric CO_2. *Tree Physiol.* 14:947–60

115. Thomas RB, Richter DD, Ye H, Heine PR, Strain BR. 1991. Nitrogen dynamics and growth of seedlings of an N-fixing tree (*Gliricidia sepium* (Jacq.) Walp.) exposed to elevated atmospheric carbon dioxide. *Oecologia* 88:415–21

116. Thomas RB, Strain BR. 1991. Root restriction as a factor in photosynthetic acclimation of cotton seedlings grown in elevated carbon dioxide. *Plant Physiol.* 96:627–34

117. Tissue DT, Thomas RB, Strain BR. 1993. Long-term effects of elevated CO_2 and nutrients on photosynthesis and rubisco in loblolly pine seedlings. *Plant, Cell and Environ.* 16:859–65

118. Tschaplinski TJ, Norby RJ, Wullschleger SD. 1993. Responses of loblolly pine seedlings to elevated CO_2 and fluctuating water supply. *Tree Physiol.* 13:283–96

119. Van Cleve K, Barney R, Schlentner R. 1981. Evidence of temperature control of production and nutrient cycling in two interior Alaska black spruce ecosystems. *Can. J. For. Res.* 11:258–73

120. Van Cleve K, Oechel WC, Hom JL. 1990. Response of black spruce (*Picea mariana*) ecosystems to soil temperature modification in interior Alaska. *Can J. For. Res.* 20:1530–35

121. van Veen JA, Liljeroth E, Lekkerkerk LJA, van de Geijn SC. 1991. Carbon fluxes in plant-soil systems at elevated atmospheric CO_2 levels. *Ecol. Appl.* 1: 175–81

121a. VEMAP Participants. 1995. Vegetation/Ecosystem Modeling and Analysis Project (VEMAP): comparing biogeography and biogeochemistry models in a continental-scale study of terrestrial ecosystem responses to climate change and CO_2 doubling. *Global Biogeochem. Cycles.* In press

122. Vitousek PM, Howarth RW. 1991. Nitrogen limitation on land and in the sea: How can it occur? *Biogeochemistry* 13: 87–115

123. Williams RS, Lincoln DE, Thomas RB. 1994. Loblolly pine grown under elevated CO_2 affects early instar pine sawfly performance. *Oecologia* 98:64–71

124. Williams WE, Garbutt K, Bazzaz FA, Vitousek PM. 1986. The response of plants to elevated CO_2 IV. Two deciduous-forest tree communities. *Oecologia* 69:454–59.

125. Wong SC. 1979. Elevated atmospheric partial pressure of CO$_2$ and plant growth. I. Interactions of nitrogen nutrition and photosynthetic capacity in C$_3$ and C$_4$ plants. *Oecologia* 44:68–74

126. Wong SC, Kriedemann PE, Farquhar GD. 1992. CO$_2$ × nitrogen interaction on seedling growth of four species of eucalypt. *Austral. J. Bot.* 40:457–72

127. Woodrow IE. 1994. Optimal acclimation of the C$_3$ photosynthetic system under enhanced CO$_2$. *Photosynth. Res.* 39:401–12.

128. Working Group I. 1990. Policymakers Summary. In *Climate Change: The IPCC Scientific Assessment,* ed. JT Houghton et al, pp. vii-xxxiv. Cambridge: Cambridge Univ. Press

129. Wulff RD, Strain BR. 1982. Effects of CO$_2$ enrichment on growth and photosynthesis in *Desmodium paniculatum. Can. J. Bot.* 60:1084–91

130. Wullschleger SD, Norby RJ. 1992. Respiratory cost of leaf growth and maintenance in white oak saplings exposed to atmospheric CO$_2$ enrichment. *Can. J. For. Res.* 22:1717–21

131. Wullschleger SD, Norby RJ, Gunderson CA. 1992. Growth and maintenance respiration in leaves of *Liriodendron tulipifera* L. exposed to long-term carbon dioxide enrichment in the field. *New Phytol.* 121:515–23.

132. Wullschleger SD, Norby RJ, Hanson PJ. 1995. Growth and maintenance respiration of *Quercus alba* after four years of CO$_2$ enrichment. *Physiol. Plant.* 93:47–54

133. Wullschleger SD, Norby RJ, Hendrix DL. 1992. Carbon exchange rates, chlorophyll content, and carbohydrate status of two forest tree species exposed to carbon dioxide enrichment. *Tree Physiol.* 10:21–31

134. Wullschleger SD, Ziska LH, Bunce JA. 1994. Respiratory responses of higher plants to atmospheric CO$_2$ enrichment. *Physiol. Plant.* 90:221–29

135. Zak DR, Pregitzer KS, Curtis PS, Teeri JA, Fogel R, Randlett DL. 1993. Elevated atmospheric CO$_2$ and feedback between carbon and nitrogen cycles. *Plant & Soil* 151:105–17

Annu. Rev. Ecol. Syst. 1995. 26:505–29

FOOD WEB ARCHITECTURE AND POPULATION DYNAMICS: Theory and Empirical Evidence

Peter J. Morin

Department of Biological Sciences, Rutgers University, New Brunswick, New Jersey 08855-1059

Sharon P. Lawler[1]

Center for Ecology, Evolution and Behavior, Biological Sciences, University of Kentucky, Lexington, Kentucky 40506-0225

KEY WORDS: complexity, food web, food chain length, omnivory, population dynamics, productivity, stability

ABSTRACT

Food web theory makes quantitative and qualitative predictions about the patterns of population dynamics to be expected in food webs with particular structures. Some of these predictions can be tested by comparing population dynamics in simple food chains of different architecture. Few studies have been designed specifically to manipulate food chain properties as a test of food web theory, but relevant information can be gleaned from studies of predator-prey dynamics in which food chain structure is known to vary in important ways. For example, comparisons of prey population dynamics in the presence or absence of a predator can be used to infer the consequences of a small change in food chain length. Common consequences of increased food chain length include greater temporal variation in abundance and a greater frequency of local extinctions. Studies that compare the impact of omnivore and nonomnivore predators are so infrequent that few conclusions can be reached. Ex-

[1]Present Address: Department of Entomology, University of California, Davis, California 95616

505

perimental studies of links between productivity and food chain length or population dynamics are also scarce. However, an emerging theme is that both increased and decreased productivity sometimes result in shorter food chains, probably for different mechanistic reasons. Studies of trophic cascades indicate that differences in the length of linear food chains have important consequences for the standing stock of species in different trophic levels, regardless of any effects on dynamics. Finally, a few studies of relations between food web complexity and the dynamics suggest that more complex systems can be less stable than simple systems, although these effects probably depend on how complexity is distributed among trophic levels.

INTRODUCTION

Food Web Theory

Food web theory remains controversial, largely because it is difficult to test its chief predictions about population dynamics in natural systems (65). Although Elton (29) sensitized ecologists to the importance of basic food web concepts during the formative years of modern ecology, the quantitative study of food webs has only recently come into its own. The recent growth of interest in food webs has been stimulated largely by theoretical work on factors that might constrain patterns of trophic connections within webs (16, 83, 117, 172). Many generalizations about the structure and function of food webs are based on the dynamic properties of relatively simple models of food webs or food chains (65, 117, 123, 172). Other generalizations about the architecture of food webs arise from the comparative study of patterns in natural food webs (16, 80, 81, 161). Indeed, Lawton & Warren (67) noted that food web theory can be divided into studies providing a population dynamic explanation for the static patterns of food webs (8, 23, 36, 47, 49, 82, 83, 102, 113–115, 117, 120, 121, 124, 131, 137) and other approaches that provide descriptive statistical or graphical generalizations about web structure, without explicit reference to the effects of web structure on dynamics (16–19, 64, 80, 81, 101, 116, 122, 147, 165–172). We are primarily interested in examining the effects of food web architecture on population dynamics, because theory suggests some clearly testable hypotheses about the population dynamics of species embedded in webs of different structure.

We focus here on a restricted set of issues selected from within the broad realm of food web theory. Those issues include: 1. factors that limit the length of food chains, including constraints imposed by population dynamics and energetics, 2. relations between the length and complexity of food chains and the existence of trophic cascades, and 3. relations between food web complexity and population dynamics. All of these issues have a rich if not entirely realistic theoretical underpinning. Often, more than one theoretical framework has been proposed to account for a particular pattern.

In contrast to the wealth of theoretical work addressing food web patterns, the amount of experimental work that can be mustered to test food web theory is surprisingly limited. This limitation reflects two factors. First, the data needed to address the predictions made by food web theory about population dynamics are exceedingly difficult to measure for long-lived organisms in natural food webs. Predictions about the relative stability of populations require data on dynamics that span at least several generations (20). When those data exist, corresponding data on food web structure are usually lacking, or vice versa. Second, many empirically oriented ecologists retain a healthy skepticism about the applicability of food web theory to natural communities (125, 126, 161). That skepticism reflects, in part, concerns about the realism of the basic models used to analyze the properties of food webs.

Renewed interest in the dynamics and architecture of food webs draws heavily on the contributions made by theoretical analyses and the comparative method, while experimental tests of food web theory have lagged far behind theoretical advances (65, 123). Our goal is to emphasize that some predictions of food web theory are empirically testable, although the tests are often more tractable in somewhat contrived or artificial systems than in nature. We present a selected overview of the kinds of experimental studies that can be used to test selected aspects of food web theory. Along the way, we emphasize the special features of experimental design needed to make inferences about food webs. Because so few experiments have been designed specifically to test food web theory, our conclusions must remain very tentative. Nonetheless, some tantalizing patterns have begun to emerge. Below we briefly review some predictions of food web theory that relate population dynamics to food chain length, omnivory, complexity, and other aspects of trophic architecture that are open to experimental tests.

Food Chain Length, Population Dynamics, and Energetics

Elton (29) observed that food chains are short. Energy often passes through only about four or five species in a chain before being passed along to decomposers. Particularly detailed descriptions of trophic interactions suggest that longer chains can be found (38, 44, 80, 125, 158). Several hypotheses have been proposed to account for the length of food chains.

Ideas about the factors that limit the length of food chains fall into two groups. One group emphasizes the constraints imposed by ecological energetics. The other group focuses on the possibility that certain food chain configurations, long chains in particular, are unlikely to persist because they are dynamically unstable, regardless of energetic constraints. Both groups of ideas present testable hypotheses. If the length of food chains is constrained by available energy, then experimental increases or decreases in productivity should cause commensurate changes in food chain length. If food chain length is constrained by population dynamics, species in longer food chains should exhibit the hallmarks of less stable dynamics when compared to the same species in shorter chains. The problem in putting these tests into action involves logistic constraints associated with the range of productivity that can be realistically manipulated in most systems, and the need to observe population dynamics over a time frame long enough to assess stability.

Population dynamic explanations for the limits to food chain length rely on the notion that long chains are locally unstable, or recover from perturbations more slowly than do shorter chains; they are therefore less likely to persist. Pimm & Lawton (120) used model food chains to argue that the complex population dynamics of trophically linked species could constrain food chain length. Their results suggest that over a short range of food chain length (2–4 levels) and a small constant number of species (4 total), populations in longer model food chains will recover from perturbations more slowly than those in shorter chains, and they are therefore less likely to persist. These conclusions have generated some debate (e.g. 137, 145), but there have been very few empirical tests of possible effects of food chain structure on population dynamics (e.g. 63).

Omnivores feed on more than one trophic level (120, 121). Lotka-Volterra models of food webs that include omnivory tend to be less stable than webs without omnivory (120). Early surveys of natural food webs suggested that omnivory was infrequent (114). However, population dynamic explanations for the relative lack of omnivory have been questioned (169), and subsequent surveys suggest that omnivory may be more common in nature than was previously thought (125, 144, 154, 158, 161).

Other models of linear food chains predict a variety of dynamic behaviors that depends on the details of the interactions. For instance, Hastings & Powell (48) showed that simple linear food chains consisting of three species can exhibit chaotic dynamics. The distinction between this finding and the local stability seen in Pimm & Lawton's analysis probably reflects differences in the ways that predators and prey are assumed to interact. A key difference is that Hastings & Powell's model incorporates nonlinear (saturating) functional responses and numerical responses, while these are linear functions in the models of Pimm & Lawton. Other similar models of three-level food chains are extensions of Monod models for population dynamics in continuous culture, and they also include nonlinear functional and numerical responses (21, 152). These models display a rich array of dynamics, ranging from stable limit cycles and stable point equilibria through unstable dynamics that result in the loss of species, depending on the values of parameters used.

Some energetic explanations for the limits to food chain length rely on the notion that energy transfer between trophic levels is inefficient, and therefore food chain length is assumed to be limited by the inefficient transfer of energy up through the chain. Lindeman (71) attributed the short length of food chains to the inefficiency of energy transfer between trophic levels. This hypothesis implies that food chains can be longer in more productive habitats (32). For similar reasons, Oksanen et al (103) predicted an increase in food chain length as productivity increased.

In contrast to the foregoing argument, many simple food chain models predict that stable predator-prey interactions will be destabilized if resource augmentation increases the prey's carrying capacity or intrinsic growth rate. This phenomenon has been termed "the paradox of enrichment" (133, reviews: 3, 24, 86). Abrams & Roth (3) point out that this prediction must somehow be reconciled with the ideas of Oksanen et al (103) that predict an increase in food chain length along a nutrient gradient. Abrams & Roth (2, 3) analyzed the responses of unstable two- and three-level model food chains to enrichment. They found that enrichment could either increase or decrease the abundance of the top predator, depending on the form of the functional responses, the level of immigration into the system, and the presence or absence of refuges. It seems likely that two opposing effects of increased energy availability interact to determine food chain length. Increased productivity creates a greater potential food chain length, but that potential is accompanied by reduced dynamic stability caused by energy input into a chain of a given length. These opposing factors may contribute to the apparent lack of correlation between productivity and food chain length noted by some authors (117).

Other hypotheses combine energetic and dynamic explanations for the limited length of food chains. Hutchinson (55) and Hastings & Conrad (49) suggested that predators feeding low in the food chain can attain large population sizes because there is more energy available at lower levels. In turn, predators with large population sizes may be less prone to extinction than species with smaller population sizes that feed higher in the web. This advantage could be offset by other factors, especially if species lower in the food chain experience a greater risk of predation than do those near the top of the chain (90, 117). Size constraints on predators may also limit food chain length. Predators are usually larger than their prey (29), and correlations between body size, home range, growth rate, and resource availability place an upper limit on feasible predator body sizes, which in turn may limit the number of trophic levels that can be assembled in a food chain (142). This hypothesis is reasonable but difficult to test. The relative sizes of predators and prey may also account for other aspects of food web architecture (67).

Relations Between the Length and Complexity of Food Chains and the Existence of Trophic Cascades

Trophic cascades occur when the addition or deletion of a higher trophic level affects the standing stock of species in lower levels. Paine (105) apparently coined the term *trophic cascade,* although the basic idea was well developed in the writings of Hairston et al (43) and Fretwell (32). Cascades reflect shifts in levels of abundance or biomass at equilibrium, rather than a change in dynamics per se. The theory of how cascades should work in simple, short, linear food chains is well developed (1, 32, 103). The problem is that natural chains are seldom linear, and even modest departures from linear food chains can prevent cascades (1). Despite this, there are many examples of trophic cascades, particularly in aquatic systems (e.g. 129).

Trophic cascade theory predicts that the biomass and population dynamics of species are determined in part by the number of trophic levels in the food chain to which that species belongs (e.g. 12, 43, 103, 134). The theory is an outgrowth of the classic paper by Hairston, Smith, & Slobodkin (43), which predicted that the particular trophic level that a species occupies will determine whether it is most likely to be regulated by competition or predation. They argued that in three-level terrestrial food chains, to the extent that carnivores decrease the abundance of herbivores, carnivores will alleviate competition among herbivores for their food, the primary producers. In turn, herbivores limited to low population densities by their predators will not be sufficiently abundant to graze producers down to levels where interplant competition is unimportant. Consequently, the abundance of trophic level two (herbivores) is controlled by predation, but the abundance of levels one (plants) and three (carnivores) should be resource limited.

Fretwell (32) and Oksanen et al (103) extended this idea, observing that predators that restrict prey biomass to low levels also reduce their own resource base. Such predators should themselves remain scarce and be unable to support a higher secondary predator. Under conditions where predators overexploit prey, additional trophic levels could be added only if increases in energy or nutrients boost per capita prey productivity (32, 103). Predator populations supplied with faster growing prey could produce sufficient biomass to support another trophic level. Upon addition of another trophic level, prey biomass in the second trophic level could increase again if additional resources became available, because predators in the third trophic level would be limited by consumers in the fourth trophic level (33, 42, 103). This scenario leads to a stepwise addition of trophic levels with increasing productivity, as might occur along a gradient of light or limiting nutrients. There is intriguing correlational evidence for a stepwise addition of trophic levels along a gradient running from oligotrophic to eutrophic lakes in Northern Europe (107). This process implies a limit to food chain length set by physiological limits to the productivity of organisms along a gradient of increasing resource availability. In cases where prey have reached the maximum productivity set by physiological constraints, additional resources will not lengthen the food chain (2, 42, 103).

Predator-prey dynamics are very important in most trophic cascade scenarios, and thus cascade theory may not be very distinct from Pimm & Lawton's (120) population dynamic explanation for why food chains are short (139). A predator that becomes scarce by overexploiting its prey would become more vulnerable to stochastic extinction (139). Examples of extreme overexploitation are uncommon in nature (146), but of course, if overexploiting predators tend to become rare, examples should be difficult to find! Hence, the nature of this process makes its prevalence impossible to ascertain by the simple inspection of natural food chains.

Food Web Complexity and Dynamics

Community ecology has a long tradition of conjectures about relations between complexity and stability (30, 40, 76, 83). Increasing complexity generally implies increasing species richness, connectance, or numbers of trophic levels in a community. Stability is well defined for systems of differential equations (10, 27, 83, 117), but the operational measures of stability applied to natural or experimental populations are less precise (20, 85, 117). Measures such as the standard deviation over time of the log of population density may only represent loose allegories for mathematical stability (138). However, these statistics are intrinsically interesting because of what they imply about the predictability of community composition. Measures of the temporal variability of population size have their own special problems related to the temporal and spatial scales of measurement (85). Despite these potential problems, they are often the only measures available for comparing the dynamics of populations.

Expectations about the stability of populations in food webs of differing complexity come from two rather different ecological traditions. The two traditions are not entirely at odds, since they focus on different things. The first tradition assumes a positive relation between stability and complexity. Some early theoretical work by MacArthur (76), and Elton's (30) comparisons of natural and artificial systems, reinforced the widely held belief that simple communities were in some way less stable than complex ones. Elton's observations (30) of organisms in natural and modified communities suggested that communities with few species tend to be unstable. In retrospect, Elton's anecdotal examples of unstable simple food webs could be explained by factors other than low complexity (40, 88, 117). MacArthur (76) suggested that increased prey species richness would enhance the stability of a predator population, because there would be more alternate pathways for energy to reach the predator population in the event that one or more of its prey populations crashed. The extent of trophic linkage between a given number of species was later formalized as the concept of connectance (34, 82), the number of realized links between species divided by the total number of possible pairwise links.

The second tradition assumes a negative association between stability and complexity. May (82) used mathematical models of food webs to argue that more complex communities should be less stable than simple communities. May's analysis did not address the situation considered by MacArthur (76), specifically whether feeding on different numbers of prey species would affect the stability of the predator population. May's conclusions about relations between complexity and the stability of the entire system depended critically on some biologically unrealistic assumptions used to generate randomly connected model webs (23, 64).

Connectance in some natural food webs apparently declines with increasing species richness, as May predicted (e.g. 8, 17). However, descriptions of natural food webs tend to omit connections, and lump species into categories at lower trophic levels, casting doubt on the robustness of some of these patterns (106, but see 131). More recent work with finely resolved food webs suggests that connectance may in fact be constant across webs of varying species richness (81).

Empirical studies of the connectance-stability relation have identified potentially interesting patterns in natural communities (8, 130), but the correlational nature of these studies makes it difficult to infer clear causal relations between stability and connectance. In an array of field studies, Pimm (117) found no consistent empirical relation between species diversity and food web stability. Few of the field studies reviewed by Pimm (117) were specifically designed to test the effects of species diversity on stability, and several studies were simply anecdotal reports of "simple" systems that seemed stable, or "diverse" systems that seemed variable. We are unaware of experiments that have demonstrated effects of connectance on population dynamics by varying connectance while holding species richness constant in webs of real organisms.

WAYS TO TEST FOOD WEB THEORY

Linking the Dynamics of Model Food Webs and Real Species

Tests of theory can be qualitative or quantitative. A qualitative test assesses the general agreement between broad dynamic patterns predicted by models and those observed in experimental systems. For instance, if a model predicts that a given shift in food chain structure will produce a shift from stable to oscillating population dynamics, and an analogous change occurs when the structure of a food chain is experimentally manipulated, then some qualitative agreement between theory and experimental evidence would seem to exist. If the predicted pattern fails to materialize, then the theory is found wanting. Of course, it is always possible that populations in model and real settings will display similar dynamics for entirely different and fortuitous reasons. For this reason, quantitative tests are needed to rule out spurious agreement between the dynamics of real food chains and their model representations. In our view, a quantitative test would involve first estimating the parameters of the model used to predict food chain dynamics. Then, the parameterized model can be used to predict the dynamics that might be observed in an experimental system. A decision about whether any model provides a reasonable fit to experimentally observed dynamics is a question ultimately settled by the statistical approach used to assess the goodness of fit between predicted and observed dynamics. For systems like food chains that are prone to chaotic behavior (48), which means that their temporal dynamics will be exquisitely sensitive to initial conditions, quantitative tests may not be a realistic goal. Clearly, quantitative tests will be much more difficult to accomplish than are qualitative tests.

Dynamic Implications of Long Return Times or Locally Unstable Equilibria

Qualitative tests of food web theory require some way of comparing the dynamics of species in different food webs with the general predictions of food web theory. One key attribute of the model food chains studied by Pimm & Lawton (120, 121) was the tendency for longer food chains to exhibit longer return times. Strictly speaking, all of the chains without omnivores were locally stable and would return to an equilibrium following a perturbation. The existence of longer return times means that when displaced from an equilibrium, populations in model chains with longer return times would display damped oscillations for a longer period of time before returning to stable equilibrium levels. One consequence of the longer return times of longer food chains should be the existence of prolonged oscillations following a perturbation. Such oscillations should translate into a statistical signature that should be measurable as an increase in temporal variation in abundance (see 150).

Models with an unstable local equilibrium can yield a diversity of dynamic behaviors, including stable limit cycles, chaos, or the extinction of one or more species (10, 27). Systems with an unstable local equilibrium point may persist indefinitely (60), but they probably exhibit higher levels of temporal variability than do systems with stable equilibria and rapid returns to those equilibria. A key point is that an unstable system need not generate the extinction of a species, but it should display greater temporal variation than would a locally stable system with a short return time. The other difference, of course, will be the shorter persistence time of species in systems where extinctions occur.

There has been much debate about whether local stability in model systems is analogous to the persistence of species in nature. Some authors have argued that only systems with stable equilibria are likely to persist in nature (117). Others have shown that systems without a locally stable equilibrium may nonetheless persist indefinitely (14, 60). Indeed, nonequilibrium metapopulation systems may persist for long periods of time, despite the presence of regular local extinctions (14). Rather than cavil over what measure of dynamic behavior is the best predictor of persistence in natural systems, we prefer to focus on whether the qualitative behaviors predicted by particular models are consistent with the observed dynamics of populations.

EMPIRICAL EVIDENCE

Food Chain Length and Dynamics

Evidence for the effects of food chain length on population dynamics mostly comes from three sources: 1. microbial food chains assembled under continuous culture or batch culture conditions, 2. arthropod predator-prey or parasitoid-host systems in laboratory microcosms, and 3. a small assortment of field systems where population dynamics can be compared among situations where food chains differ in some known way, usually because of the presence or absence of an important predator. Of these sources of information, laboratory studies of predator-prey dynamics in simple microbial food chains provide most of what we know about the consequences of increasing food chain length for population dynamics. For example, Gause (35) emphasized that his simple laboratory systems of infusoria feeding on bacteria corresponded to two-level food chains.

The problem in extracting usable information about the influence of food chain length on dynamics from many studies is that critical controls are either omitted or not reported. This appears to reflect an overriding early interest in the dynamics of predator-prey species pairs per se, rather than an interest in comparing the dynamics of prey with or without their predators. Consequently, there is much information about the dynamics of an assortment of readily culturable predator-prey pairs (46, 153), but critical information about the dynamics of the prey in the absence of the predator is much harder to come by. It is also necessary to follow dynamics for a sufficiently long period of time that reasonable measures of temporal dynamics (e.g. temporal variation in abundance, persistence times) can be compared. It is surprising how few studies meet these minimal requirements. Those that do typically vary in food chain length by only one link, which corresponds to the addition or deletion of a predator.

Experimental Manipulations of Food Chain Length

One common consequence of adding predators to prey populations in simple laboratory systems is a rapid decrease in prey abundance, followed by the extinction of both the predator and prey species, or just the predator (35, 41, 51, 74, 77, 91). Extinctions fall within the range of dynamics predicted by models with nonlinear interactions between species, but those models probably predict the observed extinctions for the wrong reasons. In most cases, extinctions in laboratory systems seem to be correlated with time lags that allow predators to overexploit their prey and drive them to extinction before starving out themselves (35, 73).

The failure of predators and prey to persist in spatially simple laboratory environments is frequently cited as evidence for the importance of spatial refuges or other metapopulation mechanisms in promoting predator-prey coexistence (35, 51). This argument probably overstates the case for the importance of spatial refuges, however, because there are numerous examples of predator-prey pairs that do manage to coexist for long periods of time in simple, spatially homogeneous, laboratory settings (153, 5, 22, 50, 57, 61–63, 74, 92, 135, 152, 155). These persisting systems provide support for the notion that relatively simple models may adequately describe the dynamics of a certain class of food chains.

Prolonged interactions among predators and prey in laboratory settings can produce a variety of dynamics. In microbial systems, one common pattern is an increase in the temporal variation in prey population size in longer food chains. This pattern holds for interactions between bacteria and phage (15, 50, 69, 70), bacteria and protists (21, 57, 152, 155), and protists and protists (6, 62, 63, 73, 92, 135).

In some cases prey dynamics may be relatively unaffected by predators. Luckinbill (75) found no obvious difference between dynamics of the prey species *Colpidium campylum* with and without the predator *Didinium nasutum*. Morin & Lawler (92) also noted a minority of cases where temporal variation in prey dynamics was unaffected by predators. None of these studies has examined the full range of food chain length (from two to four levels) explored theoretically by Pimm & Lawton (120, 121).

An assortment of studies of arthropod predator-prey interactions in laboratory settings describes either sustained oscillations or irregular fluctuations in population size over time (51, 91, 111, 112, 153). Huffaker's (51) classic study of interactions between herbivorous and predatory mites is unusual in that it does describe the dynamics of prey with and without predators. For persisting systems of differing spatial complexity, prey dynamics appeared more variable in shorter food chains without predatory mites. Mitchell et al (91) show that the population dynamics of *Drosophila* become increasingly irregular as parasitoids become more abundant.

Damped oscillations in the abundances of coevolved predators and prey, like those described by Pimentel (111) and others (15, 50, 70), can have multiple causes, including reduced attack rates by predators and increased resistance by hosts. Some studies of interactions between bacteria and viruses in chemostats offer indirect insights into links between dynamics and food chain lengths, which are inferred from well-known evolutionary changes in the bacteria and viruses that tend to decouple the predator-prey interaction. In most cases, bacterial populations initially display erratic temporal fluctuations in abundance as they interact with viruses that function as predators (15, 50, 70). After the bacteria and viruses coevolve for a time, resistant bacteria attain much higher densities and exhibit temporally stable dynamics, which can be interpreted as the dynamics typical of shorter food chains.

It is also worth pointing out that there have been very few efforts to actually estimate the parameters of simple model food chains that correspond to laboratory food chains with more than two trophic levels. Consequently, few strictly quantitative tests of food chain models exist. Some exceptions include the continuous culture studies of Tsuchiya et al (152), Curds (21), and Lenski & Levin (69), and the batch culture studies of Maly (77). Tsuchiya et al (152) concluded that a Monod model with terms for saturation kinetics in resource uptake provided a better fit to their data than did a classic Lotka-Volterra model without saturation kinetics. In practice, the Monod model would have some of the same features as the models elaborated by Pimm & Lawton (120), such as density dependence at the base of the food chain, and the Monod model would differ in the kind of nonlinear functional and numerical responses linking prey-predator dynamics.

Dynamics of predators and prey in natural nonexperimental settings also sometimes yield patterns of greater temporal variation where predators are present (45, 94). Some examples of biological control fall into this category. There are fewer examples of what happens after specialized predators of the herbivores are introduced, but predators appear to accentuate the population fluctuations of organisms as different as California red scale (96) and voles (45). Although a wealth of population dynamic data exists for different kinds of species (20, 162), it is usually impossible to determine the structure of the food chains in which such species are embedded. This makes it impossible to say much about links between observed population dynamics and food web theory in most natural systems.

Although there is a modest amount of data concerning the effects of typical top predators on prey dynamics, few analogous data exist for omnivorous top predators. This is unfortunate, because theory predicts that chains containing omnivores should be much less stable than comparable chains without omnivores (120, 121). Diehl (26) has pointed out the virtual absence of population dynamic information about the prey of omnivores, or the omnivores themselves. Lawler & Morin (63) and Morin & Lawler (92) have compared the dynamics of species in simple food chains with and without omnivorous top predators. They found no consistent effects of omnivores on prey dynamics for two omnivore species feeding on two different prey, a total of four different food chains, but they did find that omnivores often attained higher population densities than did nonomnivorous predators.

To our knowledge, there are no other comparative studies of population dynamics in natural settings for predators that differ with respect to omnivory. Diehl (26) has attempted to survey the relative net impacts of omnivores and nonomnivores in natural communities, but the data do not address dynamics per se. Other surveys of the effects of predators on prey populations, such as that of Sih et al (140), make clear that most experimental studies do not distinguish among types of predators (omnivore or not), and few studies contain the information on long-term dynamics needed to assess stability.

Productivity and Food Chain Length

A handful of researchers have tested whether enrichment destabilizes predator-prey population dynamics in aquatic microcosms and lake enclosures, with mixed results. Luckinbill (73, 74) destabilized protist predator-prey interactions by enriching the bacterial food supply of the prey species. Enrichment improved the nutrient content of the prey, which in turn delayed predator starvation and allowed predators to drive prey populations to extinction. Neill (100) found that adding nutrients to enclosures in an oligotrophic lake improved rotifer recruitment, which supported high densities of midge larvae (*Chaoborus*), which in turn overexploited their alternate crustacean prey. In a study conducted in estuarine enclosures, Björnsen et al (7) showed that correlated oscillations in bacteria and heterotrophic flagellates increased in amplitude when nutrients were added. Analogous responses to increased nutrient supply rates occur in some continuous culture systems (57). Balčiūnas & Lawler (6) enriched basal resources in a microbial food chain, which caused the extinction of the top predator, but not through the "paradox of enrichment" mechanism. Augmenting resources allowed most of the prey to achieve a size refuge from predation. McCauley & Murdoch (86) found no effect of enrichment on population cycles involving *Daphnia* feeding on algae; they suggested that the observed dynamics were driven by time lags in the response of consumers to prey abundance.

Unfortunately, comparable information about the effects of productivity of terrestrial systems is usually unavailable (24), primarily because the long generation times of most terrestrial producers and consumers make it difficult to say much about long-term population dynamics, other than those in rather unique circumstances (141).

Unlike the pattern predicted by the paradox of enrichment, the energy transfer hypothesis suggests that longer food chains should occur in more productive habitats. Pimm & Kitching (118) and Jenkins et al (56) have manipulated productivity in an effort to learn whether productivity affects the number of tropic levels supported in simple tree-hole communities. The results have been mixed. Additions of nutrients above ambient levels actually appeared to reduce the number of trophic levels in one study (118), a result loosely consistent with the paradox of enrichment. In a second study of the same system where nutrient inputs were reduced to two orders of magnitude below ambient levels, food chains again became shorter (56). As noted above, enrichment of laboratory systems tends to destabilize some interactions that are stable at lower levels of enrichment, resulting in either the extinction of species and a concomitant reduction in food chain length, or increasing temporal variation in abundance.

Although Pimm (117) found no evidence for a correlation between productivity and food chain length in his survey of systems, Persson et al (107) found that more productive lakes tended to support an additional trophic level, up to a point. Lakes with the highest productivity also had slightly shorter food chains. The uncertainty in interpreting such comparative studies comes from the inability to disentangle observed differences in productivity from other factors that might affect food chain length.

Trophic Cascades, Food Chain Length, and Population Dynamics

In evaluating the potential for predation to cause cascading effects, researchers have noted that prey are often resistant to predators, so that predators may not reduce prey biomass enough for a strong cascade to occur (e.g. 66, 68, 128, 156). We use "prey resistance" as a general term to encompass prey inedibility, physical defenses, behavioral defenses, and the use of spatial or temporal refuges. Prey resistance could also affect food chain length, but whether resistance lengthens or shortens food chains depends on the details of how predators and prey interact (2). If most prey are resistant in some way, the amount of biomass available to support predators will be low. Strong resistance consequently decreases the predator's resource base and potentially shortens the food chain (e.g, 6). However, a partial refuge that prevents the predator from overexploiting the prey may ensure sufficient stability of prey production to allow both prey and predators to attain a higher biomass, creating the potential for a longer food chain. Refuges that protect a large portion of the population may result in donor-controlled population dynamics, so called because the refuge prevents predators from strongly affecting prey dynamics, while prey dynamics still influence the predator (23). Such systems are often quite stable.

Elliott et al (28) demonstrated how the presence of a refuge can allow a potentially unstable trophic cascade to persist by preventing the overexploitation of prey. They constructed a tri-trophic food chain of algae, *Daphnia*, and fish in replicated experimental microcosms, plus the shorter food chains of algae alone, and algae plus *Daphnia*. When fish were present, *Daphnia* populations crashed. However if the fish were confined to a cage, they were prevented from overexploiting the *Daphnia*, and all three trophic levels coexisted in the microcosms. Similarly, Takahashi (149) was able to stabilize a host-parasitoid interaction by providing a spatial refuge for the host.

In other systems, cascades are truncated before they reach the lowest trophic levels (13, 104). Pace & Funke (104) found that negative effects of *Daphnia* on protists failed to enhance the abundance of bacteria in lake enclosure experiments, despite the fact that protists are important consumers of bacteria.

Reduced predator efficiency can also counteract the tendency of predators to overexploit prey (59, 97). Spatial heterogeneity can reduce predator efficiency even in the absence of an explicit spatial refuge (that is, space that is available to prey but not to predators, 51, 77, 98, 148, 157). Space per se can also dilute prey so that predator-prey encounters become infrequent, thereby allowing prey to persist (74). Interference among predators can prevent excessive attack rates, and aggregation of predators in dense patches of prey can stabilize predator-prey dynamics by reducing attacks in areas where prey are sparse (31).

The trophic cascade hypothesis has been quite controversial (e.g. 1, 11, 25, 42, 52, 126, 136, 146). Debate has centered on one main issue: Does the cascade hypothesis apply to entire, complex food webs? The answer is probably "rarely," for three reasons. First, omnivory can blur trophic levels and weaken cascading effects, and omnivores are prevalent in many systems (93, 108, 125, 126, 161). Second, strong full-web cascades are unlikely unless the prey within a trophic level are homogeneously edible and available to predators (1, 47, 68, 89, 126). Finally, interference among predators (4, 37) and abiotic forces may weaken cascades by preventing predators from reaching abundances such that they limit prey. All these factors can reduce the strength and extent of cascades in food webs. Top-down effects are evident in many communities, but they vary greatly in magnitude and in the number of food web members affected. This has led some to question the strength of an effect needed before a top-down effect can be legitimately termed a cascade, since the original theory predicted overwhelming top-down effects (25, 47). Nevertheless, few can dispute that trophic cascades can affect the population dynamics of species within single food chains embedded in larger food webs. Cascades are one of the most common and widely recognized indirect effects in nature, and they should retain a central place in ecological theory (reviews: 33, 42, 127, 163; see also 9, 39, 72, 78, 79, 84, 132, 143). Trophic cascades influence population dynamics, and population dynamics potentially determine the number of trophic levels in food chains.

Recent work also suggests that effects of enrichment on the abundance of species in different trophic levels will depend on the trophic level considered and on food chain length, as suggested by Fretwell (32) and Oksanen et al (103). Wootton & Power (164) have shown that increased algal productivity in stream systems has no effect on the abundance of grazers, but significantly increases the abundance of small predators that feed on grazers. Where food chains are one level longer, and larger predators depress the abundance of small predators, grazers become more abundant, while algal abundance is depressed.

Complexity and Dynamics

Rigorous experimental tests of effects of food web complexity on stability should manipulate species richness and/or connectance independently of other potentially correlated factors. Unfortunately, many field studies confound differences in species richness with differences in successional age, so that the factors contributing to differences among communities remain somewhat ambiguous (e.g. 53, 87, 95, 110). Important exceptions include the microcosm studies of Hairston et al (41), Tsuchiya et al (152), Luckinbill (75), and Lawler (61, 62) that directly created increases in species richness while measuring the stability of relatively simple laboratory communities of bacteria and protists. Hairston et al (41) found that increased bacterial diversity enhanced the stability of bactivorous *Paramecium* spp., but communities containing three species of *Paramecium* appeared less stable than those containing fewer species. Addition of the predators *Didinium* and *Woodruffia* also destabilized the communities, although these predators can sometimes coexist with prey for long periods under other conditions (73, 74, 135). Hairston et al could not resolve, from this study, whether diversity and stability were related, and they called for more experiments. With the exception of a handful of studies, the call went largely unheeded. Tsuchiya et al (152) found that the inclusion of a third trophic level stabilized an unstable competitive interaction between two species of bacteria. Luckinbill (75) and Lawler (62) both found that the addition of an alternate prey destabilized a previously stable interaction between different protist predator-prey pairs. In a different study, Lawler (61) assembled food webs of differing complexity (2, 4, and 8 protist species) from pairs of predators and prey that were known to be stable. Frequencies of extinctions were significantly higher in the more complex communities, suggesting that more complex communities were indeed less stable. Resurgent interest in the structure and dynamics of food webs has prompted renewed calls for experimental studies of stability and complexity (67, 88, 119). The recent experiments by Naeem et al (99) and Tilman & Downing (151) have examined various aspects of the performance of ecosystems with different degrees of species richness. Because most of the species in these terrestrial systems are relatively long-lived, these experiments have not directly addressed the issue of how complexity affects population dynamics. They have suggested that the ability of communities to absorb carbon dioxide (99) or retain biomass in the face of environmental perturbations (151) is positively related to initial species richness and food web complexity.

Some field studies of relations between complexity and stability have measured how communities of relatively long-lived organisms respond to various perturbations (e.g. 53, 54, 87, 110). Other studies estimated temporal variation in the attributes of communities of different complexity, to measure dynamic correlates of complexity (e.g. 95, 109, 159, 160). These studies have yielded interesting results, but they have their limitations. Logistic constraints common to most field studies limit replication, and experimental time scales are typically too brief (relative to the generation time of key organisms) to permit assessment of stability (20).

Complex natural food webs are reticulate in structure, so that one population can affect another through many indirect pathways. We reviewed the potential consequences of one type of indirect effect, trophic cascades, above. Other indirect effects include consumptive competition, apparent competition, indirect mutualisms, and interaction modifications. The implications of these various indirect effects for community structure and population dynamics have been reviewed recently by Wootton (163), so we will not duplicate that effort here. However, it is worth pointing out that as food webs become more species rich and connected, more potential pathways of indirect effects arise. Indirect effects complicate the predictions of population dynamics. Nevertheless, ecologists have made progress in identifying the food web structures that produce indirect effects. In many cases, empirical studies of trophic cascades, competition, and apparent competition have verified theoretical predictions about how these processes influence population dynamics (163).

SYNTHESIS

Kinds of Experiments/Studies That Are Needed

This review should make clear the pressing need for more experimental studies of the effects of food web architecture and energetics on the dynamics of populations. Population dynamic data must be collected over many generations of the focal organisms in order to address basic questions about stability. This requirement will necessarily exclude most long-lived organisms from studies that attempt to link food web structure to population dynamics. Despite that limitation, meaningful data on the dynamics of real species in food webs of known structure can be collected and compared. With somewhat more effort, it should be possible to estimate the parameters of simple food web models that correspond to the kinds of food webs that can be assembled in ecological experiments. Such efforts are critically needed to assess whether the models used to explore the dynamics of food webs and food chains depart in important ways from the dynamics of systems assembled from real species.

The length of experimental food chains examined so far is extremely limited. There seems to be little problem in building short chains with two or three levels in laboratory microcosms. Experimental studies of chains with four or more levels in laboratory microcosms are nonexistent. It is unclear whether such chains will persist long enough, in general, for studies of their long-term dynamics to be feasible, or whether the intrinsic interest in the dynamics of such systems has gone unrecognized. Long chains clearly exist in nature. What we don't know is whether those longer natural chains can only persist when embedded in a more complex food web, and if that is the case, why it might be so. Much the same case can be made for studies of how complexity affects dynamics. The range of species richness explored in these studies is very small, generally a mere handful of species. Species richness in most laboratory microcosms falls several orders of magnitude below the complexity of even the more depauperate natural communities. If effects of complexity on dynamics are a nonlinear function of species richness, the few studies completed so far are likely to yield a very uncertain estimate of relations between complexity, dynamics, and ecosystem functions. Much important work remains to be done to test food web theory. That work is needed both to expand the information base about dynamics of real species in webs of known configuration and to test the very tentative generalizations that we outline below.

Emerging Generalizations

1. *Food Chain Length and Population Dynamics* The limited evidence available suggests that population dynamics become more variable, and in some sense less stable, as food chains increase in length. There are occasional examples of trends in the opposite direction, but these seem to be cases where time lags create variable dynamics in prey populations even in the absence of predators. This pattern suggests that the precise details of the models used to represent food chains may matter very little, as long as a qualitative agreement between food chain structure and dynamics is all that is desired.

2. *Omnivory and Dynamics* Evidence is insufficient to say whether food chains with omnivores are less stable than chains without omnivores. One intriguing pattern is that omnivores appear to attain consistently higher, and occasionally more stable, population sizes than do predators that feed on a single trophic level. Much more work needs to be done to determine whether any of these patterns are general.

3. *Productivity and Food Chains* The effects of various kinds of enrichment on population dynamics and the length of food chains are complex. In the majority of laboratory studies, enrichment is destabilizing, causing a partial or

complete collapse of the food chain and a net decrease in food chain length. Such responses support the basic mechanisms underlying the paradox of enrichment, although some important exceptions do occur. Reductions in nutrient inputs in some systems also cause a reduction in the number of trophic levels, which suggests that energy availability does play some role in the setting the length of food chains. This raises a new paradox, delineated by the observation that increases or decreases in productivity appear to have the same negative effect on the length of food chains. The tentative resolution of this paradox is that food chains collapse for different reasons. Reductions in energy input make it impossible for species on higher trophic levels to obtain enough energy to persist. Increases in energy input destabilize the dynamics of populations in food chains to the point where one or more trophic levels go extinct. It is unclear whether natural food chains contain species that have coevolved to interact stably at some characteristic level of productivity, or whether natural chains persist despite any departures from stable dynamics forced by energetic constraints.

4. *Trophic Cascades* Trophic cascades are a natural consequence of interactions in simple linear food chains. In some situations, cascades are truncated after one or two trophic levels, for a variety of reasons that require further exploration. More work needs to be done to test current ideas about how departures from simple linear food chain architectures will influence the appearance and propagation of cascades in real systems.

5. *Complexity and Dynamics* Very modest increases in the complexity of simple food webs lead to decreases in stability, as is shown by the extinctions of species in more complex communities that readily persist in less complex subsets of the same communities. In at least one case, the mechanism responsible for the decreased stability is a kind of indirect effect, apparent competition. Similar mechanisms seem plausible in related systems. Models suggest that increases in the stability of community- or ecosystem-level responses with increasing complexity are not at odds with the apparent decreased stability of population dynamics (58). It will be particularly interesting to explore whether the dynamics of populations and ecosystem processes are tightly linked or only loosely correlated in systems composed of real species.

ACKNOWLEDGMENTS

We are pleased to acknowledge the support of the National Science Foundation, Grants BSR 9006462 and DEB 9220665, for our experimental studies of food webs.

Literature Cited

1. Abrams PA. 1993. Effect of increased productivity on the abundances of trophic levels. *Am. Nat.* 141:351–71
2. Abrams PA, Roth JD. 1994. The effects of enrichment of three species food chains with nonlinear functional responses. *Ecology* 75:1118–30
3. Abrams PA, Roth JD. 1994. The responses of unstable food chains to enrichment. *Evol. Ecol.* 8:150–71
4. Arditi R, Saiah H. 1992. Empirical evidence of the role for heterogeneity in ratio-dependent consumption. *Ecology* 73:1544–51
5. Ashby RE. 1976. Long term variations in a protozoan chemostat culture. *J. Exp. Mar. Biol. Ecol.* 24:227–35
6. Balčiūnas D, Lawler SP. 1995. Effects of basal resources, predation, and alternate prey in microcosm food chains. *Ecology* 76: 1327–36
7. Bjørnsen PK, Riemann B, Horsted SJ, Neilson TG, Pock-Sten J. 1988. Trophic interactions between heterotrophic nanoflagellates and bacterioplankton in manipulated seawater enclosures. *Limnol. Oceanogr.* 33:409–20
8. Briand F. 1983. Environmental control of food web structure. *Ecology* 64:253–63
9. Brönmark C, Klosiewski SP, Stein RA. 1992. Indirect effects of predation in a freshwater benthic food chain. *Ecology* 73:1662–74
10. Bulmer MG. 1994. *Theoretical Evolutionary Ecology.* Sunderland: Sinauer. 352 pp.
11. Carpenter SR, Kitchell JF. 1992. Trophic cascade and biomanipulation: interface of research and management—a reply to the comment by DeMelo et al. *Limnol. Oceanogr.* 37:208–13
12. Carpenter SR, Kitchell JF, Hodgson JR. 1985. Cascading trophic interactions and lake productivity. *BioScience* 35: 634–39
13. Carpenter SR, Kitchell JF, Hodgson JR, Cochran PA, Elser JJ, et al. 1987. Regulation of lake primary productivity by food web structure. *Ecology* 68:1863–76
14. Caswell H. 1978. Predator-mediated coexistence: a nonequilibrium model. *Am. Nat.* 112:127–54
15. Chao L, Levin BR, Stewart FM. 1977. A complex community in a simple habitat: an experimental study with bacteria and phage. *Ecology* 58:369–78
16. Cohen JE. 1978. *Food Webs and Niche Space.* Princeton: Princeton Univ. Press. 189 pp.
17. Cohen JE, Briand F. 1984. Trophic links of community food webs. *Proc. Natl. Acad. Sci. USA* 81:4105–9
18. Cohen JE, Newman CM. 1985. A stochastic theory of community food webs. I. Models and aggregated data. *Proc. R. Soc. London Ser. B* 224:421–48
19. Cohen JE, Newman CM, Briand F. 1985. A stochastic theory of community food webs. II. Individual webs. *Proc. R. Soc. London Ser. B* 224:449–61
20. Connell JH, Sousa WP. 1983. On the evidence needed to judge ecological stability or persistence. *Am. Nat.* 121:789–824
21. Curds CR. 1971. A computer-simulation study of predator-prey relationships in a single-stage continuous-culture system. *Water Res.* 5:793–812
22. Curds CR, Cockburn A. 1971. Continuous monoxenic culture of *Tetrahymena pyriformis*. *J. Gen. Microbiol.* 66:95–108
23. DeAngelis DL. 1975. Stability and connectance in food web models. *Ecology* 56:238–43
24. DeAngelis DL, Mulholland PJ, Palumbo AV, Steinman AD, Huston MA, Elwood JW. 1989. Nutrient dynamics and food-web stability. *Annu. Rev. Ecol. Syst.* 20:71–95
25. DeMelo R, France R, McQueen DJ. 1992. Biomanipulation: Hit or myth? *Limnol. Oceanogr.* 37:192–207
26. Diehl S. 1993. Relative consumer sizes and the strengths of direct and indirect interactions in omnivorous feeding relationships. *Oikos* 68:151–57
27. Edelstein-Keshet L. 1988. *Mathematical Models in Biology.* New York: Random House. 586 pp.
28. Elliott ET, Castañares G, Perlmutter D, Porter KG. 1983. Trophic-level control of production and nutrient dynamics in an experimental planktonic community. *Oikos* 41:7–16
29. Elton C. 1927. *Animal Ecology.* London: Methuen & Co.
30. Elton C. 1958. *The Ecology of Invasions by Animals and Plants.* London: Chapman & Hall
31. Free CA, Beddington JR, Lawton JH. 1977. On the inadequacy of simple models of mutual interference for parasitism and predation. *J. Anim. Ecol.* 46:543–54
32. Fretwell S. 1977. The regulation of plant communities by the food chains exploiting them. *Perspect. Biol. Med.* 20:169–85
33. Fretwell SD. 1987. Food chain dynam-

ics: the central theory of ecology? *Oikos* 50:291–301

34. Gardner MR, Ashby WR. 1970. Connectance of large dynamical (cybernetic) systems: critical values for stability. *Nature* 228:784

35. Gause GF. 1934. *The Struggle for Existence.* Reprinted 1971. New York: Dover

36. Gilpin ME. 1975. Stability of feasible predator-prey systems. *Nature* 254:137–39

37. Ginzburg L, Akcakaya HR. 1992. Consequences of ratio-dependent predation for steady-state properties of ecosystems. *Ecology* 73:1536–43

38. Goldwasser L, Roughgarden JR. 1993. Construction and analysis of a large Caribbean food web. *Ecology* 74:1216–33

39. Gómez JM, Zamora R. 1994. Top-down effects in a tritrophic system: parasitoids enhance plant fitness. *Ecology* 75:1023–30

40. Goodman D. 1975. The theory of diversity-stability relationships in ecology. *Q. Rev. Biol.* 50:237–66

41. Hairston NG, Allan JD, Colwell RK, Futuyma DJ, Howell J, et al. 1968. The relationship between species diversity and stability: an experimental approach with protozoa and bacteria. *Ecology* 49:1091–1101

42. Hairston NG Jr, Hairston NG Sr. 1993. Cause-effect relationships in energy flow, trophic structure, and interspecific interactions. *Am. Nat.* 142:379–411

43. Hairston NG, Smith FE, Slobodkin LB. 1960. Community structure, population control, and competition. *Am. Nat.* 94:421–25

44. Hall SJ, Raffaelli DG. 1991. Food web patterns: lessons from a species-rich web. *J. Anim. Ecol.* 60:823–41

45. Hanski I, Turchin P, Korpimaki E, Henttonen H. 1993. Population oscillations of boreal rodents: regulation by mustelid predators leads to chaos. *Nature* 364:232–35

46. Hassell MP. 1978. *The Dynamics of Arthropod Predator-Prey Systems.* Princeton: Princeton Univ. Press

47. Hastings A. 1988. Food web theory and stability. *Ecology* 69:1665–68

48. Hastings A, Powell T. 1991. Chaos in a three-species food chain. *Ecology* 72:896–903

49. Hastings HM, Conrad M. 1979. Length and evolutionary stability of food chains. *Nature* 282:838–39

50. Horne MT. 1970. Coevolution of *Escherichia coli* and bacteriophages in chemostat culture. *Science* 168:992–93

51. Huffaker CB. 1958. Experimental studies on predation: dispersion factors and predator-prey oscillations. *Hilgardia* 27:343–83

52. Hunter MD, Price PW. 1992. Playing chutes and ladders: heterogeneity and the relative roles of bottom-up and top-down forces in natural communities. *Ecology* 73:724–32

53. Hurd LE, Mellinger MV, Wolf LL, McNaughton SJ. 1971. Stability and diversity at three trophic levels in terrestrial successional ecosystems. *Science* 173:1134–36

54. Hurd LE, Wolf LL. 1974. Stability in relation to nutrient enrichment in arthropod consumers of old field successional ecosystems. *Ecol. Monogr.* 44:465–82

55. Hutchinson GE. 1959. Homage to Santa Rosalia; or, why are there so many kinds of animals? *Am. Nat.* 93:145–59

56. Jenkins B, Kitching RL, Pimm SL. 1992. Productivity, disturbance and food web structure at a local spatial scale in experimental container habitats. *Oikos* 65:249–55

57. Jost JL, Drake JF, Fredrickson AG, Tsuchiya HM. 1973. Interactions of *Tetrahymena pyriformis, Escherichia coli, Azotobacter vinelandii,* and glucose in a minimal medium. *J. Bacteriol.* 113:834–40

58. King AW, Pimm SL. 1983. Complexity, diversity, and stability: a reconciliation of theoretical and empirical results. *Am. Nat.* 122:229–39

59. Kuno E. 1987. Principles of predator-prey interaction in theoretical, experimental, and natural population systems. *Adv. Ecol. Res.* 16:249–337

60. Law R, Blackford JC. 1992. Self assembling food webs: a global viewpoint of coexistence of species in Lotka-Volterra communities. *Ecology* 73:567–78

61. Lawler SP. 1993. Species richness, species composition, and population dynamics of protists in experimental microcosms. *J. Anim. Ecol.* 62:711–19

62. Lawler SP. 1993. Direct and indirect effects in microcosm communities of protists. *Oecologia* 93:184–90

63. Lawler SP, Morin PJ. 1993. Food web architecture and population dynamics in laboratory microcosms of protists. *Am. Nat.* 141:675–86

64. Lawlor LE. 1978. A comment on randomly constructed model ecosystems. *Am. Nat.* 112:445–47

65. Lawton JH. 1989. Food webs. In *Ecological Concepts,* ed. JM Cherrett, pp. 43–78. Oxford: Blackwell Sci.

66. Lawton JH, McNeill S. 1979. Between

the devil and the deep blue sea: on the problem of being a herbivore. In *Population Dynamics,* ed. RM Anderson, BD Turner, LR Taylor, pp. 223–44. London: Blackwell

67. Lawton JH, Warren PH. 1988. Static and dynamic explanations for patterns in food webs. *TREE* 3:242–45

68. Leibold M. 1989. Resource edibility and the effects of predators and productivity on the outcome of trophic interactions. *Am. Nat.* 134:922–49

69. Lenski RE, Levin BR. 1985. Constraints on the coevolution of bacteria and virulent phage: a model, some experiments, and predictions for natural communities. *Am. Nat.* 125:585–602

70. Levin BR, Stewart FM, Chao L. 1977. Resource-limited growth, competition, and predation: a model and experimental studies with bacteria and bacteriophage. *Am. Nat.* 111:3–24

71. Lindeman RL. 1942. The trophic-dynamic aspect of ecology. *Ecology* 23:399–413

72. Lodge DM, Kershner MW, Aloi JE. 1994. Effects of an omnivorous crayfish (*Orconectes rusticus*) on a freshwater littoral food web. *Ecology* 75:1265–81

73. Luckinbill LS. 1973. Coexistence in laboratory populations of *Paramecium aurelia* and its predator *Didinium nasutum. Ecology* 54:1320–27

74. Luckinbill LS. 1974. The effects of space and enrichment on a predator-prey system. *Ecology* 55:1142–47

75. Luckinbill LS. 1979. Regulation, stability, and diversity in a model experimental microcosm. *Ecology* 60:1098–102

76. MacArthur R. 1955. Fluctuations of animal populations, and a measure of community stability. *Ecology* 36:533–36

77. Maly EJ. 1978. Stability of the interaction between *Didinium* and *Paramecium:* effects of dispersal and predator time lag. *Ecology* 59:733–41

78. Marquis RJ, Whelan CJ. 1994. Insectivorous birds increase growth of white oak through consumption of leaf-chewing insects. *Ecology* 75:2007–14

79. Martin TH, Crowder LB, Dumas CF, Burkholder JM. 1992. Indirect effects of fish on macrophytes in Bays Mountain Lake: evidence for a littoral trophic cascade. *Oecologia* 89:476–81

80. Martinez ND. 1991. Artifacts or attributes? Effects of resolution on the Little Rock Lake food web. *Ecol. Monogr.* 61:367–92

81. Martinez ND. 1992. Constant connectance in community food webs. *Am. Nat.* 139:1208–18

82. May RM. 1972. Will a large complex system be stable? *Nature* 238:412–13

83. May RM. 1973. *Stability and Complexity in Model Ecosystems.* Princeton: Princeton Univ. Press

84. Mazumder A. 1994. Patterns of algal biomass in dominant odd- vs. even-link lake ecosystems. *Ecology* 75:1141–49

85. McArdle BH, Gaston KJ, Lawton JH. 1990. Variation in the size of animal populations: patterns, problems, and artefacts. *J. Anim. Ecol.* 59:439–54

86. McCauley E, Murdoch WW. 1990. Predator-prey dynamics in environments rich and poor in nutrients. *Nature* 343:455–57

87. McNaughton SJ. 1977. Diversity and stability of ecological communities: a comment on the role of empiricism in ecology. *Am. Nat.* 111:515–25

88. McNaughton SJ. 1988. Diversity and stability. *Nature* 333:204–5

89. McQueen DJ, Johannes MRS, Post JR, Stewart TJ, Lean DRS. 1989. Bottom-up and top down impacts on freshwater pelagic community structure. *Ecol. Monogr.* 59:289–309

90. Menge BA, Sutherland JP. 1976. Species diversity gradients: synthesis in the roles of predation, competition, and temporal heterogeneity. *Am. Nat.* 110:351–69

91. Mitchell P, Arthur W, Farrow M. 1992. An investigation of population limitation using factorial experiments. *J. Anim. Ecol.* 61:591–98

92. Morin PJ, Lawler SP. 1995. Effects of food chain length and omnivory on population dynamics in experimental microcosms. In *Food Webs: Integration of Patterns and Dynamics,* ed. G Polis, K Winemiller. New York: Chapman & Hall

93. Murdoch WW. 1966. Community structure, population control and competition—a critique. *Am. Nat.* 100:219–26

94. Murdoch WW. 1994. Population regulation in theory and practice. *Ecology* 75:271–87

95. Murdoch WW, Evans FC, Peterson CH. 1972. Diversity and pattern in plants and insects. *Ecology* 53:819–29

96. Murdoch WW, Luck RF, Swarbrick SL, Walde S, Yu DS, Reeve JD. 1995. Regulation of an insect population under biological control. *Ecology* 76:206–17

97. Murdoch WW, Oaten A. 1975. Predation and population stability. *Adv. Ecol. Res.* 9:1–131

98. Nachman G. 1991. An acarine predator-prey metapopulation system inhabiting greenhouse cucumbers. *Biol. J. Linnean Soc.* 42:285–303

99. Naeem S, Thompson LJ, Lawler SP, Lawton JH, Woodfin RM. 1994. Declining biodiversity can alter the performance of ecosystems. *Nature* 368: 734–37

100. Neill WE. 1988. Complex interactions in oligotrophic lake food webs: responses to nutrient enrichment. In *Complex Interactions in Lake Communities*, ed. SR Carpenter, pp. 31–44. Berlin: Springer-Verlag

101. Newman CM, Cohen JE. 1986. A stochastic theory of community food webs. IV. Theory of food chain lengths in large webs. *Proc. R. Soc. London Ser. B* 228:355–77

102. Nunney L. 1980. The stability of complex model ecosystems. *Am. Nat.* 115: 639–49

103. Oksanen L, Fretwell SD, Arruda J, Niemelä P. 1981. Exploitation ecosystems in gradients of primary productivity. *Am. Nat.* 118:240–61

104. Pace ML, Funke E. 1991. Regulation of planktonic microbial communities by nutrients and herbivores. *Ecology* 72: 904–14

105. Paine RT. 1980. Food webs: linkage interaction strength and community infrastructure. *J. Anim. Ecol.* 49:667–85

106. Paine RT. 1988. Food webs: road maps of interactions or grist for theoretical development? *Ecology* 69:1648–54

107. Persson L, Diehl S, Johansson L, Andersson G, Hamrin SF. 1992. Trophic interactions in temperate lake ecosystems: a test of food chain theory. *Am. Nat.* 140:59–84

108. Peters RH. 1977. Unpredictable problems of tropho-dynamics. *Environ. Biol. Fishes* 2:97–101

109. Peterson CH. 1975. Stability of species and of community for the benthos of two lagoons. *Ecology* 56:958–65

110. Pimentel D. 1961. Species diversity and insect population outbreaks. *Ann. Entomol. Soc. Am.* 54:76–86

111. Pimentel D. 1968. Population regulation and genetic feedback. *Science* 159: 1432–37

112. Pimentel D, Nagel WP, Madden JL. 1963. Space-time structure of the environment and the survival of parasite-host systems. *Am. Nat.* 97:141–67

113. Pimm SL. 1979. The structure of food webs. *Theor. Popul. Biol.* 16:144–58

114. Pimm SL. 1980. Properties of food webs. *Ecology* 61:219–25

115. Pimm SL. 1980. Food web design and the effects of species deletion. *Oikos* 35:139–49

116. Pimm SL. 1980. Bounds on food web connectance. *Nature* 285:591

117. Pimm SL. 1982. *Food Webs*. London: Chapman & Hall

118. Pimm SL, Kitching RL. 1987. The determinants of food chain lengths. *Oikos* 50:302–7

119. Pimm SL, Kitching RL. 1988. Food web patterns: trivial flaws or the basis of an active research program? *Ecology* 69: 1669–72

120. Pimm SL, Lawton JH. 1977. Number of trophic levels in ecological communities. *Nature* 268:329–31

121. Pimm SL, Lawton JH. 1978. On feeding on more than one trophic level. *Nature* 275:542–44

122. Pimm SL, Lawton JH. 1980. Are food webs divided into compartments? *J. Anim. Ecol.* 49:879–98

123. Pimm SL, Lawton JH, Cohen JE. 1991. Food web patterns and their consequences. *Nature* 350:669–74

124. Pimm SL, Rice JC. 1987. The dynamics of multispecies, multi-life-stage models of aquatic food webs. *Theor. Popul. Biol.* 32:303–25

125. Polis GA. 1991. Complex trophic interactions in deserts: an empirical critique of food-web theory. *Am. Nat.* 138:123–55

126. Polis GA. 1994. Food webs, trophic cascades and community structure. *Aust. J. Ecol.* 19:121–36

127. Power ME. 1992. Top-down and bottom-up forces in food webs: Do plants have primacy? *Ecology* 73:733–46

128. Power ME. 1992. Habitat heterogeneity and the functional significance of fish in river food webs. *Ecology* 73:1675–88

129. Power ME, Matthews WJ, Stewart AJ. 1985. Grazing minnows, piscivorous bass, and stream algae: dynamics of a strong interaction. *Ecology* 66:1448–56

130. Redfearn A, Pimm SL. 1988. Population variability and polyphagy in herbivorous insect communities. *Ecol. Monogr.* 58: 39–55

131. Rejmánek M, Stary P. 1979. Connectance in real biotic communities and critical values for stability of model ecosystems. *Nature* 280:311–13

132. Rosemond AD, Mulholland PJ, Elwood JW. 1993. Top-down and bottom-up control of stream periphyton: effects of nutrients and herbivores. *Ecology* 74: 1264–80

133. Rosenzweig ML. 1971. The paradox of enrichment: destabilization of exploitation ecosystems in ecological time. *Science* 171:385–87

134. Rosenzweig ML. 1973. Exploitation in three trophic levels. *Am. Nat.* 107:275–94

135. Salt GW. 1967. Predation in an experimental protozoan population (*Woodrufia-Paramecium*). *Ecol. Monogr.* 37:113–44

136. Sarnelle O. 1992. Nutrient enrichment and grazer effects on phytoplankton in lakes. *Ecology* 73:551–60

137. Saunders PT. 1978. Population dynamics and the length of food chains. *Nature* 272:189–90

138. Schoener TW. 1985. Are lizard population sizes unusually constant through time? *Am. Nat.* 126:633–41

139. Schoener TW. 1989. Food webs from the small to the large. *Ecology* 70:1559–89

140. Sih A, Crowley P, McPeek M, Petranka J, Strohmeier K. 1985. Predation, competition, and prey communities: a review of field experiments. *Annu. Rev. Ecol. Syst.* 16:269–311

141. Silvertown J. 1987. Ecological stability: a test case. *Am. Nat.* 130:807–10

142. Slobodkin LB. 1961. *Growth and Regulation of Animal Populations.* New York: Holt, Rinehart, & Winston. 184 pp.

143. Spiller DA, Schoener TW. 1994. Effects of top and intermediate predators in a terrestrial food web. *Ecology* 75:182–96

144. Sprules WG, Bowerman JE. 1988. Omnivory and food chain length in zooplankton food webs. *Ecology* 69:418–26

145. Stenseth NC. 1985. The structure of food webs predicted from optimal food selection models: an alternative to Pimm's stability hypothesis. *Oikos* 42:361–64

146. Strong DR Jr. 1992. Are trophic cascades all wet? Differentiation and donor-control in speciose ecosystems. *Ecology* 73:747–54

147. Sugihara G, Schoenly K, Trombla A. 1989. Scale invariance in food web properties. *Science* 245:48–52

148. Takafuji A. 1977. The effect of the rate of successful dispersal of a phytoseiid mite, *Phytosciulus persimilis* Athias-Henriot (Acarina: Phytoseiidae) on the persistence in the interactive systems between the predators and its prey. *Res. Popul. Ecol.* 18:210–22

149. Takahashi F. 1959. An experimental study on the suppression and regulation of the host population by the action of the parasitic wasp. *Jpn. J. Ecol.* 19:225–32

150. Taylor AD. 1992. Deterministic stability analysis can predict the dynamics of some stochastic population models. *J. Anim. Ecol.* 61:241–48

151. Tilman D, Downing JA. 1994. Biodiversity and stability in grasslands. *Nature* 367:363–65

152. Tsuchiya HM, Drake JF, Jost JL, Fredrickson AG. 1972. Predator-prey interactions of *Dictyostelium discoideum* and *Escherichia coli* in continuous culture. *J. Bacteriol.* 110:1147–53

153. Utida S. 1957. Cyclic fluctuations of population density intrinsic to the host-parasite system. *Ecology* 38:442–49

154. Vadas RL. 1990. The importance of omnivory and predator regulation of prey in freshwater fish assemblages of North America. *Environ. Biol. Fishes* 27:285–302

155. van den Ende P. 1973. Predator-prey interactions in continuous culture. *Science* 181:562–64

156. Vanni MJ. 1987. Effects of nutrients and zooplankton size on the structure of a phytoplankton community. *Ecology* 68:624–35

157. Walde SJ. 1994. Immigration and the dynamics of a predator-prey interaction in biological control. *J. Anim. Ecol.* 63:337–46

158. Warren PH. 1989. Spatial and temporal variation in the structure of a freshwater food web. *Oikos* 55:299–311

159. Watt KEF. 1964. Comments on fluctuations of animal populations and measures of community stability. *Can. Entomol.* 96:1434–42

160. Watt KEF. 1965. Community stability and the strategy of biological control. *Can. Entomol.* 97:887–95

161. Winemiller KO. 1990. Spatial and temporal variation in tropical fish trophic networks. *Ecol. Monogr.* 60:331–67

162. Woiwod IP, Hanski I. 1992. Patterns of density dependence in moths and aphids. *J. Anim. Ecol.* 61:619–29

163. Wootton JT. 1994. The nature and consequences of indirect effects in ecological communities. *Annu. Rev. Ecol. Syst.* 25:443–66

164. Wootton JT, Power ME. 1993. Productivity, consumers, and the structure of a river food chain. *Proc. Natl. Acad. Sci. USA* 90:1384–87

165. Yodzis P. 1980. The connectance of real ecosystems. *Nature* 284:544–45

166. Yodzis P. 1981. The stability of real ecosystems. *Nature* 289:674–76

167. Yodzis P. 1981. The structure of assembled communities. *J. Theor. Biol.* 92:103–17

168. Yodzis P. 1982. The compartmentation

of real and assembled ecosystems. *Am. Nat.* 120:551–70

169. Yodzis P. 1984. The structure of assembled communities. II. *J. Theor. Biol.* 107:115–26

170. Yodzis P. 1984. How rare is omnivory? *Ecology* 65:321–23

171. Yodzis P. 1984. Energy flow and the vertical structure of real ecosystems. *Oecologia* 65:86–88

172. Yodzis P. 1989. *Introduction to Theoretical Ecology.* New York: Harper & Row

Annu. Rev. Ecol. Syst. 1995. 26:531–52

ARCHITECTURAL EFFECTS AND THE INTERPRETATION OF PATTERNS OF FRUIT AND SEED DEVELOPMENT

Pamela K. Diggle

Department of Environmental, Population and Organismic Biology, University of Colorado, Boulder, Colorado 80309–0334

KEY WORDS: architecture, positional effects, resource limitation, allocation, control of fruit set

ABSTRACT

The commonly observed proximal-to-distal decrease within inflorescences of fruit and/or seed maturation per flower has frequently been attributed to competition among developing fruits for resources. The research summarized in this review suggests, however, that the observed variation can also be due to architecture—that is, to sources of variation inherent in plant axes. Thus, the fate of a developing flower depends not only on the reproductive events that have preceded it during the ontogeny of the organism, but on where it occurs within the architecture of an individual. The effects of architecture are separable experimentally from the effects of differential resource allocation, and careful experimental analysis of these two factors will enhance our understanding of the physiological, developmental, and evolutionary controls of fruit and seed production in flowering plants.

INTRODUCTION

In many outcrossing hermaphroditic plants, more flowers are produced than mature and set viable seed (15, 20, 50, 52, 53). Since the seminal review by Stephenson (50) summarizing the proximate and ultimate causes of flower and fruit abortion, there has been a wealth of research, both empirical and theo-

531

retical, on limitations to plant reproduction. It is generally assumed that, where flower production exceeds successful fruit maturation, fruit and/or seed set[1] may be limited by: 1. the quantity and quality of pollen transferred (pollen limitation), 2. the amount of nutrients and photosynthate available for allocation to fruits and seeds (resource limitation), 3. herbivores, predators, and disease, and 4. agents of the physical environment such as freezing or drought (32).

The distribution of viable fruits and seeds within individual plants is typically not uniform (32). Most frequently, fruit maturation per flower (or seed number per fruit) has been shown to decrease with position from the base to the apex of individual inflorescences (reviewed in 32, 50, 56, 63; and see below). Because initiation and development of plant metamers is sequential, the production and distribution of flowers within an individual vary both spatially and temporally. Thus, the observed decline with position in fruit and/or seed set has frequently been ascribed to the abortion of distal fruits (or increased abortion of ovules within fruits) due to preemption of resources by older, basal fruits (resource competition hypothesis). It is difficult, however, to distinguish temporal from spatial effects with respect to the differential allocation of resources within an inflorescence; preferential access to photosynthate, water, and nutrients may be a consequence of temporal precedence and/or preferential location. Whatever the proximate explanation for the widely noted pattern of basal to distal decline in fruit and seed maturation, the generality of this observation has led to the hypothesis that plants have evolved a "strategy" of aborting offspring in which the least investment has been made (32, 40, 50).

What is absent from many of these discussions, however, is the recognition that quantitative variation of floral organ size within inflorescences is common and that this variation may occur independently of the fruiting status of an individual. For example, Weberling's definition of a raceme specifies an acropetal decrease in flower size as a distinguishing feature of this widespread type of inflorescence (58). Consideration of such positional variation in floral characters within inflorescences may provide additional insight into the causes of observed variation in fruit and/or seed maturation. The goal of this review is to summarize the data on positional variation among flowers and fruits within inflorescences and to examine whether these patterns are due to variation inherent in inflorescence architecture as well as to resource allocation effects.

---

[1]The term "fruit set" has been defined specifically to refer to the initiation of fruit development (50). However, I adopt the broader definition of fruit set from the ecological and evolutionary literature to mean successful fruit maturation.

PATTERNS OF FRUIT AND SEED MATURATION WITHIN INFLORESCENCES: EVIDENCE FOR COMPETITION AMONG FRUITS

A proximal-to-distal decrease within inflorescences of fruit and/or seed maturation per flower has frequently been attributed to competition among developing fruits for resources (reviewed in 32, 50). Distal flowers typically reach anthesis after proximal flowers and thus initiate fruit development at a later time. If sink strength of fruits increases with size and age, distal fruits will have lower competitive ability (with respect to resources) compared to early fruits within an inflorescence. In addition, distal flowers and fruits, located farther from the supply of resources, may therefore have less access to resources. This section reviews selected studies of temporal and/or positional variation in fruit set per flower and/or seed set per fruit within inflorescences, and it compares the observed patterns to the predictions of the resource competition hypothesis.

Observational and Statistical Examinations of Patterns of Fruit and Seed Maturation

One approach to the investigation of positional and/or temporal variation in fruit and seed maturation is the analysis of the distribution of fruit, and seeds per fruit, within inflorescences. This type of study has typically focused on unbranched inflorescences such as racemes. Nevertheless, the focal taxa are diverse and have included annuals, herbaceous perennials, shrubs, and trees, from both temperate and tropical habitats. Data are usually collected in natural populations without modification of existing levels of pollination and/or resource availability. Many studies have demonstrated an acropetal decline in fruit and/or seed set within inflorescences.

Asphodelus albus (Liliaceae) is a rhizomatous perennial that bears an indeterminately flowering raceme of 50 to 150 flowers. As a result of the 40-day flowering period of individual inflorescences, basal flowers are pollinated long before distal flowers, and the potential temporal separation between initiation of fruits in basal and distal positions is considerable. Fruit set was not evenly distributed among sectors: In the basal, middle, and distal sectors, 50%, 40%, and 10% of flowers produced fruit, consistently over two years (43). Thus, fruit in lower positions of the inflorescence matured preferentially. An analysis of variance for seed mass in *A. albus* showed that most of the observed variation occurred within plants (variation was assessed among populations, among individuals within populations, and within individual inflorescences) (42). Significant variation in both fruit and seed production occurred within inflorescences of *A. albus,* and the proportion of variation in seed mass explained by position was greater than that explained by genetic and environmental differences among individuals and among populations.

Similar results have been obtained for *Caesalpinia eriostachys* (Fabaceae, Caesalpinioidae), a tree of the tropical dry zone of Costa Rica. Analysis of the distribution of mature fruit within basal, middle, and distal portions of individual racemes showed that pods beginning to mature in the upper third were much more likely to be aborted than were those in the lower or middle portions of the inflorescence (4).

Thalaspi arvense (Brassicaceae) is a weedy, readily autogamous annual. The typical pattern of fruit, aborted fruit, and unopened flower buds within inflorescences of plants grown in the greenhouse at different densities was analyzed (36). Regardless of density, there was always a zone of aborted fruit and unopened flower buds at the distal end of each raceme. The absolute number of distal nodes bearing aborted fruits was about the same for plants grown at high and low densities, but because fewer flowers are produced per raceme under high density, aborted fruits comprised a larger proportion of the total production of reproductive structures. Thus, the number of fruit initiated, rather than the number aborted, varied with the growth conditions of individual plants.

Even when fruit maturation does not vary significantly with position within an inflorescence, seed number per fruit may vary, producing a pattern of decline in reproductive success with position. For example, initial analyses of fruit set within inflorescences of *Arisaema triphyllum* (Araceae), an herbaceous perennial, showed no differences in total fruit production among the base, middle, and distal sectors. Further analysis, however, uncovered the fact that a large proportion of distal fruits contained no seeds, and seed number per fruit decreased from the base to the top of the inflorescence (33). Similarly, in *Prunella vulgaris* (Lamiaceae), an annual to short-lived perennial, seed mass declined significantly with position within the terminal raceme, and overall, 61% of the total variation in seed mass was due to within-inflorescence variation (62). Although seed maturation per flower is never 100% in *Echium vulgare* (Boraginaceae), an herbaceous perennial, flowers at the base of the cymose inflorescences mature more seeds per fruit than distal flowers (41).

The studies summarized above (see 11, 32, 50, 56 for additional examples) demonstrate a positional decline in the proportion of flowers that successfully mature fruit, or in seed number or mass per fruit. The ubiquity of this pattern among diverse taxa with a wide diversity of morphologies and life histories is congruent with the hypothesis that the presence of developing fruit decreases the potential for fruit maturation at later, distal positions within the same inflorescence. Other studies, however, have identified contrasting patterns in which the basal positions are not associated with the greatest probability of fruit maturation.

For example, McKone (38) examined patterns of fruit set in five species of *Bromus* (Poaceae), including *B. tectorum*, a self-compatible annual weedy species, *B. inermis*, a widely cultivated, self-incompatible rhizomatous perennial that had become naturalized in the study area, and *B. kalmii, B. ciliatus*, and *B. latiglumis*, all self-compatible, native, nonrhizomatous perennials. All species have a compound inflorescence consisting of a central axis bearing spikelets (inflorescences) with florets (flowers). Fruit abortion was high in *B. inermis,* but aborted fruits were distributed evenly within spikelets, i.e. there was no observable pattern of positional variation in fruit maturation. For the remaining species, fruit production was nonuniform with respect to position. *Bromus kalmii, B. ciliatus*, and *B. latiglumis* had peak fruit set in the second or third floret from the glumes. Although *B. ciliatus* and *B. kalmii* had the same number of florets per spikelet (with 70% fruit set) and similar total dry masses, the pattern of fruit maturation for these species differed in many details. *Bromus ciliatus* produced more than 1.5 times as many florets, fruits, and spikelets per plant as did *B. kalmii*, and successful fruits were more evenly distributed within the spikelets. Fruit set in *B. ciliatus* varied from 73.65% to 78% to 66.7% in basal, middle, and distal positions, respectively. In *B. kalmii*, fruit set was 77% in basal positions, 83% in middle positions, and 41% in distal positions. Thus, for the same total plant dry weight, *B. ciliatus* had more spikelets, more flowers, and more fruits, yet had a shallower decline in fruit production within spikelets than did *B. kalmii*. Only in *B. tectorum* did fruit set decline consistently with floret position. McKone (38) argued that pollen was unlikely to limit reproduction in any of the five species, so variation in pollen availability may not provide an explanation for the observed patterns.

Fruit set in other species also showed no consistent relationship to position. *Phytolacca rivinoides* (Phytolaccaecae) is a short-lived perennial of disturbed tropical habitats. Seed number and seed mass either remained constant or increased along the length of the raceme while pulp mass per fruit declined (11). Similarly, fruit maturation in the orchid *Aspasia principissa* was independent of flower position within each six-flowered raceme (66). In *Clintonia borealis* (Liliaceae), an herbaceous understory perennial, flowering order had no effect on seed set (fruit set was not reported) (23).

In field studies, the confounding effects of variation in pollination intensity and order may hamper interpretation of fruit set patterns. Berry & Calvo (6) found that the distribution of mature fruit within racemes varied with pollination regime in *Myrosmodes cochleare* (Orchidaceae). When plants were self-pollinated, fruit set declined rapidly with position, ranging from 100% in basal positions to 0% in distal positions. In contrast to the steep acropetal decline in fruit set observed for self-pollinated plants, fruit set in open-pollinated plants was maximal in the middle portion of each inflorescence and declined both basipetally and acropetally. The authors suggested that the pattern observed in the open-pollination treatment reflected the activity of pollinators. Floral display was maximal when flowers in the middle portion of the inflorescence were in anthesis, and these flowers may have had a higher frequency of successful pollination, whereas flowers in basal positions might have been pollen-limited. Thus, the observed pattern of fruit set may reflect the patterns of pollinator activity or effectiveness.

Similar results have been obtained for two species of *Epilobium, E. dodonaei* and *E. fleischeri* (Onagraceae) (51). Abortion frequency of flower buds, flowers, fruits, and ovules by position within the racemes of each species were examined. For both species, frequency of abortion of all reproductive structures was typically greatest in the proximal and distal parts of the inflorescence. The probability of abortion of proximal buds was higher than that of buds in the middle of the inflorescence, leading the authors to conclude that proximity to resources and the absence of competing older fruits were not necessarily advantages. They further suggested that the switch from vegetative to reproductive shoot growth may involve instabilities in resource partitioning for vegetative versus reproductive structures (51). Although unreported, pollinator activity may have limited the reproductive potential of early flowers. If this interpretation is correct, within a single inflorescence, reproduction can be limited by different factors—early reproduction by pollen and late reproduction by resource competition.

As can be seen from the preceding summary, a diverse group of species have yielded results that may be inconsistent with the pattern of fruit and seed maturation expected if competition for resources among distal and basal fruits is the proximate cause of that pattern. The advantage of observational and statistical analyses of fruit and seed set is that they can be accomplished in natural populations. They chronicle the diversity of intra-inflorescence variation in fruit and seed maturation as it occurs within the context of variation among individuals and populations. However, given that the observations reported for *Myrosmodes, Epilobium, Bromus, Phytolacca, Aspasia*, and *Clintonia* were made under field conditions that may include temporal variation in pollinator activity and other biotic and abiotic factors, the cause of fruit abortion and/or lack of development cannot be readily identified (for example, aborted fruits cannot be distinguished from unpollinated ovaries). Thus, it is difficult to evaluate the significance of these patterns relative to predictions based on a hypothesis of resource competition among developing fruits. Even when a pattern of declining fruit set is demonstrated, the potential "function" of flowers at distal positions remains unknown. In order to demonstrate effects of developing fruits on subsequent fruit and/or seed maturation, it is necessary to move from the analysis of naturally occurring patterns of variation to experimental manipulation of pollination and fruit production.

Experimental Examinations of Fruit and Seed Maturation

In addition to resource competition among developing fruits, at least two other hypotheses may account for patterns of unequal fruit set among flowers within inflorescences. First, pollination may not be distributed uniformly among flowers, and second, flowers in distal positions may not be capable (when pollinated) of setting fruit, even in the absence of basal fruit. Uniform hand-pollination is used to demonstrate that fruit abortion is due to temporal or positional precedence rather than lack of, or differences in, successful pollination among flowers. Additionally, pollination of flowers in basal positions may be prevented (or fruit in these positions removed) to determine whether lower rates of maturation of distal fruits can be attributed solely to the effects of proximal fruits. If, in the absence of basal flowers or fruits, fruit and seed maturation of distal flowers does not differ from those of proximal flowers on control plants, it can be inferred that the presence of proximal fruits affects the fate of distal fruits. Differences in fruit development that persist regardless of the absence of basal fruits suggest that positional variation in the *potential* for fruit set exists, and that this variation may underlie observed patterns of declining fruit maturation.

An example of experimental analysis of patterns of fruit set is provided by Stephenson's (48, 49) studies of *Catalpa speciosa* (Bignoniaceae), a tree of riparian habitats. Inflorescences of *C. speciosa* are thyrses with an average of 27 flowers that open sequentially over a period of 5 to 6 days. Three flowers within bagged inflorescences were hand-pollinated (with outcrossed pollen): one flower on the first day after initiation of flowering, one flower on the third day, and one flower on the fifth day. Significantly more fruits began to develop from flowers pollinated on the first day than from flowers pollinated on either the third or fifth days (48), and significantly more of the fruits initiated on the first day reached maturity (49). No difference appeared in fruit maturation, however, between flowers pollinated on the third and fifth days. This experiment confirmed a pattern of unequal fruit set among flowers within inflorescences and eliminated unequal pollination as an explanation of the observed pattern.

Stephenson also provided evidence that failure of later (distal) fruits was due to the presence of early (basal) fruits rather than to some other inherent property of the inflorescence. Removal of the earliest initiated fruit from each inflorescence (presumably the basal fruit) resulted in increased fruit set of the later (distal) fruits (49). In addition, hand-pollination of single flowers within bagged inflorescences showed no significant effect of time or flower position on probability of fruit set (48). Thus, flowers at each position within the inflorescence appear to be equally capable of maturing a fruit in the absence of other fruits.

These experimentally based observations are consistent with the interpretation that the reproductive potential of distal flowers in *C. speciosa* was modified by successful fruit initiation at basal positions. The effect appears to be temporal rather than positional, because flowers at all positions are equally capable of maturing a fruit in the absence of other fruits.

Hand-pollination and fruit removal experiments also have been used to demonstrate the capacity of distal flowers to produce fruit in the absence of proximal flowers/fruits in species with racemose inflorescences. In *Lathyrus vernus* (Fabaceae), an herbaceous perennial, removal of basal flowers within inflorescences resulted in increased fruit set in distal flowers (21). Removal of basal flowers in *Prunus mahaleb* (Rosaceae), a small tree, increased the fruit set of distal flowers but did not affect the fruit set of the plant as a whole (25, 26). *Calochortus leichtlinii* is an herbaceous perennial lily that typically produces two flowers per inflorescence. Fruit set of the second flower is normally rare, but removal of the basal flower increased the probability that the second flower set fruit, and seed set of the second flower was negatively correlated with the age of the first flower when it was removed (28). In *Erythronium* and *Clintonia*, also herbaceous perennials of the Liliaceae, distal flowers had a higher seed set when proximal flowers remained unpollinated as compared to controls in which all flowers were pollinated (56).

Catalpa, *Lathyrus*, *Prunus*, *Calochortus*, *Erythronium*, and *Clintonia* all register a significant effect of basal fruit on the potential reproductive success of flowers in distal positions. Fruit initiation in basal positions decreases the probability of maturation of distal fruits. All flowers have the potential to produce a fruit, and it is the presence of other fruit that affects their success. These results are consistent with the hypothesis that the development of younger, distal fruit is precluded by competition with older, proximal fruit for resources.

Results from a few studies are difficult to reconcile with a straightforward explanation of competition among fruits based on temporal and/or spatial advantages. For example, *Cryptantha flava* (Boraginaceae) is an herbaceous perennial that, like all members of the Boraginaceae, is characterized by a four-ovulate gynoecium (two ovules in each of two carpels) and by cymose inflorescences. Although the probability of fruit set was distributed at random within a cyme, the probability of maturing at least two seed per fruit decreased distally (14). Removal of proximal flowers did not increase seed set per flower in distal flowers, however, suggesting that the greater frequency of abortion of seeds per fruit in these positions was not due to resource limitation. While these experimental studies of *C. flava* may not be consistent with direct resource competition as the proximate cause of seed abortion, the ultimate effect of these observed patterns may be one of optimal resource allocation.

In *Asclepias speciosa* (Asclepiadaceae), temporal precedence has only a transitory effect on fruit maturation (7). This species produces flowers simultaneously within umbels, and there are essentially no positional differences among flowers within an umbel. When individual flowers were pollinated within umbels at one-day intervals, no differences were observed in the proportion of fruit set between flowers pollinated on successive days. When pollinations within umbels were separated by two days, fruit set of the earlier pollinated flowers increased to a level twice that of the later pollinated flowers. These results are consistent with an interpretation of preferential maturation of the more developed fruit. However, when pollinations occurred four days apart, no difference in the proportion of fruit set between the earlier and later pollinated flowers was observed. Seed maturation per fruit followed the same pattern as fruit set. These results are consistent with the hypothesis that successful fruit maturation is affected by competition with other fruits, but they also suggest that the effects of competition may decline as the time interval between fruit initiation increases. The interval between fruit initiation by successive flowers is rarely considered and may be a crucial factor regulating competition among developing fruit.

In summary, evidence from experimental studies confirms that early/proximal fruit may significantly alter the reproductive potential of later/distal fruit in many taxa. This pattern is consistent with a "strategy" that terminates investment of resources in the youngest fruit. However, direct competition among developing fruit for access to resources may not be the complete explanation for observed patterns of variation in successful fruit and seed maturation within inflorescences. The next section explores evidence for alternative causes of positional variation in fruit and seed maturation.

ALTERNATIVE EXPLANATIONS OF VARIATION IN FRUIT AND SEED DEVELOPMENT WITHIN INFLORESCENCES

Variation in Potential Reproductive Capacity Among Flowers

Many of the experiments summarized in the preceding section demonstrated that all flowers within an inflorescence were equally capable of producing fruit in the absence of other fruit. This is not always the case. Distal flowers may have a lower capacity or efficiency of fruit set compared to basal flowers, unrelated to the fruiting status of basal positions. For example, in experiments on soybean (*Glycine max*, Fabaceae), distal flowers within racemes were capable of setting fruit if proximal flowers were removed. However, even in the absence of basal fruits, the proportion of fruit set by flowers in distal positions was never as great as the fruit set of proximal flowers (29). This pattern was examined in more detail by placing flowers in tissue culture. When distal flowers from inflorescences bearing basal flowers were cultured, they showed little increase in mass. When proximal flowers were removed from inflorescences before distal flowers were placed in culture, growth occurred, but the increase in mass of these distal flowers was never equal to that of cultured proximal flowers (29). Similarly, fruit removal experiments demonstrated that although flower position had no effect on capacity to set fruit in *Lupinus luteus* (also Fabaceae), growth of distal pods was slower than that of proximal pods, even when basal pods were removed (57).

These results suggest that the potential for fruit development can vary with flower position irrespective of the presence of other fruit. If the capacity or efficiency of fruit production varies with position, a pattern of declining fruit maturation is predicted to result. Competition for resources between distal and basal flowers and fruits, however, would not be the sole explanation of declining fruit set. Thus, careful experimental analysis of the potential for fruit maturation by flower position within an inflorescence is necessary before an effect of basal fruit on distal flowers can be demonstrated.

In addition to a quantitative decrease in the inherent capacity for fruit set by flowers in distal positions, distal flowers may in some cases be incapable of setting fruit. For example, the probability of fruit set decreased distally in *Aquilegia caerulea* (Ranunculaceae), and even when fruit set of basal flowers was prevented, distal flowers did not set fruit (9). A similar result was reported for *Myrosmodes cochleare* (Orchidaceae) (6). Thus, while morphological inspection of such species would classify them as hermaphroditic, they are functionally andromonoecious: Individuals bear both hermaphrodite and functionally staminate flowers. Clearly, female reproductive potential of morphologically uniform flowers within inflorescences may vary both quantitatively and qualitatively with position. The hypothesis that resource competition among fruits explains observed acropetal declines in fruit and/or seed maturation is based on the assumption that the reproductive potential of all hermaphrodite flowers within an inflorescence is equal. The research summarized above shows that this is not necessarily a universal biological model.

Positional Variation in Ovule Number per Flower

Seed set per fruit often declines with position (reviewed in 32), a decline attributed to increased abortion of ovules in distal fruits. Alternatively, a positional decline in ovule number per flower would also be predicted to result in declining seed number per fruit. Thus, consideration of ovule number per ovary is critical for understanding positional variation in seed maturation.

Inflorescences of liliaceous species are typically racemes (47) with acropetal initiation of floral meristems. Analyses of patterns of variation of ovule number per ovary within racemes of 15 liliaceous species showed a nearly ubiquitous decline in ovule number per ovary with flower position (56). In the species examined, flowers are preformed and ovule number per ovary is determined for all flowers within an inflorescence before flowering begins. Thus, variation in ovule number cannot be a response to earlier commitment to fruit set. If distal flowers have fewer ovules to mature as seeds, then declines in seed number per fruit may be due solely to positional variation in ovule number.

Interestingly, the acropetal declines in ovule number with position (56) occurred in seven species with a basipetal pattern of floral anthesis. Relative seed set per flower was recorded for four of these species (*Clintonia borealis*, *Calochortus leichtlinii*, *Erythronium grandiflorum*, and *Medeola virginiana*). In spite of the fact that distal flowers in these species may have been pollinated first and thus have had a potential temporal advantage in terms of developing sink strength, seed number per fruit declined acropetally. This result suggests that positionally determined commitment to ovule primordia during inflorescence development can override temporal precedence of fruit initiation on seed set per fruit.

In *Zigadenus paniculatus*, also a lily with preformed flowers, the pattern of ovule number per ovary varied with plant size (22). Ovule number per ovary increased with position in small plants but decreased with position in large plants. The number of ovules per ovary declined with position in *Lupinus luteus* (Fabaceae) (57) and in *Lycopersicon esculentum* (Solanaceae) (1), even in flowers developing in the absence of other flowers and fruits. While fruit set was independent of flower position in *Aspasia principissa* (as discussed above), fruit length declined with flower position (66). Declining fruit length may have been due to a decline in number of ovules fertilized or to an increase in ovule or seed abortion, but the decline could also have been caused by a decrease in ovary size and number of ovules per ovary with position.

Examination of ovule number per ovary within inflorescences of the closely related species *Epilobium didonaei* and *E. fleischeri* demonstrated contrasting patterns (51). In *E. fleischeri*, the position of the flower within the inflorescence had no effect on the number of ovules per ovary, while in *E. dodonaei*, the mean number of ovules per ovary decreased acropetally. In a third species of *Epilobium*, *E. montanum*, ovule number per ovary decreased from 220 at node 1 to 170 at node 10. Flower position accounted for 59% of the variation in ovule number (27).

Ovule number per ovary declines during the growing season in *Raphanus sativus* (Brassicaceae), an annual (65), and in *Diervilla lonicera* (Caprifoliaceae), a clone-forming perennial (55). Flower position was not analyzed in these studies, but it is likely that later blooming flowers are located distally within inflorescences.

These studies confirm that variation in ovule number per ovary exists among flowers within inflorescences. This pattern may result in an acropetal decline in seed number per fruit. Thus, patterns of declining seed maturation can be established by events that occur during gynoecial development, long before anthesis, fertilization, and initiation of fruit- and seed-set. Declining ovule number may also affect patterns of fruit abortion if there is a threshold number of seeds per fruit required for fruit maturation [evidence for a relationship between the number of seeds in a fruit and its probability of maturing is reviewed by Lee (32) and Stephenson (50)]. If distal flowers have fewer ovules or set fewer seeds, and seed set falls below a threshold, these fruit will be aborted. The underlying cause would be subminimal ovule number (and variation in floral development) rather than competitive interactions among fruits.

INTRA-INFLORESCENCE FLORAL VARIATION AND ARCHITECTURAL EFFECTS

Fruits, and the ovaries from which they develop, are integral components of flowers. If the sizes of organs within flowers are highly correlated (or coregulated) during development, then it is possible that variation in ovule number per ovary reported in the preceding section is indicative of a pattern of intra-inflorescence variation that potentially affects all floral organs. If floral organ size varies systematically within an inflorescence, fruit size may reflect this variation regardless of patterns of allocation to developing fruit. The term *architectural effect* is used to describe morphological variation that can be ascribed to the position of a flower (and its organs) within an inflorescence, rather than to effects of resource supply. The data available to examine this subject of architectural effects are limited. This section summarizes the few studies that provide data on variation in floral organ dimensions within inflorescences in order to address the issue of regulation of fruit and seed development.

Minimus guttatus

Mimulus guttatus (Scrophulariaceae) is an herbaceous annual to facultative perennial that produces indeterminately flowering bracteose racemes. Measurements were made of corolla size (length and width) and mass; anther size, mass, and number of viable pollen grains; ovary length, width, and mass; number of ovules; and capsule mass (35, 39, 44, 45). Within each raceme, flower size decreased distally (35, 39). The sizes of all floral organs were positively correlated, both phenotypically and genetically (34, 35), and quantitative analysis demonstrated a coordinated decline in calyx, corolla, anther, ovary, and mature fruit size and mass—a decline that was a function of flower position within inflorescences. This coordinated decrease in floral organ size and mass occurred in plants bearing fruit (every flower hand-pollinated) and plants bearing no fruit (ovaries removed at anthesis) (35, 39). The rate of decline in size and mass with position, however, was greater in plants that bore developing fruit. Thus, an acropetal diminution of all floral organs is an inherent property of *M. guttatus* inflorescences and occurs in addition to the effects of allocation to developing fruit.

If patterns of fruit set had been the only focus of study in *M. guttatus*, the decline in fruit size or seed number with position might have been interpreted as evidence of resource competition among fruits. Even the decline in ovule number per ovary could have been attributed to a decrease in resources allocated to floral development following the initiation of fruit at early (basal) positions. When variation of all floral organs on plants both with and without fruit was studied, a contrasting hypothesis emerged. Declines in fruit mass, seed mass, or seed number with position may be manifestations of a regular pattern of quantitative intra-inflorescence positional variation in floral characters—that is, of *architectural effects* on plant-reproductive characters. Distal flowers are smaller and have smaller ovaries that contain fewer ovules. This decline occurs in the absence of competition from basal fruits, so it cannot be attributed solely to competition among fruits for resources.

The *M. guttatus* data do show an effect of allocation to basal fruits on the development of distal flowers and fruits; the sizes of floral organs in successive flowers decreased more rapidly in fruit-bearing plants (35, 39). Comparison of the rate of decline in floral organs in the presence and absence of basal fruit provides an accurate measure of the actual effect of resource commitment to earlier fruit. Without such a comparison, the effect of basal fruit may be overestimated.

Although the sizes of all floral organs of *M. guttatus* differed with position, gynoecial and androecial characters showed separate patterns of variation with respect to flower position. Pollen-ovule ratios (estimated from the ratio of anther mass to ovary size) declined with flower position (33), suggesting that anther mass declined at a greater rate than ovary size. The authors suggested that if decreases in pollen-ovule ratios are an unavoidable consequence of reduced flower size in *Mimulus* species, then the low pollen-ovule ratios typical of related self-fertilizing taxa within the *M. guttatus* complex may be an epiphenomenon of the small flower size characteristic of these species rather than an adaptation per se.

The decline in pollen-ovule ratios with decreased flower size in *M. guttatus* and other members of the *M. guttatus* complex is unlikely to be the rule among other flowering plants. For example, Brunet (9, 10) found that variation in anther mass with position in *Aquilegia caerulea* (Ranunculaceae) was not significant, although floral size and ovary size decreased significantly. Thomson (56) reported that pollen-ovule ratio increased with position in *Erythronium grandiflorum* (Liliaceae).

Hydrophyllum appendiculatum

Wolfe (63) designed a study to determine whether temporal reduction in the sizes of reproductive structures was due to declining resources or to architectural effects in *Hydrophyllum appendiculatum* (Hydrophyllaceae), a biennial herb that produces multiple inflorescences throughout a season. The number of flowers per inflorescence, flower size, and individual seed weight (fruits are single seeded) were measured on successive inflorescences of plants grown with high and low levels of fruit set.

Although all three reproductive traits declined with inflorescence sequence in *H. appendiculatum*, the traits differed in their pattern of response to the fruit set treatments. The decline in number of flowers per inflorescence and mean seed weight was greater in plants with high fruit set than in plants with low fruit set. In contrast, the decline in flower size with inflorescence position did not differ between the treatments. Thus, unlike *Mimulus guttatus*, in which flower size and fruit size were highly correlated in plants both with and without fruit, in *H. appendiculatum* flower size and seed size (presumably reflecting ovary size to some extent) varied independently. Wolfe concluded that both resource limitation and architectural effects (Wolfe used the term *developmental constraint* rather than *architectural effect)* were in operation and that their effects were trait-specific.

In *H. appendiculatum*, the decline in corolla size was due entirely to flower position and occurred irrespective of the level of fruit set (the rate of decline did not vary among treatments). In contrast, seed weight and number of flowers per inflorescence were affected by both architecture and resource competition. Wolfe suggested that plants of *H. appendiculatum* maintain flower size as large as possible within the constraints of architectural/positional limitations on size. The large (or more constant) flower size was preserved at the expense of flower number and seed mass in high fruiting individuals. Several studies have shown that corolla size is more highly correlated with estimates of male reproductive success (pollen production, pollen export) than with female reproductive success (seed production) (5, 12, 24, 64). Thus, flower size may be buffered against resource limitations so that pollen export is maximized throughout the growing season (63). If this interpretation is correct, then the decline in seed mass in *H. appendiculatum* is ultimately attributable to tradeoffs between male and female function rather than to direct competition among fruits.

Solanum hirtum

In a study of *Solanum hirtum* (Solanaceae), intra-inflorescence variation in floral organ size under contrasting conditions of high fruit set (with the potential for competition between basal fruits and distal flowers) and no fruit set (no basal fruits to affect development at distal positions) of 10 clonally replicated genotypes was examined (17). *Solanum hirtum* is a woody shrub that produces sympodial raceme-like inflorescences successively along indeterminately growing branches. This species is andromonoecious and the proportions of hermaphrodite and staminate flowers borne by individuals depend, in part, on resource availability (18, 19).

Unpollinated (no fruit) plants of *S. hirtum* produced hermaphrodite flowers both basally and distally within inflorescences, but positional variation was evident among these hermaphrodite flowers (17). Anther, stigma, ovary size, and ovule number of distal hermaphrodite flowers were smaller than in basal hermaphrodite flowers (Figure 1b,c). This quantitative variation, however, was not associated with variation in potential reproductive function among these flowers; all hermaphrodite flowers were capable of producing fruit regardless of position (16).

On inflorescences of pollinated (+fruit) plants of *S. hirtum*, flowers varied both quantitatively and functionally with position (17). Basal flowers were hermaphrodite, whereas distal flowers were functionally staminate and did not produce fruit. All organs of distal staminate flowers were significantly smaller than those of basal hermaphrodite flowers, but the sources of quantitative variation differed among the floral organs. The decrease in corolla length of staminate flowers was due solely to treatment. Corolla length of staminate flowers of the +fruit treatment was significantly less than corolla length of basal and distal hermaphrodite flowers of the no fruit treatment (Figure 1a). In contrast, variation in anther length appeared to be due solely to position (architectural effect). There was no difference in anther length between (distal) staminate flowers of +fruit plants and the distal hermaphrodite flowers of unpollinated (no fruit) plants (Figure 1b). Gynoecial variation was due both to flower position and to treatment. Style length, stigma width, and ovary length and width of staminate flowers were all less than those of both basal hermaphrodite flowers within the same inflorescence and distal hermaphrodite flowers of genetically identical unpollinated plants (Figure 1c).

Variation in floral morphology and function within inflorescences of *S. hirtum* arose from flower position (basal vs. distal within each inflorescence) and treatment (fruit or no fruit). Each floral organ was affected by either or both sources. Anther size in *S. hirtum* decreased with position but was unaffected by the presence of developing fruit. Thus, variation in anther size was solely an architectural effect. Positional variation in gynoecial characters also occurred, but, ultimately, the presence of developing fruit had an additional effect: gynoecial development of distal flowers was arrested prior to anthesis, resulting in further size decreases and precluding female function (16, 19). The development of basal flowers within inflorescences was not affected by treatment, even when large numbers of fruit were present on the branch bearing that inflorescence, and other symptoms of resource depletion were evident

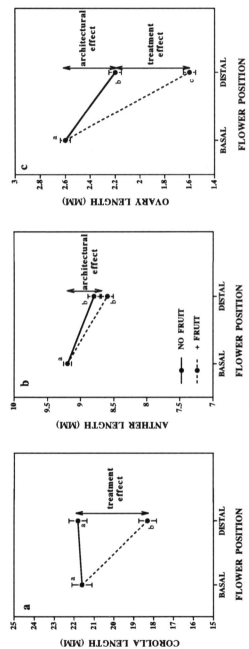

Figure 1 Mean corolla, anther, and ovary length in *Solanum hirtum* as a function of flower position (basal or distal within an inflorescence) and treatment [pollination and fruit set (+ fruit) or no pollination and no fruit set (no fruit)]. Bars indicate standard errors. Within each graph, means associated with the same letter are not significantly different (ANOVA followed by Student-Neuman-Keuls test; $P < 0.01$). *1a.* Corolla length did not vary with flower position but decreased in distal flowers following fruit set at basal positions. The decline in corolla length with position in + fruit plants is a treatment effect. *1b.* Anther length decreased in distal flowers, and fruit set had no significant effect on the decline (mean anther length of distal flowers did not differ significantly between treatments). Thus, the decline in anther length is an architectural effect. *1c.* Ovary length decreased with position in both the no fruit and + fruit treatment but the decline was significantly greater in + fruit plants. The difference between ovary lengths of basal and distal flowers of the no-fruit treatment is an architectural effect, whereas the difference between ovary lengths of distal flowers in no-fruit and + fruit plants is a treatment effect. Data are from (17).

(e.g. decline in number of flowers initiated per inflorescence—18). Thus, the effects of developing fruit can be restricted to certain organs within flowers, and to certain flowers within inflorescences.

Bawa & Webb (3) also showed that floral organs within flowers in *Muntingia calabura* (Elaeocarpaceae) differed in their response to resource manipulation. *Muntingia calabura* is a small fast-growing tree of disturbed sites in lowland tropics. In this species, flowers are produced in fascicles. Quantitative analysis showed that ovary size decreased with position and that large (basal) ovaries set fruit while small distal ones did not. In contrast, stamen number per flower increased with flower position. Removal of basal flowers caused an increase in the size of distal ovaries but no change in stamen number.

Quantitative studies on intra-inflorescence variation in floral organ dimensions are, at present, too few to permit broad generalizations. Yet, in three species in which sizes of floral organs have been analyzed by position, an *architectural effect* has been demonstrated. These results suggest that, at minimum, the interpretation of any quantitative variation in reproductive attributes must take account of potential effects of inflorescence architecture on flower and fruit development. The studies summarized in this section show that size variation in some floral organs is due primarily to differences in flower position (architectural effects). Other floral organs of these same species varied with both flower position and treatment (level of fruit set). Thus, the proximate causes of observed variation may differ among organs within flowers.

The hypothesis that the likelihood of fruit maturation at distal positions within inflorescences is affected by competition for resources with basal fruit is predicated on the expectation that flower and floral organ sizes, in the absence of resource competition, are invariant among positions within inflorescences. This implicit assumption of the resource allocation hypothesis has rarely been tested. The research summarized in the preceding sections leaves little doubt that basal fruit can affect the potential for successful fruit maturation at distal positions. The magnitude of this effect, and the importance of allocation to early fruit for the success of subsequent fruit production, however, remains uncertain. A substantial part of the oft-reported pattern of acropetal decline in fruit and seed maturation may result from architectural/positional effects rather than the history of allocation to developing fruit within the inflorescence. Architectural effects may also explain divergent patterns of fruit and seed maturation such as those reported for *Myrosmodes* (6), *Epilobium* (51), *Bromus* (38), *Phytolacca* (11), *Aspasia* (66), and *Clintonia* (23).

ARCHITECTURAL EFFECTS: INHERENT FEATURES OF INFLORESCENCES?

If intra-inflorescence variation in floral form and function is common, are these architectural effects inevitable consequences of mechanisms of plant growth and metameric construction? Winn (62) provided convincing evidence that positional variation in seed mass persists despite strong directional selection for a uniform crop of large seeds in *Prunella vulgaris*. Seed mass declined significantly with position within racemes. Yet, large seeds were clearly advantageous in this species. They had a significantly greater probability of emergence in the field in two consecutive years (61), and large seeds were able to emerge in a greater variety of microsites than were small seeds (60). Winn concluded that *P. vulgaris* experienced strong directional selection favoring large seeds and that this selection was consistent across years and microsites. Thus, it was difficult to support a hypothesis that within-individual variation in seed mass is adaptive because of spatial or temporal variation in the optimum seed mass or that weak or inconsistent selection on seed mass prevented selection against parents producing seeds of mixed size (62).

Variation in floral organ size (including ovule/seed size) might be unavoidable in racemose inflorescences. The developmental control necessary to achieve uniform seed mass in *P. vulgaris* may be unavailable despite strong selection for such control. If this interpretation is correct, then positional variation in seed mass is an unavoidable or inherent feature of the racemes of *P. vulgaris*. Comparative study of seed mass variation in related species could be used in the future to determine whether the pattern reported for *P. vulgaris* (or any other species) is invariant within a clade.

Evidence from developmental genetics suggests that gene expression also varies with flower position within inflorescences. Analyses of induced mutations in *Arabadopsis thaliana* (Brassicaceae) demonstrate that the effects of a mutation vary with flower position. In general, the effects of mutations in meristem identity genes become less severe with flower position. For example, *leafy* and *apetala-1* caused the transformation of proximal flowers to inflorescences whereas distal flowers were only partially affected (8, 59). In contrast, the effects of organ identity mutations became more severe in distal flowers within inflorescences [e.g. alleles of *ap2*, (31)]. It may be that when the normal roles of *LEAFY* or *APETALA1* are removed by mutation, underlying physiological and/or developmental variation is revealed to a greater degree in mutant as compared to wild type plants (F Hempel, personal communication).

Similarly, mutations affecting floral development in natural populations vary in expression with flower position. For example, Barrett & Harder (2) have studied self-fertilizing genotypes of the typically outcrossing *Eichhornia paniculata* (Pontideriaceae). Self-fertilization was due to increased stamen elongation. Increased stamen length, however, occurred only in flowers at specific locations within an inflorescence. The particular positions of self-fertilizing flowers varied among the genotypes examined. Thus, expression of new (mutant) phenotypes in both natural (*E. paniculata*) and artificial (*A. thaliana*) populations varied with flower position—that is, they were subject to architectural effects. If this is a typical scenario for the initial introduction of a novel floral morphology during evolution, then the occurrence of more stable or uniform differences among taxa may reflect subsequent evolution of modifying loci.

Proximate causes of architectural effects are not known. It has been suggested that the vasculature of inflorescences may regulate fruit and/or seed maturation (11, 13, 63). In contrast to the reproductive structures in basal positions, those produced distally are borne on stems of smaller diameter that contain less vascular tissue. Thus, the supply of resources to these distal structures may be lower. While the diameter of an inflorescence, and the vascular tissue within, typically declines acropetally, the requirements of transport also decrease acropetally. Furthermore, evidence from studies of xylem differentiation suggest that the presence of a fruit has the potential to stimulate the development of vascular tissue supplying it. Xylem development responds to a source of auxin production (30, 46), and seeds within fruits are a potential source of auxin (54). Thus, developing fruit are as likely to regulate the development of vascular tissue as vascular tissue is to limit fruit maturation. The distribution of vascular tissue within inflorescence axes may be yet another manifestation of architectural variation in plant characteristics.

Rather than being inherent features of plant axes, architectural effects may have a selective or adaptive explanation. Brunet & Charlesworth (10) developed an ESS model to predict the occurrence of positional variation of intrafloral allocation. Based on Brunet's analysis of *Aquilegia*, they showed that in protandrous species, the probability of pollen transfer (both receipt and donation) changed with flower position. This type of variation in the mating environments of flowers within inflorescences selects for differences in sex allocation among flowers at different positions. For protandrous species, lower relative allocation to male function in early opening flowers and lower allocation to female function in later opening flowers was predicted. Protogyny was associated with the opposite pattern. In contrast, when transfer probabilities for flowers at different positions were equal (equal pollen movement) and when the exponents to the male and female gain curves were equal, no variation in sex allocation among positions was expected, even when resources varied among positions. Thus, intra-inflorescence variation is predicted to evolve under some circumstances and to be absent in other cases.

SUMMARY

Architectural effects within plants are likely to be a fundamental component of plant phenotypes. Because plant metamers are produced sequentially and acropetally, there is always the potential for early developing structures to affect the fate of later formed structures. That is, the phenotype and function of any particular plant organ may be determined, in part, by the ontogenetic history of the organism. Within inflorescences, flowers produced early in inflorescence development are likely to have a different female reproductive potential than flowers produced later. Part of this variation among flowers is typically due to differential allocation of resources during flower and fruit development. The research summarized in this review suggests, however, that observed variation can also be due to architecture—that is, to sources of variation inherent in plant axes. Thus, the fate of a developing flower depends not only on the events that have preceded it during the ontogeny of the organism, but on where it occurs within the architecture of an individual plant. The effects of architecture are separable experimentally from the effects of differential allocation, and careful experimental analysis of these two factors will enhance our understanding of the physiological, developmental, and evolutionary controls of fruit and seed production in flowering plants.

ACKNOWLEDGMENTS

I thank William E. Friedman for stimulating discussion of the ideas presented and for careful reading of the manuscript; Beth Krizek and Frederick Hempel for discussion of intra-inflorescence variation in *Arabadopsis*; and Mimi Lam for xeroxing and entering references. Support has been provided by grants from Apple Computer, Inc. and the National Science Foundation.

Literature Cited

1. Bangerth F, Ho LC. 1984. Fruit position and fruit sequence in a truss as factors determining final fruit size of tomato fruits. *Ann. Bot.* 53:315–319

2. Barrett SCH, Harder LD. 1992. Floral variation in *Eichhornia paniculata* (Spreng.) Solms (Pontederiaceae) II. Effects of development and environment on the formation of selfing flowers. *J. Evol. Biol.* 5:83–107

3. Bawa KS, Webb CJ. 1983. Floral variation and sexual differentiation in *Muntingia calabura* (Elaeocarpaceae) a species with hermaphrodite flowers. *Evolution* 37:1271–82

4. Bawa KS, Webb CJ. 1984. Flower, fruit and seed abortion in tropical forest trees: implications for the evolution of paternal and maternal reproductive patterns. *Am. J. Bot.* 71:736–51

5. Bell G. 1985. On the function of flowers. *Proc. R. Soc. London* 224:223–65

6. Berry PE, Calvo RN. 1991. Pollinator limitation and position dependent fruit set in the high Andean orchid *Myrosmodes cochleare* (Orchidaceae). *Plant Syst. Evol.* 174:93–101

7. Bookman SS. 1983. Effects of pollination timing on fruiting in *Asclepias speciosa* Torr. (Asclepiadaceae). *Am. J. Bot.* 70:897–905

8. Bowman JL, Alvarez J, Weigel D, Meyerowitz EM, Smyth DR. 1993. Control of flower development in *Arabidopsis thaliana* by *APETALA1* and interacting genes. *Development* 119:721–43

9. Brunet J. 1990. *Gender specialization of flowers within inflorescences of hermaphroditic plants.* PhD thesis. State Univ. New York, Stony Brook

10. Brunet J, Charlesworth D. 1995. Floral sex allocation in sequentially blooming plants. *Evolution* 49:70–79

11. Byrne M, Mazer SJ. 1990. The effect of position on fruit characteristics, and relationships among components of yield in *Phytolacca rivinoides* (Phytolaccaceae). *Biotropica* 22:353–65

12. Campbell DR. 1989. Measurements of selection in a hermaphroditic plant: variation in male and female pollination success. *Evolution* 43:318–34

13. Carlquist S. 1969. Toward acceptable evolutionary interpretations of floral anatomy. *Phytomorphology* 19:332–62

14. Casper BB. 1984. On the evolution of embryo abortion in the herbaceous perennial *Cryptantha flava*. *Evolution* 38:1337–49

15. Cohen D, Dukas R. 1990. The optimal number of female flowers and the fruits-to-flowers ratio in plants under pollination and resources limitation. *Am. Nat.* 135:218–41

16. Diggle PK. 1991. Labile sex expression in andromonoecious *Solanum hirtum*: floral development and sex determination. *Am. J. Bot.* 78:377–93

17. Diggle PK. 1991. Labile sex expression in andromonoecious *Solanum hirtum*: pattern of variation in floral structure. *Can. J. Bot.* 69:2033–43

18. Diggle PK. 1993. Developmental plasticity, genetic variation, and the evolution of andromonoecy in *Solanum hirtum* (Solanaceae). *Am. J. Bot.* 80:967–73

19. Diggle PK. 1994. The expression of andromonoecy in *Solanum hirtum* (Solanaceae): phenotypic plasticity and ontogenetic contingency. *Am. J. Bot.* 81:1354–65

20. Ehrlén J. 1991. Why do plants produce surplus flowers? A reserve-ovary model. *Am. Nat.* 138:918–33

21. Ehrlén J. 1992. Proximate limits to seed production in a herbaceous perennial legume, *Lathyrus vernus*. *Ecology* 73:1820–31

22. Emms SK. 1993. Andromonoecy in *Zigadenus paniculatus* (Liliaceae): spatial and temporal patterns of sex allocation. *Am. J. Bot.* 80:914–23

23. Galen C, Plowright RC, Thomson JD. 1985. Floral biology and regulation of seed set and seed size in the lily, *Clintonia borealis*. *Am. J. Bot.* 72:1544–52

24. Galen C, Stanton ML. 1989. Bumble bee pollination and floral morphology: factors influencing pollen dispersal in the alpine sky pilot, *Polemonium viscosum* (Polemoniaceae). *Am. J. Bot.* 76:419–26

25. Guitain J. 1993. Why *Prunus mahaleb* (Rosaceae) produces more flowers than fruits. *Am. J. Bot.* 80:1305–9

26. Guitian J. 1994. Selective fruit abortion in *Prunus mahaleb* (Rosaceae). *Am. J. Bot.* 81:1555–58

27. Harper JL, Wallace HL. 1987. Control of fecundity through abortion in *Epilobium montanum* L. *Oecologia* 74:31–38

28. Holtsford T. 1985. Nonfruiting hermaphroditic flowers of *Calochortus leichtlinii* (Liliaceae): potential reproductive functions. *Am. J. Bot.* 72:1687–94

29. Huff A, Dybing CD. 1980. Factors affecting shedding of flowers in soybean (*Glycine max* (L.) Merrill). *J. Exp. Bot.* 31:751–62

30. Jacobs WP. 1952. The role of auxin in the differentiation of xylem around a wound. *Am. J. Bot.* 39:301–9

31. Kunst L, Klenz JE, Martinez-Zapater J, Haughn GW. 1989. *AP2* gene determines the identity of perianth organs in flowers of *Arabidopsis thaliana*. *Plant Cell* 1:1195–208

32. Lee TD. 1988. Patterns of fruit and seed production. In *Plant Reproductive Ecology: Patterns and Strategies*, ed. J Lovett Doust, L Lovett Doust, pp. 179–202. New York: Oxford Univ. Press

33. Lovett Doust L, Lovett Doust J, Turi K. 1986. Fecundity and size relationships in Jack-in-the-Pulpit, *Arisaema triphyllum* (Araceae). *Am. J. Bot.* 73:489–94

34. Macnair MR, Cumbes QJ. 1989. The genetic architecture of interspecific variation in *Mimulus*. *Genetics* 122:211–22

35. Macnair MR, Cumbes QJ. 1990. The pattern of sexual resource allocation in the yellow monkey flower, *Mimulus guttatus*. *Proc. R. Soc. Lond. B.* 242:100–7

36. Matthies D. 1990. Plasticity of reproductive components at different stages of development in the annual plant *Thalaspi arvense* L. *Oecologia* 83:105–16

37. Omitted in proof

38. McKone MJ. 1985. Reproductive biology of several bromegrasses (*Bromus*): breeding system, pattern of fruit maturation, and seed set. *Am. J. Bot.* 72:1334–39

39. Mossop R, Macnair MR, Robertson AW. 1994. Within-population variation in sexual resource allocation in *Mimulus guttatus*. *Funct. Ecol.* 8:410–18

40. Nakamura RR. 1986. Maternal investment and fruit abortion in *Phaseolus vulgaris*. *Am. J. Bot.* 73:1049–57

41. Nicholls MS. 1987. Spatial pattern of ovule maturation in the inflorescence of *Echium vulgare*: demography, resource

allocation and the constraints of architecture. *Biol. J. Linn. Soc.* 31:247–56

42. Obeso JR. 1993. Seed mass variation in the perennial herb *Asphodelus albus*: sources of variation and position effect. *Oecologia* 93:571–75

43. Obeso JR. 1993. Selective fruit and seed maturation in *Asphodelus albus* Miller (Liliaceae). *Oecologia* 93:564–70

44. Ritland C, Ritland K. 1989. Variation of sex allocation among eight taxa of the *Mimulus guttatus* species complex. *Am. J. Bot.* 76:1731–39

45. Robertson AW, Diaz A, Macnair MR. 1994. The quantitative genetics of floral characters in *Mimulus guttatus*. *Heredity* 72:300–11

46. Sachs T. 1981. The control of the patterned differentiation of vascular tissues. *Adv. Bot. Res.* 9:152–255

47. Smith JP Jr. 1977. *Vascular Plant Families*. Eureka, CA: Mad River Press

48. Stephenson AG. 1979. An evolutionary examination of the floral display of *Catalpa speciosa* (Bignoniaceae). *Evolution* 33:1200–9

49. Stephenson AG. 1980. Fruit set, herbivory, fruit reduction, and the fruiting strategy of *Catalpa speciosa* (Bignoniaceae). *Ecology* 6:57–64

50. Stephenson AG. 1981. Flower and fruit abortion: proximate causes and ultimate functions. *Annu. Rev. Ecol. Syst.* 12: 253–79

51. Stocklin J, Favre P. 1994. Effects of plant size and morphological constraints on variation in reproductive components in two related species of *Epilobium*. *J. Ecol.* 82:735–46

52. Sutherland S. 1986. Patterns of fruit set: what controls fruit-flower ratios in plants? *Evolution* 40:117–28

53. Sutherland S, Delph LF. 1984. On the importance of male fitness in plant plants: patterns of fruit set. *Ecology* 65: 1093–104

54. Taiz L, Zeiger E. 1991. *Plant Physiology*. Redwood City, CA: Benjamin Cummings

55. Thomson JD. 1985. Pollination and seed set in *Diervilla lonicera* (Caprifoliaceae): temporal patterns of flower and ovule deployment. *Am. J. Bot.* 72:737–40

56. Thomson JD. 1989. Deployment of ovules and pollen among flowers within inflorescences. *Evol. Trends in Plants* 3:65–68

57. Van Stevenick RFM. 1957. Factors affecting the abscission of reproductive organs in yellow lupins (*Lupinus luteus* L.). I. The effect of different patterns of flower removal. *J. Exp. Bot.* 8:373–81

58. Weberling F. 1992. *Morphology of Flowers and Inflorescences*. Cambridge: Cambridge Univ. Press

59. Weigel D, Alvarez J, Smyth DR, Ynofsky MF, Meyerowitz EM. 1992. *LEAFY* controls floral meristem identity in *Arabidopsis*. *Cell* 69:843–59

60. Winn AA. 1985. The effects of seed size and microsite on seedling emergence in four populations of *Prunella vulgaris*. *J. Ecol.* 73:831–40

61. Winn AA. 1988. Ecological and evolutionary consequences of seed size in *Prunella vulgaris*. *Ecology* 68:1224–33

62. Winn AA. 1991. Proximate and ultimate sources of within-individual variation in seed mass in *Prunella vulgaris* (Lamiaceae). *Am. J. Bot.* 78:838–44

63. Wolfe LM. 1992. Why does the size of reproductive structures decline through time in *Hydrophyllum appendiculatum* (Hydrophyllaceae)?: developmental constraints vs. resource limitation. *Am. J. Bot.* 79:1286–90

64. Young HJ, Stanton ML. 1990. Influences of floral variation on pollen removal and seed production in wild radish. *Ecology* 71:536–47

65. Young HJ, Stanton ML. 1990. Temporal patterns of gamete production within individuals of *Raphanus sativus* (Brassicaceae). *Can. J. Bot.* 68:480–86

66. Zimmerman JK, Aide TM. 1989. Patterns of fruit production in a neotropical orchid: pollinator vs. resource limitation. *Am. J. Bot.* 76:67–73

Annu. Rev. Ecol. Syst. 1995. 26:553–78

MUTATION AND ADAPTATION: The Directed Mutation Controversy in Evolutionary Perspective

P. D. Sniegowski and R. E. Lenski

Center for Microbial Ecology, Michigan State University, East Lansing, Michigan, 48824-1325

KEY WORDS: directed mutation, adaptive mutation, natural selection, evolution of mutation rates

ABSTRACT

A central tenet of evolutionary theory is that mutation is random with respect to its adaptive consequences for individual organisms; that is, the production of variation precedes and does not cause adaptation. Several recent experimental reports have challenged this tenet by suggesting that bacteria (and yeast) "may have mechanisms for choosing which mutations will occur" (6, p. 142). The phenomenon of nonrandom mutation claimed in these experiments was initially called "directed mutation" but has undergone several name changes during its brief and controversial history. The directed mutation hypothesis has not fared well; many examples of apparently directed mutation have been rejected in favor of more conventional explanations, and several reviews questioning the validity of directed mutation have appeared (53, 54, 59–61, 79, 80). Nonetheless, directed mutation has recently been reincarnated under the confusing label "adaptive mutation" (5, 23, 24, 27, 35, 74). Here we discuss the many experimental and conceptual problems with directed/adaptive mutation, and we argue that the most plausible molecular models proposed to explain "adaptive mutation" are entirely consistent with the modern Darwinian concept of adaptation by natural selection on randomly occurring variation.

In the concluding section of the paper, we discuss the importance of an informed evolutionary approach in the study of the potential adaptive significance of mutational phenomena. Knowledge of the molecular bases of muta-

0066-4162/95/1120-0553$05.00

tion is increasing rapidly, but rigorous evolutionary understanding lags behind. We note that ascribing adaptive significance to mutational phenomena (for example, "adaptive mutation") is beset with some of the same difficulties as ascribing adaptive significance to features of whole organisms (29). We consider some examples of mutational phenomena along with possible adaptive and nonadaptive explanations.

INTRODUCTION: THE HISTORICAL RELATIONSHIP BETWEEN VARIATION AND ADAPTATION

> Any heuristic can be treacherous, but a Darwinian explanation is the first I would seek in explaining a biological enigma. I do not insist that it will always last, but it has had enormous power in bringing us to our present understanding.
>
> J. Lederberg (49, p. 398)

We begin by providing a brief historical sketch of theories and evidence concerning the relationship between heritable variation and adaptation. Our purpose is not to present a formal or comprehensive historical analysis of this subject. Rather, we wish to place the directed mutation controversy in perspective by illustrating some of the ways in which variation and adaptation have repeatedly been confounded and then sorted out since Lamarck. We acknowledge that a brief historical synopsis such as this risks oversimplifying the rich history of scientific ideas and debate. Readers should consult comprehensive treatments of the history of biology (e.g. 63) for more detail.

Lamarck and Darwin

Lamarck (47) theorized that heritable adaptive variation arises in individual organisms as a consequence of needs and activities stimulated by environmental conditions. In the Lamarckian view, the origin of heritable variation and the origin of evolutionary adaptation are one and the same. Darwin, in contrast, conceived of a separation between variation and adaptation. According to Darwin, heritable variation arises continually as a result of (unknown) processes; evolutionary adaptation occurs as a consequence of natural selection acting on this heritable variation. Mayr has succinctly contrasted the evolutionary theories of Lamarck and Darwin: "The crucial difference between Darwin's and Lamarck's mechanisms of evolution is that for Lamarck the environment and its changes had priority. They produced needs and activities in the organism, and these, in turn, caused adaptational variation. For Darwin random variation was present first, and the ordering activity of the environment ('natural selection') followed afterwards" (63, p. 354). The primacy of natural selection in Darwin's theory of adaptation is illustrated by the following passage, which concludes Chapter Five of *The Origin of Species* (14, p. 170):

Whatever the cause may be for each slight difference in the offspring from their parents—and a cause for each must exist—it is the steady accumulation, through natural selection, of such differences, when beneficial to the individual, that gives rise to all the more important modifications of structure, by which the innumerable beings on the face of this earth are enabled to struggle with each other, and the best adapted to survive.

From Darwin to the Modern Synthesis

The blending mechanism of heredity widely accepted in Darwin's time posed problems for natural selection (42). Blending swamps variation, eroding both the hereditary differences among individuals that are necessary for natural selection and any heritable differences that might accumulate by natural selection. Under the assumption of blending, an enormous input of new variation is required at each generation to maintain distinct variants in an interbreeding population. Darwin argued both for the existence of variation and for the evolutionary transformation of old forms into new by natural selection, but he did not demonstrate a source of variation or a hereditary mechanism that could withstand the effects of blending. To account for the problem of blending, Darwin included roles for such Lamarckian factors as "the effects of use and disuse of parts" in the generation of new variation and in adaptation, both in the *Origin of Species* and in his pangenesis theory of inheritance (15).

Support for Lamarckian evolution, however, began to erode in the late 1880s. Weismann (93) made a forceful case against Lamarckism, arguing, among other things, that the observed separation of germline and soma in many organisms was inconsistent with the inheritance of acquired characters. Further doubt was cast on both Lamarckian evolution and blending inheritance by the rediscovery of Mendelism around 1900 (e.g. 68). Mutations in Mendelian factors were found to be infrequent and usually deleterious, contrary to the requirement of blending inheritance for tremendous inputs of variation and also inconsistent with what would be expected if mutations were to direct evolution by arising in response to adaptive need.

During the 1920s and 1930s, the emerging science of genetics and the theory of natural selection were incorporated into a comprehensive view of evolution—the modern synthesis (39). The modern synthetic theory identifies natural selection as the sole evolutionary force responsible for the adaptation of organisms to their environment. (There remain, of course, debates as to the levels of selection necessary to explain apparently adaptive phenomena; see, e.g. 94). A central tenet of the modern synthetic theory, therefore, is that mutation is random with respect to the adaptive needs of individual organisms.

As we show in the next section, direct experimental evidence indicating that mutation is random with respect to its adaptive consequences was not available prior to the publication of several classic experiments with bacteria in the

1940s and 1950s. Nonetheless, by the 1930s the randomness of mutation (in the sense given above) was widely accepted among geneticists and evolutionary theorists (e.g. 22, 88). Circumstantial evidence clearly favored random mutation; it also seems that the need to invoke an adaptive role for mutation had been effectively eliminated by the perceived potential of natural selection to explain adaptation.

Lamarckism and Bacterial Adaptation

Long after natural selection on randomly arising variation had gained wide acceptance as the mechanism of adaptation in higher organisms, debate continued over the relationship between variation and adaptation in bacteria (41, 85). In contrast to the situation in higher organisms, it was impossible to observe the origin of an individual bacterial variant in circumstances in which it was disfavored; the only way to isolate a specific bacterial variant was by altering the environment so as to favor its phenotype. Bacteria also had no separation of germline and soma as found in most higher organisms. Early experiments had shown that pure cultures of bacteria would quickly adapt to a selective agent when challenged, but it was unclear whether such adaptation should be attributed to the mass conversion of cells from one state to another or to selection on randomly occurring genetic variation. In retrospect, some of these cases may have been the result of physiological adaptation, i.e. the regulation of gene expression. But in 1934, Lamarckian inheritance in bacteria remained a definite possibility (Lewis in Ref. 57, p. 636):

> The subject of bacterial variation and heredity has reached an almost hopeless state of confusion. Almost every possible view has been set forth, and there seems no reason to hope that any uniform concensus of opinion may be reached in the immediate future. There are many advocates of the Lamarckian mode of bacterial inheritance, while others hold to the view that it is essentially Darwinian.

Indeed, it was unclear whether bacteria even had genes analogous to those of higher organisms. Julian Huxley was careful to exclude bacteria from the modern synthesis in 1942 (39, pp. 131–132).

Bacteria Enter the Modern Synthesis

In the 1940s and 1950s, elegant experiments opening the way to modern bacterial genetics provided strong support for the Darwinian view of bacterial adaptation, and, in fact, provided the first direct demonstrations, in any organism, of the random nature of mutation. The logic and methodology of these experiments have proven important in the recent debate over directed mutation, and so we describe them briefly here.

THE FLUCTUATION AND RESPREADING TESTS   In 1943, Luria & Delbrück formulated and tested two competing hypotheses to account for the appearance of cells resistant to viral infection in populations of *Escherichia coli* that were

(a)

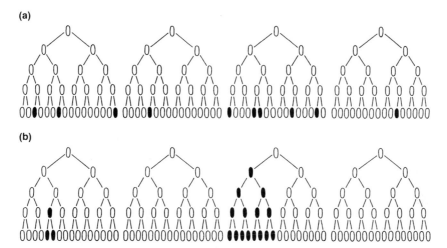

(b)

Figure 1 Schematic representation of the Luria-Delbrück fluctuation test. Distributions of mutants (filled symbols) across four populations, each founded from a single progenitor cell, expected under the hypotheses of (*a*) acquired hereditary immunity and (*b*) spontaneous mutation prior to exposure to the selective agent. The final row of cells represents the generation that is exposed to the selective agent. (Reprinted with permission from reference 85.)

previously sensitive to infection (58). The "acquired hereditary immunity" hypothesis supposed that each bacterium has a certain small probability of surviving exposure to the virus, and survival confers immunity that is inherited. In contrast, the "mutation" hypothesis supposed that each bacterium has a small probability of mutating spontaneously to viral resistance even in the absence of the virus, and that each descendant of a resistant mutant is itself resistant.

Luria & Delbrück deduced that the expected distribution of resistant mutants among independent cultures (each grown from a few sensitive cells) was markedly different under these two hypotheses. Under acquired hereditary immunity, resistance that arises with small probability per cell upon exposure to the virus should result in a Poisson distribution of resistant cells among cultures, with the expected variance equal to the mean (Figure 1a). Under the mutation hypothesis, however, occasional cultures in which resistant clones arose several generations before selection are expected to contain large numbers of resistant cells ("jackpots") compared with the average. The mutation hypothesis, therefore, predicts a clumped distribution of mutants among cultures, with variance greater than the Poisson expectation (Figure 1b). By spreading many cultures on agar plates containing the virus, Luria & Delbrück observed that resistant mutants were in fact distributed in jackpot fashion. This result was consistent with the mutation hypothesis but not with the hypothesis of acquired hereditary immunity.

A related test was presented by Newcombe in 1949 (71). In the respreading test, thousands of sensitive bacteria were allowed to grow from single cells into nearly confluent lawns of microcolonies on agar plates. Control plates were sprayed with the selective virus without disrupting the colonies, while other plates were sprayed with the virus and then the colonies were respread around the plate. Because respreading does not change the number of cells present on a plate, the hypothesis of acquired hereditary immunity predicted that control and respread plates would show equal numbers of resistant colonies. However, Newcombe observed a large increase in the numbers of resistant colonies on respread plates relative to controls. This result indicated that clones of resistant mutants had arisen spontaneously during growth on the plate before exposure to the virus and had then been dispersed around the plate by respreading.

REPLICA PLATING AND SIB SELECTION   The fluctuation test and the respreading test relied on quantitative reasoning to demonstrate the preexistence of bacterial variants resistant to selection. Neither test actually enabled an investigator to isolate variants without first exposing bacteria to a selective agent. Skeptics were not convinced (e.g. 37). In the 1950s, two additional tests—replica plating and sib selection—succeeded in demonstrating resistant variants in bacterial cultures never exposed to viruses or antibiotics.

In the replica plating test reported in 1952 by J. and E. Lederberg (50), cells were grown into a nearly confluent lawn of microcolonies on a nonselective agar plate. A piece of velveteen was then used to transfer a spatially ordered inoculum of cells from this "master plate" to several "replica plates" containing the selective agent, on which only resistant cells could form colonies (Figure 2). The Lederbergs observed a striking correspondence in the locations of resistant colonies on replica plates made from the same master plate, indicating that resistant cells had arisen by spontaneous mutation and increased in number by clonal growth on the master prior to selection. Furthermore, by pursuing a succession of master and replica plates with cells from suspected locations of resistant clones on each successive master plate, the Lederbergs were able to establish pure cultures of resistant bacteria without ever exposing them to the selective agent.

Cavalli-Sforza and J. Lederberg presented the related method of sib selection by limit sampling in 1956 (7). In this method, a primary culture containing a small number of presumptive mutants resistant to an antibiotic was divided into several equal subcultures, resulting in a chance increase in the proportion of resistant mutants in some subcultures. In essence, this procedure employs random genetic drift (i.e. founder effect) to increase nonselectively the proportion of mutants in certain subcultures. For example, if a primary culture containing one mutant is divided into ten equal subcultures, the subculture that receives the single mutant has a proportion of mutant cells that is tenfold higher than the original proportion in the primary culture. Cavalli-Sforza & Lederberg recognized that this increased proportion of mutants should be roughly maintained upon regrowth of this subculture in fresh medium, provided the growth rate of the mutants in nonselective medium is comparable to that of the nonmutant cells. Selective plating of samples from each regrown subculture allowed Cavalli-Sforza & Lederberg to determine retrospectively which subculture contained the increased proportion of mutants. This regrown subculture was then subjected to a new round of subculturing and regrowth. By repeating this cycle several times, Cavalli-Sforza & Lederberg were able to isolate pure cultures of antibiotic-resistant bacteria from cells that had never been exposed to antibiotics.

CONCLUSIONS FROM THE CLASSIC EXPERIMENTS   The experiments of Luria & Delbrück, Newcombe, Cavalli-Sforza, and J. and E. Lederberg showed that

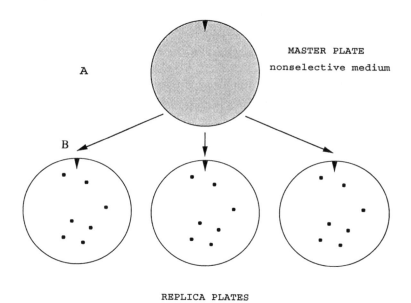

Figure 2   The logic of replica plating. (A) A large number of cells is grown into a nearly confluent lawn on a plate containing nonselective agar. A piece of velveteen (not shown) is then used to transfer spatially structured inocula from this master plate to replica plates containing selective agar. (B) The correspondence in location of colonies resistant to the selective agent on the replica plates indicates the spontaneous origin and clonal growth of resistance mutants on the master plate prior to selection. Sampling from the master plate using the spatial information in the replica plates allows the isolation of mutants that were never exposed to the selective agent.

heritable variants resistant to lethal agents could arise in bacterial populations before selection was applied; selection, therefore, could not have caused the occurrence of such variants. This finding suggested in turn that bacteria, as well as higher organisms, possess stable hereditary factors—genes—and that evolutionary adaptation in bacteria also proceeds via the action of natural selection on spontaneously arising genetic variation.

In retrospect, these experiments provided the first *direct* demonstrations of the random nature of mutation in any organism. These demonstrations were made possible by the ability of bacteriologists to manipulate and quantify vast clonal populations under controlled environmental circumstances, features not available to students of higher organisms. Bacteriology, the last major stronghold of Lamarckism, provided the textbook examples against Lamarckism.

THE DIRECTED MUTATION CONTROVERSY

Origin of the Controversy

In 1988, Cairns et al argued that the "classical experiments could not have detected (and certainly did not exclude) the existence of a non-random, possibly product-oriented form of mutation" (6, p. 142). They maintained that, because the classic experiments had employed lethal selective agents (viruses and antibiotics), the possibility that bacteria might adapt to nonlethal selective agents by some directed mutational process had been ignored. To that end, Cairns et al investigated cases "where the selective pressure rewards mutants by letting them multiply but allows all the other, non-mutant cells to survive so that they can at least have the opportunity to perform directed mutation" (6, p. 142). They concluded that the most plausible explanations for their experimental results resided in mechanisms that would confer on cells the capacity to adapt through the "inheritance of acquired characteristics" (6, p. 145). This and subsequent claims of "directed mutation" challenged the generality of the classic experiments demonstrating spontaneous mutation and raised a new controversy over the possibility of non-Darwinian adaptation.

Certain geneticists (e.g. 21, 83) seem to have found the evidence for directed mutation convincing. After all, claims of directed mutation emerged from the same powerful experimental system as the original demonstrations of random mutation. It is important to emphasize, however, that two significant aspects of the classic experiments have largely been overlooked in the directed mutation controversy. First, the authors of the classic experiments were careful about the assumptions of their tests. For example: in its simplest form, the sib selection experiment assumes that putative mutants and their progenitors grow at equal rates (are equally fit) in the absence of a selective agent. If, instead, mutants grow more slowly (are less fit), then the results of this experiment will deviate from randomness in a manner suggestive of directed mutation. Rather than immediately invoke directed mutation on such evidence, Cavalli-Sforza & Lederberg (7) considered and quantitatively tested the alternative hypothesis of differential growth rate. By contrast, the failure to consider and test alternative hypotheses led to harsh criticisms of the recent experiments that claim to demonstrate directed mutation.

Second, the observation that some mutations occur after cells are exposed to a selective agent does not indicate that those mutations are caused by selection. To imply that postselection mutations per se challenge the Darwinian view of adaptation (46) is to confuse the method of the classic experiments (showing that variation arises before the imposition of selection) with the logical interpretation of their results (variation is not caused by selection).

We discuss the important evidence and ideas in the directed mutation controversy in the remainder of this section. As we show, several apparent cases of directed mutation have been undermined by subsequent demonstrations that experimental problems gave rise to artifactual results. Furthermore, the most plausible mechanisms proposed to explain remaining cases of apparently directed mutation are entirely consistent with the modern Darwinian view that genetic variation arises without regard to adaptive need; that is, variation precedes adaptation.

Initial Claims of Directed Mutation Advanced by Cairns et al

"POISSON-LIKE" DISTRIBUTIONS OF LAC+ REVERTANTS IN FLUCTUATION TESTS The first modern case of apparently directed mutation involved the appearance of Lac^+ revertants in cultures of a Lac^- ($lacZ_{am}$ $uvrB$) strain of $E.$ $coli$ starving in a medium containing only lactose as a potential carbon source (lactose minimal medium). Cairns et al (6) reported that when numerous cultures of this strain were grown under permissive conditions and subsequently plated onto lactose minimal medium, the observed distribution of Lac^+ mutants per culture was markedly different from the jackpot distribution expected in a fluctuation test if mutants arose only before plating. In particular, substantial numbers of mutants appeared some days after plating on the lactose minimal medium, giving rise to a hybrid, "Poisson-like" distribution of mutants; such a distribution might be expected if these late-arising mutants occurred during starvation specifically in response to lactose. Cairns et al tested for the dependence of the late-arising mutants on lactose by plating cells onto medium containing no carbon source and adding lactose later; they found that Lac^+ revertants did not begin accumulating until after lactose was added. Furthermore, Cairns et al observed that mutants to a phenotype (valine resistance) unrelated to the lactose selection did not appear during starvation on lactose minimal plates. To Cairns et al this result indicated that mutation rates were not generally elevated in the starving cultures.

Cairns et al argued that the occurrence of Poisson-like distributions, the appearance of late-arising Lac+ mutants only after the addition of lactose, and the lack of increased mutation at an unselected locus all were consistent with the hypothesis of directed mutation to Lac+ in the presence of lactose. However, numerous authors subsequently noted that these results were also consistent with spontaneous mutation. Many questioned the appropriateness of valine-resistance mutations as a control for elevation of the general mutation rate (13, 21, 38, 53, 54, 59, 61). Mutations to valine resistance can arise in several loci and by many types of sequence alteration, and they therefore may not be comparable to the reversion or suppression of an amber mutation in *lacZ*. MacPhee (60) showed that the assay conditions Cairns et al had used to detect mutations to valine resistance in starving cells actually suppressed the occurrence of those mutations. Many authors also pointed out that Poisson-like distributions can result from violations of various assumptions of the fluctuation test; hence, the appearance of Poisson-like distributions of mutants need not indicate directed mutation (10, 54, 55, 86, 87, 89). Indeed, a number of earlier authors had noted that discrepancies between fluctuation test results and the predicted jackpot distribution of mutants were not sufficient to reject the hypothesis of spontaneous mutation (44, 48, 75). For example, if mutants grow more slowly than nonmutants before exposure to the selective agent, then the distribution of mutants observed when the replicate cultures are plated on selective medium will be less variable than the jackpot expectation. Several authors noted that among the late-arising Lac+ phenotypes observed by Cairns et al were many amber suppressor mutants, which are likely to grow slowly in permissive medium as a consequence of altered transcription (10, 38, 54, 55). Indeed, Cairns et al noted that these suppressor mutants produced characteristically small colonies on permissive agar plates (6).

No further experimental evidence has appeared for or against directed mutation to Lac+ in the *lacZ*$_{am}$ *uvrB* strain investigated by Cairns et al, and this case must thus be regarded as unresolved. However, Cairns acknowledged the potential problems with the case, noting that "if these had been the only experiments, the [1988] paper would not have been written" (4, p. 527). Ironically, the case that Cairns regarded as stronger evidence for directed mutation has fared much worse.

A CASE OF DIRECTED MUTATION THAT SEEMED PARTICULARLY STRONG IS REJECTED: EXCISION OF PROPHAGE MU   In *E. coli* strain MCS2 (76), part of the *ara* operon including a regulatory region has been joined to structural genes from the *lac* operon by bacteriophage Mu DNA containing transcription terminating signals. With this prophage intact, MCS2 cannot grow on either lactose or arabinose. However, upon excision of the prophage in a suitable reading frame, MCS2 is phenotypically Lac(Ara)+; it can grow on lactose if arabinose is present as an inducer. Shapiro (76) had noted that Lac(Ara)+ excision mutants almost never arise in MCS2 cultures that are actively growing on glucose or glycerol, but that substantial numbers of Lac(Ara)+ excision mutants appear in MCS2 cultures that have been starved for several days on medium containing only lactose and arabinose as potential carbon sources. In their 1988 paper (6), Cairns et al reported further experiments in which they were unable to recover Lac(Ara)+ mutants from cultures starving on media not containing lactose and arabinose. These results led them to conclude that Lac(Ara)+ mutants arose only when MCS2 cells were starving in the presence of lactose and arabinose, so that the occurrence of Mu excisions in MCS2 seemed a particularly clear case of directed mutation.

Mittler & Lenski (66) confirmed Shapiro's observations that Lac(Ara)+ mutants almost never occur in growing cultures but do occur at high frequency when cells are starved on medium containing lactose and arabinose. However, in contrast to Cairns et al, Mittler & Lenski found that Lac(Ara)+ mutants also occur in starving cultures on media that do not contain lactose and arabinose. The latter result suggested that Lac(Ara)+ mutants are not directed by the presence of lactose and arabinose but instead are induced by starvation. Mittler & Lenski further showed that the frequency of Lac(Ara)+ mutants detected in cultures of MCS2 starved without lactose and arabinose is stable when those cultures are regrown in glucose (66). This result clearly did not support the existence during starvation of unstable Mu excision intermediates that rapidly convert to the Lac(Ara)+ phenotype only upon exposure to lactose and arabinose. Nonetheless, some proponents of directed mutation were skeptical of Mittler & Lenski's results (23, 78). Foster (23) implied that use of the classical methods of detecting preexisting mutations was necessary to confirm or reject directed mutation in MCS2.

Fluctuation analysis, sib selection, and replica plating have now all been used to test the directed mutation hypothesis in MCS2. All three approaches uphold Mittler & Lenski's finding that Lac(Ara)+ mutations occur in starving cultures regardless of whether lactose and arabinose are present. Foster & Cairns (26) employed the fluctuation test to show that a jackpot distribution of mutants was obtained when replicate MCS2 cultures starved in liquid medium without lactose and arabinose were regrown and plated on medium containing lactose and arabinose. This result implies the existence of Lac(Ara)+ excision mutants before the exposure of cultures to lactose and arbinose. Maenhaut-Michel & Shapiro (62) used sib selection to enrich the proportion of Lac(Ara)+ excision mutants in starved MCS2 cultures, and they obtained pure cultures of Lac(Ara)+ mutants without ever exposing the progenitor cells to lactose and arabinose. Finally, Sniegowski (81) used replica plating to show that nearly all the Lac(Ara)$^+$ mutants detected when a starved MCS2 culture was exposed to lactose and arabinose were preexisting.

The Many Potential Flaws in Claims of Directed Mutation

In an earlier review, Lenski & Mittler (54) identified several effects that have the potential to mislead experimenters into concluding that directed mutation is occurring when, in fact, it is not. The $lacZ_{am}$ and Mu cases indeed illustrate two of these effects; others will be brought out below when we discuss subsequent cases of apparently directed mutation. The "Poisson-like" distributions of $lacZ_{am}$ revertants that Cairns et al observed in fluctuation tests were quite plausibly due to slow growth of some revertants, particularly amber suppressors, prior to selective plating. In the case of Mu excision, starvation and the presence of selective substrates were confounded; the observed discrepancy between rates of mutation to Lac(Ara)$^+$ during growth and during starvation on lactose-arabinose medium was a nonspecific consequence of starvation rather than a specific response to the presence of lactose and arabinose. Indeed, the case of Mu excision illustrates the general point, made earlier, that the occurrence of mutations after the imposition of a selective agent does not demonstrate that the selective agent is the cause of those mutations.

Subsequent Cases of Apparently Directed Mutation

We next consider several cases of apparently directed mutation that were reported after the 1988 paper by Cairns et al. We focus upon cases for which detailed experimental reevaluation has supported alternative explanations consistent with the modern Darwinian view of adaptation. We acknowledge that not all cases of apparently directed mutation have been so examined (see, e.g., 33, 84). Given the general nature of the potentially misleading effects in directed mutation experiments, some of these other cases may have explanations similar to those described below. We do not speculate here on such possible alternative explanations, except to note that no case of apparently directed mutation has received a full mechanistic explanation that supports a non-Darwinian process of adaptation. At the end of this section, we describe the most studied remaining case of apparently directed mutation, the so-called "adaptive" reversion of a *lac* frameshift in *E. coli*. After discussing recent results in this case, we consider the molecular models that have been invoked to explain it and other cases of apparently directed mutation. We stress that mechanisms in the most plausible models, though inherently fascinating and potentially important to the study of mutagenesis, are consistent with the modern Darwinian view that variation precedes adaptation.

EVENTS GIVING RISE TO DOUBLE MUTANTS IN THE *Bgl* OPERON: ANTICIPATORY
MUTATION?   Hall (31) studied an *E. coli* K12 strain in which two mutations
in the *bgl* operon are apparently required for growth on salicin: excision of an
insertion sequence, *IS*150, from a structural gene, *bglF*, and a mutation in a
regulatory region, *bglR*. Hall observed that *IS*150 excision almost never occurred in growing cultures of this strain; consequently, salicin-utilizing (Sal⁺)
double mutants did not arise at detectable frequencies during growth on some
other substrate. However, Hall detected large numbers of Sal⁺ cells in cultures
subjected to prolonged incubation on agar supplemented with salicin as the
only available growth substrate. Hall reported that *IS*150 excision-mutant intermediates accumulated in these cultures before the appearance of Sal⁺ double
mutants, but that an excision mutant clone was incapable of growth on salicin
without the second mutation in *bglR*. On this basis, Hall argued that the
observed increase in the frequency of excision-mutant intermediates was the
result of anticipatory directed mutation to produce a population of cells large
enough to acquire the second, random mutation in *bglR* that would allow
growth on salicin.

Hall's extraordinary claim of anticipatory directed mutation was challenged
by Mittler & Lenski (67). Contrary to Hall's claim, these authors found that
many excision mutants, including the one tested by Hall, are in fact capable
of some growth on salicin. The growth of these excision-mutant intermediates
increases the expected number of fully Sal⁺ double mutants on selective salicin
agar by many orders of magnitude, such that there is no need to invoke
anticipatory directed mutation.

Hall has acknowledged that some excision mutants are capable of growth
on salicin without the second mutation in *bglR*, and that such growth can
explain his previous results in the *bgl* system without the need to invoke
anticipatory directed mutation (34a). At the same time, however, Hall has made
a further claim that *IS*150 excision is nonetheless directed in a genetic background in which no other mutations are required for full utilization of salicin
(34a). To date, this new claim has not been challenged experimentally.

ENHANCED RATE OF APPEARANCE OF TRP⁺ CELLS DURING STARVATION OF A
trpA trpB DOUBLE MUTANT FOR TRYPTOPHAN   Hall (32) also claimed that the
appearance of Trp⁺ cells in cultures of a *trpA trpB* strain of *E. coli* starved for
tryptophan is "selection-induced," in that *trpA⁺ trpB⁺* cells arise at far higher
rates than expected from the product of the reversion rates of single *trpA* and
trpB mutants in similar circumstances. Foster, however, suggested that single-
mutant *trpA trpB⁺* intermediates might be able to grow on indole, a tryptophan
precursor that can accumulate in medium without tryptophan as a result of
excretion by *trpA⁺ trpB* intermediates or breakdown of indoleglycerol phosphate excreted by the *trpA trpB* progenitor (25). As in the *bgl* case, the

accumulation, by growth, of an intermediate genotype could explain the increased occurrence of double mutants in starving *trpA trpB* populations without the need to invoke directed mutation. Further experiments by Hall have in fact revealed substantial growth of *trpA trpB+* cells in mixed culture with *trpA trpB* cells on selective medium, and Hall now acknowledges that selective enrichment of *trpA trpB+* intermediates may explain the increase in *trpA+ trpB+* double revertants on medium without tryptophan (34).

BIASED RECOVERY OF DEX$^+$ MUTANTS   Benson et al (2) examined mutation in an *E. coli* strain that lacks the LamB outer membrane protein and thus is unable to grow on large maltodextrins (Dex$^-$). Mutations in genes for two other membrane proteins, OmpC and OmpF, can give rise to Dex$^+$ phenotypes in this strain. Benson et al observed that when Dex$^-$ populations were starved on a medium containing only maltodextrins as a potential carbon source, OmpF$^+$ mutations apparently occurred at a much higher frequency than did OmpC$^+$ mutations (2), as though a process of directed mutation were taking place at the ompF locus. Upon further investigation, however, Benson et al discovered that OmpF$^+$ mutants overgrew their Dex$^-$ progenitors much more quickly than did OmpC$^+$ mutants, leading to a bias in the recovery, rather than the occurrence, of the OmpF$^+$ mutation (1).

REVERSION TO LEUCINE PROTOTROPHY IN *SALMONELLA TYPHIMURIUM*   Dijkmans et al (18) observed Poisson-like distributions of Leu$^+$ revertants in fluctuation tests with a Leu$^-$ strain of *S. typhimurium*. The growth rates of Leu$^+$ mutants on permissive media prior to selective plating were similar to that of the Leu$^-$ progenitor, and this seemed to rule out one frequently suggested alternative to directed mutation. However, many Leu$^+$ clones consisted of cells that were 10- to 100-fold larger than nonmutant cells. Dijkmans et al postulated that the transition from nonmutant Leu$^-$ to much larger Leu$^+$ mutant cells is likely to involve a substantial initial delay in cell division as mutant daughter cells increase in size; this delay appears to be responsible for the observed Poisson-like distribution of seemingly directed, late-arising Leu$^+$ mutants on selective plates. Consistent with this hypothesis, Dijkmans et al observed that Leu$^+$ mutants that gave rise to jackpots on selective plates had normal cell sizes, in contrast to the late-arising mutants (18).

The Case of "Adaptive Mutation"

Despite the setbacks in the cases described above, directed mutation has recently garnered renewed publicity (27, 35, 74) under the guise of "adaptive mutation," a term that sits uneasily between Lamarckian and Darwinian connotations. DNA sequence data have suggested the involvement of known molecular mechanisms in this case. In a later section, we argue that the suggested mechanistic basis for the phenomenon of "adaptive mutation" is entirely consistent with the modern Darwinian view that adaptation is a consequence of natural selection, not mutation. However, we first describe the important features in this case.

In 1991, Cairns & Foster reported that a strain of *E. coli* unable to grow on lactose because of a *lacI* frameshift polar on the *lacZ* region would revert to Lac+ during prolonged incubation on lactose minimal medium (5). Unlike the case of Mu excision described above, this case was not a clearcut candidate for directed mutation; some Lac+ mutations occurred in populations of this strain growing in permissive (nonselective) medium. However, Cairns & Foster showed that Lac+ revertants did not accumulate in this strain during starvation when lactose was absent or when lactose was present but another growth requirement was unfulfilled. (The latter finding implies that lactose per se is not sufficient to promote recovery of the *lac* frameshift revertants. This observation is critical when it comes to considering the mutational mechanisms that may be involved and their implications, as we discuss further below.) Foster (24) has examined and apparently rejected many potential artifactual explanations similar to those we have described in conjunction with other cases of apparently directed mutation and concluded that the presence of lactose is necessary for Lac+ mutations to occur during starvation.

DNA sequencing of Lac+ revertants recovered during starvation of the *lac* frameshift strain on lactose minimal medium revealed that the majority of these are the result of single-base deletions in short mononucleotide repeats (27, 74). In contrast, sequencing of revertants recovered from growing cultures indicates a broader spectrum of mutational events, including duplications, deletions, and insertions that are many nucleotides in length. This change in the relative frequencies of recovered mutations suggests that certain mutational events occur more frequently in *lac* frameshift cells starving in the presence of lactose than in growing cells. (Artifactual explanations, however, have not been completely ruled out. For example, selection may favor certain mutants over others, as in the case of biased recovery of Dex+ mutants.) Very recently, it has unexpectedly been shown that replication and possibly conjugal transfer of the plasmid carrying the defective *lac* gene may be involved in "adaptive mutation" (27a, 73a). These findings, while intriguing, further illustrate the lack of a clear understanding of the molecular mechanisms and population dynamics underlying apparently directed mutation in this system.

MOLECULAR MODELS PROPOSED TO EXPLAIN APPARENTLY DIRECTED MUTATION

... only a vitalist Pangloss would consider that the genes know how and when it is good for them to mutate.

Th. Dobzhansky (19, p. 92)

Here we consider several molecular models that have been proposed to explain "adaptive mutation" and other cases of apparently directed mutation. We identify two major categories of model: 1. neo-Lamarckian models in which individual cells are postulated to possess the capacity to monitor their own fitness and somehow increase the probability of mutations conferring higher fitness; and 2. non-Lamarckian models. We argue that the mechanisms invoked in the second category of model are the more plausible, but we caution that no model has received experimental confirmation.

Neo-Lamarckian Models

SPECIFIC REVERSE TRANSCRIPTION OF mRNAS ENCODING SUCCESSFUL PROTEINS In conjunction with the original claims of directed mutation, Cairns et al suggested that "the cell could produce a highly variable set of mRNA molecules and then reverse-transcribe the one that made the best protein" (6, p. 145). In other words, if a cell could somehow monitor protein variants and reverse transcribe the specific message that encoded the most successful one, then the result truly would be directed mutation. It is a fact that a single allele yields variable mRNA molecules as a consequence of transcription errors; upon translation, such variable mRNAs can produce variable proteins. Also, reverse transcriptase is present in some *E. coli* strains. However, the specific reverse transcription model postulates the existence of a heretofore unknown cellular component that can somehow monitor the effect of variant proteins on fitness and choose the appropriate mRNA for reverse transcription.

Foster (23) has argued that the specific reverse transcription model incorporates selection, in that cells first generate variable mRNAs and proteins and only successful ones are reverse transcribed. Clearly, however, the model invokes a non-Darwinian process in which an individual cell somehow assesses its own fitness and selects the appropriate protein, RNA, and ultimately mutation. It seems very unlikely that a cell can assess its own fitness dependably in this manner. Fitness is the cross-generational product of survival and reproductive success, and it need not correlate predictably (monotonically) with the activity of specific proteins or with the ability of cells to utilize specific growth substrates. How is an organism, in this case a cell, to assess its own fitness?

There is little current support for the specific reverse transcription model. Reverse transcriptase has not been discovered in the particular *E. coli* K12 strains used to study directed mutation. In addition, suppressor mutations that occur outside the transcribed gene also accumulate when Lac⁻ cells are incubated on media containing lactose, a phenomenon not predicted by the model (25). Foster & Cairns have acknowledged the lack of evidence supporting the specific reverse transcription model, concluding that "selective condition does not play an 'instructional' role in determining which DNA sequence changes arise" (25, p. 785).

NONRANDOM AMPLIFICATION OF BENEFICIAL MUTANT GENES   Cairns & Foster (5) showed that RecA function is required for "adaptive" reversion of a *lac* frameshift mutant in *E. coli*, and they suggested that this finding implicates gene amplification as part of a mechanism for directed mutation. In essence, gene amplification could create a large target for mutation, consisting of an array of many copies of the relevant gene. Any gene in the array that garnered a mutation conferring growth would allow a cell to escape starvation, after which the amplified region could be resolved by a RecA-dependent process.

The problem with this model is that it can produce a high bias in favor of beneficial mutations only if such mutations can be identified by the cell and preferentially amplified further within an array. Otherwise, any favorable mutant sequence within an array has as high a probability as any other sequence of being lost when the array is resolved back to a single copy by random recombinational processes. Stahl (83) has noted that there is no known process that would allow a cell to treat one sequence in an array differently from another. In fact, the preferential amplification model suffers from the same general difficulty as that described above for the reverse transcriptase model: There is no known molecular mechanism that would allow a cell to assess its own fitness and preferentially generate or retain the mutation it needs.

Non-Lamarckian Models

A number of models have invoked mechanisms in which a cell need not assess the effects of different genetic variants on its own fitness. Instead, these models propose to explain apparently directed mutation either as a consequence of increased random mutation at specific loci induced by the selective agent (mutagenic transcription model) or of differential proliferation of genetic variants arising during limited DNA replication in nongrowing cells (incipient mutation models). In these non-Lamarckian models, as in the neo-Lamarkian models, the initial variation is proposed to occur at random. There is a crucial distinction between these two categories of models, however. The neo-Lamarckian models require that an individual cell somehow be able to scrutinize variants and select the most appropriate mutation. The non-Lamarckian models, on the other hand, simply assume that randomly arising genetic variants proliferate differentially as a consequence of their fitness effects. These models, therefore, actually invoke natural selection to explain apparently directed mutation (56, 80).

MUTAGENIC TRANSCRIPTION   Davis (16) hypothesized that transcription might be mutagenic, such that the presence of a selective substrate (e.g. lactose) that induces transcription increases the mutation rate at the selected locus. Such a mechanism could give the appearance of directed mutation, because beneficial mutations would arise at a higher rate in the presence of the substrate. However, the model also predicts that nonbeneficial (misdirected) mutations should arise at a higher rate in the presence of the substrate, and so it does not imply that mutation would be systematically beneficial at the selected locus (54).

Experimental evidence published to date has not supported the transcriptional mutagenesis model as an explanation for apparently directed mutation. Although addition of the gratuitous inducer IPTG to growing populations of an inducible Lac⁻ strain does slightly increase the number of Lac⁺ mutants observed, this effect is absent when IPTG is added to starving Lac⁻ populations, contrary to the prediction of the model (16). In addition, the model does not explain why starving cells that constitutively transcribe the *lac* operon apparently accumulate mutations only in the presence of lactose (5, 6, 25).

INCIPIENT MUTATION MODELS   Earlier, we alluded to an important finding in studies of "adaptive mutation": revertants of a Lac⁻ frameshift mutant do not accumulate in the presence of lactose when there is a second, unfulfilled growth requirement (5). Evidently, lactose is not sufficient to promote apparently directed mutations in this system. The requirement for cell growth offers support for another category of molecular models for "adaptive mutation" first proposed by Stahl (82). These models invoke random DNA sequence alterations in starving cells, which we call "incipient mutations," to explain apparently directed mutation (Figure 3). If a coding strand in a starving cell should be altered so as to encode a variant sequence that can be transcribed and translated, and if the resulting protein allows the cell to grow and replicate its DNA, then one of the two daughter cells could possess a mutation at the site of the sequence alteration in the parent cell. If, however, the incipient mutation does not allow replication and cell growth (e.g. if there is a second, unfulfilled growth requirement), then nonmutagenic mismatch repair may eventually restore the original sequence, or the cell may die as a consequence of unrepaired damage. In either of the latter cases, a mutation will not be detected.

In response to the initial Cairns et al paper, Stahl (82) and Boe (3) proposed that the methyl-directed mismatch repair system might act more slowly during starvation than during growth, allowing unrepaired sequence alterations to be made permanent by chromosome replication if they enable cells to grow. Mismatch repair-deficient strains do show elevated mutation rates under selective conditions (3). However, the slow repair model predicts that uncorrected mismatch mutations should accumulate when a mismatch repair-deficient strain is starved, regardless of whether selection is applied. This is not the case: Lac⁺ mutants apparently do not accumulate in Lac⁻ strains deficient in repair when lactose is absent (25).

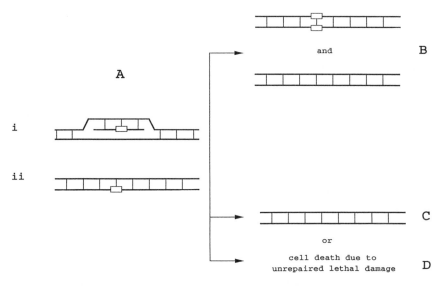

Figure 3 Hypothetical "incipient mutation" model to explain why certain mutations might be recovered only in environments in which they are advantageous. Polymerase error during limited DNA synthesis (A_i) or slow repair of DNA damage (A_{ii}) alters the genetic sequence originally present in a cell. By chance, the altered sequence encodes a functional gene (such as *lacZ*); before the altered sequence can be (correctly) repaired, it is transcribed and translated. (*B*) In an environment where the gene product allows the cell to grow and replicate its DNA (in the case of *lacZ*, where lactose is the sole carbon source), one of the daughter cells could inherit a mutation at the site where the original sequence alteration occurred. (*C, D*) In an environment where the gene product is superfluous (lactose is absent) or insufficient for cell growth (some other nutritional requirement is unfulfilled), then either nonmutagenic mismatch repair may restore the original sequence (*C*) or the cell may die as a consequence of unrepaired damage (*D*).

The discovery by Foster & Cairns that "adaptive mutation" is RecA dependent (25) led Stahl subsequently to suggest a second incipient mutation model. This model invokes a form of DNA synthesis, called "stable DNA replication" (17), which occurs in nondividing cells and is RecA dependent (95). Stahl suggested that such replication might ordinarily halt at the D-loop stage during starvation (Figure 3, part A_i), with subsequent degradation of the incipient strand. However, if a growth-enabling sequence change on the incipient strand could be transcribed and translated, then a full replication fork might form and the useful mutation could be transmitted to a daughter cell (Figure 3B).

We emphasize that there is presently no evidence to confirm any incipient mutation model. Nonetheless, mutations associated with limited DNA replication are implicated by some recent results on "adaptive" mutation (27, 74). As mentioned above, revertants recovered during starvation of the *lac* frameshift strain on lactose minimal medium are mostly the result of single-base deletions in short mononucleotide repeats. Such sequence changes implicate polymerase errors, possibly associated with strand slippage during recombination, repair, or replication, as the cause of these mutations (27, 74).

It will be interesting to see whether an incipient mutation model such as those described above can be experimentally confirmed. It seems unlikely, however, that such a finding would alter evolutionary theory. As we have argued repeatedly in this paper, the key feature of the modern Darwinian theory of adaptation is that genetic variation arises at random with regard to its effects on fitness, such that adaptation occurs solely as a consequence of natural selection on this variation. According to the incipient mutation models, discrepancies between the two DNA strands arise at random with respect to their adaptive utility; the systematic difference in the proliferation of variant strands that these models invoke is due to natural selection. In contrast to the neo-Lamarckian models, the individual cell does not select, choose, or instruct anything in the incipient mutation models.

MUTATION AND ADAPTATION IN EVOLUTIONARY PERSPECTIVE: EVALUATING THE POTENTIAL ADAPTIVENESS OF MUTATIONAL PHENOMENA

> Beneath the imposing building called 'Heredity' there has been a dingy basement called 'Mutation.' Lately the searchlight of genetic analysis has thrown a flood of illumination into many of the dark recesses there, revealing some of them as ordinary rooms in no wise different from those upstairs, that merely need to have their blinds flung back, while others are seen to be subterranean passageways of a quite different type.
>
> H. J. Muller, 1921 (70, p. 106)

Throughout this chapter, we have argued that the evidence for directed mutation does not warrant a revision or qualification of the modern Darwinian theory that evolutionary adaptation occurs solely as a consequence of natural selection acting on randomly occurring variation. We have shown that many purported examples of directed mutation have alternative explanations of a more conventional nature. In addition, we have argued that the most plausible molecular models proposed to explain the current incarnation of directed mutation ("adaptive mutation") are fully consistent with the modern Darwinian theory of adaptation.

By arguing that mutation is random, we have not meant to imply that mutation occurs at equal rates at all loci or in all environments, or that mutations do not have definable, proximate causes. Rather, we argue that environmental factors (proximate causes) do not induce specifically those mutations that are beneficial. In this final section, we shift our focus and briefly consider

Table 1 Several mutational phenomena, their hypothesized adaptive significance, and possible alternative explanations.

Mutational phenomenon	Hypothesized adaptive significance	Alternative explanation
Starvation-induced mutagenesis	May occasionally allow an organism that is physiologically stressed, and which presumably has little to lose, to acquire the ability to use some available resource (e.g. 32).	See text.
Transposon activity	May promote complex variation not accessible by point mutation (64, 77).	Mutation may be an indifferent consequence of the activities of selfish DNA (9, 20, 73); causes mostly deleterious mutations
Transcription-induced mutagenesis	May allow an organism to improve particular genes under specific ecological conditions where the gene product is required for growth (16).	May be an unavoidable consequence of mechanistic constraints during transcription; may increase deleterious mutations in essential genes (54).
Hypermutable loci	May allow an organism to increase variation in certain "contingency" genes without increasing load of deleterious mutations in essential "housekeeping" genes (69).	Variation in rates among loci may have arisen for reasons unrelated to postulated adaptive value. Requires confirmation using comparative and experimental methods (69).
Mutation rate disparity between leading and lagging strands during DNA replication	May provide a balance between novelty and conservatism superior to what can be achieved by having both strands equally mutable (92).	May be an unavoidable consequence of DNA replication machinery (92).

the possibility that various mutational phenomena are nonetheless adaptive in the sense that they have been "designed" or maintained by natural selection (ultimate causes) because the random variation they produce increases evolutionary flexibility. As more is known about mechanisms causing mutation, speculation increases about the possible adaptive significance of these mechanisms as sources of variation (see Table 1 for some examples). Here we emphasize that notions about the adaptive significance of mutational phenomena must be regarded as evolutionary hypotheses, which require rigorous testing and independent confirmation.

The process of adaptation by natural selection is the cornerstone of modern evolutionary theory, and so it is natural to look for adaptive explanations for organismal traits. But not all traits are the result of adaptation by natural selection. Students of morphology and behavior were given a sharp reminder of this in 1979 when Gould & Lewontin (29) labelled the uncritical invocation of adaptive explanations for various organismal traits as "adaptationism." Gould & Lewontin provided numerous alternative explanations for the existence of any particular trait, including the random fixation of alleles by genetic drift, developmental or mechanistic correlations among traits, phylogenetic inertia and constraints, and so on.

In the context of this paper, we suggest that, while a given mutation may sometimes have beneficial effects (e.g. the Lac$^+$ mutant that can grow in an environment where the cell would otherwise starve), it is not the case that the mechanism that causes mutation is necessarily adaptive. One finding from studies of the directed mutation phenomenon that is well supported is that certain mutagenic processes (for example, Mu excision) are increased in starving bacterial cells. One might assume that a starving cell has nothing to lose, and that an elevated rate of mutation during starvation is adaptive (beneficial) because a cell might thereby stumble on a good mutation that allowed it to grow on a substrate that happened to be available. But there is also an evolutionary downside, which is the possibility that a cell might acquire a deleterious mutation that would prevent it from growing in the event that the environment later became more favorable for growth (prior to death by starvation). In addition, the very mechanism of mutagenesis itself could involve the risk of cell death through unrepaired DNA damage (35).

Perhaps increased mutation in response to starvation is not an adaptation at all, but rather is symptomatic of a cell that is falling apart and losing control over its genetic integrity. (Indeed, in the case of Mu excision, a plausible evolutionary hypothesis is that the Mu bacteriophage has evolved the capacity to detect when its host is dying, and, as a consequence, to leave in search of a new host (65).) Consider the SOS response, in which an elevated mutation rate is induced by environmental stresses, such as UV irradiation, that cause damage to DNA (45). The increased mutations result from the action of enzymes that bypass DNA damage (e.g. pyrimidine dimers) that would otherwise block replication. This replicative bypass introduces mutations, and these mutations might occasionally have beneficial consequences. Even if the vast majority of mutations are detrimental, however, it is clearly more evolutionarily advantageous to repair the damage mutagenically than not to repair it at all, since the alternative is failure to replicate. Perhaps, then, the mutagenic effects of the SOS response are the best that can be made of a bad situation (e.g. see 27a, p. 510).

Given the above criticisms, some might throw up their hands at the apparent difficulty of determining whether a mutational phenomenon is or is not adaptive. Certainly the task is not likely to be easy. Several approaches, however, may allow this question to be addressed (69). One is theoretical analysis, which examines the costs and benefits of one evolutionary strategy relative to another. Such an approach can establish the conditions under which an adaptive explanation is feasible, and it may suggest variables that could be measured to shed further light on this feasibility. For example, in the case of starvation-induced mutation, the feasibility of the adaptive explanation may hinge on the relative rates of death due to starvation and environmental change that relieves starvation. There already exists a substantial theoretical literature on the evolution of mutation rates (e.g. 28, 40, 43, 51, 52), which may provide a framework for further analyses to address specific issues. A second approach is comparative. In essence, one tests the correlation between organismal traits and features of their environments. Although the comparative approach is very old, important methodological advances have recently been made that reflect the importance of phylogenetic considerations in developing appropriate statistical criteria for accepting or rejecting an association (36). A third approach, for which bacteria are particularly well suited, is experimental. The idea here is to devise selective regimes that would be expected to favor, for example, an increase in the trait of interest under one hypothesis but not under an alternative. Several experimental studies have examined the evolutionary adjustment of mutation rates (8, 11, 12, 72, 90, 91); the methodology of these experiments can provide a foundation for future research. Of course, the most compelling cases of adaptation are those that can be supported by careful theoretical, comparative, and experimental analyses.

Our point in criticizing adaptive explanations for various mutational mechanisms and phenomena is not to imply that these explanations are wrong or implausible. We believe, however, that such explanations should be regarded as evolutionary hypotheses until sufficient evidence is provided to corroborate or refute them. The rapidly advancing field of molecular genetics is sure to provide more intriguing possibilities of the kind listed in Table 1. We suggest that studies reflecting an informed evolutionary perspective will be essential to a comprehensive understanding of such phenomena. Such studies may further enrich the modern Darwinian perspective on mutation and adaptation.

Acknowledgments

We are grateful to John E. Mittler for calling our attention to the relevant passage from I. M. Lewis, and we are indebted to him for his fundamental contributions toward the resolution of this controversy. In addition, we thank J. Cairns, P. L. Foster, B. G. Hall and J. A. Shapiro for their open exchange of ideas, even during periods of intense disagreement. Financial support was provided by NSF grant DEB-9421237 to R. E. Lenski and by the NSF Center for Microbial Ecology (BIR-9120006) at Michigan State University.

Literature Cited

1. Benson SA, Decloux AM, Munro J. 1991. Mutant bias in nonlethal selections results from selective recovery of mutants. *Genetics* 129:647–58
2. Benson SA, Occi JL, Sampson BA. 1988. Mutations that alter the pore function of the OmpF porin of *Escherichia coli* K12. *J. Mol. Biol.* 203:961–70
3. Boe L. 1990. Mechanism for induction of adaptive mutations in *Escherichia coli. Mol. Microbiol.* 4:597–601
4. Cairns J. 1988. Origin of mutants disputed. *Nature* 336:527–58
5. Cairns J, Foster PL. 1991. Adaptive reversion of a frameshift mutation in *Escherichia coli. Genetics* 128:695–701
6. Cairns J, Overbaugh J, Miller S. 1988. The origin of mutants. *Nature* 335:142–45
7. Cavalli-Sforza LL, Lederberg J. 1956. Isolation of preadaptive mutants in bacteria by sib selection. *Genetics* 41:367–81
8. Chao L, Cox EC. 1983. Competition between high and low mutating strains of *Escherichia coli. Evolution* 37:125–34
9. Charlesworth B, Sniegowski PD, Stephan W. 1994. The evolutionary dynamics of repetitive DNA in eukaryotes. *Nature* 371:215–20
10. Charlesworth D, Charlesworth B, Bull JJ. 1988. Origin of mutants disputed. *Nature* 336:525
11. Cox EC. 1976. Bacterial mutator genes and the control of spontaneous mutation. *Annu. Rev. Genet.* 10:135–56
12. Cox EC, Gibson TC. 1974. Selection for high mutation rates in chemostats. *Genetics* 77:169–84
13. Danchin A. 1988. Origin of mutants disputed. *Nature* 336:527
14. Darwin CR. 1859. *The Origin of Species.* London: John Murray
15. Darwin CR. 1868. *Variation of Animals and Plants Under Domestication.* London: John Murray
16. Davis BD. 1989. Transcriptional bias: a non-Lamarckian mechanism for substrate-induced mutations. *Proc. Natl. Acad. Sci. USA* 86:5005–9
17. Demassey B, Fayet O, Kogoma T. 1984. Multiple origin usage for DNA replication in *sdr* (*rnh*) mutants of *Escherichia coli* K12: initiation in the absence of *oriC. J. Mol. Biol.* 128:227–36
18. Dijkmans R, Kreps S, Mergeay M. 1994. Poisson-like fluctuation patterns of revertants of leucine auxotrophy (*leu-500*) in *Salmonella typhimurium* caused by delay in mutant cell division. *Genetics* 137:353–59
19. Dobzhansky T. 1970. *Genetics of the Evolutionary Process.* New York: Columbia Univ. Press
20. Doolittle WF, Sapienza CS. 1980. Selfish genes, the phenotype paradigm, and genome evolution. *Nature* 284:601–7
21. Drake JW. 1991. Spontaneous mutation. *Annu. Rev. Genet.* 25:125–46
22. Fisher RA. 1930. *The Genetical Theory of Natural Selection.* Oxford: Clarendon
23. Foster PL. 1993. Adaptive mutation: the uses of adversity. *Annu. Rev. Microbiol.* 47:467–504
24. Foster PL. 1994. Population dynamics of a Lac⁻ strain of *Escherichia coli* during selection for lactose utilization. *Genetics* 138:253–61
25. Foster PL, Cairns J. 1992. Mechanisms of directed mutation. *Genetics* 131:783–89
26. Foster PL, Cairns J. 1994. The occurrence of heritable *Mu* excisions in starving cells of *Escherichia coli. EMBO J.* 13:5240–44
27. Foster PL, Trimarchi JM. 1994. Adaptive reversion of a frameshift mutation in *Escherichia coli* by simple base deletions in homopolymeric runs. *Science* 265:407–9
27a. Friedberg EC, Walker GC, Siede W. 1995. *DNA Repair and Mutagenesis.* Washington, DC: Am. Soc. Microbiol.
27b. Galitsky T, Roth JR. 1995. Evidence that F plasmid transfer replication underlies apparent adaptive mutation. *Science* 268:421–23
28. Gillespie JH. 1981. Mutation modification in a random environment. *Evolution* 35:468–76
29. Gould SJ, Lewontin RC. 1979. The spandrels of San Marco and the Panglossian paradigm: a critique of the adaptationist programme. *Proc. Roy. Soc. Lond. B* 205:581–98
30. Deleted in proof
31. Hall BG. 1988. Adaptive evolution that

requires multiple spontaneous mutations. I. Mutations involving an insertion sequence. *Genetics* 120:887–97

32. Hall BG. 1990. Spontaneous point mutations that occur more often when advantageous than when neutral. *Genetics* 126:5–16

33. Deleted in proof

34. Hall BG. 1993. The role of single-mutant intermediates in the generation of *trpAB* double revertants during prolonged selection. *J. Bacteriol.* 175: 6411–14

34a. Hall BG. 1994. On alternatives to selection-induced mutation in the *bgl* operon of *Escherichia coli*. *Mol. Biol. Evol.* 11:159–68

35. Harris RS, Longerich S, Rosenberg SM. 1994. Recombination in adaptive mutation. *Science* 264:258–60

36. Harvey PH, Pagel MD. 1991. *The Comparative Method in Evolutionary Biology*. Oxford: Oxford Univ. Press

37. Hinshelwood CN. 1950. Chemistry and bacteria. *Nature* 166:1089–92

38. Holliday R, Rosenberger RF. 1988. Origin of mutants disputed. *Nature* 336:526

39. Huxley J. 1942. *Evolution: The Modern Synthesis*. New York: Harper

40. Ishii K, Matsuda H, Iwasa Y, Sasaki A. 1989. Evolutionarily stable mutation rate in a periodically changing environment. *Genetics* 121:163–74

41. Jacob F, Wollmann EL. 1971. *Sexuality and the Genetics of Bacteria*. New York: Academic

42. Jenkin F. 1867. "The Origin of Species." *North British Rev.* 46:149–71

43. Kimura M. 1967. On the evolutionary adjustment of spontaneous mutation rates. *Genet. Res. Camb.* 9:23–34

44. Koch AL. 1982. Mutation and growth rates from Luria-Delbrück fluctuation tests. *Mutation Res.* 95:129–43

45. Kornberg A, Baker T. 1992. *DNA Replication*. New York: Freeman

46. Krawiec S. 1994. Misdirected controversy? *Am. Sci.* 82:3–4

47. Lamarck J-B. 1809. *The Zoological Philosophy*. Transl. H Elliot, 1963. London: Macmillan

48. Lea DE, Coulson CA. 1949. The distribution of the number of mutants in bacterial populations. *J. Genet.* 49:264–85

49. Lederberg J. 1989. Replica plating and indirect selection of bacterial mutants: isolation of preadaptive mutants in bacteria by sib selection. *Genetics* 121:395–99

50. Lederberg J, Lederberg EM. 1952. Replica plating and indirect selection of

bacterial mutants. *J. Bacteriol.* 63:399–406

51. Leigh EG. 1970. Natural selection and mutability. *Am. Nat.* 104:301–5

52. Leigh EG. 1973. The evolution of mutation rates. *Genet. Suppl.* 73:1–18

53. Lenski RE. 1989. Are some mutations directed? *Trends Ecol. Evol.* 4:148–50

54. Lenski RE, Mittler JE. 1993. The directed mutation controversy and neo-Darwinism. *Science* 259:188–94

55. Lenski RE, Slatkin M, Ayala FJ. 1989. Another alternative to directed mutation. *Nature* 337:123–24

56. Lenski RE, Sniegowski PD. 1995. Directed mutations slip-sliding away? *Curr. Biol.* 5:97–99

57. Lewis IM. 1934. Bacterial variation with special reference to behavior of some mutable strains of colon bacteria in synthetic media. *J. Bacteriol.* 28:619–38

58. Luria SE, Delbrück M. 1943. Mutations of bacteria from virus sensitivity to virus resistance. *Genetics* 28:491–511

59. MacPhee D. 1993. Directed evolution reconsidered. *Am. Sci.* 81:554–61

60. MacPhee DG. 1993. Directed mutation: paradigm postponed. *Mutation Res.* 285: 109–16

61. MacPhee DG. 1993. Is there evidence for directed mutation in bacteria? *Mutagenesis* 8:3–5

62. Maenhaut-Michel G, Shapiro JA. 1994. The roles of starvation and selective substrates in the emergence of *araB-lacZ* fusion clones. *EMBO J.* 13:5229–39

63. Mayr E. 1982. *The Growth of Biological Thought*. Cambridge, Mass.: Belknap

64. McDonald JF. 1993. Evolution and consequences of transposable elements. *Curr. Opin. Genet. Devel.* 3:855–64

65. Mittler JE, Lenski RE. 1990. Causes of mutation and *Mu* excision. *Nature* 345: 213

66. Mittler JE, Lenski RE. 1990. New data on excisions of *Mu* from *E. coli* MCS2 cast doubt on directed mutation hypothesis. *Nature* 344:173–75

67. Mittler JE, Lenski RE. 1992. Experimental evidence for an alternative to directed mutation in the *bgl* operon. *Nature* 356:446–48

68. Morgan TH. 1903. *Evolution and Adaptation*. New York: Macmillan

69. Moxon ER, Rainey PB, Nowak MA, Lenski RE. 1994. Adaptive evolution of highly mutable loci in pathogenic bacteria. *Curr. Biol.* 4:24–33

70. Muller HJ. 1923. Mutation. *Eugenics, Genetics and the Family.* 1:106–12

71. Newcombe HB. 1949. Origin of bacterial mutations. *Nature* 164:150–51

72. Nöthel H. 1987. Adaptation of *Drosophila melanogaster* populations to high mutation pressure: evolutionary adjustment of mutation rates. *Proc. Natl. Acad. Sci. USA* 84:1045–49

73. Orgel LE, Crick FHC. 1980. Selfish DNA: the ultimate parasite. *Nature* 284:604–7

73a. Radicella JP, Park PU, Fox MS. 1995. Adaptive mutation in *Escherichia coli*: a role for conjugation. *Science* 268:418–20

74. Rosenberg SM, Longerich S, Gee P, Harris RS. 1994. Adaptive mutation by deletions in small mononucleotide repeats. *Science* 265:405–7

75. Ryan FJ. 1952. Distribution of numbers of mutant bacteria in replicate cultures. *Nature* 169:882–83

76. Shapiro JA. 1984. Observations on the formation of clones containing *araB-lacZ* cistron fusions. *Mol. Gen. Genet.* 194:79–90

77. Shapiro JA. 1992. Natural genetic engineering in evolution. *Genetica* 86:99–111

78. Shapiro JA, Leach D. 1990. Action of a transposable element in coding sequence fusions. *Genetics* 126:293–99

79. Smith KG. 1992. Spontaneous mutagenesis: experimental, genetic and other factors. *Mutation Res.* 277:139–62

80. Sniegowski PD. 1995. The origin of adaptive mutants: random or nonrandom? *J. Mol. Evol.* 40:94–101

81. Sniegowski PD. 1995. A test of the directed mutation hypothesis in *Escherichia coli* MCS2 using replica plating. *J. Bacteriol.* 177:1119–20

82. Stahl FW. 1988. A unicorn in the garden. *Nature* 355:112–13

83. Stahl FW. 1992. Unicorns revisited. *Genetics* 132:865–67

84. Steele F, Jinks-Robertson S. 1992. An examination of adaptive reversion in *Saccharomyces cerevisiae*. *Genetics* 132:9–21

85. Stent GS. 1971. *Molecular Genetics.* New York: WH Freeman

86. Stewart FM. 1994. Fluctuation tests: How reliable are the estimates of mutation rates? *Genetics* 137:1139–46

87. Stewart FM, Gordon DM, Levin BR. 1990. Fluctuation analysis: the probability distribution of the number of mutants under different conditions. *Genetics* 124:175–85

88. Sturtevant AH. 1937. Essays on evolution. I. On the effects of selection on mutation rate. *Q. Rev. Biol.* 12:467–77

89. Tessman I. 1988. Origin of mutants disputed. *Nature* 336:527

90. Tröbner W, Piechocki R. 1984. Competition between isogenic *mutS* and *mut⁺* populations of *Escherichia coli K12* in continuously growing cultures. *Mol. Gen. Genet.* 198:175–76

91. Tröbner W, Piechocki R. 1984. Selection against hypermutability in *Escherichia coli* during long term evolution. *Mol. Gen. Genet.* 198:177–78

92. Wada K-N, Doi H, Tanaka S-I, Wada Y, Furosawa M. 1993. A neo-Darwinian algorithm: asymmetrical mutations due to semiconservative DNA-type replication promote evolution. *Proc. Natl. Acad. Sci. USA* 90:11934–38

93. Weismann A. 1889. *Essays Upon Heredity.* Oxford: Clarendon

94. Williams GC. 1992. *Natural Selection: Domains, Levels, Challenges.* Oxford: Oxford Univ. Press

95. Witkin E, Kogoma T. 1984. Involvement of the activated form of RecA protein in SOS mutagenesis and stable DNA replication in *Escherichia coli*. *Proc. Natl. Acad. Sci. USA* 81:7539–43

Annu. Rev. Ecol. Syst. 1995. 26:579–600

SPECIATION IN EASTERN NORTH AMERICAN SALAMANDERS OF THE GENUS *PLETHODON*

Richard Highton

Department of Zoology, University of Maryland, College Park, Maryland 20742

KEY WORDS: allozymes, electrophoresis, *Plethodon*, plethodontid salamanders, molecular evolution, speciation

---

ABSTRACT

Studies on allozyme variation in eastern plethodontid salamanders of the woodland genus *Plethodon* have revealed a large number of cryptic species. Genetic variation within and among species reveals patterns of speciation. In the late Miocene and early Pliocene there appear to have been only five clades of eastern North American *Plethodon* that still survive today. A burst of speciation took place in the Pliocene, so there are now at least 35 species in four species groups in eastern North America. It is hypothesized that the arid climates of the Pliocene isolated many populations in wetter forested mountainous areas where in isolation allopatric speciation occurred. With the return of wetter climates, many of these species dispersed and a large number of parapatric contact zones may now be studied where closely related species are interacting. Hybridization between closely related species is frequent in these contact zones.

INTRODUCTION

The woodland salamanders, genus *Plethodon*, with 43 recognized species, occur in forested areas of eastern and western North America (Table 1). This review concentrates on the eastern section of the genus, which contains some species with very broad ranges, while others are restricted to only a few km² on a single mountain in the Appalachian or Ouachita mountains. Species of *Plethodon* are

579

Table 1 Species of the genus *Plethodon*

Western *Plethodon*	*P. wehrlei* species group
P. vehiculum species group	*punctatus*
dunni	*wehrlei*
vehiculum	*P. glutinosus* species group
P. elongatus species group	*Petraeus*
elongatus	*yonahlossee*
stormi	*P. ouachitae* complex
P. vandykei species group	*caddoensis*
idahoensis	*fourchensis*
vandykei	*ouachitae*
P. neomexicanus species group	*P. jordani* complex
larselli	*jordani*
neomexicanus	*P. glutinosus* complex
Eastern *Plethodon*	*albagula*
P. cinereus species group	*aureolus*
cinereus	*chattahoochee*
hoffmani	*chlorobryonis*
hubrichti	*cylindraceus*
nettingi	*glutinosus*
richmondi	*grobmani*
serratus	*kentucki*
shenandoah	*kiamichi*
P. welleri species group	*kisatchie*
angusticlavius	*mississippi*
dorsalis	*ocmulgee*
websteri	*savannah*
welleri	*sequoyah*
	teyahalee
	variolatus

mostly terrestrial and deposit their direct-developing eggs on land, thus omitting the aquatic larval stage characteristic of many amphibians. Thus they are not restricted to aquatic habitats for reproduction and dispersal. Species of *Plethodon* are sometimes the most abundant terrestrial vertebrates in forests of eastern North America, with recorded densities of thousands of individuals per hectare (6, 71). Up to five species of *Plethodon* may occur in the same habitat in the southern Appalachians. The scientific name of the genus refers to the large number of teeth in these salamanders, but the prefix *pleth* also appropriately describes population densities of these salamanders, which provide a plethora of opportunities for studies on their behavior (3, 11–13, 52–54, 64–66) and ecology (16, 17, 23, 24, 48–51, 55, 100), as well as their evolution.

Through my extensive field studies (29, 30) in the middle Atlantic states and in the southern Appalachian Mountains and studies based on large museum collections (21, 28) as well as distribution maps in the field guides and many state herpetologies, the ranges of most eastern species are now probably accurately determined. Many sites are known where ecological and genetic interactions between species may be studied. I have obtained allozyme data on most of the approximately 140 contacts between pairs of eastern species of *Plethodon*. These data indicate the presence of reproductive isolation or reveal the amount of interspecific hybridization at each site. Fixed or complete genetic differences at several loci are usually found between sympatric species. However, at contacts between closely related parapatric species, natural hybridization is common. Allozyme data permit a quantitative analysis of the extent of natural hybridization and the amount and distance of gene flow, especially between parapatric species with narrow contact zones. An example of the interactions of nine forms of the *P. glutinosus* and *P. jordani* complexes in the southern Appalachians is given.

Before the advent of molecular genetic studies, the genus *Plethodon* had been revised by systematists three times in this century (15, 21, 28). The first species of *Plethodon* were discovered in the early nineteenth century, but as recently as 1962, only 16 of the 43 currently named species had been recognized. Taxonomists had difficulty identifying many of the species of *Plethodon* because speciation often has been decoupled from the evolution of morphological novelties in these salamanders (59, 60), so that far more species exist than can be recognized on the basis of morphological differentiation. The primary subdivision of the genus *Plethodon* is into a western group of eight species (in four species groups) in the Pacific northwest and the northern and southern Rocky Mountains, and an eastern group of 35 presently recognized species (in four species groups) continuously distributed in forested areas in most of the eastern United States and southeastern Canada (Table 1, Figure 1). *Plethodon* may be paraphyletic: The five species of its closest relative, *Aneides*, may be part of the western *Plethodon* radiation (37, 62).

Herpetologists have been aware that *Plethodon* is a difficult group because of the relatively few taxonomic characters available for systematic analysis. Molecular methods have provided the data necessary to clarify the patterns of inter- and intraspecific variation and have resulted in the recognition of many additional species and in clarifying the phylogeny of the genus (27, 37, 38, 41, 62, 67). At least 10 undescribed eastern species revealed by allozyme studies are currently under study.

RELATIONSHIP BETWEEN MORPHOLOGICAL AND ALLOZYME DIVERGENCE

Larson (59, 60) and Wake (97) have excellent reviews on this topic for plethodontid salamanders. I review the same topic for *Plethodon*. As in many other well-known vertebrate groups, the taxonomy of *Plethodon* has passed

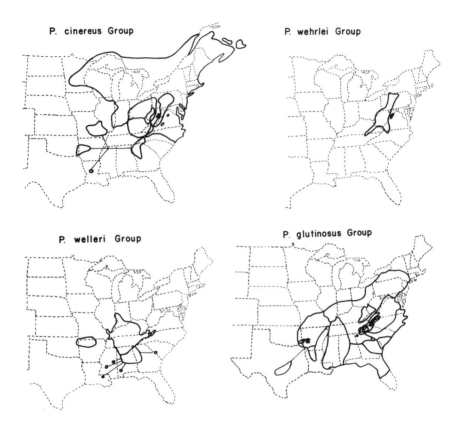

Figure 1 Distribution of the 35 species in each of the four species groups of eastern *Plethodon*. The Appalachian, Ozark, or Ouachita mountains are included in part of the range of all but six southern species of the *P. glutinosus* complex.

through several stages. During the nineteenth century, some morphotypes were named as full species by naturalists who used the then-current typological species concept. Thus, the unstriped and striped morphs of *P. cinereus* were recognized for almost 100 years as different species (*P. cinereus* and *P. erythronotus*, respectively). During the first 60 years of the twentieth century some geographic variants were described as subspecies (2, 21, 22, 40, 74, 81); some of these are now recognized as full species. At the present time, no recognized subspecies remain, primarily because within-species patterns of geographic variation in most morphological characters are discordant (28). By 1962 the taxonomy appeared stable, with 16 species recognized in the genus. Since then, five previously unknown, morphologically distinct species have been discovered (30, 39, 44, 102), but the remaining species (over half of the presently known species in the genus) have been revealed by allozyme studies. There are at least three different causes for the failure to recognize so many species by morphological analysis: 1. stasis, parallelism and convergence; 2. speciation while retaining ancestral color patterns, polymorphisms or ontogenetic trajectories; and 3. failure of independent characters that vary among species or populations to vary concordantly.

Stasis seems common throughout the genus. Most surprising was the discovery of *P. websteri* (32, 61), a species that so resembles *P. dorsalis* that it had never even been recognized as a different subspecies. These species have a Nei (77) genetic distance of 1.3–2.1, depending on which populations are compared and which proteins are examined (41, 61). According to molecular clock estimates by Maxson & Maxson (68), who suggested it takes about 14 Myr to evolve a Nei genetic distance of 1.0 in *Plethodon*, *P. dorsalis* and *P. websteri* probably separated at least 18 Myr ago. They are sympatric in a narrow contact zone in Jefferson County, Alabama, where no hybridization was detected (34). D. Wake and R. Highton (unpublished data) have found differences in the size of the feet in these two species, so at least one morphological difference is known, but the extremely slow rate of morphological evolution during the long period of divergence of *P. dorsalis* and *P. websteri* is remarkable, even in this slowly evolving genus (59, 97, 98).

Several species of *Plethodon* (*vehiculum, cinereus, serratus, shenandoah, dorsalis, websteri*) have a similar dorsal color polymorphism (a red-striped morph and a dark brown, unstriped morph) and several other species are monomorphic for one or the other morph. This polymorphism is usually present from hatching throughout the life of the salamander, but in *P. elongatus* the striped morph usually occurs only in juveniles. The adaptive significance of this color polymorphism is unknown. The two most common and wide-ranging western and eastern species of *Plethodon* (*vehiculum* and *cinereus*, respectively) are small elongate species with the two color morphs. They also have similar venters that are mottled black and white. However, the close resemblance between these two species is superficial; their average number of trunk vertebrae differs by three, their trunk vertebrae are differently proportioned (a major morphological difference between the eastern and western groups; 28), and the amount of nuclear DNA is about twice as much in the western as in the eastern species (75, 86), resulting in a large difference in cell size. Allozymic differences at most loci (41) and immunological distances (67, 68) between eastern and western *Plethodon* indicate that the two groups have been diverging for over 40 Myr.

Black bodies with small dorsal iridophore spots that vary in size and range in color from white to brassy in appearance characterize the 16 species of the *P. glutinosus* complex as well as several other species (*Aneides flavipunctatus*; *P. punctatus* of the *P. wehrlei* species group; a local population of *P. yonah-*

lossee in the vicinity of Bat Cave, Rutherford County, North Carolina (1); and the isolates of *P. ouachitae* on Winding Stair and Kiamichi mountains, Oklahoma). No one has speculated on the possible adaptive significance of this color pattern. All of the species of eastern large *Plethodon* possess tails that exude a great deal of slime when they are disturbed. They have many amphibian, reptile, bird, and mammal predators. Huheey (45), working with *P. jordani,* showed that *Plethodon* slime may stick to the feathers around the eyes of a bird predator (shrike), interfering with its vision by gluing the eyelids shut. Species of *Plethodon* may be found under leaf litter both day and night; they often forage on the forest floor on wet nights and occasionally are also out during the day. In contrast to many species of salamanders (e.g. other species of *Plethodon,* as well as other genera such as *Desmognathus*) with cryptic dorsal color patterns, it may be that the black bodies present in many of the *P. glutinosus* group species are very conspicuous to bird predators, especially against the background of brown leaves on the forest floor. This may serve as a warning to avian predators of their sliminess. In contrast to birds, many amphibian, reptile, and mammal predators do not seem to be deterred by the tail slime.

Although Bishop (5) included *P. glutinosus* as one of several North American salamanders that is likely "a complex of species or subspecies requiring additional study before its components can be properly delineated," and Highton (30) mapped the distribution of forms of *P. glutinosus* in the middle Atlantic states that have dorsal spots of different size and color, it was not until allozyme analysis that 16 genetically divergent forms in the *P. glutinosus* complex were identified (35). All 16 are recognized as taxonomic species because there are four cases of sympatry without hybridization between pairs of species within this complex as well as large genetic differences between all parapatric forms. Thus, there have been numerous speciation events with little associated morphological divergence. The color pattern has probably been retained from a common ancestor (or evolved by parallelism or convergence) in a number of related species within the *P. glutinosus* species group, in other species groups of *Plethodon* and also in one species of *Aneides.*

A probable example of convergence in the color pattern of eastern large *Plethodon* is the similar dorsal coloration of three large species of the *P. glutinosus* group: *P. yonahlossee, P. petraeus,* and many individuals in one of the five isolates of the subdivided species *P. ouachitae* (Rich Mountain form). None are closely related genetically (14, 35; R. Highton, unpublished data), but all have a dorsum with abundant red or yellowish-red pigment.

Some young of five species of eastern large *Plethodon* have paired or alternating red dorsal spots (*wehrlei, jordani, yonahlossee, petraeus, ouachitae*). In *P. wehrlei* the red (yellow in southwestern populations) spots are often retained throughout life; in *P. jordani* they usually disappear before maturity;

and in the last three species they enlarge to form the dorsal stripe of adults. If the dorsal spots are primitive, they have been lost in most of the other species of the group. *Plethodon fourchensis* also has large dorsal spots throughout its life, but they are white in color. Thus, retention of ancestral characters may result in morphological similarities that may not be indicative of close relationships.

In spite of large differences in proportions, the genus *Plethodon* is osteologically quite uniform (96). The modal number of trunk vertebrae varies among species from 15 to 23, but there is little indication of phylogenetic pattern in variation in this character. There is a difference in the shape of the trunk vertebrae between most of the eastern as opposed to western groups, but *P. neomexicanus* is an exception, resembling the eastern group (28). Most osteological features have remained very similar during the approximately 40 Myr of evolution of the genus. Although there is not a good fossil record in *Plethodon*, speciation in these salamanders would be difficult or impossible to study using fossil material.

Carr (7) did a morphometric analysis of variation in 14 of the 16 species of the *P. glutinosus* complex. Using multivariate statistics and discriminant function analysis, he found that 48–90% (mean 63%) of adult salamanders from a single sample of each species could be allocated correctly to its population. However, additional populations of some species also could be distinguished from different conspecific populations, sometimes at the same level of discrimination as from samples of other species. Multiple populations of the same species sometimes did not cluster together in phenograms. Thus, some local populations within species of *Plethodon* are as morphologically differentiated from some conspecific populations as they are from populations of different species. This may be the result of genetic adaptation to local environments or of ecophenotypic variation. Therefore, even sophisticated morphometric analysis may not distinguish individuals of different species.

The systematic relationships of *Plethodon* species, even those that are morphologically easy to distinguish, have proven difficult to determine. Each taxonomist who has tried to estimate the relationships of eastern species has suggested quite different phylogenies (15, 21, 28, 92). The reasons are (*a*) few variable morphological characters and (*b*) disagreement over which morphological characters are useful in phylogenetic reconstruction. The application of molecular techniques (allozymes: 41; immunology: 27, 67; and DNA hybridization: 75) has led to general agreement about the relationships among the eastern species. Relationships based on morphological similarity do not always agree with these molecular results. For example, the *P. welleri* and the *P. cinereus* groups were always considered closely related on the basis of their small size, short legs, and elongate bodies. The other two groups (*P. glutinosus* and *P. wehrlei* groups) are much larger in size, have proportionally longer legs, and are less elongate than all small eastern species except *P. welleri.* Yet the molecular data show that the *P. cinereus* group is a sister group to the other three groups. It may be that the ancestral eastern *Plethodon* was a small elongate species, similar to the western *P. vehiculum,* and that the *cinereus* and *welleri* groups have retained this primitive morphology, while the eastern large groups are derived. Molecular data indicate that western *Plethodon* represent a distinct clade, so it is likely that parallel and/or convergent interspecific variation in size, body proportions, and number of vertebrae has occurred in the eastern and western groups.

Smaller body size seems to be characteristic of all Coastal Plain populations of species in the *P. glutinosus* complex, while all but two non-Coastal Plain species are larger in size (7, 35). Yet there is little indication of genetic similarity uniting species in either of these size groups into clades.

All of the species of *Plethodon* that have been studied have the same diploid 28 chromosome number. Cytologists have generally pursued studies on the chromosomes of other salamander genera that have evolved exceedingly interesting chromosome variation (56, 76, 85–87). However, modern banding techniques have not yet been applied to analyze *Plethodon* chromosome evolution.

In summary, it is unwise to rely very heavily on known variable morphological characters, including coloration, in attempting to reconstruct the phylogeny of *Plethodon.* This is because of the few taxonomic characters available to the systematist as well as differential rates in the evolution of morphological characters. Because of the numerous remarkable cases of evolutionary stasis as well as probable adaptive convergences, without molecular data, it is sometimes nearly impossible even to distinguish species of *Plethodon.*

SPECIES CONCEPTS IN PLETHODON

Until recently, the biological species concept of Mayr (groups of naturally or potentially interbreeding natural populations that are reproductively isolated from other such groups—69) was almost universally applied to most vertebrate groups. It was generally believed that the same number of species would be recognized in a group no matter what criteria (morphological, genetic, ethological) were employed.

The biological species concept cannot be applied objectively in determining species boundaries in asexually reproducing groups, in fossil lineages, and in geographically isolated populations. Peripheral speciation may sometimes result in paraphyletic species, which are not favored by systematists employing cladistic methodology (all taxa must be monophyletic). Some cladists also regard reproductive compatibility as an ancestral character, which they believe is not appropriate for use in phylogenetic reconstruction. The importance of reproductive isolation in the taxonomic recognition of species should not be dismissed because this factor determines whether previously isolated populations will merge or continue to diverge. Reproductive isolation develops gradually by the accumulation of mutations throughout the genome (9, 78), and it therefore increases in a manner similar to the increase in genetic distance through time (36). Criticisms of the biological species concept have resulted in proposals for new species concepts [e.g. recognition, evolutionary, phylogenetic, and cohesion; see reviews by Templeton (90) and Avise (4)]. The application of some of these substitute species concepts would make little difference in the number of species recognized in some animal groups, but in others it would make it difficult for the practicing taxonomist to decide on species boundaries and the number of species to be recognized. During the last 50 years, most taxonomists studying *Plethodon* have applied the biological species concept. Because there is a poor fossil record of this genus and there are no known parthenogenetic species, the major problem in applying the biological species concept prior to the development of molecular methods was the taxonomic treatment of isolated allopatric forms.

The application of molecular methods has shown that the traditional method for determining species limits in this genus, the analysis of morphological variation, failed to distinguish many distinct, sympatric, and reproductively isolated species. That *Plethodon* is not unusual in this regard is shown by studies on other genera of salamanders in which morphologically similar cryptic species have been revealed by molecular data (e.g. 10, 19, 25, 47, 58, 93, and others summarized in 58, 59). Thus far, few comprehensive allozymic studies on species of North American salamanders from throughout the range of a species have failed to discover cryptic species (e.g. 72, 73).

In the case of sympatric forms with several proteins that show fixed or complete differences, there is no problem in recognizing taxonomically distinct species because there is obviously no natural hybridization. An example discussed above is *P. websteri* and *P. dorsalis* (32, 61). Another example is the pair of sympatric species *P. glutinosus* and *P. kentucki,* in which morphometric analysis by Clay et al (8) failed to detect differences. Highton & MacGregor (42) later reported differences in color pattern and mental gland shape, and Carr (7) found significant morphometric differences between the two species. As often occurs in such cases, once genetic differences are detected so that specimens can be identified correctly, morphological characters with consistent differences that distinguish the species are revealed. The characters that distinguish *P. kentucki* from sympatric *P. glutinosus* (shape of the male mental gland, lighter chin, and color and size of the dorsal spots) are variable in other allopatric *P. glutinosus* complex populations, at one time thought to be a single species, so their importance was not recognized.

The ability of electrophoretic analysis of allozymes to sort out populations of *Plethodon* into species has been remarkable. In six geographic variation studies that involve multispecies comparisons of eastern *Plethodon* (14, 33,

35, 42, 43, 61), only one of 264 populations failed to cluster on UPGMA phenograms within a monophyletic group of all other populations of its own species. In four of these studies, however, species were recognized on the basis of the genetic data, so some circularity is involved. The above studies also include many groupings of populations that had been previously recognized as distinct species on the basis of morphology, and all were sorted out to their correct species on the basis of data from 22–26 protein loci.

All currently recognized species in the genus *Plethodon*, originally described on the basis of morphological characters, have Nei genetic distances above 0.15 from their closest relatives. This is also the case in the species recognized on the basis of allozyme studies. How successful have allozyme studies been in detecting all species of *Plethodon* as well as of other groups of salamanders? It would be surprising if some species have not been missed, considering only 20–30 protein loci were analyzed, and electrophoretic analysis detects only about 40% of real protein variation (70, 82). These might be detected by analyzing additional allozymes or by using newer molecular techniques such as sequential or two-dimensional electrophoresis, restriction site analysis, microsatellite analysis, or DNA sequencing. Few parapatric contact zones between genetically different populations of *Plethodon* that have Nei distances below 0.15 have been studied. One such zone is the contact between genetically different populations of *P. cinereus* on the Del-Mar-Va peninsula that have a Nei genetic distance of about 0.1 (31, 101). Evidence from five transects across the peninsula indicates considerable interbreeding between a northern and a southern form, with variation in the estimates of cline widths ranging from 1 to 16 km. At two esterase loci, a linkage disequilibrium was detected in the intergrade zone (101), suggesting a secondary contact. Other contacts between very closely related forms are currently being studied in my laboratory. If any should represent unrecognized species with unusually low genetic distances, we may still be underestimating the number of biological species in *Plethodon*.

There is a problem of how to treat allopatric forms that differ genetically but are not differentiated morphologically. Some taxonomists would not recognize them as different species in the absence of morphological differentiation (e.g. 97). Others (e.g. 35) believe that the buildup of genetic distance to a level as great as that found in over 98% of interspecific comparisons in vertebrates other than birds ($D = 0.15$) (91) is sufficient evidence to recognize a pair of allopatric or parapatric forms as different taxonomic species, after consideration of other evidence (if available) such as morphological variation, patterns of geographic genetic variation, and genetic interactions in contact zones. Good et al (20) suggested a method for detecting gaps in gene flow that may be very helpful in these cases, and it was applied successfully by Good & Wake (19) in recognizing sibling species of the salamander *Rhyacotriton*.

Frost & Hillis (18) point out that different mixes of fast, intermediate, and slowly evolving proteins (84) would be expected to yield different D values for the same comparison. Thus, for example, the genetic distances obtained by Wake's laboratory in studies on *Ensatina* (46, 99) would be higher if fast evolving proteins such as albumin, esterase, and transferrin had been included. Such fast evolving proteins often account for a substantial portion of the genetic distance between populations of *Plethodon* analyzed in my laboratory. Moreover, statistical sampling errors would be expected if different sets of proteins are used to compare two populations, even if the same mix of fast and slowly evolving proteins is used. One comparison could easily be higher than an arbitrarily defined species divergence level, while another could be below it. On the other hand, comparing the matrices of genetic identities for 28 species of plethodonine salamanders in 55 randomly divided sets of 14 and 15 protein loci, from the total of 29 analyzed (41), Highton (37) found that the mean correlation coefficient between the two independent data sets was high ($r = 0.86$). Taxonomic conclusions based on a larger number of loci are more likely to be valid (38, 77).

The species level taxonomy is not only important to systematists, but it makes a profound difference in the interpretation of other types of data. For example, Tilley et al (94) compared genetic distance with ethological reproductive isolation among seven populations of the plethodontid "species" *Desmognathus ochrophaeus* in the southern Appalachian mountains. In only 2 of their 27 comparisons of genetic distance with ethological isolation was the genetic distance below 0.15. It is likely that several species were used in their study, some with genetic distances over 0.5 (see also 93). In interpreting data on ethological isolation, it is important to know whether the populations studied are (*a*) of the same continuously distributed species with gene flow throughout its range, (*b*) between long-isolated species that have been diverging from one another for millions of years, or (*c*) a mix of the two. Correlations between geographic distance and ethological differentiation might be high in (*a*), but low in (*b*), especially if extensive migration of populations occurred after the geographic barriers to interbreeding between the species were removed. The taxonomic status of the various populations is also important because of the rarity of reports in the literature of ethological isolation within continuously distributed species.

Another example of the importance of a knowledge of species-level taxonomy involves the use of F_{ST} statistics as a measure of subdivision of populations. The Wahlund variance is often used to measure the rate of gene flow among populations. High F_{ST} values are sometimes interpreted as indications of reduced gene flow within species of plethodontid salamanders (61, 88; see also 63). These results seem to conflict with estimates of migration rates in several species of *Plethodon* that appear to have migrated northward for hundreds of km, including the crossing of major rivers, within the last few thousand years (21, 35, 43). If migration of such great distances into unoccupied territory has been possible since the end of the Wisconsin continental glaciation, why should it be so difficult for individuals to migrate short distances to exchange genes with nearby conspecific populations? And why do so many populations of closely related *Plethodon* on opposite sides of smaller rivers than those crossed by the above mentioned post-glacial migrating forms have fixed differences at several loci? The obvious explanation is that there is considerable gene flow within species, but little or none between cryptic parapatric species. When the taxonomy has not been clarified, samples of two or more species are erroneously considered conspecific, and it appears that there is little gene flow among some populations of a "species" that is, in reality, a species complex of more than one species. The ranges of *P. cinereus* and *P. hoffmani* in Washington County, Maryland, are separated by Licking Creek (30). This creek is so small that surely these salamanders can walk across it on fallen trees without entering the water. Since these same species apparently have crossed some major rivers, probably by rafting inside rotting logs, it is likely that competitive exclusion rather than the stream is the barrier to dispersal of these two species.

The extremely high F_{ST} statistics for some salamander species, e.g. populations 4–16 of *Plethodon dorsalis,* reported by Larson & Highton (61), become much lower if their samples 4–10 are regarded as coming from a different species than samples 11–16, and this is indeed probably the case (R Highton, unpublished data). High F_{ST} statistics are an indication of subdivision and reduced gene flow, but it makes a major difference in interpretation if one is considering conspecific populations or populations of different species.

Detailed geographic allozyme analyses should therefore precede ethological, ecological, and other studies, so that the number of species is clearly indicated before interpretations of other kinds of biological data are attempted. This often is not done; indeed, some nonsystematists appear to have little concern with the taxonomy of the group they are studying. In salamanders it is especially important to know the number of biological species included in a study because of the large number of cryptic species that have been revealed by allozyme studies.

SPECIATION

If molecular clock estimates of the time of divergence are applied to the genetic distances obtained in the allozyme study of Highton & Larson (41), there appears to have been a burst of speciation in all four species groups of eastern species as well as in two of the four western species groups of *Plethodon* in the Pliocene Epoch. The genetic distances (> 0.15) between all currently recognized species of *Plethodon* indicate that speciation was initiated before the Pleistocene Epoch began. Thus climatic changes due to Pleistocene glaciation events of the last 2 Myr do not appear to be responsible for the original subdivision of the ancestral forms of each species group, although Pleistocene climates may have played a continuing role in the interruption of gene flow among the many incipient species. Recent data (57) indicate that glaciation began in Greenland as early as the Late Miocene; and thus climatic effects initiating speciation in *Plethodon* may have been related to these earlier glaciation events.

The results of allozyme studies have increased the number of recognized species in the *P. glutinosus* group from five in 1962 to 22 at present time (14, 33, 35, 42, 102). All but two (*fourchensis* and *petraeus*) of the newly recognized species are sibling species of the *P. glutinosus* complex. In addition, the *P. jordani* complex, as analyzed by Peabody (79), was found to include at least five different parapatric and allopatric forms, based on allozyme variation.

The genetic distances among the 22 currently recognized species of the *P. glutinosus* group indicate that all had a common ancestor in the Pliocene. This is also true of the seven species of the *P. cinereus* group, the two species of the *P. wehrlei* group, and three (*P. welleri, P. dorsalis,* and *P. angusticlavius*) of four species of the *P. welleri* group. Thus, the only species of eastern *Plethodon* that does not have sister species with a common ancestor in the Pliocene is *P. websteri*. Although the six southern species of the *P. glutinosus* complex (*grobmani, savannah, ocmulgee, variolatus, mississippi* and *kisatchie*) have entirely Coastal Plain distributions; the present distributions of the remaining 27 species of eastern *Plethodon* all include mountainous areas in the Appalachian, Ozark, or Ouachita mountains.

During the Pliocene periods of glaciation and aridity occurred in North America (57, 83, 89, 95). *Plethodon* is generally absent from drier areas such as the prairie regions of North America, suggesting an explanation for the subdivision of the ranges of eastern *Plethodon* at that time. If lower elevation habitats were largely grasslands for long periods during the Pliocene, this may have isolated many local populations of *Plethodon* in the relict forests of upland areas, thus initiating the divergence that led to so many speciation events in these woodland-adapted salamanders. If arid climates were responsible for the initial isolation of many forms, these dry conditions probably lasted for long periods—an estimation based upon the degree of genetic differentiation between the many living species of *Plethodon*. With the return of wetter climates and the spread of forests to the lowlands, some of the differentiated forms have been able to expand their ranges, but others have remained restricted to their mountain isolates. That species of eastern *Plethodon* can rapidly expand their ranges into uninhabited areas is illustrated by the northern spread of *P. cinereus* into glaciated territory during the last 12,000 years (21, 43). Three fourths of its present large range was under the last continental ice sheet at its maximum 21,000 years ago (80).

In eastern North America, only five phylogenetic lines with surviving descendants were present in the early Pliocene. Two of these are in the *P. welleri* group: *P. websteri* and the ancestor of the three remaining species in that group. The other three surviving lines represent the single ancestor of each of the remaining three species groups. If other lineages were present in eastern North America at that time, they apparently have become extinct. Harvey et al (26) have calculated a high rate of extinction in *Plethodon*. Some of the isolated mountain regions where speciation was initiated were probably occupied by coexisting species of *Plethodon* living in different ecological niches, and these species likely were differentiated in size as are the members of some different species groups in eastern North America today. In eastern North America it is common to find a large species of the *P. wehrlei* or *P. glutinosus* groups sympatric with a small species of the *P. cinereus* or *P. welleri* groups (21). Thus, the ancestors of two or more species of *Plethodon*, particularly if they were members of different species groups, might have been isolated together in a single mountain range. With the spread of newly differentiated species that arose during the burst of speciation in the Pliocene, many contacts among species have taken place in eastern North America, thus providing an opportunity to study many contact zones. When completed, my allozyme studies will be able to correlate the amount of genetic distance that has evolved between each pair of species with the extent of reproductive isolation that occurs in nature, the geographic and ecological overlap that has occurred, the amount of hybridization (if any) that is now taking place, and the extent of gene flow from one form to the other as a result of that hybridization. The amount of divergence between many species of eastern large *Plethodon* has not been sufficient to result in complete reproductive isolation between many recently differentiated species; examples of hybridization are given in the next section.

The hypothesis that explains the burst of speciation in *Plethodon* during the Pliocene implies that geographic speciation took place during long periods of isolation and is consistent with the allopatric model of speciation. There is evidence for the hybrid origin of one species, *P. teyahalee* (35). Even in this case the allopatric model is not ruled out, for it may merely mean that the isolated population that gave rise to *P. teyahalee* was of hybrid origin. *Plethodon teyahalee* is a large member of the *P. glutinosus* complex found in the southern Appalachians west of the French Broad River. It differs from most members of the *P. glutinosus* and *P. jordani* complexes at six allozyme loci. At two of these loci it has unique alleles, but at the other four loci it possesses alleles that are common in two other species, two in the parapatric *P. cylindraceus* (a large, white-spotted species of the *P. glutinosus* complex), and two in the smaller, parapatric unspotted, red-legged member of the *P. jordani* complex, once considered a separate species, *P. shermani,* or subspecies, *P. j. shermani. Plethodon teyahalee* is also somewhat intermediate in color pattern between these two species, being the only member of the *P. glutinosus* complex that sometimes has small red spots on its legs. Its dorsal white spots are smaller than in most other species of the *P. glutinosus* complex. It is genetically homogeneous throughout its range (35), except when it is involved in hybridization with other southern Appalachian species, and it may therefore be of Pleistocene origin.

NATURAL HYBRIDIZATION AMONG SPECIES

From my analysis of the distributional and genetic interactions of the many newly discovered species of eastern *Plethodon*, it is clear that many pairs have parapatric distributions (Figure 1) (35). It also is apparent that some species hybridize extensively, mostly in narrow contact zones. Most of these have Nei genetic distances below 0.4. Figure 2 shows the distributions of nine forms of the *P. jordani* (A-D) and *P. glutinosus* (E-I) complexes in the southwestern Appalachian Mountains between 34°45′ and 36°N and 83° and 84°45′W. The genetic distances among all nine forms are greater than 0.15. Studies on geographic variation in allozymes in these forms have been made (33, 35, 79), and in this particular group of forms, color pattern differences are so consistent that distributional data from Highton (29) may be used to show the details of the ranges of all nine forms in Figure 2.

The following pairs of forms are broadly sympatric with little or no evidence of hybridization in all or most of their range overlaps: A,E; B,E; D,E; E,F. D and H are widely sympatric in Virginia, to the northeast of this region, without known hybridization, but they do not occur sympatrically in the area included in Figure 2. No hybrids between E and F are known although they are sympatric throughout the range of F. B and E are sympatric throughout the range of B, but hybridization is extremely rare (R Highton, unpublished data). In the remaining pairs of sympatric forms, hybridization is frequent in only a portion of the area of sympatry or contact. D and E hybridize extensively at only one locality in the Cowee Mountains (29). A and E are widely sympatric at intermediate elevations in most of the Great Smoky Mountains, but they replace each other altitudinally within a narrow contact zone at the northeastern edge of the range of A, an area in which they hybridize extensively (29, 79).

The remaining pairs, all with parapatric distributions, hybridize in contact zones: A,D; C,E; C,F; C,I; D,G; E,G; E,I; G,I (29, 33, 79; R Highton, unpublished data). Forms E and H do not hybridize at one site in Polk County, Tennessee (33), but transects between the two forms in two other areas to the northeast indicate parapatric hybridation (R Highton, unpublished data). Data

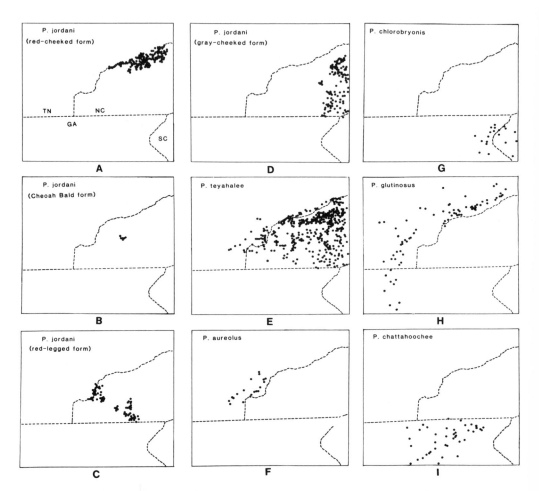

Figure 2 The distribution of nine forms of the *P. jordani* (A-D) and *P. glutinosus* (E-I) complexes in the southern Appalachian Mountains between 34°45′ and 36°N and 83° and 84°45′W.

from a transect between H and I have not yet been analyzed. The widths of these hybrid zones are variable. Highton (29) suggested that a former population of C in the Snowbird mountains of North Carolina has been swamped by hybridization with form E.

Extensive parapatric hybridization occurs between five pairs of the same complex (A,D; E,G; E,H; E,I; G,I) and between four pairs of different complexes (C,E; C,F; C,I; D,G). Only one case of sympatry with no hybridization occurs between species of the same complex (E,F), while three cases (A,E; B,E; D,E) occur between pairs of different complexes. Clearly, the classical recognition of the two complexes as only two species, *P. glutinosus* and *P. jordani* (22, 28, 29), is not correct. Most or all of the nine forms should be regarded as separate species. Although the members of each complex are usually easily recognized as distinct by coloration and size, there is little evidence that either of these complexes is monophyletic (79; R Highton and RB Peabody, unpublished data).

All except group B hybridize extensively with at least one other form, and *P. teyahalee* (E), itself probably of hybrid origin, hybridizes with six other forms. Yet five of the forms (A, B, D, E, F) in this area and one (H) in another region are widely sympatric with at least one other form of the two complexes without hybridization. In these cases, the forms clearly have reached the species level of divergence.

The parapatric hybrid zones should be regarded as hybrid zones between closely related species with incomplete reproductive isolation, rather than as intergrade zones between subspecies, in spite of the fact that most hybrid zones resemble the traditional view of the latter. Most of the hybrid zones are relatively narrow in comparison with the ranges of the forms, and often there is little indication of introgression beyond a limited area near the contact zone. If, on the basis of parapatric hybridization involving all the forms except B, the nine forms were regarded taxonomically as a single species, an impossible taxonomy would result. It would imply that 1. the widely sympatric, morphologically distinct, and usually noninterbreeding *P. jordani* and *P. glutinosus* complexes were the same biological species; 2. several circular overlaps ("rings") occur within a single species; and 3. noninterbreeding sympatric forms with up to six fixed differences and genetic distances up to 0.5 were all members of the same biological species. It also would place groups E, F, and H in the same taxonomic species although all three have been found sympatrically at a site in Polk County, Tennessee, without evidence of hybridization (32). The suggested taxonomic arrangement is to regard most or all of these forms as separate species but to recognize in this group that reproductive isolation may take several million years to evolve and has not yet been completed between a number of pairs of species.

CONCLUSIONS

Salamanders of the lungless terrestrial genus *Plethodon* are often the most abundant vertebrate in forests of eastern North America. From five ancestral lineages that were present in the late Miocene or early Pliocene, a burst of speciation has resulted in 35 recognized species in four species groups. There are two pre-Pliocene clades in the *P. welleri* group (*P. websteri* and another clade that has given rise to at least three existing species, *P. dorsalis, P. angusticlavius* and *P. welleri*). The remaining three pre-Pliocene clades each gave rise to a separate living species group (*P. cinereus* group [7 sp.]; *P. glutinosus* group [22 sp.] and the *P. wehrlei* group [2 sp.]). At least part of the range of most eastern species is in an upland area. The dry periods of the Pliocene probably isolated many populations of these lineages because of the reduction of forests at lower elevations, which thus led to allopatric speciation. In *Plethodon*, and probably in many other salamander genera as well, speciation has often been decoupled from the evolution of morphological novelties, with far more speciation events than number of new morphotypes evolved. This has resulted in numerous cryptic species that were not discovered until allozyme studies revealed their genetic distinctness.

When wetter climates returned, a number of species survived only in their restricted mountain habitats, and these now have very small ranges. Others were able to expand their ranges and at least 140 contact and overlap zones have resulted between species of eastern *Plethodon*. In many contacts, reproductive isolating mechanisms either already existed or were subsequently evolved so that many cases of wide sympatry now occur. At many parapatric contact zones, however, hybridization between closely related species occurred. Many of these hybrid zones still exist and are now under study.

ACKNOWLEDGMENTS

I wish to thank the National Science Foundation, and the General Research Board and the Computer Science Center of the University of Maryland, for their many years of support of my research on *Plethodon*, as well as all the students and colleagues who aided in the work in countless ways. Bretton W Kent and Geerat J Vermeij provided helpful information. David E Carr, Douglas E Gill, Arnold B Grobman, Carla A Hass, S Blair Hedges, Richard L Hoffman, Allan Larson, Stephen G Tilley, David B Wake, and Addison H Wynn read the manuscript and provided numerous suggestions for which I am most grateful.

Literature Cited

1. Adler KK, Dennis DM. 1962. *Plethodon longicrus*, a new salamander (Amphibia: Plethodontidae) from North Carolina. *Ohio Herp. Soc., Sp. Pub.* 4:1–14
2. Allen ER, Neill WT. 1949. A new sub-
species of salamander (genus *Plethodon*) from Florida and Georgia. *Herpetologica* 5:112–14
3. Arnold SJ. 1976. Sexual behavior, sexual interference and sexual defense in the salamanders *Ambystoma macula-*

tum, *Ambystoma tigrinum* and *Plethodon jordani*. *Z. Tierpsychol.* 42:247–300

4. Avise JC. 1994. *Molecular Markers, Natural History and Evolution.* New York: Chapman & Hall. 511 pp.

5. Bishop SC. 1943. *Handbook of Salamanders.* Ithaca, NY: Comstock. 555 pp.

6. Burton TM, Likens GE. 1975. Salamander populations and biomass in the Hubbard Brook Experimental Forest, New Hampshire. *Copeia* 1975:541–46

7. Carr DE. 1995. Morphological variation among species and populations of salamanders of the *Plethodon glutinosus* complex. *Herpetologica*. In press

8. Clay WM, Case RB, Cunningham R. 1955. On the taxonomic status of the slimy salamander, *Plethodon glutinosus* (Green), in southeastern Kentucky. *Trans. Acad. Sci.* 16:57–65

9. Coyne JA, Orr HA. 1989. Patterns of speciation in *Drosophila*. *Evolution* 43:362–81

10. Darda DM. 1994. Allozyme variation and morphological evolution among Mexican salamanders of the genus *Chiropterotriton* (Caudata: Plethodontidae). *Herpetologica* 50:164–87

11. Dawley EM. 1984. Recognition of individual, sex and species odours by salamanders of the *Plethodon glutinosus–P. jordani* complex. *Anim. Behav.* 32:353–61

12. Dawley EM. 1986. Behavioral isolating mechanisms in sympatric terrestrial salamanders. *Herpetologica* 42:156–64

13. Dawley EM. 1987. Species discrimination between hybridizing and non-hybridizing terrestrial salamanders. *Copeia* 1987:924–31

14. Duncan R, Highton R. 1979. Genetic relationships of the eastern large Plethodon of the Ouachita Mountains. *Copeia* 1979:96–110

15. Dunn ER. 1926. *The Salamanders of the Family Plethodontidae.* Southampton: Smith College Ann. Pub. 441 pp.

16. Fraser DF. 1976. Coexistence of salamanders of the genus *Plethodon:* a variation of the Santa Rosalia theme. *Ecology* 57:238–51

17. Fraser DF. 1976. Empirical evaluation of the hypothesis of food competition in salamanders of the genus *Plethodon*. *Ecology* 57:459–71

18. Frost DR, Hillis DM. 1990. Species in concept and practice: herpetological applications. *Herpetologica* 46:87–104

19. Good DA, Wake DB. 1992. Geographic variation and speciation in the torrent salamanders of the genus *Rhyacotriton*

(Caudata: Rhyacotritonidae). *Univ. Calif. Pub. Zool.* 126:1–91

20. Good DA, Wurst GZ, Wake DB. 1987. Patterns of geographic variation in allozymes of the Olympic salamander, *Rhyacotriton olympicus* (Caudata: Dicamptodontidae). *Fieldiana* 32:1–15

21. Grobman AB. 1944. The distribution of the salamanders of the genus *Plethodon* in eastern United States and Canada. *Ann. NY Acad. Sci.* 45:261–316

22. Hairston NG. 1950. Intergradation in Appalachian salamanders of the genus *Plethodon*. *Copeia* 1950:262–73

23. Hairston NG. 1980. Evolution under interspecific competition: field experiments on terrestrial salamanders. *Evolution* 34:409–20

24. Hairston NG. 1980. The experimental test of an analysis of field distributions: competition in terrestrial salamanders. *Ecology* 61:817–26

25. Hanken J. 1983. Genetic variation in a dwarfed lineage, the Mexican salamander genus *Thorius* (Amphibia: Plethodontidae): taxonomic, ecologic and evolutionary implications. *Copeia* 1983:1051–73

26. Harvey PH, Holmes EC, Nee S. 1994. Model phylogenies to explain the real world. *BioEssays* 16:767–70

27. Hass CA, Highton R, Maxson LR. 1992. Relationships among the eastern *Plethodon:* evidence from immunology. *J. Herpetol.* 26:137–41

28. Highton R. 1962. Revision of North American salamanders of the genus *Plethodon*. *Bull. Fl. State Mus.* 6:235–367

29. Highton R. 1970. Evolutionary interactions between species of North American salamanders of the genus *Plethodon*. Part 1. Genetic and ecological relationships of *Plethodon jordani* and *P. glutinosus* in the southern Appalachian Mountains. *Evol. Biol.* 4:211–41

30. Highton R. 1972. Distributional interactions among eastern North American salamanders of the genus *Plethodon*. In *The Distributional History of the Biota of the Southern Appalachians.* Part III: *Vertebrates*, ed. PC Holt, pp. 139–88. Blacksburg, VA: Res. Div. Monogr. Va. Poly. Univ.

31. Highton R. 1977. Comparison of microgeographic variation in morphological and electrophoretic traits. *Evol. Biol.* 10:397–436

32. Highton R. 1979. A new cryptic species of salamander of the genus *Plethodon* from the southeastern United States (Amphibia: Plethodontidae). *Brimleyana* 1:30–36

33. Highton R. 1984. A new species of woodland salamander of the *Plethodon glutinosus* group from the southern Appalachian Mountains. *Brimleyana* 9:1–20

34. Highton R. 1985. The width of the contact zone between *Plethodon dorsalis* and *P. websteri* in Jefferson County, Alabama. *J. Herpetol.* 19:544–46

35. Highton R. 1989. Biochemical evolution in the slimy salamanders of the *Plethodon glutinosus* complex in the eastern United States. Part 1. Geographic protein variation. *Ill. Biol. Monogr.* 57:1–78

36. Highton R. 1990. Taxonomic treatment of genetically differentiated populations. *Herpetologica* 46:114–21

37. Highton R. 1991. Molecular phylogeny of plethondonine salamanders and hylid frogs: statistical analysis of protein comparisons. *Mol. Biol. Evol.* 8:796–818

38. Highton R. 1993. The relationship between the number of loci and the statistical support for the topology of UPGMA trees obtained from genetic distance data. *Mol. Phylogen. Evol.* 2:337–43

39. Highton R, Brame A. 1965. *Plethodon stormi* species nov. *Pilot Reg. Zool.* 20:1–2

40. Highton R, Grobman AB. 1956. Two new salamanders of the genus *Plethodon* from the southeastern United States. *Herpetologica* 12:185–88

41. Highton R, Larson A. 1979. The genetic relationships of the salamanders of the genus *Plethodon*. *Syst. Zool.* 28:579–99

42. Highton R, MacGregor JR. 1983. *Plethodon kentucki* 1Mittleman: a valid species of Cumberland Plateau woodland salamander. *Herpetologica* 39: 189–200

43. Highton R, Webster TP. 1976. Geographic protein variation and divergence in populations of the salamander *Plethodon cinereus*. *Evolution* 30:33–45

44. Highton R, Worthington RD. 1967. A new salamander of the genus *Plethodon* from Virginia. *Copeia* 1967:617–26

45. Huheey JE. 1960. Mimicry in the color pattern of certain Appalachian salamanders. *J. Elisha Mitchell Sci. Soc.* 76:246–51

46. Jackman TR, Wake DB. 1994. Evolutionary and historical analysis of protein variation in the blotched forms of salamanders of the Ensatina complex (Amphibia: Plethodontidae). *Evolution* 48:876–97

47. Jacobs JF. 1987. A preliminary investigation of geographic genetic variation and systematics of the two-lined salamander, *Eurycea bislineata* (Green). *Herpetologica* 43:423–46

48. Jaeger RG. 1971. Moisture as a limiting factor influencing the distributions of two species of terrestrial salamanders. *Oecologia* 6:191–207

49. Jaeger RG. 1971. Competitive exclusion as a factor influencing the distributions of two species of terrestrial salamanders. *Ecology* 52:632–37

50. Jaeger RG. 1972. Food as a limited resource in competition between two species of terrestrial salamanders. *Ecology* 53:535–46

51. Jaeger RG. 1980. Density-dependent and density-independent causes of extinction of a salamander population. *Evolution* 34:617–21

52. Jaeger RG. 1981. Dear enemy recognition and the costs of aggression between salamanders. *Am. Nat.* 177:962–74

53. Jaeger RG. 1986. Pheromonal markers as territorial advertisement by terrestrial salamanders. In *Chemical Signals in Vertebrates*, ed. D. Duvall, D Muller-Schwarze, R Silverstein, 4:191–203. Plenum

54. Jaeger RG, Gergits WF. 1979. Intra- and interspecies communication in salamanders through chemical signals on the substrate. *Anim. Behav.* 27:150–56

55. Jaeger RG, Lucas J. 1990. On evaluation of foraging strategies through estimates of reproductive success. In *Behavioral Mechanisms of Food Selection*, ed. RN Hughes. NATO ASI Ser., Subser. G. *Ecol. Sci.* :83–94. Heidelberg: Springer-Verlag

56. Kezer J, Sessions SK, León P. 1989. The meiotic structure and behavior of the strongly heteromorphic X/Y sex chromosomes of neotropical plethodontid salamanders of the genus *Oedipina*. *Chromosoma* 98:433–42

57. Larsen HC, Saunders AD, Clift PD, Beget J, Wei W, Spezzaferri S, ODP Leg Scientific Party. 1994. Seven million years of glaciation in Greenland. *Science* 264:952–55

58. Larson A. 1983. A molecular phylogenetic perspective on the origins of a lowland tropical salamander fauna. I. Phylogenetic inferences from protein comparisons. *Herpetologica* 39:85–99

59. Larson A. 1984. Neontological inferences of evolutionary pattern and process in the salamander family Plethodontidae. *Evol. Biol.* 17:119–217

60. Larson A. 1989. The relationship between speciation and morphological evolution. In *Speciation and Its Consequences*, ed. D Otte, Endler JA, pp. 579–98. Sunderland, MA: Sinauer

61. Larson A, Highton R. 1978. Geographic protein variation and divergence in the salamanders of the *Plethodon welleri* group (Amphibia: Plethodontidae). *Syst. Zool.* 27:431–48

62. Larson A, Wake DB, Maxson LR, Highton R. 1981. A molecular phylogenetic perspective on the origins of morphological novelties in the salamanders of the tribe Plethodontini (Amphibia, Plethodontidae). *Evolution* 35:405–22

63. Larson A, Wake DB, Yanev KP. 1984. Measuring gene flow among populations having high levels of genetic fragmentation. *Genetics* 106:293–308

64. Madison DM. 1969. Homing behavior of the red-cheeked salamander, *Plethodon jordani*. *Anim. Behav.* 17:25–39

65. Madison DM. 1970. Homing behavior, orientation, and home range of salamanders tagged with Tantalum-182. *Science* 168:1484–87

66. Madison DM. 1972. Homing orientation in salamanders: a mechanism involving chemical cues. In *Animal Orientation and Navigation*, ed. SR Galler, K Schmidt-Koenig, G Jacobs, RE Belleville. *NASA Sp. Pub.* 262:485–98

67. Maxson LR, Highton R, Wake DB. 1979. Albumin evolution and its phylogenetic implications in the plethodontid salamander genera *Plethodon* and *Ensatina*. *Copeia* 1979:502–8

68. Maxson LR, Maxson RD. 1979. Comparative albumin and biochemical evolution in plethodontid salamanders. *Evolution* 33:1057–62

69. Mayr E. 1942. *Systematics and the Origin of Species.* New York: Columbia Univ. Press. 334 pp.

70. McClellan T. 1984. Molecular charge and electrophoretic mobility in Cetacean myoglobins of known sequence. *Biochem. Genet.* 22:181–200

71. Merchant H. 1972. Estimated population size and home range of the salamanders *Plethodon jordani* and *Plethodon glutinosus*. *J. Wash. Acad. Sci.* 62:248–57

72. Merkle DA, Guttman SI. 1977. Geographic variation in the cave salamander *Eurycea lucifuga*. *Herpetologica* 33:313–21

73. Merkle DA, Guttman SI, Nickerson MA. 1977. Genetic uniformity throughout the range of the hellbender, *Cryptobranchus alleganiensis*. *Copeia* 1977:549–53

74. Mittleman MB. 1951. American Caudata. VII. Two new salamanders of the genus *Plethodon*. *Herpetologica* 7:105–12

75. Mizuno S, Macgregor HC. 1974. Chromosomes, DNA sequences, and evolution in salamanders of the genus *Plethodon*. *Chromosoma* 48:239–96

76. Moler PE, Kezer J. 1993. Karyology and systematics of the salamander genus *Pseudobronchus*. *Copeia* 1993:39–47

77. Nei M. 1972. Genetic distance between populations. *Am. Nat.* 106:283–92

78. Orr HA, Coyne JA. 1989. The genetics of postzygotic isolation in the *Drosophila virilis* group. *Genetics* 121:527–37

79. Peabody RB. 1978. *Electrophoretic analysis of geographic variation of two Appalachian salamanders,* Plethodon jordani *and* Plethodon glutinosus. PhD thesis, Univ. Maryland, College Park

80. Peltier WR. 1994. Ice age paleotopography. *Science* 265:195–201

81. Pope CH, Hairston NG. 1948. Two new subspecies of the salamander *Plethodon shermani*. *Copeia* 1948:106–7

82. Ramshaw JAM, Coyne JA, Lewonton RC. 1979. The sensitivity of gel electrophoresis as a detector of genetic variation. *Genetics* 93:1019–37

83. Rea DK. 1994. The paleoclimatic record provided by eolian deposition in the deep sea: the geologic history of wind. *Rev. Geophys.* 32:159–95

84. Sarich VM. 1977, Rates, samples sizes, and the neutrality hypothesis for electrophoresis in evolutionary studies. *Nature* 265:24–28

85. Sessions SK, Kezer J. 1987. Cytogenetic evolution in the plethodontid salamander genus *Aneides*. *Chromosoma* 95:17–30

86. Sessions SK, Larson A. 1987. Developmental correlates of genome size in plethodontid salamanders and their implications for genome evolution. *Evolution* 41:1239–51

87. Sessions SK, Wiley JE. 1985. Chromosome evolution in salamanders of the genus *Necturus*. *Brimleyana* 10:37–52

88. Slatkin M. 1981. Estimating levels of gene flow in natural populations. *Genetics* 99:323–35

89. Stanley SM. 1989. *Earth and Life Through Time.* New York: WH Freeman. 689 pp.

90. Templeton AR. 1989. The meaning of species and speciation: a genetic perspective. In *Speciation and its Consequences*, ed. D Otte, JA Endler, pp. 2–38. Sunderland, MA: Sinauer

91. Thorpe JP. 1982. The molecular clock hypothesis: biochemical evolution, genetic differentiation and systematics. *Annu. Rev. Ecol. Syst.* 13:139–68

92. Thurow GR. 1968. On the small black

Plethodon problem. *W. Ill. Univ. Ser. Biol. Sci.* 6:1–48

93. Tilley SG, Merritt RB, Wu B, Highton R. 1978. Genetic differentiation in salamanders of the *Desmognathus ochrophaeus* complex (Plethodontidae). *Evolution* 32:93–115

94. Tilley SG, Verrell PA, Arnold SJ. 1990. Correspondence between sexual isolation and allozyme differentiation: a test in the salamander *Desmognathus ochrophaeus*. *Proc. Natl. Acad. Sci. USA* 87: 2715–19

95. Van Valkenburgh B, Janis CM. 1993. Historical diversity patterns in North American large herbivores and carnivores. In *Species Diversity in Ecological Communities,* ed. RR Ricklefs, D Schluter, pp. 330–40. Chicago: Univ. Chicago Press

96. Wake DB. 1966. Comparative osteology and evolution of the lungless salamanders, family Plethodontidae. *Mem. S. Cal. Acad. Sci.* 4:1–111

97. Wake DB. 1981. The application of allozyme evidence to problems in the evolution of morphology. *Proc. Second Int. Congr. Syst. Evol. Biol.*:257–270

98. Wake DB, Roth G, Wake MH. 1983. On the problem of stasis in organismal evolution. *J. Theor. Biol.* 101:211–24

99. Wake DB, Yanev KP. 1986. Geographic variation in allozymes in a "ring species," the plethodontid salamander *Ensatina eschscholtzii* of western North America. *Evolution* 40:702–15

100. Wrobel DJ, Gergits WF, Jaeger RG. 1980. An experimental study of interference competition among terrestrial salamanders. *Ecology* 61:1034–39

101. Wynn AH. 1986. Linkage disequilibrium and a contact zone in *Plethodon cinereus* on the Del-Mar-Va peninsula. *Evolution* 40:44–54

102. Wynn AH, Highton R, Jacobs JF. 1988. A new species of rock-crevice dwelling *Plethodon* from Pigeon Mountain, Georgia. *Herpetologica* 44:135–43

Annu. Rev. Ecol. Syst. 1995. 26:601–29

MULTIPLE FITNESS PEAKS AND EPISTASIS

Michael C. Whitlock

Institute for Cell, Animal, and Population Genetics, University of Edinburgh, Edinburgh, EH9 3JT Scotland[1]

Patrick C. Phillips

Department of Biology, University of Texas, Arlington, Texas 76019

Francisco B.-G. Moore and Stephen J. Tonsor[2]

Kellogg Biological Station, Michigan State University, Hickory Corners, Michigan 49060

KEY WORDS: adaptive landscape, epistasis, quantitative genetics, natural selection, shifting-balance theory

---

Abstract

The importance of genetic interactions in the evolutionary process has been debated for more than half a century. Genetic interactions such as underdominance and epistasis (the interaction among genetic loci in their effects on phenotypes or fitness) can play a special role in the evolutionary process because they can create multiple fitness optima (adaptive peaks) separated by fitness minima (adaptive valleys). The valleys prevent deterministic evolution from one peak to another. We review the evidence that genetic interaction is a common phenomenon in natural populations. Some studies give strong circumstantial evidence for multiple fitness peaks, although the mapping of epistatic interactions onto fitness surfaces remains incompletely explored, and absolute proof that multiple peaks exist can be shown to be empirically im-

---

[1]Address as of 1 July 1995: Department of Zoology, University of British Columbia, Vancouver, British Columbia V6T 1Z4, Canada

[2]Address as of 1 May 1995: Department of Biological Sciences, University of Pittsburgh, Pittsburgh, PA 15260

601

possible. We show that there are many reasons that epistatic polymorphism is very difficult to find, even when interactions are an extremely important part of the genetic system. When polymorphism results in the presence of multiple fitness peaks within a group of interbreeding populations, one fitness peak will quickly be nearly fixed within all interbreeding populations, but when epistatic or underdominant loci are nearly fixed, there will be no direct evidence of genetic interaction. Thus when complex landscapes are evolutionarily most important, evidence for alternative high fitness genetic combinations will be most ephemeral. Genetic interactions have been most clearly demonstrated in wide crosses within species and among closely related species. This evidence suggests that genetic interactions may play an important role in taxonomic diversification and species-level constraints. Population genetic analyses linked with new approaches in metabolic and molecular genetic research are likely to provide exciting new insights into the role of gene interactions in the evolutionary process.

INTRODUCTION

The ecological, developmental, and metabolic requirements of life pose problems to be answered by evolution. These problems most often have more than one genotypic or phenotypic solution; however, not all these solutions are necessarily equally fit. Some fit types may be separated from more fit forms by intermediates of lower fitness. With these "adaptive valleys" intervening, a species cannot evolve from one "fitness peak" to another by the deterministic process of natural selection. Significant controversy has surrounded the arguments concerning the mechanisms for adaptive change from one fitness peak to another, particularly those dealing with Sewall Wright's "shifting balance" process. Most of the controversy has centered around the relative roles of selection and drift in evolution and on the likelihood of population structures capable of producing peak shifts in nature. A more fundamental issue, however, is often not addressed. There is little consensus about the extent or importance of genetic interactions and about whether actual fitness surfaces have multiple adaptive peaks. In other words, it is an open question whether peak shift models are needed at all. In this review we assess the empirical and theoretical evidence on the need for a peak shift model; that is, we review the evidence that fitness functions have multiple peaks.

In this review we argue that multiple-peaked fitness functions exist in nature and that ruggedness may be a common feature of adaptive landscapes. We argue this from three perspectives: that genetic interactions among loci (i.e. epistasis) are common and important in the mapping between genotypes and fitness, that underdominance can sometimes create multiple peaks from interactions within loci, and that phenotypic fitness functions are also rugged with multiple peaks. After some introductory sections to clarify some of the concepts and terminology, this review consists of essentially three parts: The first two sections focus on the reasons to expect epistatic and dominance interactions that can cause complex landscapes, and also review some of the evidence that such interactions are in fact common. We briefly review the evidence that epistatic interactions affect phenotypic expression, then we concentrate on the genetic interactions that can be shown to affect fitness. We then discuss some ecological and quantitative genetic reasons to expect disruptive selection on hypothetical adaptive landscapes, and we review the evidence for these multiple peaks on a phenotypic scale. Finally we discuss in some detail the reasons that epistasis is extremely difficult to detect, such that the examples observed represent a very small fraction of the cases where epistasis and multiple peaks are important components of evolution.

This paper essentially makes two points. First that there is substantial evidence that adaptive landscapes are not smooth, but there are interactions within and between genetic loci in determining fitness. The evidence for this comes from a variety of studies, many of which are difficult to do (and therefore are rarely done,) and which are biased against finding evidence for these interactions, so the studies we have must represent the tip of the iceberg. Second, we show that proving these rugged landscapes have true multiple peaks is impossible, but we maintain that the fact that landscapes are rugged per se should motivate more studies of evolution on complex adaptive landscapes.

Some of the literature on epistatic interactions has been reviewed previously (1, 51, 88). Because of the limitations of space, we have not been able to cite all of the literature that bears on this broad topic; instead we have attempted to choose representative examples.

Genotypic Versus Phenotypic Fitness Functions

The adaptive landscape can be drawn at two distinct scales: mapping genotypes to fitness or mapping phenotypes to fitness. Ultimately what matters to evolution is the fitness of genotypes, of course, so landscapes must be uneven at the genotypic scale for genetic interactions to affect the course of evolution. Often, however, we know little about the genetic basis of phenotypes and more about the relationship between the phenotype and fitness. For traits determined by many loci that interact additively, complex phenotypic fitness functions imply complex genotypic fitness functions. This paper examines both scales of fitness function.

The Definition of Epistasis and of Multiple Peak Systems

Using Sewall Wright's familiar metaphor, the mean fitness of a population can be viewed as a function of the gene frequencies of the population (135, 136) if one is focusing on the genetic scale, or as a function of the frequencies of particular phenotypic classes (68, 69, 112) if one is focusing on the phenotypic scale. The shape of this mean fitness function, or adaptive landscape, to a large degree specifies the evolutionary direction of the population under the deterministic influence of natural selection. Natural selection can therefore be seen to drive populations to local peaks on the adaptive landscape. (Even if the adaptive landscape metaphor does not strictly hold, there are stable states, perhaps even strange attractors, analogous to adaptive peaks towards which populations will evolve—5). An adaptive landscape thus serves to visualize dynamical systems in which there are multiple stable equilibria.

As originally envisaged by Wright, these multiple adaptive peaks are generated by epistatic interactions between genes (135). The term epistasis actually has two distinct usages in the genetic literature (129). Epistasis was originally coined (8) to describe the interaction between genes in which the action of one gene was blocked by the action of another gene (epistasis literally means "standing above"—44). This is the standard usage of the term today in molecular and developmental biology. Other forms of interaction were also given names such as "cumulative" and "mutually supplementary" (77), but these descriptions seem to have fallen out of favor. Evolutionary geneticists tend to use epistasis to describe any form of gene interaction. This usage stems from Fisher (38), who described all genetic variance not attributable to additive and dominance effects as caused by "epistacy." The evolutionary genetics literature generally has continued to focus on the variance definition of epistasis (129), but we make a strict distinction between gene action, the statistical "effects" attributed to that action, and the genetic variance that arises from the multilocus composition of individuals within a population. As we show in the next section, there can be substantial epistatic gene action, but little manifestation of this in the form of epistatic genetic variance within a population. Indeed, one of the major conclusions of this paper is that this is a common state of natural populations.

Following the evolutionary genetic usage, we use epistasis to describe any form of gene interaction, but we focus on gene interactions that generate multiple fitness peaks. In the simplest two-locus case, multiple adaptive peaks can be generated when two mutually exclusive sets of alleles at each locus produce high fitness genotypes, but any mixing between the sets results in genotypes of lowered fitness (Table 1). For multiple peak epistasis, it is important to make a distinction between epistasis for an arbitrary character and epistasis for the particular character that is fitness. To a large extent, epistasis is often viewed as a nuisance that arises from measuring a character on a particular scale. Historically, quantitative geneticists have tried to eliminate epistatic variance via transformation, although it is not always clear how to proceed (98), and transformations may simply shift the context in which epistasis is observed (54). In this regard, fitness is unique because it does not

Table 1 A simple epistatic fitness function.

	AA	Aa	aa
BB	1 + i	1	1 − i
Bb	1	1	1
bb	1 − i	1	1 + i

exist on an arbitrary scale and should not be transformed (72). Thus, epistasis for fitness will always be epistasis and is not an artifact of scale.

In determining whether a particular fitness function generates a fitness landscape with multiple peaks, it is very important to distinguish between the fitness of individuals and the mean fitness of the population. Mean fitness functions are always less peaked than the individual fitness functions of which they are composed. If an individual fitness function has multiple peaks, but there is a lot of phenotypic variance in the population which is evolving on that landscape, the mean fitness function need not and often does not have multiple stable equilibria (this relationship is even more complex in the presence of frequency dependent selection and other factors) (31).

Epistatic Variance versus Epistatic Genetic Effects

There is an important distinction between epistatic genetic variance and the genetic effects due to epistasis. Particular alleles at different loci may interact strongly to produce radically different phenotypes relative to when they are paired with other alleles, but if these alleles are rare, then there will be very little epistatic variance. Epistatic variance is also a function of the allele frequencies (proportional to $p_A q_A p_B q_B$, where p_A and p_B are the allele frequencies at loci A and B and the q's are equal to $1 - p$), whereas the potential effect of an allelic combination depends only on the actual phenotype produced by that combination. Fitness functions are defined in terms of the fitness of allelic combinations; hence, the evolutionary potential of allelic combinations depends on these epistatic effects, rather than the epistatic variance.

A Simple Model of Gene Action

Many of the issues involved in the estimation of the potential for epistatic gene action can best be discussed with a simple model of gene interaction. The essential attribute of such a model for our purposes must be that it generates more than one fit genotype. Figure 1a presents a simple two-locus model with interaction for fitness for the loci. This sort of model closely parallels the kinds of fitness interactions we might expect from simple stabilizing selection or a canalized metabolic pathway. Notice that the mean fitness function has two peaks, at opposite corners of the distribution, with a valley between, which,

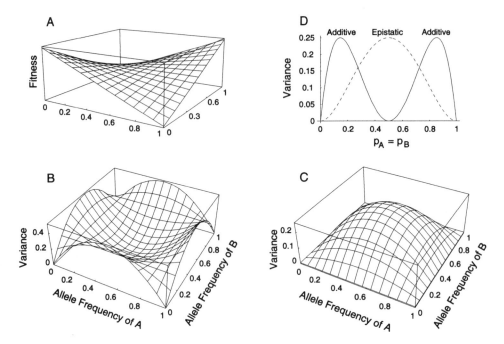

Figure 1 An epistatic fitness landscape and the variance components for fitness. The fitness function listed in Table 1 translates into the mean fitness landscape illustrated in *a*. The two horizontal axes represent the allele frequencies of A and B; the vertical axis represents the mean fitness of a population that has those allele frequencies. In *b*, the additive variance for fitness is represented as a function of these allele frequencies. Notice that the additive genetic variance is at a minimum at the saddle on the mean fitness landscape. Figure *c* represents the epistatic genetic variance as a function of the same allele frequencies. Epistatic variance reaches a maximum at the saddle point but increases very slowly near the fixation points. Near the stable equilibrium allele frequencies (near fixation for AB or ab), most of the genetic variance for fitness is additive, not epistatic, despite the fact that the interaction of genes in this system is entirely synergistic. This is shown in *d*, where the diagonal across gene frequency space connecting the two peaks is plotted (i.e. the x-axis represents the allele frequency $p_A = p_B$) against the variance components for fitness.

in this case, is deepest at $p_A = p_B = 0.5$. Populations at equilibrium will be near one or the other of these peaks. The gene action of these alleles is clearly epistatic for fitness. Yet if we examine the variance components for fitness with this model, we see that most of the genetic variance for fitness near the stable points is additive, not epistatic (see Figure 1b–d). In fact, the only point in genotype space with a large proportion of epistatic variance is near the saddle, at $p_A = p_B = 0.5$. This relationship can be formalized by remembering the least squares definition of additive variance—the additive variance for fitness components corresponds to the slope of the mean fitness function. The slope of the fitness function is 0 at maxima and minima, so their additive variance goes to zero. Any genetic variance at these points must therefore be nonadditive. In our example the stable points correspond to fixation points, which will always be the case without some form of overdominance. When equilibria are at fixed states, nonzero allele frequencies are expected by mutation-selection balance, which causes small allele frequencies precisely in the range where the ratio of epistatic to additive genetic variance is small. We therefore expect to see large amounts of additive variance, or in the case of overdominance, dominance variance, in natural systems even in systems with a large degree of epistatic gene action.

The Importance of Epistasis and Multiple Peak Systems

The critical problem posed by rugged fitness landscapes is: How do species pass through valleys to reach other, higher fitness peaks? Wright differed from Fisher in his answer to this question. Fisher thought the problem to be reasonably unimportant, thinking that there is always genetic variation for fitness available to a species, largely because the environment was constantly changing (in other words, that fitness landscapes are like the surface of a turbulent ocean). Wright, however, viewed the question of adaptive valleys as central to the study of evolution. In response to this difficulty, Wright created his famous Shifting Balance theory (135, 136), which hypothesizes that populations can move from one peak to another by random genetic drift caused by periods of small population size. Following this drift to the domain of attraction of a new peak, a population would be selected uphill to the new peak. The new peak would then be "exported" to other populations by differential migration and/or extinction. This theory has proven to be one of the more intriguing and controversial in the history of population genetics.

There are many other models of peak shifts that may also explain evolution between adaptive peaks. It is possible that macromutations, either in the sense of hopeful monsters (43) or merely in the sense of mutations of large effect (14), can cause the transition between peaks on the phenotypic fitness function. Similarly, change in the biotic or abiotic environment can cause selection pressures to change sufficiently to allow transitions (5, 63, 133), as can changes in the phenotypic variance of populations due to inbreeding (133) or in the genetic correlations among traits (100).

Whatever the mechanism, it is critical that we know the extent to which adaptive landscapes are peaked in nature, in order that we can know the importance of peaks shifts in general. If landscapes are rugged and stable through time, then many traditional models of evolution will need to be reevaluated. The response to selection and adaptation to local peaks, such as that described by Fisher's Fundamental Theorem of Natural Selection, depends on additive gene action (39, 91); if epistatic deviations from additivity are common, then the particular processes involved in transitions between peaks will have to play a more central role in evolutionary thought.

There are many other processes in evolutionary biology that have been shown to be significantly changed by the presence of epistasis for fitness. The evolution of sex (65), diploid life cycles (67), and mating systems (66) are all significantly influenced by interactions among loci. Inbreeding depression and heterosis (52, 54) as well as outbreeding depression (130) may be influenced by epistatic interactions among loci. The evolution of reproductive isolation (27) also depends critically on epistatic interactions. The importance of epistatic interactions and complex landscapes extends well beyond the initial interest motivated by Wright.

EPISTASIS IN THE MAPPING BETWEEN GENOTYPE AND FITNESS

An important way that genetic landscapes can have more than one adaptive peak is if there is significant interaction among alleles at different loci. For epistasis to act as a genetic constraint (such as in the shifting balance models), there have to be significant genetic interactions for fitness. In other words, genotypes must combine nonadditively to determine the reproductive success of individuals; for some purposes, the interactions have to be severe enough to cause multiple peaks on the adaptive landscape. This section reviews some of the evidence for epistatic interactions for fitness.

Direct Measures of Fitness Effects

Classical genetic analysis has determined that there is epistasis for fitness components (e.g. 123) or for strongly selected traits (e.g. 19, 20) in many cases. The tightly linked loci within messenger RNA are selected on their ability to create an appropriate secondary structure; this results in strong disequilibrium among species in predictable allele combinations (119). Genes also are known to interact strongly with sex-determining factors to determine fitness (104).

Studies of the phenotypic effects of particular proteins often indicate epistatic interactions. Examples come from examination of the phenotypic and fitness effects of electrophoretic variants of enzymes (e.g. 13, 86). In the tidepool copepod *Tigriopus californicus,* osmoregulation is accomplished by manipulation of cellular free amino acid content. Osmoregulation is differentially influenced by production of alanine and proline. Sharing of substrates between the Krebs cycle and the alanine and proline pathways means that genes for cellular energetics are epistatic with osmoregulatory genes (45). Alternative alleles for 6-phosphogluconate dehydrogenase in *E. coli* show fitness differences dependent on genetic backgrounds that differ in the presence or absence of an alternative pathway for the metabolism of 6-phosphogluconate. However, they showed no fitness differences in their "normal," i.e. co-evolved, background (35).

Epistasis for Phenotypes

Many types of evidence indicate the existence of epistatic interactions in the mapping between genotype and phenotype. The fitness of organisms depends on the phenotypes, and therefore epistatic interactions for phenotypic traits imply the strong possibility of rugged adaptive landscapes. Epistatic interactions among loci have been investigated using a variety of techniques, ranging from classical genetic analyses to modern QTL mapping, from molecular biology to field ecology. Epistatic interactions have been observed directly, with the genotype and phenotype both known, and they have been inferred by the patterns of genetic variation present in populations. In fact, it was Wright's early investigations of guinea pig coloration patterns (138) that initially led him to consider the importance of epistasis in the evolutionary process (101).

VARIANCE COMPONENT ANALYSES   As we have seen, it is possible to have pronounced epistatic interactions among alleles, without that trait expressing much epistatic variance at all. The opposite is not true, however; epistatic variance does imply the presence of epistatic interactions at the allelic level. The epistatic variance component is extremely difficult to measure (see below), but the presence of epistasis has been tested and found a number of times, particularly in agriculturally important species (41, 54, 102). Many other examples can be drawn from the literature. It is worth noting, however, that the proportion of variance explained by epistatic effects is normally rather small (< 30%) even when significant. Later we explain why this is expected even if there are large interactions among alleles.

QUANTITATIVE TRAIT LOCI   Recent advances in the types of polymorphic genetic markers available to evolutionary geneticists and breeders have enabled the location of the major (and some of the minor) loci responsible for observed phenotypic variation in polygenic traits. Tanksley (122) and Cheverud & Routman (17) review the identification of quantitative trait loci (QTL) through the use of polymorphic DNA markers. These authors also review some of the advantages and drawbacks of estimating genetic interactions in this way. The general picture that emerges from the limited data on QTL for crop species is that few epistatic interactions are important for determining the phenotypes of interest.

Long et al (79) recently completed a study of a synthetic population formed by hybridizing lines selected for high- and low-bristle number in *Drosophila melanogaster,* one of the classical examples of an additive polygenic trait. They produced a highly saturated map of chromosome 3 and the X chromosome, with markers approximately every 4 cM. Thirteen of 60 tests for two-factor epistasis were significant; the probability of this occurring by chance is less than 0.00001. The epistatic effects were of the same order of magnitude as the average allelic effects of the QTL, as were the sex-by-QTL interactions. Thus, although the variation they detected behaved overall as though it were additive, when two-way interactions were examined, 20–25% of the loci showed important epistatic effects. Moreno shows similar results for bristle number mutations acting in concert (88).

THE GENETIC INTERACTIONS OF METABOLIC PATHWAYS   The properties of metabolic pathways imply that both dominance and epistasis are inevitable consequences of the sequential processing of substrates and products in metabolic chains. Kacser & Burns (58) introduced the control coefficient to quantify the extent to which flux through the pathway depends on the catalytic properties of any one enzyme. The introduction of an allele that changes the kinetic properties of any rate-limiting enzyme in the pathway necessarily changes the control coefficients of all the enzymes in the pathway. Thus, the phenotypic effect of one locus is modified by the alleles present at another locus. In spite of the prevalence of such epistatic interactions, the epistatic variance in these systems is expected to be small (60).

If selection instead optimizes flux at an intermediate value, the fitness of alleles at a locus will depend on their relative kinetic properties, the control coefficient for that enzyme, and the flux rate through the pathway as a whole. As the flux rate for the pathway as a whole is moved above and below the optimum, the fitness effects of increases in velocity through any one enzyme will change from negative to positive, respectively (120). Metabolic pathways under selection for intermediate fluxes are therefore very much like any other trait undergoing optimizing selection: There is diffuse epistasis for fitness among the loci affecting the trait (see the section on stabilizing selection below).

DEVELOPMENTAL PROCESSES AND EPISTASIS   Genes must interact via developmental pathways, in much the same way as genes interact in metabolic pathways. Decisions made early in development affect the phenotypic manifestation of genes expressed later. Homeotic mutants in both animals (see 88) and plants (see 22) give clear examples. The segregating alleles at loci polymorphic for homeotic mutations influence the phenotypic expression of whole suites of genes during organ development. There are many examples of gross and widespread epistatic interaction during development varying from the earliest events, i.e. meiosis (107) and mitosis (92), through embryogenesis (110), metamorphosis (105), final height (53), and reproduction (e.g. 22). See

(88) for a review of the literature on epistatic interactions in *Drosophila* development.

Mutation Accumulation

There is not sufficient data on the interactions of new mutations affecting fitness and phenotypes. Some evidence suggests that new mutations are likely to interact predominantly additively (11). Moreno (88) argues that mutations whose singular effect is small and additive can have large synergistic effects with other mutations. New mutations conferring T4 phage resistance in *Escherichia coli* have negative effects on fitness through other pathways but rapidly evolve compensatory epistatic modifiers (74). Similar results in other organisms (cited in 74) suggest that epistatic interactions can be both primitive and evolving traits of genes, but more studies are required.

Recombination Load

Recombination load is the loss in fitness because recombination breaks up associations between beneficial combinations of interacting alleles (16). If recombination load could be precisely measured, it would give the difference between the fitness at the highest peak on the landscape produced by existing genotypes and the mean fitness of the population as it is now. Unfortunately, this is not possible, but we can ascertain the fitness costs of a single generation of recombination, at least in organisms without recombination in one sex or in those like *D. melanogaster* for which we have many genetic markers. Dobzhansky and his co-workers first measured the loss of fitness in recombinant chromosomes, which have lower fitness than nonrecombinant controls (34, 115–117). Chromosomes derived from wild-caught male flies (which do not have recombination) have higher fitness than those derived from females from the same collections (which do recombine) (16, 89). The empirical literature on recombination load has been reviewed briefly in a theoretical context by Charlesworth & Barton (15). The absolute magnitude of this load, as measured by these experiments, is not great but strongly implicates epistatic interactions in determining fitness.

F_2 Breakdown

As we discuss further below, the nature of genetic variation within species is restrained by selection to be near adaptive peaks, and this is likely to constrain epistatic variance. If alleles are maintained in different populations or different species, however, this constraint is lifted, and we might begin to get a less biased perspective on the ruggedness of adaptive landscapes. Thus, one of the best available ways to investigate the importance of epistatic interactions for fitness is by measuring the fitness of the first and second generation offspring of hybrid crosses. If the second generation (the F_2) has an average fitness less than the average of the parents and the F_1, then this is referred to as F_2 breakdown, which can only be attributed to epistasis for fitness. Reductions in fitness in the F_1 (outbreeding depression) can also be attributed to epistasis or dominance and (in the absence of environmental differences) may imply that the adaptive landscape is uneven (see the later section on underdominance).

F_2 breakdown is frequently observed in intraspecific crosses between lines or distant populations (10, 49, 62, 126) (see reviews in 41, 139), particularly in crosses between more divergent lines or populations (41). Hybrid breakdown also occurs in interspecific crosses (46, 118). Loss of fitness, or diminishment of selected traits, has often been observed in crosses between selected lines and unselected controls, showing that genetic interactions can be important in evolution on a very short time scale (106).

Speciation Genetics

The genetic studies of the results of inter-specific crosses have revealed at least two relatively general patterns (30): genes responsible for the loss of fitness in hybrids between species are likely to be on the X-chromosomes (the "large X-effect"), and if only one sex is affected, that sex is likely to be the heterogametic sex (i.e. the sex with only one sex chromosome, like male mammals—this fitness pattern is called Haldane's Rule—47). That alleles on the X chromosomes are fit in their own species but fail in the genetic context of other species' genes (even in a hemizygous state) implies that epistasis between X-linked loci and the autosomes determine the fitness of hybrids (27, 33). For some species of *Drosophila*, even in a homozygous state, X-linked alleles in the presence of autosomes from another species can cause low fitness (94).

We would like to know whether this sort of epistasis implies multiple fitness peaks, but unfortunately this information is unavailable and indeed unknowable. We do not know the full fitness function for any of the speciation genes studied. We do know that the fitness of same-species homozygotes is high, and that the fitness of cross-species heterozygotes is low. However, we cannot know the fitness of all cross-species homozygotes, i.e. individuals homozygous for alleles from the same species at each locus, but with different loci having different sources. For the very reason that the phenotype being studied is loss of fitness and certain hybrids are inviable or sterile, all of these genotypes cannot be constructed. That being so, it is impossible to reconstruct the fitness functions.

Even if the fitness functions for a two-locus pair were measured, this still would not tell us if the fitness landscape had true valleys. It is always possible that the state of a third locus determines the precise contour of the two-locus landscape, and that some unexamined state of that third locus might allow deterministic evolution over the two-locus space. Furthermore, it is also possible that within one of the loci, there has been a sequential substitution of alleles. For example, let A_1, A_2, and A_3 represent alleles at a locus A, and B_1 and B_2 at locus B, where A_1 and B_1 are the ancestral alleles. If A_1 and A_2 (but not A_3) interact well with B_1, and A_2 and A_3 (but not A_1) interact well with B_2, then in an evolving species A_2 can be substituted first for A_1, and then to B_2, and then from A_2 to A_3. The population need never pass through an adaptive valley, but the derived species would be reproductively isolated from the ancestral species in a way that would implicate epistasis. It is impossible to demonstrate that an allele such as A_2 did not ever exist, and therefore we cannot show that a landscape is truly peaked. Similar problems haunt our studies of F_2 breakdown of among-population crosses, recombination load, and linkage disequilibrium (below).

Recent, more precise maps of isolating factors have shown that epistasis influences fitness at a very small chromosomal scale as well. A series of papers from Wu and colleagues have shown epistatic interactions in a variety of regions in crosses between *Drosophila simulans* and its close relatives, *D. mauritiana* and *D. sechellia* (12, 32, 96). Factors that seemed to indicate single locus effects, on more detailed investigation turned out to be closely linked sets of loci that interact epistatically to affect fitness in male hybrids (12, 96); in each case, two closely linked factors were required to get the deleterious effects. These kinds of interactions are of the type described in the previous paragraph: Interaction with a third locus determines the fitness consequences of a two-way interaction. Isolating mechanisms in *Mimulus* also show epistatic effects (82).

Stabilizing Selection and the Adaptive Landscape

A presumably common form of selection on quantitative characters is stabilizing (or optimizing) selection, where selection favors intermediate trait values in preference to either extreme. Stabilizing selection on polygenic traits, even when the genetic effects and variance for the phenotype are additive or involve dominance, generates epistasis for fitness (3, 4, 137). This is because the effect on fitness of an allele at a segregating locus depends on whether the allele finds itself in an individual with a slightly high or slightly low genetic value for the trait in question. If the mean of the population is at the optimal phenotype (i.e. there is no directional component of selection), then approximately equal numbers of the individuals in the population will be slightly above or slightly below that optimum. The value of the allele on fitness, then, depends on the state of all other alleles in the individual, and is therefore strongly epistatic (see Figure 2). Note that, as above, the prevalence of epistatic effects does not mean that the variance for fitness should be epistatic; the

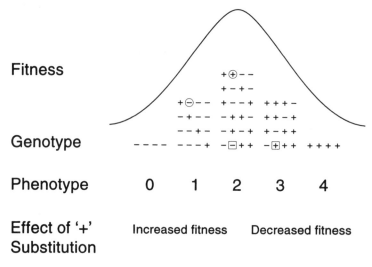

Fitness

Genotype

Phenotype   0   1   2   3   4

**Effect of '+'   Increased fitness   Decreased fitness
Substitution**

Figure 2 Stabilizing selection on an additive trait generates epistasis for fitness. The pluses and minuses represent the relative effects of alleles that affect the trait. The value of an allele for fitness depends upon the genetic background on which it finds itself: a + allele is favored in a background which would give a phenotype too small, but selected against if it is with too many other plus alleles. For example, the substitution from − to + is favored in the genetic background circled, but disfavored in the genotypes with boxes around them. Further, there are many different equally fit genotypes (e.g. −+−+ and Σ− −++) that correspond to multiple peaks in genotypic space.

degree of epistatic variance for fitness depends critically on the distribution of allelic effects (121).

There are many pitfalls in demonstrating the existence of stabilizing selection (124), including many other processes that can lead to the appearance of optimizing selection but that do not share the property of epistatic effects for fitness. For example, a trait under direct and directional selection toward one extreme, and under indirect selection toward the other extreme due to correlated response to selection on another character, can appear to be under stabilizing selection. This generates epistasis for fitness only temporarily, however, until selection can act on the component of variation for fitness that is not antagonistic. Even though in this example the hypothetical equilibrium does not necessitate epistasis for fitness, reaching this equilibrium may take a very long time, and there can be great epistasis for fitness in the meantime. The real problem is distinguishing these latter types of processes from true optimizing or stabilizing selection. Nevertheless, Travis's review shows several solid studies that demonstrate optimizing selection (124).

Selection in Subdivided Populations

As a consequence of Wright's shifting balance model, the prediction that subdivision of populations may increase the response to selection has been tested several times. This result would hold if there were strong epistatic interaction factors segregating in the population, but also if recessive alleles were sufficiently important to response to selection. Many of these studies, unfortunately, involve very high rates of mixing between subpopulations (59, 83, 103), which should eliminate the expected selection response even with strong epistasis (87). There are a few studies without too strong migration that do demonstrate stronger (36, 128) or divergent (24, 62, 75) responses to selection.

Linkage Disequilibrium in Selected Populations

As a consequence of selection on complex landscapes, selection tends to build up associations of alleles that function well together, leaving poor allele combinations in lower frequencies. This linkage disequilibrium is, of course, counterbalanced by various processes that lead to recombination, which breaks up these associations. As a result, many loci shown to interact epistatically are tightly linked, or in "co-adapted gene complexes" (see 51 for review). Linkage disequilibrium is often thought to be rare in natural populations (51), but new, more powerful techniques have demonstrated that there is more disequilibrium than formally thought (140).

The presence in natural or experimental populations of linkage disequilibrium has often been taken as good evidence of epistatic interactions for fitness, but this evidence must be taken with a few caveats. There are several studies that claim to have found disequilibria corresponding to epistatic selection (1, 21, 51, 64, 76, 119). Lewontin & White have shown that there are interactions among different grasshopper chromosomes in fitness, such that there are multiple fitness peaks that are apparently stable over time (76). Similarly, two experimental populations of cultivated barley demonstrate similar levels of disequilibrium, with the same chromosome pairs being common (21).

Unfortunately, while these results are consistent with multiple-peak epistasis, there are alternative explanations. The interactions of inversion types to produce overdominance have often been listed as evidence of coadaptation (e.g. 99), but in fact associative overdominance, due to the accumulation of different recessive deleterious alleles on different inversions, can create the same pattern without epistasis. Similarly, it is important to distinguish the effects of genetic drift or hitchhiking, which can also generate disequilibrium (93), particularly if the study is done at a geographic scale such that there is nonrandom mating.

Particularly good examples of disequilibrium in nature are given by *Cepaea nemoralis* and *C. hortensis* (57). In these snails, many of the multiple loci that control various shell patterns and shell color are in strong disequilibrium, packed into a "supergene" with low recombination among loci. These genes have been studied extensively in the field and have been shown to interact strongly to determine fitness in an environment-specific way.

MULTIPLE PEAKS FROM UNDERDOMINANCE

Underdominance at a single locus is probably the simplest type of multiple peak system. The term underdominance implies that heterozygotes are less fit than either homozygous type. With underdominance, the adaptive landscape is U-shaped, with peaks at fixation for either allele. Largely because it is a single locus system, peak shifts with underdominance have received the most complete theoretical investigation of any multiple peak system (see 6 and references cited there). One of the main results of these theoretical studies is that the underdominant valley can be traversed only if population (or neighborhood) sizes are small (6, 70), although the rate of fixation in these circumstances could be appreciable in subdivided populations, even with strong selection (1, 6, 7).

Underdominance has been studied primarily to explain the existence of the fixation of alternative chromosomal inversions in different populations and species (131). Chromosomal inversions can reduce the fitness of heterozygotes through the segregation of aneuploid gametes. Major chromosomal rearrangements of this type are widespread among animals (131, 132) and are thought to be fixed in populations at the rate of roughly 10^{-6} to 10^{-7} per generation in vertebrates (69). The existence of large numbers of cases of chromosomal inversions suggests that multiple peaks via underdominance may be a common occurrence, and this point has been used as the major evidence for the importance of genetic drift in natural populations, especially during speciation (40, 76, 132). Recent work on second- and third-chromosome pericentric inversions in *Drosophila* suggests that these conclusions may be premature, however (28, 29). Coyne and co-workers have demonstrated that some types of chromosomal inversions thought to lead to underdominance do not necessarily cause a reduction in heterozygote fitness (29), and these types have been found segregating in natural populations (28). Although most pericentric inversions are likely to be underdominant, some subsets may not be, with the obvious implication that the non-underdominant types are those fixed between populations (29). The underdominance of chromosomal rearrangements must therefore be demonstrated rather than assumed. Such a demonstration has not been performed for any polymorphic pericentric inversion (28).

Given the uncertain status of chromosomal inversions, the only demonstrable case of underdominance is shell coiling in snails (56, 78). Many, often closely related, gastropod species differ in the direction of coiling, some coiling dextrally and some sinistrally (42). Variation in coiling direction exists within some populations of the snail *Partula* (56, 78) and is known to be caused by variation at a single locus (90). Since snails with different coiling directions find mating difficult (78) or impossible (42), this is a solid example of under-dominance for fitness. It has been suggested that coiling differences may lead to rapid speciation through peak-shifts (42, 95). The importance of a peak-shift is somewhat mitigated, however, by the fact that maternal inheritance of this character reduces (but does not eliminate) the depth of the valley (95).

MULTIPLE FITNESS PEAKS ON PHENOTYPIC LANDSCAPES

As discussed in the introduction, fitness functions can also be (and often are) drawn to show the relationship between phenotypes and fitness. If these phe-notypes are genetically determined by genes that interact additively, then the genetic landscape will be complex as well. This section describes some of the evidence that the fitness functions of phenotypes can be complex and possibly multiple-peaked.

Disruptive Selection in Natural Populations

While it is very difficult to demonstrate multiple peaks in the fitness functions of natural populations for reasons outlined in the next section, several examples exist in which researchers have shown just that. For example, in the African *Pyrenestes* finches, there are two morphs of bill width, one that corresponds to a shape appropriate to cracking and handling one type of seed, and another that corresponds to another seed type. The phenotypic values between these forms are much less fit in certain life stages than are the forms at the adaptive peaks (113). Functional morphological and ecological analyses have demon-strated that the Galápagos finches have different species that correspond simi-larly to discrete niches determined by seed size and hardness (109). The analysis of the Galápagos finches was sufficiently successful to predict the presence of extinct members of the clade. Disruptive selection, caused by the presence of discrete niches in resource use, can cause multiple adaptive peaks to exist in nature.

Characters sometimes demonstrate adaptive peaks when selected in concert with other characters. One particularly good example of this is *Thamnophis* garter snakes, which have multiple color pattern morphs and multiple behav-ioral strategies for escape from predators. Snakes that are striped are easy to see when still, but give little indication of ground speed when racing away from a predator. On the other hand, spotted snakes are more likely to be cryptic

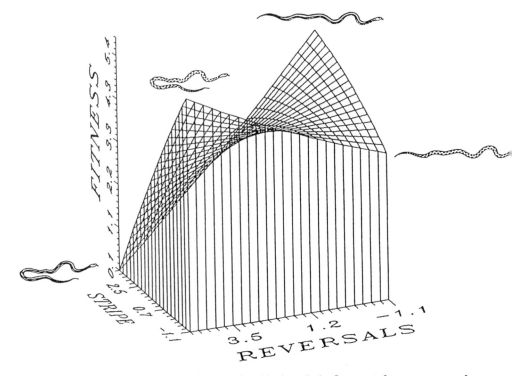

Figure 3 Brodie (9) measured the survivorship through the first year of young garter snakes (*Thamnophis ordinoides*), a highly polymorphic species. Snakes with longitudinal stripes survive better if they run straight away from predators, but snakes with blotched patterns are better able to avoid predators by reversing the direction of flight and attempting to hide cryptically. This is the presumed reason for the saddle in the fitness function of striping and the tendency to reverse direction. [From (9)] © 1992 Society for the Study of Evolution. Reprinted with the permission of the author and publisher.

when stationary but are more easily seen when moving. Brodie has shown that the stripedness of snakes is negatively correlated with the extent to which snakes reverse direction when fleeing a predator (9). Furthermore, the snakes that have the appropriate behavior to match their color pattern type are much more likely to survive through their first year (9) (see Figure 3). Similar results have been demonstrated with cryptic coloration patterns and resting site choice in *Biston betularia* (61).

Selfing rates provide another example of this type of disruptive selection. The evolution of selfing and close inbreeding in plants is facilitated if the plant is not very subject to inbreeding depression. Outcrossing is likely to be favored if the level of inbreeding depression is high. Inbreeding depression itself is a genetic property of a species, which can respond to selection; close inbreeding tends to decrease the marginal costs of inbreeding, because deleterious recessive alleles have been selected out of the population. As a result of these processes, the stable genetic states for mating systems are thought to be outcrossing (with a relatively high degree of inbreeding depression) or close inbreeding (with a relatively small associated cost to fitness) (73). This pattern has been arguably demonstrated in a collection of natural populations (108). Thus mating systems and degree of inbreeding depression are two genetically controlled characters which determine an adaptive landscape with a saddle and two peaks.

Multiple Phenotypic Solutions to the Same Evolutionary Problem

Another kind of phenotypic analysis providing evidence for the ruggedness of adaptive peaks comes from studies of genetic redundancy for fitness; there are many different ways in which different populations can solve the same environmental challenge. For example, some *Drosophila melanogaster* populations have avoided alcohol toxicity by avoiding foods that contain ethanol, while others have evolved stronger physiological mechanisms to metabolize the alcohols (23, 25, 26). Similarly, different populations of the same species of butterflies have evolved different phenotypes to mimic different forms of other toxic butterflies: hybrids between these forms are distinctly less fit (84, 85). This sort of polymorphism is extremely common in butterfly mimicry systems (125). Many other examples of this sort of genetic redundancy are known (23, 26, 134). As will be discussed further in the next section, multiple solutions to the same problem do not necessarily imply multiple fitness peaks. Only if there is a fitness cost to combining multiple solutions (as in the mimicry examples) will there be more than one peak on the adaptive surface.

LIMITATIONS TO THE STUDY OF EPISTASIS AND MULTIPLE FITNESS PEAKS

We have seen that there are many examples of epistatic interaction both at the phenotypic and genetic scale in natural populations. We also, however, have seen that there are many examples where evidence of epistasis is lacking. This section explores the overt consequences of epistatic interactions and multiple peaks, and asks, what are the strengths and weaknesses of various possible tests of the prevalence of epistatic interactions and multiple peaks?

The Problem with Epistatic Genetic Variance as a Measure of Epistasis

Because epistatic variance is rarely found in breeding experiments, it is often assumed that epistatic interactions are of little importance in creating phenotypes from genotypes. As we have seen earlier (see Figure 1), even pervasive epistatic interactions on peaked fitness landscapes are expected to demonstrate little epistatic genetic variance at equilibrium. Epistatic variance is a poor measure of the ruggedness of an adaptive landscape. There are also several reasons why the epistatic variance that does exist is very difficult to demonstrate statistically.

First, the analysis of variance techniques used for these experiments are biased against finding epistatic effects (127). The least squares analysis implicitly removes as much variation as possible to be attributed to main effects, leaving little available for attribution to interaction terms. Because correlations between variance components tend to be large (18), the hierarchical fitting of effects minimizes estimates of epistatic components. Further, since many genotypes containing epistatic alleles will be rare, estimates of the main (additive) effects of alleles are confounded with the true underlying interactions in a manner analogous to estimates arising from an unbalanced ANOVA (127). While this is appropriate for understanding the short-term response to selection (37), it is misleading if we want to understand the real genetic architecture of traits.

Second, confidence limits on genetic variance components are generally large, and this is particularly true for the epistatic variance (18, 48, 81). The magnitude of the confidence limits derives in part from the small coefficients associated with the epistatic effects in the covariances among relatives. For example, one half of the additive genetic variance contributes to the variance among full sibs, whereas one fourth of the additive-by-additive epistatic variance does so, and only one sixteenth of the dominance-by-dominance epistatic variance (37). Unfortunately, the error for each component is relatively constant, so there is one half to one eighth the power to detect the epistatic components compared to the additive variance. Experiments seeking reliable estimates of epistatic variance must therefore be quite large (80). As a result these experiments are rarely performed.

Third, in experiments conducted in artificial environments, interactions between the genotype and the environment in producing phenotypes can obscure the true nature of genetic variation in natural environments (111). Many evolutionary genetic experiments that aim to estimate nonadditive genetic variance are conducted in artificial environments and may be extremely misleading. Epistatic variance can vary strongly from one environment to another (2, 50, 55).

Finally, linkage disequilibria can hide epistasis. Tightly linked, interacting genes will appear to be additive in effect, because the alleles will, in the course of an experiment, rarely be separated. Because many examples of epistatic interactions also involve tight linkage (so-called coadapted gene complexes), the bias associated with linkage can be extreme.

The Difficulties of Estimating Epistatic Gene Effects

The problems associated with measuring epistatic genetic effects are also significant. One source of bias is that those who study genetic effects directly rarely also study fitness per se, and even more rarely are the fitness effects of gene combinations examined. A fundamental difficulty in understanding the potential for epistatic interactions is that we should expect variants that interact to affect fitness to be rare (see "simple model" section above), and therefore that they only infrequently come to the attention of biologists studying these systems.

Fortunately, there are ways of studying fitness landscapes that escape these difficulties, although they have up to now been applied only infrequently. The difficulties of estimating epistatic interactions stem from the constraints within populations to be at stable equilibria. When gene flow is restricted between two different breeding pools, the genetic systems are no longer constrained to function well together. Therefore, crosses between distant populations or closely related species should give a less biased picture of the ruggedness of genetic landscapes. In other words, the nature of genetic variation within species or populations is constrained to be at a fitness peak, and therefore biased, but the effects of genes combined across species are less constrained in this way. The genotype that occurs as a result of the combination of two distinct gene pools that have evolved completely in allopatry lands to a small extent in a random part of the adaptive landscape and gives us a less biased view of what unselected genotypes might be like. (We say "to a small extent" because even this analysis is predominantly biased: Most genes are shared by species related closely enough to allow viable offspring. The very existence of reproductive isolation between species that have had no direct selection to evolve reproductive isolation gives very strong evidence for the extremely rugged nature of adaptive landscapes.)

Difficulties in Demonstrating Disruptive Selection

The difficulties in determining the extent of disruptive selection or the density of valleys on a phenotypic adaptive landscape are very similar to the reasons why discovering epistasis on the genetic level is difficult. Sexual species are usually constrained to be near single adaptive peaks (71), and this constrains the range of phenotypes available for study. We cannot measure the fitness effects of phenotypes that do not exist; therefore we are usually limited to measuring the effects of phenotypes clustered around single peaks. The few examples we have discussed above are unusual, not because they demonstrate disruptive selection, but rather because there is some unusual situation that allows variation to be maintained in the population. In the case of the Py-renestes finches, for example, the genetic variance of the trait is essentially due to a single locus (114). It is much easier to maintain variance with single macromutations than with a polygenic basis for trait variation. Other examples also are maintained by likely frequency dependent selection or recent intergradation of populations (9). The landscape for other examples is inferred across species (108). Studies based on interspecific comparisons, interpopulational transplants, phenotypic modification, and other experimental techniques should be more promising avenues for truly understanding the nature of adaptive landscapes than is possible with the strictly observational approach. Further, adaptive landscapes on the phenotypic scale to some extent reflect the functional relationship between fitness and phenotype, such that expectations based on functional morphology may point the way to the existence of multiple peaks (e.g. 109).

The Impossibility of Finding True Valleys

As has been briefly discussed in the section on speciation genetics, it is difficult to know whether epistasis for fitness translates into multiple fitness peaks. Alternatively, the multidimensional fitness function may be simply a rugged landscape with ridges connecting all high spots and no true valleys, merely depressions around which selection must detour. We face this problem not only because it is difficult to reconstruct all pertinent genotypes from the alleles which segregate within a population (and therefore the "local" landscape is difficult to measure), but also because we do not have access to all possible alleles. The number of possible genotypes in a *Drosophila* genome, for example, is close to 10 raised to the 200 millionth power, which is the sort of number that makes us glad for scientific notation but despair of ever testing even a tiny subset of the possibilities. Alleles at one locus can change the fitness function of other two-locus pairs (12, 32, 96) and eliminate the apparent adaptive valley between certain combinations. Furthermore, it is often quite difficult to reconstruct all corners of the genotype-to-fitness matrix; we most often simply do not know what trans-species homozygotes are like, for example. It is therefore impossible to ever rigorously claim that any particular landscape is necessarily multiply peaked; we can only know that there is epistasis for fitness.

Underdominance systems have the same difficulties. Like all of the cases of multiple peak systems discussed above, underdominance at a single locus is a very unstable state. We might therefore expect to see variation for underdominant alleles far more often between populations than within populations, an expectation that does match the usual pattern of chromosomal inversions (131). Indeed, there are several examples of hybrid zones in which one chromosomal inversion is on one side of the tension zone and another on the other side (132).

These problems, while substantial, do not negate the importance of studying landscapes. If alleles that allow deterministic change do not yet exist, then populations will experience constraints to adaptation until those alleles appear in the population, which could be a very long time (97). If the slopes of the fitness function are very slight in the neighborhood of the current allele frequencies, then evolution may not follow the deterministic path; mechanisms of shifts across local valleys may yet be important. Finally, adaptive landscapes may in fact be peaked; if so the only way we can test the hypothesis is by attempts at falsification. The evolutionary consequences of genetic interactions for fitness are too great not to investigate these issues as fully as we can.

CONCLUSIONS

Many examples of epistatic interactions, both for phenotypic traits and fitness, have been reported in the genetics literature. Some of these examples report multiple adaptive peaks, many others demonstrate rugged adaptive landscapes, and many show that the evolution of traits need not follow an additive model. There are many other traits, however, that do not show this kind of interaction: The additive model is adequate for many of the characters that have been studied. This review has focused on showing that this need not be the case, but rather, genetic interactions can be an important part of evolution.

There are many reasons why traditional approaches to the study of epistatic interactions and multiple adaptive peaks are biased strongly against finding evidence of these phenomena. In spite of these difficulties, there are many examples in the literature of epistasis and of multiple peaks on a phenotypic scale, which is amazing in the face of the statistical (and sociological) difficulties of studying the phenomenon. This argues that epistasis plays a much larger role in determining phenotypes and fitness than is generally thought. Furthermore, we have been looking in the wrong place; interspecific comparisons are critical to determining the ruggedness of adaptive landscapes. When these comparisons are made, epistasis (or underdominance) is most often found. The importance of epistasis in evolutionary genetics is not in the generation of epistatic variance per se, but in the particular interactions of specific loci, and the way in which these interactions structure the set of possible evolutionary outcomes. Our interpretation of the relative importance of multiple fitness peaks in the evolutionary process will depend on how prevalent these interactions turn out to be and how carefully we endeavor to find them.

A biologically significant contrast has struck us from our reading of the literature, although there is little hope of reliably testing this with the data we have now. The level of epistasis for fitness tends to increase with phylogenetic distance. Crosses performed within populations show little epistasis for fitness, but crosses from distantly related populations or between closely related species almost universally show such interactions for fitness. To a certain extent this may seem tautological: Speciation can only occur if there are gene combinations that do not function well together but do perform well with other combinations. The data strongly demonstrate this pattern. We need to lift the constraints imposed by selection and recombination, which limit the frequencies of alleles that do interact with other alleles in affecting fitness, in order to see what adaptive landscapes truly are like. Each species cross represents a random sample of the landscape. The simple fact that there are species implies that adaptive landscapes are rugged indeed.

A fundamental problem of studying adaptive landscapes, as pointed out above, is that we can never truly know whether there are ridges connecting "peaks." We cannot measure all possible genotypes, and we do not have access to all genotypes which may have existed in the history of the divergence of natural populations. We must, then, be satisfied to know that landscapes are rugged, without knowing that they are peaked, or else we must study fine-scale differentiation in much more detail, for example in laboratory experiments. Furthermore, fitness landscapes are intrinsically linked to the environments in which the species evolves; it is clear that the landscape can change as a function of environmental changes. We must learn more about the permanence of landscapes over time. Until such work is performed, we can never truly know whether peak shift models are necessary, and we run the risk, in the absence of such proof, of ignoring the genetic interactions that may be an extremely important part of the evolutionary process.

ACKNOWLEDGMENTS

We would like to thank Michael Wade for the role he has played for each of us in our thinking about epistasis. We are grateful to Nick Barton, Norman Johnson, Peter Keightley, Sally Otto, Joe Travis, and Mike Wade for criticism of the manuscript, and to Jim Crow, Dolph Schluter, Tony Long, Andy Clark, Bill Hill, Rich Lenski, Charlie Fenster, Charles Goodnight, Tom Getty, Mike Travisino, and Lin Chao for useful discussions and ideas. This work was supported by a grant from SERC and NSF grant 8906956 to SJT. This is Kellogg Biological Station contribution #791.

Literature Cited

1. Barker JSF. 1979. Inter-locus interactions: a review of experimental evidence. *Theor. Pop. Biol.* 16:323–46
2. Barnes PT, Holland B, Courreges V. 1989. Genotype-by-environment and epistatic interactions in *Drosophila melanogaster:* the effects of *Gpdh* allozymes, genetic background and rearing temperature on larval developmental time and viability. *Genetics* 122:859–68

3. Barton NH. 1986. The maintenance of polygenic variation through a balance between mutation and stabilising selection. *Genet. Res., Camb.* 47:209–16

4. Barton NH. 1989. The divergence of a polygenic system under stabilising selection, mutation and drift. *Genet. Res., Camb.* 54:59–77

5. Barton N, Charlesworth B. 1984. Genetic revolutions, founder effects, and speciation. *Annu. Rev. Ecol. Syst.* 15: 133–64

6. Barton NH, Rouhani S. 1991. The probability of fixation of a new karyotype in a continuous population. *Evolution* 45:499–517

7. Barton NH, Rouhani S. 1993. Adaptation and the 'shifting balance'. *Genet. Res.* 61:57–74

8. Bateson W. 1909. *Mendel's Principles of Heredity.* Cambridge: Cambridge Univ. Press

9. Brodie ED III. 1992. Correlational selection for color pattern and antipredator behavior in the garter snake *Thamnophis ordinoides*. *Evolution* 46:1284–98

10. Burton RS. 1990. Hybrid breakdown in developmental time in the copepod *Tigriopus californicus*. *Evolution* 44: 1814–22

11. Caballero A, Toro MA, Lopez-Fanjul C. 1991. The response to artificial selection from new mutations in *Drosophila melanogaster*. *Genetics* 128: 89–102

12. Cabot EL, Davis AW, Johnson NA, Wu CI. 1994. Genetics of reproductive isolation in the *Drosophila simulans* clade: complex epistasis underlying hybrid male sterility. *Genetics* 137:175–89

13. Cavener DR, Clegg MT. 1981. Multigenic response to ethanol in *Drosophila melanogaster*. *Evolution* 35:1–10

14. Charlesworth B. 1990. The evolutionary genetics of adaptation. In *Evolutionary Innovations*, ed. M. Nitecki, pp. 47–68. Chicago: Univ. Chicago Press

15. Charlesworth B, Barton NH. 1995. Recombination load due to selection for increased recombination. In press

16. Charlesworth B, Charlesworth D. 1975. An experiment on recombination load in *Drosophila melanogaster*. *Genet. Res., Camb.* 25:267–74

17. Cheverud JM, Routman E. 1993. Quantitative trait loci: individual gene effects on quantitative characters. *J. Evol. Biol.* 6:463–80

18. Chi RK, Eberhart SA, Penny LH. 1969. Covariances among relatives in a maize variety (*Zea mays* L.). *Genetics* 63:511–20

19. Clarke CA, Sheppard PM. 1971. Further studies on the genetics of the mimetic butterfly *Papilio memnon* L. *Phil. Trans. R. Soc. London Ser.* 847:35–65

20. Clarke CA, Sheppard PM, Thornton IWB. 1968. The genetics of the mimetic butterfly *Papilio memnon* L. *Phil. Trans. R. Soc. London B Ser.* 254B. 791:37–89

21. Clegg MT, Allard RW, Kahler AL. 1972. Is the gene the unit of selection? Evidence from two experimental plant populations. *Proc. Natl. Acad. Sci. USA* 69:2474–8

22. Coen ES, Meyerowitz EM. 1991. The war of the whorls: genetic interactions controlling flower development. *Nature* 353:31–37

23. Cohan FM. 1984. Can uniform selection retard random genetic divergence between isolated conspecific populations? *Evolution* 38:495–504

24. Cohan FM. 1984. Genetic divergence under uniform selection. I. Similarity among populations of *Drosophila melanogaster* in their responses to artificial selection for modifiers of ci^D. *Evolution* 38:55–71

25. Cohan FM, Hoffmann AA. 1986. Genetic divergence under uniform selection. II. Different responses to selection for knockdown resistance to ethanol among *D. melanogaster* populations and their replicates. *Genetics* 114:145–63

26. Cohan FM, Hoffmann AA. 1989. Uniform selection as a diversifying force in evolution: evidence from *Drosophila*. *Am. Nat.* 134:613–37

27. Coyne JA. 1992. The genetics of speciation. *Nature* 355:511–5

28. Coyne JA, Aulard S, Berry A. 1991. Lack of underdominance in a naturally occurring pericentric inversion in *Drosophila melanogaster* and its implications for chromosome evolution. *Genetics* 129:791–802

29. Coyne JA, Meyers W, Crittenden AP, Sniegowski P. 1993. The fertility effects of pericentric inversions in *Drosophila melanogaster*. *Genetics* 134:487–96

30. Coyne JA, Orr HA. 198. Two rules of speciation. In *Speciation and its Consequences*, ed. D. Otte, J. Endler, pp. 180–207. Sunderland, MA: Sinauer

31. Curtsinger JW. 1984. Evolutionary landscapes for complex selection. *Evolution* 38:359–67

32. Davis AW, Noonberg EG, Wu CI. 1994. Evidence for complex genic interactions between conspecific chromosomes underlying hybrid female sterility in the *Drosophila simulans* clade. *Genetics* 137:191–99

33. Dobzhansky Th. 1937. *Genetics and the*

Origin of Species. New York: Columbia Univ. Press

34. Dobzhansky T, Levene H, Spassky B, Spassky N. 1959. Release of genetic variability through recombination. III. *Drosophila prosaltans. Genetics* 44:75–92

35. Dykhuizen D, Hartl DL. 1980. Selective neutrality of 6PGD allozymes in *E. coli* and the effects of genetic background. *Genetics* 96:801–17

36. Enfield FD. 1977. Selection experiments in Tribolium designed to look at gene action issues. In *Proc. Int. Conf. on Quant. Genet.*, ed. E Pollack, O Kempthorne, JTB Bailey, pp. 177–190. Ames: Iowa State Press

37. Falconer DS. 1981. *Introduction to Quantitative Genetics.* New York: Longman

38. Fisher RA. 1918. The correlation between relatives on the supposition of Mendelian inheritance. *Trans. R. Soc. Edinburgh* 3:399–433

39. Fisher RA . 1958. *The Genetical Theory of Natural Selection.* New York: Dover

40. Futuyma DJ, Mayer GC. 1980. Non-allopatric speciation in animals. *Syst. Zool.* 29:254–71

41. Geiger HH. 1988. Epistasis and heterosis. In *Proc. 2nd Int. Cong. Quant. Genetics,* ed. B Weir, E Eisen, M Goodman, G Namkoong, pp. 395–99. New York: Sinauer

42. Gittenberger E. 1988. Sympatric speciation in snails; a largely neglected model. *Evolution* 42:826–28

43. Goldschmidt RB. 1940. *The Material Basis of Evolution.* New Haven, CT: Yale Univ. Press

44. Goldschmidt RB. 1952. *Understanding Heredity: An Introduction to Genetics.* New York: Wiley

45. Goolish EM, Burton RS. 1989. Energetics of osmoregulation in an intertidal copepod: effects of anoxia and lipid reserves on the pattern of free amino acid accumulation. *Funct. Ecol.* 3:81–9

46. Grant V. 1975. *Genetics of Flowering Plants.* New York: Columbia Univ. Press

47. Haldane JBS. 1922. Sex ratio and unisexual sterility in hybrid animals. *J. Genet.* 12:101–9

48. Hallauer AR, Miranda JB. 1988. *Quantitative Genetics in Maize Breeding.* Ames, IA: Iowa State Univ. Press. 2nd ed.

49. Hard JJ, Bradshaw WE, Holzapfel CM. 1992. Epistasis and the genetic divergence of photoperiodism between populations of the pitcher-plant mosquito, *Wyeomyia smithii. Genetics* 131:389–96

50. Hayman BI. 1958. The separation of epistatic from additive and dominance variation in generation means. *Heredity* 12:371–90

51. Hedrick P, Jain S, Holden L. 1978. Multilocus systems in evolution. *Evol. Biol.* 11:101–84

52. Hill WG. 1982. Dominance and epistasis as components of heterosis. *Sonderdruck aus Zeitschrift für Tierzüchtungsbiologie* 99:161–68

53. Inai S, Ishikawa K, Nunomura O, Ikehashi H. 1993. Genetic analysis of stunted growth by nuclear-cytoplasmic interaction in interspecific hybrids of *Capsicum* by using RAPD markers. *Theor. Appl. Genet.* 87:416–22.

54. Jinks JL. 1955. A survey of the genetical basis of heterosis in a variety of diallel crosses. *Heredity* 9:223–38

55. Jinks JL, Perkins JM, Pooni S. 1973. The incidence of epistasis in normal and extreme environments. *Heredity* 31: 263–69

56. Johnson MS. 1982. Polymorphism for direction of coil in *Partula suturalis*: behavioural isolation and positive frequency dependent selection. *Heredity* 49:145–51

57. Jones JS, Leith BH, Rawlings P. 1977. Polymorphism in *Cepaea*: a problem with too many solutions? *Annu. Rev. Ecol. Syst.* 8:109–43

58. Kacser H, Burns JA. 1973. Rate control of biological processes. *Symp. Soc. Exp. Biol.* 27:65–104

59. Katz AJ, Enfield FD. 1977. Response to selection for increased pupal weight in *Tribolium castaneum* as related to population structure. *Genet. Res., Camb.* 30:237–46

60. Keightley P. 1989. Models of quantitative variation of flux in metabolic pathways. *Genetics* 121:869–76

61. Kettlewell HBD. 1955. Recognition of appropriate background by the pale and black of Lepidoptera. *Nature* 175:943–44

62. King JC. 1955. Evidence for the integration of the gene pool from studies of DDT resistance in *Drosophila. Cold Spring Harbor Symp. Quant. Biol.* 20: 311–17

63. Kirkpatrick M. 1982. Quantum evolution and punctuated equilibrium in continuous genetic characters. *Am. Nat.* 119:833–48

64. Klitz W, Thomson G. 1987. Disequilibrium pattern analysis. II. Application to Danish HLA A and B locus data. *Genetics* 116:633–43

65. Kondrashov AS. 1982. Selection against harmful mutations in large sexual and

asexual populations. *Genet. Res., Camb.* 40:325–32

66. Kondrashov AS. 1985. Deleterious mutations as an evolutionary factor. II. Facultative apomixis and selfing. *Genetics* 111:635–53
67. Kondrashov AS, Crow JF. 1991. Haploidy or diploidy: Which is better? *Nature* 351:314–15
68. Lande R. 1976. Natural selection and random genetic drift in phenotypic evolution. *Evolution* 30:314–34
69. Lande R. 1979. Quantitative analysis of multivariate evolution, applied to brain: body size allometry. *Evolution* 33:402–16
70. Lande R. 1984. The expected fixation rate of chromosomal inversions. *Evolution* 48:743–52
71. Lande R. 1986. The dynamics of peak shifts and the pattern of morphological evolution. *Paleobiology* 12:343–54
72. Lande R, Arnold SJ. 1983. The measurement of selection on correlated characters. *Evolution* 37:1210–26
73. Lande R, Schemske DW. 1985. The evolution of self-fertilization and inbreeding depression in plants 1. Genetic models. *Evolution* 39:24–40
74. Lenski RE. 1988. Experimental studies of pleiotrophy and epistasis in *Escherichia coli.* II. Compensation for maladaptive effects associated with resistance to T4. *Evolution* 42:433–40
75. Lenski RE, Travisano M. 1994. Dynamics of adaptation and diversification: a 10,000 generation experiment with bacterial populations. *Proc. Natl. Acad. Sci. USA* 91:6808–14
76. Lewontin R, White MJD. 1960. Interaction between inversion polymorphisms of two chromosome pairs in the grasshopper, *Moraba scurra. Evolution* 14:116–29
77. Lindsey AW. 1932. *A Textbook of Genetics.* New York: MacMillan
78. Lipton CS, Murray J. 1979. Courtship of land snail of the genus *Partula. Malacologia* 19:129–46
79. Long AD, Mullaney SL, Reid LA, Fry JD, Langley CH, Mackay TFC. 1995. High resolution mapping of genetic factors affecting abdominal bristle number in *Drosophila melanogaster. Genetics.* 139:1273–91
80. Lynch M. 1988. Design and analysis of experiments on random drift and inbreeding depression. *Genetics* 120:791–807
81. Lynch M. 1991. The genetic interpretation of inbreeding depression and outbreeding depression. *Evolution* 45:622–29

82. Macnair MR, Cumbes QJ. 1989. The genetic architecture of interspecific variation in *Mimulus. Genetics* 122:211–22
83. Madalena FE, Robertson A. 1975. Population structure in artificial selection: studies with *Drosophila melanogaster. Genet. Res.* 24:113–26
84. Mallet J. 1989. The genetics of warning color in hybrid zones of *Heliconius erato* and *H. melpomene. Proc. R. Soc. London B* 236:163–85
85. Mallet J, Barton N, Lamas G, Santisteban J, Muedas M, Eeley H. 1990. Estimates of selection and gene flow from measures of cline width and linkage disequilibrium in *Heliconius* hybrid zones. *Genetics* 124:921–36
86. McKechnie SW, Geer BW. 1988. The epistasis of *Adh* and *Gpdh* allozymes and variation in the ethanol tolerance of *Drosophila melanogaster* larvae. *Genet. Res., Camb.* 52:179–84
87. Moore FB-G, Tonsor SJ. 1994. A simulation of Wright's shifting-balance process: migration and the three phases. *Evolution* 48:69–80
88. Moreno G. 1994. Genetic architecture, genetic behavior, and character evolution. *Annu. Rev. Ecol. Syst.* 25:31–44
89. Mukai T, Yamaguchi O. 1974. The genetic structure of natural populations of *D. melanogaster.* XI Genetic variability in a local population. *Genetics* 76:339–66
90. Murray J, Clarke B. 1966. The inheritance of polymorphic shell characters in *Partula* (Gastropoda). *Genetics* 54:1261–77
91. Nagylaki T. 1992. *Introduction to Theoretical Population Genetics.* Berlin: Springer-Verlag
92. Neiman AM, Chang R, Komachi K, Herskowitz I. 1990. CDC36 and CDC39 are negative elements in the signal transduction pathway of yeast. *Cell Regul.* 1:391–401
93. Ohta T, Kimura M. 1969. Linkage disequilibrium due to random genetic drift. *Genet. Res.* 13:543–63
94. Orr HA. 1993. Haldane's Rule has multiple genetic causes. *Nature* 361:532–33
95. Orr HA. 1991. Is single-gene speciation possible? *Evolution* 45:764–69
96. Palapoli MF, Wu CI. 1994. Genetics of hybrid male sterility between *Drosophila* sibling species: A complex web of epistasis is revealed in interspecific studies. *Genetics* 138:329–41
97. Phillips PC. 1995. Waiting for compensatory mutation: phase zero of the shifting-balance process. *Genet. Res.* In press
98. Powers PL. 1950. Determining scales and the use of transformation in studies

of weight per locule of tomato fruit. *Biometrics* 6:145–63

99. Prakash S, Lewontin RC. 1968. A molecular approach to the study of genic heterozygosity in natural populations. III. Direct evidence of coadaptation in gene arrangements of *Drosophila*. *Proc. Natl. Acad. Sci. USA* 59:398–405

100. Price T, Turelli M, Slatkin M. 1993. Peak shifts produced by correlated response to selection. *Evolution* 47:280–90

101. Provine WB. 1986. *Sewall Wright and Evolutionary Biology*. Chicago: Univ. Chicago Press

102. Ram T, Singh J, Singh RM. 1989. Genetics and order effects of seed weight in rice: a triallel analysis. *J. Genet. Breed.* 44:53–58

103. Rathie KA, Nicholas FW. 1980. Artificial selection with differing population structures. *Genet. Res., Camb.* 36:117–31

104. Rice WR. 1992. Sexually antagonistic genes: experimental evidence. *Science* 256:1436–39

105. Riddle DL, Swanson MM, Albert PS. 1981. Interacting genes in a nematode dauer larva formation. *Nature* 290:668–71

106. Robertson FW, Reeve ECR. 1953. Studies in quantitative inheritance. IV. The effects of substituting chromosomes from selected strains in different genetic backgrounds. *J. Genet.* 51:586–610

107. Rockmill B, Roeder GS. 1990. Meiosis in asynaptic yeast. *Genetics* 126:563–74

108. Schemske DW, Lande R. 1985. The evolution of self-fertilization and inbreeding depression in plants. 2. Empirical observations. *Evolution* 39:41–52

109. Schluter D, Grant PR. 1984. Determinants of morphological patterns in communities of Darwin's finches. *Am. Nat.* 123:175–96

110. Schupbach T. 1987. Germ line and soma cooperate during oogenesis to establish the dorsoventral pattern of egg shell and embryo in *Drosophila melanogaster*. *Cell* 49:699–707

111. Service PM, Rose MR. 1985. Genetic covariation among life-history components: the effect of novel environments. *Evolution* 22:406–21

112. Simpson GG. 1953. *The Major Features of Evolution*. New York: Columbia Univ. Press

113. Smith TB. 1990. Natural selection on bill characters in the two bill morphs of the African finch *Pyrenestes ostrinus*. *Evolution* 44:832–42

114. Smith TB. 1993. Disruptive selection and the genetic basis of bill size polymorphism in the African finch *Pyrenestes*. *Nature* 363:618–20

115. Spassky B, Spassky N, Levene H, Dobzhansky T. 1958. Release of genetic variability through recombination. I. *Drosophila pseudoobscura*. *Genetics* 43:844–65

116. Spiess EB. 1958. The effect of genetic recombination on viability in *Drosophila*. *CSHQB* 23:239–50

117. Spiess EB. 1958. Release of genetic variability through recombination. II. *Drosophila persimilis*. *Genetics* 44:43–58

118. Stebbins GL. 1950. *Variation and Evolution in Plants*. New York: Columbia Univ. Press

119. Stephan W, Kirby DA. 1993. RNA folding in *Drosophila* shows a distance effect for compensatory fitness interactions. *Genetics* 135:97–103

120. Szathmary E. 1993. Do deleterious mutations act synergistically? Metabolic control theory provides a partial answer. *Genetics* 133:127–32

121. Tachida H, Cockerham CC. 1988. Variance components of fitness under stabilizing selection. *Genet. Res.* 51:47–53

122. Tanksley SD. 1993. Mapping polygenes. *Annu. Rev. Genet.* 27:205–33

123. Templeton AR, Hollocher H, Johnston JS. 1993. The molecular through ecological genetics of abnormal abdomen in *Drosophila mercatorum*: V. Female phenotypic expression on natural genetic backgrounds and in natural environments. *Genetics* 134:475–85

124. Travis J. 1989. The role of optimizing selection in natural populations. *Annu. Rev. Ecol. Syst.* 20:279–96

125. Turner JRG. 1977. Butterfly mimicry: the genetical evolution of an adaptation. *Evol. Biol.* 10:163–206

126. Vetukhiv M. 1956. Fecundity of hybrids between geographical populations of *Drosophila pseudoobscura*. *Evolution* 10:139–46

127. Wade MJ. 1992. Sewall Wright: Gene interaction and the Shifting Balance Theory. *Oxford Surveys in Evolutionary Biol.* 8:35–62

128. Wade MJ, Goodnight CJ. 1991. Wright's shifting balance theory: an experimental study. *Science* 253:1015–8

129. Wade MJ. 1992. Epistasis. In *Keywords in Evolutionary Biology*, ed. EF Keller, EA Lloyd, pp. 87–91. Cambridge, MA: Harvard Univ. Press

130. Waser NM, Price MV. 1994. Crossing-distance effects in *Delphinium nelsonii*: outbreeding and inbreeding depression in progeny fitness. *Evolution* 48:842–52

131. White MJD. 1973. *Animal Cytology and Evolution.* London: William Clowes

132. White MJD. 1978. *Modes of Speciation.* San Francisco: Freeman

133. Whitlock MC. 1995. Variance-induced peak shifts. *Evolution.* In press

134. Wilkens H. 1971. Genetic interpretation of regressive evolutionary processes: studies on hybrid eyes of two *Astyanax* cave populations (Characidae, Pisces). *Evolution* 25:530–44

135. Wright S. 1931. Evolution in Mendelian populations. *Genetics* 16:97–159

136. Wright S. 1932. The roles of mutation, inbreeding, crossbreeding, and selection in evolution. *Proc. 6th Int. Cong. Genet.* 1:356–66

137. Wright S. 1935. Evolution in populations in approximate equilibrium. *J. Genet.* 30:257–66

138. Wright S. 1968. *Evolution and the Genetics of Populations.* Vol. 1. *Genetics and Biometric Foundations.* Chicago: Univ. Chicago Press

139. Wright S. 1978. *Evolution and the Genetics of Populations.* Vol. 4. *Variability Within and Among Natural Populations.* Chicago: Univ. Chicago Press

140. Zapata C, Alvarez G. 1992. The detection of gametic disequilibrium between allozyme loci in natural populations of *Drosophila. Evolution* 46:1900–17

Annu. Rev. Ecol. Syst. 1995. 26:631–56

ECOLOGY AND EVOLUTION OF SOCIAL ORGANIZATION:
Insights from Fire Ants and Other Highly Eusocial Insects

Kenneth G. Ross

Department of Entomology, University of Georgia, Athens, Georgia 30602-2603

Laurent Keller

Institut de Zoologie et d'Ecologie Animale, Université de Lausanne, Bâtiment de Biologie, 1015 Lausanne, and Zoologisches Institut, Universität Bern, Ethologische Station Hasli, Wohlenstrasse 50a, CH-3032 Hinterkappelen, Switzerland

KEY WORDS: monogyny, polygyny, social evolution, gene flow, polymorphism

ABSTRACT

Social organisms exhibit conspicuous intraspecific variation in all facets of their social organization. A prominent example of such variation in the highly eusocial Hymenoptera is differences in the number of reproductive queens per colony. Differences in queen number in ants are associated with differences in a host of reproductive and social traits, including queen phenotype and breeding strategy, mode of colony reproduction, and pattern of sex allocation. We examine the causes and consequences of changes in colony queen number and associated traits using the fire ant *Solenopsis invicta* as a principal model. Ecological constraints on mode of colony founding may act as important selective forces causing the evolution of queen number in this and many other ants, with social organization generally perpetuated across generations by means of the social environment molding appropriate queen phenotypes and reproductive strategies. Shifts in colony queen number have profound effects on genetic structure within nests and may also influence genetic structure at higher levels (aggregations of nests or local demes) because of the association of queen number with particular mating and dispersal habits. Divergence of breeding habits between populations with different social organizations has

631

0066-4162/95/1120-0631$05.00

the potential to promote genetic differentiation between these social variants. Thus, evolution of social organization can be important in generating intrinsic selective regimes that channel subsequent social evolution and in initiating the development of significant population genetic structure, including barriers to gene flow important in cladogenesis.

INTRODUCTION

Social organization refers to the number of individuals in a social group, their behavioral and genetic relationships, and the way in which reproduction is partitioned among them. Social organization comprises the most fundamental defining features of animal societies, forged by ecological and social selection acting over the history of a population, and creating social and genetic environments that govern the course of subsequent social evolution (1, 80, 97, 143, 177).

It has become increasingly evident over the past decade that social organization varies not only among species but also within species or even populations. Striking variation in group composition and in the partitioning of reproduction among group members is proving to be the rule rather than the exception in a wide array of social vertebrate and insect species (74a, 75, 148, 150, 183). Such variation raises two sets of issues with broad ecological and evolutionary implications. The first set relates to the underlying causes. Variation in social organization is the product of diverse extrinsic selection pressures generated by the local ecology that interact with intrinsic selection pressures related to competitive and cooperative interactions among group members (1, 75, 121, 123). Thus, studies of variation in social organization may shed light on how ecological and social factors jointly influence the course of social evolution.

The second set of issues relates to the consequences of variation in social organization for the evolution of individual phenotypes and patterns of gene flow. Alternative phenotypes associated with variation in social organization differ in fitness according to ecological and social context. The mechanism for regulating their production may include a genetic component (18, 51, 182), or it may involve phenotypic plasticity (151), with the social environment inducing expression of the appropriate phenotype (77, 78, 107). In either case, differing reproductive habits of the alternative phenotypes may incidentally inhibit interbreeding between them, allowing accumulating genetic differentiation and the eventual incorporation of a genetic basis to phenotypic expression. Thus, studies of variation in social organization are relevant to understanding how phenotypic responses to ecological and social selection mediate population genetic divergence (176, 178, 179).

Eusocial insects exhibit enormous variation in virtually all elements of social organization (11, 35, 37a, 74a, 80, 148, 180) and thus are ideal subjects for studying the causes and consequences of its evolution. Variation in colony queen number, one conspicuous element of social organization, assumes special significance with respect to the issues raised above. Queen number apparently responds to variation in local ecological selection pressures and, in turn, influences the intrinsic selective milieu of a society by shaping within-nest genetic structure (e.g. 7, 11, 12, 68, 75, 120, 123, 143, 157). Moreover, variation in queen number often is associated with parallel variation in other reproductive traits, illustrating concerted phenotypic responses to ecological and social selection (66). For instance, in ants, the presence of a single queen per colony (*monogyny*) typically is associated with colony reproduction by the emission of sexuals, mating away from the nest following flight, extensive queen dispersal, and independent colony founding by queens (no assistance from workers). In contrast, the presence of several queens per colony (*polygyny*) often coincides with loss of the mating flight (with mating sometimes occurring in the nest), limited dispersal, and dependent colony founding (workers assist queens) (11, 12, 61, 65, 72, 73). These different reproductive strategies are associated with particular queen phenotypes. Monogyne ant queens tend to have larger body sizes (70), higher nutrient reserves (69, 102), longer life spans (71), higher fecundity (55, 64, 88, 170), and a later first age of sexual production (71) than do their polygyne counterparts.

Colony queen number and associated elements of social organization can vary dramatically at a hierarchy of levels. Queen number often is labile across populations, among colonies within a population, and even during the ontogeny of single colonies in many ants and wasps (31, 34, 44, 58, 67, 74, 117, 125, 129–131). Similarly, the way in which reproduction is divided among nestmate queens (15, 54, 62) and the strategies employed by queens to become active reproductives (44, 125) can vary geographically or even within local populations in these insects.

In this review we discuss the ecological and social factors that promote shifts in social organization and the consequences of such shifts for reproductive strategies, phenotypic evolution, and genetic structure. We focus on the "highly eusocial" Hymenoptera, which generally have clear morphological differences between queens and workers and form large perennial colonies (35). We use the fire ant *Solenopsis invicta* as a principal subject because variation in its social organization has been studied extensively. We focus on colony queen number because it is central to the variation in social organization found in fire ants and many other eusocial insects.

VARIATION IN COLONY QUEEN NUMBER

Association of Colony Queen Number with Other Properties of Social Organization

The biology of *Solenopsis invicta* has been studied extensively in the southeastern United States, where the species was introduced from South America

Table 1 Major differences between the monogyne (M) and polygyne (P) social forms of the fire ant *Solenopsis invicta* in the United States.

	Monogyne (M) form	Polygyne (P) form
Number of wingless (reproductive) queens per nest[a]	1	2 to more than 200
Relative fecundity of wingless (reproductive) queens[b]	High	Low
Unmated wingless (reproductive) queens in nests[c]	Absent (or rare)	Frequent
Mode of colony founding[d]	Independent	Dependent
Relative weight of winged (nonreproductive) queens[e]	High	Low
Relative worker size[f]	Large	Small
Relative next density[g]	Low	High
Connections between nests[h]	No	Yes
Genetic relatedness of nestmate workers[i]	Close to 0.75	Variable and often close to 0
Diploid males[j]	Absent	Frequent
Relative number of new sexuals produced[k]	Many	Few

[a] Includes mated and unmated queens in the P form. See section on "Within-nest genetic structure."
[b] (34, 166, 167, 170).
[c] See section on "Within-nest genetic structure."
[d] M queens start new colonies independently after a mating flight by raising brood without the help of workers; P queens return to or remain within an established P nest after shedding their wings and initiating oogenesis (41, 77, 86, 108, 114, 165). P colonies reproduce dependently through fissioning, a process in which wingless queens leave their nest with workers to initiate a new nest in close proximity to and in continuing contract with the parent nest (172).
[e] Mature winged queens of the M form weight on average 48% more than their counterparts in the P form, mainly as a result of their greater fat reserves (77, 79).
[f] (42).
[g] (109, 111).
[h] P workers and brood are transported from one nest to another (172), and P workers have a high tolerance of non-nestmate workers (92). M colonies are territorial (181), and M workers have a low tolerance of non-nestmate workers (92).
[i] (129, 133, 141).
[j] See section on "Production of diploid males."
[k] See section on "Sex allocation."

earlier this century (81). Two distinctive social forms exist in *S. invicta*: the monogyne (M) form, in which colonies have a single egg-laying queen, and the polygyne (P) form, in which colonies have multiple queens. The two forms often occur in distinct populations in which nests of one form predominate (83, 129, 135, 139), yet the P form is normally found in geographical association with the M form (42, 109, 110, 139). The forms differ in a number of traits besides colony queen number (Table 1); most important among these are the breeding habits of queens. Monogyne queens mate in aerial swarms and found new nests without the help of workers, that is, independently (86, 165), whereas P queens mate in the nest or during swarming but always initiate egg-laying in established P nests, that is, dependently (41, 77, 108). Polygyne nests multiply by fissioning (172) (Table 1).

Colonization of the United States by *S. invicta* has resulted in a unique opportunity to view selection in progress (41a) by comparing the social biology of this ant in the introduced range with that in its native range. Although the biology of the ant in South America is not yet well known, available data indicate that both social forms occur there and that, as in the United States, queen number is associated with other properties of the social organization. The native P form is known to differ from this form in the United States in some important social attributes, however, a point discussed more fully below.

Differences in properties of social organization associated with variation in queen number have been reported in several ant species and tend to be similar to those found in *S. invicta*. For example, queens and workers produced in polygyne colonies are smaller than those produced in monogyne colonies in *Formica truncorum* (158) and several *Myrmica* species (27). In the fire ant *S. geminata*, polygyne colonies produce fewer new queens, rear smaller workers, contain more unmated reproductive queens, and occur at higher densities than monogyne colonies (84, 168). Lower fecundity of queens in polygyne than in monogyne nests has been reported for a considerable number of species (55, 64, 88, 168), as has a tendency for polygyne colonies to reproduce through fissioning, while monogyne colonies produce queens that initiate nests independently (12, 65, 174). Finally, polygyny and colony reproduction via fissioning commonly are associated with persistent connections between nests (11, 12, 157).

Intraspecific variation in colony queen number in *S. invicta* and other ants, and associated differences in other properties of social organization, often parallel the variation seen between related monogyne and polygyne ant species. This suggests that variation at both levels stems from similar causes and that identification of the ecological and social factors influencing queen number within species may shed light on some features of cladogenesis in these insects.

Causes of Variation in Colony Queen Number

Solenopsis invicta has spread rapidly throughout the southeastern United States during the 60 years since its introduction (81). Entomologists have studied colonies throughout this period, yet it was not until 1973 that polygyny was first reported (39). Since this initial report, the frequency of polygyny apparently has increased throughout the introduced range, with many areas originally colonized by the M form now occupied by the P form (109, 110). Indeed, polygyny is now more common than monogyny in some large regions, such as east-central Texas (111).

The distributions of the two forms apparently are not associated with specific habitat types in the United States (111), and colonies of the two forms are intermingled at very localized scales in Argentina (KG Ross, EL Vargo, & L Keller, unpublished). This suggests that variation in queen number in fire ants is not a response to habitat differences, as has been suggested for other ants (60), and that another explanation for the gradual increase in the incidence of polygyny in the introduced range must be sought. Nonacs (91) suggested that polygyny arises as newly colonized habitats become increasingly saturated. Suitable habitats are widely available for colonization when *S. invicta* first enters an area, so the independent-founding strategy of the M form is highly successful at this point (114, 161). However, the success of dispersing M queens probably decreases as appropriate nesting sites become filled, and as predation and brood raiding by workers from existing colonies increase (162, 163). The diminished founding success of dispersing queens may generate strong selection on queens to attempt to remain in or reenter their natal colony (e.g. 12, 58, 124, 125), a behavior commonly associated with polygyny. Workers may be selected to accept such queens, some of which may be their sisters, once the probability of successful independent founding falls below a certain threshold (68, 144). Moreover, fissioning, the mode of colony reproduction associated with polygyny, may be more successful than independent colony founding in saturated habitats because queens are tended by workers during this process and thus are less vulnerable to predation and raiding (83, 162). Suggestive evidence for a link between a low success rate for independent founding and the inception of polygyny comes from the apparent correspondence of high nest densities in the monogyne form with high frequencies of occurrence of polygyny in Texas (111).

Constraints on independent founding may select for polygyny in other ant species as well (68). Herbers (56) showed that the availability of empty nest sites and degree of polygyny are inversely correlated across populations of *Leptothorax longispinosus,* and Bourke & Heinze (12) concluded from a comparative study of lepothoracine ants that polygyny is associated with limited nest sites, cold climate, and patchy habitats, factors that decrease the success of queen dispersal and independent founding.

A significant and unresolved issue is whether shifts from monogyny to polygyny generally are accompanied by important genetic changes (e.g. 178). If the alternative reproductive phenotypes have some genetic basis, then genes encoding the "polygyne phenotype" will increase in frequency when polygyny is favored, given some limitations on gene flow from the monogyne form. If alternative phenotypes are "inherited" by virtue of the social environment in which an individual is reared (as a result of phenotypic plasticity), then the polygyne phenotype will increase in frequency when and where polygyny becomes favored regardless of patterns of gene flow. Current evidence suggests that such "cultural inheritance" underlies the distinctive reproductive phenotypes of the two forms of *S. invicta.*

Maintenance of Variation in Colony Queen Number

A geographically extensive survey by Porter (110) showed that most *S. invicta* populations in the United States remained predominantly monogyne or polygyne over a three-year period. This stability can be explained by the fact that the social environment in which queens mature influences their phenotype and reproductive opportunities, biasing them toward living in a colony with the same social organization as that in which they were reared. Newly emerged queens of the two forms are indistinguishable in weight, an important component of reproductive phenotype, yet M queens become 48% heavier than P queens during the period of adult maturation, mainly because of greater fat accumulation (77). Cross-fostering experiments showed that these differences are largely dependent on the number of reproductive queens present in the nest (77, 79). Young queens maturing in the presence of a single egg-laying queen develop the M phenotype, whereas those maturing in the presence of multiple queens develop the P phenotype, no matter which type of nest they developed in as brood.

In addition to this effect of social environment, maturation of P queens is influenced by their genotype at a single gene, designated *Pgm-3* (78, 79). Winged queens with the homozygous genotype *Pgm-3$^{a/a}$* become 26% heavier than queens with the alternate genotypes in the P form. Significantly, there is no difference in weight among queens with different *Pgm-3* genotypes in the M form. The result of this interaction between genotype and social environment is the production of three phenotypic classes of queens: queens with extensive energy reserves (M queens), queens with minimal energy reserves (P queens with genotypes other than *Pgm-3$^{a/a}$*), and an intermediate class (P queens with genotype *Pgm-3$^{a/a}$*).

The phenotypic differences that develop between these three types of queens affect their reproductive options. The two types produced in P colonies apparently have insufficient reserves to found colonies independently. This conclusion is supported by the finding that most P queens attempting to found colonies alone in the laboratory produce either no workers or too few to give the colony a reasonable chance of survival (114). In contrast, queens from M nests have high success in producing the requisite number of workers to ensure colony survival (113, 164). Queens that cannot found colonies independently must be adopted into existing P colonies. Only small queens produced in P colonies are accepted by P workers; large P queens (all those with genotype *Pgm-3$^{a/a}$*) invariably are destroyed by workers during their attempts to become egg layers (78, 128). Understandably, P workers also seem not to accept newly mated M queens (137), which are even heavier than P queens with the genotype *Pgm-3$^{a/a}$* (78, 79, 171).

Social organization in *S. invicta* thus is "culturally inherited" because the type of social environment in which a queen matures specifies her phenotypic characteristics, which, in turn, dictate the type of society in which she can survive and reproduce. Because of this mode of inheritance, populations retain their characteristic social organization in spite of extensive gene flow between the social forms (see below).

Occasional changeovers from one social form to the other have been documented in some fire ant populations (110). Polygyne nests, which multiply and expand their territory by fissioning, may replace M nests by outcompeting them or taking over their territories after their single queen dies (43). Monogyne nests may replace P nests when the latter are eliminated from an area by some catastrophe such as flooding (110), the vacant habitats thus created being colonized most readily by dispersing M queens.

Intraspecific variation in queen phenotypes associated with differences in social organization is known in other ants and can be of three types. First, queens from colonies differing in queen number can display modest but consistent phenotypic differences, as in fire ants. For example, queens of *Formica truncorum* produced in polygyne colonies are smaller and have less fat reserves than queens produced in monogyne colonies (158). These differences, which arise during adult maturation, are coupled with tendencies for monogyne queens to disperse on the wing and polygyne queens to mate and shed their wings in the nest. Thus, as in *S. invicta*, the social environment in which young *F. truncorum* queens mature affects their reproductive options.

The second type of variation involves distinct queen phenotypes (with size the principal difference), but the different morphs are not strictly associated with different social organizations (10, 87). As an example, *Myrmica ruginodis* has both "macrogyne" (normal) and "microgyne" (miniature) queen morphs. Colonies headed by macrogynes generally are monogyne and reproduce by independent founding, whereas those headed by microgynes are polygyne and reproduce by fissioning (13).

The third and most distinctive type of queen variation is similar to the second except that the smaller queens are wingless and resemble workers in other respects as well (5, 18, 48, 50, 53, 184). This pronounced dimorphism possibly has a simple genetic basis in some species (18), but in others the morphs appear to be environmentally induced (e.g. 184). Queens of the winged morph typically disperse and initiate new colonies independently, whereas wingless queens mate in the vicinity of their natal nest, in which they then remain. In one *Leptothorax* species, most wingless queens are accepted as reproductives by their natal colony, whereas winged queens generally are destroyed by workers if they fail to disperse (48).

In summary, variation in colony queen number in *S. invicta* and other ants frequently is associated with other important differences in the social biology, including distinctive queen behaviors and morphologies. We next review the consequences of these different syndromes for population genetic structure, production of diploid males, and sex ratios. We then consider how the suite of changes accompanying shifts in queen number may affect gene flow between socially divergent populations.

Consequences of Variation in Colony Queen Number

GENETIC STRUCTURE   Changes in queen number affect genetic structure within the nest and at higher population levels, altering the focus of selection (3) and possibly setting in motion further social evolution (97). For instance, a shift from monogyny to polygyny alters the patterns of genetic relatedness that determine the extent of genetic conflict among nestmates, thus influencing such social traits as nepotism, the partitioning of reproduction, and sex allocation (7, 19, 75, 97, 118, 120, 123). At larger scales, modified breeding habits linked to a shift in queen number may impart local genetic structure within polygyne populations or even initiate genetic divergence between variant social forms.

Within-nest genetic structure   The number of reproductive queens is a basic determinant of patterns of relatedness within nests, but its effects depend on other factors such as the apportionment of maternity among nestmate queens, their relatedness to one another, and numbers of matings by each (115, 129). Thus it is important to consider all of these features when evaluating how differences in queen number influence within-nest genetic structure.

Nests of the M form of *S. invicta* generally possess only a single reproductive female (86, 133, 141, 149, 165). Workers are incapable of laying eggs (35), and eggs laid by winged (virgin) queens in queenright colonies seem not to be viable (33, 133, 140). The number of wingless reproductive queens in P nests of *S. invicta* has been studied by direct counts and by indirect estimates based on within-nest genetic variation (relatedness). The number of mated queens per nest is highly variable, ranging from 2 to almost 200 in a well-studied Georgia population (Figure 1), with queen number highly predictive of the genetic variation measured among workers within a nest (129). The effective number of queens per nest in this population ($N_e = 4.9$), estimated indirectly from nestmate relatedness values, corresponds well to the harmonic mean number of queens actually collected from these nests ($N_h = 4.5$), as expected from the mathematical theory underlying relatedness estimation (115, 129, 173a). As is also expected when queen number varies substantially among nests, N_e is considerably less than the arithmetic mean number of queens per nest ($N_a = 25$). In native P populations, direct counts and indirect estimates both suggest that the number of mated queens per nest generally is lower than in introduced populations. For instance, estimates of N_e range from 4.0 to 6.1

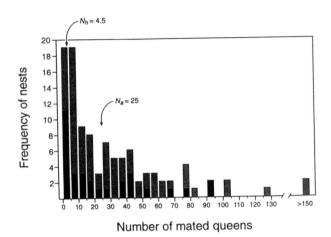

Figure 1 Frequency distribution of number of mated reproductive queens per nest in polygyne *S. invicta* from northern Georgia, United States (data from 129; EL Vargo, unpublished). The subset of 31 nests for which genetic data for estimating relatedness are available is indicated by black shading. The harmonic mean (N_h) and arithmetic mean (N_a) numbers of queens per nest are indicated for this subset.

for confirmed P nests in two Argentine populations, but estimates exceed 13 in the Georgia P population when nests with single queens are excluded (KG Ross, EL Vargo, L Keller, unpublished).

A large proportion of reproductive queens in P nests (25–40%) are unmated in introduced populations (114, 135, 170); although reproductively active, these queens produce few viable eggs (170, 173). The proportion of unmated queens is lower in native P populations (2–12%), presumably because operational sex ratios are less female-biased than in introduced populations (see below).

Empirical measurements of within-nest relatedness also have been used to obtain reliable estimates of N_e in other ants (2, 9, 49, 59, 101, 126, 145, 146, 153, 156). For instance, Seppä (146) and Banschbach & Herbers (2) estimated N_e at 2.3 and 1.8 for polygyne *Myrmica ruginodis* and *M. punctiventris,* respectively, in good agreement with N_h values of 3.2 and 1.4. Sundström (156) estimated N_e at 16 in a population of *Formica truncorum* in which nests regularly contained more than 20 queens. The only other group of highly eusocial polygyne insects in which queen number and genetic structure have been studied jointly is the epiponine wasps. Observed colony queen numbers in four epiponine species were found to be variable and often low, with N_h between 3 and 10 (38, 116, 154, 155). Measured relatedness values in these wasps yielded estimates of N_e close to N_h in each case.

Fire ant queens of both social forms invariably mate only once (133, 140, 141, 149). Effective numbers of matings are low in most other ants (8a, 126) and in epiponine wasps (116) as well, with single matings apparently more common in polygyne than monogyne ant queens (76). Thus, variation in queen mating frequency plays a relatively minor role in altering within-nest genetic structure in most highly eusocial Hymenoptera with variable colony queen number, compared to the variation in queen number.

Variation in maternity apportionment among nestmate queens (reproductive skew; 75, 123) has been studied in the P form of *S. invicta* in the laboratory and field. Considerable skew was observed in the laboratory (127), with the result that relatedness of worker offspring was twice the value expected had all queens reproduced equally. Furthermore, greater skew was documented with respect to the maternity of new queens than the maternity of workers, resulting in consistently higher relatedness of queen offspring to one another than of worker offspring to one another. This difference in maternity apportionment according to offspring caste, which has been detected also in field studies (129), emphasizes that production of worker offspring may not always serve as a reliable indicator of skew in actual reproductive success in highly eusocial insects (66).

Few studies of other polygyne Hymenoptera have used genetic markers to quantify reproductive skew (see 47, 106, 127, 131 for behavioral studies). Pamilo & Seppä (100) obtained higher relatedness values for queen than for worker pupae in *Formica sanguinea*, consistent with greater skew in the maternity of sexual than of worker offspring. Seppä (146) concluded that skew among *Myrmica ruginodis* queens is minimal because worker relatedness values were such as would be expected had all captured nestmate queens reproduced equally. Skew appears minimal in epiponine wasps judging from oocyte counts (38, 116, 154, 155), a view supported by similarities between offspring relatedness values measured and those expected if equal maternity apportionment is assumed. Studies such as these can be useful not only because they illuminate one factor influencing within-nest genetic structure, but also because they can test theoretical predictions of how nestmate reproduction is partitioned in particular ecological and social settings (12, 75, 122, 123).

Relatedness of reproductive nestmate queens has been investigated in P populations of *S. invicta* in Argentina and the United States. Queens are closely related in native populations ($r = 0.46$; KG Ross, EL Vargo & L Keller, unpublished), implying that they usually are sisters recruited as egg-layers into their natal colony. In contrast, queen relatedness is essentially zero in the introduced ants (129, 133), which accords with mark-recapture evidence that at least some queens here are recruited from foreign nests after making mating flights (41, 108). The apparent breakdown in adoption of a colony's daughter queens as well as the increase in colony queen number in the United States may stem from the higher population densities characteristic of introduced populations (112) and consequent increased selection on young queens to seek adoption into existing colonies.

Nestmate queen relatedness in other highly eusocial species varies across the spectrum of possible values (58, 68, 126, 131). In those groups in which queens are moderately or highly related (2, 26, 38, 94, 99, 104, 116, 146, 153, 155), most new queens likely are derived from the parent nest. Low values reported for several ants (63, 104, 156) presumably result from recruitment of queens from foreign nests, acceptance of daughter queens in nests in which queen number consistently is high, and/or acceptance of daughter queens over several generations (68, 97).

Relatedness of mating pairs (inbreeding) interacts with colony queen number and other elements of social organization in complicated ways to determine within-nest relatedness patterns (89, 96). Extensive genetic studies of introduced and native *S. invicta* of both social forms employing many polymorphic allozyme loci indicate that these ants invariably are outbred (133, 137–139, 141). Similarly, genetic studies of other highly eusocial Hymenoptera seldom have found evidence of pronounced inbreeding (2, 9, 38, 63, 101, 116, 126, 131, 146, 153–156). Inbreeding might be anticipated in polygyne ants because of the association of restricted dispersal with polygyny. Although it has been detected in a few such ants (8, 59, 145), in others it has not (63, 99, 152, 153, 156), presumably because brood is moved between nests or adult males disperse widely (103).

Higher-level population genetic structure   This refers to relatedness between nearby nests as well as genetic differentiation among very localized demes. Such structure may arise from mating near the natal nest, adoption of daughter queens, colony fissioning, and persistent interconnections between nests (23, 46, 94, 95, 131, 147), some or all of which are associated with polygyny in various highly eusocial Hymenoptera.

Higher-level genetic structure has been studied at scales relevant to the above processes in *S. invicta* from the native range only (KG Ross, EL Vargo & L Keller, unpublished). Both P but neither of the M populations studied in Argentina were significantly structured at scales of 10 m and 5 km. These results suggest more restricted dispersal in the P than the M form, consistent with the notion that queens are adopted by their natal colony and that colony reproduction occurs by fissioning in the native P populations. Among other highly eusocial Hymenoptera, higher-level structure has been detected in polygyne ants from six genera (2, 8, 24, 25, 45, 59, 95, 99, 147, 152, 156). Particularly interesting are cases in which it exists in the absence of local inbreeding (25, 99, 152, 156), presumably as a result of substantial dispersal by males but not queens or of single colonies occupying multiple nests. Pronounced higher-level structure in polygyne but not monogyne populations of single species or related species has been reported in *Formica* and *Myrmica* (99, 147, 156), as well as fire ants, and highlights the distinctive mating and dispersal strategies of the alternative forms. A major unresolved issue is whether these distinctive strategies commonly drive genetic differentiation between sympatric social forms (see below).

PRODUCTION OF DIPLOID MALES   A unique consequence of variation in colony queen number in fire ants is the different distributions of diploid males in the two social forms. Although social Hymenoptera are characterized by a male-haploid genetic system, diploid (2N) males can be produced in some circumstances. Sex appears to be determined by heterozygosity at one or a few loci, with individuals heterozygous at such loci becoming females and individuals either hemizygous (haploid) or homozygous (diploid) becoming males (21). Diploid males have low viability and/or fertility and are produced at the expense of workers or fertile sexuals, so they constitute a cost to a colony. The resulting selection against their production acts to maintain many alleles and high heterozygosity at sex-determining loci. Production of 2N males is affected not only by sex-locus variation but also by the mating system (e.g. extent of inbreeding) and, importantly, by colony queen number (93, 101, 141).

Diploid males are common in P populations of *S. invicta* in the United States, representing 73–100% of all males produced (132, 141). In contrast, 2N males are absent from mature M nests. This pattern is not the result of greater inbreeding or smaller effective population sizes in the P than in the M form (132, 137, 139, 141), nor is it the result of M workers possessing and P workers lacking an ability to recognize and destroy 2N males early in development (134). Rather, it results from differential mortality of nests producing such males in the two forms. Colonies founded independently by M queens that produce 2N males invariably die early in development because resources are diverted from production of workers (the labor of which is crucial to early colony growth and survival) to production of males (132, 134). This source of mortality does not afflict the P form because colonies are founded by fissioning, during which a sizable worker force already is present to ensure sustained colony growth.

Production of 2N males constitutes a significant genetic load on *S. invicta* populations in the United States, with 15–20% of all newly founded colonies of the M form succumbing to this mortality factor. It is unlikely that such a load could persist in ecologically balanced communities, and indeed, male diploidy is uncommon in native populations (141). Diploid male production has increased in introduced populations because sex alleles were lost during a population bottleneck early in the colonization process (138, 141).

SEX ALLOCATION   As is true for solitary organisms, colonies of social insects face the decision of how to allocate resources between growth (worker production) and reproduction (production of sexuals). However, eusocial Hymenoptera are unique because this decision depends on two parties, workers and queens, that have conflicting interests over such allocation of colony resources (98) as well as over the investment in each sex within the fraction allocated to reproduction (24a, 160). This conflict arises because of the differential relatedness between workers and brood of each sex that stems from male haploidy. Variation in colony queen number can affect this relatedness asymmetry and, thus, the preferred sex-investment ratio of workers (7, 98). Different dispersal strategies associated with variation in queen number also affect the preferred ratio, but for both parties (7, 20, 36, 91, 96a, 98).

The M and P forms of *S. invicta* differ in the amount of resources they allocate to production of workers and sexuals, with M colonies producing a higher proportion of sexuals among their offspring than do P colonies (114, 169, 169a). This difference in caste resource allocation reflects the different modes of colony reproduction of the two forms. Fissioning, characteristic of the P form, requires that a substantial number of workers accompany each queen inhabiting a new nest, whereas independent founding, characteristic of the M form, occurs in the absence of workers.

Estimates of the numerical sex ratio in the United States reveal a modest male bias in the M form (0.7 female:1 male) and female bias in the P form (1.4:1) (169). However, a high proportion of males produced in the P form are diploid and infertile (see above), and exclusion of such males results in an even stronger female bias (6.2:1) in the numerical ratio in the P form (only haploid males are present in the M form). This is important because it affects the operational sex ratio (OSR), the relative abundance of members of each sex available for mating. The OSR is equivalent to the numerical sex ratio for fertile sexuals in *S. invicta* because both sexes mate only once (see 129). Thus, the OSR is strongly female biased in the P form but moderately male biased in the M form in the United States, which may help explain why many P queens here remain permanently unmated, why, among those P queens that do mate, most mate with M males (see below), and why virtually all M queens become mated (40, 134).

Fire ant queens are larger and thus presumably require more investment than males. The M form produces larger queens and smaller males than does the P form, so that the sex-investment ratios estimated from the numerical ratios turn out to be equivalently biased toward females (1.5:1) for the two forms in the United States (169). Such female-biased investment, while lower than expected based on observed relatedness patterns (6, 90, 160), is in the direction predicted if there is at least partial worker control over colony investment. The identical investment by the two forms may be linked to the similar asymmetries in relatedness between workers and brood of each sex (which are similar between the forms because of low queen relatedness in the P form) and the correspondingly similar worker-preferred sex-investment ratios (7). This scenario assumes that workers regard 2N males as normal males, and indeed, the sex-investment patterns may be taken as evidence that this is the case. Investment ratios in native P populations, where nestmate queens are closely related and 2N males are rare, are expected to be less biased toward females than in the M form and to vary predictably with colony queen number, according to this scenario.

Variation in queen number is known to influence sex allocation in other highly eusocial Hymenoptera. Lower investment in reproductives in polygyne than in monogyne nests has been documented in several ant species (20, 157, 175), where it probably is associated with colony fissioning (20). Relative investment in each sex also commonly varies according to queen number. Polygyne colonies of several ants (20, 29, 31a, 55, 57, 157, 175) and epiponine wasps (117) are characterized by less female-biased sex-investment ratios than occur in conspecific monogyne colonies. This may be explained by the lower asymmetry of relatedness between workers and brood of each sex in polygyne colonies containing related queens compared to monogyne colonies. Female brood in such polygyne colonies are more closely related to workers than are male brood, but the difference diminishes with increasing queen number, so workers should prefer less female-biased sex ratios as queen number increases (7). Additionally, local resource competition may occur between daughter colonies when parent colonies adopt related queens and reproduce by fissioning, thus devaluing investment in queens and contributing to a less female-biased sex ratio in polygyne than in monogyne nests (7, 20, 36, 91, 96a, 97).

Differences in operational sex ratios between conspecific social forms such as occur in introduced *S. invicta* may influence the types of matings that occur between the forms, as discussed below. Therefore, their study can be valuable for understanding how shifts in social organization affect the routes and magnitude of interform gene flow.

Gene Flow between Alternative Social Forms

A major theme of this review is that shifts from monogyny to polygyny in ants often are accompanied by changes in the breeding biology, such as the site of mating, context in which queens begin reproducing, and mode of colony founding. These correlated changes raise questions as to whether interbreeding between populations differing in colony queen number ever is impeded and, if so, by what route existing interform gene flow occurs. Finally, the related question of whether any such restricted interbreeding can develop into reproductive isolation also must be considered.

The M and P forms of *S. invicta* are conspecific (159). Evidence of their close relationship includes the following: (i) They are more similar genetically than are any two recognized species in their species complex (138, 139); (ii) the P form always is associated geographically with the M form (42, 109, 110, 139); (iii) the forms are not distinguishable by any taxonomically informative morphological characters (JC Trager, unpublished); and (iv) egg-laying queens can be adopted into queenless nests of the alternate form (32, 134). Several studies have sought evidence for interform gene flow and attempted to determine its mode and magnitude, with the following results. Introduced populations typically are more similar genetically to adjacent populations of the alternate form than to distant populations of the same form, based on data from 26 protein markers (139). Adjacent M and P populations in Georgia, United States, possess the identical alleles at similar frequencies at 12 polymorphic protein loci, as well as similar numbers of sex-determining alleles (137). Most significantly, the allele $Pgm-3_a$, which occurs at high frequency in M populations, is common also in the P form despite being under intense negative selection in this form (78, 128). Taken together, these results indicate substantial gene flow between the social forms in the United States (128, 135, 137).

Laboratory and field studies using $Pgm-3$ as a marker to estimate the magnitude of interform matings suggest that 80–100% of mated P queens are inseminated by M males (128, 135, 137). Such matings may represent the predominant or sole route of interform gene flow in introduced *S. invicta*. Fertile (haploid) males are rare in P nests and thus are not expected to disperse and mate with M queens to any appreciable extent. Moreover, queens of neither form are likely to establish themselves as reproductives in nests of the alternate form. Newly mated M queens attempting to enter P nests likely are destroyed, because P workers are intolerant of the M queen phenotype (see above). Winged P queens may disperse into neighboring M populations, but they also are unlikely to affect interform gene flow. Their meager energy reserves make it doubtful that they can found colonies independently (77, 114), and they are not normally accepted into queenright M colonies (32). Gene flow between the social forms in the United States thus may be largely unidirectional and mediated by males, apparently as a result of the unique suites of reproductive traits distinguishing the two forms.

Patterns of interform gene flow in native populations may differ substantially from those in the United States. Assuming that the rarity of sterile diploid males in native P populations results in less female-biased operational sex ratios than in introduced P populations, gene flow via males is less likely to occur in the direction M→P but more likely to occur in the direction P→M (queens still are unlikely to mediate such gene flow). Although the route of interform gene flow thus may differ between the native and introduced ants, there is no reason to expect differences in the magnitude of gene flow. Nonetheless, significant genetic differentiation has been detected between M and P nests at one location in Argentina (KG Ross, EL Vargo, L Keller, unpublished), suggesting more substantial barriers to interform gene flow at this site than at any sites in the United States (137, 139). Such restricted interbreeding plausibly arises from site-specific assortative mating in native populations, with P sexuals mating primarily at the nest and M sexuals frequenting mating swarms. Mating at the nest in native P populations is strongly suggested by the observed high relatedness of mated nestmate queens.

Gene flow between related social forms seldom has been investigated in other ants. Elmes (30) studied mating swarms of *Myrmica ruginodis* to determine whether assortative mating occurred in the "microgyne" and "macrogyne" forms. No evidence of this was found, but a mating advantage of macrogyne males in swarms, combined with a possible tendency for macrogyne queens to mate in swarms and microgyne queens to mate at the nest, hint that some barriers to interbreeding between the forms may exist. Genetic surveys typically have revealed that monogyne and polygyne forms of single nominal species share identical alleles at similar frequencies (2, 126), as is expected with ongoing gene flow or recent isolation. Conspecific social forms within two separate *Rhytidoponera* species are not differentiated genetically, even though females of the alternate forms probably mate at different sites (174); interform gene flow may be mediated through males in these species, as it is in fire ants. Genetic studies of European *Formica* have failed to reveal significant differentiation between conspecific social forms, despite evidence for occasionally extensive local structure in some polygyne populations (99, 156). Clearly, additional genetic data from diverse taxa are needed to answer the question of whether the derived breeding habits that create local structure in polygyne ants also can restrict interbreeding between sympatric social forms.

ALTERNATIVE SOCIAL ORGANIZATIONS: PATHWAYS TO SPECIATION?

The potential for restricted interbreeding between conspecific social forms raises the question of whether reproductive isolation ever develops between them and, thus, whether evolution of social organization can be important in promoting speciation in ants (10, 13, 22, 139, 176). Possible steps in this process are:

(i) Some queens in an ancestral monogyne population forgo independent founding to be adopted in their natal colonies, inducing polygyny.

(ii) Polygyne colonies produce small queens with low reserves that specialize on the dependent strategy, leading to the coexistence of two queen reproductive strategies and phenotypes.

(iii) A shift in queen mating site to the area of the natal nest occurs in the novel type, increasing the probability that queens are adopted into their own colony and further diminishing the need for reserves to support flight and independent founding.

(iv) Sexual selection acts on males to track the novel mating behavior of polygyne queens.

(v) Positive assortative mating by site and continued disruptive selection on queen (and perhaps male) phenotypes leads to decreased interbreeding and increased differentiation between the ancestral monogyne and derived polygyne forms, culminating in reproductive isolation.

Evidence for such a mode of speciation, such as it exists, is highly inferential. Polygyny seems often to arise secondarily within essentially monogyne taxa (5, 13, 22, 37, 60, 124), resulting both in clusters of closely related species that differ in colony queen number and in frequent intraspecific variation in this trait (4, 5, 8, 13, 60, 142). In some cases the species status of the social variants is difficult to decide (50, 105, 142), possibly because these socially polymorphic clusters are actively radiating. Two predictions concerning the phylogenetic distribution of queen number across species in such clusters have been generated to test this hypothesis of speciation (176): Sister species should differ frequently in social organization, and species-rich clades should display more variation in social organization than do their species-poor sister clades. Unfortunately, the phylogeny and social biology of the relevant taxa seldom are sufficiently well understood to search for either pattern (but see 176). Furthermore, frequent shifts in social organization relative to the pace of cladogenesis (36a) as well as extinction may obscure the predicted pattern of fixed alternative social organizations in sister species.

Divergence in breeding behaviors observed between conspecific social forms constitutes the best available evidence for this proposed route of speciation. Restrictions on gene flow may be caused by differences in mating sites between the forms (13, 30, 99, 174), as well as by an inability of queens to establish themselves in populations of the alternate type due to worker intolerance (13, 137) or phenotypes ill-suited to the alternate breeding system (5, 12, 48, 65, 77, 79, 114). Pronounced queen morphological differentiation may accompany continued divergence in breeding strategies, an extreme result of which is a stable queen dimorphism featuring a winged morph specialized for independent founding and a miniaturized wingless morph specialized for adoption (see above). Disruptive sexual selection on males to track queens of both types may lead to the evolution of alternative male mating habits and morphologies as well (52). Both queen morphs may be maintained in some cases, presumably due to balancing selection associated with environmental heterogeneity and consequent variation in the success of independent founding (12, 50, 53, 124; also 174). In other cases, the queen morph specialized for adoption may replace the independent-founding morph (5, 13, 53). This latter pattern generally has been attributed to anagenesis but could result from a polygyne daughter species replacing its monogyne parent after reproductive isolation has been achieved.

Virtually no comprehensive genetic studies have been undertaken in the relevant species clusters to quantify the magnitude of differentiation that can develop between social variants. The studies cited above suggest that interform gene flow is not restricted sufficiently to permit substantial differentiation in sympatry in *Formica* and *Rhytidoponera,* although the significant genetic divergence found between the M and P forms of *S. invicta* in one native habitat hints at such a process. A frequently cited example of divergence between the "microgyne" and "macrogyne" forms of *Myrmica rubra* (105) is, unfortunately, based on only a single genetic marker.

Additional evidence that incipient barriers to gene flow between conspecific social forms can progress to complete reproductive isolation may come from combining population genetic and phylogenetic studies. Genetic analyses can resolve the species status of populations, quantify the extent and route of gene flow between socially divergent populations, and provide data for phylogeny reconstruction. With well-supported phylogenies, social organization in ancestral groups can be inferred (85) to determine whether speciation events typically coincide phylogenetically with shifts in social organization, and species richness of sister groups can be compared to learn whether significant radiations are associated with pronounced variation in social organization (14, 176).

Although questions remain about whether the inception of polygyny can drive speciation in the manner described above, there is growing consensus that polygyny plays an important role in ant speciation through its link to social parasitism (10, 13, 16, 17, 28, 176, 178). True social parasites in ants establish new colonies dependently by means of queens infiltrating nests of closely related polygyne host species, suggesting that they arose from populations in which unrelated foreign queens sometimes were adopted. Such "preparasitic" queens likely experienced selection to specialize in production of sexuals, to the detriment of the colony and their unrelated nestmates (reduced importance of workers is characteristic of social parasitism). Changes in parasite queen phenotypes facilitating more effective colony entry likely contributed to disruptive selection for alternative reproductive strategies in parasite and host subpopulations, analogous to the selection driving divergence between conspecific monogyne and polygyne social forms. Temporal or spatial isolation arising from changes in the incipient parasites' mating behaviors (perhaps associated with their diminutive size) and subsequent sexual selection on males may have further restricted interbreeding between parasites and hosts (10). The evolution of polygyny thus may set the stage for a subsequent dramatic innovation in social organization, social parasitism, that appears to be prominently involved in cladogenesis in ants.

ACKNOWLEDGMENTS

We thank AFG Bourke, M Chapuisat, PW Sherman, and EL Vargo for comments on an earlier draft of this paper. Our work is supported by grants from the National Geographic Society and the Swiss National Science Foundation.

Literature Cited

1. Alexander RD, Noonan KM, Crespi BJ. 1991. The evolution of eusociality. In *The Biology of the Naked Mole-Rat*, ed. PW Sherman, JUM Jarvis, RD Alexander, pp. 3–44. Princeton: Princeton Univ. Press

2. Banschbach VS, Herbers JM. 1995. Complex colony structure in social insects: I. Ecological determinants and genetic consequences. *Evolution.* In press

3. Barton N, Clark A. 1990. Population structure and process in evolution. In *Population Biology; Ecological and Evolutionary Viewpoints*, ed. K Wöhrmann, SK Jain, pp. 115–73. Berlin: Springer-Verlag

4. Bennett B. 1987. Ecological differences between monogynous and polygynous sibling ant species (Hymenoptera: Formicidae). *Sociobiology* 13:249–70

5. Bolton B. 1986. Apterous females and shift of dispersal strategy in the *Monomorium salomonis*-group (Hymenoptera: Formicidae). *J. Nat. Hist.* 20: 267–72

6. Boomsma JJ. 1989. Sex-investment ratios in ants: Has female bias been systematically overestimated? *Am. Nat.* 133:517–32

7. Boomsma JJ. 1993. See Ref. 67, pp. 86–109

8. Boomsma JJ, Brouwer AH, van Loon AJ. 1990. A new polygynous *Lasius* species (Hymenoptera: Formicidae) from central Europe. II. Allozymatic confirmation of species status and social structure. *Insectes Soc.* 37:363–75

8a. Boomsma JJ, Ratnieks FLW. 1995. Unpublished ms.

9. Boomsma JJ, Wright PJ, Brouwer AH. 1993. Social structure in the ant *Lasius flavus*: multi-queen nests or multi-nest mounds? *Ecol. Entomol.* 18: 47–53

10. Bourke AFG, Franks NR. 1991. Alternative adaptations, sympatric speciation and the evolution of parasitic, inquiline ants. *Biol. J. Linn. Soc.* 43:157–78

11. Bourke AFG, Franks NR. 1995. *Social Evolution in Ants*. Princeton: Princeton Univ. Press

12. Bourke AFG, Heinze J. 1994. The ecology of communal breeding: the case of multiple-queen leptothoracine ants. *Philos. Trans. R. Soc. London Ser. B* 345: 359–72

13. Brian MV. 1983. *Social Insects: Ecology and Behavioural Biology*. London: Chapman & Hall

14. Brooks DR, McLennan DA. 1993. Comparative study of adaptive radiations with an example using parasitic flatworms (Platyhelminthes: Cercomeria). *Am. Nat.* 142:755–78

15. Buschinger A. 1968. Mono- und Polygynie bei Arten der Gattung *Leptothorax* Mayr (Hymenoptera, Formicidae). *Insectes Soc.* 15:217–26

16. Buschinger A. 1986. Evolution of social parasitism in ants. *Trends Ecol. Evol.* 1:155–60

17. Buschinger A. 1990. Sympatric speciation and radiative evolution of socially parasitic ants—heretic hypotheses and their factual background. *Z. Zool. Syst. Evolut.-Forsch.* 28:241–60

18. Buschinger A, Heinze J. 1992. Polymorphism of female reproductives in ants. In *Biology and Evolution of Social Insects*, ed. J Billen, pp. 11–23. Leuven: Leuven Univ. Press

19. Carlin NF, Reeve HK, Cover SP. 1993. See Ref. 67, pp. 362–401

20. Chan GL, Bourke AFG. 1994. Split sex ratios in a multiple-queen ant population. *Proc. R. Soc. London Ser. B* 258: 261–66

21. Cook JM, Crozier RH. 1995. Sex determination and population biology in the Hymenoptera. *Trends Ecol. Evol.* 10:281–86

22. Crozier RH. 1977. Evolutionary genetics of the Hymenoptera. *Annu. Rev. Entomol.* 22:263–88

23. Crozier RH. 1980. Genetical structure of social insect populations. In *Evolution of Social Behavior: Hypotheses and Empirical Tests,* ed. H Markl, pp. 129–46. Weinheim: Chemie

24. Crozier RH, Pamilo P. 1986. Relatedness within and between colonies of a queenless ant species of the genus *Rhytidoponera* (Hymenoptera: Formicidae). *Entomol. Gener.* 11:113–17

24a. Crozier RH, Pamilo P. 1995. *Kin Selection and Sex Allocation in Social Insects.* Oxford: Oxford Univ. Press

25. Crozier RH, Pamilo P, Crozier YC. 1984. Relatedness and microgeographic genetic variation in *Rhytidoponera mayri,* an Australian arid-zone ant. *Behav. Ecol. Sociobiol.* 15:143–50

26. Douwes P, Sivusaari L, Niklasson M, Stille B. 1987. Relatedness among queens in polygynous nests of the ant *Leptothorax acervorum. Genetica* 75:23–29

27. Elmes GW. 1974. The effect of colony population on caste size in three species of *Myrmica* (Hymenoptera: Formicidae). *Insectes Soc.* 21:213–30

28. Elmes GW. 1978. A morphometric comparison of three closely related species of *Myrmica* (Formicidae), including a new species from England. *Syst. Entomol.* 3:131–45

29. Elmes GW. 1987. Temporal variation in colony populations of the ant *Myrmica sulcinodis.* 2. Sexual production and sex ratios. *J. Anim. Ecol.* 56:559–71

30. Elmes GW. 1991. Mating strategy and isolation between the two forms, macrogyna and microgyna, of *Myrmica ruginodis* (Hym. Formicidae). *Ecol. Entomol.* 16:411–23

31. Elmes GW, Keller L. 1993. See Ref. 67, pp. 294–307

31a. Evans JD. 1995. Relatedness threshold for the production of female sexuals in colonies of a polygynous ant, *Myrmica tahoensis,* as revealed by microsatellite DNA analysis. *Proc. Natl. Acad. Sci. USA* 92:6514–17

32. Fletcher DJC, Blum MS. 1983. Regulation of queen number by workers in colonies of social insects. *Science* 219:312–14

33. Fletcher DJC, Blum MS. 1983. The inhibitory pheromone of queen fire ants: effects of disinhibition on dealation and oviposition by virgin queens. *J. Comp. Physiol. A* 153:467–75

34. Fletcher DJC, Blum MS, Whitt TV,

Temple N. 1980. Monogyny and polygyny in the fire ant *Solenopsis invicta. Ann. Entomol. Soc. Am.* 73:658–61

35. Fletcher DJC, Ross KG. 1985. Regulation of reproduction in eusocial Hymenoptera. *Annu. Rev. Entomol.* 30:319–43

36. Frank SA. 1987. Variable sex ratio among colonies of ants. *Behav. Ecol. Sociobiol.* 20:195–201

36a. Frumhoff PC, Reeve HK. 1994. Using phylogenies to test hypotheses of adaptation: a critique of some current proposals. *Evolution* 48:172–80

37. Frumhoff PC, Ward PS. 1992. Individual-level selection, colony-level selection, and the association between polygyny and worker monomorphism in ants. *Am. Nat.* 139:559–90

37a. Gadagkar R. 1994. Why the definition of eusociality is not helpful to understand its evolution and what should we do about it. *Oikos* 70:485–88

38. Gastreich KR, Strassmann JE, Queller DC. 1993. Determinants of high genetic relatedness in the swarm-founding wasp, *Protopolybia exigua. Ethol. Ecol. Evol.* 5:529–39

39. Glancey BM, Craig CH, Stringer CE, Bishop PM. 1973. Multiple fertile queens in colonies of the imported fire ant, *Solenopsis invicta. J. Georgia Entomol. Soc.* 8:237–38

40. Glancey BM, Lofgren CS. 1985. Spermatozoon counts in males and inseminated queens of the imported fire ants, *Solenopsis invicta* and *Solenopsis richteri* (Hymenoptera: Formicidae). *Fla. Entomol.* 68:162–68

41. Glancey BM, Lofgren CS. 1988. Adoption of newly-mated queens: a mechanism for proliferation and perpetuation of polygynous red imported fire ants, *Solenopsis invicta* Buren. *Fla. Entomol.* 71:581–87

41a. Grafen A. 1988. On the uses of data on lifetime reproductive success. In *Reproductive Success: Studies of Individual Variation in Contrasting Breeding Systems,* ed. TH Clutton-Brock, pp. 454–71. Chicago: Univ. Chicago Press

42. Greenberg L, Fletcher DJC, Vinson SB. 1985. Differences in worker size and mound distribution in monogynous and polygynous colonies of the fire ant *Solenopsis invicta* Buren. *J. Kansas Entomol. Soc.* 58:9–18

43. Greenberg L, Vinson SB, Ellison S. 1992. Nine-year study of a field containing both monogyne and polygyne red imported fire ants (Hymenoptera: Formicidae). *Ann. Entomol. Soc. Am.* 85:686–95

44. Greene A. 1991. See Ref. 136, pp. 263–305

45. Halliday RB. 1983. Social organization of meat ants *Iridomyrmex purpureus* analysed by gel electrophoresis of enzymes. *Insectes Soc.* 30:45–56

46. Hamilton WD. 1972. Altruism and related phenomena, mainly in social insects. *Annu. Rev. Ecol. Syst.* 3:193–232

47. Heinze J. 1993. See Ref. 67, pp. 334–61

48. Heinze J. 1993. Habitat structure, dispersal strategies and queen number in two boreal *Leptothorax* ants. *Oecologia* 96:32–39

49. Heinze J. 1994. Genetic colony and population structure of the ant *Leptothorax* cf *canadensis*. *Can. J. Zool.* 72:1477–80

50. Heinze J, Buschinger A. 1987. Queen polymorphism in a non-parasitic *Leptothorax* species (Hymenoptera, Formicidae). *Insectes Soc.* 34:28–43

51. Heinze J, Buschinger A. 1989. Queen polymorphism in *Leptothorax* spec. A: its genetic and ecological background (Hymenoptera: Formicidae). *Insectes Soc.* 36:139–55

52. Heinze J, Hölldobler B. 1993. Fighting for a harem of queens: physiology of reproduction in *Cardiocondyla* male ants. *Proc. Natl. Acad. Sci. USA* 90:8412–14

53. Heinze J, Hölldobler B, Cover SP. 1992. Queen polymorphism in the North American harvester ant, *Ephebomyrmex imberbiculus*. *Insectes Soc.* 39:267–73

54. Heinze J, Lipski N, Hölldobler B, Bourke AFG. 1995. Geographic variation in the social and genetic structure of the ant *Leptothorax acervorum*. *Zoology.* 98:127–35

55. Herbers JM. 1984. Queen-worker conflict and eusocial evolution in a polygynous ant species. *Evolution* 38:631–43

56. Herbers JM. 1986. Nest site limitation and facultative polygyny in the ant *Leptothorax longispinosus*. *Behav. Ecol. Sociobiol.* 19:115–22

57. Herbers JM. 1990. Reproductive investment and allocation ratios for the ant *Leptothorax longispinosus*: sorting out the variation. *Am. Nat.* 136:178–208

58. Herbers JM. 1993. See Ref. 67, pp. 262–93

59. Herbers JM, Grieco S. 1994. Population structure of *Leptothorax ambiguus*, a facultatively polygynous and polydomous ant species. *J. Evol. Biol.* 7:581–98

60. Hölldobler B, Wilson EO. 1977. The number of queens: an important trait in ant evolution. *Naturwissenschaften* 64:8–15

61. Hölldobler B, Wilson EO. 1990. *The Ants.* Cambridge: Harvard Univ. Press

62. Ito F. 1990. Functional monogyny of *Leptothorax acervorum* in Northern Japan. *Psyche* 97:203–11

63. Kaufmann B, Boomsma JJ, Passera L, Petersen KN. 1992. Relatedness and inbreeding in a French population of the unicolonial ant *Iridomyrmex humilis* (Mayr). *Insectes Soc.* 39:195–213

64. Keller L. 1988. Evolutionary implications of polygyny in the Argentine ant, *Iridomyrmex humilis* (Mayr) (Hymenoptera: Formicidae): an experimental study. *Anim. Behav.* 36:159–65

65. Keller L. 1991. Queen number, mode of colony founding, and queen reproductive success in ants (Hymenoptera Formicidae). *Ethol. Ecol. Evol.* 3:307–16

66. Keller L. 1993. The assessment of reproductive success of queens in ants and other social insects. *Oikos* 67:177–80

67. Keller L, ed. 1993. *Queen Number and Sociality in Insects.* New York: Oxford Univ. Press

68. Keller L. 1995. Social life: the paradox of multiple-queen colonies. *Trends Ecol. Evol.* In press

69. Keller L, Passera L. 1989. Size and fat content of gynes in relation to the mode of colony founding in ants (Hymenoptera; Formicidae). *Oecologia* 80:236–40

70. Keller L, Passera L. 1990. Queen number, social structure, reproductive strategies and their correlates in ants. In *Social Insects and the Environment,* ed. GK Veeresh, B Mallick, CA Viraktamath, pp. 236–37. New Delhi: Oxford/IBH

71. Keller L, Passera L. 1990. Fecundity of ant queens in relation to their age and the mode of colony founding. *Insectes Soc.* 37:116–30

72. Keller L, Passera L. 1992. Mating system, optimal number of matings, and sperm transfer in the Argentine ant *Iridomyrmex humilis*. *Behav. Ecol. Sociobiol.* 31:359–66

73. Keller L, Passera L. 1993. Incest avoidance, fluctuating asymmetry, and the consequences of inbreeding in *Iridomyrmex humilis*, an ant with multiple queen colonies. *Behav. Ecol. Sociobiol.* 33:191–99

74. Keller L, Passera L, Suzzoni JP. 1989. Queen execution in the Argentine ant *Iridomyrmex humilis* (Mayr). *Physiol. Entomol.* 14:157–63

74a. Keller L, Perrin N. 1995. Quantifying the level of eusociality. *Proc. R. Soc. London Ser. B.* In press

75. Keller L, Reeve HK. 1994. Partitioning of reproduction in animal societies. *Trends Ecol. Evol.* 9:98–102

76. Keller L, Reeve HK. 1994. Genetic variability, queen number, and polyandry in social Hymenoptera. *Evolution* 38:694–704

77. Keller L, Ross KG. 1993. Phenotypic plasticity and "cultural" transmission of alternative social organizations in the fire ant *Solenopsis invicta. Behav. Ecol. Sociobiol.* 33:121–29

78. Keller L, Ross KG. 1993. Phenotypic basis of reproductive success in a social insect: genetic and social determinants. *Science* 260:1107–10

79. Keller L, Ross KG. 1995. Gene by environment interaction: effects of a single gene and social environment on reproductive phenotypes of fire ant queens. *Funct. Ecol.* In press

80. Keller L, Vargo EL. 1993. See Ref. 67, pp. 16–44

81. Lofgren CS. 1986. See Ref. 82, pp. 36–47

82. Lofgren CS, Vander Meer RK, ed. 1986. *Fire Ants and Leaf-Cutting Ants: Biology and Management.* Boulder: Westview

83. MacKay WP, Greenberg L, Vinson SB. 1991. Survivorship of founding queens of *Solenopsis invicta* (Hymenoptera: Formicidae) in areas with monogynous and polygynous nests. *Sociobiology* 19:293–304

84. MacKay WP, Porter S, Gonzalez D, Rodriguez A, Armendedo H, et al. 1990. A comparison of monogyne and polygyne populations of the tropical fire ant, *Solenopsis geminata* (Hymenoptera: Formicidae), in Mexico. *J. Kansas Entomol. Soc.* 63:611–15

85. Maddison DR. 1994. Phylogenetic methods for inferring the evolutionary history and processes of change in discretely valued characters. *Annu. Rev. Entomol.* 39:267–92

86. Markin GP, Collins HL, Dillier JH. 1972. Colony founding by queens of the red imported fire ant, *Solenopsis invicta. Ann. Entomol. Soc. Am.* 65:1053–58

87. McInnes DA, Tschinkel WR. 1995. Queen dimorphism and reproductive strategies in the fire ant *Solenopsis geminata* (Hymenoptera: Formicidae). *Behav. Ecol. Sociobiol.* 36:367–75

88. Mercier B, Passera L, Suzzoni JP. 1985. Etude de la polygynie chez la fourmi *Plagiolepis pygmaea* Latr. (Hymenoptera: Formicidae). II - La fécondité des reines en condition expérimentale polygyne. *Insectes Soc.* 32:349–62

89. Michod RE. 1993. Inbreeding and the evolution of social behavior. In *The Natural History of Inbreeding and Outbreeding: Theoretical and Empirical Perspectives,* ed. NW Thornhill, pp. 74–96. Chicago: Univ. Chicago Press

90. Nonacs P. 1986. Ant reproductive strategies and sex allocation theory. *Q. Rev. Biol.* 61:1–21

91. Nonacs P. 1993. See Ref. 67, pp. 110–31

92. Obin MS, Morel L, Vander Meer RK. 1993. Unexpected, well-developed nestmate recognition in laboratory colonies of polygyne imported fire ants (Hymenoptera, Formicidae). *J. Insect Behav.* 6:579–87

93. Page RE. 1980. The evolution of multiple mating behavior by honey bee queens (*Apis mellifera* L.). *Genetics* 96:263–73

94. Pamilo P. 1982. Genetic population structure in polygynous *Formica* ants. *Heredity* 48:95–106

95. Pamilo P. 1983. Genetic differentiation within subdivided populations of *Formica* ants. *Evolution* 37:1010–22

96. Pamilo P. 1985. Effect of inbreeding on genetic relatedness. *Hereditas* 103:195–200

96a. Pamilo P. 1990. Sex allocation and queen-worker conflict in polygynous ants. *Behav. Ecol. Sociobiol.* 27:31–36

97. Pamilo P. 1991. Evolution of colony characteristics in social insects. II. Number of reproductive individuals. *Am. Nat.* 138:412–33

98. Pamilo P. 1991. Evolution of colony characteristics in social insects. I. Sex allocation. *Am. Nat.* 137:83–107

99. Pamilo P, Rosengren R. 1984. Evolution of nesting strategies of ants: genetic evidence from different population types of *Formica* ants. *Biol. J. Linn. Soc.* 21:331–48

100. Pamilo P, Seppä P. 1994. Reproductive competition and conflicts in colonies of the ant *Formica sanguinea. Anim. Behav.* 48:1201–6

101. Pamilo P, Sundström L, Fortelius W, Rosengren R. 1994. Diploid males and colony-level selection in *Formica* ants. *Ethol. Ecol. Evol.* 6:221–35

102. Passera L, Keller L. 1990. Loss of mating flight and shift in the pattern of carbohydrate storage in sexuals of ants (Hymenoptera, Formicidae). *J. Comp. Physiol.* B 160:207–11

103. Passera L, Keller L. 1994. Mate availability and male dispersal in the Argentine ant *Linepithema humile. Anim. Behav.* 48:361–69

104. Pearson B. 1982. Relatedness of normal queens (macrogynes) in nests of the

polygynous ant *Myrmica rubra* Latreille. *Evolution* 36:107–12

105. Pearson B, Child AR. 1980. The distribution of an esterase polymorphism in macrogynes and microgynes of *Myrmica rubra* Latreille. *Evolution* 34:105–9

106. Peeters C. 1993. See Ref. 67, pp. 234–61

107. Pfennig DW, Collins JP. 1993. Kinship affects morphogenesis in cannibalistic salamanders. *Nature* 362:836–38

108. Porter SD. 1991. Origins of new queens in polygyne red imported fire ant colonies (Hymenoptera: Formicidae). *J. Entomol. Sci.* 26:474–78

109. Porter SD. 1992. Frequency and distribution of polygyne fire ants (Hymenoptera: Formicidae) in Florida. *Fla. Entomol.* 75:248–57

110. Porter SD. 1993. Stability of polygyne and monogyne fire ant populations (Hymenoptera: Formicidae: *Solenopsis invicta*) in the United States. *J. Econ. Entomol.* 86:1344–47

111. Porter SD, Bhatkar A, Mulder R, Vinson SB, Clair DJ. 1991. Distribution and density of polygyne fire ants (Hymenoptera: Formicidae) in Texas. *J. Econ. Entomol.* 84:866–74

112. Porter SD, Fowler HG, MacKay WP. 1992. Fire ant mound densities in the United States and Brazil (Hymenoptera: Formicidae). *J. Econ. Entomol.* 85: 1154–61

113. Porter SD, Tschinkel WR. 1986. Adaptive value of nanitic workers in incipient fire ant colonies. *Ann. Entomol. Soc. Am.* 79:723–26

114. Porter SD, Van Eimeren B, Gilbert LE. 1988. Invasion of red imported fire ants (Hymenoptera: Formicidae): microgeography of competitive replacement. *Ann. Entomol. Soc. Am.* 81: 913–18

115. Queller DC. 1993. See Ref. 67, pp. 132–52

116. Queller DC, Negrón-Sotomayor JA, Strassmann JE, Hughes CR. 1993. Queen number and genetic relatedness in a neotropical wasp, *Polybia occidentalis*. *Behav. Ecol.* 4:7–13

117. Queller DC, Strassmann JE, Solís CR, Hughes CR, Deloach DM. 1993. A selfish strategy of social insect workers that promotes social cohesion. *Nature* 365: 639–41

118. Ratnieks FLW. 1988. Reproductive harmony via mutual policing by workers in eusocial Hymenoptera. *Am. Nat.* 132: 217–36

119. Deleted in proof

120. Ratnieks FLW, Reeve HK. 1992. Conflict in single-queen hymenopteran societies: the structure of conflict and processes that reduce conflict in advanced eusocial species. *J. Theor. Biol.* 158:33–65

121. Reeve HK. 1991. See Ref. 136, pp. 99–148

122. Reeve HK, Keller L. 1995. Partitioning of reproduction in mother-daughter versus sibling associations: a test of optimal skew theory. *Am. Nat.* 145:119–32

123. Reeve HK, Ratnieks FLW. 1993. See Ref. 67, pp. 45–85

124. Rosengren R, Pamilo P. 1983. The evolution of polygyny and polydomy in mound-building *Formica* ants. *Acta Entomol. Fennica* 42:65–77

125. Rosengren R, Sundström L, Fortelius W. 1993. See Ref. 67, pp. 308–33

126. Ross KG. 1988. Population and colony-level genetic studies of ants. In *Advances in Myrmecology*, ed. JC Trager, pp. 189–215. New York: Brill

127. Ross KG. 1988. Differential reproduction in multiple-queen colonies of the fire ant *Solenopsis invicta* (Hymenoptera: Formicidae). *Behav. Ecol. Sociobiol.* 23:341–55

128. Ross KG. 1992. Strong selection on a gene that influences reproductive competition in a social insect. *Nature* 355: 347–49

129. Ross KG. 1993. The breeding system of the fire ant *Solenopsis invicta*: effects on colony genetic structure. *Am. Nat.* 141:554–76

130. Ross KG, Carpenter JM. 1991. Phylogenetic analysis and the evolution of queen number in eusocial Hymenoptera. *J. Evol. Biol.* 4:117–30

131. Ross KG, Carpenter JM. 1991. See Ref. 136, pp. 451–79

132. Ross KG, Fletcher DJC. 1985. Genetic origin of male diploidy in the fire ant, *Solenopsis invicta* (Hymenoptera: Formicidae), and its evolutionary significance. *Evolution* 39:888–903

133. Ross KG, Fletcher DJC. 1985. Comparative study of genetic and social structure in two forms of the fire ant, *Solenopsis invicta* (Hymenoptera: Formicidae). *Behav. Ecol. Sociobiol.* 17: 349–56

134. Ross KG, Fletcher DJC. 1986. Diploid male production—a significant colony mortality factor in the fire ant *Solenopsis invicta* (Hymenoptera: Formicidae). *Behav. Ecol. Sociobiol.* 19:283–91

135. Ross KG, Keller L. 1995. Joint influence of gene flow and selection on a reproductively important genetic polymorphism in the fire ant *Solenopsis invicta*. *Am. Nat.* 146:325–48

136. Ross KG, Matthews RW, ed. 1991. *The*

Social Biology of Wasps. Ithaca: Cornell Univ. Press

137. Ross KG, Shoemaker DD. 1993. An unusual pattern of gene flow between the two social forms of the fire ant *Solenopsis invicta. Evolution* 47:1595–605

138. Ross KG, Trager JC. 1990. Systematics and population genetics of fire ants (*Solenopsis saevissima* complex) from Argentina. *Evolution* 44:2113–34

139. Ross KG, Vargo EL, Fletcher DJC. 1987. Comparative biochemical genetics of three fire ant species in North America, with special reference to the two social forms of *Solenopsis invicta* (Hymenoptera: Formicidae). *Evolution* 41:979–90

140. Ross KG, Vargo EL, Fletcher DJC. 1988. Colony genetic structure and queen mating frequency in fire ants of the subgenus *Solenopsis* (Hymenoptera: Formicidae). *Biol. J. Linn. Soc.* 34:105–17

141. Ross KG, Vargo EL, Keller L, Trager JC. 1993. Effect of a founder event on variation in the genetic sex-determining system of the fire ant *Solenopsis invicta. Genetics* 135:843–54

142. Satoh T. 1989. Comparisons between two apparently distinct forms of *Camponotus nawai* Ito (Hymenoptera: Formicidae). *Insectes Soc.* 36:277–92

143. Seger J. 1991. Cooperation and conflict in social insects. In *Behavioural Ecology: An Evolutionary Approach*, ed. JR Krebs, NB Davies, pp. 338–73. Oxford: Blackwell

144. Seger J. 1993. See Ref. 67, pp. 1–15

145. Seppä P. 1992. Genetic relatedness of worker nestmates in *Myrmica ruginodis* (Hymenoptera: Formicidae) populations. *Behav. Ecol. Sociobiol.* 30:253–60

146. Seppä P. 1994. Sociogenetic organization of the ants *Myrmica ruginodis* and *Myrmica lobicornis*: number, relatedness and longevity of reproducing individuals. *J. Evol. Biol.* 7:71–95

147. Seppä P, Pamilo P. 1995. Gene flow and population viscosity in *Myrmica* ants. *Heredity* 74:200–9

148. Sherman PW, Lacey EA, Reeve HK, Keller L. 1995. The eusociality continuum. *Behav. Ecol.* 6:102–8

149. Shoemaker DD, Costa JT, Ross KG. 1992. Estimates of heterozygosity in two social insects using a large number of electrophoretic markers. *Heredity* 69:573–82

150. Stacey PB, Koenig WD, ed. 1991. *Cooperative Breeding in Birds*. Cambridge: Cambridge Univ. Press

151. Stearns SC. 1992. *The Evolution of Life Histories*. Oxford: Oxford Univ. Press

152. Stille M, Stille B. 1993. Intrapopulation nest clusters of maternal mtDNA lineages in the polygynous ant *Leptothorax acervorum* (Hymenoptera: Formicidae). *Insect Mol. Biol.* 1:117–21

153. Stille M, Stille B, Douwes P. 1991. Polygyny, relatedness and nest founding in the polygynous myrmicine ant *Leptothorax acervorum* (Hymenoptera; Formicidae). *Behav. Ecol. Sociobiol.* 28:91–96

154. Strassmann JE, Gastreich KR, Queller DC, Hughes CR. 1992. Demographic and genetic evidence for cyclical changes in queen number in a neotropical wasp, *Polybia emaciata. Am. Nat.* 140:363–72

155. Strassmann JE, Queller DC, Solís CR, Hughes CR. 1991. Relatedness and queen number in the neotropical wasp, *Parachartergus colobopterus. Anim. Behav.* 42:461–70

156. Sundström L. 1993. Genetic population structure and sociogenetic organisation in *Formica truncorum* (Hymenoptera; Formicidae). *Behav. Ecol. Sociobiol.* 33:345–54

157. Sundström L. 1995. Sex allocation and colony maintenance in monogyne and polygyne colonies of *Formica truncorum* (Hymenoptera: Formicidae); the impact of kinship and mating structure. *Am. Nat.* In press

158. Sundström L. 1995. Dispersal polymorphism and physiological condition of males and females in the ant *Formica truncorum. Behav. Ecol.* 6:132–39

159. Trager JC. 1991. A revision of the fire ants, *Solenopsis geminata* group (Hymenoptera: Formicidae: Myrmicinae). *J. New York Entomol. Soc.* 99:141–98

160. Trivers RL, Hare H. 1976. Haplodiploidy and the evolution of the social insects. *Science* 191:249–63

161. Tschinkel WR. 1986. See Ref. 82, pp. 72–87

162. Tschinkel WR. 1992. Brood raiding and the population dynamics of founding and incipient colonies of the fire ant, *Solenopsis invicta. Ecol. Entomol.* 17:179–88

163. Tschinkel WR. 1993. Sociometry and sociogenesis of colonies of the fire ant *Solenopsis invicta* during one annual cycle. *Ecol. Monogr.* 63:425–57

164. Tschinkel WR. 1993. Resource allocation, brood production and cannibalism during colony founding in the fire ant, *Solenopsis invicta. Behav. Ecol. Sociobiol.* 33:209–23

165. Tschinkel WR, Howard DF. 1983. Colony founding by pleometrosis in the fire ant, *Solenopsis invicta*. *Behav. Ecol. Sociobiol.* 12:103–13

166. Vander Meer RK, Morel L, Lofgren CS. 1992. A comparison of queen oviposition rates from monogyne and polygyne fire ants, *Solenopsis invicta*. *Physiol. Entomol.* 17:384–90

167. Vargo EL. 1992. Mutual pheromonal inhibition among queens in polygyne colonies of the fire ant *Solenopsis invicta*. *Behav. Ecol. Sociobiol.* 31:205–10

168. Vargo EL. 1993. Colony reproductive structure in a polygyne population of *Solenopsis geminata* (Hymenoptera: Formicidae). *Ann. Entomol. Soc. Am.* 86:441–49

169. Vargo EL. 1995. Unpublished

169a. Vargo EL, Fletcher DJC. 1987. Effect of queen number on the production of sexuals in natural populations of the fire ant, *Solenopsis invicta*. *Physiol. Entomol.* 12:109–16

170. Vargo EL, Fletcher DJC. 1989. On the relationship between queen number and fecundity in polygyne colonies of the fire ant, *Solenopsis invicta*. *Physiol. Entomol.* 14:223–32

171. Vargo EL, Laurel M. 1994. Studies on the mode of action of a queen primer pheromone of the fire ant *Solenopsis invicta*. *J. Insect Physiol.* 40:601–10

172. Vargo EL, Porter SD. 1989. Colony reproduction by budding in the polygyne form of the fire ant, *Solenopsis invicta* (Hymenoptera: Formicidae). *Ann. Entomol. Soc. Am.* 82:307–13

173. Vargo EL, Ross KG. 1989. Differential viability of eggs laid by queens in polygyne colonies of the fire ant, *Solenopsis invicta*. *J. Insect Physiol.* 35:587–93

173a. Wade MJ. 1985. The influence of multiple inseminations and multiple foundresses on social evolution. *J. Theor. Biol.* 112:109–21

174. Ward PS. 1983. Genetic relatedness and colony organization in a species complex of ponerine ants. I. Phenotypic and genotypic composition of colonies. *Behav. Ecol. Sociobiol.* 12:285–99

175. Ward PS. 1983. Genetic relatedness and colony organization in a species complex of ponerine ants. II. Patterns of sex ratio investment. *Behav. Ecol. Sociobiol.* 12:301–7

176. Ward PS. 1989. Genetic and social changes associated with ant speciation. In *The Genetics of Social Evolution*, ed. MD Breed, RE Page, pp. 123–48. Boulder: Westview

177. West-Eberhard MJ. 1983. Sexual selection, social competition, and speciation. *Q. Rev. Biol.* 58:155–83

178. West-Eberhard MJ. 1986. Alternative adaptations, speciation, and phylogeny (a review). *Proc. Natl. Acad. Sci. USA* 83:1388–92

179. West-Eberhard MJ. 1989. Phenotypic plasticity and the origins of diversity. *Annu. Rev. Ecol. Syst.* 20:249–78

180. Wilson EO. 1971. *The Insect Societies.* Cambridge: Harvard Univ. Press

181. Wilson NL, Dillier JH, Markin GP. 1971. Foraging territories of imported fire ants. *Ann. Entomol. Soc. Am.* 64:660–65

182. Winter U, Buschinger A. 1986. Genetically mediated queen polymorphism and caste determination in the slave making ant, *Harpagoxenus sublaevis* (Hymenoptera: Formicidae). *Entomol. Gener.* 11:125–37

183. Woodroffe R. 1993. Alloparental behaviour in the European badger. *Anim. Behav.* 46:413–15

184. Yamauchi K, Furukawa T, Kinomura K, Takamine H, Tsuji K. 1991. Secondary polygyny by inbred wingless sexuals in the dolichoderine ant *Technomyrmex albipes*. *Behav. Ecol. Sociobiol.* 29:313–19

Annu. Rev. Ecol. Syst. 1995. 26: 657–81

SEPARATE VERSUS COMBINED ANALYSIS OF PHYLOGENETIC EVIDENCE

Alan de Queiroz[1]

Department of Ecology and Evolutionary Biology, University of Arizona, Tucson, Arizona 85721

Michael J. Donoghue

Department of Organismic and Evolutionary Biology, Harvard University, Cambridge, Massachusetts 02138

Junhyong Kim[2]

Department of Ecology and Evolutionary Biology, University of Arizona, Tucson, Arizona 85721

KEY WORDS: phylogeny, taxonomic congruence, character congruence, consensus, total evidence

---

ABSTRACT

There has been much discussion in the recent systematic literature over whether different data sets bearing on phylogenetic relationships should be analyzed separately or combined and analyzed simultaneously. We review arguments in favor of each of these views. Assuming that the goal is to uncover the true phylogeny of the entities in question, arguments for combining data based on the notions that one should use the "total evidence" available, or that the combined analysis gives the tree with the greatest descriptive and explanatory power, are not compelling. However, combining data sets can enhance detection of real phylogenetic groups. On the other hand, if there is heterogeneity

[1]Present address: University Museum and Department of Environmental, Population and Organismic Biology, University of Colorado, Boulder, Colorado 80309

[2]Present address: Department of Biology, Yale University, New Haven, Connecticut 06511

657

0066-4162/95/1120-0657$05.00

among data sets with respect to some property that affects phylogeny estimation, then combining the data can give misleading results. Thus, there are reasonable arguments on both sides of the debate.

We present a conceptual framework based on the reasons that different data sets may give conflicting estimates of phylogeny. The framework illustrates the point that the precise nature of the difference among data sets is critical in the choice of a method of analysis. In particular, very different approaches are necessary to deal with data sets that differ in processes of character change compared to ones that differ in branching histories. We highlight several recently developed methods designed to deal with these different situations. All of these methods avoid the loss of information that is likely to be associated with summarizing data sets as trees in an intermediate step (an advantage of typical combined analyses), while taking into account heterogeneity among data sets (an advantage of separate analyses). We suggest that the recognition and further development of such methods will help depolarize the debate over combined and separate analysis.

INTRODUCTION

The availability of a variety of sources of evidence on phylogenetic relationships has focused attention on a fundamental question: Should different kinds of data bearing on a given phylogenetic problem be analyzed separately or combined and analyzed simultaneously? Our aim is to review arguments for and against separate and combined analyses and to provide a general conceptual framework within which to explore the basic but sometimes subtle issues associated with the problem. We hope to show that both kinds of analyses are useful in estimating phylogenetic relationships, but that the standard forms of these alternatives by no means exhaust the possible solutions to the problem.

Some additional clarification of our goals is in order, especially to say what we do not intend to provide. First, although we need to refer to particular examples in order to clarify arguments, we do not provide a thorough review of studies in which separate and/or combined analyses have actually been carried out. In addition, although much emphasis has been placed on molecular versus morphological evidence (26, 29, 38, 45, 84, 85, 108), that contrast is not our specific concern here. The partitioning of data into molecular and morphological subsets is only one of many divisions that may be relevant to the problem at hand.

It is also critical to clarify our general perspective from the outset. We assume that the ultimate goals of phylogenetic analysis are to discover the true phylogeny of the entities under investigation, and to understand evolutionary processes. Although this review is written from that perspective, we recognize that certain arguments that we criticize may be valid given different basic goals (see, in particular, the arguments in favor of combined analysis based on the principle of "total evidence" and on maximizing descriptive and explanatory power). It is also important to appreciate that different sorts of interrelated entities have their own histories, which may or may not coincide with one another. For example, the branching history of a particular gene or organellar genome may not coincide exactly with the branching history of the populations or species of organisms in which it resides (e.g. 3, 29, 109). Consequently, the objects of study must be clearly specified. For the purposes of this paper, we assume that the goal is to estimate relationships among taxa. (Many of our arguments apply equally well to estimation of relationships among other entities—e.g. genes—but may have to be transformed slightly for that purpose.) Finally, we recognize that the "best" method of analysis in a given instance may depend on the relative importance given to resolving power versus avoidance of error (e.g. see 20, 105).

PREVIOUS VIEWS

Much of the discussion in the literature has revolved around a contrast between the "consensus" of trees derived from separately analyzed data sets ("taxonomic congruence" sensu Mickevich, 68; see 50, 57) and what has been called the "total evidence" approach ("character congruence" sensu Mickevich, 68; 56; see below). In general, we consider a method to be a consensus method if the characters in two (or more) data sets are not allowed to interact directly with one another in a single analysis, but instead interact only through the trees derived from them. Given this definition, consensus includes methods such as Brooks parsimony analysis (113, 115) in which two or more trees derived from individual data sets are coded for parsimony analysis as a set of characters that reflect the underlying tree structure (also see 7, 29, 89a, 89b). It is important to recognize that the contrast with which we are primarily concerned—between separate and combined analysis—is not identical to this standard distinction between consensus and the "total evidence" approach. Choosing to analyze data sets separately does not necessitate the use of consensus trees, and proponents of separate analysis have not always condoned the use of consensus techniques (e.g. 10). Instead, separate analyses may be seen as a means of exploring possible disagreements among data sets. Similarly, a combined approach does not necessarily imply incorporation of all character data in a single analysis. Although our discussion of previous views is constrained somewhat by the particular methods that the authors were addressing, ultimately we consider alternatives beyond those typically discussed in the literature (see the "Framework" section).

The idea that one might want to perform separate analyses on subsets of the available data relies on the existence of different classes of evidence with respect to phylogeny estimation. To qualify as a distinct class of evidence, characters in a data set must, in a statistical sense, be more similar to each other than they are to characters in other data sets with respect to some property that affects phylogeny estimation by the given method (10; see also 19, 55, 96, 105).

Recently, Kluge & Wolf (57; also see 50) have questioned whether such classes of data actually exist. They urged cladists to "question artificial subdivisions of evidence because there is no reason to believe those definitions have discoverable boundaries" (p. 190). This is an important issue to address before proceeding further because, if classes of evidence do not exist, the justification for analyzing data sets separately disappears (72). Several studies suggest that some traditional distinctions (e.g. between molecular, morphological, and behavioral characters; 21, 22, 26, 27, 95) may not be relevant with respect to estimating phylogenies. Nonetheless, molecular studies have made it increasingly clear that distinct, identifiable classes of evidence do exist. The case for the existence of such classes has been articulated most strongly by Miyamoto & Fitch (72; also see 10, 22, 105), who use as an example sequences of the γ^1-globin and 12S rRNA genes, which differ in a number of properties (e.g. substitution patterns, overall rate of evolution, and frequencies of recombination, gene conversion, and gene duplication) that are likely to affect their behavior as indicators of phylogeny. Many other molecular examples could be given. For example, in cases where gene trees may differ from each other and from the overall species tree (see below), data from any pair of unlinked genes may be considered different classes of evidence. As entities with distinct locations in the genome, these different sequences have real, discoverable boundaries. (In stating that the boundaries are real and discoverable we are not suggesting that they are always clean; for example, since linkage varies continuously, classes of characters that are based on linkage relationships may have fuzzy boundaries.) We conclude that the argument of Kluge & Wolf is not universally valid.

Arguments in Favor of Separate Analysis

Bull et al (10; see also 49) presented a persuasive general argument against combining and in favor of separate analysis in some circumstances, which rests on the view that any estimate of phylogeny assumes a model of evolution. If, under the chosen method of estimation, data sets give phylogenetic estimates that are too different to be ascribed to sampling error (due to the limited number of characters and/or taxa sampled), then they must have been governed by different evolutionary rules. The significant difference between the phylogenetic estimates further indicates that the data sets differ in whether or how they violate the assumptions of the method (i.e. either one data set violates the assumptions and the other does not, or both violate the assumptions but in different ways). Bull et al argued that, if this is the case, the data sets should not be combined unless the method can be changed to account for the difference. In support of this argument they pointed out that assessing heterogeneity prior to combining data is an accepted procedure in science in general, applied, for instance, through analysis of variance and contingency tables.

One argument for separate analysis concerns simply the ability to quickly detect such heterogeneity in the form of areas of agreement and disagreement, which might highlight conflicts caused by natural selection, differential rates of evolution, hybridization, horizontal transfer, or lineage sorting. Comparison of separately analyzed trees has been seen as especially useful in identifying hybrids, wherein, for example, one may see conflicts between uniparentally inherited genomes (most mitochondria and chloroplasts) and nuclear genes and/or morphological characters (e.g. 90, 92). It has been argued that combined analysis may obscure significant patterns of congruence or conflict among characters (10, 19, 105), and it is true that we currently lack efficient methods for keeping track of the behavior of whole suites of characters in combined analyses. However, as noted by Chippindale & Wiens (14), proponents of combined analysis generally also carry out separate analyses to explore such possibilities.

Two arguments that have been made against combining data sets can be viewed as special cases of Bull et al's general argument. One concerns the impact of putting together a "bad" data set with a "good" one, where bad and good refer to the ability to accurately reflect true phylogenetic relationships (10). Within the framework of Bull et al, this can be viewed as an argument against combining a data set that violates the assumptions of the estimation method with one that does not (or at least violates the assumptions less drastically). The idea is that combining "bad" with "good" may actually give a less accurate estimate than using the "good" data by themselves. In the extreme this has led some authors to dismiss large classes of data. For example, some proponents of molecular approaches have written off morphology on the grounds that morphological features are subject to natural selection and therefore may be misleading due to convergent evolution (e.g. 98). As indicated above, there is no compelling evidence that molecular characters are in general better than morphological characters for estimating phylogeny, and the same can be said for some other traditional distinctions (21, 22, 27, 95).

Nonetheless, there do seem to be identifiable "good" and "bad" classes of evidence. For example, certain genes or gene regions may evolve much more rapidly than others; for some levels of divergence, the distribution of nucleotides among taxa for these rapidly evolving characters may be essentially random, whereas genes/gene regions that evolve more slowly may retain phylogenetic information (42, 59, 61, 73, 74). Bull et al (10) explored a similar case in a series of simulations that show circumstances under which the combined analysis is less likely to recover the true tree. In particular, they examined cases in which characters in one data set evolved at a significantly faster rate than did those in another, and they demonstrated instances in which the best results were obtained from the slowly evolving characters alone. They showed that this result can be obtained even when the estimation method is consistent for both data sets, i.e. converges on the truth as the number of characters is increased (35).

Barrett et al (5) suggested that in such cases characters might be weighted to reflect differences in evolutionary rate and then combined, a possibility also noted by Bull et al (10). Chippindale & Wiens (14) showed that, in the cases examined by Bull et al, such a weighting scheme would indeed render the combined analysis equal or superior to either of the individual analyses in recovering the true tree. However, in this example the true tree was known and various weighting schemes were applied to determine which worked best. Left open is the very real problem of determining an appropriate weighting scheme at the outset of an analysis (see "Framework" section).

The second special case of Bull et al's argument is the concern that one data set may have an inordinately great influence on an analysis, simply by virtue of having a larger number of characters (29, 45, 54). This argument hinges on the possibility that the larger data set might be misleading in some way that the smaller one is not; thus, like the previous argument, it relies on the potential for differential violation of the assumptions of the analysis. Donoghue & Sanderson (26) pointed out that the addition of even a small number of characters can have a significant impact on the outcome, and in practice it often emerges that the smaller of two data sets does have a substantial impact on the resulting trees (e.g. 28). However, the fact that the smaller data set might have an impact on the combined analysis does not in itself indicate lack of any swamping, and it seems certain that swamping must occur in at least some cases.

A number of authors have pointed to independence between data sets (and explicitly or implicitly, nonindependence within data sets) as the basis for arguments in favor of separate analysis (19, 55, 58, 72, 75, 96, 105). In this context, nonindependence within data sets does not necessarily imply functional or physical linkage, but only that characters within a data set are more likely to share some property relevant for phylogeny estimation than are characters in different data sets. We view this as an alternative way of expressing the idea of heterogeneity; if characters within a data set are less independent than characters in different data sets, then there is heterogeneity among data sets. Using the support for conflict among trees from different data sets as a means of assessing such independence (19, 96) can be seen as a test of heterogeneity. In general, we prefer the construction of Bull et al (10) because the term "nonindependence" conjures up functional or physical links between characters and is thus somewhat misleading.

Focusing on the idea of independence, however, does help to highlight important aspects of the debate. The possibility that different data sets give independent estimates of phylogeny underlies perhaps the most common argument in favor of consensus, namely, that areas of agreement among trees from separate analyses are especially likely to be true and are therefore conservative estimates of phylogeny (16, 19, 45, 55, 68, 72, 75, 86, 87, 96, 105). Because of their lack of independence, characters within a data set might as a whole tend to give misleading results. The same might be true of other data sets, but if there is independence among data sets, they should in general not mislead in the same way. Thus, areas of agreement are likely to represent real groups. [Of course the nature of the data sets must be considered here; data sets that are independent of each other in some ways might still share misleading properties (14).] The same reasoning has been applied in vicariance biogeographic studies (77, 80, 94, 99). Others have wondered whether there really are special advantages derived from assessing confidence through consensus (5, 50, 57), as opposed, for example, to combining the data in one analysis and performing bootstrap and decay analyses with the combined data set. However, for at least some of these latter authors (i.e. 50, 57), this criticism is tied to the idea that "evidence is evidence" (57, p. 190), i.e. that classes of evidence do not exist. The special advantages of consensus derive from the notion that such classes do exist.

The feeling that consensus trees might be safe estimates of phylogeny may have motivated Hillis's (45) suggestion that these be used in formulating classifications, where stability is often a concern (23). Barrett et al (5) challenged this belief by presenting hypothetical data for which a clade supported by strict consensus does not appear in an analysis of the combined data. Although this shows that the consensus result might not be sanctioned by all of the data analyzed together, resolving which approach (if either) is more likely to result in the true tree requires additional arguments (see 6, 14, 49, 76). The frequency of such occurrences with real data sets remains to be examined. It may be that consensus trees—at least those that include only clades found in more than one of the original trees (19)—will tend to contain fewer incorrect clades than do combined trees, at least in part because they generally make fewer claims about relationships.

Kluge & Wolf (57) criticized the argument that consensus analyses may be preferred due to their conservative nature by suggesting that safety in classification is not the goal of cladistics (13, 33). They pointed out that a completely unresolved tree would be maximally conservative; the absurdity of desiring such a tree is apparently meant to imply that conservatism cannot be a compelling criterion in constructing phylogenetic hypotheses. Instead they suggested, citing Popper (88, 89), that completely resolved hypotheses are to be preferred because they are bolder. Counter to the point regarding the completely unresolved tree, it can be argued that conservatism could influence one's choice of a method without leading to a maximally conservative tree. Specifically, one might want a tree in which all clades have received a certain level of support. The idea that bold hypotheses are to be preferred is undoubtedly valid in some contexts, but one can argue that in phylogenetic studies, particularly those in which phylogenies are used as assumptions of an analysis, what one wants are well-supported hypotheses (105).

Chippindale & Wiens (14) questioned the idea that independence between and nonindependence within data sets might favor separate analysis and consensus over combined analysis. They suggested (p. 280) that examples of this sort "do not involve weaknesses unique to data combination; rather they are cases in which the fundamental assumptions of parsimony analysis are violated" (for example, independence of characters and lineages). However, this criticism ignores important differences in the assumptions of consensus versus combined analysis. For example, to estimate relationships among species from several gene sequences (which may have different histories), separate analyses assume that the characters are independent estimators of the gene trees, not of the species tree, and consensus assumes that the different gene trees are not likely to differ from the species history in the same way. A combined analysis of such data, on the other hand, would assume that the characters are independent estimators of the species tree. It is precisely this kind of difference in the assumptions made by a consensus versus a combined analysis that may justify consensus in some circumstances.

Although advocates of separate analysis agree that independence among data sets is important, there is disagreement about the evidence that should compel one not to combine. Some authors suggest that data sets are combinable unless one can show that there is significant conflict among the phylogenies estimated from them (10, 19, 96). This view of combining as the default strategy may be motivated by the potential benefits of combining (see below). Miyamoto & Fitch (72), however, contended that, if there are biological reasons for believing that there is heterogeneity among data sets, then they should not be combined, regardless of the level of disagreement among the phylogenetic estimates. These latter authors placed great emphasis on corroboration of phylogenetic hypotheses by independent data.

A final positive argument for consensus is simply that certain techniques preclude combining some data sets (5, 30, 58, 97). For example, there are no methods for combining DNA hybridization or immunological distance data with a set of morphological characters or molecular sequences. In such cases, consensus is the only option available if one wishes to present a single estimate (5).

Arguments in Favor of Combined Analysis

Arguments for combining data can be divided into five categories:

1. A philosophical argument based on the idea of "total evidence."
2. Objections to arbitrariness in consensus methods.
3. The difficulty of choosing a scheme of partitioning.
4. The greater descriptive and explanatory power of phylogenetic hypotheses generated from the combined data.
5. The greater ability of combined analyses to uncover real phylogenetic groups.

Kluge (56; also see 5, 50, 57) argued in favor of combining data based on the philosophical principle that one should use the total evidence available (12, 41, 43). Conclusions based on all of the relevant evidence are certainly to be preferred. However, general admonitions of this sort are of limited value in choosing among the very particular alternatives in the case of phylogenetic analysis. Probably no current method of analysis takes into account all of the relevant evidence. Consensus methods may lose information in the intermediate step of summarizing individual data sets as trees. However, the standard "total evidence" approach ignores both the problem of data sets being systematically misleading (19), and any trees generated from distance data (58, 89a). The argument for combining based on total evidence stems from the goal of minimizing ad hoc assumptions of homoplasy counted on a character-by-character basis (56, 57). However, if instead the goal of phylogenetic analysis is to uncover the true phylogeny, then evidence beyond what is required to construct the most parsimonious combined tree may argue for separate analysis (19).

The second category of arguments for combining points out that combined analysis of all the data avoids the arbitrariness inherent in consensus analyses. One argument, presented by Kluge (56, also see 57), is that combined analysis circumvents the need to choose among the various methods of consensus, a choice that is characterized as fundamentally arbitrary. A related argument (57) points to the arbitrary nature of deciding how to summarize congruence among data sets when each one may result in two or more best trees. As noted above, one course of action would be to carry out separate analyses without proceeding to a consensus solution (cf 10). However, if the goal is to achieve an estimate of the true phylogeny, rather than simply to explore the conflicts among data sets, then such arguments need to be addressed. It may be possible to defend the choice of a particular consensus method based on the goals of a particular study (as opposed to a more general defense), although such arguments have rarely been articulated in practice. Thus, it might be argued that strict consensus (81, 105) provides the most conservative assessment of the agreement between trees, or that Adams consensus trees (1) are best at identifying taxa whose position is at odds in two or more trees (39, 45).

A related argument against consensus points out that these methods entail an arbitrary weighting of characters, because the individual trees are accorded equal weight in forming the consensus, regardless of the total number of characters that underlie them, or the number that support particular branches (17, 70; also see 26, 45). That is, the characters in a tree based on more characters will effectively be downweighted in comparison to those in a smaller data set. It might be argued that differential weighting of the characters in different data sets is warranted, but one would then need a further defense of weights that effectively reflect the number of characters in a data set. As Barrett et al (5) argued, weighting decisions should be explicitly defended, rather than being a passive and arbitrary outcome of the method of analysis. An alternative response is possible from those who argue that characters from different sources should not be combined under certain circumstances, namely, that in such instances one wishes to examine trees as bits of evidence rather than characters. For example, if one had confidence in several gene trees, one might wish to use these trees to infer a species phylogeny (83, 114). At this point, one has chosen to ignore individual characters, so their weights become irrelevant.

The third category of arguments is that there are many ways to partition all the data, and it is unclear how a particular scheme of partitioning can be justified (14). For example, DNA sequence data might be partitioned into separate genes or by position in the codon, and morphological data might be partitioned into larval versus adult, cranial versus post-cranial, or soft versus hard anatomy. A general response to this argument is that multiple partitions should be investigated to the extent that this is practical. One can draw an analogy with a multiple regression analysis: Factors should be added to the model if they add significantly to its accuracy. Nonetheless, exactly how one should examine the effects of multiple partitions in phylogenetic analysis remains problematic in many cases. However, with certain methods of analysis, the recognition of distinct classes will not always call for separate analyses, thus simplifying the general problem; for example maximum likelihood, neighbor-joining, and weighted parsimony may account for certain distinctions among classes of characters in a single analysis.

The fourth category of arguments in favor of combined analysis involves criticism of the efficacy of consensus methods as a means of producing phylogenetic hypotheses with descriptive and explanatory power. Miyamoto (70) highlighted the fact that a consensus of trees produced by separate analysis of each data set can be less parsimonious than the tree(s) from a combined analysis of the data. He argued that the consensus approach fails to take into account the underlying evidential support for the fundamental trees (i.e. the trees from separate analyses) and that consensus trees do not represent the best summary of the character information. He therefore recommended that consensus trees not be used in studies of evolution or as a basis for classification. Instead, the tree from the combined analysis is to be preferred as the most efficient summary of the available evidence. This message appears to have been widely appreciated, and consensus trees are now seldom used to portray character evolution. However, one can argue (105) that polytomies in consensus trees resulting from conflict among the fundamental trees, are "soft" polytomies (62), representing the various possible resolutions of the tree with their attendant character optimizations. Under this view, the only situation in which a consensus tree may be considered less parsimonious than a combined tree is when the consensus actually conflicts with the combined tree (as in the example in 5). Nonetheless, the consensus tree might still be considered a less efficient summary of the evidence if it includes more ambiguity than does the combined tree, as is likely (see below).

In a similar vein, Kluge & Wolf (57) pointed to the greater explanatory power of combined trees. Because consensus trees tend to be less resolved than combined trees, the argument goes, the former are worse at explaining the data than the latter. Here, explanatory power is judged by the ability of the phylogenetic hypothesis to explain shared character states as homologies or, equivalently, to avoid ad hoc assumptions of homoplasy (13, 33, 34; see 101 for objections to this characterization of explanatory power).

From our perspective these arguments based on maximizing descriptive and explanatory power suffer from the same deficiency as the "total evidence" argument. Specifically, they assume that maximizing potential instances of homology counted on a character-by-character basis is the sole criterion of descriptive or explanatory power. The best explanation of the data viewed in this narrow sense may not be the best explanation when all evidence relevant to estimating the phylogeny is considered. For example, consider a case in which separate analyses of several unlinked genes all strongly support the same tree, while one gene gives a different tree. A combined parsimony analysis could give the latter tree (if, for example, the one dissenting gene was larger than the other genes combined); however, although requiring the fewest ad hoc assumptions of character homoplasy, this tree would require ad hoc assumptions to explain why nearly all the genes give the same wrong estimate of phylogeny.

The argument in favor of combining that has perhaps received the most attention recently concerns the ability to uncover real phylogenetic groups. Hillis (45) was concerned with the possibility that two data sets/trees might not be positively at odds, yet standard consensus methods (such as strict consensus) might still yield an unresolved tree. He showed several examples involving differential resolution of various parts of the tree by different data sets (due perhaps to differences in the rate of evolution). As a means of circumventing this problem, he presented a method of consensus that was subsequently formalized by Bremer (8) as "combinable component consensus" (see 105) and is referred to in PAUP (106) as "semi-strict" consensus. The combinable component consensus is a tree that contains any clade found in any fundamental tree that is not contradicted by another fundamental tree. This consensus is thus always at least as resolved as the strict consensus, and often, more resolved. However, we note that the combinable component consensus lacks a property of both strict and majority-rule consensus that may be desirable in certain circumstances, namely, that any clade in the consensus tree must be found in at least two of the fundamental trees, thus reflecting agreement by (presumably) independent sources of data (19).

In the examples used by Hillis (45), it is assumed that the trees from the separate analyses show real and uncontradicted phylogenetic groups that are then hidden by the use of some consensus methods (e.g. strict consensus). However, it is also possible for a combined analysis to resolve conflicts among trees from separate analyses or even to reveal real groups not present in any of the separate trees. The underlying argument is that with an increasing number of characters the phylogenetic signal is more likely to assert itself over the noise, resulting in a more accurate estimate of the true phylogeny (5, 19). In essence, one is reducing sampling error by increasing the number of data points (see "Framework" section below). Simulation studies have shown that a greater number of characters translates into greater accuracy under a wide variety of circumstances (48).

Chippindale & Wiens (14) summarized a variety of cases in which novel phylogenetic results have been obtained in combined analyses, suggestive of this process. A particularly striking example is provided by an analysis of the angiosperm family Solanaceae by Olmstead & Sweere (79), based on three chloroplast DNA data sets. Each of the data sets results in a tree with some elements not seen in the trees derived from the other two data sets. The combined analysis of any two of the data sets yields a tree that has at least one of the unique elements found in the third data set (Figure 1). This implies that there is indeed signal for this arrangement present in these data, but that the signal is masked in some of the individual data sets and not recovered until they are combined. A similar example using artificial data was presented by Barrett et al (5). Another phenomenon consistent with the notion of enhanced

----

→

Figure 1 Strict consensus trees for 17 species of Solanaceae from parsimony analyses of each combination of two data sets from a total of three (*ndh*F and *rbc*L gene sequences, and restriction sites for the entire chloroplast genome). Relationships indicated in boldface were not found through separate analyses of either data set, but were found through analysis of the third data set. Modified from Olmstead & Sweere (79). See text for further explanation.

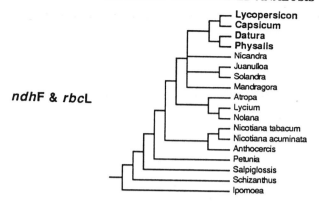

*ndh*F & *rbc*L

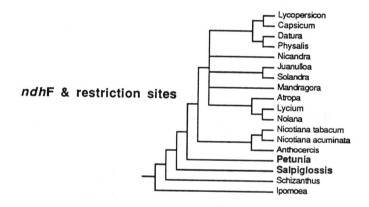

*ndh*F & restriction sites

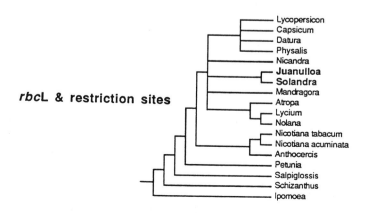

*rbc*L & restriction sites

signal that is observed frequently is a surprising increase in bootstrap support for a particular branch in a combined analysis. Such a case was highlighted in an analysis of seed plant relationships by Doyle et al (28) based on ribosomal sequences and morphological characters. A "eudicot" clade appeared in the trees based on morphology, but not in those based on ribosomal sequences. This clade was found in trees based on the combined data, but with even higher bootstrap and decay index values than were seen in the morphological trees. This suggests that signal is present for this clade in the ribosomal data, but that it is masked until the data sets are combined.

It should be noted that consensus methods that involve recoding trees as characters (7, 29, 89a, 89b, 113, 115) can also result in such "signal enhancement." However, because of the probable loss of information associated with summarizing a data set as a tree, it seems likely that the effect will be less frequent than when data sets are combined. Given that our goal is to uncover phylogenetic relationships, this loss of information may be the most general argument against consensus and in favor of combining data sets. To date there has been little attempt to examine the effect of this loss of information (beyond noting that consensus trees are often less resolved than combined trees).

Conclusions from Previous Views

Given that our goal in conducting phylogenetic analysis is to discover the true relationships among the entities in question, arguments against separate analyses based on the principle of total evidence (56, 57), or on preferences for trees that are bold (13, 57) or efficient descriptions and explanations of data on a character-by-character basis (57, 70), are not compelling. However, even given the above goal, legitimate arguments exist on both sides of the debate. The following points seem especially critical: 1. Combining data sets can give misleading results if there is heterogeneity among data sets. 2. Combining data sets can enhance the detection of real groups. In a given case, both combined and separate analyses can have advantages. Coupled with the fact that investigators differ in the emphasis they place on high resolution versus avoiding error, this gives much room for disagreement. Except in extreme cases—where the benefits of one of the approaches disappear—it is difficult to say what one should do. This is an important message that seems to have been lost on most participants in the debate.

Given the potential benefits of both approaches, one "solution" would be to perform both kinds of analyses (19, 60). However, a better approach may be to employ nontraditional methods that incorporate the advantages of combined and separate analyses simultaneously. In the next section, we highlight a number of such methods that attempt to include "the best of both worlds."

A CONCEPTUAL FRAMEWORK

The empirical result underlying most arguments in favor of separate analyses is the observation that phylogenetic estimates derived from different sources of data often disagree. As indicated above, Bull et al (10) and Huelsenbeck et al (49) argued that if this disagreement is greater than one would expect from sampling error, then combining the data sets is inappropriate unless the source of the conflict can be identified and accounted for by changing the method of estimation. In this section, we expand upon this framework and argue that the reason for conflict is critical in devising solutions. Although the problem of distinguishing among sources of conflict is an area in need of development (see below), here we assume that one can actually identify the reason(s) for conflict.

Reasons for Conflicting Phylogenies and Possible Solutions

We begin with the assumption that each data set is a sample of the results of a stochastic process that can be specified by a particular tree topology and character change model. The term "character change model" is used broadly here. Examples of differences in the model include overall rate differences, whether different types of character state changes have different probabilities [e.g. Jukes-Cantor (51) vs Kimura (53) models], and differences in branch lengths. Although such models have generally been applied only to molecular data, they could, in theory, apply to other kinds of data as well (e.g. morphological, behavioral, physiological). Under this construction, there are three general, not mutually exclusive, reasons why trees estimated from two or more data sets might differ: 1. sampling error; 2. different stochastic processes acting on the characters; 3. different branching histories.

SAMPLING ERROR   Data sets may be samples from the same tree topology and the same stochastic process and yet give different estimates of phylogeny due purely to sampling error. If sampling error is the only problem, then the data sets should be combined (10). By combining the data, one is increasing the sample size and therefore generally tending to reduce the error around the estimate of phylogeny (48). There seems to be little disagreement on the appropriate course of action in this case (but see 72 for a dissenting view).

This argument assumes that the data are drawn from a distribution such that the estimation method is consistent. If the method is not consistent, then increasing the amount of data may decrease the likelihood of obtaining the correct tree. However, this should not be taken as an argument against combining. The problem here is inconsistency, regardless of whether the data sets are combined or analyzed separately (14), and the solution is to change the method of estimation.

DIFFERENT STOCHASTIC PROCESSES   Data sets may be samples from the same tree topology, but the characters in each set might be affected by distinct stochastic processes (e.g. tend to have different rates of change). It seems clear that if the method of estimation is consistent for one data set and inconsistent for another, then combining the data may lead to problems. Perhaps less obvious is the fact that problems can arise even if the method is consistent for all the data sets. JT Chang (submitted) has shown that with data drawn from different stochastic process distributions, representing what is called a mixture model, even a generally consistent method such as maximum likelihood can be inconsistent. In fact, Steel et al (104) have shown that with general mixture models, for some parameter values every tree will have the same expected data set, which implies that estimation can be impossible regardless of the estimation method. Finally, the sampling error associated with stochastic models over a tree is dependent on the parameter values. Therefore, data drawn from different stochastic process distributions may have different sampling errors. Here the important point is that the sampling error for the combined data may actually be larger than that for some subset of the data (see 10).

A reasonable general strategy in this case is to find the tree (or set of trees) for which the fit combined over all the data sets analyzed separately is the greatest (11). For methods that use a quantitative optimality criterion (e.g. parsimony, maximum likelihood, least squares minimum evolution), the fit of the tree to the data can be assessed using the value of this criterion. The objective function for such methods takes the form $f(T,D)$, where T is the tree topology and D is the data set. The value of this function is the measure of goodness of fit of the tree to the data. For data sets $D1, D2,...,Dn$ the problem is to find the tree such that the combination of the objective functions $f(T,D1)$, $f(T,D2),..., f(T,Dn)$ is optimal.

This general framework leaves open the problem of properly combining the values of this criterion over all of the data sets. For maximum likelihood there is a straightforward solution: The combined optimality criterion should be the sum of the log likelihoods of all the data sets given the tree in question (11). The tree for which this sum is maximized has a straightforward meaning; it is the tree topology that maximizes the joint probability of all data sets.

Cao et al (11) introduced a heuristic use of this approach to estimate phylogenetic relationships among orders of Eutherian mammals. They obtained maximum likelihood trees for each of 13 mitochondrial DNA genes analyzed separately. The model parameter values were allowed to vary among the genes, which is critical because otherwise the analysis does not account for differences in the character change model. For each topology that was the ML estimate for any single gene, the log likelihood for each data set was computed. The tree for which the sum of the 13 log likelihoods was greatest was chosen as the best estimate of the phylogeny. We emphasize that the method is heuristic because it is possible for some topology other than those that were the ML estimates for single genes to have the greatest summed log likelihood (although not in the specific example analyzed in 11). Ideally one would want to consider all possible topologies.

The approach of Cao et al (see also 112 for a method with some similar properties), with the modification suggested above, has advantages over simply combining the data or using some form of consensus. The method takes into account differences in the character change model, which a simple combined analysis would not do (but see the discussion of weighting below). Differences in the reliability of the data sets should be reflected in their contributions to the overall sum of log likelihoods. The advantage of this approach over consensus is that it does not involve summarizing each data set as a tree, with the probable loss of information that such a summary entails.

A similar approach can be envisioned using parsimony. Under parsimony, the optimality criterion is to minimize tree length, i.e. the number of character steps required. An approach paralleling the maximum likelihood approach above would then be to find the tree(s) for which the length summed over all the data sets is a minimum. Under simple parsimony, length is computed for each character independently and in the same manner for all characters. Thus, the tree that minimizes the sum of lengths of two data sets is also the tree with minimum length for the combined data set. Therefore, it would seem that to achieve the same end as the maximum likelihood approach, we should simply combine the data.

The problem with this parsimony approach is that it does not account for differences in the stochastic processes affecting the characters. When these processes differ it is doubtful whether simple addition of tree lengths is an appropriate procedure. For example, if one data set contains only characters that evolve slowly and show no homoplasy, whereas a second data set contains rapidly evolving, highly homoplasious characters, the meaning of an estimated step in the two data sets is not equivalent. This brings us to what may be a reasonable parsimony solution when stochastic processes differ, namely, differential character weighting (5, 10, 14). Currently, the problem with this procedure under parsimony is that the theoretical justification for any particular scheme of weights has not been well developed (but see, e.g. 32, 36, 40 for some reasonable attempts).

DIFFERENT HISTORIES   Data sets may differ not only in their character change models but in the sequence of branching events they have experienced. In other words, the true tree topology might be different for each data set. Different histories can be the result of differential lineage sorting of ancestral polymorphisms (3, 109; Figure 2a) and/or hybridization/horizontal transfer between taxa (e.g. 52, 66, 100; Figure 2b). Lineage sorting is probably a very

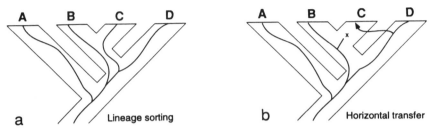

Figure 2 Two processes giving disagreement between gene trees and population trees. The curved lines represent the history of gene lineages and the straight lines represent the history of populations. (a) In the case of lineage sorting, an ancestral polymorphism persists through more than one speciation event, leading to a mismatch between gene and population trees. In this case the gene phylogeny links C and D as sister taxa whereas in the overall population history B and C are sister taxa. (b) In the case of horizontal transfer, discordance results from the transfer of a gene lineage from one population to another. The "x" indicates extinction of a gene lineage. Modified from Maddison (63). See text for further explanation.

common problem for closely related taxa (3, 74, 109), while hybridization/horizontal transfer, although perhaps less common, may be frequent in certain taxa (e.g. many plant groups—103). One of the main points of this review is that conflicts due to different histories present very different analytical problems than those due to different processes of character change. This point has generally been ignored or glossed over in previous considerations of the issue of combined vs separate analysis (e.g. 5, 10, 14, 19, 56).

If data sets have different branching histories, simply combining the data does not solve the problem. That approach is meant to reduce the error around the estimate of a single branching history by compiling as much information as possible from characters that have experienced that history. Allowing for different processes acting on different sets of characters will not solve the problem either; this can account for different expected patterns of character state distribution, but one is still assuming that all characters have experienced the same history. One can imagine a combined analysis of all the data that keeps track of the assignment of each character to its original data set and estimates a tree accounting for the possibility of different histories. However, the development of a generally applicable method of this kind does not appear to be imminent.

When faced with different histories there are at least three obvious ways in which one might represent phylogeny: 1. as the collection of individual histories, 2. as a reticulating tree, 3. as a non-reticulating (i.e. constantly diverging) tree representing the single dominant pattern among data sets. The dominant pattern might be interpreted either as the history of populations/species or just as a description of the central tendency of the distribution of all the histories (64). If there is no correlation among the different histories (i.e.

their topologies are no more similar than randomly chosen topologies), then perhaps only the first option makes sense. However, in most cases data set histories will be largely correlated, and 2. and 3. will be viable options. Here we focus only on the problem of obtaining the single dominant tree. With regard to reticulate trees, some promising methods have been developed (2, 4, 82, 91), but deciding on the number of reticulate connections to allow remains a major difficulty.

To formulate an approach to obtaining the dominant tree, we need to consider the nature of the disagreement among data sets. Specifically, it is important to know whether disagreements are confined to a few taxa or are spread over the tree. The taxon excision and pairwise outlier excision methods described below are designed to deal, respectively, with these two cases. The motivation behind both methods is to allow a single combined analysis while removing data that are misleading with respect to estimating the dominant tree.

Differences that are localized to a small part of the tree can occur, for example, because of lineage sorting associated with a particularly short internal branch near the tips of the tree or through hybridization between a small number of terminal taxa. In such cases, a reasonable procedure may be to excise the taxa involved in the conflict, an idea suggested by Rodrigo et al (93) with a slightly different method than that described here. [Funk (39) and Wagner (110) suggested excising possible hybrids in plants, although not in the context of combining data sets.] A major difficulty is determining which taxa to eliminate. One would begin by testing for significant incongruence among the data sets using a test of heterogeneity (see "Conclusions"). If there were significant incongruence, one would excise the taxon for which a different placement in the estimated trees was most strongly supported, as indicated, for example, by bootstrapping (37) or T-PTP testing (31). One would then cycle back to the heterogeneity test, continuing to excise taxa until there was no significant incongruence among the data sets. The resulting tree from a combined analysis would ideally represent only those taxa for which all data sets had experienced the same history.

Conflicts due to different histories may also be spread widely over the tree. This might occur, for example, if many internal branches were short enough to produce discordances due to lineage sorting, or in situations involving frequent horizontal transfer (e.g. 15). In such cases, excising the "offending" taxa may not be practical since many taxa would have to be eliminated. However, if we have multiple data sets, we can use an approach that relies on the assumption that the relationship between any two taxa will be distorted for only a minority of the data sets. D. Dykhuizen (personal communication to J. Kim) suggested such a method that involves computing distances between pairs of taxa separately for each data set. If a small number of data sets were involved in a branch rearrangement along the path connecting the two taxa, the distances for those data sets should be outliers in the distribution of distances for all the data sets. The data for such outliers could be eliminated in various ways. If one were using a distance method to estimate the tree, a combined distance for each pair of taxa could be computed excluding the outlier values. Similarly, using a character-based method, one could set the character states in a data set to missing in those taxa for which the data set was an outlier.

Implications

An important implication of this framework is that there is no simple answer to the question of whether to combine data sets for phylogenetic analysis. In some cases, combining all the data, with either equal or unequal weighting of characters, may be appropriate. However, as we have argued above, combining the data implicitly assumes that all data sets are products of the same branching history; when this assumption does not hold, simple combination of all the data may not be the best approach.

Another message of this section is that polarizing the debate into "combining all the data" versus consensus methods obscures some possible solutions. For example, both the taxon excision and pairwise outlier excision methods involve combining information from different data sets, but do not use all the data in the final analysis. Thus they avoid the potential loss of information associated with an intermediate step of summarizing data sets as trees (an advantage of a typical combined analysis), while mitigating the effect of misleading information (an advantage of some consensus methods). The maximum likelihood approach of Cao et al also avoids this loss of information, yet keeps the data sets separate and thus can account for different evolutionary processes acting on them. These methods demonstrate the possibility of incorporating some of the best aspects of traditional combined and separate analyses in one analysis. Nonetheless it is important to note that these methods need to be rigorously tested; it would be premature at this point to claim that they are necessarily superior to more traditional combined or consensus analyses.

A final point concerns the applicability of the framework to nonmolecular data. As suggested above, we believe that, in theory, the framework applies to these kinds of data as well. However, we currently have little idea about what divisions to make in such data. Some traditional divisions may have little relevance in the context of phylogeny estimation (14). An inability to identify relevant classes will clearly limit the kinds of approaches one can take. Our argument, however, is not that one must always partition data sets. Rather the point is that, when faced with biologically defensible divisions, the nature of these partitions should be taken into account in the choice of analytical methods.

CONCLUDING REMARKS

As we have emphasized, there are reasonable arguments in favor of combined analysis of data sets, as well as for separate analysis in some cases. Our goal has been to provide a general conceptual framework within which these options can be better evaluated. This exercise highlights the need for an expanded set of methods to cope with the variety of circumstances that may cause phylogenetically significant heterogeneity among data sets. As indicated above, such methods can incorporate the benefits of traditional combined and separate analyses. Much more effort is also needed to develop and validate specific tests for heterogeneity among data sets (see 60, 69, 93, 105, 107)—tests that are able to pinpoint whether the cause of heterogeneity is, for example, a global difference in evolutionary rate or a highly localized evolutionary event such as hybridization.

In the context of the framework presented above, biogeographic data and data from parasites of the focal organisms may be viewed as peculiar subsets of the problem of diverse data sets. In some cases these data may be products of the exact same branching history as the taxa under study. However, the fact that they may also represent histories somewhat different from the focal taxa should not automatically exclude them from analysis, since methods are available that can mitigate this problem.

A variety of other problems in phylogenetic systematics appear to have a fundamentally similar structure, in that they revolve around the inclusion or exclusion of data from an analysis. For example, there has been controversy over whether (and under what circumstances) fossils should be included in phylogenetic analyses along with information on living organisms (e.g. 24, 47, 78). Likewise there has been disagreement over whether to include the characters being investigated by ecologists and evolutionary biologists using comparative methods (e.g. 9, 18, 65; K de Queiroz, submitted), and whether to include distantly related outgroups (25, 67, 111). Some of the issues raised above are echoed in these controversies. For example, fossils and distant outgroups have been presented as causing more problems than they solve, akin to the idea that a "bad" data set may increase the total sampling error. The exclusion of characters under investigation by comparative biologists has been viewed as conservative, whereas it has been argued that including the characters gives the best estimate of the phylogeny; this distinction mirrors arguments for consensus and combined analyses, respectively. The similarity of these various problems suggests the possibility of a general conceptual framework that would clarify many of the issues involved.

ACKNOWLEDGMENTS

We are grateful to J Gatesy, C von Dohlen, and JJ Wiens for many useful comments on the manuscript. We thank B Baldwin, D Cannatella, J Silva, K de Queiroz, JA Doyle, T Eriksson, D Hillis, W Maddison, M Sanderson, D Swofford, A Yoder, the Phylogenetic Discussion Group at the University of Arizona, and N Moran's lab group for helpful discussion. A de Queiroz was supported by post-doctoral fellowships from the National Science Foundation and from the NSF-funded Research Training Group in the Analysis of Biological Diversification at the University of Arizona. MJ Donoghue was supported by an NSF grant (DEB-9318325) and a Mellon Foundation Fellowship from the Smithsonian Institution. J Kim was supported by an NSF grant (DEB-9119349) to M Kidwell.

Literature Cited

1. Adams EN. 1972. Consensus techniques and the comparison of taxonomic trees. *Syst. Zool.* 21:390–97
2. Alroy J. 1995. Continuous track analysis: a new phylogenetic and biogeographic method. *Syst. Biol.* 44:152–78
3. Avise JC, Shapira JF, Daniel SW, Aquadro CF, Lansman RA. 1983. Mitochondrial DNA differentiation during the speciation process in *Peromyscus. Mol. Biol. Evol.* 1:38–56
4. Bandelt HJ, Dress AWM. 1989. Weak hierarchies associated with similarity measures—an additive clustering technique. *Bull. Math. Biol.* 51:133–66
5. Barrett M, Donoghue MJ, Sober E. 1991. Against consensus. *Syst. Zool.* 40:486–93
6. Barrett M, Donoghue MJ, Sober E. 1993. Crusade? A response to Nelson. *Syst. Biol.* 42:216–17
7. Baum BR. 1992. Combining trees as a way of combining data sets for phylogenetic inference, and the desirability of combining gene trees. *Taxon* 41:3–10
8. Bremer K. 1990. Combinable component consensus. *Cladistics* 6:369–72
9. Brooks DR, McLennan DA. 1991. *Phylogeny, Ecology, and Behavior.* Chicago: Univ. Chicago Press
10. Bull JJ, Huelsenbeck JP, Cunningham CW, Swofford DL, Waddell PJ. 1993. Partitioning and combining data in phylogenetic analysis. *Syst. Biol.* 42:384–97
11. Cao Y, Adachi J, Janke A, Paabo S, Hasegawa M. 1994. Phylogenetic relationships among Eutherian orders estimated from inferred sequences of mitochondrial proteins: instability of a tree based on a single gene. *J. Mol. Evol.* 39:519–27
12. Carnap, R. 1950. *Logical Foundations of Probability.* Chicago: Univ. Chicago Press
13. Carpenter JM. 1988. Choosing among multiple equally parsimonious cladograms. *Cladistics* 4:291–96
14. Chippindale PT, Wiens JJ. 1994. Weighting, partitioning, and combining characters in phylogenetic analysis. *Syst. Biol.* 43:278–87
15. Clark JB, Maddison WP, Kidwell MG. 1994. Phylogenetic analysis supports horizontal transfer of P transposable elements. *Mol. Biol. Evol.* 11:40–50
16. Cracraft J, Helm-Bychowski K. 1991. Parsimony and phylogenetic inference using DNA sequences: some methodological strategies. See Ref. 71, pp. 184–220
17. Cracraft J, Mindell DP. 1989. The early history of modern birds: a comparison of molecular and morphological evidence. See Ref. 38, pp. 389–403
18. Deleporte P. 1993. Characters, attributes and tests of evolutionary scenarios. *Cladistics* 9:427–32
19. de Queiroz A. 1993. For consensus (sometimes). *Syst. Biol.* 42:368–72
20. de Queiroz A, Lawson R. 1994. Phylogenetic relationships of the garter snakes based on DNA sequence and allozyme variation. *Biol. J. Linn. Soc.* 53:209–29
21. de Queiroz A, Wimberger PH. 1993. The usefulness of behavior for phylogeny estimation: levels of homoplasy in

behavioral and morphological characters. *Evolution.* 47:46–60

22. de Queiroz A, Wimberger PH. 1995. Comparisons of behavioral and morphological characters as indicators of phylogeny. In *Phylogenies and the Comparative Method in Animal Behavior,* ed. E Martins. New York: Oxford Univ. Press. In press

23. de Queiroz K, Gauthier J. 1992. Phylogenetic taxonomy. *Annu. Rev. Ecol. Syst.* 23:449–80

24. Donoghue MJ, Doyle JA, Gauthier J, Kluge AG, Rowe T. 1989. The importance of fossils in phylogeny reconstruction. *Annu. Rev. Ecol. Syst.* 20:431–60

25. Donoghue MJ, Maddison WP. 1986. Polarity assessment in phylogenetic systematics: a response to Meacham. *Taxon* 35:534–38

26. Donoghue MJ, Sanderson MJ. 1992. The suitability of molecular and morphological evidence in reconstructing plant phylogeny. See Ref. 102, pp. 340–68

27. Donoghue MJ, Sanderson MJ. 1994. Complexity and homology in plants. In *Homology: The Hierarchical Basis of Comparative Biology,* ed. B Hall, pp. 393–421. San Diego: Academic

28. Doyle JA, Donoghue MJ, Zimmer EA. 1994. Integration of morphological and rRNA data on the origin of angiosperms. *Ann. Mo. Bot. Gard.* 81:419–50

29. Doyle JJ. 1992. Gene trees and species trees: molecular systematics as one-character taxonomy. *Syst. Bot.* 17:144–63

30. Eernisse DJ, Kluge AG. 1993. Taxonomic congruence versus total evidence, and amniote phylogeny inferred from fossils, molecules and morphology. *Mol. Biol. Evol.* 10:1170–95

31. Faith DP. 1991. Cladistic permutation tests for monophyly and nonmonophyly. *Syst. Zool.* 40:366–75

32. Farris JS. 1977. Phylogenetic analysis under Dollo's law. *Syst. Zool.* 26:77–88

33. Farris JS. 1979. The information content of the phylogenetic system. *Syst. Zool.* 28:483–519

34. Farris JS. 1983. The logical basis of phylogenetic analysis. In *Advances in Cladistics,* ed. NI Platnick, VA Funk, 2:7–36. New York: Columbia Univ. Press

35. Felsenstein J. 1978. Cases in which parsimony or compatibility methods will be positively misleading. *Syst. Zool.* 27:401–10

36. Felsenstein J. 1981. A likelihood approach to character weighting and what

it tells us about parsimony and compatibility. *Biol. J. Linn. Soc.* 16:183–96

37. Felsenstein J. 1985. Confidence limits on phylogenies: an approach using the bootstrap. *Evolution* 39:783–91

38. Fernholm B, Bremer K, Jornvall H. 1989. *The Hierarchy of Life: Molecules and Morphology in Phylogenetic Analysis.* Amsterdam: Elsevier

39. Funk VA. 1985. Phylogenetic patterns and hybridization. *Ann. Mo. Bot. Gard.* 72:681–715

40. Goloboff PA. 1993. Estimating character weights during tree search. *Cladistics* 9:83–91

41. Good IJ. 1983. *Good Thinking: The Foundations of Probability and its Applications.* Minneapolis: Univ. Minn. Press

42. Graybeal A. 1994. Evaluating the phylogenetic utility of genes: a search for genes informative about deep divergences among vertebrates. *Syst. Biol.* 43:174–93

43. Hempel CG. 1965. *Aspects of Scientific Explanation and Other Essays in the Philosophy of Science.* New York: Free Press

44. Hennig W. 1966. *Phylogenetic Systematics.* Urbana: Univ. Ill. Press

45. Hillis DM. 1987. Molecular versus morphological approaches to systematics. *Annu. Rev. Ecol. Syst.* 18:23–42

46. Hoch PC, Stephenson AG, ed. 1995. *Experimental and Molecular Approaches to Plant Biosystematics.* St. Louis: Monogr. Syst., Mo. Bot. Gard.

47. Huelsenbeck JP. 1991. When are fossils better than extant taxa in phylogenetic analysis? *Syst. Zool.* 40:458–69

48. Huelsenbeck JP, Hillis DM. 1993. Success of phylogenetic methods in the four-taxon case. *Syst. Biol.* 42:247–64

49. Huelsenbeck JP, Swofford DL, Cunningham CW, Bull JJ, Waddell PJ. 1994. Is character weighting a panacea for the problem of data heterogeneity in phylogenetic analysis? *Syst. Biol.* 43:288–91

50. Jones TR, Kluge AG, Wolf AJ. 1993. When theories and methodologies clash: a phylogenetic reanalysis of the North American ambystomatid salamanders (Caudata: Ambystomatidae). *Syst. Biol.* 42:92–102

51. Jukes TH, Cantor CR. 1969. Evolution of protein molecules. In *Mammalian Protein Metabolism,* ed. HN Munro, pp. 21–132. New York: Academic

52. Kidwell MG. 1993. Lateral transfer in natural populations of eukaryotes. *Annu. Rev. Genet.* 27:235–56

53. Kimura M. 1980. A simple method for

estimating evolutionary rate of base substitution through comparative studies of nucleotide sequences. *J. Mol. Evol.* 16: 111–20

54. Kluge AG. 1983. Cladistics and the classification of the great apes. In *New Interpretations of Ape and Human Ancestry*, ed. RL Ciochon, RS Corruccini, pp. 151–77. New York: Plenum

55. Kluge AG. 1984. The relevance of parsimony to phylogenetic inference. In *Cladistics: Perspectives on the Reconstruction of Evolutionary History*, ed. T Duncan, TF Stuessy, pp. 24–38. New York: Columbia Univ. Press

56. Kluge AG. 1989. A concern for evidence and a phylogenetic hypothesis of relationships among Epicrates (Boidae, Serpentes). *Syst. Zool.* 38:7–25

57. Kluge AG, Wolf AJ. 1993. Cladistics: What's in a word? *Cladistics* 9:183–99

58. Lanyon SM. 1993. Phylogenetic frameworks: towards a firmer foundation for the comparative approach. *Biol. J. Linn. Soc.* 49:45–61

59. Larson A. 1991. Evolutionary analysis of length variable sequences: divergent domains of ribosomal DNA. See Ref. 71, pp. 221–48

60. Larson A. 1994. The comparison of morphological and molecular data in phylogenetic systematics. In *Molecular Ecology and Evolution: Approaches and Applications*, ed. B Schierwater, B Streit, GP Wagner, R DeSalle, pp. 371–90. Basel, Switzerland: Birkhauser Verlag

61. Li W-H, Graur D. 1991. *Fundamentals of Molecular Evolution*. Sunderland, Mass: Sinauer

62. Maddison WP. 1989. Reconstructing character evolution on polytomous cladograms. *Cladistics* 5:365–77

63. Maddison WP. 1995. The growth of phylogenetic biology. In *Molecular Approaches to Zoology and Evolution*, ed. J Ferraris. New York: Wiley In press

64. Maddison WP. 1995. Phylogenetic histories within and among species. See Ref. 46, pp. 273–87

65. Maddison WP, Maddison DR. 1992. *MacClade: Interactive Analysis of Phylogeny and Character Evolution, Version 3.0*. Sunderland, Mass: Sinauer

66. McDade LA. 1992. Hybrids and phylogenetic systematics. II. The impact of hybrids on cladistic analysis. *Evolution* 46:1329–46

67. Meacham CA. 1986. More about directed characters: a reply to Donoghue and Maddison. *Taxon* 35:538–40

68. Mickevich MF. 1978. Taxonomic congruence. *Syst. Zool.* 27:143–58

69. Mickevich MF, Farris JS. 1981. The implications of congruence in Menidia. *Syst. Zool.* 30:351–70

70. Miyamoto MM. 1985. Consensus cladograms and general classifications. *Cladistics* 1:186–89

71. Miyamoto MM, Cracraft J, ed. 1991. *Phylogenetic Analysis of DNA Sequences*. New York: Oxford Univ. Press

72. Miyamoto MM, Fitch WM. 1995. Testing species phylogenies and phylogenetic methods with congruence. *Syst. Biol.* 44:64–76

73. Moritz C, Dowling TE, Brown WM. 1987. Evolution of animal mitochondrial DNA: relevance for population biology and systematics. *Annu. Rev. Ecol. Syst.* 18:269–92

74. Nei M. 1987. *Molecular Evolutionary Genetics*. New York: Columbia Univ. Press

75. Nelson G. 1979. Cladistic analysis and synthesis: principles and definitions, with a historical note on Adanson's Familles Des Plantes (1763–1764). *Syst. Zool.* 28:1–21

76. Nelson G. 1993. Why crusade against consensus? A reply to Barrett, Donoghue, and Sober. *Syst. Biol.* 42:215–16

77. Nelson G, Platnick NI. 1981. *Systematics and Biogeography: Cladistics and Vicariance*. New York: Columbia Univ. Press

78. Novacek MJ. 1992. Fossils as critical data for phylogeny. In *Extinction and Phylogeny*, ed. MJ Novacek, QD Wheeler, pp. 46–88. New York: Columbia Univ. Press

79. Olmstead RG, Sweere JA. 1994. Combining data in phylogenetic systematics: an empirical approach using three molecular data sets in the Solanaceae. *Syst. Biol.* 43:467–81

80. Page RDM. 1988. Quantitative cladistic biogeography: constructing and comparing area cladograms. *Syst. Zool.* 37:254–70

81. Page RDM. 1989. Comments on component-compatibility in historical biogeography. *Cladistics* 5:167–82

82. Page RDM. 1994. Maps between trees and cladistic analysis of historical associations among genes, organisms, and areas. *Syst. Biol.* 43:58–77

83. Pamilo P, Nei M. 1988. Relationships between gene trees and species trees. *Mol. Biol. Evol.* 5:568–83

84. Patterson C, ed. 1987. *Molecules and Morphology in Evolution: Conflict or Compromise?* Cambridge: Cambridge Univ. Press

85. Patterson C, Williams DM, Humphries CJ. 1993. Congruence between molecu-

lar and morphological phylogenies. *Annu. Rev. Ecol. Syst.* 24:153–88

86. Penny D, Foulds LR, Hendy MD. 1982. Testing the theory of evolution by comparing phylogenetic trees constructed from five different protein sequences. *Nature* 297:197–200

87. Penny D, Hendy MD. 1986. Estimating the reliability of evolutionary trees. *Mol. Biol. Evol.* 3:403–17

88. Popper KR. 1965. *Conjectures and Refutations: The Growth of Scientific Knowledge*. New York: Harper & Row

89. Popper KR. 1968. *The Logic of Scientific Discovery*. New York: Harper & Row

89a. Purvis A. 1995. A modification to Baum and Ragan's method for combining phylogenetic trees. *Syst. Biol.* 44:251–55

89b. Ragan MA. 1992. Phylogenetic inference based on matrix representation of trees. *Mol. Phylog. Evol.* 1:53–58

90. Rieseberg LH, Brunsfeld S. 1992. Molecular evidence and plant introgression. See Ref. 102, pp. 151–76

91. Rieseberg LH, Morefield JD. 1995. Character expression, phylogenetic reconstruction, and the detection of reticulate evolution. See Ref. 46, pp. 333–53

92. Rieseberg LH, Soltis DE. 1991. Phylogenetic consequences of cytoplasmic gene flow in plants. *Evol. Trends Plants* 5:65–84

93. Rodrigo AG, Kelly-Borges M, Bergquist PR, Bergquist PL. 1993. A randomization test of the null hypothesis that two cladograms are sample estimates of a parametric phylogenetic tree. *NZ J. Bot.* 31:257–68

94. Rosen DE. 1978. Vicariant patterns and historical biogeography. *Syst. Zool.* 27:159–88

95. Sanderson MJ, Donoghue MJ. 1989. Patterns of variation in levels of homoplasy. *Evolution* 43:1781–95

96. Shaffer HB, Clark JM, Kraus F. 1991. When molecules and morphology clash: a phylogenetic analysis of North American ambystomatid salamanders (Caudata: Ambystomatidae). *Syst. Zool.* 40:284–303

97. Sheldon FH, Bledsoe AH. 1993. Avian molecular systematics, 1970s to 1990s. *Annu. Rev. Ecol. Syst.* 24:243–78

98. Sibley CG, Ahlquist JE. 1987. Avian phylogeny reconstructed from comparisons of the genetic material, DNA. See Ref. 84, pp. 95–121

99. Simberloff D. 1987. Calculating probabilities that cladograms match: a method of biogeographic inference. *Syst. Zool.* 36:175–95

100. Smith MW, Feng D-F, Doolittle RF. 1992. Evolution by acquisition: the case for horizontal gene transfers. *Trends Biochem. Sci.* 17:489–93

101. Sober E. 1988. *Reconstructing the Past: Parsimony, Evolution, and Inference*. Cambridge, Mass: MIT Press

102. Soltis PS, Soltis DE, Doyle JJ, ed. 1992. *Molecular Systematics in Plants*. New York: Chapman & Hall

103. Stebbins GL. 1950. *Variation and Evolution in Plants*. New York: Columbia Univ. Press

104. Steel MA, Szekely LA, Hendy MD. 1994. Reconstructing trees when sequence sites evolve at variable rates. *J. Comp. Biol.* 1:153–63.

105. Swofford DL. 1991. When are phylogeny estimates from molecular and morphological data incongruent? See Ref. 71, pp. 295–333

106. Swofford DL. 1993. PAUP: *Phylogenetic Analysis Using Parsimony, Version 3.1*. Champaign: Illinois Nat. Hist. Surv.

107. Swofford DL. 1994. *Inferring phylogenies from combined data. Look before you leap*. Presented at Annu. Meet. Soc. Syst. Biol., Athens, Georgia

108. Sytsma KJ. 1990. DNA and morphology: inference of plant phylogeny. *Trends Ecol. Evol.* 5:104–10

109. Tajima F. 1983. Evolutionary relationships of DNA sequences in finite populations. *Genetics* 105:437–60

110. Wagner WH. 1980. Origin and philosophy of the groundplan-divergence method of cladistics. *Syst. Bot.* 5:173–93

111. Wheeler WC. 1990. Nucleic acid sequence phylogeny and random outgroups. *Cladistics* 6:363–67

112. Wheeler WC. 1991. Congruence among data sets: a Bayesian approach. See Ref. 71, pp. 334–46

113. Wiley EO. 1987. Methods in vicariance biogeography. In *Systematics and Evolution: A Matter of Diversity*, ed. P Hovenkamp, pp. 283–306. Utrecht: Inst. Syst. Bot.

114. Wu C-I. 1991. Inferences of species phylogeny in relation to segregation of ancient polymorphisms. *Genetics* 127:429–35

115. Zandee M, Roos MC. 1987. Component-compatibility in historical biogeography. *Cladistics* 3:305–32

Annu. Rev. Ecol. Syst. 1995. 26:683–704

ANTARCTIC TERRESTRIAL ECOSYSTEM RESPONSE TO GLOBAL ENVIRONMENTAL CHANGE

Andrew D. Kennedy

British Antarctic Survey, Natural Environment Research Council, High Cross, Madingley Road, Cambridge, CB3 OET, United Kingdom

KEY WORDS: Antarctica, climate change, polar ecology, ice recession, biogeochemical feedback

ABSTRACT

Geographical isolation and climatic constraints are responsible for the low biodiversity and structural simplicity of the antarctic terrestrial ecosystem. Under projected scenarios of global change, both limiting factors may be released. Alien species immigration is likely to be facilitated as modified ocean and atmospheric circulation introduce exotic water- and air-borne propagules from neighboring continents. Elevated temperature, UV radiation, CO_2, and precipitation will combine additively and synergistically to favor new trajectories of community development. It can be predicted that existing patterns of colonization, recruitment, succession, phenology and mortality will be perturbed with concomitant effects for ecosystem function through changes in biomass, trophodynamics, nutrient cycling, and resource partitioning. Soil propagule banks will play an important role through founder effects. Uniquely in Antarctica, many of the short-term consequences of global change will depend on the ecophysiological relationships of cryptogamic plants. However, in the long term, climatic warming will favor an increase in phanerogamic biomass since these species are currently excluded by the low cumulative degree-days $> 0°C$. It has been suggested that antarctic communities may be particularly vulnerable to global change: Their slow rate of development and restricted gene flow limit response to new conditions. However, vulnerability must be defined with respect to both the di-

0066-4162/95/1120-0683$05.00

rection and rate of change and it is likely that some perturbations will enhance the complexity and productivity of the biota, with negative feedback to the global carbon cycle. The chapter concludes with a discussion of institutional issues surrounding this topic.

INTRODUCTION

Antarctica has traditionally played a pivotal role in discussions of global environmental change (GEC). By virtue of its geographical isolation and unique meteorological conditions, this most southerly of continents provides unparalleled opportunities for monitoring globally integrated geophysical processes. The discovery of the stratospheric ozone hole in 1985 (42) best exemplifies this potential. Less popularized, but just as significant, were 160,000-year $\delta^{18}O$, CO_2, and CH_4 profiles drilled from the Antarctic Ice Cap that revealed the close relationship between climate and the chemical composition of the atmosphere (84). More recently, the rapid disintegration of the Wordie and Larsen ice shelves in response to atmospheric forcing has received much attention (35).

Against this background of high profile geophysical research, it is disappointing to learn that the biological consequences of global change in Antarctica have failed to attract much attention. The southern polar ecosystem has been dismissed on account of the low abundance, sparse distribution, limited diversity, and small total biomass of its communities (100), and for the lack of economic incentives (114).

This paper sets out to redress this imbalance by reviewing what is known and what is postulated about antarctic terrestrial ecosystem response to global environmental change. It is shown that the dismissive approach to Antarctica derives from a "static" view of the contemporary biota that is inappropriate in the context of environmental perturbations. Climatic constraints are largely responsible for the present-day ecosystem's diminutive properties; biotic interactions such as predation and competition play only a minor role (11). Consequently, any amelioration of ambient temperature, water availability, or light regime will encourage the development of new associations with concomitant effects on ecosystem function. The exotic communities growing under conditions of elevated temperature and humidity around antarctic volcanic fumaroles (107) demonstrate clearly the potential for change.

Here, the limited ecophysiological data available from Antarctica are combined with relevant arctic and cool-temperate research to explore ecosystem response to elevated temperature, UV radiation, atmospheric CO_2, and water balance. Emphasis is placed on describing "dynamic" responses to GEC that derive from the unique form and function of the continent's biota. Special sections are devoted to alien species introductions and biogeochemical feedbacks because these are likely to have long-term significance for the biosphere's equilibrium.

THE ANTARCTIC TERRESTRIAL ECOSYSTEM

For the purpose of this discussion, "Antarctica" is defined as the region south of the Antarctic Convergence.[1] Potential terrestrial habitats in this area cover 14×10^6 km^2, comprising one tenth of the Earth's land surface. However, at the present stage of the Holocene Interglacial, just over 98% of this land mass is covered by the Antarctic Ice Sheet and is largely unavailable for colonization (116). The remaining $\sim 2 \times 10^5$ km^2 of ice-free ground is restricted to coastal regions, islands, nunataks, and inland dry valleys.

Antarctica is a harsh continent by any frame of reference. It is the highest (mean elevation = 2,300 m), coldest (minimum temperature = $-89,6°C$), and geographically most isolated land mass on Earth. In winter, the land surface is essentially devoid of free water (59, 69). During summer, katabatic winds, frequent freeze-thaw cycles, low humidity, and inorganic soils combine to limit survival (77, 127).

The severe nature of Antarctica's climate, together with the continent's isolation from sources of immigrant propagules, explains many of the characteristics of its terrestrial biota. Communities consist of species typically preadapted for dispersal and able to exploit the low-energy environment. Terrestrial plants are predominantly cryptogamic and include algae, fungi, mosses, liverworts, and lichens, while the terrestrial fauna is limited to nematodes, tardigrades, rotifers, and protozoa, with a small number of microarthropods (10, 108). Only two flowering plants occur, the hairgrass *Deschampsia antarctica* and the pearlwort *Colobanthus quitensis*, both restricted to the west coast of the Antarctic Peninsula to 68°42'S. These taxa possess adaptations to survive the hostile conditions, including the sequestering of cryoprotectant substances to prevent freezing, the mobilization of chemical reactions at low temperatures, and the ability to adjust life cycles to exploit the ephemeral growth conditions (12, 18, 21, 64). Collectively, these traits demonstrate "A-selection" (53), defined as "favouring the conservation of adaptations to consistently and predictably adverse environments".

---

[1]The Antarctic Convergence is now technically referred to as the Polar Frontal Zone (PFZ). This term describes the marked discontinuity where cold southern waters meet the upper layers of the relatively warm northern oceans. South of the PFZ, terrestrial habitats are conventionally separated into three biogeographical regions: the sub-, maritime, and continental Antarctic [see (108) for delineation]. However, in the interest of brevity, all three regions are collectively considered here as "Antarctica."

At the landscape scale, antarctic terrestrial communities vary in their extent, structure, and floristic composition. At one end of the environmental spectrum are the moss carpet and turf formations that occur whenever meltwater and shelter coincide [see the sociations of Gimingham & Smith (50a).] At the other end, in areas of extreme stress such as the McMurdo Desert and Vestfold Hills, organisms rely upon strategies of avoidance and opportunistic growth: Cryptoendolithic communities exploit the protection from desiccation, wind-blown particles and freeze-thaw cycles provided within the matrix of porous sandstone rocks (47a), while cryoconite holes support cyanobacteria, notably *Phormidium* and *Nostoc,* together with occasional green algae and diatoms.

The severe nature of Antarctica's climate plays a controlling role in ecosystem function. Rates of production in closed bryophyte communities are limited to between 5 and 100 g m^{-2} yr^{-1} at xeric continental sites and around 300–650 g m^{-2} yr^{-1} for more favorable oceanic locations (108). Food webs are characterized by the absence of macroherbivores, so that most of the energy and materials assimilated by primary production enter a detritus, rather than a grazing, trophic pathway (32, 57). On the Antarctic Peninsula, the highest consumer level is occupied by the mesostigmatid mite *Gamasellus racovitzai* (79). Considerable energy loss occurs at the interfaces between these trophic platforms, and nutrients are removed at an early stage by freeze-thaw damage to plant cell membranes (63).

ECOSYSTEM RESPONSE TO GLOBAL CHANGE SCENARIOS

Global Warming

The Intergovernmental Panel on Climate Change (IPCC) predicts that within the next 100 years significant changes in the Earth's climate will be caused by anthropogenic emissions to the atmosphere. Radiant energy trapped by greenhouse gases (GHGs) such as carbon dioxide, methane, chlorofluorocarbons and nitrous oxide will induce a global temperature rise averaging 0.3°C per decade (range 0.2–0.5°C dec^{-1}) (60, 61).

How climate warming will translate to Antarctica is uncertain. The high albedo of snow and ice, together with meridional patterns of heat flux through oceanic and atmospheric circulation, suggests that polar regions are likely to experience a winter temperature rise of 2–2.4 times the global average and a summer temperature rise of 0.5–0.7 times that average (88). Less warming is expected in the Southern Hemisphere than in the Northern Hemisphere, a function of the smaller land surface available to respond to changes in radiative forcing (60). However, radiative scattering by anthropogenic sulfate aerosols over northern industrialized regions (116a) may partially offset this difference.

Recent meteorological observations from Antarctica suggest that global warming may already be tangible. Between 1955 and 1985, the Southern Hemisphere warmed by 0.3°C (60). In the subantarctic zone, mean daily temperatures have increased by ~1°C at Macquarie Island since 1949, a rate of 0.026°C yr^{-1} (1). On the Antarctic Continent, significant warming trends have been recorded at Casey and Dumont d'Urville Stations (2), while an increase of ~4°C in decadal mid-winter temperatures between 1947 and 1990 has been reported from Faraday Station on the Antarctic Peninsula (115). However, it is uncertain whether these disparate data reflect the enhanced greenhouse effect or natural climate variability (74).

ECOSYSTEM CONSEQUENCES

Ice retreat, habitats, and soil formation   The primary factor limiting the present-day distribution of Antarctic terrestrial communities is the scarcity of ice-free terrain. [The ice cap covers ~13.8 × 10^6 km^2 of Antarctica's ~14 × 10^6 km^2 surface area, and the resulting shortage of stable substrata restricts opportunities for colonization.] Although in the short term, global warming may lead to ice-sheet growth through enhanced snow accumulation (93), the major long-term effect of a warmer climate is likely to be the diminution of this ice cover (54). Greatest melting is likely to occur at northern latitudes of the maritime Antarctic where the relatively low density ice is sensitive to even small changes in summer temperature: An increase of >1°C may alter the balance between ice accumulation and ice loss. In this region, localized reduction in ice cover of up to 35% over the past 40 years has been reported (109). In the longer term, ice retreat is likely to occur around continental glaciers and coastal margins as well as inland around nunataks, dry valleys, and oases.

Melting of low-density snow pack and retraction of ice fields will alter both the rate and manner of rock weathering. The underlying bedrock will be exposed to a variety of mechanical and chemical weathering processes, including thermal and mechanical stresses, abrasion by wind- and water-borne particles, hydration-dehydration effects, atmospheric oxidation, hydrolysis, and solubilization of salts. Hall & Walton (54) suggest that two physical processes will play a particularly important role under ameliorated climate conditions: The formation of rock microfractures will facilitate the ingression of water, while the thawing of permafrost will increase the soil active depth. A substantial increase in the rate of soil formation may thus be expected.

The lithosols formed by frost shattering of rocks and fluvial-glacial sorting provide the basic substrata for primary colonization by a range of opportunistic taxa (125). Cyanobacteria, algae, yeasts, and microfungi stabilize and enrich this matrix by the secretion of mucilage, concentration of nutrients, and formation of polysaccharide meshes (128). Because it is biogenic, this modification process is limited by the rate of growth of microbial populations (127). Comparison of growth rates in culture (29) with measurements of temperature in fellfield soils (31) indicates that low temperature plays a limiting role. Global warming should thus accelerate soil development to create conditions favorable for the establishment of secondary communities.

Biogeographical distributions   A unifying principle of global biogeographical research is that boundaries to the distribution of species often coincide with isometric lines of climatological variables (52). In Antarctica, this relationship is reflected in the decreasing gradient of diversity that exists along a north-south latitudinal transect. Phanerogams cease at 68°42′S, mosses at 84°S, and the southernmost record for lichens is 86°S. Psychrophilic bacteria and yeasts extend to at least 87°21′S. Low summer temperatures, often acting through secondary factors such as water availability and freeze-thaw stress, are believed responsible for this latitudinal gradient (82, 108, 113). Climate warming can thus be expected to modify the spatial pattern of Antarctica's biota; species having environmental intolerance that currently limits them to the relatively benign coastal margins may extend their ranges latitudinally southward and altitudinally upward. Exotic species that cannot survive the contemporary environment may, subject to overcoming the constraints of immigration, also become established.

Evidence that global warming is affecting antarctic species distributions has been reported for taxa as diverse as bryophytes (112), phanerogams (45, 104, 109), and penguins (117). On Signy Island, expansion of lichens and mosses onto previously uncolonized glacial moraines has been interpreted in terms of milder summer temperatures (109). In the Argentine Islands, populations of the two native phanerogams, *Colobanthus quitensis* and *Deschampsia antarctica*, increased by ~5× and ~25× respectively between 1964 and 1990 (45). In the subantarctic zone, *Rumex crispus*, *Galium antarcticum*, and *Anthoxanthum odoratum* have all recently been recorded for the first time on Macquarie Island, while the ranges of *Poa annua*, *Ranunculus biternatus*, and *Montia fontana* have expanded on Heard Island (104).

The consequences of global warming for species distributions depend not only on the absolute range of thermal elevation but also on the rapidity of change. The rate of warming predicted by general circulation models (GCMs) is more than an order of magnitude faster than the most rapid climate changes of the recent geological past (62). Under such an accelerated transition, natural dispersal mechanisms may be unable to keep up with the shift in climate zones. Mosses and lichens possess aerodynamic dispersal stages that permit migration over greater distances than higher plants (19, 122), and in this respect the antarctic flora may be less affected than elsewhere. However, the apterous habit of antarctic invertebrates will restrict their dispersal, potentially leading to the formation of unique nonfaunal associations on isolated ground.

Community development   The vertical structure of antarctic plant communities is likely to develop heterogeneity under ameliorated thermal conditions. The contemporary vegetation is exclusively of low-growth form, a property that minimizes heat and moisture loss and maximizes the absorption of solar radiation. Lichens are generally crustose and dwarf foliose, mosses grow adpressed to the substratum, and of the two flowering plants, the hairgrass *D. antarctica* forms low mats while the pearlwort *C. quitensis* grows in compact cushions. Warmer temperatures should release plants from these constraints. Dwarf, cushion, and prostrate growth forms will give way to tall foliose, canopy, and hummock-forming habits.

The reproductive strategies of antarctic plants are likely to be affected by warmer conditions. At present, most antarctic cryptogams reproduce asexually by means of thallus fragments, bulbils, rhizoid buds, tubers, and gemmae. Sexual reproduction occurs in less than 25% of maritime antarctic bryophytes (notably the genus Schistidium) and in only one species of continental antarctic moss (108). However, warmer temperatures (particularly the cumulative degree-days $> 0°C$) have been shown in laboratory experiments to increase the production of sporangia and the germination of spores (82, 83, 109, 112). A comparable situation exists in antarctic phanerogams, for which increasingly successful seed production, germination, and seedling survival have been observed (38, 54, 113), with 15°–30°C optima for a range of seeds from the Kerguelen Archipelago (36). Recruitment should thus be promoted by warmer conditions.

Elevated temperature, acting as an environmental cue, may exert an adverse effect upon the life cycles of a variety of plants and invertebrates (19). Many antarctic species survive the winter in a desiccated inactive state, resuming metabolism and growth only when favorable conditions return (81). Thermoperiods provide temporal information from which to trigger such physiological changes (31). For example, germination of indigenous phanerogamic seeds occurs only after exposure to mild temperatures, a mechanism that Frenot & Gloaguen (47) suggest prevents young seedlings from being exposed to low temperatures during winter. Synthesis of cryoprotectants in *Alaskozetes antarcticus* is triggered by the repeated frost exposures that characterize the start of winter (21). Furthermore, the development of male and female gametophytes in the Bryophyta may require specific thermal budgets, so that warmer temperatures may yield a localized sexual imbalance (82).

An insight into the additive and synergistic consequences of climatic warming for antarctic community development has been gained by deploying greenhouses over barren fellfield soils (see e.g. 70, 112, 113, 128). Biological response to this heuristic treatment is dramatic. In one instance, moss cover of the substratum increased to 40% after 2 years while microbial crusts, comprising predominantly filamentous cyanobacteria, covered 74% of the soil after 3 years compared to only 5% coverage of controls (128). Soil invertebrate populations were similarly enhanced (70). Although the precise conformity of greenhouse microclimate to global change scenarios is unknown (71), the results clearly demonstrate the potential magnitude and direction of community response to elevated temperatures. Of particular significance is the colonization potential of the soil propagule bank (112): Plant development beneath sealed greenhouses indicates that a substantial reservoir of spores lies dormant in antarctic soils, ready to germinate given suitable climate conditions.

Ecosystem function   Global warming is likely to have special significance for the functioning of antarctic ecosystems. Under contemporary climate conditions, many vegetated habitats have a mean annual temperature <0°C and experience only a short summer growing season >0°C. This low cumulative heat sum severely limits rates of production, decomposition, and nutrient cycling. Further limitation is imposed by the transition of water from liquid (biologically available) to ice (largely biologically unavailable) occurring at ~0°C.

In common with species from other cold environments (9), a temperature increase of <5°C is likely to promote the net productivity of antarctic plants by stimulating photosynthesis more than respiration. The optimum temperature for carbon fixation is above that usually experienced under ambient field conditions. For example, on Signy Island (mean summer temperature = 0.8°C) positive net photosynthesis of *Deschampsia antarctica* and *Colobanthus quitensis* occurs between −5°C and 35°C with optima of 13°C and 19°C, respectively (40). At the same site, maximum net photosynthesis of *Umbilicaria antarctica*, *Drepanocladus uncinatus*, and *Polytrichum alpestre* is attained at 13°C, 15°C, and 5–10°C, respectively (26, 56). Reviewing lichen photosynthetic data from the Antarctic continent, Kappen (64) writes, "even at local noontime of warm summer days, thallus temperatures in the field were at least 2°C lower than the optimum temperature for photosynthetic rates." Regional warming of 1.5–4.5°C, as predicted by the IPCC, should thus increase net carbon assimilation.

Under conditions of greater productivity, the low nutrient content of antarctic fellfield soils may become limiting. This effect will be partially offset by increased rates of decomposition of soil organic matter; the microbial processes that mediate the transfer of organically bound nutrient elements to inorganic forms respond positively and often exponentially to temperature (16, 44). However, soil hydrological status also plays a limiting role, and at oceanic sites, waterlogging may slow nutrient mineralization through loss of soil aeration (91, 94, 114).

The effects of global warming on trophodynamics will not be confined to primary producers but will also extend to herbivores. In Antarctica these are entirely ectothermic; a positive correlation between consumption and temperature can thus be expected. For example, the oribatid mite *Alaskozetes antarcticus* has a sigmoid activity curve over the range –4 to 24°C with an optimum temperature between 12 and 16°C (12). For the isotomid collembolan *Cryptopygus antarcticus*, thermal elevation from 0 to 5°C increases food consumption by both juvenile and mature individuals (17). Few antarctic invertebrates feed directly upon bryophytes (11), so a potential increase in phanerogam abundance may shift the emphasis away from the existing plant-microbiota-invertebrate foodchain toward the plant-invertebrate pathway more characteristic of low latitudes.

Ultraviolet Radiation

The discovery of the Antarctic ozone hole in 1985 (42) provided the first real evidence for the globally pervasive nature of anthropogenic pollution. Chlorofluorocarbons emitted by industrialized nations in the Northern Hemisphere were seen to be capable of dispersing to the southern tip of the earth where, under the influence of low temperature and sunlight, their chlorine radicals caused the catalytic destruction of stratospheric ozone (5). Recent satellite measurements over Antarctica suggest that, from values >300 Dobson Units (DU) in the late 1950s, total column ozone had dropped to <100 DU by October 1993 (72).

Ozone is the only major atmospheric gas to absorb radiation at wavelengths <300 nm. Accordingly, it regulates the amount of ultraviolet radiation (UVR) reaching the earth's surface. The efficiency of absorption varies with the wavelength being considered: At 315 nm, an ozone reduction from 315 to 110 DU is accompanied by a factor of 2.2 increase in surface irradiance whereas at 305 nm, the corresponding enhancement is a factor of 14 (46). The biological UV dose (an estimate of UVR damage) is also wavelength dependent and is defined as the integrated product of the action spectrum of the particular biological component being studied (DNA, proteins, etc) with the incident spectral irradiance (27).

ECOSYSTEM CONSEQUENCES

Enhanced UVR resulting from ozone depletion has been described as a potential threat to all antarctic ecosystems (129). Absorption of UV-B (280–315 nm) and UV-C (100–280 nm) by DNA, RNA, and proteins causes, in experiments on lower latitude organisms, mutagenesis and changes in membrane structure and interferes with normal metabolic functioning. Long-term biological consequences include morphogenetic aberrations, impaired growth, restricted mobility, a decrease in chlorophyll, lipid, and protein content of leaves, and an increase in stomatal resistance [see Young et al (131) and references therein]. However, the lack of research into the photobiology of endemic antarctic species makes prediction of the polar consequences of ozone depletion difficult: Even slight variations in susceptibility may alter the effects of UVR in as yet undetermined ways.

Changes in the taxonomic composition of antarctic terrestrial communities are likely to be the most noticeable consequence of increased UVR. Susceptibility to UV damage is species-specific (123), so dominance patterns will change as UV-tolerant taxa are favored by the new light conditions. The degree of UV sensitivity relates to the efficiency of a species' repair mechanisms (photoreactivation, excision, and recombination) and to the existence of avoidance strategies (behavioral migration and pigment production) (67). A number of antarctic taxa synthesize protective pigments; these are either extracellular, as in the case of scytonemin in the sheath matrix of *Nostoc commune* (48), or intracellular, as in the case of the mycosporine-like amino acids in the cytoplasm of *Gleocapsa* sp. (49). Extracellular pigments act as sunscreens filtering out UVR, whereas intracellular pigments are thought to function as quenching agents, dissipating excess energy that would otherwise generate cytotoxic singlet oxygen (129). The relative pigment-synthesizing abilities of antarctic taxa may exert a structuring role in future community succession.

An increase in UVR is likely to modify ecosystem function. Primary producers depend on photosynthetically active radiation (PAR: 400–700 nm) for carbohydrate synthesis and so are confined to the euphotic zone where UVR receipt is greatest. Located at the base of the food chain, they play an important role in trophodynamics. In Antarctica, photoinhibition may already be a factor influencing the net productivity of such taxa (3). Growth impairment of the cyanobacteria *Phormidium murayi* and *Oscillatoria priestleyi* by UVB has been demonstrated experimentally (123), while growth of *Chlorogloeopsis* sp. resumes only after accumulation of scytonemin in the cellular envelopes (50). Comparatively little research has been performed on antarctic cryptogams, but these are likely to be particularly sensitive on account of their adaptation to shade (19). Photoinhibition has been demonstrated in *Grimmia antarctici* and *Schistidium antarctici* growing on the Antarctic continent (65, 99) while a decrease in the photosynthetic rate of *Usnea sphacelata* has been recorded under an ambient photon flux density of 600 μmol m^{-2} s^{-1} PAR (64). In comparable arctic habitats, *Sphagnum* species grow optimally in less than full sunlight (55).

Any decrease in productivity caused by UVR is likely to feed through to higher trophic levels. Consumers will face reduced food resources, and this will counteract the stimulatory effects of global warming. Trophic relationships may also be influenced directly by the effect of UV radiation on secondary metabolism (89): Increased flavonoid synthesis is often accompanied by increased production of unpalatable terpenoids, resulting in reduced invertebrate feeding upon plant hosts. Furthermore, increased UVR has been shown to alter competitive interactions between plants and to change levels of allelochemicals in plant tissues (76, 119).

Increased Atmospheric CO_2

The Antarctic Ice Cap provides an excellent record of changes in atmospheric composition during the recent geological past. Analysis of gas bubbles occluded at different depths reveals that CO_2 increased rapidly from ~200 ppmv at the end of the Wisconsin glaciation to ~260 ppmv in the Holocene (33). During the Industrial Revolution, CO_2 concentrations increased further, from 280 ppmv around the year 1750 to 345 ppmv in 1984 (96). Under a stabilized emissions scenario, CO_2 concentrations are expected to reach 415–480 ppmv by the year 2050 and 460–560 ppmv by the year 2100. A "business as usual" scenario suggests an increase to 830 ppmv over the same period (61).

ECOSYSTEM CONSEQUENCES
Elevated CO_2 exerts both direct and indirect effects upon ecosystem structure and function. A range of responses at the individual level (metabolism, stress tolerance, resource allocation, water-use efficiency, phenology) and community level (species composition, carbon and nutrient cycling, competitive and plant-herbivore interactions) has been reported from forest, salt marsh, prairie, and agricultural ecosystems [see Bazzaz (8) and references therein]. However, minimal data for antarctic species exist, and to what extent lower latitude observations may be extrapolated is unknown.

The best-documented consequence of elevated CO_2 is the "CO_2 fertilization effect" whereby net photosynthetic capacity is enhanced, leading to increased growth and primary production. The mechanism for this response involves reduced competition from O_2 for the primary carboxylase Rubisco, thereby increasing its activation and suppressing photorespiration. Although the traditional dogma that C_3 species are favored at the expense of C_4 species has recently been challenged (98), the bulk of experimental evidence still suggests that increased CO_2 will benefit the more highly conserved mechanisms of carboxylation and oxygenation found in the C_3 pathway. Antarctic plants fix carbon entirely by C_3 photosynthesis (2), and on this basis, elevated atmospheric CO_2 should lead to increased productivity.

Realization of the CO_2 fertilization effect will depend upon water, nutrients, light, and other growth resources being nonlimiting. Plants cannot accelerate their rate of tissue synthesis without simultaneously using greater amounts of these resources. In polar nutrient-poor soils, increased growth in one year may decrease nutrient availability the next year, thus limiting any long-term increase in productivity (43). However, warmer temperatures may offset this limitation by stimulating the release of nutrients through microbial decomposition and rock weathering (16, 54, 94).

Changes in resource partitioning from above- to below-ground biomass have been reported for a variety of plants grown at low latitudes (75). The increased root:shoot ratio may be explained on the basis of optimal allocation of resources: With carbon easier to obtain, the plant invests more materials in root tissue to enhance nutrient foraging (98). In Antarctica, increased root growth may occur in the phanerogams *D. antarctica* and *C. quitensis*, but this effect cannot be translated to cryptogams, which lack true root systems. Potentially of greater significance for such taxa will be changes in resource allocation favoring reproductive output: CO_2 modifications involving flower phenology and longevity, and seed number, size, and nutrient content have been reported from higher plants (8), including components of arctic tundra (120), but no data for antarctic cryptogams are available.

At the community level, a variety of changes may result from increased atmospheric CO_2. Prominent among these will be tissue-chemistry inspired modifications in interspecific interactions between plants and their herbivores, pathogens, and symbionts (7). Although high CO_2 concentrations are unlikely to influence invertebrates directly (19), increased plant C:N ratios will force herbivores to consume a greater volume of plant material to gain the same amount of protein (41, 87). In Antarctica, the quantity of primary production routed to a grazing (as opposed to detrital) trophic pathway may thus increase. The net effect upon invertebrates will be a reduction in growth rates, increased larval mortality, and a lowering of overall fitness (41). Infection by parasitic and symbiotic organisms may also increase, yielding increased N_2 fixation and accelerated plant growth (19).

At the ecosystem level, changes in plant tissue chemistry are likely to affect rates of organic matter decomposition. Synthesis of carbon-based secondary metabolites such as lignin, terpenes, and condensed tannins is believed to be promoted by the high carbohydrate content of plants grown at elevated CO_2 (44). These compounds slow the breakdown and mineralization of nutrients by detrivores and soil microbiota (7, 58). In Antarctica, the potential for this limitation will depend on the chemical composition of secondary metabolites synthesized by cryptogams in response to elevated CO_2.

Precipitation and Water Balance

Atmospheric greenhouse gas concentrations exert an effect not only on surface temperature but also on precipitation. Cold air is incapable of supporting a high moisture content so that for a doubling of CO_2, the global concentration of water vapor is expected to increase by between 20% and 33% (121).

Penetration of this moisture-rich air into high latitude regions will be facilitated by the retraction of sea ice. On this basis it has been estimated that, for every 1°C gained, precipitation in polar regions may increase by 5% to 20% (102). Rain and snowfall over the Antarctic continent currently average a water equivalent of between 146 and 192 mm yr^{-1} (101): An increase in the range 7.3 to 38.4 mm yr^{-1} $°C^{-1}$ may thus be expected.

ECOSYSTEM CONSEQUENCES   The availability of water is a fundamental factor limiting the distribution and composition of life in Antarctica (69). Under contemporary climate conditions, katabatic winds compressed and heated adiabatically during the descent from the polar plateau depress relative humidity, causing water lost by ablation to exceed water gained by precipitation; water activities [a_w, sensu Horowitz (59)] across much of the continent are among the lowest recorded on Earth. In oceanic regions, meltwater ponds and drainage creeks lack sufficient spatial and temporal permanence to satisfy the growth requirements of all but the most opportunistic taxa. The biological consequences are visible throughout Antarctica, where fine-scale distributions correlate closely to gradients in meltwater, seepage, and upwelling (78).

An increase in water availability is likely to promote the establishment of increasingly complex communities. Liverworts, mosses, and other dehydration-susceptible taxa will be favored at the expense of xerophytic lichens, yeasts, and algae. Smith (109) provided a flow diagram showing trajectories of community development under contrasting moisture conditions: Moss carpets and turves prevail at the hydric end of the scale, grading into ephemeral mosses and algae under unstable conditions, and microlichens at the xeric end. Global change can be expected to direct succession along these axes. A contemporary analogue exists beside the Canada Glacier in Southern Victoria Land: *Bryum argenteum* occurs in areas of flowing water and seepage, while *Pottia heimii* dominates the edge of the meltwater zone; beyond these sites the cyanobacteria *Nostoc* grows opportunistically wherever traces of moisture exist (103). On the Antarctic Peninsula, the phanerogams *Deschampsia antarctica* and *Colobanthus quitensis* are themselves associated with moister soils (111).

Changes in moisture availability are likely to modify ecosystem function. Plant productivity, soil respiration, organic matter, and nutrient cycling at high latitudes are all sensitive to soil water content (57). In arctic tundra, decomposition rates are slow and independent of temperature under conditions of low moisture (≤ 20% of dry mass) but become increasingly temperature dependent as soil water increases. Under saturated conditions, soil aeration becomes a limiting factor (94). In Antarctica, although dry and wet moss systems have similar ecological efficiencies and levels of net primary production, the turnover of dry matter is completely different. Decomposition and leaching proceed more quickly in wet habitats, with little accumulation of dead organic matter (32).

From these observations, it may be predicted that increased moisture availability will have a net stimulatory effect on antarctic ecosystem structure and function. However, three negative impacts may disrupt the simplicity of this relationship. First, increased groundwater in the presence of low temperatures will facilitate soil cryoturbation, causing the formation of periglacial features such as frost polygons, stripes, and solifluction lobes (88). Soil instability is already an important factor limiting the succession from initial pioneer species to climax vegetation (111), and under wetter conditions, this limitation may be enhanced. Second, excess water may cause the waterlogging of antarctic soils, inflicting direct mortality upon the biota: Such a phenomenon has been witnessed on Signy Island where Davey (30) recorded a marked decrease in soil chlorophyll content at the time of spring thaw. Third, increased water availability is likely to reduce the cold hardiness of a variety of cryptogams and invertebrates (22, 68) through cryoprotectant dilution.

It must be emphasized that greenhouse gas–inspired climate change will not necessarily lead to greater water availability at all sites in Antarctica. Retreating glaciers and snow fields may yield locally xeric conditions (116). Where melting of permafrost leads to the exposure of the underlying substrata, soil permeability will be increased, facilitating drainage and promoting desertification (19). Lacking roots or internal vascular structure, the bryophytes that comprise the bulk of the antarctic vegetation will be particularly vulnerable to desiccation stress (82, 118).

SPECIES IMMIGRATION

The Antarctic Continent has been isolated from other landmasses for over 25 million years. During this period, the immigration of alien species has been restricted by the geophysical barriers of the Southern Ocean and South Polar Air Vortex (66). Ice scouring during a succession of glacial maxima makes it unlikely that many species have survived from what the fossil record suggests was a rich and diverse preglacial epoch (125). Consequently, geographical isolation has probably played an important role in structuring the composition of the present-day biota.

The advent of polar exploration in the eighteenth century opened new opportunities for species immigration. Whale and seal hunters imported a variety of exotic organisms to the subantarctic islands, including angiosperms, cryptogams, invertebrates, mammals, and some land birds (14, 124). The long-term impacts of these introductions can be seen on the Kerguelen and Crozet Islands where cats, reindeer, and rats have exploited the lack of competitors, predators, and disease to radically modify the terrestrial communities

(24). In the maritime and continental Antarctic, alien species have been introduced both intentionally (39) and accidentally (13): Several species of grass, the chironomid midge *Eretmoptera murphyi*, and a range of microorganisms have all been established in the recent past (13, 20, 95). However, the adverse environment has prevented these organisms from expanding their ranges to displace the indigenous biota.

There is a genuine concern that global change will further disrupt natural barriers to alien species immigration. Changing oceanic and atmospheric circulation, together with the growing volume of air and sea transport, are likely to provide new opportunities for dispersal (66, 104, 110, 114). Once established, climate warming, increased precipitation, and greater habitat availability will encourage population growth so that alien taxa, which hitherto would not have survived the adverse environment, will no longer be so constrained [see Chown & Language (25) for a recent example involving Diptera and Lepidoptera]. The net result is that endemic antarctic species may face exotic predators and competitors for the first time.

Comparison with lower latitude ecosystems demonstrates the dramatic effect that immigrant species may have on existing community structure (see e.g. 28, 37). Local extinction of species, fluctuations in relative abundance, and evolution in plant macrostructure are three impacts of alien polyphagous herbivores recorded from the subantarctic islands (24). On Marion Island, *Agrostis stolonifera* and *Sagina apetala* have displaced endemic taxa through competition, while herbivory by the moth *Plutella xylostella* has modified vegetation morphology (104). The simplicity of antarctic ecosystems renders them acutely vulnerable to such introductions: the absence of predators, pathogens, and competitive taxa means that natural regulatory mechanisms are weak. Furthermore, once alien species are established, artificial control and eradication measures will be complicated by the continent's size and remote location.

FEEDBACK TO GLOBAL CHANGE

Any change in the spatial area of the Antarctic Ice Cap is likely to exert positive feedback on GEC. The polar plateau reflects 40–80% of incident radiation, and this high albedo acts to sustain the low temperature environment (85). Diminution of the Antarctic Ice Cap, with exposure of low albedo snow and rock, will lead to greater absorption of solar irradiance and positive feedback to climate warming. Conversely, an increase in ice cover, as predicted by some climatologists (93), will promote surface cooling. Some buffering of change may be caused by the latent energy needed to convert water between the solid and liquid states (melting 1% of terrestrial ice in 100 years is estimated to require a global average of 0.06 W m^{-2}). However, this input is minimal compared to CO_2-induced heat fluxes and will not significantly offset the positive feedback caused by changes in surface albedo (116).

At the biological level, positive feedback between the biosphere and geosphere may result from the release of GHGs from melting permafrost (51). In the Arctic, CO_2 emissions from this source have recently been estimated at $53 - 286$ g C m^{-2} yr^{-1} (97). CH_4, with a greenhouse potential 21 times that of CO_2, is also being released by the melting of methane gas hydrates within permafrost and by the action of methanogenic bacteria; combined outgassing of CH_4 from northern tundra and boreal wetlands is estimated at 40 Tg yr^{-1} (91). In Antarctica, data for GHG emissions from fellfield soils are inadequate to evaluate whether the continent is a net source or sink of atmospheric carbon. Yarrington & Wynn-Williams (130) recorded mean summer methane emissions of 1.24 mg C m^{-2} day^{-1} (range: 0.02–15.2 mg C m^{-2} day^{-1}) from a wet moss-carpet on Signy Island, while Hall & Walton (54) suggest that, in the maritime Antarctic, up to 1 m depth of permafrost may already have melted in response to warmer temperatures. However, the balance between these losses and photosynthetic carbon fixation is unknown.

Terrestrial ecosystems play an important role in the exchange of moisture between land surfaces and the lower atmosphere. Changes in the structure and abundance of vegetation can affect hydrological processes through boundary layer dynamics, circulation patterns, convective activity, and rainfall (105). In Antarctica, any increase in the stature of fellfield communities will amplify local surface roughness, leading to increased rates of water loss and negative feedback to the hydrological cycle (90). Any expansion of their spatial distribution will have the same effect. However, vegetation also absorbs solar radiation and insulates against heat loss, so any increase in above- and below-ground biomass may also promote soil saturation through permafrost melting.

INSTITUTIONAL ISSUES

Antarctica is unique among continents in being controlled by a single institutional framework, the Antarctic Treaty (AT). This agreement, signed in Washington, DC in December 1959, freezes territorial claims and devotes all land south of 60°S to peaceful purposes. Although environmental protection is not referred to in the AT per se, the document cites the "preservation and conservation of living resources" as a subject about which the signatories should consult. Since 1961, various environmental resolutions have been passed, including the Agreed Measures for the Conservation of Antarctic Fauna and Flora (1964), a Recommendation for the designation of Sites of Special Scientific Interest (1975), and a Recommendation on Environmental Impact Assessment (1987). Hypothetically, these protocols provide a comprehensive system for controlling human impacts on the antarctic environment.

Unfortunately, the enthusiasm with which such measures have been drafted has not always been reflected at the implementation level. Auburn (6) and Manheim (86) outline some of the discrepancies. A fundamental weakness of the AT is that member nations are self-regulating: breaches of conduct are unlikely to be detected except by the perpetrating nation itself. To quote Bonner (15):

> The problem was that the (treaty) regulations were too often ignored. Famously, a station was built in one of the first Specially Protected Areas. Even quite recently, buildings have been put up without the circulation, or even preparation, of an environmental evaluation. Birds have been introduced and released in the Antarctic to celebrate the opening of a station. Thoughtless and pointless road-building through fragile vegetated areas has been permitted by a Party which was one of the main proponents for a more rigorous environmental convention....

In response to public calls for greater environmental responsibility, the AT Parties held a Special Consultative Meeting in Madrid in October 1991, where a Protocol on Environmental Protection was drafted to supersede all previous legislation [see (4) for summary]. Although not yet fully ratified, the protocol declares Antarctica a natural reserve and commits signatories to the "comprehensive protection of the antarctic environment and its dependent and associated ecosystems." Specific regulations cover mineral exploitation, waste disposal, alien species introductions, and endangered species. Summing up the significance of these measures, Bonner (15) concludes that "no equivalent part of the earth is better protected, in theory and in practice, than Antarctica."

While Bonner's statement is undoubtedly true for local pollution sources, nothing in either the original 1959 treaty or in the 1991 environmental protocol protects the continent from anthropogenic pressures originating north of 60°S. In this respect, Antarctica remains totally at risk. Approximately 95% of industrial CO_2 emissions originate in the Northern Hemisphere while an even greater proportion of CFCs are manufactured there (61). In the long term, it may be argued that these substances pose a far greater threat to antarctic ecology than do local pollution sources. The best-designed AT legislation will be ineffectual if global environmental change causes the collapse of terrestrial food chains and biodiversity loss.

ACKNOWLEDGMENTS

I wish to thank my colleagues at the British Antarctic Survey for their support and encouragement in embarking upon this review, and the CSIRO Marine Laboratories in Western Australia for providing facilities to write whilst overseas.

Literature Cited

1. Adamson DA, Whetton P, Selkirk PM. 1988. An analysis of air temperature records for Macquarie Island: decadal warming, ENSO cooling and Southern Hemisphere circulation patterns. *Pap. Proc. R. Soc. Tasman.* 122:107–12
2. Adamson H, Adamson E. 1992. Possible effects of global climate change on antarctic terrestrial vegetation. See Ref. 34, pp. 52–62
3. Adamson H, Wilson M, Selkirk PM, Seppelt RD. 1988. Photoinhibition in antarctic mosses. *Polarforschung* 58:103–11
4. Antarctic Journal of the United States. 1991. Protocol on Environmental Protection to the Antarctic Treaty. *Antarct. J. US* 26:5–11
5. Anderson JG, Toohey DW, Brune WH. 1991. Free radicals within the antarctic vortex: the role of CFCs in antarctic ozone loss. *Science* 251:39–46
6. Auburn FM. 1981. The antarctic environment. In *Year Book of World Affairs*, pp. 248–65. London: Stevens & Sons
7. Ayres MP. 1993. Plant defense, herbivory, and climate change. In *Biotic Interactions and Global Change*, ed. PM Kareiva, JG Kingsolver, RB Huey, pp. 75–94. Sunderland, MA: Sinauer
8. Bazzaz FA. 1990. The response of natural ecosystems to the rising global CO_2 levels. *Annu. Rev. Ecol. Syst.* 21:167–96
9. Berry J, Björkman O. 1980. Photosynthetic response and adaptation to temperature in higher plants. *Annu. Rev. Plant Physiol.* 31:491–543
10. Block W. 1984. Terrestrial microbiology, invertebrates and ecosystems. See Ref. 77, pp. 163–236
11. Block W. 1985. Arthropod interactions in an antarctic terrestrial community. See Ref. 106, pp. 614–19
12. Block W. 1990. Cold tolerance of insects and other arthropods. *Phil. Trans. R. Soc. Lond. B.* 326:613–33
13. Block W, Burn AJ, Richard KJ. 1984. An insect introduction to the maritime Antarctic. *Biol. J. Linn. Soc.* 23:33–39
14. Bonner WN. 1984. Introduced mammals. See Ref. 77, pp. 237–78
15. Bonner WN. 1994. Antarctic conservation and management. *Polar Biol.* 14:301–5
16. Bouwmann AF, ed. 1990. *Soils of a Warmer Earth.* Chichester: Wiley
17. Burn AJ. 1984. Energy partitioning in the antarctic collembolan *Cryptopygus antarcticus. Ecol. Entomol.* 9:11–21
18. Burn AJ. 1984. Life cycle strategies in two antarctic Collembola. *Oecologia* 64:223–29
19. Callaghan TV, Sonesson M, Sømme L. 1992. Responses of terrestrial plants and invertebrates to environmental change at high latitudes. *Phil. Trans. R. Soc. Lond. B.* 338:279–88
20. Cameron RE, Morelli FA, Honour RC. 1973. Aerobiological monitoring of Dry Valley drilling sites. *Antarct. J. US* 8:211–14
21. Cannon RJC, Block W. 1988. Cold tolerance of microarthropods. *Biol. Rev.* 63:23–77
22. Cannon RJC, Block W, Collett GD. 1985. Loss of supercooling ability in *Cryptopygus antarcticus* (Collembola: Isotomidae) associated with water uptake. *Cryo-Letters* 6:73–80
23. Chapin FS III, Jefferies RL, Reynolds JF, Shaver GR, Svoboda J, Chu EW, eds. 1992. *Arctic Ecosystems in a Changing Climate: an Ecophysiological Perspective.* London: Academic
24. Chapuis JL, Boussès P, Barnaud G. 1994. Alien mammals, impact and management in the French subantarctic islands. *Biol. Conserv.* 67:97–106
25. Chown SL, Language K. 1994. Recently established Diptera and Lepidoptera on sub-antarctic Marion Island. *Afr. Entomol.* 2:57–76
26. Collins NJ. 1977. The growth of mosses in two contrasting communities in the maritime Antarctic: measurement and prediction of net annual production. See Ref. 80, pp. 921–33
27. Dahlback A, Henriksen T, Larsen SHH, Stamnes K. 1989. Biological UV-doses and effect of an ozone layer depletion. *Photochem. Photobiol.* 49:621–25
28. D'Antonio CM, Vitousek PM. 1992. Biological invasions by exotic grasses, the grass/fire cycle, and global change. *Annu. Rev. Ecol. Syst.* 23:63–87

29. Davey MC. 1991. Effects of physical factors on the survival and growth of antarctic terrestrial algae. *Br. Phycol. J.* 26:315–25

30. Davey MC. 1991. The seasonal periodicity of algae on antarctic fellfield soils. *Holarct. Ecol.* 14:112–20

31. Davey MC, Pickup J, Block W. 1992. Temperature variation and its biological significance in fellfield habitats on a maritime Antarctic island. *Antarct. Sci.* 4:383–88

32. Davis RC. 1981. Structure and function of two antarctic terrestrial moss communities. *Ecol. Monogr.* 51:125–43

33. Delmas RJ, Ascencio J, Legrand M. 1980. Polar ice evidence that atmospheric CO_2 20,000 yr BP was 50% of present. *Nature* 284:155–57

34. Department of the Arts, Sport, the Environment and Territories. 1992. *Impact of Climate Change on Antarctica - Australia.* Canberra: Aust. Govt. Publ. Serv.

35. Doake CSM, Vaughan DG. 1991. Rapid disintegration of the Wordie Ice Shelf in response to atmospheric forcing. *Nature* 350:328–30

36. Dorne A-J. 1977. Analysis of the germination under laboratory and field conditions of seeds collected in the Kerguelen Archipelago. See Ref. 80, pp. 1003–13

37. Drake JA, Mooney HA, di Castri F, Groves RH, Kuger FJ, et al. 1989. *Biological Invasions. A Global Perspective.* Chichester: Wiley

38. Edwards JA. 1974. Studies in *Colobanthus quitensis* (Kunth) Bartl. and *Deschampsia antarctica* Desv.: VI. Reproductive performance on Signy Island. *Br. Antarct. Surv. Bull.* 39:67–86

39. Edwards JA. 1980. An experimental introduction of vascular plants from South Georgia to the maritime Antarctic. *Br. Antarct. Surv. Bull.* 49:73–80

40. Edwards JA, Smith RIL. 1988. Photosynthesis and respiration of *Colobanthus quitensis* and *Deschampsia antarctica* from the maritime Antarctic. *Br. Antarct. Surv. Bull.* 81:43–63

41. Fajer ED, Bowers MD, Bazzaz FA. 1989. The effects of enriched carbon dioxide atmospheres on plant-insect herbivore interactions. *Science* 243:1198–1200

42. Farman JC, Gardiner BG, Shanklin JD. 1985. Large losses of total ozone in Antarctica reveal seasonal ClO_x/NO_x interaction. *Nature* 315:207–10

43. Field CB. 1994. Arctic chill for CO_2 uptake. *Nature* 371:472–73

44. Field CB, Chapin FS III, Matson PA, Mooney HA. 1992. Responses of terrestrial ecosystems to the changing atmosphere: a resource-based approach. *Annu. Rev. Ecol. Syst.* 23:201–35

45. Fowbert JA, Smith RIL. 1994. Rapid population increases in native vascular plants in the Argentine Islands, Antarctic Peninsula. *Arct. Alp. Res.* 26:290–97

46. Frederick JE, Snell HE. 1988. Ultraviolet radiation levels during the antarctic spring. *Science* 241:438–39

47. Frenot Y, Gloaguen JC. 1994. Reproductive performance of native and alien colonizing phanerogams on a glacier foreland, Iles Kerguelen. *Polar Biol.* 14:473–81

47a. Friedmann EI. 1982. Endolithic microorganisms in the antarctic cold desert. *Science* 215:1045–53

48. Garcia-Pichel F, Castenholz RW. 1991. Characterization and biological implications of scytonemin, a cyanobacterial sheath pigment. *J. Phycol.* 27:395–409

49. Garcia-Pichel F, Castenholz RW. 1993. Occurrence of UV-absorbing, mycosporine-like compounds among cyanobacterial isolates and an estimate of their screening capacity. *Appl. Environ. Microbiol.* 59:163–69

50. Garcia-Pichel F, Sherry ND, Castenholz RW. 1992. Evidence for an ultra-violet sunscreen role of the extra-cellular pigment scytomenin in the terrestrial cyanobacterium *Chlorogloeopsis* sp. *Photochem. Photobiol.* 56:17–23

50a. Gimingham CH, Smith RIL. 1970. Bryophyte and lichen communities in the maritime Antarctic. In *Antarctic Ecology*, ed. MW Holdgate, 1:752–85. London: Academic

51. Gorham E. 1991. Northern peatlands: role in the carbon cycle and probable responses to climatic warming. *Ecol. Appl.* 1:182–95

52. Grace J. 1987. Climatic tolerance and the distribution of plants. *New Phytol.* 106:113–30

53. Greenslade PJM. 1983. Adversity selection and the habitat templet. *Am. Nat.* 122:352–65

54. Hall KJ, Walton DWH. 1992. Rock weathering, soil development and colonisation under a changing climate. *Phil. Trans. R. Soc. Lond. B.* 338:269

55. Harley PC, Tenhunen JD, Murray KJ, Beyers J. 1989. Irradiance and temperature effects on photosynthesis of tussock tundra *Sphagnum* mosses from the foothills of the Phillip Smith Mountains, Alaska. *Oecologia* 79:251–59

56. Harrisson PM, Walton DWH, Rothery P. 1989. The effects of temperature and moisture on CO_2 uptake and total resistance to water loss in the antarctic fo-

liose lichen *Umbilicaria antarctica.*
New Phytol. 111:673–82

57. Heal OW, Block W. 1987. Soil biological processes in the North and South. *Ecol. Bull.* 38:47–57

58. Horner JD, Gosz JR, Cates RG. 1988. The role of carbon-based plant secondary metabolites in decomposition in terrestrial ecosystems. *Am. Nat.* 132: 869–83

59. Horowitz NH. 1979. Biological water requirements. In *Strategies of Microbial Life in Extreme Environments*, ed. M Shilo, pp. 15–27. Berlin: Dahlem Konferenzen

60. Houghton JT, Callander BA, Varney SK, eds. 1992. *Climate Change 1992—Suppl. Rep. IPCC Scie. Assess.* Cambridge: Cambridge Univ. Press

61. Houghton JT, Jenkins GJ, Ephraums JJ, eds. 1990. *Climate Change—the IPCC Scientific Assessment.* Cambridge: Cambridge Univ. Press

62. Huntley B. 1991. How plants respond to climate change: migration rates, individualism and the consequences for plant communities. *Ann. Bot. (Lond.)* 67:15–22 (Suppl. 1)

63. Hurst JL, Pugh GJF, Walton DWH. 1985. The effects of freeze-thaw cycles and leaching on the loss of soluble carbohydrates from leaf material of two subantarctic plants. *Polar Biol.* 4:27–31

64. Kappen L. 1993. Lichens in the Antarctic region. In *Antarctic Microbiology*, ed. EI Friedmann, pp. 433–90. New York: Wiley-Liss

65. Kappen L, Smith RIL, Meyer M. 1989. Carbon dioxide exchange of two ecodemes of *Schistidium antarctici* in continental Antarctica. *Polar Biol.* 9:415–22

66. Kappen L, Straka H. 1988. Pollen and spores transport into the Antarctic. *Polar Biol.* 8:173–80

67. Karentz D. 1991. Ecological considerations of Antarctic ozone depletion. *Antarct. Sci.* 3:3–11

68. Kennedy AD. 1993. Photosynthetic response of the antarctic moss *Polytrichum alpestre* Hoppe to low temperatures and freeze-thaw stress. *Polar Biol.* 13:271–79

69. Kennedy AD. 1993. Water as a limiting factor in the antarctic terrestrial environment: a biogeographical synthesis. *Arct. Alp. Res.* 25:308–15

70. Kennedy AD. 1994. Simulated climate change: a field manipulation study of polar microarthropod community response to global warming. *Ecography* 17:131–40

71. Kennedy AD. 1995. Simulated climate change: Are passive greenhouses a valid microcosm for testing the biological effects of environmental perturbations? *Global Change Biol.* 1:29–42

72. Kerr RA. 1994. Antarctic ozone hole fails to recover. *Science* 266:217

73. Kerry KR, Hempel G, eds. 1990. *Antarctic Ecosystems. Ecological Change and Conservation.* Berlin: Springer-Verlag

74. King JC. 1994. Recent climate variability in the vicinity of the Antarctic Peninsula. *Int. J. Climatol.* 14:357–69

75. Larigauderie A, Hilbert DW, Oechel WC. 1988. Effect of CO_2 enrichment and nitrogen availability on resource acquisition and resource allocation in a grass, *Bromus mollis. Oecologia* 77: 544–49

76. Larson RA, Garrison WJ, Carlson RW. 1990. Differential response of alpine and non-alpine *Aquilegia* species to increased ultraviolet-B radiation. *Plant Cell Environ.* 13:983–87

77. Laws RM, ed. 1984. *Antarctic Ecology.* London: Academic

78. Light JJ, Heywood RB. 1975. Is the vegetation of continental Antarctica predominantly aquatic? *Nature* 256:199–200

79. Lister A, Block W, Usher MB. 1988. Arthropod predation in an antarctic terrestrial community. *J. Anim. Ecol.* 57: 957–71

80. Llano GA, ed. 1977. *Adaptations within Antarctic Ecosystems.* Washington DC: Smithsonian

81. Longton RE. 1985. Terrestrial habitats—vegetation. In *Key Environments: Antarctica*, ed. WN Bonner, DWH Walton, pp. 73–105. Oxford: Pergamon

82. Longton RE. 1988. *The Biology of Polar Bryophytes and Lichens.* Cambridge: Cambridge Univ. Press

83. Longton RE. 1990. Sexual reproduction in bryophytes in relation to physical factors of the environment. In *Bryophyte Development: Physiology and Biochemistry*, ed. RN Chopra, SC Bhatla, pp. 139–66. Boca Raton, FL: CRC

84. Lorius C, Jouzel J, Ritz C, Merlivat L, Barkov NI, et al. 1985. A 150,000-year climate record from antarctic ice. *Nature* 316:591–96

85. Manabe S, Wetherald RT. 1980. On the distribution of climate change resulting from an increase in carbon content of the atmosphere. *J. Atmos. Sci.* 37:99–118

86. Manheim BS Jr. 1992. The failure of the National Science Foundation to protect Antarctica. *Mar. Pollut. Bull.* 25: 253–54

87. Mattson WT. 1980. Herbivory in rela-

tion to plant nitrogen content. *Annu. Rev. Ecol. Syst.* 11:119–61
88. Maxwell JB, Barrie LA. 1989. Atmospheric and climatic change in the Arctic and Antarctic. *Ambio* 18:42–49
89. McCloud ES, Berenbaum MR. 1994. Stratospheric ozone depletion and plant-insect interactions: effects of UVB radiation on foliage quality of *Citrus jambhiri* for *Trichoplusia ni. J. Chem. Ecol.* 20:525–39
90. McNaughton KG, Jarvis PG. 1983. Predicting effects of vegetation changes on transpiration and evaporation. In *Water Deficits and Plant Growth*, ed. TT Kozlowski, 7:1–47. New York: Academic
91. Melillo JM, Callaghan TV, Woodward FI, Salati E, Sinha SK. 1990. Effects on ecosystems. See Ref. 61, pp. 285–310
92. Miles J, Walton DWH, eds. 1993. *Primary Succession on Land*, Oxford: Blackwell
93. Morgan VI, Goodwin ID, Etheridge DM, Wookey CW. 1991. Evidence from antarctic ice cores for recent increases in snow accumulation. *Nature* 354:58–60
94. Nadelhoffer KJ, Giblin AE, Shaver GR, Linkins AE. 1992. Microbial processes and plant nutrient availability in arctic soils. See Ref. 23, pp. 281–300
95. Nedwell DB, Russell NJ, Cresswell-Maynard T. 1994. Long-term survival of microorganisms in frozen material from early antarctic base camps at McMurdo Sound. *Antarct. Sci.* 6:67–68
96. Neftel A, Moor E, Oeschger H, Stauffer B. 1985. Evidence from polar ice cores for the increase in atmospheric CO_2 in the past two centuries. *Nature* 315:45–47
97. Oechel WC, Hastings SJ, Vourlitis G, Jenkins M, Riechers G, Grulke N. 1993. Recent change of arctic tundra ecosystems from a net carbon dioxide sink to a source. *Nature* 361:520–23
98. Pitelka LF. 1994. Ecosystem response to elevated CO_2. *Trends Ecol. Evol.* 9:204–07
99. Post A, Adamson E, Adamson H. 1990. Photoinhibition and recovery of photosynthesis in antarctic bryophytes under field conditions. In *Current Research in Photosynthesis*. Vol. IV, ed. M Baltscheffsky, pp. 635–38. Dordrecht, Netherlands: Kluwer
100. Roberts L. 1989. Does the ozone hole threaten antarctic life? *Science* 244:288–89
101. Rubin MJ. 1965. Antarctic climatology. In *Biogeography and Ecology in Ant-arctica*, ed. J Van Mieghem, P Van Oye, pp. 72–96. The Hague: Junk
102. Schlesinger ME, Mitchell PFB. 1987. Climate model simulations of the equilibrium climate response of increased carbon. *Rev. Geophys.* 25:760–98
103. Schwartz AMJ, Green TGA, Seppelt RD. 1992. Terrestrial vegetation at Canada Glacier, Southern Victoria Land, Antarctica. *Polar Biol.* 12:397–404
104. Selkirk PM. 1992. Climate change and the Subantarctic. See Ref. 34, pp. 43–51
105. Shukla J, Mintz Y. 1982. Influence of land-surface evapotranspiration on the earth's climate. *Science* 215:1498–501
106. Siegfried WR, Condy PR, Laws RM, eds. 1985. *Antarctic Nutrient Cycles and Food Webs.* Berlin: Springer-Verlag
107. Smith RIL. 1984. Colonization and recovery by cryptogams following recent volcanic activity on Deception Island, South Shetland Islands. *Br. Antarct. Surv. Bull.* 62:25–51
108. Smith RIL. 1984. Terrestrial plant biology of the Subantarctic and Antarctic. See Ref. 77, pp. 61–162
109. Smith RIL. 1990. Signy Island as a paradigm of biological and environmental change in antarctic terrestrial ecosystems. See Ref. 73, pp. 32–49
110. Smith RIL. 1991. Exotic sporomorpha as indicators of potential immigrant colonists in Antarctica. *Grana* 30:313–24
111. Smith RIL. 1993. Dry coastal ecosystems of Antarctica. In *Ecosystems of the World. 2A. Dry Coastal Ecosystems—Polar Regions and Europe.* ed. E van der Maarel, pp. 51–71. London: Elsevier
112. Smith RIL. 1993. The role of bryophyte propagule banks in primary succession: case-study of an antarctic fellfield soil. See Ref. 92, pp. 55–78
113. Smith RIL. 1994. Vascular plants as bioindicators of regional warming in Antarctica. *Oecologia* 99:322–28
114. Smith VR, Steenkamp M. 1990. Climatic change and its ecological implications at a subantarctic island. *Oecologia* 85:14–24
115. Stark P. 1994. Climatic warming in the central Antarctic Peninsula area. *Weather* 49:215–20
116. Street RB, Melnikov PI. 1990. Seasonal snow cover, ice and permafrost. In *Climate Change—The IPCC Impacts Assessment*, ed. WJ McG Tegart, GW Sheldon, DC Griffiths, 33:1–7. Canberra: Aust. Govt. Publ. Serv.
116a. Taylor KE, Penner JE. 1994. Response of the climate system to atmospheric aerosols and greenhouse gases. *Nature* 369:734–37

117. Taylor RH, Wilson PR. 1990. Recent increase and southern expansion of Adélie Penguin populations in the Ross Sea region related to climatic warming. *NZ J. Ecol.* 14:25–29

118. Tenhunen JD, Lange OL, Hahn S, Siegwolf R, Oberbauer SF. 1992. The ecosystem role of poikilohydric tundra plants. See Ref. 23, pp. 213–37

119. Teramura AH. 1990. Implications of stratospheric ozone depletion upon plant production. *Hortic. Sci.* 25:1557–60

120. Tissue DT, Oechel WC. 1987. Response of *Eriophorum vaginatum* to elevated CO_2 and temperature in the Alaskan tussock tundra. *Ecology* 68:401–10

121. Trenberth KE, Christy JR, Olson JG. 1987. Global atmospheric mass, surface pressure and water vapour variations. *J. Geophys. Res.* 92:14815–26

122. van Zanten BO, Pocs T. 1981. Distribution and dispersal of bryophytes. *Adv. Bryol.* 1:479–562

123. Vincent WF, Quesada A. 1994. Ultraviolet radiation effects on cyanobacteria: implications for antarctic microbial ecosystems. See Ref. 126, pp. 111–24

124. Walton DWH. 1975. European weeds and other alien species in the sub-Antarctic. *Weed Res.* 15:271–82

125. Walton DWH. 1990. Colonization of terrestrial habitats: organisms, opportunities and occurrence. See Ref. 73, pp. 51–60

126. Weiler CS, Penhale PA, eds. 1994. *Ultraviolet Radiation in Antarctica: Measurements and Biological Effects.* Washington DC: Am. Geophys. Union

127. Wynn-Williams DD. 1990. Ecological aspects of Antarctic microbiology. *Adv. Microb. Ecol.* 11:71–146

128. Wynn-Williams DD. 1993. Microbial processes and initial stabilization of fellfield soil. See Ref. 92, pp. 17–32

129. Wynn-Williams DD. 1994. Potential effects of ultraviolet radiation on antarctic primary terrestrial colonizers: cyanobacteria, algae, and cryptogams. See Ref. 126, pp. 243–57

130. Yarrington MR, Wynn-Williams DD. 1985. Methanogenesis and the anaerobic micro-biology of a wet moss community at Signy Island. See Ref. 106, pp. 229–233

131. Young AR, Björn LO, Moan J, Nultsch W, eds. 1993. *Environmental UV Photobiology.* New York: Plenum

Annu. Rev. Ecol. Syst. 1995. 26:705–27

PLANT-VERTEBRATE SEED DISPERSAL SYSTEMS IN THE MEDITERRANEAN: Ecological, Evolutionary, and Historical Determinants

Carlos M. Herrera

Estación Biológica de Doñana, Consejo Superior de Investigaciones Científicas, Apartado 1056, E-41080 Sevilla, Spain

KEY WORDS: coevolution, frugivorous birds, mutualism, seed dispersal

ABSTRACT

Investigations on vertebrate seed dispersal systems in the Mediterranean show that extremely efficient plant-disperser mutualisms do not require, and thus are not evidence for, mutual evolutionary adjustments of participants. Current Mediterranean dispersal systems have apparently been shaped by means of 1. trophic and behavioral adaptations of birds morphologically preadapted to pre-existing plant resources, and 2. disperser-mediated processes of habitat-shaping occurring at an ecological time scale. These processes depend on differential recruitment of plant species as a function of disperser preferences, rather than on adjustments based on evolutionary processes. On the plant side, there is a prevalence of historical and phylogenetic effects, which reflects a series of ecological limitations inherent to the interactions between plants and dispersal agents that constrain plant adaptation to dispersers. To test adaptive hypotheses and explanations, future investigations on Mediterranean plant-disperser systems should concentrate more on the animal than on the plant side of the interaction.

INTRODUCTION

By acting as seed vectors, frugivorous animals play an essential role in the reproductive cycle of their food plants. This circumstance, and the mutualistic

705

nature of the relationship, have given rise to considerable interest in the evolutionary consequences of frugivory and seed dispersal for plants and animals (44, 46). Although botanists have long been interested in the natural history of seed dispersal by animals (141, 169), evolutionary ecologists turned their attention to the subject rather recently, following seminal contributions by Snow (149, 150), McKey (125), and Howe & Estabrook (92). The current interest in plant-disperser mutualisms contrasts with an older research tradition that studied aspects of plant reproduction such as pollination, sex expression, and breeding system, which has been part of mainstream evolutionary biology since its Darwinian inception (24, 25).

Interest in plant-disperser systems, and most earlier evolutionary interpretations, was motivated by investigations conducted in tropical forests, habitats that have continued to contribute decisively to our knowledge in this field (44, 46). Away from the tropics, most studies on plant-disperser mutualisms have been in American temperate habitats and in the Mediterranean Basin. This review focuses on the latter region (see 171, 172 for partial reviews of frugivory and seed dispersal by vertebrates in temperate America). Two major reasons justify restricting the geographical scope to the Mediterranean region. On formal grounds, studies on the evolutionary ecology of plant-disperser interactions in the Mediterranean, although numerous, have been not reviewed previously. One further, conceptual reason is that in no other large and ecologically well-defined geographic area have most facets of plant-disperser interactions been so thoroughly investigated as to provide a comprehensive picture of the relative importance of ecological, evolutionary, and historical factors in shaping plant-disperser systems. Studies in the Mediterranean have revealed that current plant-disperser interactions not only may be shaped by adaptive evolution of participants, but may also reflect important limitations imposed by regional history and the phylogeny of the taxa involved (73, 74, 82, 83).

The Mediterranean Environment

In this review I adhere to the delineation of the Mediterranean region adopted by di Castri (42). The Mediterranean climate is a transitional regime between temperate and dry tropical climates, characterized by a concentration of rainfall in winter, occurrence of a distinct summer drought of variable length, warm-to-hot summers, and cool-to-cold winters (42). These climatic characteristics, together with the original features of the plants populating the present-day Mediterranean region, are key elements to understanding the peculiarities of Mediterranean ecosystems. The Mediterranean climate is very young in geological terms. It first appeared in the Pliocene, approximately 3.2 million years ago (155) and, as discussed below, most of the fleshy-fruited plant taxa living at present in the Mediterranean existed prior to the appearance of this climatic type.

Due to a long history of human-induced perturbations, original Mediterranean habitats either were extirpated long ago or, in the relatively few places where they still persist, have almost invariably been subject to some degree of disturbance (158, 161). Plant-disperser interactions are seriously distorted in heavily disturbed Mediterranean habitats (38, 81, 139, 143), hence investigations conducted on extensively human-modified habitats are only rarely considered in this review.

THE PARTICIPANTS

Vertebrate-Dispersed Plants

In European temperate forests, fleshy-fruited plants are most abundant in clearings and forest edges but become scarce in the interior of mature forests, which are dominated by nut- or cone-producing trees (136). In lowland and mid-elevation Mediterranean habitats, in contrast, fruit-bearing plants generally replace earlier successional species that are not vertebrate-dispersed (e.g. Cistaceae, Labiatae, Leguminosae; 88, 89), and such plants may eventually dominate the vegetation in undisturbed or lightly disturbed woodlands and shrublands (54, 86, 90). Vertebrate-dispersed species account for 32–64% of local woody species richness and 20–95% of woody plant cover in Mediterranean habitats (64, 86, 104, 114). Their importance is greatest in lowland vegetation on fertile soils, and it declines with elevation, increasing aridity, decreasing soil fertility, and severity of habitat disturbances such as nitrification and fire (6, 54, 64, 66, 86, 88, 89). Mediterranean shrublands and woodlands are intermediate between tropical and temperate forests in fruit production (70). Annual fruit production for some southern Spanish habitats ranges between 60 and $1,400 \times 10^3$ ripe fruits/ha, representing 6–100 kg dry mass/ha (114).

In contrast to temperate habitats, where most fleshy fruit–producing species belong to relatively few plant families (mostly Rosaceae and Caprifoliaceae in Europe and America; 136, 171), local assemblages of vertebrate-dispersed plants in the Mediterranean are taxonomically diverse at the familial level, particularly in the warmer lowlands. This is largely due to the widespread occurrence of several families with current distributions centered on tropical and subtropical regions and northernmost distributional limits in the Mediterranean (e.g. Anacardiaceae, Oleaceae, Santalaceae, Myrtaceae, Lauraceae, Palmae). Most of these families are locally represented by only 1–2 species in single genera (e.g. *Pistacia* in Anacardiaceae, *Olea* in Oleaceae, *Osyris* in Santalaceae, *Chamaerops* in Palmae; 28, 64, 96, 104). However, as shown later, some play a prominent role in the maintenance of Mediterranean plant-disperser systems owing to their abundance and the nutritional characteristics of their fruits.

Nonavian Dispersers

Ants may act as secondary dispersers of the seeds of some fleshy-fruited Mediterranean plants (5), but further studies are needed to assess the generality of the phenomenon. The role of reptiles as dispersal agents has been not studied in detail, but it seems quantitatively unimportant except in some insular situations (167, 168). Several species of mammals belonging to the order Carnivora are seasonally frugivorous and disperse the seeds of many plants in undisturbed or lightly disturbed habitats (18, 37, 81, 135). In a southeastern Spanish region, three species of carnivores disperse seeds of 27 species, representing 40% of the regional fleshy-fruited flora (81). Other mammalian groups such as ungulates (52), rabbits (128), and hedgehogs (56) also sporadically ingest fleshy fruits and disperse seeds, but their importance as dispersers is probably local. The same applies to the Brown Bear (*Ursus arctos*) and the Barbary Macaque (*Macaca sylvanus*), which feed on fleshy fruits and disperse the seeds of many plants, in the few regions where they still survive (81, 126, 127).

The vast majority of Mediterranean fleshy-fruited plants are dispersed either by birds alone or by some combination of birds and carnivorous mammals (37, 81). Indirect evidence based on concurrent studies of the frugivorous diet of birds and mammals in the same region (64, 81, 130) indicates that birds are the main dispersers of most plants with mixed dispersal. As a group, therefore, birds are by far the most important vertebrate seed dispersers in the Mediterranean, and I focus on bird-plant relationships for the rest of this review.

Avian Dispersers

Birds of many species eat fleshy fruits in Mediterranean shrublands and woodlands. Not all of these, however, are legitimate seed dispersers, as some feed only on pulp or seeds without effecting dispersal. These "fruit predators" (65) typically are small- and medium-sized finches (Fringillidae) and titmice (Paridae) (55, 64, 65, 85, 112, 116, 163). The distinction between fruit predators and legitimate dispersers, however, is somewhat context-dependent. Fruit predators may sometimes act as dispersers when feeding on small-seeded fruits (e.g. titmice feeding on blackberries; 48, 103), and dispersers may act as predators when feeding on large-seeded fruits (e.g. warblers feeding on olive fruits; 138).

Most legitimate avian dispersers are small- to medium-sized (body mass range = 10–110 g) passerines in the families Turdidae (thrushes), Sylviidae (old World warblers), and Muscicapidae (old World flycatchers). In all Mediterranean habitats so far studied, the most important seed dispersers are some combination of species in the genera *Sylvia, Turdus,* and *Erithacus* (39, 50, 64, 85, 96, 103, 104, 116, 165). Species in these genera are strong seasonal frugivores that feed on many fruit species (*Sylvia*: 30, 96, 108, 109, 118; *Turdus*: 32, 33, 53, 60, 115, 131, 151, 173; *Erithacus*: 19, 34, 58, 111). In four southern Spanish localities, *Sylvia atricapilla* consumed the fruits of 29 plant species in the period October-March (118). *Erithacus rubecula* fed on 21 fruit species in southern France (34), and thrushes consume locally the fruits of 5–18 species (32, 53, 64, 131). Species of Sturnidae (47) and Corvidae (116, 152, 150a) often feed on fleshy fruits but do not seem to be important frugivores anywhere except in the Canary Islands, where *Corvus corax* is a major disperser for at least 16 plant species (129). Bulbuls (Pycnonotidae) are important dispersers in the Middle East (3, 96, 98, 99) and, most likely, in northern Africa as well (21).

With only minor exceptions (e.g. *Sylvia melanocephala, Pycnonotus* spp.), Mediterranean avian dispersers are medium- and long-distance migrants that breed in central and northern Europe (21, 22). Some species overwinter in tropical and subtropical Africa, appearing in Mediterranean habitats only during spring and autumn migrations. As a rule, these trans-Saharan migrants eat fleshy fruits only in autumn passage (96), but some species are also frugivorous in spring if ripe fruits are available (78). Other species of dispersers overwinter in the Mediterranean Basin from October-March, when they feed almost entirely on fleshy fruits (see below). For most of these species, the Mediterranean Basin is the major or exclusive wintering area in the western Palaearctic (21, 22; see 124, 156, for reviews of bird migration in the Mediterranean). Due to their abundance, extended permanence in the region, and extensive frugivory, overwintering species are the most important and genuine avian dispersers of Mediterranean plants.

CONSEQUENCES OF PLANT-BIRD INTERACTIONS

The Plant Side

The efficiency of the relationship of the fleshy-fruited plants with dispersal agents may be assessed by considering the success of fruit removal by legitimate dispersers and the patterns of postdispersal seed deposition. On these grounds, the interaction with Mediterranean frugivorous birds is remarkably efficient for most species, particularly in lowland habitats.

Predispersal seed predation by invertebrates on fleshy-fruited Mediterranean taxa is lower, on average, than among coexisting dry-fruited species (88) or on European temperate fleshy-fruited species (67). Except in some montane (132, 163, 164, 165) or disturbed (121) habitats, fruit losses to herbivores and fruit predators are negligible for most species (68, 88, 107, 110, 112). Crops of ripe fruits are thoroughly depleted by legitimate frugivores, particularly in the lowlands (Table 1). On average, species in well-preserved lowland habitats

Table 1   Proportion of ripe fruit crops of Mediterranean fleshy-fruited plants removed by legitimate avian seed dispersers. Species are arranged in decreasing order of fruit removal rate.

Species	Ripe fruits removed[a] (%)	References
Asparagus aphyllus	100, 100	1, 96
Pistacia lentiscus	100, 99, 99, 91	64, 96, 110
Smilax aspera	100, 91, 86	64, 96, 104
Phillyrea angustifolia	99, 96[b], 93, 83, 72[b]	64, 104, 162
Rhamnus lycioides	98, 98, 97	64, 104
Osyris alba	98, 76	104
Daphne gnidium	97, 92	104
Myrtus communis	95, 95, 89	64, 104
Osyris quadripartita	94	64
Olea europaea	94, 52[d]	107
Rhamnus alaternus	93, 61[b]	64, 130
Rubus ulmifolius	90, 92, 88, 80, 86[b], 43[c]	103, 104, 130
Lonicera periclymenum	86, 61	104
Lonicera etrusca	84	96
Rhamnus palaestinus	84	96
Rubia tenuifolia	82	96
Phillyrea latifolia	78[b,d], 32[b,d]	85
Viburnum tinus	75, 51[b]	157, CM Herrera, unpubl.
Berberis hispanica	71[b], 52[b]	130, 132
Prunus mahaleb	68[b], 61[b], 53[b], 50[b]	116
Osyris alba	68	96
Cornus sanguinea	49[c], 36[c], 36[c]	121
Pistacia terebinthus	28[b]	166

[a] Mean values of fruit removal rates obtained for individual plants. Different entries for the same species correspond to different localities or years.
[b] Highland habitat ($\geq$ 1000 m elevation). Entries without this superscript are from lowland habitats.
[c] Disturbed habitat.
[d] Estimated during a season of unusually large fruit crop.

have 90.2 ± 9.4 (SD)% of their ripe fruits consumed by legitimate avian dispersers, which is significantly greater ($P \ll 0.001$, Kruskal-Wallis ANOVA) than the 62.1 ± 17.7% exhibited by highland species. Fruit removal success declines in disturbed habitats or when disperser populations become satiated during seasons with unusually large fruit crops (Table 1) (85, 107, 143).

Depending on the combination of plant and disperser species, seed ingestion by Mediterranean frugivorous birds may enhance germination or not, but there is no documented instance either of detrimental effects of bird ingestion on seed germination or of seeds that obligately require bird ingestion to germinate

(9, 10, 27, 98). The consequences of seed dispersal by birds for seed survival, seedling ecology, and population recruitment of Mediterranean fleshy-fruited plants are still little known. The limited evidence available suggests that the demographic implications of dispersal are very site- and species-specific (69, 85, 99, 119, 144), so few generalizations can be drawn. Different species of frugivores tend to generate characteristic seed shadows, depending on foraging behavior, seed retention times, patterns of fruit selection, and response to the structure of the vegetation (84, 99, 105, 144). The postdispersal seed deposition pattern of a given plant species will therefore depend on detailed aspects of habitat structure and composition of the disperser assemblage.

At the between-habitat scale, bird-dispersed species have a distinct advantage over dry-fruited ones in the colonization of newly available habitat patches (29, 40). Seed dissemination, however, is essentially a within-habitat process, as most dispersed seeds travel relatively short distances from the mother plant (35, 41, 85). In *Phillyrea latifolia*, postdispersal seed predation was not greater under conspecifics, but recruitment was much reduced there. Seed dissemination from mother plants, although rather restricted spatially, was clearly advantageous in this species (85). Within habitats, no evidence exists for birds preferentially dispersing seeds to particularly favorable germination or survival microsites, and evidence exists of conflicts in the quality of microsites for seeds and seedlings (69, 85, 119).

Avian frugivores generally feed simultaneously on fruits of several species, and fruiting plants act as focal points for frugivores' foraging, leading to a predictable concentration of dispersed seeds beneath fleshy fruit–producing species or, in the case of dioecious taxa, under fruit-bearing females (35, 41, 69, 85, 99). This effect may influence the distribution of fruiting plants in the habitat ("habitat shaping"; 70). The hemiparasitic shrub *Osyris quadripartita* mostly parasitizes species that, like itself, are bird-dispersed, in proportions correlated with the frequency of co-occurrence of their seeds in the meals of its main disperser (79, 80). By producing multispecific seed shadows, dispersers may also influence the absolute and relative abundances of their food plants in local plant communities. The recruitment of seedlings of *P. latifolia* in certain microhabitats is limited by seed rain (85, 119), and local percent cover of individual fleshy-fruited species is, at least in some plant communities, directly correlated with their area-specific intensity of fruit production (64).

The Bird Side

Regardless of whether they are transient fall migrant or overwintering species, major avian dispersers are all extensively frugivorous while inhabiting Mediterranean habitats (Table 2; also 30, 32–34, 45, 159). On average, fruits usually contribute > 75%, and often > 90%, of the food for these species.

Contrary to earlier suggestions (12), recent laboratory studies have shown

Table 2 Contribution of fleshy fruits to the diet of major Mediterranean avian seed dispersers. Different entries for the same species correspond to different localities. Average values are shown for localities with data for different years.

Species and overall mean[a]	Mean % fruit volume	Period[c]	References
Sylvia atricapilla	98.3	FM	102
86.4	93.3	FM + OW	64
	93.0	FM + OW	64, CM Herrera unpubl.
	92.4	FM + OW	CM Herrera unpubl.
	86.1	FM + OW	108
	82.5	FM + OW	51
	79.0	FM	50
	76.3	OW	138
	60.2	FM	118
Sylvia borin	94.7	FM	102
90.3	92.4	FM	108
	91.1	FM	64
	84.1	FM	51
	84.0	FM	50
Sylvia communis	95.8	FM	102
74.1	76.8	FM	64
	70.9	FM	108
	54.2	FM	51
Sylvia melanocephala	83.5	R	64, CM Herrera unpubl.
70.6	73.9	R	108
	66.6	R	64
	59.6	OW	138
Turdus iliacus	100.0	OW	CM Herrera unpubl.
86.8	98.0	OW	85
	85.7[b]	OW	151
Turdus merula	89.6	OW	64, CM Herrera unpubl.
80.1	88.2	FM + OW	CM Herrera unpubl.
	88.1	FM	102
	83.0[b]	FM + OW	153
	79.4	R + OW	130
	75.0	OW	50
	72.8	FM + OW	P Jordano unpubl.
	67.3	FM + OW	64
Turdus philomelos	97.2	OW	P Jordano unpubl.
91.6	92.9	OW	CM Hedrrera unpubl.
	91.0	OW	50
	78.6	OW	138
Turdus torquatus	96.2	OW	173
Turdus viscivorus	92.8	R	131
Erithacus rubecula	81.2	FM + OW	CM Herrera unpubl.
64.8	70.2	OW	58, 64, CM Herrera unpubl.
	69.0	FM + OW	P Jordano unpubl.
	60.5	OW	64
	46.9	OW	138
	43.0	FM + OW	50
	45.1	FM	102

[a] Computed using sample sizes as weighting factors.
[b] Based on analyses of stomach contents. Figures without this superscript are based on analyses of fecal and/or gastrointestinal samples.
[c] FM, fall migration (August–September); OW, overwintering period (October–March); R, year-round resident species, data for the autumn–winter period.

that extensive consumption of fruits by trans-Saharan migrants at Mediterranean stopover sites may play an important role in migratory fat deposition (7, 148), and field evidence confirms this. Migrating garden warblers (*Sylvia borin*) feeding on fruits are significantly heavier than conspecifics feeding on insects alone (109, 159). As premigratory fat deposition is an essential requisite for successful migration across the Saharan Desert (8, 16), extensive frugivory at Mediterranean stopover localities has obvious short-term survival value for fall migrants.

Overwintering frugivores accumulate only small-to-moderate amounts of fat during their October-March stay in Mediterranean habitats, but frugivory is also important for survival. In *Sylvia atricapilla* and *Erithacus rubecula*, body mass increases significantly over daytime due to fat storage, which is depleted during the long winter nights (23, 93). Lipid-rich fruits (*Pistacia, Olea, Viburnum*, see below) invariably predominate in the diets of overwintering frugivores (Refs. in Table 2), and they must play an essential role in the daily rebuilding of fat stores. In overwintering *S. atricapilla*, fat accumulation is directly related to the extent of frugivory (109), and fatty acid composition of accumulated fat coincides closely with that of lipids extracted from the fruits that predominate in its diet (63). In the less heavily frugivorous *Erithacus rubecula* (Table 2), a direct relationship between body mass and fruit consumption has been found among birds overwintering in southern France (34), but not in southern Spain (111).

FACTORS SHAPING THE INTERACTION

As shown above, the interaction between fleshy-fruited plants and major avian dispersers in the Mediterranean is remarkably efficient for most plants and highly beneficial for the birds. The main proximate and ultimate factors contributing to give rise to these patterns are examined in this section.

Plant Traits

FRUITING PHENOLOGY   Production of fleshy fruits in Mediterranean habitats is seasonal. In the highlands, most species fruit in summer and early autumn, while in lowland and mid-elevation habitats, fruiting peaks during autumn or early winter (28, 39, 64, 87, 96, 109, 130). The shrub *Osyris quadripartita* (Santalaceae) ripens fruits throughout the year in southwestern Spanish lowlands, but even in this rather atypical species, most individuals ripen most fruits in autumn-winter (71, 78). A close match generally exists locally between the seasonal curves of avian disperser abundance and fleshy fruit production or availability (39, 64, 96, 106). Frugivorous birds form an important part of autumn-winter bird communities in lowland and mid-elevation habitats, constituting 25–50% of all birds (39, 64, 106). From summer to winter, the density of frugivores increases locally 6- to 15-fold, as a consequence of the abundant influx of overwintering species (124).

Bird-fruit phenological matching may reflect adaptive tuning of the ripening season of individual plant species to the time of greatest disperser availability (150, 160). Fuentes (49) critically examined this hypothesis for a latitudinal gradient in western Europe including temperate and Mediterranean habitats. At the plant community level, seasonal variation in the total number and biomass of fruits matched variation in abundance of avian frugivores, yet individual species showed no evidence of phenological adjustment to variations in the seasonal pattern of disperser abundance. Fuentes concluded that fruiting patterns at the community level match disperser abundance not because of adaptations by component species, but because of the greater relative abundance (and hence greater contribution to total fruit production) in each locality of those species that fruit when birds are most abundant. These species may have achieved a demographic advantage by getting more seeds dispersed than do species that ripen in other seasons. Debussche & Isenmann (39) similarly concluded that latitudinal shifts in ripening season occurring in western European plant communities mainly reflect climatic constraints on plants rather than selective pressures from dispersers. In the autumn-fruiting *Olea europea* and *Arbutus unedo*, adequate amounts of autumn rainfall are necessary for successful fruit ripening (20, 107). Cambial activity in *A. unedo* is also restricted to the autumn and closely controlled by moisture conditions (4). Severe water stress thus most likely precludes fruit ripening during the hot-dry Mediterranean summer (26), particularly in the warmer lowlands, and prevailing autumn ripening may be a response to predictable rainfall characteristic of the Mediterranean-type climate.

FRUIT SIZE   There is a close correlation across habitats between means of fruit diameter of local species and body size of dispersers (70). Small-sized frugivores and small-fruited species predominate in lowland habitats, while larger-sized frugivores (mostly *Turdus* species) and larger-fruited plants predominate in highlands (64, 70, 77, 96, 102, 103, 106, 130, 173). Altitudinal segregation of different-sized frugivores is unlikely to represent a response to variation in fruit size, as proportional contribution of small species to bird communities generally declines with elevation regardless of feeding habits (2, 123). This decline is probably a consequence of size-related differential ability to cope with adverse thermal environments. There is, however, evidence of selection by small birds in lowland habitats against large-fruited plant species and, within species, against large-fruited phenotypes. Percentage of fruit crops removed by birds declines significantly with increasing fruit diameter both among species (64) and among individuals within species (78, 107, 110). These two levels of selection, operating concurrently, have presumably shaped fruit size distributions in lowlands via long-term processes, both demographic (consistent dispersal advantages of small-fruited over large-fruited species) and evolutionary (adaptive changes within species). The latter mechanism is supported by the observation that, in several plant genera with species that segregate altitudinally (e.g. *Pistacia, Daphne, Phillyrea, Lonicera*), lowland species invariably have smaller fruits than do highland congeners (70). In the comparative analysis of angiosperm fleshy fruits by Jordano (117), traits related to fruit size were the least influenced by phylogenetic correlations and the most closely correlated with type of dispersal agent.

NUTRITIONAL CHARACTERISTICS OF FRUITS   Extensive studies have been conducted on the structural and nutritional characteristics of Mediterranean fleshy fruits (36, 57, 75, 94, 95, 97). The potential profitability of fruits to dispersal agents in a sample of southern Spanish species was shown by Herrera (57) not to differ significantly from that exhibited by a set of tropical species. Furthermore, almost all plant genera that have very lipid-rich fruits belong to plant families of tropical affinities (Oleaceae, Anacardiaceae, Lauraceae). In the genus *Pistacia*, for example, high lipid content has been found consistently in species from Israel (58% of pulp dry mass) (97), southern France (50–61%) (36), and southern Spain (56–59%) (75, 110). Other genera with lipid-rich fruits include *Laurus, Olea, Rhus, Juniperus, Rubia*, and *Viburnum*. As noted earlier, these fruits are essential for the maintenance of frugivory among overwintering dispersers. A high protein content has been also reported for some species (36, 75, 97), but this should be interpreted with caution, because abundant nonprotein nitrogen compounds in fruits of many Mediterranean species may have led to inflated protein content estimates (95).

Nutritional characteristics of fruits are related to season of ripening. Average water content of pulp decreases, and lipid content increases, from summer-through autumn- to winter-ripening species (36, 61). Species ripening in summer produce fruits characterized by water- and carbohydrate-rich pulps, while lipid-rich fruit pulps are predominantly found among autumn- and winter-fruiting ones. Energy-rich fruits are thus most abundant precisely at a time of year when the energy demands of dispersers are probably highest, while those providing most water are more frequent during the summer, when water availability to birds is at its yearly minimum. This seasonal matching between fruit attributes and disperser requirements was interpreted by Herrera (61) in terms of ("diffuse") coevolution between plants and birds. Results of later investigations, however, cast doubt on this interpretation. Characteristics of Mediterranean fleshy fruits, and especially nutritional ones, are closely correlated with phylogeny (75, 82, 117), and these traits have apparently undergone little evolutionary change over extended geological periods (74, 78). Autumn-winter ripening plants like *Pistacia, Olea,* and *Laurus* thus most likely have very lipid-rich fruits not because of their particular ripening seasons in the Mediterranean, but because the lineages involved intrinsically have fruits with these characteristics, regardless of habitat type or fruiting phenology. The importance of phylogenetic correlations as determinants of the characteristics of Mediterranean fruits has been demonstrated by Herrera (82). Interspecific variation in fruit shape among vertebrate-dispersed plants of the Iberian Peninsula did not depart significantly from that predicted by an allometry-based null hypothesis, and deviations of individual species from the allometric relationship were unrelated to dispersal mode and originated from genus and species-specific variation in fruit shape (see also 117).

Bird Traits

FRUGIVORE ABUNDANCE AND RESOURCE TRACKING   High rates of fruit removal experienced by most Mediterranean plants are a consequence of extensive frugivory by dispersers, but also of disperser abundance and ability to track spatio-temporal variations in fruit abundance. Variation among habitats in success of fruit removal, particularly with elevation (Table 1), are due to concurrent variation in absolute and relative abundances of dispersers and fruit predators (55, 64, 70, 85, 116). During autumn-winter, dispersers were 3.5 times more abundant in lowland than highland southern Spanish habitats, accounting for 79% of frugivorous bird populations in the former but only 52% in the latter (64). In lowland habitats, species of overwintering dispersers may reach extraordinary densities—up to 154 birds/10 ha (106).

Fruit production is smaller and fluctuates more across years in highland than lowland habitats (76, 85, 106, 107, 113, 115). In highlands, between-year variation in abundance of small-sized dispersers (*Sylvia, Erithacus*) does not track variation in fruit supply (76), but abundance of larger-sized dispersers (*Turdus*) generally does (115). Although there is also some annual variation, fruit production is more predictable in lowlands (48, 106, 107, 139, 142). There, abundance of both small- and larger-sized dispersers parallels annual changes in fruit supply (106, 142). Although individual frugivorous birds tend to return to the same winter quarters in consecutive years (124), their nomadic behavior enables them to track spatio-temporal variation in fruit availability at a regional scale (31, 140). Individuals of *Sylvia atricapilla* overwintering in the Mediterranean Basin wander over hundreds of kilometers during the same winter season (120).

TROPHIC ADAPTATIONS   Mediterranean avian dispersers are seasonal frugivores that shift from an insect-dominated diet in spring-summer to a fruit-dominated one in autumn-winter. At least in some species, this seasonal dietary shift is not a short-term response to increased availability of fruits but is controlled by an endogenous rhythm of food preferences (8, 12, 148). Garden (*Sylvia borin*) and Blackcap (*S. atricapilla*) warblers kept under controlled laboratory conditions with unlimited availability of various food items exhibited seasonal changes in food preference, shifting spontaneously from insectivory to frugivory in late summer (8, 12). In the Garden Warbler, the shift to frugivory is accompanied by hyperphagia, increased assimilation efficiency, food selection–mediated compensation for specific nutrient deficiencies of fruits, and regulation of daily food intake (7).

Faced with the same abundant fruit supply, major seed dispersers (*Sylvia, Erithacus, Turdus*) rely much more heavily on fruits for food locally than do coexisting fruit predators (finches and titmice; 64, 85). This difference is partly attributable to morphological preadaptations of dispersers that enable them to handle and swallow whole fruits efficiently. Dispersers tend to have flatter and broader bills than do fruit predators, and they have a wider mouth relative to bill width (65, 108). Functional digestive adaptations, however, seem to play the main role in allowing sustained and heavy frugivory by major dispersers (65, 108). Gut passage time of legitimate dispersers is considerably shorter than that of fruit predators of similar body size (65, 104, 108). The dry mass of nutritious material obtainable per mass unit of fresh whole fruit ingested is very low, as it is "diluted" by high water content and indigestible seeds. An ability to process fruits rapidly is thus essential for dispersers to exploit extensively these superabundant but, individually, minimally rewarding items. The shorter gut passage times of Mediterranean dispersers reflect faster food passage rates, not shorter intestine lengths. In fact, dispersers have proportionally longer intestines than do nonfrugivores and fruit predators (108), a characteristic that probably enhances assimilatory efficiency (147). In *Pycnonotus leucogenys*, intestine length varies seasonally and is directly correlated with the proportion of fruits in the diet (3).

Another factor that may contribute to the extensive frugivory of major dispersers is their tolerance of secondary compounds in the pulp of ripe fruits (11, 62, 95). Seuter (145) found tolerance to the alkaloid atropine (occurring in *Atropa* spp. fruits) to be 10^3 times greater in *Turdus merula* than in humans. Fruit predators only rarely include toxic fruits in their diet, while legitimate dispersers consume them abundantly; and a close correlation exists across disperser species between degree of frugivory and occurrence of toxic fruits in the diet (72). Furthermore, degree of frugivory is correlated with the frequency of chemically defended, aposematic insects in the diet (72), which also suggests the existence of enhanced tolerance to toxic metabolites among Mediterranean dispersers.

LIMITS ON PLANT ADAPTATION TO DISPERSERS

Mediterranean frugivorous birds are able to discriminate among co-occurring fruit species, individual fruiting plants within species, and fruits within individual fruit crops (59, 64, 78, 107). Of these three selection levels, the second is apt to have evolutionary consequences for the plant species involved (133). Furthermore, disperser species vary in the quantity and quality of dispersal effected (84, 99, 144), and this variation provides the raw material for plants to evolve adaptations to the one or few taxa providing the best dispersal services. Why, then, are adaptations of Mediterranean plants to their current dispersers so infrequent, in contrast with the situation occurring among birds? Limitations on plant adaptation to dispersers that are inherent to most or all plant-disperser systems have been reviewed elsewhere (73, 74, 91, 170). I briefly consider two further reasons for Mediterranean systems.

With only minor exceptions (*Viburnum tinus* and highland junipers, which rely for dispersal on *E. rubecula* and *Turdus torquatus*, respectively; 34, 64, 115, 173), Mediterranean plants have their seeds dispersed by an array of bird species, and degrees of dependence on particular dispersers fluctuate among years and sites (37, 50, 55, 64, 96, 107, 116). Variation at different temporal and spatial scales in the composition and relative abundance of the dispersers and fruit predators that interact with a given plant, along with incongruence between the geographical distributions of plant and bird species (55, 77, 115, 116), limit the possibilities of selection by particular dispersers on dispersal-related traits.

Most studies focusing on patterns of phenotypic selection exerted by Mediterranean dispersers on dispersal-related plant traits have shown that, although individual variation in these traits actually translates into differential seed dispersal success, such variation has only a minor influence on final reproductive output of the plants (78, 85, 107, 110, 166). Individual variation in fruit traits explains only 1.8%, $\approx$ 2%, and 0.5–2.3% of the variance in reproductive output in *Pistacia lentiscus, Osyris quadripartita*, and *Phillyrea latifolia*, respectively (78, 85, 110). The overwhelming influence on reproductive output of differences in flower production and crop size, and the effects of multiple interactions with vertebrate and invertebrate fruit predators, are the main reasons for the negligible influence of selection by dispersers on the realized fecundity of Mediterranean plants. Under this regime of extremely low selection intensities by avian dispersers, responses of plants to selection on fruit or fruiting characteristics are unlikely, and strong phylogenetic correlations of fruit traits are not surprising.

HISTORY OF BIRD DISPERSAL IN THE MEDITERRANEAN

The contemporary flora of the Mediterranean Basin represents a historically heterogeneous assortment of lineages with varied origins in time and space

(83, 134, 137). Some plant genera were present in the region well before the initiation of the Mediterranean-type climatic conditions in the Pliocene, while others appeared much more recently. The first appearances in the fossil record of genuinely Mediterranean fleshy-fruited genera are spread over the Eocene (e.g. *Chamaerops, Smilax*), Oligocene (e.g. *Arbutus, Olea*), and Miocene (e.g. *Pistacia, Phillyrea*) (see 134 for review).

Lineage age of extant Mediterranean woody plants is correlated with several aspects of their reproductive biology, including seed dispersal (83). The fleshy fruit–producing habit is extremely infrequent among those lineages that arose in, or immigrated to, truly Mediterranean-climate scenarios. In this group of young lineages, only 5.6% of genera produce fleshy fruits and are bird-dispersed. Other taxa, in contrast, are the remnants of the tropical and subtropical flora that was widespread in west-central and southern Europe when the region was still tropical (Refs. in 83). In this group of old lineages, 94.4% of genera produce fleshy fruits. Most old, fruit-producing lineages presently occur in lowland and mid-elevation habitats, where they are responsible for increased familial diversity of local fleshy-fruited species assemblages. Furthermore, the group of old lineages includes all the distinctive lipid-rich, autumn-winter fruiting plants that are essential in maintaining the frugivorous diet of major avian seed dispersers (*Pistacia, Olea*).

Differences between pre-Mediterranean lineages in seed dispersal method seem to have led to differential extinction probabilities from the initiation of the Mediterranean-type climate until present. Among woody plant genera that occurred in the Mediterranean region in the Pliocene, 47.4% of those producing fleshy fruits have become extinct, compared to 69.2% extinction among genera with other seed dispersal methods ($P = 0.061$, χ^2 test; computations use data from Ref. 83). This trend suggests that fleshy-fruited lineages have some demographic advantage over dry-fruited ones that has allowed them to persist longer in the region despite changes in environmental conditions. These possible advantages, however, are not attributable to the action of current dispersal agents. In comparison with their food plants, both the species of Mediterranean avian dispersers and their migratory behavior are young. Avian fossils referable to recent species of frugivores (including *Turdus* and *Sylvia* spp.) do not appear in Europe until the Pleistocene (122), and this recent origin is consistent with biogeographical interpretations (17) and estimates of lineage age obtained from DNA hybridization studies (146). In addition, the present global pattern of seasonal migrations must have been established even more recently, during the 15,000–20,000 years since the peak of the last glaciation (124). The migratory behavior of overwintering species such as *Sylvia atricapilla* and *Erithacus rubecula* has a genetic basis (13–15), and microevolutionary changes may take place rapidly in response to modifications in geographical patterns of food availability. Populations of *S. atricapilla* breeding in western Europe have evolved a new, genetically based migratory behavior in only three decades in response to improved wintering conditions in the new British wintering areas (14).

CONCLUSION

While there is substantial evidence that Mediterranean avian seed dispersers have evolved physiological, digestive, and behavioral adaptations to take advantage of the abundant and profitable fruit supply in the region, analogous evidence is lacking for fruiting plants. Interspecific variation in fruiting phenology, fruit shape, nutritional composition of fruits, and structural fruit characteristics of Mediterranean plants are more closely tied to phylogeny or to the abiotic environment than to the current disperser/dispersal environment (49, 75, 82, 117). Mediterranean fleshy-fruited plants and their current avian dispersers lack a common history of interaction; the birds presumably evolved their present trophic and migratory adaptations in response to a previous ecological scenario characterized by mild winters and abundant and energetically profitable fruits.

It once seemed intuitive to expect that, in ecological interactions in which counterparts mutually benefit from participating in the interaction, reciprocal adaptations enhancing these benefits should evolve ("coevolution"; 100). The earliest evolutionary treatments of plant-animal seed dispersal systems were imbued with this view (92, 125, 150), but later studies showed that, for a variety of reasons, plant-disperser coevolution was unlikely to occur (73, 74, 91, 170). Studies on Mediterranean seed dispersal systems not only support this latter view, they also indicate that extremely efficient plant-disperser mutualisms do not require, and thus are not evidence for, mutual evolutionary adjustments of participants.

In ecological and evolutionary time, current Mediterranean dispersal systems seem to have been shaped through 1) actual adaptations of morphologically and behaviorally preadapted dispersers to preexisting plant resources, and 2) disperser-mediated processes occurring at an ecological time scale and based on differential recruitment of plant species as a function of disperser preferences. These adjustments are based on ecological, not evolutionary, processes and are best described as instances of "ecological fitting" (101) or "habitat shaping" (70). On the plant side, the prevalence of historical effects (long-term persistence of traits evolved in temporally and ecologically distant scenarios; 83) apparently reflects a series of ecological limitations inherent to the interaction between plants and dispersal agents that constrain the adaptation of plants to dispersers. A practical corollary following from this review is that future investigations on the evolutionary ecology of Mediterranean plant-disperser systems should concentrate more on the animal than on the plant side of the interaction, for it is among dispersers that adaptive explanations related to present-day environments seem both most straightforward and best justified.

ACKNOWLEDGMENTS

I am indebted to Pedro Jordano for valuable discussion, assistance with the literature, and unpublished data; to him, Conchita Alonso, Juan A. Amat, José L. Yela, and an anonymous reviewer for critical comments on the manuscript; and to Javier Guitián, Pedro Rey, and Anna Traveset for generously supplying preprints or unpublished information. While preparing this paper, I was funded by DGICYT grant PB91–0114.

Literature Cited

1. Adar M, Safriel UN, Izhaki I. 1992. Removal of the green fruit of *Asparagus aphyllus* by birds—seed predation or dispersal? In *Plant-Animal Interactions in Mediterranean-Type Ecosystems*, ed. CA Thanos, pp. 281–86. Athens: Univ. Athens
2. Affre G. 1980. Distribution altitudinale des oiseaux dans l'est des Pyrenees françaises. *Oiseau R.F.O.* 50:1–22
3. Al-Dabbagh KY, Jiad JH, Waheed IN. 1987. The influence of diet on the intestine length of the white-cheeked bulbul. *Ornis Scand.* 18:150–52
4. Arianoutsou-Faraggitaki M, Psaras G, Christodoulakis N. 1984. The annual rhythm of cambial activity in two woody species of the Greek "Maquis." *Flora* 175:221–29
5. Aronne C, Wilcock CC. 1994. First evidence of myrmecochory in fleshy-fruited shrubs of the Mediterranean region. *New Phytol.* 127:781–88
6. Arroyo J, Marañón T. 1990. Community ecology and distributional spectra of Mediterranean shrublands and heathlands in southern Spain. *J. Biogeogr.* 17:163–76
7. Bairlein F. 1991. Nutritional adaptations to fat deposition in the long-distance migratory garden warbler *Sylvia borin*. *Acta 20th Congr. Int. Ornithol., Christchurch*, pp. 2149–58
8. Bairlein F, Gwinner E. 1994. Nutritional mechanisms and temporal control of migratory energy accumulation in birds. *Annu. Rev. Nutr.* 14:187–215

9. Barnea A, Yom-Tov Y, Friedman J. 1990. Differential germination of two closely related species of *Solanum* in response to bird ingestion. *Oikos* 57:222–28
10. Barnea A, Yom-Tov Y, Friedman J. 1991. Does ingestion by birds affect seed germination? *Funct. Ecol.* 5:394–402
11. Barnea A, Harborne JB, Pannell C. 1993. What parts of fleshy fruits contain secondary compounds toxic to birds and why? *Biochem. Syst. Ecol.* 21:421–29
12. Berthold P. 1976. Animalische und vegetabilische Ernährung omnivorer Singvogelarten: Nahrungsbevorzugung, Jahresperiodik der Nahrungswahl, physiologische und ökologische Bedeutung. *J. Ornithol.* 117:145–209
13. Berthold P, Querner U. 1981. Genetic basis of migratory behavior in European warblers. *Science* 212:77–79
14. Berthold P, Helbig AJ, Mohr G, Querner U. 1992. Rapid microevolution of migratory behaviour in a wild bird species. *Nature* 360:668–69
15. Biebach H. 1983. Genetic determination of partial migration in the European Robin (*Erithacus rubecula*). *Auk* 100:601–6
16. Biebach H, Friedrich W, Heine G. 1986. Interaction of body-mass, fat, foraging and stopover period in trans-Sahara migrating passerine birds. *Oecologia* 69:370–79
17. Blondel J. 1982. Caractérisation et mise

en place des avifaunes dans le bassin méditerranéen. *Ecol. Medit.* 8:253–72

18. Calisti M, Ciampalini B, Lovari S, Lucherini M. 1990. Food habits and trophic niche variation of the red fox *Vulpes vulpes* (L., 1758) in a Mediterranean coastal area. *Rev. Ecol. (Terre Vie)* 45:309–20

19. Calvario E, Fraticelli F. 1986. *Rubia peregrina* berries in the winter food of Robins *Erithacus rubecula*. *Avocetta* 10:115–18

20. Chiarucci A, Pacini E, Loppi S. 1993. Influence of temperature and rainfall on fruit and seed production of *Arbutus unedo* L. *Bot. J. Linn. Soc.* 111:71–82

21. Cramp S, ed. 1988. *Handbook of the Birds of Europe, the Middle East and North Africa.* Vol. 5. Tyrant Flycatchers to Thrushes. Oxford: Oxford Univ. Press. 1063 pp.

22. Cramp S, ed. 1992. *Handbook of the Birds of Europe, the Middle East and North Africa.* Vol. 6. *Warblers.* Oxford: Oxford Univ. Press. 728 pp.

23. Cuadrado M, Rodríguez M, Arjona S. 1989. Fat and weight variations of Blackcaps wintering in southern Spain. *Ring. & Migr.* 10:89–97

24. Darwin C. 1862. *On the Various Contrivances by Which British and Foreign Orchids Are Fertilised by Insects.* London: Murray. 300 pp.

25. Darwin C. 1877. *The Different Forms of Flowers on Plants of the Same Species.* London: Murray. 352 pp.

26. de Lillis M, Fontanella A. 1992. Comparative phenology and growth in different species of the Mediterranean maquis of central Italy. *Vegetatio* 99:83–96

27. Debussche M. 1985. Rôle des oiseaux disséminateurs dans la germination des graines de plantes à fruits charnus en région méditerranéenne. *Acta Oecol., Oecol. Plant.* 6:365–74

28. Debussche M. 1988. La diversité morphologique des fruits charnus en Languedoc méditerranéen: relations avec les caractéristiques biologiques et la distribution des plantes, et avec les disséminateurs. *Acta Oecol., Oecol. Gener.* 9:37–52

29. Debussche M, Escarré J, Lepart J. 1982. Ornithochory and plant succession in mediterranean abandoned orchards. *Vegetatio* 48:255–66

30. Debussche M, Isenmann P. 1983. La consommation des fruits chez quelques fauvettes méditerranéennes (*Sylvia melanocephala, S. cantillans, S. hortensis* et *S. undata*) dans la région de Montpellier (France). *Alauda* 51:302–8

31. Debussche M, Isenmann P. 1984. Origine et nomadisme des fauvettes à tête noire (*Sylvia atricapilla*) hivernant en zone méditerranéenne française. *Oiseau R.F.O.* 54:101–7

32. Debussche M, Isenmann P. 1985. Le régime alimentaire de la Grive musicienne (*Turdus philomelos*) en automne et en hiver dans les garrigues de Montpellier (France méditerranénne) et ses relations avec l'ornithochorie. *Rev. Ecol. (Terre Vie)* 40:379–88

33. Debussche M, Isenmann P. 1985. An example of Redwing diet in a Mediterranean wintering area. *Bird Study* 32:152–53

34. Debussche M, Isenmann P. 1985. Frugivory of transient and wintering European robins *Erithacus rubecula* in a Mediterranean region and its relationship with ornithochory. *Holarctic Ecol.* 8:157–63

35. Debussche M, Lepart J, Molina J. 1985. La dissémination des plantes à fruits charnus par les oiseaux: rôle de la structure de la végétation et impact sur la succession en région méditerranéenne. *Acta Oecol., Oecol. Gener.* 6:65–80

36. Debussche M, Cortez J, Rimbault I. 1987. Variation in fleshy fruit composition in the Mediterranean region: the importance of ripening season, lifeform, fruit type and geographical distribution. *Oikos* 49:244–52

37. Debussche M, Isenmann P. 1989. Fleshy fruit characters and the choices of bird and mammal seed dispersers in a Mediterranean region. *Oikos* 56:327–38

38. Debussche M, Isenmann P. 1990. Introduced and cultivated fleshy-fruited plants: consequences for a mutualistic Mediterranean plant-bird system. In *Biological Invasions in Europe and the Mediterranean Basin*, ed. F di Castri, AJ Hansen, M Debussche, pp. 399–416. Dordrecht: Kluwer

39. Debussche M, Isenmann P. 1992. A Mediterranean bird disperser assemblage: composition and phenology in relation to fruit availability. *Rev. Ecol. (Terre Vie)* 47:411–32

40. Debussche M, Lepart J. 1992. Establishment of woody plants in mediterranean old fields: opportunity in space and time. *Landscape Ecol.* 6:133–45

41. Debussche M, Isenmann P. 1994. Bird-dispersed seed rain and seedling establishment in patchy Mediterranean vegetation. *Oikos* 69:414–26

42. di Castri F. 1981. Mediterranean-type shrublands of the world. In Ref. 43, pp. 1–52

43. di Castri F, Goodall DW, Specht RL,

eds. 1981. *Ecosystems of the World.* Vol. 11. *Mediterranean-Type Shrublands.* Amsterdam: Elsevier. 643 pp.

44. Estrada A, Fleming TH, eds. 1986. *Frugivores and Seed Dispersal.* Dordrecht: Junk. 392 pp.

45. Ferns PN. 1975. Feeding behaviour of autumn passage migrants in north east Portugal. *Ring. & Migr.* 1:3–11

46. Fleming TH, Estrada A, eds. 1993. *Frugivory and Seed Dispersal: Ecological and Evolutionary Aspects.* Dordrecht: Kluwer. 392 pp.

47. Fortuna P. 1991. Studio sull'alimentazione della popolazione di Storni svernante nella città di Roma. *Avocetta* 15:25–31

48. Fraticelli F, Gustin M. 1986. Blackberries, *Rubus ulmifolius,* in the autumnal feeding of blue tits, *Parus caeruleus. Riv. Ital. Orn.* 56:114–16

49. Fuentes M. 1992. Latitudinal and elevational variation in fruiting phenology among western European bird-dispersed plants. *Ecography* 15:177–83

50. Fuentes M. 1994. Diets of fruit-eating birds: What are the causes of interspecific differences? *Oecologia* 97:134–42

51. Gardiazábal A. 1990. *Untersuchungen zur Ökologie rastender Kleinvögel im Nationalpark von Doñana, (Spanien): Ernährung, Fettdeposition, Zugstrategie.* PhD thesis. Univ. Köln. 158 pp.

52. Genard M, Lescourret F. 1985. Le sanglier (*Sus scrofa scrofa* L.) et les diaspores dans le sud de la France. *Rev. Ecol. (Terre Vie*) 40:343–53

53. González-Solís J, Ruiz X. 1990. Alimentación de *Turdus philomelos* en los olivares mediterráneos ibéricos durante la migración otoñal. *Misc. Zool.* 14:195–206

54. Guitián J, Sánchez JM. 1992. Seed dispersal spectra of plant communities in the Iberian Peninsula. *Vegetatio* 98:157–64

55. Guitián J, Fuentes M, Bermejo T, López B. 1992. Spatial variation in the interactions between *Prunus mahaleb* and frugivorous birds. *Oikos* 63:125–30

56. Hernández A. 1990. Observaciones sobre el papel del lagarto ocelado (*Lacerta lepida* Daudin), el erizo (*Erinaceus europaeus* L.) y el tejón (*Meles meles* L.) en la dispersión de semillas. *Doñana Acta Vert.* 17:235–42

57. Herrera CM. 1981. Are tropical fruits more rewarding to dispersers than temperate ones? *Am. Nat.* 118:896–907

58. Herrera CM. 1981. Fruit food of Robins wintering in southern Spanish mediterranean scrubland. *Bird Study* 28:115–22

59. Herrera CM. 1981. Fruit variation and competition for dispersers in natural populations of *Smilax aspera. Oikos* 36:51–58

60. Herrera CM. 1982. Datos sobre la dieta frugívora del Mirlo (*Turdus merula*) en dos localidades del sur de España. *Doñana Acta Vert.* 8:306–10

61. Herrera CM. 1982. Seasonal variation in the quality of fruits and diffuse coevolution between plants and avian dispersers. *Ecology* 63:773–85

62. Herrera CM. 1982. Defense of ripe fruits from pests: its significance in relation to plant-disperser interactions. *Am. Nat.* 120:218–41

63. Herrera CM. 1983. Coevolución de plantas y frugívoros: la invernada mediterránea de algunos paseriformes. *Alytes* 1:177–90

64. Herrera CM. 1984. A study of avian frugivores, bird-dispersed plants, and their interaction in Mediterranean scrublands. *Ecol. Monogr.* 54:1–23

65. Herrera CM. 1984. Adaptation to frugivory of Mediterranean avian seed dispersers. *Ecology* 65:609–17

66. Herrera CM. 1984. Tipos morfológicos y funcionales en plantas del matorral mediterráneo del sur de España. *Studia Oecol.* 5:7–33

67. Herrera CM. 1984. Avian interference of insect frugivory: an exploration into the plant-bird-fruit pest evolutionary triad. *Oikos* 42:203–10

68. Herrera CM. 1984. Selective pressures on fruit seediness: differential predation of fly larvae on the fruits of *Berberis hispanica. Oikos* 42:166–70

69. Herrera CM. 1984. Seed dispersal and fitness determinants in wild rose: combined effects of hawthorn, birds, mice, and browsing ungulates. *Oecologia* 63:386–93

70. Herrera CM. 1985. Habitat-consumer interactions in frugivorous birds. In *Habitat Selection in Birds,* ed. ML Cody, pp. 341–65. New York: Academic

71. Herrera CM. 1985. Predispersal reproductive biology of female *Osyris quadripartita* (Santalaceae), a hemiparasitic dioecious shrub of Mediterranean scrublands. *Bot. J. Linn. Soc.* 90:113–27

72. Herrera CM. 1985. Aposematic insects as six-legged fruits: incidental short-circuiting of their defense by frugivorous birds. *Am. Nat.* 126:286–93

73. Herrera CM. 1985. Determinants of plant-animal coevolution: the case of mutualistic dispersal of seeds by vertebrates. *Oikos* 44:132–41

74. Herrera CM. 1986. Vertebrate-dispersed plants: Why they don't behave the way they should. In Ref. 44, pp. 5–18

75. Herrera CM. 1987. Vertebrate-dispersed plants of the Iberian Peninsula: a study of fruit characteristics. *Ecol. Monogr.* 57:305–31

76. Herrera CM. 1988. Variaciones anuales en las poblaciones de pájaros frugívoros y su relación con la abundancia de frutos. *Ardeola* 35:135–42

77. Herrera CM. 1988. Avian frugivory and seed dispersal in Mediterranean habitats: regional variation in plant-animal interaction. In *Acta XIX Congressus Internationalis Ornithologici*, Vol. 1, ed. H Ouellet, pp. 509–17. Ottawa: Univ. Ottawa Press

78. Herrera CM. 1988. The fruiting ecology of *Osyris quadripartita*: individual variation and evolutionary potential. *Ecology* 69:233–49

79. Herrera CM. 1988. Habitat-shaping, host plant use by a hemiparasitic shrub, and the importance of gut fellows. *Oikos* 51:383–86

80. Herrera CM. 1988. Plant size, spacing patterns, and host-plant selection in *Osyris quadripartita*, a hemiparasitic dioecious shrub. *J. Ecol.* 76:995–1006

81. Herrera CM. 1989. Frugivory and seed dispersal by carnivorous mammals, and associated fruit characteristics, in undisturbed Mediterranean habitats. *Oikos* 55:250–62

82. Herrera CM. 1992. Interspecific variation in fruit shape: allometry, phylogeny, and adaptation to dispersal agents. *Ecology* 73:1832–41

83. Herrera CM. 1992. Historical effects and sorting processes as explanations for contemporary ecological patterns: character syndromes in Mediterranean woody plants. *Am. Nat.* 140:421–46

84. Herrera CM, Jordano P. 1981. *Prunus mahaleb* and birds: the high-efficiency seed dispersal system of a temperate fruiting tree. *Ecol. Monogr.* 51:203–18

85. Herrera CM, Jordano P, López-Soria L, Amat JA. 1994. Recruitment of a mast-fruiting, bird-dispersed tree: bridging frugivore activity and seedling establishment. *Ecol. Monogr.* 64:315–44

86. Herrera J. 1984. Vegetación del Valle del Guadahornillos (Sierra de Cazorla, Jaén). *Studia Oecol.* 5:77–96

87. Herrera J. 1986. Flowering and fruiting phenology in the coastal shrublands of Doñana, south Spain. *Vegetatio* 68: 91–98

88. Herrera J. 1987. Flower and fruit biology in southern Spanish Mediterranean shrublands. *Ann. Miss. Bot. Gard.* 74: 69–78

89. Herrera J. 1987. Biología reproductiva de algunas especies del matorral de Doñana. *Anal. Jard. Bot. Madrid* 44: 483–97

90. Houssard C, Escarré J, Romane F. 1980. Development of species diversity in some mediterranean plant communities. *Vegetatio* 43:59–72

91. Howe HF. 1984. Constraints on the evolution of mutualisms. *Am. Nat.* 123: 764–77

92. Howe HF, Estabrook GF. 1977. On intraspecific competition for avian dispersers in tropical trees. *Am. Nat.* 111: 817–32

93. Ioale P, Benvenuti S. 1982. Seasonal and diurnal variation of weight in four passeriformes in autumn and winter. *Avocetta* 6:63–74

94. Izhaki I. 1992. A comparative analysis of the nutritional quality of mixed and exclusive fruit diets for yellow-vented bulbuls. *Condor* 94:912–23

95. Izhaki I. 1993. Influence of nonprotein nitrogen on estimation of protein from total nitrogen in fleshy fruits. *J. Chem. Ecol.* 19:2605–15

96. Izhaki I, Safriel UN. 1985. Why do fleshy-fruit plants of the mediterranean scrub intercept fall- but not spring-passage of seed-dispersing migratory birds? *Oecologia* 67:40–43

97. Izhaki I, Safriel UN. 1989. Why are there so few exclusively frugivorous birds? Experiments on fruit digestibility. *Oikos* 54:23–32

98. Izhaki I, Safriel UN. 1990. The effect of some Mediterranean scrubland frugivores upon germination pattern. *J. Ecol.* 78:56–65

99. Izhaki I, Walton PB, Safriel UN. 1991. Seed shadow generated by frugivorous birds in an eastern Mediterranean scrub. *J. Ecol.* 79:575–90

100. Janzen DH. 1980. When is it coevolution? *Evolution* 34:611–12

101. Janzen DH. 1985. On ecological fitting. *Oikos* 45:308–10

102. Jordano P. 1981. Alimentación y relaciones tróficas entre los passeriformes en paso otoñal por una localidad de Andalucía central. *Doñana Acta Vert.* 8:103–24

103. Jordano P. 1982. Migrant birds are the main seed dispersers of blackberries in southern Spain. *Oikos* 38:183–93

104. Jordano P. 1984. *Relaciones entre plantas y aves frugívoras en el matorral mediterráneo del área de Doñana.* PhD thesis. Univ. Sevilla. 284 pp.

105. Jordano P. 1984. Seed weight variation

and differential avian dispersal in blackberries *Rubus ulmifolius*. *Oikos* 43:149–53

106. Jordano P. 1985. El ciclo anual de los paseriformes frugívoros en el matorral mediterráneo del sur de España: importancia de su invernada y variaciones interanuales. *Ardeola* 32:69–94

107. Jordano P. 1987. Avian fruit removal: effects of fruit variation, crop size, and insect damage. *Ecology* 68:1711–23

108. Jordano P. 1987. Frugivory, external morphology and digestive system in mediterranean sylviid warblers *Sylvia* spp. *Ibis* 129:175–89

109. Jordano P. 1988. Diet, fruit choice and variation in body condition of frugivorous warblers in Mediterranean scrubland. *Ardea* 76:193–209

110. Jordano P. 1989. Predispersal biology of *Pistacia lentiscus* (Anacardiaceae): cumulative effects on seed removal by birds. *Oikos* 55:375–86

111. Jordano P. 1989. Variación de la dieta frugívora otoño-invernal del Petirrojo (*Erithacus rubecula*): efectos sobre la condición corporal. *Ardeola* 36:161–83

112. Jordano P. 1990. Utilización de los frutos de *Pistacia lentiscus* (Anacardiaceae) por el verderón común (*Carduelis chloris*). In *Principios en etología*, ed. L Arias, P Recuerda, T Redondo, pp. 145–53. Córdoba: Monte Piedad

113. Jordano P. 1991. Gender variation and expression of monoecy in *Juniperus phoenicea* (L.) (Cupressaceae). *Bot. Gaz.* 152: 76–485

114. Jordano P. 1992. Fruits and frugivory. In *Seeds. The Ecology of Regeneration in Plant Communities*, ed. M Fenner, pp. 105–56. London: CAB Int.

115. Jordano P. 1993. Geographical ecology and variation of plant-seed disperser interactions: southern Spanish junipers and frugivorous thrushes. In Ref. 46, pp. 85–104

116. Jordano P. 1994. Spatial and temporal variation in the avian-frugivore assemblage of *Prunus mahaleb*: patterns and consequences. *Oikos* 71:479–91

117. Jordano P. 1995. Angiosperm fleshy fruits and seed dispersers: a comparative analysis of adaptation and constraints in plant-animal interactions. *Am. Nat.* 145: 163–91

118. Jordano P, Herrera CM. 1981. The frugivorous diet of Blackcap populations *Sylvia atricapilla* wintering in southern Spain. *Ibis* 123:502–7

119. Jordano P, Herrera CM. 1995. Shuffling the offspring: uncoupling and spatial discordance of multiple stages in vertebrate seed dispersal. *Ecoscience.* In press

120. Klein H, Berthold P, Gwinner E. 1973. Der Zug europäischer Garten und Mönchsgrasmücken (*Sylvia borin* und *S. atricapilla*). *Vogelwarte* 27:73–134

121. Krüsi BO, Debussche M. 1988. The fate of flowers and fruits of *Cornus sanguinea* L. in three contrasting Mediterranean habitats. *Oecologia* 74:592–600

122. Lambrecht K. 1964. *Handbuch der Palaeornithologie.* Amsterdam: Asher. 1024 pp.

123. Lebreton P, Broyer J. 1981. Contribution à l'étude des relations avifaune/altitude. I. Au niveau de la région Rhône-Alpes. *Oiseau R.F.O.* 51:265–85

124. Lövei GL. 1989. Passerine migration between the Palaearctic and Africa. *Current Ornithol.* 6:143–74

125. McKey D. 1975. The ecology of co-evolved seed dispersal systems. In *Coevolution of Animals and Plants*, ed. LE Gilbert, PH Raven, pp. 159–91. Austin: Univ. Texas Press

126. Ménard N. 1985. Le régime alimentaire de *Macaca sylvanus* dans differents habitats d'Algerie. I. Régime en chênaie decidue. *Rev. Ecol. (Terre Vie)* 40:451–69

127. Ménard N, Vallet D. 1986. Le régime alimentaire de *Macaca sylvanus* dans differents habitats d'Algerie: II. Régime en forêt sempervirente et sur les sommets rocheux. *Rev. Ecol. (Terre Vie)* 41:173–92

128. Muñoz Reinoso JC. 1993. Consumo de gálbulos de sabina (*Juniperus phoenicea* ssp. *turbinata* Guss, 1891) y dispersión de semillas por el conejo (*Oryctolagus cuniculus* L.) en el Parque Nacional de Doñana. *Doñana Acta Vert.* 20:49–58

129. Nogales M, Hernández EC. 1994. Interinsular variations in the spring and summer diet of the Raven *Corvus corax* in the Canary Islands. *Ibis* 136:441–47

130. Obeso JR. 1985. *Comunidades de paseriformes y frugivorismo en altitudes medias de la Sierra de Cazorla.* PhD thesis. Univ. Oviedo. 475 pp.

131. Obeso JR. 1986. Alimentación del zorzal charlo (*Turdus viscivorus*) en la Sierra de Cazorla. *Doñana Acta Vert.* 13:95–102

132. Obeso JR. 1989. Fruit removal and potential seed dispersal in a southern Spanish population of *Berberis vulgaris* ssp. *australis* (Berberidaceae). *Acta Oecol., Oecol. Plant.* 10:321–28

133. Obeso JR, Herrera CM. 1994. Inter- and intraspecific variation in fruit traits

in co-occurring vertebrate-dispersed plants. *Int. J. Plant Sci.* 155:382–87

134. Palamarev E. 1989. Paleobotanical evidences of the Tertiary history and origin of the Mediterranean sclerophyll dendroflora. *Plant Syst. Evol.* 162:93–107

135. Pigozzi G. 1992. Frugivory and seed dispersal by the European badger in a mediterranean habitat. *J. Mamm.* 73:630–39

136. Polunin O, Walters M. 1985. *A Guide to the Vegetation of Britain and Europe.* Oxford: Oxford Univ. Press. 238 pp.

137. Pons A. 1981. The history of the Mediterranean shrublands. In Ref. 43, pp. 131–38

138. Rey PJ. 1992. *Preadaptación de la avifauna frugívora invernante al cultivo del olivar.* PhD thesis. Univ. Granada. 231 pp.

139. Rey PJ. 1993. The role of olive orchards in the wintering of frugivorous birds in Spain. *Ardea* 81:151–60

140. Rey PJ. 1995. Spatio-temporal variation in fruit availability and frugivorous bird abundance in Mediterranean olive orchards. *Ecology* In press

141. Ridley HN. 1930. *The Dispersal of Plants Throughout the World.* Ashford: Reeve. 744 pp.

142. Rodríguez M, Cuadrado M, Arjona S. 1986. Variation in the abundance of Blackcaps (*Sylvia atricapilla*) wintering in an olive (*Olea europaea*) orchard in southern Spain. *Bird Study* 33:81–86

143. Santos T, Tellería JL. 1994. Influence of forest fragmentation on seed consumption and dispersal of Spanish juniper *Juniperus thurifera*. *Biol. Conserv.* 70:129–34

144. Schupp EW. 1993. Quantity, quality and the effectiveness of seed dispersal by animals. In Ref. 46, pp. 15–29

145. Seuter F. 1970. Ist eine endozoische Verbreitung der Tollkirsche durch Amsel und Star möglich? *Zool. Jahrb. Physiol.* 75:342–59

146. Sibley CG, Ahlquist JE. 1990. *Phylogeny and Classification of Birds.* New Haven: Yale Univ. Press. 976 pp.

147. Sibly RM. 1981. Strategies of digestion and defecation. In *Physiological Ecology,* ed. CR Towsend, P Calow, pp. 109–39. London: Blackwell

148. Simons D, Bairlein F. 1990. Neue Aspekte zur zugzeitlichen Frugivorie der Gartengrasmücke (*Sylvia borin*). *J. Ornithol.* 131:381–401

149. Snow DW. 1965. A possible selective factor in the evolution of fruiting seasons in tropical forest. *Oikos* 15:274–81

150. Snow DW. 1971. Evolutionary aspects of fruit-eating by birds. *Ibis* 113:194–202

150a. Soler JJ, Soler M. 1991. Análisis comparado del régimen alimenticio durante el período otoño-invierno de tres especies de córvidos en un área de simpatría. *Ardeola* 38:69–89

151. Soler M, Pérez-González JA, Tejero E, Camacho I. 1988. Alimentación del zorzal alirrojo (*Turdus iliacus*) durante su invernada en olivares de Jaén (sur de España). *Ardeola* 35:183–96

152. Soler M, Alcalá N, Soler JJ. 1990. Alimentación de la grajilla *Corvus monedula* en tres zonas del sur de España. *Doñana Acta Vert.* 17:17–48

153. Soler M, Pérez-González JA, Soler JJ. 1991. Régimen alimenticio del mirlo comuacn (*Turdus merula*) en el sureste de la Península Ibérica durante el período otoño-invierno. *Doñana Acta Vert.* 18:133–48

154. Deleted in proof

155. Suc JP. 1984. Origin and evolution of the Mediterranean vegetation and climate in Europe. *Nature* 307:429–32

156. Tellería JL, ed. 1988. *Invernada de aves en la Península Ibérica.* Madrid: Soc. Esp. Ornitol. 208 pp.

157. Thébaud C, Debussche M. 1992. A field test of the effects of infructescence size on fruit removal by birds in *Viburnum tinus. Oikos* 65:391–94

158. Thirgood JV. 1981. *Man and the Mediterranean Forest.* London: Academic Press. 194 pp.

159. Thomas DK. 1979. Figs as a food source of migrating Garden Warblers in southern Portugal. *Bird Study* 26:187–91

160. Thomson JN, Willson MF. 1979. Evolution of temperate fruit/bird interactions: phenological strategies. *Evolution* 33:973–82

161. Tomaselli R. 1976. La dégradation du maquis méditerranéenne. *Not. Tech. MAB* 2:35–76

162. Traveset A. 1992. Production of galls in *Phillyrea angustifolia* induced by cecidomyiid flies. In *Plant-Animal Interactions in Mediterranean-Type Ecosystems*, ed. CA Thanos, pp. 198–204. Athens: Univ. Athens

163. Traveset A. 1993. Weak interactions between avian and insect frugivores: the case of *Pistacia terebinthus* L. (Anacardiaceae). In Ref. 46, pp. 191–203

164. Traveset A. 1994. The effect of *Agonoscena targionii* (Licht.) (Homoptera: Psylloidea) on seed production by *Pistacia terebinthus* L. *Oecologia* 98:72–75

165. Traveset A. 1994. Influence of type

of avian frugivory on the fitness of *Pistacia terebinthus* L. *Evol. Ecol.* 8:618–27

166. Traveset A. 1994. Cumulative effects on the reproductive output of *Pistacia terebinthus* (Anacardiaceae). *Oikos* 71: 152–62

167. Traveset A. 1995. Seed dispersal of *Cneorum tricoccon* L. (Cneoraceae) by lizards and mammals in the Balearic Islands. *Acta Oecol.* In press

168. Valido A, Nogales M. 1994. Frugivory and seed dispersal by the lizard *Gallotia galloti* (Lacertidae) in a xeric habitat of the Canary Islands. *Oikos* 70: 403–11

169. van der Pijl L. 1972. *Principles of Dispersal in Higher Plants.* Berlin: Springer. 162 pp. 2nd ed.

170. Wheelwright NT, Orians GH. 1982. Seed dispersal by animals: contrasts with pollen dispersal, problems of terminology, and constraints on coevolution. *Am. Nat.* 119:402–13

171. Willson MF. 1986. Avian frugivory and seed dispersal in eastern North America. *Curr. Ornithol.* 3:223–79

172. Willson MF. 1993. Mammals as seed-dispersal mutualists in North America. *Oikos* 67:159–76

173. Zamora R. 1990. The fruit diet of Ringouzels (*Turdus torquatus*) wintering in the Sierra Nevada (south-east Spain). *Alauda* 58:67–70

SUBJECT INDEX

A

Abalone
  sperm lysins in
    protein divergence in, 413
Abir-Am, P., 340
Abrams, P. A., 509
ACA
  See Association of Collegiate
    Alumnae
ACEER
  See Amazon Center for Environmental Education and Research
Acer pennsylvanicum
  leaf litter of
    nitrogen concentration in, 494
Acer pseudoplantanus
  leaf litter of
    nitrogen concentration in, 494
Acer rubrum
  leaf litter of
    nitrogen concentration in, 494
Acer saccharum
  respiration rate in
    atmospheric carbon dioxide and, 487
Acnathaster planci
  outbreak on Great Barrier Reef, 290
Acquired hereditary immunity, 557
Adaptation, 553–75
  bacterial
    Lamarckism and, 556
  evolutionary perspective and, 572–75
  vestigialization and, 251
Adaptationism, 574
Adaptive landscape
  natural selection and, 604
  stabilizing selection and, 613–14
Adaptive management, 56–57
Adrenergic receptor
  accelerated evolution and, 411
Aegricorpus, 205
Aerobic respiration
  carbon loss from plant tissues and, 476
African origin hypothesis
  origin of *Homo sapiens* and, 362
Agenda 21, 227–28, 231

Aggarwal, M., 242
Agricultural crops
  root exudation in, 73
Agricultural productivity
  management of soil organic matter and, 71
Agriculture
  alternative, 201–17
    integration in, 212–14
    pest control in, 203–6
    problem of conversion in, 214–16
    soil management in, 206–12
  biological resource base of, 78–79
  holistic, 202
  mycorrhizas in, 75
  organic
    mulching and, 211
  sustainable
    biological basis of, 69–87
    low input, 201
    resource conservation and, 202
    traits under development in, 79–83
Agroecosystems
  added nitrogen interaction effect in, 209
  ecology of diseases in, 205
  green manure in, 211
  soil functions in, 33–34
Agronomic systems
  soil temperature and, 28
Agrostis stolonifera
  establishment in Antarctic terrestrial ecosystem, 697
Ahlquist, J. E., 349
Ahmand, Y., 15
Air pollution
  excessive burning of fossil fuels and, 241
Akashi, H., 403–19
Alaskozetes antarcticus
  sigmoid activity curve for, 691
  synthesis of cryoprotectants in, 689
Alberts, J. J., 240
Alchian, A., 123
Alcorn, J. B., 235
Alexander, A. M., 330
Allelic genealogy, 348–49
Allelochemicals
  plant nutrient uptake and, 73
Allocation
  atmospheric carbon dioxide and, 492–93, 495–96

Allopatry
  phylogenetic biogeography and, 380
Allozyme divergence
  in *Plethodon*, 581–86
Allozyme polymorphism
  balancing selection and, 409–11
  DNA polymorphism vs., 405–6
Allozyme variation
  evolution and, 404–6
Alluvial soils, 26
Alnus rubra
  leaf nitrogen concentration in
    fertilization and, 477
Alpha diversity, 96
Alstad, D. N., 435
Alternative agriculture, 201–17
  integration in, 212–14
  pest control in, 203–6
  problem of conversion in, 214–16
  soil management in, 206–12
Altieri, M. A., 240
Altschul, S. F., 98
Aluminum toxicity
  downward root growth and, 30
AM
  See Arbuscular mycorrhizae
Amazon Center for Environmental Education and Research (ACEER), 169
Amboseli basin
  bone frequencies of major species in, 275
  vertebrate assemblages in
    live:dead data on, 281–82
Ambystoma maculatum, 254
American Association for the Advancement of Science, 328, 332
American Museum of Natural History, 330, 339
American Nature Study Society, 338
Amerinidians
  natural resource utilization by, 127
Amino acid racemization geochronology
  modern death assemblages and, 284
Ammonium
  storage in cells, 209
Ampelocera macrocarpa
  seed dispersers/predators and, 159

729

CUMULATIVE INDEXES

CONTRIBUTING AUTHORS, VOLUMES 22–26

CHAPTER TITLES, VOLUMES 22–26

ANNUAL REVIEWS

a nonprofit scientific publisher
4139 El Camino Way
P.O. Box 10139
Palo Alto, CA 94303-0139 • USA

Annual Reviews publications may be ordered directly from our office; through stockists, booksellers and subscription agents, worldwide; and through participating professional societies. **Prices are subject to change without notice. We do not ship on approval.**

- **Individuals:** Prepayment required on new accounts. in US dollars, checks drawn on a US bank.

- **Institutional Buyers:** Include purchase order.   Calif. Corp. #161041 • ARI Fed. I.D. #94-1156476

- **Students / Recent Graduates:** $10.00 discount from retail price, per volume. *Requirements:* 1. be a degree candidate at, or a graduate within the past three years from, an accredited institution;   2. present proof of status (photocopy of your student I.D. or proof of date of graduation);   3. Order direct from Annual Reviews;   4. prepay. This discount **does not** apply to standing orders, *Index on Diskette,* Special Publications, ARPR, or institutional buyers.

- **Professional Society Members:** Many Societies offer *Annual Reviews* to members at reduced rates. Check with your society or contact our office for a list of participating societies.

- **California orders** add applicable sales tax. • **Canadian orders** add 7% GST. Registration #R 121 449-029.

- **Postage paid** by Annual Reviews (4th class bookrate/surface mail). UPS ground service is available at $2.00 extra per book within the contiguous 48 states only. UPS air service or US airmail is available to any location at actual cost. UPS requires a street address. P.O. Box, APO, FPO, not acceptable.

- **Standing Orders:** Set up a standing order and the new volume in series is sent automatically each year upon publication. Each year you can save 10% by prepayment of prerelease invoices sent 90 days prior to the publication date. Cancellation may be made at any time.

- **Prepublication Orders:** Advance orders may be placed for any volume and will be charged to your account upon receipt. Volumes not yet published will be shipped during month of publication indicated.

N O T E	For copies of individual articles from any *Annual Review,* or copies of any article cited in an *Annual Review,* call **Annual Reviews Preprints and Reprints (ARPR)** toll free 1-800-347-8007 (fax toll free 1-800-347-8008) from the USA or Canada. From elsewhere call 1-415-259-5017.

ANNUAL REVIEWS SERIES *Volumes not listed are no longer in print*	**Prices, postpaid, per volume.** **USA/other countries**	Regular Order Please send Volume(s):	Standing Order Begin with Volume:
❏ *Annual Review of* **ANTHROPOLOGY**			
Vols. 1-20	(1972-91) ....$41 / $46		
Vols. 21-22	(1992-93) ....$44 / $49		
Vol. 23-24	(1994 and Oct. 1995) ....$47 / $52	Vol(s). _____	Vol. _____
❏ *Annual Review of* **ASTRONOMY AND ASTROPHYSICS**			
Vols. 1, 5-14, 16-29	(1963, 67-76, 78-91) ....$53 / $58		
Vols. 30-31	(1992-93) ....$57 / $62		
Vol. 32-33	(1994 and Sept. 1995) ....$60 / $65	Vol(s). _____	Vol. _____
❏ *Annual Review of* **BIOCHEMISTRY**			
Vols. 31-34, 36-60	(1962-65,67-91).........................$41 / $47		
Vols. 61-62	(1992-93) ....$46 / $52		
Vol. 63-64	(1994 and July 1995) ....$49 / $55	Vol(s). _____	Vol. _____
❏ *Annual Review of* **BIOPHYSICS AND BIOMOLECULAR STRUCTURE**			
Vols. 1-20	(1972-91) ....$55 / $60		
Vols. 21-22	(1992-93) ....$59 / $64		
Vol. 23-24	(1994 and June 1995) ....$62 / $67	Vol(s). _____	Vol. _____

❑ *Annual Review of* **CELL AND DEVELOPMENTAL BIOLOGY** (new title beginning with volume 11)
Vols. 1-7 (1985-91) ....$41 / $46
Vols. 8-9 (1992-93) ....$46 / $51
Vol. 10-11 (1994 and Nov. 1995) ....$49 / $54   Vol(s). _____ Vol. _____

❑ *Annual Review of* **COMPUTER SCIENCE** (Series suspended)
Vols. 1-2 (1986-87) ....$41 / $46
Vols. 3-4 (1988-89/90) ....$47 / $52   Vol(s). _____
Special package price for
Vols. 1-4 (if ordered together) ....$100 / $115 ❑ Send all four volumes.

❑ *Annual Review of* **EARTH AND PLANETARY SCIENCES**
Vols. 1-6, 8-19 (1973-78, 80-91) ....$55 / $60
Vols. 20-21 (1992-93) ....$59 / $64
Vol. 22-23 (1994 and May 1995) ....$62 / $67   Vol(s). _____ Vol. _____

❑ *Annual Review of* **ECOLOGY AND SYSTEMATICS**
Vols. 2-12, 14-17, 19-22..(1971-81, 83-86, 88-91) ..$40 / $45
Vols. 23-24 (1992-93) ....$44 / $49
Vol. 25-26 (1994 and Nov. 1995) ....$47 / $52   Vol(s). _____ Vol. _____

❑ *Annual Review of* **ENERGY AND THE ENVIRONMENT**
Vols. 1-16 (1976-91) ....$64 / $69
Vols. 17-18 (1992-93) ....$68 / $73
Vol. 19-20 (1994 and Oct. 1995) ....$71 / $76   Vol(s). _____ Vol. _____

❑ *Annual Review of* **ENTOMOLOGY**
Vols. 10-16, 18, 20-36 (1965-71, 73, 75-91) ....$40 / $45
Vols. 37-38 (1992-93) ....$44 / $49
Vol. 39-40 (1994 and Jan. 1995) ....$47 / $52   Vol(s). _____ Vol. _____

❑ *Annual Review of* **FLUID MECHANICS**
Vols. 2-4, 7 (1970-72, 75)
9-11, 16-23 (1977-79, 84-91) ....$40 / $45
Vols. 24-25 (1992-93) ....$44 / $49
Vol. 26-27 (1994 and Jan. 1995).... ....$47 / $52   Vol(s). _____ Vol. _____

❑ *Annual Review of* **GENETICS**
Vols. 1-12, 14-25 (1967-78, 80-91) ....$40 / $45
Vols. 26-27 (1992-93) ....$44 / $49
Vol. 28-29 (1994 and Dec. 1995) ....$47 / $52   Vol(s). _____ Vol. _____

❑ *Annual Review of* **IMMUNOLOGY**
Vols. 1-9 (1983-91) ....$41 / $46
Vols. 10-11 (1992-93) ....$45 / $50
Vol. 12-13 (1994 and April 1995) ....$48 / $53   Vol(s). _____ Vol. _____

❑ *Annual Review of* **MATERIALS SCIENCE**
Vols. 1, 3-19 (1971, 73-89) ....$68 / $73
Vols. 20-23 (1990-93) ....$72 / $77
Vol. 24-25 (1994 and Aug. 1995) ....$75 / $80   Vol(s). _____ Vol. _____

❑ *Annual Review of* **MEDICINE: Selected Topics in the Clinical Sciences**
Vols. 9, 11-15, 17-42 (1958, 60-64, 66-42) ....$40 / $45
Vols. 43-44 (1992-93) ....$44 / $49
Vol. 45-46 (1994 and April 1995) ....$47 / $52   Vol(s). _____ Vol. _____